Spectroscopy of Semiconductor Microstructures

NATO ASI Series

Advanced Science Institutes Series

A series presenting the results of activities sponsored by the NATO Science Committee, which aims at the dissemination of advanced scientific and technological knowledge, with a view to strengthening links between scientific communities.

The series is published by an international board of publishers in conjunction with the NATO Scientific Affairs Division

A	**Life Sciences**	Plenum Publishing Corporation
B	**Physics**	New York and London
C	**Mathematical and Physical Sciences**	Kluwer Academic Publishers Dordrecht, Boston, and London
D	**Behavioral and Social Sciences**	
E	**Applied Sciences**	
F	**Computer and Systems Sciences**	Springer-Verlag
G	**Ecological Sciences**	Berlin, Heidelberg, New York, London,
H	**Cell Biology**	Paris, and Tokyo

Recent Volumes in this Series

Series B: Physics

Spectroscopy of Semiconductor Microstructures

Edited by

Gerhard Fasol

University of Cambridge
Cambridge, United Kingdom

Annalisa Fasolino

International School for Advanced Studies (SISSA)
Trieste, Italy

and

Paolo Lugli

University of Rome II
Rome, Italy

Springer Science+Business Media, LLC

Proceedings of a NATO Advanced Research Workshop on
Spectroscopy of Semiconductor Microstructures,
held May 9–13, 1989,
in Venice, Italy

Library of Congress Cataloging in Publication Data

NATO Advanced Research Workshop on Spectroscopy of Semiconductor Micro-
structures (1989: Venice, Italy)
    Spectroscopy of semiconductor microstructures / edited by Gerhard Fasol,
Annalisa Fasolino, and Paolo Lugli.
        p.   cm.—(NATO ASI series, Series B, Physics; vol. 206)
    Proceedings of a NATO Advanced Research Workshop on Spectroscopy of
Semiconductor Microstructures, held May 9–13, 1989, in Venice, Italy.
    Includes bibliographical references.
    ISBN 978-1-4757-6567-0   ISBN 978-1-4757-6565-6 (eBook)
    DOI 10.1007/978-1-4757-6565-6
    1. Semiconductors—Surfaces—Optical properties—Congresses. 2. Micro-
structure—Congresses. 3. Quantum wells—Congresses. 4. Superlattices as
materials—Congresses. I. Fasol, Gerhard. II. Fasolino, Annalisa. III. Lugli, Paolo.
IV. Title. V. Series: NATO advanced science institutes series. Series B, Physics; v.
206.
QC611.6.S9M35   1989   89-23015
537.6′22—dc20   CI

© 1989 Springer Science+Business Media New York
Originally published by Plenum Press, New York in 1989
Softcover reprint of the hardcover 1st edition 1989

SPECIAL PROGRAM ON CONDENSED SYSTEMS OF LOW DIMENSIONALITY

This book contains the proceedings of a NATO Advanced Research Workshop held within the program of activities of the NATO Special Program on Condensed Systems of Low Dimensionality, running from 1983 to 1988 as part of the activities of the NATO Science Committee.

Other books previously published as a result of the activities of the Special Program are:

INTRODUCTION

G. Fasol*, A. Fasolino**, and P. Lugli***

* Cavendish Laboratory, Madingley Road, Cambridge CB3 0HE, United Kingdom
** SISSA, Strada Costiera 11, I-34100 Trieste, Italy
***Dipt. di Ingegneria Meccanica, Università di Roma II, "Tor Vergata", I-00173 Roma, Italy

Semiconductor crystal growth techniques, including recent off-angle epitaxial techniques, are approaching the limit where structures can be grown with a control down to the dimensions of one single atom in all three directions. This development is possible due to large investments of ingenuity, efforts, man power and money and is supported strongly by intensive international collaboration and the international exchange of ideas. These investments are driven by now established commercial applications: high electron mobility transistors (HEMT), e.g. for low noise input stages of satellite television receiver dishes, quantum well laser diodes, integrated repeaters, detectors and modulators for data communication systems, are just some examples. Although not necessarily always profitable, III-V materials and quantum wells have become an established part of modern technology, used or investigated for use by several sectors, including the consumer electronics, telecommunication, computer and car sectors. Other areas, such as strained layer superlattices, Silicon / Germanium superlattices, one-dimensional systems, for example, have much promise to become very important, too. The development of commercial devices takes place parallel with on-going fundamental research. It is characteristic for the field of epitaxial semiconductor structures, that growth techniques, devices, characterization methods and theoretical methods are all developed at the same time in strong international collaboration. The development of commercial devices is strongly linked to fundamental research and we can observe a very active feedback loop between characterisation measurements, theory (band structure calculations, device and process modelling, etc.) and device development. This area cuts horizontally through many disciplines (growth and characterisation techniques, solid state physics, chemistry, engineering, optics...). And people move quite frequently between sectors in different stages of their careers.

The fascination comes from the aim to achieve the ultimate limit of what is physically possible: chemists / crystal growers aim to achieve single atomic layer control in the growth direction - atomically smooth interfaces, quantum wires, quantum boxes, or the highest mobilities. Spectroscopists look for the shortest possible time scales using femto-second light pulses, where the light pulses are only micrometers long, they characterize single sheets of electrons, or single layers of atoms as in δ-function doping using optical signals. Device developers try to make single quantum wells lase. Semiconductor microstructures are studied in extreme physical conditions: quasi one- or zero-dimensionality, high magnetic fields, low temperatures, high optical excitation, high electrical fields. This fundamental search accompanies the development of ultrafast transistors, ultrafast lasers and modulators and other devices. Neither would happen without the other. In fact, the classification of a particular project in the field of semiconductor microstructures as "fundamental" research, "applied" research or device development has become dependent on the country and on the time-scales in which people in different countries and cultures are conditioned to think. The present book is biased towards the European way of working, but several US, Canadian and Japanese contributions broaden the view.

The aim of this book is to explain some of the frontiers of today's work on semiconductor microstructures. The book therefore contains extended articles which join novel results with a detailed and illustrative presentation. All contributors present their *own* ongoing work in this volume. Both groups represent the most active workers, who are in the middle of fascinating research programs. They usually do not have the time to present their work in monographs, but they do so in short conference proceedings and "letter-type" journal articles which do not leave space for detailed explanations. The articles in the present book are on purpose a lot longer, so that authors can give explanations and details which are usually left out from articles for research journals. Thus this book should prove useful for the new generation of research workers and of course also for all those workers well established in semiconductor research who need to know details about neighbouring fields.

A typical example for this approach are the results on phonons, presented in the present book: phonons in quantum wells are extremely complicated. Phonons via the electron-phonon interaction have dominating influence on noise of quantum well lasers, the mobility, relaxation times in femto-second processes. Up to now, a large number of theoretical and experimental studies approximated the correct quantum well phonon modes either by bulk phonon modes, and sometimes even by quite wrong models. This book gives a detailed explanation of phonons in superlattices to people concentrating on other aspects of quantum wells but who see the need to understand phonons more fully. This is one typical example for the way we hope that the present volume contributes to an exchange of information between different areas.

A special feature of the present book is that in addition to presenting up-to-date extended review articles, it also contains much previously unpublished work!

The development of spectroscopical techniques advances in collaboration with the developments of growth, physical understanding, devices and theoretical methods. Therefore the present book has a very broad approach covering spectroscopic techniques and many surrounding areas. Reflecting the strong international interactions and the interdisciplinary character, many articles cover several different topics and represent international collaborations.

The book covers work at the frontiers of the following areas:

- spectroscopic techniques
  - magneto-optics
  - electro-optics and nonlinear optics
- growth techniques in interaction with applications and characterisation
  - molecular beam epitaxy (MBE)
  - metal organic chemical vapour deposition (MOCVD)
- quantum wires and quantum dots
- ultrafast spectroscopy
  - femtosecond experiments
  - electro-optic sampling
  - Monte-Carlo calculation techniques
- phonons
  - fundamental properties
  - use for materials characterisation
- electronic structure, theoretical methods

The present book results from a NATO Advanced Research Workshop on "Spectroscopy of Semiconductor Microstructures", which took place in Venezia - Lido between 9 and 13 May 1989. As most contributors had several months to organise the contributions to the conference and to this book, there is a good coherence of the contributions, and many contain previously unpublished results.

It has not been an easy task to organise the papers under different headings, since most of the papers involve more than one topic. Therefore the reader will find some overlap between sections, which reflects the interdisciplinary character of this field. The motivation for most work is twofold: on one hand a better understanding of physics in extreme conditions, and on the other hand the achievement of progress towards more efficient devices.

We gratefully acknowledge the generous support of the NATO Scientific Affairs Division, of the Italian Consiglio Nazionale delle Ricerche (CNR) and of DILOR SA (Lille, France), which made the workshop and this resulting book possible. We also express our gratitude to all those who have contributed to this book and to the workshop, and to Plenum Press for the efficient publication. In particular, G. F. expresses his deep gratitude to Klaus Ploog, Detlev Heitmann, Hiroyuki Sakaki and Roland Fasol for much advice, encouragement and support in planning the program and the organisation of this book and the workshop and to Gill Renshaw for much help in type-setting and organisational matters.

Gerhard Fasol
Annalisa Fasolino
Paolo Lugli

CONTENTS

GROWTH OF SEMICONDUCTOR MICROSTRUCTURES

QUANTUM WIRES AND QUANTUM DOTS

PHONONS IN III-V, Si / Ge AND FIBONACCI SUPERLATTICES

OPTICAL PROBING OF BAND STATES AND WAVEFUNCTIONS

ELECTRIC FIELDS AND NON-LINEAR OPTICS

MAGNETO-OPTICS

FEMTOSECOND SPECTROSCOPY AND MONTE-CARLO CALCULATIONS

MBE GROWTH OF CUSTOM-DESIGNED III-V SEMICONDUCTOR MICROSTRUCTURES SCALED TO THE PHYSICAL LIMIT: ULTRATHIN-LAYER SUPERLATTICES AND MONOLAYER DOPING

Klaus Ploog

Max-Planck-Institut für Festkörperforschung
D-7000 Stuttgart 80, FR Germany

1. INTRODUCTION

Molecular beam epitaxy (MBE) of custom-designed microstructures has reached a status where monolayer dimensions in artificially layered semi-conductors are being routinely controlled to form a new class of materials with accurately tailored electrical and optical properties[1,2]. The unique capabilities of molecular beam epitaxy in terms of spatially resolved ma-terials synthesis has stimulated the inspiration of device engineers to design a whole new generation of electronic and photonic devices based on the concept of band gap engineering[3,4]. This concept, also called wavefunc-tion or density-of-states engineering[5,6], respectively, relies on the ar-bitrary modulation of band-edge potentials in semiconductors through the abrupt change of composition (e.g. GaAs/AlAs, GaSb/InAs, Si/Ge, etc.) or of dopant concentration. The microscopic structuring or engineering of se-miconducting solids to within atomic dimensions is thus achieved by the incorporation of interfaces (consisting of abrupt homo- or heterojunctions) into a crystal in well-defined geometrical and spatial arrangements. The electrical and optical properties are then defined locally, and phenomena related to extremely small dimensions ("quantum size effects") become more important than the actual chemical properties of the materials involved.

In this article two prototype artificially layered semiconductor struc-tures are presented in which the concept of microscopical structuring of solids is scaled to its ultimate physical limit normal to the crystal sur-face, i.e. $(GaAs)_m/(AlAs)_n$ ultrathin-layer superlattices (UTLS) and mono-layer (or delta) doping (MD) in GaAs and $Al_xGa_{1-x}As$. Each constituent layer in the $(GaAs)_1/(AlAs)_1$ monolayer superlattice and also the narrow buried MD channel in, e.g. GaAs have a spatial extent normal to the surface of less than the lattice constant of the respective bulk material. Both the $(GaAs)_m/(AlAs)_n$ UTLS as well as the MD $GaAs/Al_xGa_{1-x}As$ heterostructures have recently become very important for fundamental studies and for appli-cation in advanced semiconductor device concepts.

2. FUNDAMENTALS OF MOLECULAR BEAM EPITAXY

The fabrication of custom-designed microstructures, in which the de-sired potential differences are defined locally by the accurate positioning of homo- or heterojunctions, requires advanced epitaxial crystal growth techniques, such as molecular beam epitaxy (MBE). The particular merits

1

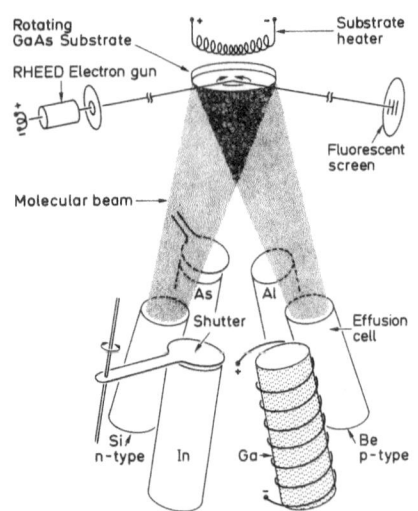

Fig. 1. Schematic illustration of evaporation and deposition process during molecular beam epitaxy of III-V compounds.

of MBE are that ultrathin films can be grown with precise control over thickness, composition, and doping level[5-8]. The technique allows atomic layer-by-layer deposition in a 2D growth process, so that crystalline materials in alternating layers of arbitrary composition and only a few atomic layers thick can be formed. The conventional MBE process, schematically shown in Fig. 1, consists of a continuous co-evaporation of the constituent solid elements (Al, Ga, In, P, As, Sb) of the epitaxial layer and of solid dopants (mainly Si for n-type and Be for p-type doping) onto a heated crystalline substrate where they react chemically under ultrahigh-vacuum (UHV) conditions. The composition of the layer and its doping level depend on the relative arrival rates of the constituent elements which in turn depend on the evaporation rates of the appropriate sources. Accurately controlled temperatures (to within $\pm$ 0.1° at 1000 °C) thus have a direct controllable effect upon the growth process.

The group III elements are always supplied as monomers by evaporation from the respective liquified element and have a unity sticking coefficient over most of the substrate temperature range used for film growth (e.g. 500 - 650 °C for GaAs). The group V elements, on the other hand, can be supplied as tetramers (P_4, As_4, Sb_4) by sublimation from the respective solid element or as dimers (P_2, As_2, Sb_2) by dissociating the tetrameric molecules in a two-zone furnace. A low growth rate of typically 0.5 - 1.5 μm/h is chosen so that extensive migration of the impinging species on the growing surface to the appropriate lattice sites is ensured and incorporation of crystalline defects is thus avoided. Simple mechanical shutters in front of the evaporation sources are used to interrupt the beam fluxes to start and stop deposition and doping in a time scale of the order of 0.1 s. Due to the selected slow growth rate of about 1 monolayer/s, changes in composition and doping can thus be abrupt on an atomic scale.

The stoichiometry of most III-V semiconductors during molecular beam epitaxy is self-regulating as long as excess group V element molecules are continuously impinging on the growing surface. The excess group V species do not stick on the heated substrate surface. The growth rate is thus essentially determined by the arrival rates of group III elements. The simplicity of the MBE process allows composition control from x = 0 to x = 1 in $Al_xGa_{1-x}As$, $Ga_xIn_{1-x}As$ etc. with a precision of $\pm$ 0.001 and doping control, both n- and p-type, from the 10^{14} cm^{-3} to the 10^{19} cm^{-3} range with a precision of a few percent. The accuracy is largely determined by the care with which the growth rate and doping level were precisely calibrated in test layers. This independent and accurate control of the individual beam sources allows the reproducible fabrication of artificially layered semiconductor structures of various complexity by MBE.

Most of the advanced MBE systems consist of several basic UHV building blocks (the growth chamber, the sample preparation chamber, and the load-lock chamber), which are separately pumped and interconnected via large diameter channels and isolation valves. High-quality layered semiconductor structures require background vacuums in the low 10^{-11} Torr range to avoid incorporation of impurities into the growing layers. Therefore, extensive LN_2 cryoshrouds are used around the substrate to achieve locally much lower background pressures of condensible species. The starting materials for the growth of III-V semiconductors are evaporated in resistively heated effusion cells made of pyrolytic BN which operate at temperatures up to 1400 °C. Most of the functions important for the MBE growth process are controlled by a computer.

Molecular beam epitaxy of III-V semiconductors is mostly performed on (001) oriented substrate slices about 300 - 500 μm thick. The preparation of the growth face of the substrate from the polishing stage to the in-situ cleaning stage in the MBE system is of crucial importance for epitaxial growth of ultrathin layers and heterostructures with high purity and crystal perfection and with accurately controlled interfaces on an atomic scale. The substrate surface should be free of crystallographic defects and clean on an atomic scale. Various cleaning methods have been described for GaAs and InP, which are the most important substrate materials for deposition of III-V semiconductors. The first step always involves chemical etching, which leaves the surface covered with some kind of a protective oxide. After insertion in the MBE system this oxide is removed by heating under UHV conditions carried out in a beam of the group-V-element.

Reflection high-energy electron diffraction (RHEED) operated at 10 - 30 keV in the small glancing angle reflection mode (Fig. 2) is the most important method to monitor surface crystallography and growth kinetics in-situ during MBE growth. The diffraction pattern projected on the fluorescent screen, mostly taken in the [110] and [$\bar{1}$10] azimuths of (001) oriented III-V semiconductors, contains information from the topmost layers of the deposited material, and it can thus be related to the topography and structure of the growing surface. The diffraction spots are elongated to characteristic streaks normal to the shadow edge. Additional features in the RHEED pattern at fractional intervals between the bulk diffraction streaks manifest the existence of specific surface reconstructions, which are correlated to the surface stoichiometry and thus directly to the MBE growth conditions (substrate temperature, molecular beam flux ratio, etc.).

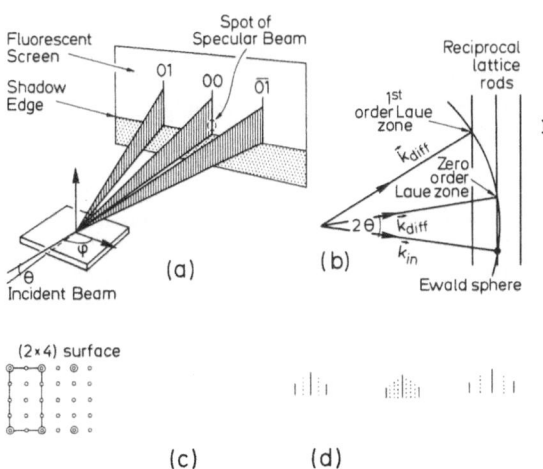

Fig. 2. Schematic illustration of electron diffraction with grazing-angle incidence (RHEED) during molecular beam epitaxy (a); Ewald construction to interpret the diffraction conditions (b); surface unit cell (c) and diffraction patterns of reconstructed (2x4) surface (d).

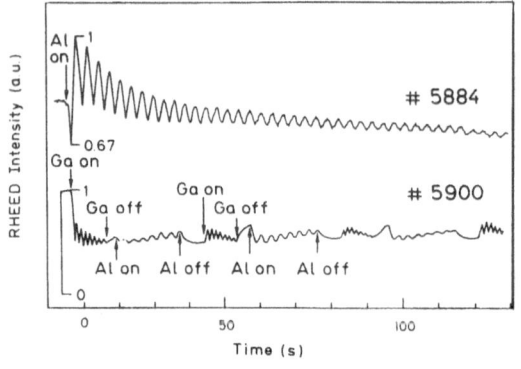

Fig. 3. Periodic intensity oscillations of specularly reflected electron beam in the
RHEED pattern as a function
of time during growth of
GaAs/AlAs layers.

Another characteristic feature of the RHEED pattern is the existence
of pronounced periodic intensity oscillations of the specularly reflected
and of the diffracted beams during MBE growth (Fig. 3)[9,10]. The period of
these oscillations corresponds exactly to the time required to deposit a
monolayer of GaAs (or AlAs, $Al_xGa_{1-x}As$, etc.) on the (001) surface. To
explain this peculiarity we can to a first approximation assume that the
amplitude of the intensity oscillation reaches its maximum when the monolayer is just completed (maximum reflectivity). The formation of the following lattice plane starts with statistically distributed 2D islands having
the height of one GaAs monolayer ($\cong 2.8$ Å). The intensity of the diffracted (or reflected) electron beam decreases with increasing size of the islands. The minimum reflectivity occurs at half-layer coverage ($\Theta = 0.5$).
When the coverage is further increased the islands coalesce more and more,
and the reflectivity finally reaches a maximum again at $\Theta = 1$. The observed intensity oscillations in the RHEED pattern provide direct evidence
that MBE growth occurs predominantly in a 2D layer-by-layer growth mode,
and this method is now widely used to calibrate and to monitor absolute
growth rates in real time with monolayer resolution. However, it is important to note that the interpretation of RHEED intensity data may be erroneous because in general both elastically and diffusely scattered electrons
contribute to the recorded intensity. Diffraction conditions where elastic
scattering dominates the intensity of the specularly reflected beam are
most easily obtained under off-azimuth conditions. The other important aspect is the damping of the RHEED intensity oscillations observed during
growth (Fig. 3) and the characteristic recovery once growth is interrupted.
This damping has been ascribed[11] to an increase of the surface step density
(i.e. an increasing number of individual surface domains probed by the
electron beam are no longer in phase). The recovery of the RHEED intensity
following interruption of growth has been identified with an expansion of
the mean terrace width of the surface and hence a reduction of the surface
step density (i.e. a few large domains on the growing surface are formed).
The damping of the RHEED intensity oscillations can be eliminated continuously when the surface migration of the Ga or Al atoms is enhanced by short
synchronized cessations of the arsenic flux[12].

The existence of 2D growth islands of one monolayer height, having
different lateral extensions (i.e. terrace widths) , leads to material interfaces between, e.g. GaAs/AlAs or $GaAs/Al_xGa_{1-x}As$, which are "rough" and
not smooth and abrupt on an atomic scale. In addition, the strong difference
in surface migration rate of Ga and Al atoms on the growing surface results
in substantial differences of the structural and thus also electronic properties of $GaAs/Al_xGa_{1-x}As$ interfaces depending on the sequence of layer
growth (binary-to-ternary or ternary-to-binary). To a certain extent, a
short interruption of growth at the interface under continuous arsenic
supply allows the surface to relax, resulting in "smoother" interfaces in
terms of larger growth islands. However, the time for growth interruption

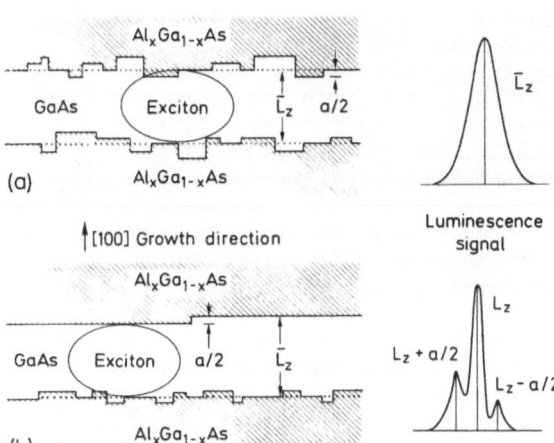

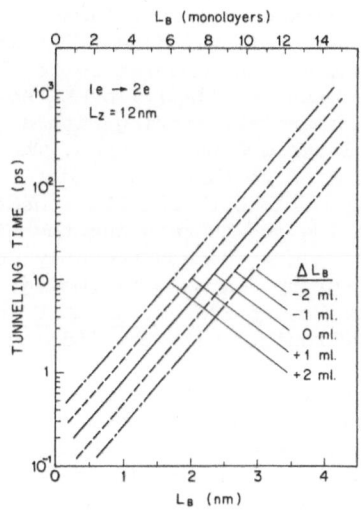

Fig. 4. Influence of lateral size of the 2D growth islands at the interface on the effective well width L_z and on the localization of excitons.

Fig. 5. Influence of the barrier width fluctuation ΔL_B on the resonant tunneling time in GaAs/AlAs superlattices ($L_z = 120$ Å).

should not exceed several tens of seconds to avoid contamination of the growing surface by residual gas impurities. The interface roughness induces fluctuations in the well width as well as in the barrier width of GaAs/ $Al_xGa_{1-x}As$ quantum wells and superlattices, and it thus strongly affects the excitonic properties (Fig. 4) and the resonant tunneling (vertical transport) of electrons (Fig. 5). When the size of the growth islands at the interface becomes larger than twice the exciton Bohr radius (> 200 Å), the excitons in the quantum well experience different confinement energies, and up to three distinct excitonic transitions are observed in the low-temperature photoluminescence (PL) spectra. For smaller-size growth islands, the excitons sense an averaged confinement energy and a spectral broadening of the PL signal occurs. We have recently shown[13] that the actual population of the three different island regions in GaAs/$Al_xGa_{1-x}As$ quantum wells depends strongly on the excitation conditions (laser wavelength and intensity) of the luminescence. The area ratio of the island regions and the exciton transfer rates can be derived from luminescence measurements only under very specific excitation conditions. Otherwise erroneous results are obtained. As shown in Fig. 5, in GaAs/AlAs superlattices the monolayer fluctuations of the barrier width has a significant effect on the resonant tunneling time[14]. As for 120 Å thick GaAs wells, the tunneling time changes by a factor of two when the barrier width fluctuates by $\pm$ 1 monolayer. This example illustrates the stringent requirements for MBE growth control imposed upon the successful fabrication of double-barrier resonant-tunneling diode structures.

In the last two years, the original MBE technique has encountered important new developments to expand its application particularly to low-temperature and heteroepitaxial (lattice-mismatched) growth. While in conventional MBE the molecular beams impinge continuously onto the substrate surface for the growth of a homogeneous layer (e.g. GaAs), alternating or modulated beams synchronized with the layer-by-layer growth mode by properly actuating the shutters are now used in migration-enhanced epitaxy (MEE)[15] and in modulated molecular beam epitaxy (MMBE)[16]. To a certain extent these new techniques artificially induce an <u>atomic</u> layer-by-layer growth sequence on the (001) surface. Their advantages are the superior smooth morphology of surfaces and interfaces (even for lattice-mismatched systems), the suppression of oval-defect formation, and the substantial lowering of the optimum growth temperature for high-quality material (by more than 200 $^{\circ}$C for GaAs and AlAs). In MEE both the group III and the group V molecular beams are alternating with a periodicity corresponding to the atomic layer-by-layer growth sequence. The group III dose per cycle has to be carefully adjusted to that required for completion of one atomic layer by using RHEED intensity oscillations as reference. The disadvantages of MEE are the low growth rate and the high mechanical load for the mechanical shutters. In the MMBE method recently developed by Briones et al.[16], on the other hand, the group III beam is impinging continuously onto the substrate while periodic short pulses of group V species are supplied with a repetition rate synchronized with the monolayer growth rate. Also in this case the atomic layer-by-layer growth sequence is accomplished, provided that (i) the group V repetition rate coincides accurately with the growth rate in monolayer/s, (ii) the group V pulses are short, and (iii) their intensity is adequate to saturate the surface at each cycle. The advantages of MMBE are thus a higher growth rate (comparable to conventional MBE) and less mechanical load on the shutters. The growth mechanism of both MEE and MMBE is based on an instantaneous group III enrichment of the (001) surface during each cycle, which can be monitored by a dramatic change of the surface reconstruction via the RHEED pattern. The growing surface thus alternates between a group III completed and a group V saturated state. While Horikoshi et al.[15] assume that the growth mechanism of MEE involves the enhancement of the surface migration of the group III species when supplied separately, Briones et al.[16] suggest a mechanism of enhanced nucleation and forced 2D growth for MMBE which is prompted by the synchronous separation of the surface from equilibrium. Probably valid for both techniques is that the periodic changes of surface stoichiometry induce a 2D growth mechanism by enhanced layer nucleation.

3. $(GaAs)_m/(AlAs)_n$ ULTRATHIN-LAYER SUPERLATTICES

The research activities on $(GaAs)_m/(AlAs)_n$ ultrathin-layer superlattices (UTLS) were initiated by some detrimental structural, electrical, and optical properties of the ternary alloy $Al_xGa_{1-x}As$. First, the interface roughness for growth of binary GaAs or AlAs on the ternary alloy[10,17] yields inferior excitonic and transport properties. Second, the electrical properties of n-type $Al_xGa_{1-x}As$ for x > 0.2 are controlled by a deep donor ("DX center") in addition to the hydrogen-like shallow donor due to the peculiar band structure of the alloy[18]. Third, the X-minimum of the conduction band becomes the lowest one when x > 0.4[19], and thus an indirect bandgap of the alloy results. The first investigations of ultrathin-layer $(GaAs)_m$ /$(AlAs)_n$ superlattices with (m, n) from 10 down to 1 were motivated by the possibilities to shift the confined-particle states of the Γ, L and X valleys of the conduction band (and also for Γ of the valence band) to high enough energy to create radiative size-determined "direct-indirect" transitions exceeding in energy the bulk direct-indirect transitions of the ternary $Al_xGa_{1-x}As$ alloy with x = 0.43[20]. The all-binary $(GaAs)_m/(AlAs)_n$ UTLS are therefore considered as possible substitutes for the random ternary alloy in advanced device structures. In addition, the electronic properties of these superlattices, which are in the transition region between the extremes of quantum well behaviour for a period length > 80 $\overset{\circ}{A}$ [i.e. (m, n) > 15] and of a possible alloy-like behaviour of monolayer (m = n = 1) superlattices, are not fully understood. Of particular importance for application in photonic devices is the question, whether the lowest energy transitions are direct, indirect or pseudodirect,and how do they change as a function of layer thickness.

The recent progress in the control of interface formation during MBE using RHEED intensity oscillations and growth interruption has led to the

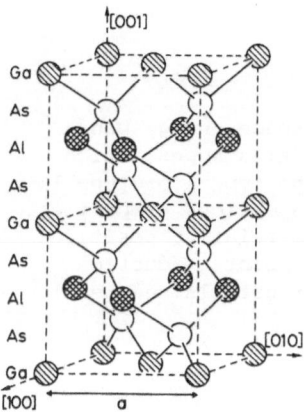

Fig. 6. Schematic arrangement of constituent atoms in $(GaAs)_1/$
$(AlAs)_1$ superlattice ("monolayer alloy") which represents
an ordered ternary alloy of
composition $Al_{0.5}Ga_{0.5}As$.

successful fabrication of $(GaAs)_m/(AlAs)_n$ UTLS with $(m,n) < 15$[21]. Growth of
these superlattices by conventional MBE was achieved[22] by monitoring each
deposited GaAs and AlAs monolayer from the RHEED oscillation period, inter-
rupting the group III beam at (m, n) = 1, 2, 3 ... and allowing the RHEED
intensity to recover almost to its initial value, and then depositing the
next constituent layer. The growth temperature and the growth rate were
lowered to prevent interdiffusion of Ga and Al across the interface and to
facilitate the build-up of the crystal atom-by-atom, respectively. The
well-ordered periodic layer-by-layer arrangement of Ga and Al atoms on the
appropriate lattice sites in [001] direction manifests itself in the appearance of distinct satellite peaks around the (004) reflection of the X-ray
diffraction patterns[22]. In the $(GaAs)_1/(AlAs)_1$ monolayer superlattice (Fig.
6), which is composed of alternating (001) GaAs and (001) AlAs monolayers,
each constituent layer has a spatial extent normal to the surface of less
than the lattice constant of the respective bulk material. The artificially
layered material has the same composition as the random ternary alloy
$Al_{0.5}Ga_{0.5}As$, and it can thus be considered as an ordered alloy.

The $(GaAs)_m/(AlAs)_n$ UTLS with low (m, n) values are expected to exhibit not the well-described confined particle states associated with quantum
wells but minibands parallel to the layer sequence within which carriers
propagate normal to the quantum well plane[23,24]. Due to the indirect-gap
nature of the AlAs barriers, however, the gradual transition of $(GaAs)_m/$
$(AlAs)_n$ superlattices from the well-understood quantum well behaviour
for (m, n) > 15 to the $(GaAs)_1/(AlAs)_1$ monolayer superlattice can produce
either type I superlatttices, where the electrons and holes are confined
in the same (GaAs)layer, or type II superlattices, characterized by the
spatial separation of electrons and holes with electrons confined in the
AlAs "barrier" region and holes in the GaAs well (see Fig. 7 for illustration)[25]. The type II behaviour was first observed in GaSb/InAs quantum
wells where it originates from the peculiar arrangement of the conduction
and valence band discontinuities[24]. In the GaAs/AlAs system the essential
conditions are that the AlAs barrier material is indirect and that the
X-point conduction band minima in the barriers lie at lower energy than
the Γ minima in the wells, which can be achieved by increasing the Γ confinement energy in very narrow wells. Note, however, that the simultaneous
reduction of the barrier width also leads to an increase of the confinement
of the AlAs X states and thus pushes them up in energy. For m = n the transition from type I to type II behaviour in $(GaAs)_m/(AlAs)_n$ superlattices
was found to occur at a critical value of 10 < (m, n) < 13[26,27]. For m ≠ n
and (m, n) < 15, on the other hand, the average Al mole fraction of the
superlattice serves as a rough guideline for determining the type I or
type II behaviour, similar to the ternary $Al_xGa_{1-x}As$ alloy. If the average
Al mole fraction is smaller than x = 0.4, the superlattice exhibits type I
behaviour, and in all other cases it behaves as a type II (indirect-gap)

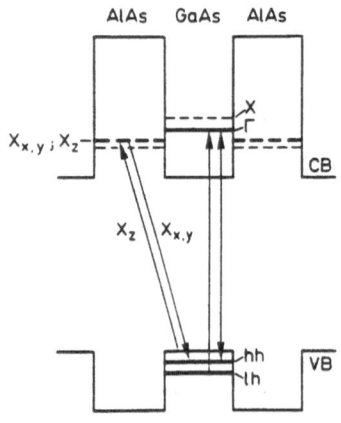

Fig. 7. Schematic real-space energy band diagram of GaAs/AlAs superlattice to illustrate the type I or type II behaviour of the optical properties, depending on the relative position of the Γ- and X-point conduction band minima (for details see text).

material[25]. Therefore, the $(GaAs)_2/(AlAs)_1$ and $(GaAs)_3/(AlAs)_1$ UTLS are of type I, while the $(GaAs)_2/(AlAs)_2$, $(GaAs)_2/(AlAs)_3$ and $(GaAs)_1/(AlAs)_2$ UTLS are of type II behaviour.

At present, there is a wide spread of spectroscopic data obtained from $(GaAs)_m/(AlAs)_n$ UTLS with $(m, n) < 15$ and of the assignment of distinct electronic properties to specific superlattice configurations given by various authors. Therefore, a clear need exists for an independent and improved structural characterization of the as-grown samples, e.g. by means of double-crystal X-ray diffraction (DCXRD)[28]. In Fig. 8 we show the X-ray diffraction patterns obtained from $(GaAs)_m/(AlAs)_n$ UTLS of different periodicity taken in the vicinity of the (004) reflection. The average Al mole fraction of the entire superlattice is deduced from the angular spacing between the substrate peak (S) and the main (zero-order) peak of the epitaxial layer ("E"). The angular distance $(\Delta\Theta)_{\pm1}$ between the zero-order peak and the satellite peak "+1" (or "-1") yields the superlattice periodicity according to the relation $T = \lambda/[\Delta\Theta_{\pm1} \sin(2\Theta_B)]$, where λ is the X-ray wavelength and Θ_B is the diffraction angle of the zero-order peak. From these data also the thicknesses of the constituent GaAs and AlAs layers can accurately be determined. Detailed information about thickness fluctuation of the constituent layers, inhomogeneity of composition, and interface quality can be extracted from the halfwidths and intensities of the satellite peaks.

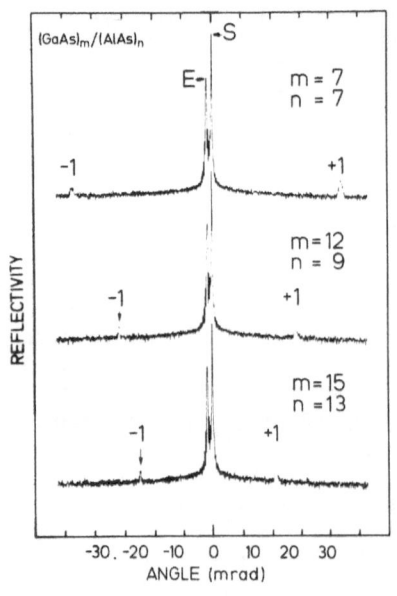

Fig. 8. Double-crystal X-ray diffraction pattern of three $(GaAs)_m/(AlAs)_n$ UTLS with different periodicity taken with $CuK_{\alpha1}$ radiation in the vicinity of the (004) reflection.

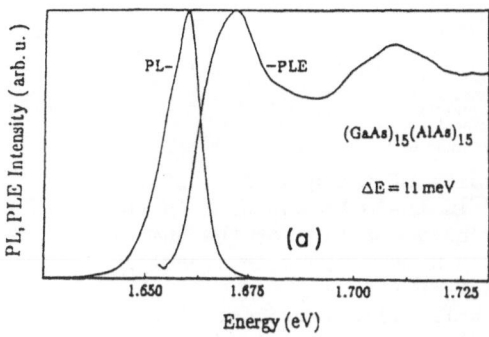 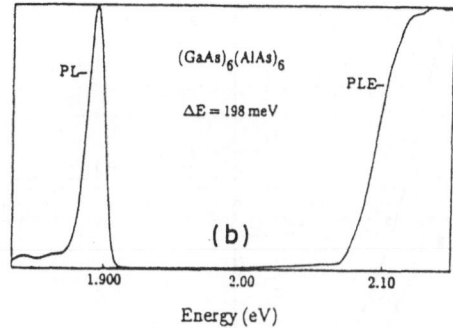

Fig. 9. Luminescence and excitation spectra of $(GaAs)_m/(AlAs)_n$ UTLS of type I (left) and type II (right) behaviour.

For this purpose we have made theoretical fits of the experimental diffraction curves based on a semi-kinematical approach of the dynamical theory of X-ray diffraction for distorted crystals[28]. The existence of interface disorder (roughness) in the GaAs-AlAs system manifests itself in an increase of the halfwidths and a decrease of the intensities of the satellite peaks. The analysis of possible changes in the lattice constant of the epitaxial layer parallel to the (001) substrate surface requires the recording of reflection curves from a lattice plane inclined to the (001) surface, e.g. from the (115) plane.

The assignment of either type I or type II behaviour to the respective superlattice configuration can be made by careful evaluation of the low-temperature photoluminescence (PL) and photoluminescence excitation (PLE) spectra[29]. In Fig. 9 we show representative PL and PLE spectra from both types of $(GaAs)_m/(AlAs)_n$ UTLS. In the type I $(GaAs)_{15}/(AlAs)_{15}$ UTLS the lowest confined GaAs Γ-conduction band state is lower in energy than the lowest confined conduction band states which originate from the X-conduction band minima of the AlAs barriers. The PLE spectrum of Fig. 9a exhibits a low-energy onset at about 1.66 eV, and we observe the well-known structure corresponding to the heavy- and light-hole exciton. The photoluminescence appears at the low-energy tail of the PLE spectrum, and the Stokes shift between the PL and the PLE maxima amounts to 11 meV. Time-resolved PL measurements on type I $(GaAs)_m/(AlAs)_n$ UTLS with different (m, n) values[29] reveal a continuous decrease of the luminescence decay time with decreasing (m, n) down to values of a few hundred picoseconds for the $(GaAs)_3/(AlAs)_1$ UTLS. This phenomenon is caused by the increase of the radiative (and non-radiative) recombination rates due to localization in agreement with previous measurements on $GaAs/Al_xGa_{1-x}As$ multi-quantum wells.

As an example of the characteristic type II behaviour we show in Fig. 9b the PL and PLE spectra of a $(GaAs)_6/(AlAs)_6$ UTLS[29]. The main PL peak is located at about 1.9 eV, and weaker bands arising from phonon assisted transitions (phonon replica) appear at lower energy. The photoluminescence is attributed to recombination of excitons formed from electrons at X_z in the AlAs "barriers" and holes at Γ in the GaAs wells (see Fig. 7 for illustration). The PLE spectrum shows a steep increase at about 2.08 eV which is attributed to the onset of direct absorption involving electron and hole states both at Γ in the GaAs well. In the range between 1.9 and 2.08 eV the PLE signal is weak, reflecting the weak strength of the "spatially" indirect transitions of electrons from a valence band state at Γ in the GaAs into a conduction band state in the AlAs which originates from the X conduction band minima. The features related to light and heavy hole excitons of the direct Γ-Γ transitions (around 2.1 eV in Fig. 9b) are always strongly damped in type-II superlattices due to the short lifetime

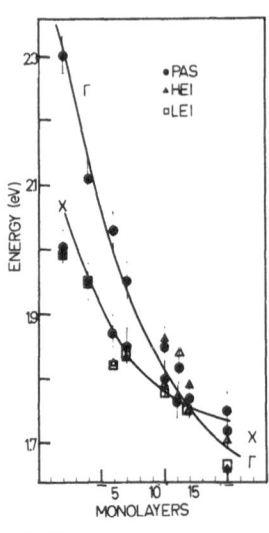

Fig. 10. Energies of the quantized states at Γ in the GaAs and at X in the AlAs as a function of the number of m = n monolayers in $(GaAs)_m$/$(AlAs)_n$ UTLS.
The solid line represents the calculated energy values.

of these excitons. The photoluminescence decay in type II superlattices is slower by orders of magnitude. In addition, we observe very fast high-energy luminescence transitions in the type II structures due to the re-combination of nonthermalized electrons in the GaAs layers. However, this PL feature rapidly disappears and thus indicates an extremely fast scatter-ing of electrons from Γ-like conduction band states in the GaAs into X-like states in the AlAs with characteristic time constants considerably smaller than 20 ps. It is finally important to note that the type II $(GaAs)_2$/$(AlAs)_2$ UTLS, although of average composition $Al_{0.5}Ga_{0.5}As$, exhibits electronic properties totally different from the random ternary alloy[25,29].

The interesting question, at which superlattice configuration the transition from type-I to type-II behaviour in $(GaAs)_m$/$(AlAs)_n$ UTLS with m = n actually occurs has recently been addressed by various spectroscopic[26] techniques with the samples under normal and under hydrostatic pressure[27]. The results which we obtained by application of photoacoustic spectroscopy combined with low (LEI) and high excitation intensity (HEI) PL measurements are depicted in Fig. 10 and compared with calculations of the energies of the quantized states at Γ in the GaAs and at X in the AlAs as a function of the m = n monolayers in the UTLS[26]. These data clearly indicate that the experimentally determined transition from type-I to type-II behaviour occurs at m = n = 12 monolayers. This important result is confirmed by PL measure-ments with the UTLS samples subject to hydrostatic pressure ranging up to 50 kbar[27]. As the energies of the Γ-like and X-like conduction band states have different pressure coefficients (E^Γ shows a strong blue shift and E^X a slight red shift with pressure enhancement), the dependence of the energy separation between Γ-like and X-like confined states on the actual UTLS configuration can be determined with high accuracy, independent of the re-lative position of these levels with respect to each other. The other im-portant result deduced from the PL intensity measurements under hydrostatic pressure is that the mixing between Γ-like and X-like states in these UTLS is rather small. As a consequence, the transition probability ratio of elec-trons in the Γ- and X-like states to heavy hole states (P^Γ/P^X) increases only slightly from 1.4×10^{-4} for (m, n) = 17 (type I) to 4.6×10^{-3} for (m, n) = 6 (type II). This result is in good agreement with the theoretical calculations of Xia[30].

The $(GaAs)_m$/$(AlAs)_n$ UTLS with (m, n) values close to the type-I/type II transition (i.e. having a small energy separation between the AlAs X-like and the GaAs Γ-like states of less than 100 meV) exhibit a number of inte-resting phenomena related to a switching between direct and indirect beha-viour in a given UTLS configuration, when (i) the excitation intensity or

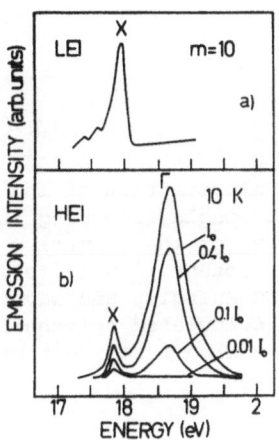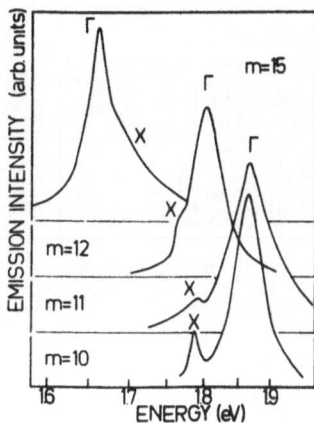

Fig. 11. PL spectra obtained from a (m, n) = 10 UTLS under different excitation power densities (left) and from UTLS of different (m, n)
values under high power density (right).

the sample temperature is varied or when (ii) an external bias or hydrostatic
pressure is applied. In Fig. 11 we show the results of PL measurements on
different UTLS configurations under increased (pulsed) excitation power
densities up to I_o = 300 kW/cm^2 ($\cong 10^{12}$ photogenerated carrier per cm^2) which
clearly reveal the strong competition between the type-I and the type-II
recombination channels[26]: In the (m, n) = 10 type-II UTLS the type-I emission (Γ-like) becomes clearly observable at increased excitation power densities and finally dominates, because the Γ-like emission grows superline-
arly with excitation power and the X-like emission saturates, as expected
from the different relaxation times of these two recombination processes[29]
The PL measurements under high-excitation power (HEI) therefore allow us
to directly observe both the direct and indirect emission in UTLS of 10 $\leq$
(m, n) $\leq$ 15, as indicated in Fig. 11, and to determine the crossover between type-I and type-II behaviour as well as even to switch between these
two states. In the (m, n) = 10 UTLS an enhancement of the sample temperature
from 10 to 40 K also favours the population of the closely spaced quasi-
resonant Γ-like state, and the type-II recombination is thus quenched on
the expense of the type-I recombination[26]. On a much faster time scale and
in a more elegant way the type-I/type-II switching in (GaAs)$_m$/(AlAs)$_n$ UTLS
can be obtained by the application of an electrical field perpendicular to
the layers[31] or by the application of hydrostatic pressure[27], because the
energies of the respective Γ-like and X-like states shift in opposite direction as a function of these external forces.

In the (GaAs)$_m$/(AlAs)$_n$ UTLS the minority carrier lifetimes can be tailored within the range of several hundred ps to several μs or even ms simply
by selecting the appropriate superlattice configuration during MBE growth[29].
This unique feature opens up a new field of application of these artificially layered semiconductor structures in nonlinear photonic and optoelectronic
devices. In addition, in properly designed type-II (GaAs)$_m$/(AlAs)$_n$ UTLS the
real-space transfer times associated with the Γ-X intervalley scattering of
electrons can directly be measured by femtosecond optical pump and probe
spectroscopy[32]. These measurements clearly reveal that in contrast to bulk
material the scattering rate for this process connecting electron states in
different slabs of the artificially layered material is determined by the
spatial overlap of the electron states in the different satellite minima.

4. MONOLAYER DOPING IN GaAs LAYERS AND IN GaAs/Al$_x$Ga$_{1-x}$As HETEROSTRUCTURES

The concept of monolayer (or delta) doping allows the deliberate positioning of Si donors and Be acceptors in precise numbers and with atomic

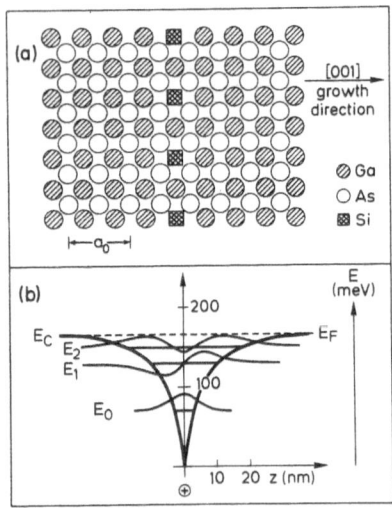

Fig. 12. Schematic illustration of the concept of Si monolayer doping in
GaAs (a) and of the formation of
a V-shaped potential well (b).
The subband energies and wavefunctions are calculated selfconsistently assuming a uniform sheet of positive charge[34].

layer precision during MBE growth of III-V semiconductors[33]. As indicated
in Fig. 12, the confinement of the dopant atoms to one (001) lattice plane
of the host crystal during epitaxial growth gives rise to the formation of
a V-shaped potential well with a quasi-2D electron (or hole) gas. The ionized
Si donors of density N_D^{2D} provide a continuous sheet of positive charge
(screened selfconsistently by the subband charge of $n_s \sim N_D^{2D}$ electrons)
which produces the V-shaped well. Direct evidence for the formation of
quasi-2D subbands in Si-monolayer-doped GaAs comes from Shubnikov-de Haas
oscillations observed during magnetotransport measurements with the magnetic field perpendicular to the (001) plane[34]. In contrast to other 2D carrier
systems, the electrons in monolayer-doped (MD) systems populate several
excited subbands in the V-shaped well even at easily accessible moderate
donor densities (e.g. four subbands at $N_D^{2D} \sim 4 \times 10^{12}$ cm^{-2}). In MD GaAs
a mobility enhancement as compared to similarly doped bulk material is in
general not observed due to the proximity of the electrons to their parent
ionized donors confined to the (001) plane. However, the narrow buried doping
channel in MD GaAs layers and especially in selectively doped $Al_xGa_{1-x}As$ heterostructures leads to a significant modification and improvement of the electrical and optical properties and to intriguing new phenomena important for
fundamental studies and also for device applications. A few aspects of the
large variety of phenomena will be discussed here.

In conventional MBE the monolayer doping profiles with Si and Be are
obtained through interruption of the growth of the GaAs or $Al_xGa_{1-x}As$ host
crystal by closing the Ga (or Ga plus Al) shutter, leaving the As shutter
open, and opening the respective dopant effusion cell for a certain time
interval. In this dopant growth mode under As-stabilized conditions a small
fraction of a monolayer of Si (or Be) atoms is deposited and incorporated
on Ga sites. To continue growth of the host crystal, the dopant shutter is
closed and the Ga (or Ga and Al) shutter is opened again. When sufficiently
low growth temperatures are applied, the dopant atoms incorporated on Ga
sites are buried within the (001) lattice plane and their spreading along
the growth direction is negligible. The measured sheet doping concentration
n_s is in good agreement with the number of supplied dopant atoms up to high
doping densities of about 8×10^{12} cm^{-2} [33].

Recently, there has been a controversial discussion about the actual
width dz of the narrow buried channel of ionized donors in Si-monolayer-doped
GaAs. On the basis of capacitance-voltage (C-V) profiling measurements
and selfconsistent calculations, Schubert et al.[35] assumed a localization
of the Si atoms in MD GaAs on the length scale of the lattice constant.
The authors even reported C-V profiles which were considerable narrower

than theoretically predicted. On the other hand, based on magnetotransport measurement under hydrostatic pressure and more sophisticated calculations, Zrenner et al.[38] demonstrated a substantial broadening of the Si atoms in MD GaAs up to 200 Å width, particularly when growth was performed at substrate temperatures higher than 600 °C. In addition, Zrenner[37] recently showed in a systematic study that for $N_D^{2D} > 4 \times 10^{12}$ cm^{-2} the existence of intrinsic deep donor states (DX centers), about 200 meV above the Γ conduction band edge, does in fact reduce the width of the measured C-V profiles considerably. Besides these two methods, which require selfconsistent model calculations to interpret the measured data, two additional techniques are available to examine the spread of the Si atoms normal to the (001) growth plane, namely secondary ion mass spectroscopy (SIMS) profiling[38] and Raman scattering by local vibrational modes[39]. While the SIMS profiling technique is limited in its depth resolution, the local vibrational mode spectroscopy, which has probably the best depth resolution, has been developed only recently. At present, the results of the four methods to determine the width of Si doping profiles in MBE grown MD GaAs can be summarized as follows: (i) The localization of the Si atoms within one (001) lattice plane can be achieved only at low growth temperatures (< 500 °C). At higher growth temperatures a marked broadening of the Si donors exists which is asymmetric towards the growing surface. (ii) The Si atoms are incorporated on Ga sites but not on As sites. At high doping densities ($N_D^{2D} > 8 \times 10^{12}$ cm^{-2}), Si-X complexes acting as acceptors can be formed. (iii) In true Si-monolayer-doped GaAs the free-electron concentration saturates at 7×10^{12} cm^{-2} because of the occupation of deep-donor levels. It is finally important to note that the electronic properties of Si MD GaAs in terms of energy, population and charge distribution of the subband scheme are not changed by the spread of the Si atoms up to dz $\sim$ 30 Å[36].

With the continuous efforts to reduce the gate length in metal-semiconductor field-effect transistors (MESFETs) for higher switching speed in microwave devices, a confinement of the electrons in the active region to a quasi-2D channel and a narrow distance between the gate electrode and the electron channel becomes increasingly important. The intentional positioning of a narrow channel of high electron density close to the crystal surface during monolayer doping has thus a great potential for device applications. The concept of the monolayer-doped (δ-) FET[40], which is schematically illustrated in Fig. 13, has the following advantages: (i) high concentration (up to 6×10^{12} cm^{-2}) of electrons in the quasi-2D channel, (ii) narrow distance (less than 100 Å) of electron channel from the gate electrode, and (iii) high breakdown voltage of the Schottky gate. These peculiarities result in high transconductances for short gate lengths and to an improved high-frequency linearity of the device. Ishibashi et al.[41] have recently fabricated monolayer-doped GaAs MESFETs with an intrinsic transconductance of 400 mS/mm.

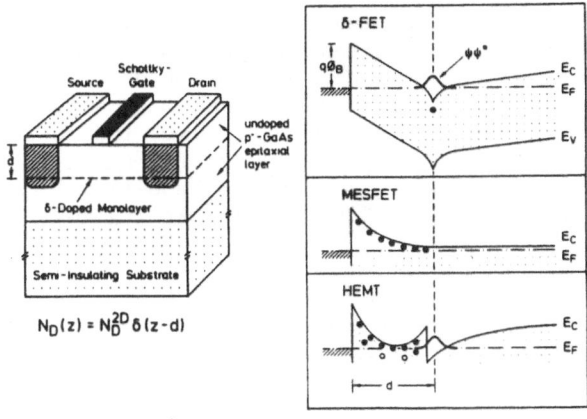

$N_D(z) = N_D^{2D} \delta(z-d)$

Fig. 13. Schematic layer sequence of a monolayer-doped (δ-) FET (left) and comparison of the conduction band diagram (right) with a MESFET and a high electron mobility transistor (HEMT).

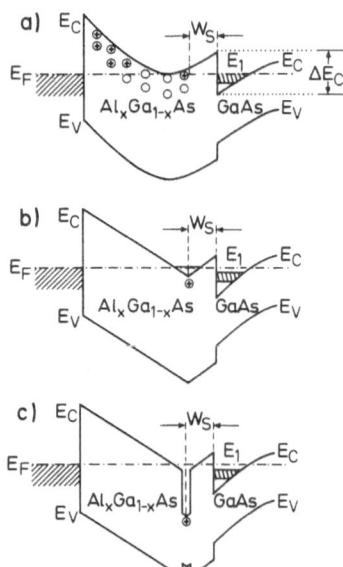

Fig. 14. Schematic real-space energy band diagram of (a) conventional selectively doped GaAs/Al$_x$Ga$_{1-x}$As heterostructure (W$_S$ is the spacer thickness, E$_C$ and E$_V$ are the conduction and valence band edges, resp., ΔE$_C$ is the conduction band discontinuity at the GaAs/Al$_x$Ga$_{1-x}$As interface, E$_F$ is the Fermi level, E$_1$ is the first subband in the triangular potential well at the interface), (b) selectively doped structure in which the Al$_x$Ga$_{1-x}$As is monolayer-doped at a distance W$_S$ from the interface, (c) structure as in (b) where the doped layer in the Al$_x$Ga$_{1-x}$As is embedded in the centre of a 20Å GaAs quantum well.

Due to the close proximity of ionized impurities and free carriers, the Hall mobilities of 2D electrons in monolayer-doped GaAs are comparable to the homogeneously doped material of equivalent doping concentration. Therefore, the concept of monolayer-doping has to be combined with the concept of selectively doped (SD) heterostructures for an effective spatial separation of the 2D electron gas (2DEG) from the atomic plane of ionized impurities, in order to achieve high carrier mobilities[33]. The concept of confinement of donors to atomic (001) planes in selectively doped GaAs/Al$_x$Ga$_{1-x}$As heterostructures is now used for three purposes, i.e. (i) to achieve high carrier mobilities at high 2D carrier densities in a simple heterostructure (recently even the record mobility of $\mu \geq 1 \times 10^7$ cm^2/Vs[44] at very low electron densities has been achieved with this configuration), (ii) to reduce the undesired persistent photoconductivity effect caused by deep donors in homogeneously doped n-Al$_x$Ga$_{1-x}$As[18], and (iii) to improve the transconductance of HEMTs by minimizing the thickness between the 2DEG and the crystal surface with the gate electrode. We discuss the new concept by means of the real-space energy band diagrams depicted in Fig. 14. When we move the doping layer away from the heterojunction at z = 0 into the high-gap material, the overlap between the 2DEG and the sheet of ionized donors is reduced. A spatial separation W$_S$ of only 60 Å is sufficient to produce a strong mobility enhancement of the 2DEG at the heterointerface[35]. In contrast to conventional selectively doped GaAs/Al$_x$Ga$_{1-x}$As heterostructures, the carrier freeze-out in these new structures is very small. It is important to note that, although the doping layer and the heterojunction are located at different z, they still form a coupled pair of quantum mechanical systems (note, e.g. that the real space extent of the lowest subband in the doping potential well is larger than 50 Å). As all the donor atoms in these heterostructures are confined in the (001) plane, the band bending is linear, as indicated in Fig. 14b, and we can thus expect that HEMTs fabricated from this structure exhibit a very constant transconductance and gate-to-source capacitance over a wide bias range. In addition, we easily achieve charge densities of more than 1.5 x 10^{12} cm^{-2} in the 2D channel at a single heterointerface having a strongly enhanced mobility even at room temperature[33]. The product of mobility and density has thus reached a value never reported before for <u>single</u> heterojunctions made of the GaAs/Al$_x$Ga$_{1-x}$As material system. Combined with the narrow distance between the gate electrode and the high-mobility 2DEG this feature will certainly improve the current driving capability of single heterojunction HEMTs.

A considerable improvement of the transport properties of these new selectively monolayer-doped heterostructures is achieved, when the doping plane in the $Al_xGa_{1-x}As$ is surrounded on each side by 10 Å GaAs so that it is actually embedded in the center of a 20-Å wide GaAs quantum well (see Fig. 14c for illustration). Using this modified structure, we were able to obtain mobilities of more than 4×10^5 cm^2/Vs with a spacer as narrow as 60 Å[33]. This strong mobility enhancement of the 2DEG even for a narrow spacer arises from the different nature of the impurity screening as compared to conventional SD heterostructures. In addition, the undesired persistent photoconductivity (ppc) due to deep donors in the ternary n-type $Al_xGa_{1-x}As$ with $x > 0.2$[18] is significantly reduced, because the $Al_xGa_{1-x}As$ barriers in the MD heterostructures remain essentially undoped. The quantum-mechanical coupling between the two potential wells from the monolayer-doping and from the band bending at the heterointerface might give rise to some remaining electron density in the MD channel and thus to a conduction in two parallel channels which is deleterious for some applications. This phenomenon can be eliminated when the narrow MD GaAs quantum well is designed to be asymmetric[42]. In such asymmetric GaAs quantum wells no bound states are formed if the well widths L_z are smaller than a critical value (for $Al_{0.4}Ga_{0.6}As$ and $Al_{0.1}Ga_{0.9}As$ barriers the critical value is $L_z = 25$ Å[42]. When such a narrow asymmetric well is doped, the electrons from the respective donors are not confined in the well. They are instead transferred and form the 2DEG at the adjacent heterojunction. We have obtained very high electron mobilities for monolayer-doped asymmetric quantum well heterostructures with AlAs spacer, where the electrons have to tunnel through a 100-Å barrier. The measured peak mobility of more than 10^6 cm^2/Vs is the highest ever reported for a selectively doped GaAs/$Al_xGa_{1-x}As$ heterostructure with a spacer of only 100 Å. In addition, it is important to note that the mobility enhancement by a factor of ten observed in this new heterostructure after illumination is much more pronounced than in conventional n-GaAs/$Al_xGa_{1-x}As$ heterostructures[42].

In the monolayer-doped GaAs/$Al_xGa_{1-x}As$ heterostructure depicted in Fig. 14b and 15a the doping density of the (001) plane at distance W_s from the heterointerface has to be rather high. After being taken away by the states which pin the Fermi level at the surface, enough free carriers should remain to be transferred to the 2DEG. As the monolayer-doping density at the heterointerface has to be increased, the impurity scattering also increases and thus the mobility of the 2DEG deteriorates. An improvement over this structure is to place an additional doped monolayer closer to the

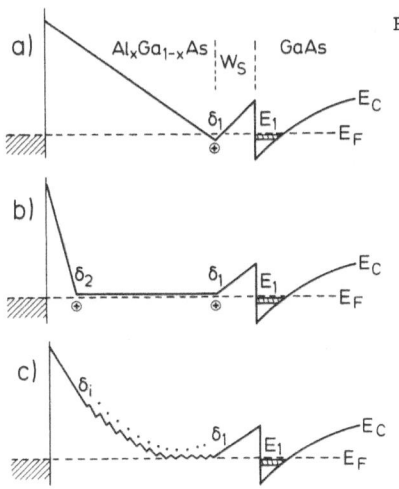

Fig. 15. Energy band diagram of (a) selectively doped heterostructure in which the $Al_xGa_{1-x}As$ is MD (marked δ_1) at a distance W_s from the interface. (b) Selectively doped structure in which the $Al_xGa_{1-x}As$ contains two MD layers. The MD layer marked δ_1 is at a distance W_s from the heterointerface and provides the carriers to the 2D electron gas. The MD layer marked δ_2 near the surface is to provide electrons to the surface states which pin the Fermi level at the surface. (c) Selectively doped structure in which the $Al_xGa_{1-x}As$ contains several MD layers. This structure gives the highest mobility values reported so far.

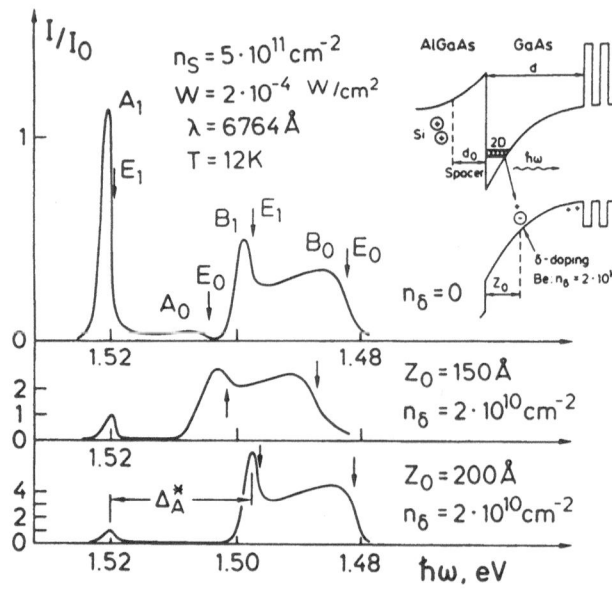

Fig. 16. Spectra of radiative recombination of 2D electrons with photoexcited holes measured in Be-monolayer-doped n-type single heterojunctions whose real-space energy band diagram is shown in the inset.

surface (marked δ_2 in Fig. 15b) to supply electrons to the surface states. The doped monolayer (marked δ_1), which contributes electrons to the 2DEG, can now be lightly doped ($\lesssim 5 \times 10^{11}$ cm^{-2}) and hence very high carrier mobilities can be obtained[43]. Further improvements with respect to the mobility can be made by placing a periodic sequence of doped monolayers in the $Al_x Ga_{1-x} As$ region as shown in Fig. 15c. For this heterostructure values of the electron mobility at low temperatures in excess of 1×10^7 cm^2/Vs have recently been reported[44].

We proceed to introduce another type of selectively monolayer-doped GaAs/$Al_x Ga_{1-x} As$ heterostructures in which the donors (or acceptors) are confined to a (001) lattice plane in the lower-gap material (i.e. GaAs) close to the heterointerface (see inset of Fig. 16). In this structure we are able to accurately position a plane of repulsive negatively charged Be or of attractive positively charged Si impurities for a controlled interaction between the high-mobility 2DEG and these charged impurities. The influence of repulsive and attractive scattering centers on the magneto-transport properties of the 2DEG is important for a quantitative interpretation of precision measurements of the quantum Hall effect[45]. The additional monolayer-doping of the GaAs near the 2DEG with donor or acceptor impurities shifts the position of the quantum Hall plateaus and of the corresponding ρ_{xx} minima to opposite directions. These shifts are interpreted as a consequence of strong non-Bornian scattering of electrons by individual impurities. The n-type single GaAs/$Al_x Ga_{1-x} As$ heterojunction depicted in Fig. 16, having a monolayer of acceptors positioned at a well defined distance z_o from the interface in the GaAs buffer layer, offers the unique possibility to study the radiative recombination of 2D electrons with photo-generated holes bound to these acceptors[46]. The luminescence spectra of Fig. 16 obtained from three samples of this type show three lines A_1, B_1, and B_o which originate from the recombination of 2D electrons with free holes (A-lines) and holes bound to the acceptors (B-lines). The assignment was derived from the observed line shifts in a perpendicular magnetic field, and the index corresponds to the number of the 2D subband. It is important to note that the broadening of the luminescence lines arising from the holes is very small and we are therefore able to directly investigate the density of states of the 2D electrons[47].

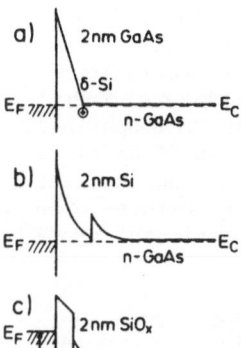

a) 2nm GaAs
δ-Si
E_F ///// n-GaAs E_C

b) 2nm Si
E_F ///// n-GaAs E_C

c) E_F ///// 2nm SiO$_x$
V<0
n-GaAs E_C E_F E_V

d) 2nm SiO$_x$
p-GaAs E_C E_F E_V
V>0

Fig. 17. Schematic conduction band diagrams of
different configurations of Si-monolayer
incorporation into GaAs, (a) for nonalloyed ohmic contact, (b) for metal-
semiconductor structures with adjustable
barrier height, (c) and (d) for metal-
insulator-semiconductor structures with
a low density of states at the insulator-
semiconductor interface.

Finally, we demonstrate how monolayer doping can be utilized to alter
the surface properties of GaAs. First we discuss the formation of low-resistance ohmic contacts to GaAs which is an important step in the device
fabrication process. The conduction band diagram of n-type GaAs with a Si-
doped monolayer 20 Å beneath the surface is shown in Fig. 17a. In order
that a metal contact on this surface becomes ohmic it is essential that
the height of the tunneling barrier and its width be reduced. We have
achieved this by placing the Si doped monolayer of $N_D^{2D} \geq 1 \times 10^{13}$ cm^{-2} plus
an additional set of 5 equally spaced Si doping sheets beneath very close
to the surface. The evaporation of Cr/Au dots on such a structure results
in nonalloyed ohmic contacts having a specific contact resistance of 10^{-6}
Ohm/cm^2 [33]. The effective Schottky barrier height of metal/GaAs heterostructures can be varied over a wide range by inserting several monolayers of
either n- or p-doped Si, as shown in Fig. 17b. For reproducibility the
fabrication of this metal/Si/GaAs heterostructure must be performed under
ultrahigh vacuum conditions. The Si effusion cell has to be operated at
very high temperatures, in order to obtain a sufficiently high flux of Si
atoms for the formation of the epitaxial Si fibres. The specially designed
Si cell[48] uses an elemental pure Si rod through which a high current is
passed for heating. The final heterostructure to be discussed here is a
metal-oxide-semiconductor (MOS) structure on GaAs with a low density of
states at the insulator-semiconductor interface (Fig. 17c and d). After
the epitaxial growth of a few monolayers of Si on the (001) GaAs surface
the Si is oxidized in-situ by an activated oxygen source. Then, an additional layer of SiO$_2$ and the metal is deposited. MOSFETs fabricated from
such oxides on GaAs show excellent characteristics[49]. The density of interface states in the structures can be reduced to 10^{12} cm^{-2} eV^{-1}.

5. CONCLUDING REMARKS

Two protype artificially layered semiconductor structures have been
presented in which the concept of microscopical structuring of solids is
scaled to its ultimate physical limit normal to the crystal surface. In
both the (GaAs)$_1$/(AlAs)$_1$ monolayer superlattice and in the monolayer (or
delta)-doped GaAs/Al$_x$Ga$_{1-x}$As structures, which have been fabricated by
molecular beam epitaxy (MBE), the characteristic material lengths have
reached a spatial extent normal to the surface of less than the lattice
constant. The (GaAs)$_m$/(GaAs)$_n$ ultrathin-layer superlattices exhibit novel
optical properties due to the indirect-gap nature of the constituent AlAs

layers. The minority-carrier lifetimes can be tailored over four orders
of magnitude by appropriately designing the superlattice configuration.
In superlattice configurations designed to have small energy separations
between the Γ and X-minima of the constituent GaAs and AlAs layers, resp.,
the artificially layered material can be switched externally from spatially
direct (type I) to spatially indirect (type II) by application of an electric field, by application of hydrostatic pressure, or by varying the excitation power density or the sample temperature. These unique features open
up new fields of application in lasers and in nonlinear photonic and optoelectronic devices.

The narrow buried doping channel in monolayer-doped GaAs layers and
in $GaAs/Al_xGa_{1-x}As$ structures leads to a significant modification and improvement of the electrical properties. Based on this concept, nonalloyed
ohmic contacts, field-effect transistors with very high transconductance,
unpinned GaAs surfaces for MOS devices, and electron mobilities as high as
10^7 cm^2/Vs have been achieved. In addition, monolayer-doping is now used
to investigate in detail the metal-insulator transition in GaAs and to modify intentionally the band offset of a given heterojunction by a dipole
layer. Finally, the exact positioning of an additional dopant monolayer in
the lower-gap material at a certain distance from the heterojunction in
selectively doped heterojunctions allows us to directly study the density
of states in 2D carrier systems by means of magnetotransport, cyclotron
resonance and magnetoluminescence measurements.

ACKNOWLEDGEMENTS

The work presented here is the outcome of our group's recent efforts.
I gratefully acknowledge the active contribution of the many of my colleagues
cited in the references. The work was sponsored by the Bundesministerium
für Forschung und Technologie of the Federal Republic of Germany.

REFERENCES

1. A.Y. Cho and J.R. Arthur, Progr. Solid State Chem. 10, 157 (1975)

2. K. Ploog, Angew. Chem. Int. Ed. Engl. 27, 593 (1988)

3. H. Sakaki, Proc. Int. Symp. Foundations of Quantum Mechanics,
   Tokyo 1983, p.94

4. F. Capasso, Science 235, 172 (1987)

5. A.C. Gossard, Treat. Mater. Sci. Technol. 24, 13 (1981)

6. B.A. Joyce, Rep. Progr. Phys. 48, 1637 (1985)

7. L.L. Chang and K. Ploog (Eds.), Molecular Beam Epitaxy and Heterostructures (Martinus Nijhoff, Dordrecht, 1985) NATO Adv. Sci. Inst.
   Ser. E 87 (1985)

8. E.H.C. Parker (Ed.), The Technology and Physics of Molecular Beam Epitaxy (Plenum Press, New York, 1985)

9. T. Sakamoto, H. Funabashi, K. Ohta, T. Nakagawa, N.J. Kawai, T. Kojima,
   and K. Bando, Superlatt. Microstruct. 1, 347 (1985)

10. B.A. Joyce, P.J. Dobson, J.H. Neave, K. Woodbridge, J. Zhang, P.K.
    Larsen, and B. Bölger, Surf. Sci. 168, 423 (1986)

11. B.A. Joyce, J. Zhang, J.H. Neave, and P.J. Dobson, Appl. Phys. A45,
    255 (1988)

12. F. Briones, D. Golmayo, L. Gonzales, and A. Ruiz, J. Cryst. Growth 81,
    19 (1987)

13. M. Kohl, D. Heitmann, S. Tarucha, K. Leo, and K. Ploog, Phys. Rev. B 39, 7736 (1989)

14. S. Tarucha and K. Ploog, Phys. Rev. B 39, 5353 (1989)

15. Y. Horikoshi and M. Kawashima, J. Cryst. Growth 95, 17 (1989)

16. F. Briones, L. Gonzales, and A. Ruiz, Appl. Phys. A 49,   (1989)

17. Y. Suzuki and H. Okamoto, J. Appl. Phys. 58, 3456 (1985)

18. E.F. Schubert and K. Ploog, Phys. Rev. B 30, 7021 (1984)

19. R. Dingle, R.A. Logan, and J.R. Arthur, Inst. Phys. Conf. Ser. 33a, 210 (1977)

20. M.D. Camras, N. Holonyak, K. Hess, J.J. Coleman, R.D. Burnham, and D.R. Scifres, Appl. Phys. Lett. 41, 317 (1982)

21. M. Nakayama, K. Kubota, H. Kato, S. Chika, and N. Sano, Solid State Commun. 53, 493 (1985)

22. T. Isu, D.S. Jiang, and K. Ploog, Appl. Phys. A 43, 75 (1987)

23. L. Esaki and R. Tsu, IBM J. Res. Develop. 14, 61 (1970)

24. G. Bastard, Wavemechanics Applied to Semiconductor Heterostructures (Les Editions de Physique, Les Ulis, France, 1988)

25. J. Nagle, M. Garriga, W. Stolz, T. Isu, and K. Ploog, J. Physique 48, Colloque C5, C5-495 (1987)

26. R. Cingolani, M. Ferrara, L. Baldassarre, M. Lugara, and K. Ploog, Phys. Rev. B 40,   (1989)

27. G.H. Li, D.S. Jiang, H.X. Han, Z.P. Wang, and K. Ploog, Phys. Rev. B 40,   (1989)

28. L. Tapfer and K. Ploog, Phys. Rev. B 33, 5565 (1986)

29. G. Peter, E.O. Göbel, W.W. Rühle, J. Nagle, and K. Ploog, Superlatt. Microstruct. 5, 197 (1989)

30. J.B. Xia, Phys. Rev. B 38, 8358 (1988)

31. M.H. Meynadier, R.E. Nahory, J.M. Worlock, M.C. Tamargo, J.L. de Miguel, and M.D. Sturge, Phys. Rev. Lett. 60, 1338 (1988)

32. J. Feldmann, R. Sattmann, E.O. Göbel, J. Kuhl, J. Hebling, K. Ploog, R. Muralidharan, P. Dawson, and C.T. Foxon, Phys. Rev. Lett. 61, 1892 (1989)

33. K. Ploog, M. Hauser, and A. Fischer, Appl. Phys. A 45, 233 (1988)

34. A. Zrenner, H. Reisinger, F. Koch, and K. Ploog, Proc. 17th Int. Conf. Phys. Semicond., Eds. J.D. Chadi and W.A. Harrison (Springer, New York, 1985) p. 325

35. E.F. Schubert and K. Ploog, Jpn. J. Appl. Phys. 25, 966 (1985); E.F. Schubert, J.B. Stark, B. Ullrich, and J.E. Cunningham, Appl. Phys. Lett. 52, 1508 (1988)

36. A. Zrenner, F. Koch, R.L. Williams, R.A. Stradling, K. Ploog, and G. Weimann, Semicond. Sci. Technol. 3, 1203 (1988)

37. A. Zrenner, Appl. Phys. Lett. 54,   (1989)

38. R.B. Beall, J.B. Clegg, and J.J. Harris, Semicond. Sci. Technol. 3, 612 (1988)

39. M. Ramsteiner, J. Wagner, H. Ennen, and M. Maier, Phys. Rev. B 38, 10669 (1988); J. Wagner, M. Ramsteiner, W. Stolz, M. Hauser, and K. Ploog, Appl. Phys. Lett. 55,   (1989)

40. E.F. Schubert, A. Fischer, and K. Ploog, IEEE Trans. Electron Devices ED-33, 625 (1986)

41. A. Ishibashi, K. Funato, and Y. Mori, Electron. Lett. 24, 1034 (1988)

42. Y. Horikoshi, A. Fischer, E.F. Schubert, and K. Ploog, Jpn. J. Appl. Phys. 26, 263 (1987)

43. B. Etienne and E. Paris, J. Physique 48, 2049 (1987)

44. J.P. Eisenstein, H.L. Störmer, L Pfeiffer, and K.W. West, Phys. Rev. Lett. 62, 1540 (1989); L. Pfeiffer, unpublished results (1988)

45. R.J. Haug, R.R. Gerhardts, K. von Klitzing, and K. Ploog, Phys. Rev. Lett. 59, 1349 (1987)

46. I.V. Kukushkin, K. von Klitzing, K. Ploog, and V.B. Timofeev, Phys. Rev. B 40,   (1989)

47. I.V. Kukushkin, V.B. Timofeev, K. von Klitzing, and K. Ploog, Festkörperprobleme (Adv. Solid State Phys.) 28, Ed. U. Rössler (Vieweg, Braunschweig, 1988) p. 21

48. K. Eberl, M. Bichler, and G. Abstreiter, paper presented during EURO MBE 89, 5 - 8 March 1989, Grainau (FRG).

49. S. Tiwari, S.L. Wright, and J. Batey, IEEE Electron Device Lett. EDL-9, 488 (1988)

MOCVD-GROWN ATOMIC LAYER SUPERLATTICES

Akira Ishibashi

SONY Corporation Research Center
174 Fujitsuka-cho, Hodogaya-ku
Yokohama 240, Japan

ABSTRACT

We have grown atomic layer superlattices, i.e., ultrathin-layer superlattices and double delta-doped structures, based on metalorganic chemical vapor deposition. The $(AlAs)_m (GaAs)_n$ ultrathin layer superlattices have been characterized by photoluminescence and Raman scattering experiment. The ultrathin-layer superlattice is revealed to be a system of quasi-three dimensional electrons and quasi-two dimensional LO phonons. The lowest conduction band in the superlattice is indicated to be a zone-folding-induced mixed-state of X and Γ bands. An idea of the isotope superlattices is proposed. We have used the double delta-doped structure to fabricate nano-structure devices with the aid of electron-beam-induced resist process, demonstrating a potential interest of universal field-effect-transistor.

INTRODUCTION

The concept of the semiconductor superlattice[1] and a great deal of investigations stimulated by this concept have urged the epitaxy technique to have good resolution in the growth direction. Also in the lateral directions, the dimensions continue to be reduced with accelerating levels of integration in ULSI devices. The dimensions in all three directions will soon become less than the de Broglie wavelength of electrons; we are proceeding to a frontier of three-dimensional superstructures. Since solid state materials are described basically by only three items of elements, the photon, the phonon, and the electron, the states of those three are of potential interest as the object of dimensional modulation in the superlattices. The wavelength of photons, except X-rays, is so large that we do not go into the details of the confinement of photons here. Instead, we shall extend the novel idea of Esaki and Tsu to the concept that we control the dimensionality of the LO phonon as well as that of the electron by taking the superstructures.

In $(AlAs)_m (GaAs)_n$ superlattices, the confinement takes place both for the electron[1] and for the LO phonon[2-4] due, respectively, to the difference in the electron affinity and to that in the reduced mass of the anion and the cation. So picture of well and barrier is adequate for electrons and for LO phonons. As a function of well and barrier widths, we show the characteristics of the electron and the LO phonon in the superlattice, in Fig. 1 (a) and 1 (b), respectively. Here L_{1e} (L_{1LO}) is the first characteristic length where a cross-over from three dimensions to quasi-two dimensions occurs for the

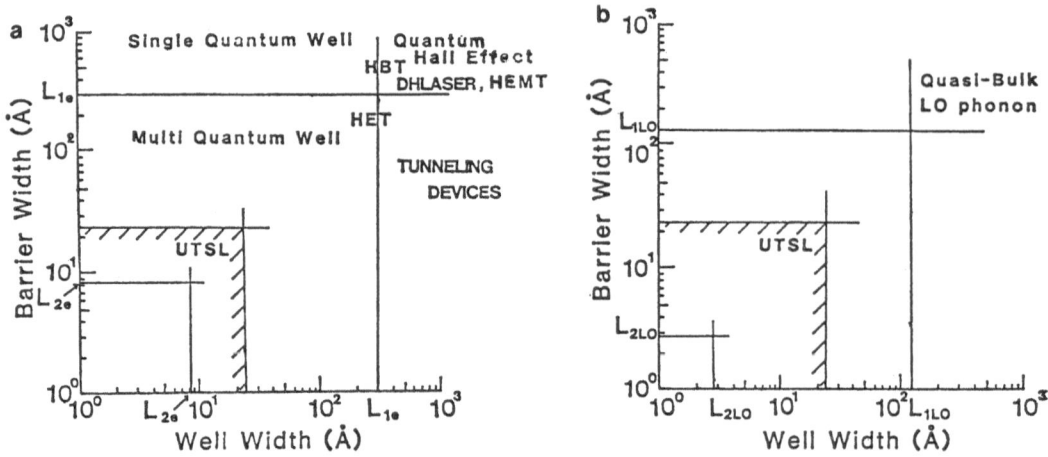

Fig. 1. (a) Electronic characteristics and (b) LO phonon characteristics, of superlattices seen as function of well and barrier widths.

electron (LO phonon). Likewise L_{2e} (L_{2LO}) is the second characteristic length where a counter-crossover from quasi-two dimensions to quasi-three dimensions takes place for the electron (LO phonon). It is estimated that $L_{1e} \sim$ de Broglie wavelength $\sim 300A$ and that $L_{1LO} \sim (\hbar/2m^*w_{LO})^{1/2} \sim 100A$, m^* being the electron effective mass and w_{LO} the LO phonon frequency. But L_{2e} and L_{2LO} have not been experimentally determined yet, to our knowledge. A mismatch in the characteristic lengths will make the electron and the LO phonon behave quite differently, when the slab-thickness is changed from infinity down to inter-atomic spacing. So it is important to determine L_{2e} and L_{2LO}. Further, the region of L_{2e} and L_{2LO} is exactly the place where the possibility of a new synthetic material has been suggested.[5-7] The Brillouin-zone-folding caused by the short superperiodicity will induce mixing of electronic bands with different characters and also generates new branches in the phonon disper-sion relation. Those facts motivate us to study the atomic layer superlat-tices, i.e., ultrathin-layer superlattice (UTSL)[8-14] and the double delta-doped structure (DDDS).[15] For the growth of atomic layer superlattices, we first establish the epitaxy-technique based on metalorganic chemical vapor deposition (MOCVD). Then we will make an application of atomic layer superlat-tice, reducing the lateral length with electron beam induced resist (EBIR).[16] The EBIR is developed based on the materials-deposition technique that util-izes the electron beam.[17-19] The DDDS is used for fabrication of ultra-short gated FETs.

MOCVD FOR ATOMIC LAYER SUPERLATTICES

We aim at controlling the slab thickness and heterointerfacial abrupt-ness to one monolayer based on MOCVD.[20] The MOCVD uses organometallic com-pounds as source materials for growth of epitaxial layers. The net reaction of the MOCVD growth of III-V compound semiconductor is given by,

$$(C_nH_{2n+1})_3M_{III} + M_VH_3 \longrightarrow M_{III}M_V + 3C_nH_{2n+2},$$

where $(C_nH_{2n+1})_3M_{III}$ denotes an organometallic compound, (C_nH_{2n+1}) an alkyl, and M_{III} the group III metal, M_VH_3 a hydride of group V element, and $M_{III}M_V$ the III-V compound semiconductor. The MOCVD growth is more chemical and less physical than MBE growth. The MOCVD consists of many processes; diffusion in the boundary layer, cracking in the ambient H_2 gas, adsorption on the

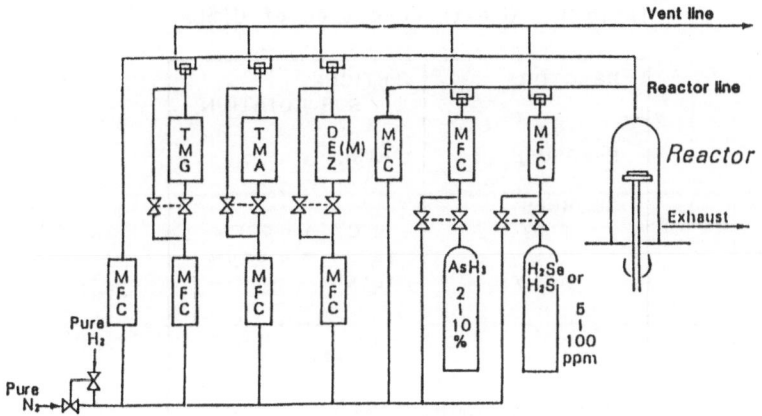

Fig. 2. Schematic diagram of our MOCVD system.

substrate, migration on the surface, and final incorporation in the substrate. Those complicated processes are, however, no fundamental demerits of MOCVD. They are advantages with which we can put ample features in the crystal growth. An example is the atomic layer epitaxy,[21] which utilizes the self-limiting mechanism of the organometallic compounds at the growth interface. For growth of the $(AlAs)_m(GaAs)_n$ UTSLs, trimethyl-aluminum (TMA) and trimethyl-gallium (TMG) are used as the sources of the group III elements. As the sources of the group V elements, hydrides are usually used. For As, we use arsine (AsH_3) as the source. We use Si_2H_6 to make the Si delta-doped layers. Figure 2 shows the schematic diagram of our MOCVD system. The organometallic compound is kept at preset temperature in a thermostatic bath, so that its vapor pressure is adequate for the epitaxial growth. Purified hydrogen gas flows bubbling through the metalorganic liquid, takes up certain concentration of the group III elements, and finally goes into the reactor as shown in Fig. 2. The amount of the carrier hydrogen gas is controlled by a mass-flow-controller (MFC). Since the hydride of the group V element is diluted in hydrogen gas, they are fed directly into the reactor, being controlled by the MFC. Those source gases are quickly turned on/off with the electrically controlled pneumatic valves placed in the pipelines. In the reactor the source gases react with each other, resulting in epitaxial layers on the GaAs substrate located on the graphite susceptor. The substrate is heated up to 600 °C ~ 800 °C by the Joule heat due to the current inducted by a radio-frequency-coil surrounding the reactor. Since the growth rate of the epitaxial layer is limited by the feed rate of the group III materials, the growth rate can be precisely set at the designed value by controlling the MFCs.

For a successful growth of UTSLs, the transition layer between the GaAs and AlAs must be made thin enough. In Table 1 the growth condition we have used to grow UTSLs is listed. First, we have used as large a gas-flow-rate as possible. The source gases of GaAs and AlAs are changed to each other in about 0.1 sec. Second, we have used a slow growth rate. It takes 2.5 sec for one monolayer thick GaAs or AlAs to grow on the substrate. Note that the growth time spent in transition across the hetero-interface is only 1/25 of the growth time of the one monolayer of GaAs or AlAs. Third, we have made the dead space in the pipeline so small that there is no delay in switching from AlAs growth to GaAs growth or vice versa.[22] Forth, the on-off control of the growth is made not by turning on/off the reactor-line but by switching the gas-flow from vent-line to reactor-line or vice versa. Fifth, we make the pressure in the vent line and in the reactor line well-balanced. The forth and the fifth recipes have been introduced to prevent the transient change in the feed of the source materials.[22] These recipes are taken to make the growth time of the interface much shorter than the time needed for monolayer growth

Table 1.   Growth condition of UTSLs.

REACTOR	VERTICAL 　　　　SUB. ROTATION
PRESSURE	1 ATM.
SOURCES 　　TMG 　　TMA 　　AsH_3 / H_2	 -10 ℃　1.4 SCCM 18 ℃　3.6 SCCM 10 %
TOTAL H_2	12 SLM
GROWTH TEMP.	750 ℃
V / III RATIO	280
GROWTH RATE	1.15 Å/sec

and also to make the growth rate quite reproducible. Under these regulations, the thickness of each slab is precisely controlled by the growth time. The growth of the superlattices have been controlled by a computer on which the growth sequence is programmed beforehand. The $(AlAs)_m(GaAs)_n$ (m,n=1~24) super-lattices have been grown on the (001) GaAs substrates. The AlAs and GaAs slabs are successively grown without intervals between the growth of two adjacent slabs. The layer thickness is uniform over the two-inch wafer except the edge. The superlattice samples consist of several tens to hundreds of periods of $(AlAs)_m(GaAs)_n$ alternating layers with total thickness of about 4200A (1400 monolayers).

EXAMINATION OF $(AlAs)_m(GaAs)_n$ UTSL STRUCTURE

Measurement of Photoluminescence from Quantum Wells[8]

Since the capability of making an abrupt heterointerface is a necessary condition to grow the UTSLs, needed is the evaluation of the abruptness of the

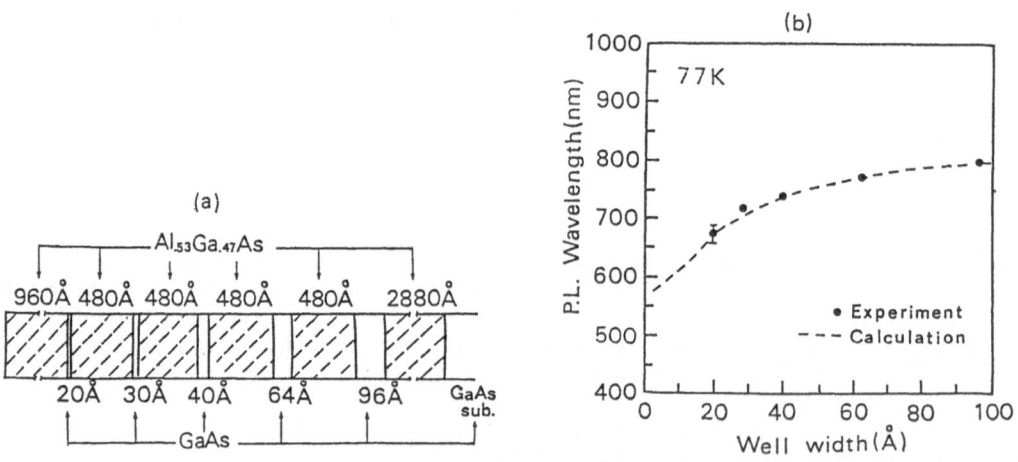

Fig. 3. (a) Structure of quantum wells and (b) wavelength of photoluminescence as a function the well width.

GaAs/AlGaAs heterointerface.[23-25] We have grown a sample having four quantum wells with different widths as shown in Fig. 3(a). Here each well is separated by AlGaAs barrier layer which is thick enough to isolate the GaAs wells. The growth condition is identical to the condition used for the growth of the UTSLs. Then we have measured the photoluminescence from those quantum wells at 77K. In Fig. 3(b) we show the observed photoluminescence wavelength as a function of the well width. The narrower the well width, the shorter the wavelength because of the increase in the ground state energy of excitons. The broken line shows the result of calculation.[10] The observed photoluminescence wave length is in good agreement with the calculation. The error bar of the datum for the narrowest well denotes the deviation that would be seen if the well width were different from the designed value by one monolayer. The observed wave length is well within the error bar. Those results indicate that the heterointerfacial abruptness is of one monolayer along the growth direction. Now it is the heterointerfacial structure in x and y directions, i.e., the lateral extent of the island that is left to be evaluated. The lateral extent of the island in our MOCVD has already been investigated[26] to be a few tens to hundreds of angstroms depending on the growth temperature.

Transmission Electron Microscopy[26]

Our MOCVD system can grow two atomically abrupt heterointerfaces at

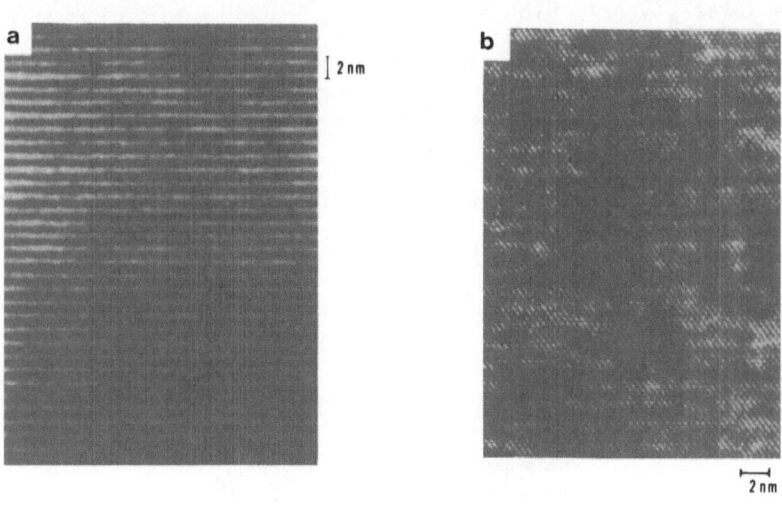

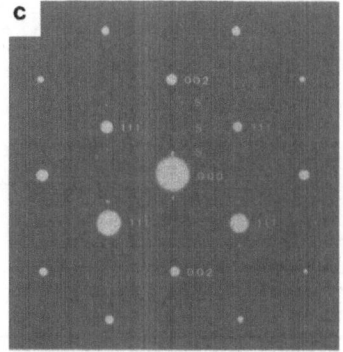

Fig. 4. (a) Dark field TEM image, (b) lattice image, and (c) electron diffraction pattern, of $(AlAs)_2(GaAs)_2$ UTSL.

least with the spacing of 20 A. The next question is whether this spacing can be reduced to the inter-atomic spacing. This is identical to the question whether our MOCVD system has the capability of controlling the absolute slab-thickness to one monolayer. We have tried to grow the $(AlAs)_2(GaAs)_2$ UTSL, for the lack of that capability will easily leads the UTSL to an $Al_{0.5}Ga_{0.5}As$ alloy due to the erosion of the heterointerfaces occurring at both sides of the slabs. We have analyzed the UTSL structure with transmission electron microscopy (TEM). Figure 4(a) shows a dark field TEM image. We can see clear stripes of AlAs and GaAs, which is a clear demonstration of the successful growth of the UTSL structure. Next question is whether the slab thickness is two monolayers as designed, or not. Using the diffracted spots we have reconstructed the lattice image. Figure 4(b) shows the lattice image of the $(AlAs)_2(GaAs)_2$ UTSL. Two lattice points are observed in each stripe. No dislocation runs through the superlattice. Merging of two AlAs or GaAs layers are not observed either. Figure 4(c) shows the transmission electron diffraction-pattern of the $(AlAs)_2(GaAs)_2$ UTSL. The pattern shows satellite spots on each side of the main diffraction spots, (000), (002), and (111) family, along the [001] growth direction. One of those satellite spots is denoted by "s". Note that the spot "s" quarters the segment between (000) and (002) which corresponds to the one monolayer thickness in real space. Thus the spot "s" manifests the superperiodicity four times as large as one monolayer thickness, i.e., the superperiodicity of the $(AlAs)_2(GaAs)_2$ UTSL. Figures 4(a), 4(b), and 4(c) verify that our MOCVD system can control both the heterointerfacial abruptness and the absolute slab-thickness to one monolayer.

LO PHONONS IN $(AlAs)_m(GaAs)_n$ UTSL'S

The effective charge that scatters the electron is given by the divergence of the polarization caused by the phonon,[27] and the polarization is proportional to a relative displacement of two adjacent atoms. Thus the main scatterer of the electron is the LO phonon, and we investigate the LO phonon in the $(AlAs)_m(GaAs)_n$ UTSLs. We performed Raman spectroscopy on the UTSLs using the 488nm and 514.5 nm lines of Ar^+ ion laser as the excitation source.

Raman Frequency in $(AlAs)_n(GaAs)_n$ UTSLs[8]

First, we investigate the LO phonons in the $(AlAs)_n(GaAs)_n$ superlattices in which the slab thickness of AlAs equals that of GaAs. Here n, the number

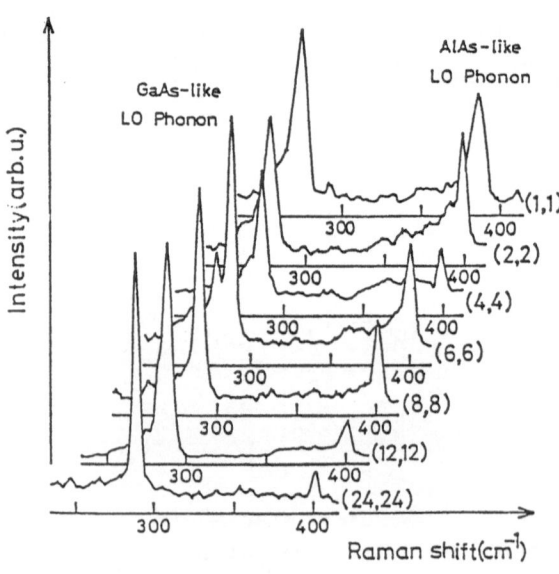

Fig. 5. Raman spectra of the $(AlAs)_n(GaAs)_n$ superlattices.

of monolayers per slab, runs from 1 through 24. We cover it from the UTSL region to the conventional superlattice region. Figure 5 plots the Raman spectra of the $(AlAs)_n(GaAs)_n$ superlattices. We can see a systematic shift in the LO phonon frequency with decreasing slab thickness. We plot the frequency in Figs. 6(a) and 6(b) respectively for GaAs-like and AlAs-like LO phonon, as a function of n. For both the GaAs-like and AlAs-like LO phonons, the frequency increases monotonously with n and then converges to the frequency of the bulk material. Using MBE-grown superlattices with n=1,2, and 4, Barker et al.[7] reported the shift in LO phonon frequency which is in congruence with our results. Although the random element isodisplacement (REI) model[28] predicts that the LO phonon frequency changes as a function of Al mole fraction in $Al_xGa_{1-x}As$ alloy, the shift in the LO phonon frequency can never be ascribed to the variation of Al content, because we have confirmed that the average Al content is 0.5 +/- 0.02 in those UTSLs by measuring Al content with electron probe microanalysis (EPMA). Note that observed GaAs-like LO phonon frequency for n=1 is 274 cm^{-1} which is different from the GaAs-like LO phonon frequency in $Al_{0.5}Ga_{0.5}As$ by 4 cm^{-1}. Also for the AlAs-like LO phonon, we observe the difference in the LO phonon frequency between the $(AlAs)_1(GaAs)_1$ UTSL and the $Al_{0.5}Ga_{0.5}As$. Those differences mean that the $(AlAs)_1(GaAs)_1$ UTSL grown by our MOCVD system is certainly different from the $Al_{0.5}Ga_{0.5}As$ alloy.

The solid line in Figs 6(a) and 6(b) is the frequency calculated from the linear chain model. In the linear chain model the dispersion relation of GaAs and AlAs is given by

$$\cos(k^G a) = [(M^G w^2 - 2K^G)(Mw^2 - 2K^G) - 2K^{G2}]/2K^{G2} \tag{1}$$

$$\cos(k^A a) = [(M^A w^2 - 2K^A)(Mw^2 - 2K^A) - 2K^{A2}]/2K^{A2}, \tag{2}$$

Here a is half of the lattice constant a_o, w the frequency, M the mass of As, k^G and k^A the confined wave vectors, M^G and M^A the mass of group III elements, and K^G and K^A the force constants, of GaAs and AlAs, respectivery. The wave vector of GaAs-like LO phonon has both imaginary and real parts in AlAs. The GaAs-like LO phonon dampens periodically in adjacent AlAs slabs. The imaginary part varies with w but the real part is constant, i.e., π/a. The GaAs-like LO phonon frequency of the $(AlAs)_m(GaAs)_n$ UTSL is given by solving

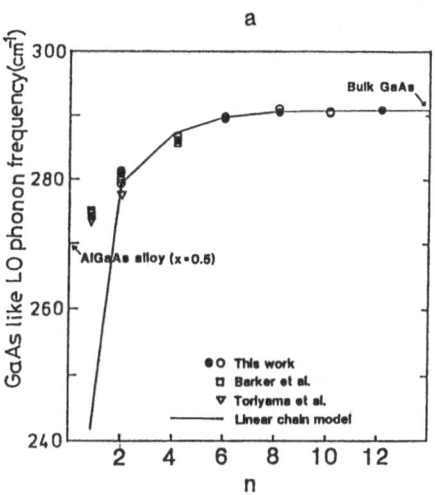 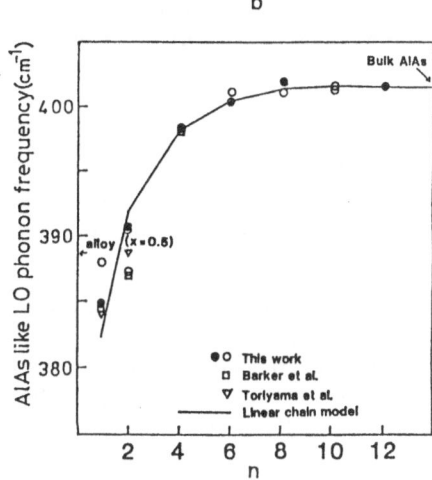

Fig. 6 Frequency (a) of GaAs-like LO phonon and (b) of AlAs-like LO phonon in $(AlAs)_n(GaAs)_n$ as a function of n.

Eqs. (1) and (2) together with the following equation:

$$\cos[q(d^G + d^A)] = \cos(k^G d^G)(-1)^{d^A/a} \cosh(\operatorname{Im}(k^A)d^A)$$
$$+ C' \sin(k^G d^G)(-1)^{d^A/a} \sinh(\operatorname{Im}(k^A)d^A), \quad (3)$$

where

$$C' = \frac{1}{2}\left| \frac{(Mw^2 - 2K^A)}{(Mw^2 - 2K^G)} \frac{\{1 + \cos(k^G a)\}\{1 - \gamma\cosh(\operatorname{Im}(k^A)a)\}}{\gamma \sin(k^G a)\sinh(\operatorname{Im}(k^A)a)} + \right.$$
$$\left. \frac{(Mw^2 - 2K^G)}{(Mw^2 - 2K^A)} \frac{\{1 - \cos(k^G a)\}\{1 + \gamma\cosh(\operatorname{Im}(k^A)a)\}}{\gamma \sin(k^G a)\sinh(\operatorname{Im}(k^A)a)} \right|.$$

with $\gamma = \cos(\pi) = -1$. Here $d^G(=na)$ and $d^A(=ma)$ are layer thickness of GaAs and AlAs, respectively, and q the superlattice wave vector, Note that the RHS of Eq. (3) changes its sign depending on whether d^A/a is odd or even. On the other hand, the wave vector of the AlAs-like LO phonon is purely imaginary in GaAs. Accordingly, the AlAs-like LO phonon dampens monotonously in adjacent GaAs slabs. For the AlAs-like LO phonon we use, instead of Eq. (3),

$$\cos[q(d^G + d^A)] = \cosh(\operatorname{Im}(k^G)d^G)\cos(k^A d^A) + C'' \sinh(\operatorname{Im}(k^G)d^G)\sin(k^A d^A), \quad (4)$$

where

$$C'' = -\frac{1}{2}\left| \frac{(Mw^2 - 2K^A)}{(Mw^2 - 2K^G)} \frac{\{1 + \cosh(\operatorname{Im}(k^G)a)\}\{1 - \cos(k^A a)\}}{\sinh(\operatorname{Im}(k^G)a)\sin(k^A a)} + \right.$$
$$\left. \frac{(Mw^2 - 2K^G)}{(Mw^2 - 2K^A)} \frac{\{1 - \cosh(\operatorname{Im}(k^G)a)\}\{1 + \cos(k^A a)\}}{\sinh(\operatorname{Im}(k^G)a)\sin(k^A a)} \right|.$$

The observed frequency is in good agreement with the calculation. This result shows that the shift in the LO phonon frequency is caused by the change in the superlattice structure. However, when we closely see Fig. 6(a), we see that the agreement is indeed very good for n≥2 but there is discrepancy for the GaAs-like LO phonon for n=1. Since the heterointerfacial abruptness of one monolayer has been proved for our MOCVD, this discrepancy suggests that the present linear chain model breaks down for n=1 especially for GaAs. Recent result by Toriyama et al.[29] is shown in Figs. 6 (a) and 6 (b). Their result explains the observed GaAs-like LO phonon frequency for n=1. They have taken into account a long-range Coulomb interaction and also the interaction up to the third nearest neighbor. The simple linear chain model, however, does not break for AlAs even for m=1. Since the second characteristic length is the dimension upon which the extrapolation based on the bulk properties breaks down, this result indicates that the L_{2LO} is about 1 monolayer for the GaAs-like LO phonon but is less than 1 monolayer for the AlAs-like LO phonon.

Raman frequency in (AlAs)$_m$(GaAs)$_n$ UTSLs[11]

In order to investigate the barrier- (or well-) thickness dependence of the LO-phonon frequency, we have changed m and n independently in the growth of the (AlAs)$_m$(GaAs)$_n$ superlattices. Raman spectra of the GaAs-like and AlAs-like LO phonons are shown in Figs. 7 (a) and 7 (b), respectively. We observe a systematic shift in frequency in accordance with the change in the UTSL structure. We plot in Figs. 8 (a) and 8 (b) respectively the frequencies of the GaAs-like LO phonon and the AlAs-like LO phonon, as a function of m+n. The LO phonon frequency scatters on the dashed line that shows n+m=constant. The LO phonon frequency is not constant at all even if the superperiodicity is identical. Thus, the LO phonon frequency is not a function of the super-periodicity, and is not determined by the straight-forward zone folding of 1/n+m. This result is in marked contrast to acoustic phonons, for which the frequency is given by the zone-folding of 1/m+n to good approximation. Then

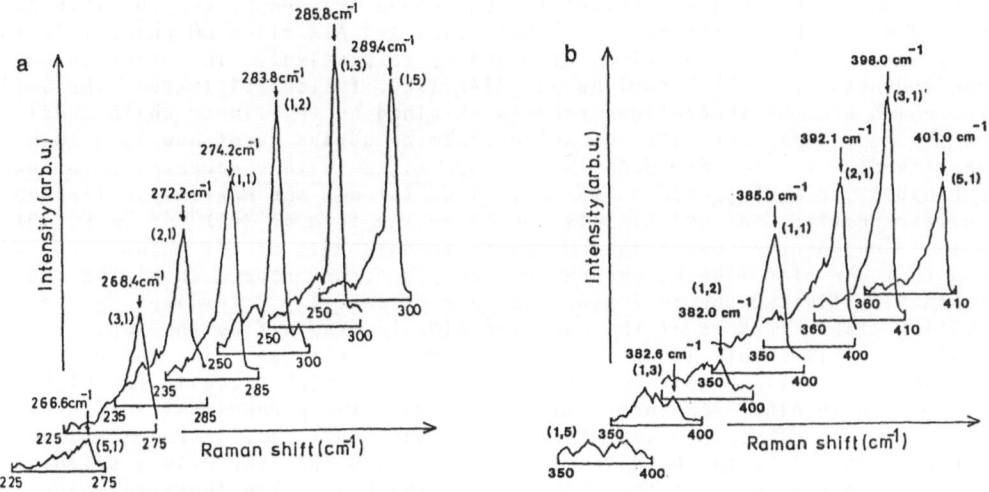

Fig. 7. Raman spectra (a) of GaAs-like LO phonon and (b) of AlAs-like LO phonon in $(AlAs)_m(GaAs)_n$ UTSL with $1 \leq m, n \leq 5$.

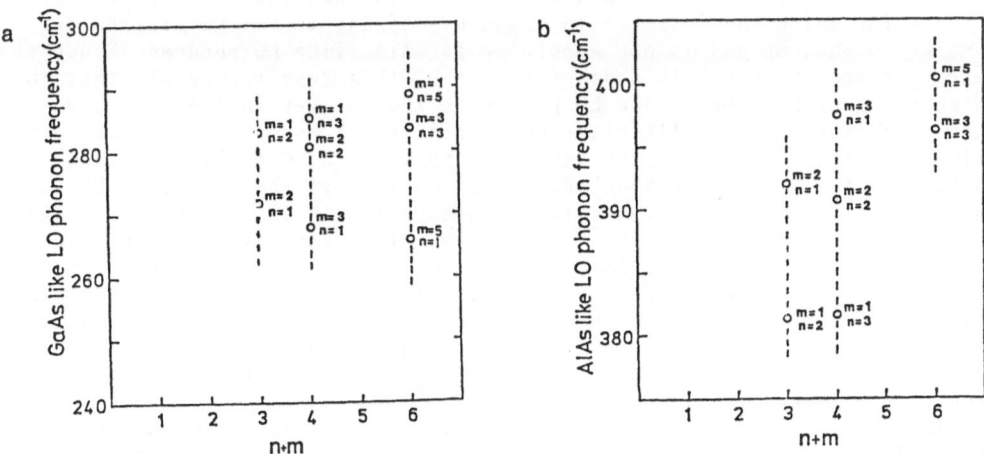

Fig. 8 Frequency (a) of GaAs-like LO phonon and (b) of AlAs-like LO phonon plotted as a function of m+n.

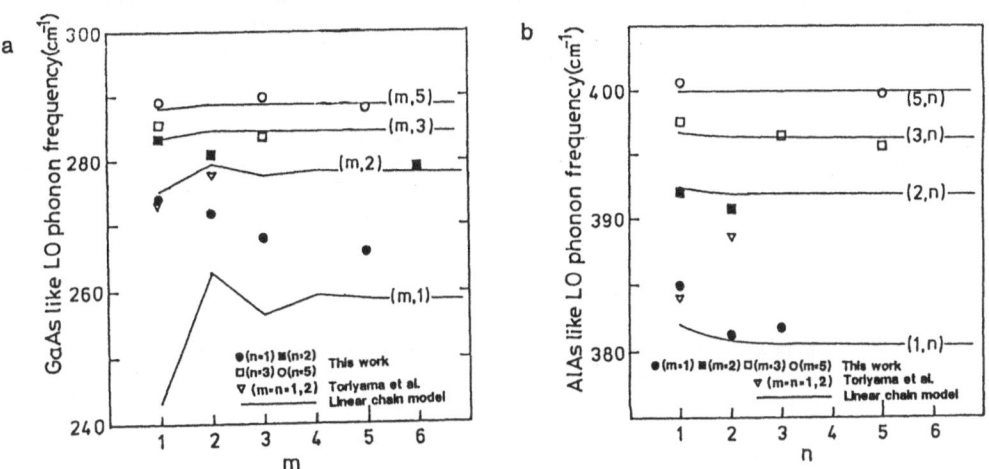

Fig. 9 Frequency (a) of GaAs-like LO phonon and (b) of AlAs-like LO phonon plotted as a function of m and n, respectively.

the question arises what determines the LO phonon frequency. We plot in Figs. 9 (a) and 9 (b) those frequencies of GaAs-like and AlAs-like LO phonons in the $(AlAs)_m (GaAs)_n$ UTSLs as functions of m and n, respectively. The observed LO phonon frequencies, both GaAs-like and AlAs-like, follow fairly well the solid lines, which are the theoretical results obtained by the linear chain model using the Eqs. (1-4). For AlAs the solid lines go upward monotonously with n decreasing down to one. For GaAs they change zigzag with m decreasing to one. This behavior is most prominent for n=1, less for n=2 and negligible for n≥3. The difference for GaAs and AlAs is caused by the term of $(-1)^{d^A/a}$ in Eq. (3). Note that the observed GaAs-like LO phonon frequency is almost independent of m, and that the AlAs-like LO phonon frequency does not depend on n. The GaAs-like and AlAs-like LO phonon frequencies are essentially determined by the respective slab-thickness of the GaAs and AlAs layers, not by the super-periodicity. This result leads to the consequence that GaAs-like and AlAs-like LO phonons are confined in respective GaAs and AlAs slabs even in the UTSLs. The LO phonon in AlAs/GaAs superlattices is quasi-two dimensional due to confinement even in UTSLs. However, the evanescence length is, in reality, finite in the adjacent slab, and both in Figs. 9 (a) and 9 (b), there is a slight increase in frequency against the thin barrier thickness. The increase shows that the confinement is a little bit relaxed, because the LO phonon frequency increases against the decrease in the wave vector in the dispersion relation. When we closely see Fig. 9 (a) and 9 (b), we note that the critical barrier thickness for which the increase in frequency begins is two monolayers for GaAs-like LO phonons and is one monolayer for AlAs-like LO phonons. Since the penetration occurs from both side of the slab, this result suggests that the evanescence length of GaAs-like LO phonon is 1 monolayer in AlAs and the evanescence length of the AlAs-like LO phonon is 0.5 monolayer in GaAs. From Eq. (1-4), on the other hand, the imaginary part of the GaAs-like LO phonon in AlAs is calculated to be about $0.3 \pi/a$, and that of the AlAs-like LO phonon in GaAs π/a.[30,31] The theoretical evanescence length is, being the reciprocal of those imaginary part, about 1 monolayer for GaAs-like LO phonon in AlAs, and about 1/3 monolayer for AlAs-like LO phonon in GaAs. These are in good agreement with the experiment. The AlAs-like LO phonon is more tightly confined than the GaAs-like LO phonon, which we can predict based on the flat dispersion relation obtained from the linear chain model. The evanescence length estimated here is in agreement with L_{2LO} estimated from the break down of the simple linear chain model.

The fact that the evanescence length is as short as, or even shorter than the inter-atomic spacing is due to the large difference in the reduced mass of the anion and the cation between AlAs and GaAs in $(AlAs)_m (GaAs)_n$ superlattices. Because of this very short evanescence length, we can obtain the structural information, i.e., the individual slab thickness of a superlattice from the frequency of the LO phonon with good accuracy, provided that the phonon dispersion relation is known. If the slab thickness is known, instead, then we can obtain the dispersion relation[4,11]. This is important, because we have now obtained a method to study the dispersion relation for such materials that are chemically too active to investigate in the air, for example, with neutron scattering. The Raman scattering of the UTSLs offers an method to get the frequency of the LO phonon with a large wave vector, in marked contrast to the Raman scattering of bulk materials giving only the dispersion relation at the zone center.

$(^{69}GaAs)_m (^{71}GaAs)_n$ Superlattices

As seen in the above section, if the reduced mass is different from each other for the constituents of the superlattice, then we have a mismatch in the frequency of the LO phonon branches. It is worthwhile calling attention to the superlattice in which the constituents are isotopes. We call this kind of superlattice an isotope superlattice. The more is the atomic number, the more is the tolerance for number of neutrons in the nucleus. We do not have a stable isotope for Al nor As, but we do have isotopes for Ga, namely, ^{69}Ga and ^{71}Ga

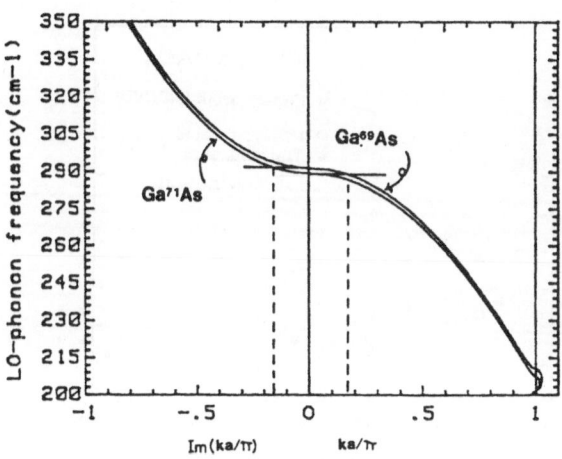

Fig. 10. LO phonon dispersion-relation for ^{69}GaAs and ^{71}GaAs.

with almost the same abundance ratio. If we make a superlattices from ^{69}GaAs and ^{71}GaAs then we have a mismatch in the LO phonon branches as shown in Fig. 10. In this calculation we can use the same force constant for ^{71}GaAs as for ^{69}GaAs, since the elastic constants, which is calculated from changes in the total energy under distortion, have nothing to do with the number of neutrons in the nucleus. Indeed the mismatch is not for the whole k-space because of the tiny difference in the mass numbers (in this sense II-VI (Se) compound semiconductors are less demanding), but the ^{69}GaAs-like LO phonon dampens strongly in ^{71}GaAs because the dw/dk is almost zero in the neighborhood of the zone center. Because of the same reason, the ^{71}GaAs-like LO phonon has in ^{69}GaAs a large wave vector which is almost 1/5 of the zone-edge wave-vector. Thus the LO phonon in the $(^{69}GaAs)_m(^{71}GaAs)_n$ superlattices is sizably modulated. It is essential that in this superlattice the electronic structure is exactly the same as the GaAs. The $(^{69}GaAs)_m(^{71}GaAs)_n$ is no superlattice for the electron, but it is for the LO phonon. This structure is important as an ideal system to investigate the effect of LO phonon modulation, because in this system we are free from the parasitic effects seen in $(AlAs)_m(GaAs)_n$ superlattices; the scattering by impurities captured in the chemically active AlAs, and the energy fluctuation in electron subbands caused by the heterointerfacial fluctuation, which always disturbs the analysis of the electron-phonon interaction.

ELECTRONIC PROPERTIES IN ULTRATHIN-LAYER SUPERLATTICES

We proceed to study the properties of another fundamental element, the electron, in the $(AlAs)_n(GaAs)_n$ UTSLs.

Analysis Based on the Raman Intensity[12]

Electronic properties of superlattices can be studied by Raman scattering, since the LO phonon couples with (virtual) excitons in the Raman scattering process.[32] Manuel et al.[33] observed that the intensity of the resonant Raman scattering is in agreement with the joint density of state of the $Al_xGa_{1-x}As/GaAs$ superlattice. Zucker et al.[34,35] used the resonant Raman scattering to reveal the delocalized exciton state in $Al_xGa_{1-x}As/$ GaAs superlattices. Here we exploit Raman scattering to infer the electronic states in the $(AlAs)_n(GaAs)_n$ UTSLs. While the Raman frequency informs us of the behavior of the atom vibrations, the Raman intensity, which is directly obtainable from the Raman spectra, reflects the cross section of the Raman process. Indeed the

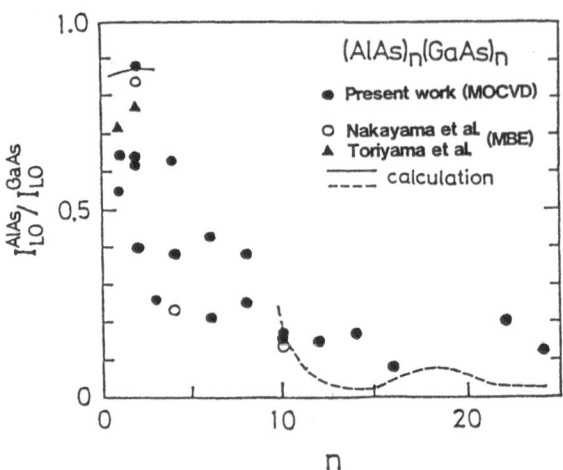

Fig. 11. Ratio of the intensity of AlAs-like LO phonon (I^A) to that of GaAs-like LO phonon (I^G) as a function of n.

absolute intensity is affected strongly by miscellaneous factors such as surface morphology, epitaxial quality, the thickness of native oxide at the surface and so on. but those factors are canceled out when we take the ratio of the LO phonon intensity in AlAs to that in GaAs. In the case of UTSLs, the slab is two orders of magnitude thinner than the penetration depth of excitation beam, and the incident photons are distributed equally in GaAs-slabs and in AlAs-slabs. Since the number of incoming photons are identical, the ratio of the numbers of outgoing photons, the ratio of Raman intensities, i.e., is exactly the ratio of the Raman scattering cross-sections in AlAs and in GaAs.

We now see Fig. 5 once again. The intensity of the AlAs-like LO phonon increases with n decreasing from 24 to 1, while that of the GaAs-like LO phonon stays almost constant. The ratio of intensity of the AlAs-like LO phonon to that of the GaAs-like LO phonon increases as n goes down from 24 to 1. Figure 11 plots the intensity ratio as a function of n. Each solid circle denotes a different UTSL sample. The scatter seen even for the same n is considered due to the fluctuation in the slab-thickness. Also the intensity ratios derived from the Raman spectra of superlattices grown by MBE are shown with blank circles[3,36] and with solid triangles[29] for comparison. No growth-technique dependence is seen. For n≥10, the intensity ratio is small, around 0.1, but is near unity for n around 2. A transition region seems to exist for n between around 2 and 8. The first order Raman scattering process is described by a Feynman diagram having three vertices, namely, an incoming photon-exciton vertex, the exciton-LO phonon vertex, and the exciton-outgoing photon vertex. The cross section of this scattering process. i.e., the Raman intensity., is given by[37,38]

$$I = s_0 \frac{m w_2 n_0^2}{2 \pi \hbar e^2} V N^{1/2} \left| \sum_{a,b} \frac{\langle f|H_{eR}|b\rangle \langle b|H_{eL}|a\rangle \langle a|H_{eR}|i\rangle}{(E_b - \hbar w_2 - i\zeta)(E_a - \hbar w_1 - i\zeta)} \right|^2 \qquad (5)$$

where s_0 is the cross-section calculated from the classical electron radius, N the number of unit cells, V the unit cell volume, and n_0 the index of refraction. H_{eL} and H_{eR} the Hamiltonians of electron-phonon and electron-photon interactions, E_a and E_b the energies of intermediate states a and b, w_1 and w_2 the frequencies of incoming and outgoing photons, and ζ the damping factor. The matrix elements $\langle f|H_{eR}|b\rangle$, $\langle b|H_{eL}|a\rangle$, and $\langle a|H_{er}|i\rangle$ correspond to the vertices and the energy denominators can be regarded as the propagators of the excitons. In general H_{eL} contains Fröehlich[39] interaction and deformation potential scattering.[40,41] The former takes place in the configuration of

$z(xx)\bar{z}$ and the latter in the configuration of $z(xy)\bar{z}$[4,42] by the selection rule. We have performed Raman spectroscopy in the configuration of $z(xy)\bar{z}$. We can rewrite Eq. (5) as follows.[34]

$$I \propto M_{DP}^2 \left| \sum_{a,b} \frac{1}{(E_b - \hbar w_2 - i\zeta)(E_a - \hbar w_1 - i\zeta)} + \frac{1}{(E_b - \hbar w_1 - i\zeta)(E_a - \hbar w_2 - i\zeta)} \right|^2, \tag{6}$$

where M_{DP} is the matrix element of the deformation potential interaction. We replace H_{eL} with M_{DP} because of the selection rule mentioned above. We take E_a as the energy gap between an electron subband and a heavy hole subband and E_b as the energy gap between an electron subband and a light hole subband with the same quantum number, i.e., the same confined wave vector. The deformation potential scattering has the matrix element,[40,41]

$$M_{DP}^2 = \hbar D_0^2 (n_p + 1)/(2Vr_{ho}w), \tag{7}$$

where n_p is the number of phonons, D_0 the optical deformation potential, r_{ho} the density, and w the angular frequency of the optical phonon. D_0 is the shift in the band-state energy per unit relative displacement of the sublattices. Poetz and Vogl[43] calculated the optical deformation potential at the point using linear combination of atomic orbital (LCAO) model. They have shown that $D_0 = 37.0$ eV and 37.7 eV for GaAs and AlAs, respectively. We now consider on the energy denominator in Eq. (6). First we consider the region $n \gtrsim 10$. We use Kronig-Penney model[44] to calculate the exciton energy in GaAs. The validity of the use of Kronig-Penney model for $n \gtrsim 10$ will be certified later in the next section. For the exciton energy in AlAs-slab we simply use the band gap energy, since the penetration depth of the electron wave function into the AlAs layer is negligible against $n \gtrsim 10$. Insertion of those energies into Eq. (6) gives the broken line in Fig. 11, the ratio of the LO phonon intensity in AlAs-slab to that in GaAs-slab. The experimental result is in the same order of magnitude as this calculation. The excitons are localized in the GaAs-slab showing the two dimensionality. Now we proceed to $n \lesssim 10$. For this region the Kronig-Penney model is not applicable. Especially for $n \sim 2$ the slab-thickness is so thin that the energy denominators cannot be considered separately for AlAs and GaAs. The energy denominator in Eq. (6) will not produce any difference in the Raman intensity for GaAs-like and AlAs-like LO phonons. Neither the deformation energy nor the refraction constant would produce a difference, since they will be defined over the UTSLs. Those assumption is equivalent to that the electronic states are quasi-three dimensional. In this case, from Eq. (6) via Eq. (7), the intensity ratio is, to a good approximation, given by

$$I_A/I_G = (n_p^A + 1)r_{ho}^G w^G/(n_p^G + 1)r_{ho}^A w^A, \tag{8}$$

where I is the LO phonon intensity and superscripts A and G denote AlAs and GaAs, respectively. Putting the above-obtained GaAs-like and AlAs-like LO phonon frequencies in Bose-Einstein statistics, we have the solid line in Fig. 11. The calculation agrees well with the observed result for $n \sim 2$. The assumption of three dimensional electronic states is proved to be true for $n \sim 2$. The excitons are also in the AlAs slab as in the GaAs, i.e., the electrons are not confined in the GaAs layer. The second characteristic length for electron is thus about 2 monolayers. This experimental result is in agreement with the theoretical estimation of Schulman et al.[45,46] and Sanches-Dehesa and Tejedor.[47] For $8 \gtrsim n \gtrsim 2$, there exists a transition region from quasi-two dimensions to quasi-three dimensions. The recovery of the electronic states to quasi-three dimensionality is in marked contrast to the case of the LO phonon which is still quasi-two dimensional even for the monolayer superlattice of the $(AlAs)_1(GaAs)_1$ as discussed in the preceding. This is the characteristic of the AlAs/GaAs system.

Analysis Based on the Photoluminescence[8]

The electronic states in heterostructures have been intensively and also extensively studied especially for AlGaAs/GaAs quantum wells and superlattices by photoluminescence[48] optical absorption[49-51] and transmission spectra.[23] Pioneering work for $(AlAs)_1(GaAs)_1$ and $(AlAs)_2(GaAs)_2$ UTSLs has been done by Gossard et al.[52] and also by van der Ziel et al.[53] In the preceding section, the electronic states in the $(AlAs)_n(GaAs)_n$ UTSL with $n \sim 2$ are found to recover the quasi-three dimensionality. Now we go further on to study the electronic states with photoluminescence measurement.

Energy Gap.[8] In Fig. 12(a) and 12(b) we plot the photoluminescence spectra of an $(AlAs)_2(GaAs)_2$ superlattice at different temperatures. the ordinate of the Fig. 12(a) is about two orders of magnitude enlarged compared to that in Fig. 12(b). In the photoluminescence measurement, the $(AlAs)_n(GaAs)_n$ UTSL sample is irradiated by 5145-A-line of Ar^+ laser. Figures 13(a) and 13(b) show the measured energy gap of the $(AlAs)_n(GaAs)_n$ superlattice for n running from 1 to 24 at 300K and at 4.2K, respectively. The energy gap becomes larger as n decreases. This result is qualitatively explained by the quantum size-effect which states that the tighter the confinement is, the higher the energy is. We find that the increase of the energy gap per unit reduction of the slab-thickness, i.e., $dE_g/d(-n)$, is small for $n \sim 2$. This result can be explained by the fact that the delocalization of the electron shown in the above weakens the quantum size-effect. In Fig. 13(b) are shown the pioneering experimental results obtained by Gossard et al.[52] and van der Ziel et al.[53] on MBE-grown samples. The solid line is a result of Schulman and McGill's calculation[54] and the dashed line the Kronig-Penney model[44]. In Fig. 13 (a), the Schulman and McGill's result is modified into the value at 300K using the temperature dependence of the band gap of GaAs as a zeroth approximation. The experimental results are in good agreement with the Kronig-Penney model for $n \gtrsim 10$, but not for $n \lesssim 8$, for which Schulman and McGill's calculation is instead in good agreement. The success of Kronig-Penney model means that the extrapolation of bulk properties is valid for $n \gtrsim 10$ in the $(AlAs)_n(GaAs)_n$ superlattice. This result justifies our usage of Kronig-Penney model in the preceding section. For $n \lesssim 8$, on the other hand, we need other theories beyond the

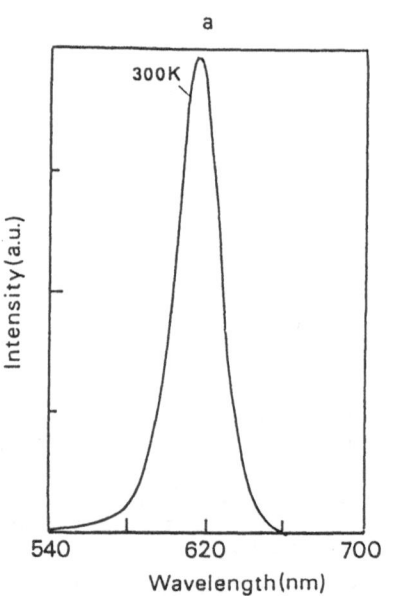
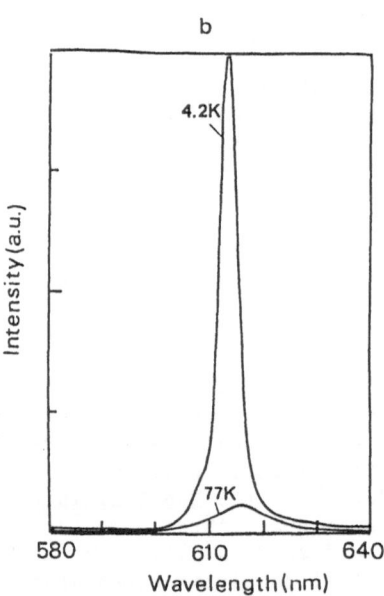

Fig. 12. Photoluminescence spectra of $(AlAs)_2(GaAs)_2$ measured (a) at 300K and (b) at 77K and 4.2K.

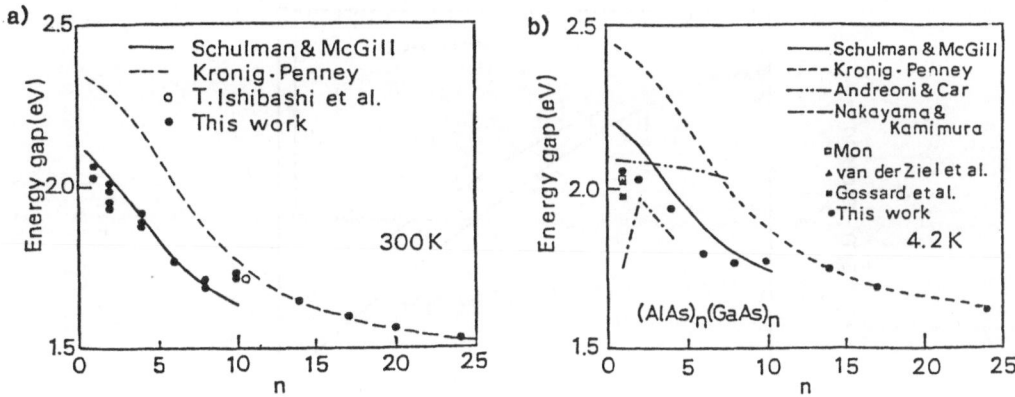

Fig. 13. Energy gap (a) at 300K and (b) at 4.2K of $(AlAs)_n(GaAs)_n$ superlattices as a function of n.

envelope function approximation. In addition to the pioneering work,[54,55,56], there have been an increasing number of theoretical calculations; pseudopotential calculations by Nakayama and Kamimura[57-59], and by Gell et al.[60,61] tight binding calculations by Yamaguchi,[62] and full potential linearized augmented plane wave method by Hamada et al.[63,64]. The observed energy gap is on the whole in conformity with those calculations. However, for n=1 we note that there exists discrepancy between the theories and our result, among the theories, and among the experiments. Our result does not show the decrease in energy gap for n=1. Garriga et al.,[65] using spectroscopic ellipsometry on MBE-grown $(AlAs)_n(GaAs)_n$ superlattices, recently reported the decrease for n=1, although the absolute energy gap itself was not in quantitative agreement with theory. As for theories, the decrease has been predicted by Nakayama and Kamimura[57-59] and also by Gell et al.[60,61] but has not been predicted by Schulman and McGill,[54] Andreoni and Car,[55] Yamaguchi,[62] nor by Hamada et al.[63,64] Raman frequency of our $(AlAs)_1(GaAs)_1$ sample is clearly different from the frequency of $Al_{0.5}Ga_{0.5}As$ alloy, and is in good agreement with the theory.[29] Therefore, we believe, the structure is grown as designed at least locally within the island extension, and the observed energy gap is not of $Al_{0.5}Ga_{0.5}As$ but of the $(AlAs)_1(GaAs)_1$ UTSL. As far as experiments are concerned, we should take the following into account. The island height, which is estimated to be one monolayer[26] is equal to the slab-thickness of the $(AlAs)_1(GaAs)_1$ UTSL. Crucial is in this case not only the superlattice structure along the z-axis, but also the lateral structure in the xy plane. The island being estimated to be a few tens to hundreds of angstroms,[26] the additional confinement in xy plane will increase the energy gap. The lateral structure seems to have strong growth-technique dependence and growth-condition dependence.[26,66] The theoretical calculation that takes into account the lateral structure of the UTSL has not yet been made, but this is an important step to make a foundation to proceed to the three dimensional superlattices.

Temperature Dependence. We plot in Fig. 14 the temperature dependence of the energy gap of the $(AlAs)_2(GaAs)_2$ UTSL. The bar shows the full width at the half maximum (FWHM) of the corresponding spectrum, where the contribution of the thermal energy has already been subtracted. The shift of the peak is very small against the temperature. For comparison we plot the temperature dependence of the Γ valley of GaAs and that of the X-valley of the AlAs.[67] The observed temperature dependece is closer to the temperature dependence of AlAs X-band rather than to that of GaAs Γ band. Although quantitative agreement has not been obtained, this is an indication that in the UTSL the conduction band has X valley component due to the zone folding effect.[57-59] Another indication of the mixing of the X-valley component is the plateau observed around

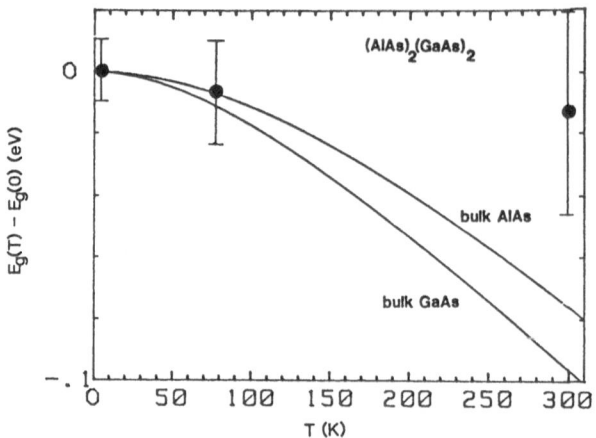

Fig. 14. Temperature dependence of the energy gap of $(AlAs)_2(GaAs)_2$ UTSL.

n=8~10 in Figs.13(a) and 13(b).[61] We may suggest that the plateau indicates the band in the UTSL is not composed of pure Γ band nor pure X band. If the band was composed of pure Γ band, the energy gap would increase monotonously without any plateau. If the band is composed of pure X band, on the other hand, then we would not see strong photoluminescence. The strong photolumines-cence observed even for small n and also the plateau at n ~ 10, thus, indicate that the lowest band of the UTSL is composed partly of Γ band and partly of X band components.

Photoluminescence Intensity.[8] The intensity of photoluminescence in a direct gap semiconductor is inherently limited by the energy separation be-tween the direct and the indirect conduction band minima.[68] With the separa-tion decreasing, an increasing number of electrons begin to populate also in the indirect valley where electrons and holes recombine non-radiatively. The population in the indirect band reduces the intensity of the photoluminescence which is the radiative recombination of electrons and holes in the Γ valley. Thus the intensity of the photoluminescence can give us informations about the

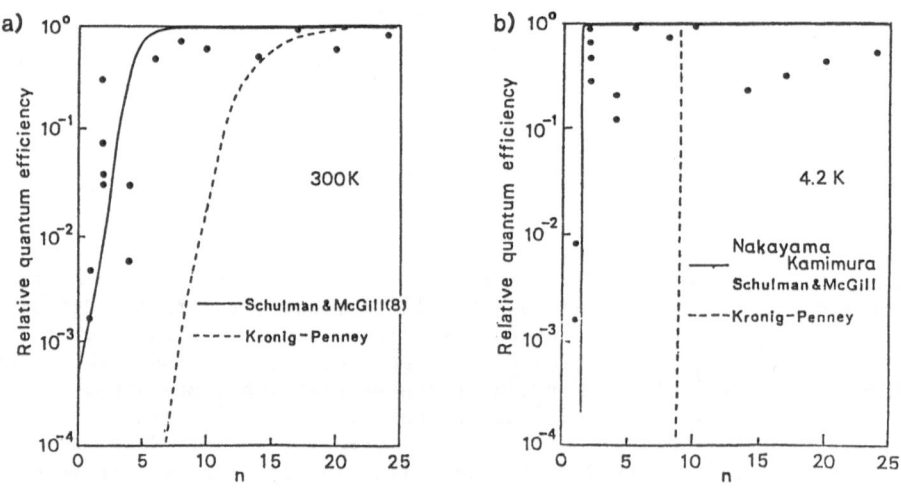

Fig. 15. Relative quantum efficiency (a) at 300K and (b) at 4.2K, of $(AlAs)_n(GaAs)_n$ superlattices as a function of n.

change in the band characters in the lowest conduction band.[69,70] Figure 15(a) and 15(b) show the relative photoluminescence intensity observed at room temperature and at 4.2K, respectively. The strongest intensity is normalized to unity, and the photoluminescence intensity of other samples are plotted relative to the strongest. The absolute intensity is about two orders of magnitude stronger than that observed at room temperature. The absolute photoluminescence intensity is affected by quality of the epitaxy, chemical treatments of the substrate before the growth, surface morphology, thickness of native oxide at the surface, etc., as discussed in the section of Raman intensity. It is inevitable to have some scatters in the intensity even for the samples whose structures are exactly identical.[70] Thus we discuss here the intensity change of more than one order of magnitude. In Figs. 15 (a) and 15 (b), we observe a drastic change in intensity as a function of the slab thickness. The large magnitude of change is considered to be due to the change in band characters. The photoluminescence intensity is proportional to an internal quantum efficiency. We assume that an h part of the recombinations in the Γ valley is radiative, and we also assume an equilibrium Boltzmann distribution among the valleys. Then the quantum efficiency is given by,[68,69]

$$g = \frac{h}{1 + (M_c^i t_d / M_c^d t_i)(m^i/m^d)^{3/2} \exp\{(E_\Gamma - E_{L(or\ X)})/kT\}} , \qquad (9)$$

where M_c^d, M_c^i; t_d, t_i; m^d, m^i; E_Γ, E_L, and E_X are numbers of equivalent minima, lifetimes, effective mass, in the direct and indirect valley, and energies of the Γ, L, and X bands, respectively. In Eq. (9) we have assumed for simplicity that $t_d \sim t_i$, and that the indirect recombination is completely nonradiative. We call g/h the relative quantum efficiency. The relative quantum efficiency is calculated with the band energies given by the Kronig-Penney model (dashed line) and the Schulman and McGill's calculation (solid line). As seen in Fig. 15 (b) the relative quantum efficiency calculated based on the Kronig-Penney model does not reproduce the experimental results. On the other hand, the Schulman and McGill's and Kamimura and Nakayama's calculations have predicted that the $(AlAs)_n(GaAs)_n$ still holds a direct energy gap for n down to two and is in agreement with our experimental result. The quantum efficiency at 4.2K changes much more abruptly, just like a step function at the place where the energy gap changes from direct to indirect, because the exponent in the denominator of Eq. (9) change from 1 to + ∞ there. We observed a drastic change in intensity between n=1 and n=2. This large magnitude of change is considered to be the step we have expected as the result of the change in the dominant band characters. We have observed intense photoluminescence for n down to two, but two orders of magnitude weaker one for n=1. The experimental result indicates that the $(AlAs)_n(GaAs)_n$ has a direct energy-gap for n down to two. The drop of relative quantum efficiency for n=1 suggests that $(AlAs)_1(GaAs)_1$ has an indirect gap, since capability of growing the $(AlAs)_1(GaAs)_1$ is indicated by TEM analysis, and the realization of the $(AlAs)_1(GaAs)_1$ UTSL is suggested by Raman scattering experiment. As discussed there, however, the electronic states in the $(AlAs)_1(GaAs)_1$ UTSL with monolayer-high islands are unknown. Further study will be necessary. Recently Hamada et al.[63,64] predicted theoretically that the $(AlAs)_n(GaAs)_n$ is direct and indirect for even n and odd n, respectively, due to the contribution from the L-valley. Our result is in conformity with their theoretical result.

Recently Finkman et al. have reported that in the $(AlAs)_3(GaAs)_3$ and the $(AlAs)_7(GaAs)_7$ superlattices X-point is the lowest in conduction band based on the result that they have observed long-lived excitons.[71] This result is in agreement with the Yamaguchi's result that the energy gap in the UTSL is indirect.[62] However we should take into consideration the fact that the dielectric function in the UTSL is small, as pointed out by Kamimura and Nakayama.[59] The long life of the exciton can also be ascribed to this effect. We believe based on the result of the photoluminescence intensity that the $(AlAs)_n(GaAs)_n$ superlattice has a pseudo-direct energy gap for n down to two,

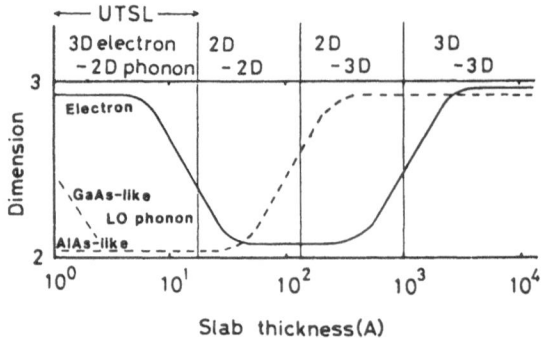

Fig. 16. Dimensions of electrons and LO phonons in $(AlAs)_n(GaAs)_n$ superlattices as a function of the slab thickness.

where 'pseudo' means that it contains also the X-band character.

We now summarize in Fig. 16 the dimensionality of the electron and the LO phonon in the $(AlAs)_n(GaAs)_n$ superlattice as a function of the slab-thickness.[13,14] With the slab-thickness decreasing, the dimensions of the electron change from three to two, and then back to three again at the slab-thickness of L_{2e} (~2 monolayers). The LO phonons change from three dimensions to two dimensions, and remain in two dimensions even for one monolayer-thick slabs, since $L_{2LO} \lesssim 1$ monolayer. In marked contrast to the bulk materials, the UTSL is a system of quasi-three dimensional electron and the quasi-two dimensional LO phonon.

DOUBLE DELTA DOPED STRUCTURES[15]

Now we make an application of the atomic layer superlattice. We have grown GaAs with delta-doped layers[72,73] by MOCVD with the same recipes as in the growth of UTSLs. In Fig. 17 (a), solid circles show the sheet carrier concentration of the single delta-doped layer as a function of the amount of Si_2H_6. We see two regions; one where n_s shows good linearity. and the other where n_s is independent of the amount of the Si_2H_6. Here n_s has been measured using a Hall measurement. The saturated n_s is about $1.5 \times 10^{13} cm^{-3}$. The FWHM measured by capacitance-voltage technique is 50A for the sample with n_s of $4 \times 10^{12} cm^{-3}$. The mobility (open circle) is almost constant for the saturated region. We use the saturated region to have good reproducibility. Next we measured n_s in the delta doped layers as a function of the depth (d) of the delta-doped layer from the surface under the condition denoted by the arrow in Fig. 17 (a). In Fig. 17 (b) we plot n_s against d. The observe n_s decreases for small d owing to the surface depletion. Under the approximation of a two-parallel plain capacitor, n_s is given as a function of d by

$$n_s(d) = n_{so} - \varepsilon\phi/d \qquad (10)$$

where ε is the dielectric constant and ϕ is the Schottky barrier height. In Fig. 17 (b) the solid line plots $n_s(d)$. In the calculation we put the value observed for saturated n_s into n_{so}, and have assumed that ϕ is independent of d for simplicity. The calculation reproduces the experimental result pretty well. To maintain the saturated n_s we need d of 200A. The critical depth at which the single delta-doped layer becomes completely insulating is about 30 A. However, when we insert a second delta-doped layer, supplying electrons to

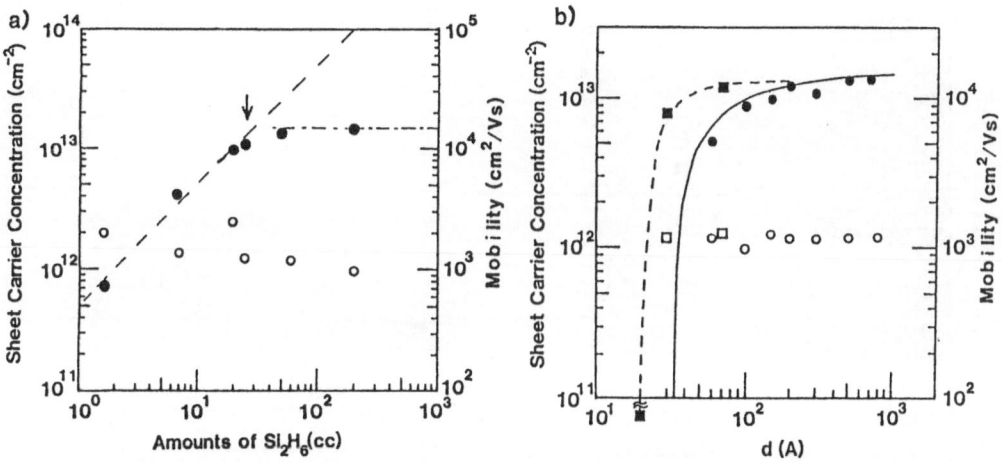

Fig. 17. (a) n_s and mobility (open circle) vs. amount of Si_2H_6. (b) n_s and mobility vs. d. Circles (squares) are for single (double) delta-doped layers. Solid (open) symbols show n_s (mobility). For the case of DDDS, d is the depth of the first delta-doped layer.

the surface state, we see a drastic change. Insertion of the second delta-doped layer at 10A depth enables a high n_s at small d, as denoted by the solid squares. The critical depth of the first delta-doped layer is much reduced, as seen in Fig. 17 (b). For d=70 A a saturated n_s is safely obtained. Even for d=30 A we have observed n_s of $8 \times 10^{12} cm^{-2}$.

Electron Beam Induced Resist (EBIR)[16]

Since the very shallow channel is obtained with ultimate control in the growth direction, we proceed to a reduction of the lateral dimensions in the device fabrication. The EBIR has been developed based on the electron beam in-duced deposition.[17-19] In the EBIR, an electron beam is swept on the sub-strate on introduction of the resist source in gas-phase. The source material changes into the solid-phase resist directly at the beam spot on the sub-strate, and the gate pattern is formed along the swept line. We can complete any pattern using the EBIR as a mask. The EBIR enables us to make finer pat-terns than with the conventional electron-beam lithography where the electrons struggle through several hundreds to thousands of angstroms of precoated resist. We used an alkyl naphthalene as the resist source. The beam spot size is about 80A, acceleration voltage 6 KV, the beam current 2×10^{-11} A, and the gas pressure about 1×10^{-6} Torr, typically. We may call EBIR a 'three in one' because resist-coating, exposure, and development are accomplished in only one process. The shortest gate length ever achieved is about 200A.[16]

Device Fabrication and Characterization[15,16]

First the channel is mesa-isolated, then the source/drain contacts are formed with conventional optical lithography. The ohmic metals, Ni/AuGe, are annealed at 450 °C. Then the gate metal, W, is deposited up to 1000 A over all of the sample. For defining the ultrashort gate, we used the EBIR process. Finally we drive the W slightly inside with thermal process to have a good Schottky contact.

Figure 18 (a) shows the overall view of the short-gated MESFET fabricated from the DDDS. The inset shows the enlarged view of the gate electrode. We show the room-temperature characteristics of the device in Fig. 18 (b). The observed transconductance is about 270 mS/mm. The intrinsic transconductance is 400mS/mm. On the other hand, the theoretical g_m is given by[72,74,75]

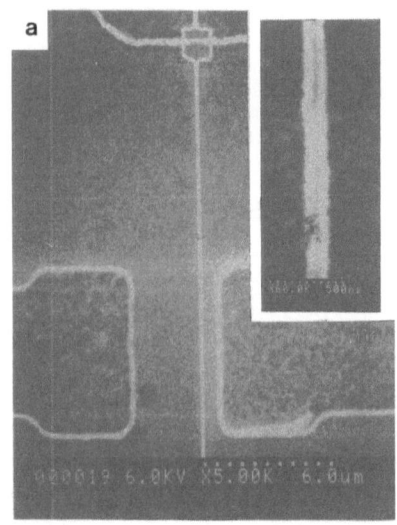
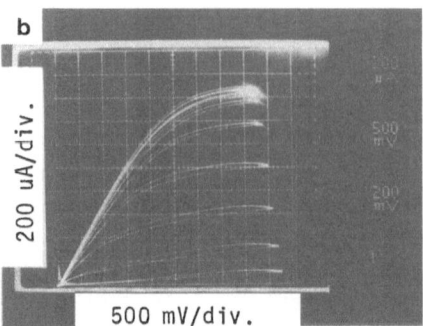

Fig. 18. (a) Overall view of the DDDS MESFET. The inset scale is 500A/div.
(b) Characteristics of the DDDS MESFET. The Gate voltage is in
200 mV step.

$$g_m = \frac{e\mu W_g n_s}{L_g} \left(1 + \left(\frac{e\mu n_s D}{\varepsilon v_s L_g} \right)^2 \right)^{-1/2} \tag{10}$$

where e is unit charge, μ the mobility, W_g the gate width, N_s the sheet carrier density, L_g the gate length, ε the dielectric constant, and D the channel
depth. When $e\mu N_s D / \varepsilon v_s L_g \gg 1$, g_m converges to $\varepsilon v_s W_g / D$ that depends only on
v_s. For Lg~200A, this condition is satisfied even for μ~100cm^2/Vs. Note that
v_s is neither very dependent on what the material is nor whether the carrier
is an electron or a hole.[76] A thin channel being needed for small L_g,[16, 76]
the success of the ultrashort-gated FET demonstrates the possibility of a
'universal' FET whose g_m (hence f_t, the cutoff frequency also) is as large as
theoretically possible, regardless of the materials or the carriers. The observed g_m is one-third of the theoretical value that the Eq. (10) predicts,
but the optimization of the device structure will lead to the theoretically
maximum g_m for such an ultrathin-chanelled short-gated FET.

SUMMARY AND CONCLUSION

We have reviewed our recent results of characterization and application
of atomic layer superlattices, i.e., UTSLs and DDDS's. By controlling the
abruptness and slab-thickness to one monolayer, we have established a technique for growing atomic layer superlattices based on MOCVD. The dimensionality of electron and LO phonon can be controlled by adjusting the slab-
thickness of the superlattice, knowing the second characteristic lengths that
can be designed by selecting adequate mismatch in the electron affinity as
well as in the reduced mass. Those facts may open up the extended concept that
the superlattice is a tool to control separately the dimensionality of the
electron and the LO phonon, the most important elements in compound semiconductors. Of potential interest is the isotope superlattice, in which optical
phonon states are quasi-two dimensional with the electronic states unchanged.
This superlattice, being free from electronically parasitic effects, is unique

object to study the interaction between the electron and the confined optical phonons. As an application of the atomic layer superlattice, we have fabricated nano-structure FET from a DDDS. With ultrathin-channeled short-gated FETs we have pointed out the possibility of the 'universal' FET. The atomic layer superlattices is of great importance for fundamental physics and also for future devices.

REFERENCES

1. L. Esaki and R. Tsu, Superlattic and Negative Differential Conductivity in Semiconductors, IBM J. Res. Develop. 14;61 (1970)
2. B. Jusserand, D. Paquet, and A. Regreny, Folded Optical Phonons in GaAs/AlGaAs Superlattices, Phys. Rev. B30;6245 (1984)
3. M. Nakayama, K. Kubota, H. Kato, S. Chika, and N. Sano, Raman Scattering from GaAs-AlAs Monolayer-Controlled Superlattices, Solid State Commun. 53:493 (1985)
4. A. K. Sood, J. Menendez, M. Cardona, and K. Ploog, Resonance Raman Scattering by Confined LO and TO phonons in GaAs-AlAs Superlattices, Phys. Rev. Lett. 54:2111 (1985)
5. U. Gnutzmann and K. Clauseker, Theory of Optical Transitions in an Optical Indirect Semiconductor with a Superlattice Structure, Appl. Phys. 3:9 (1974)
6. A. Maduhukar, Modulated Semiconductor Structures: An Overview of Some Basic Considerations for Growth and Desired Electronic Structure, J. Vac. Sci. Technol. 20:149 (1982)
7. A. S. Barker, J. L. Merz, and A. C. Gossard, Study of Zone-Folfing effects on Phonons in Alternating monolayers of GaAs-AlAs, Phys. Rev. B17: 3181 (1978)
8. A. Ishibashi, Y. Mori, M. Itabashi, N. Watanabe, Optical Properties of $(AlAs)_m(GaAs)_n$ Superlattices Grown by Metalorganic Chemical Vapor Deposition, J. Appl. Phys. 58:2691 (1985)
9. A. Ishibashi, Y, Mori, F. Nakamura, and N. Watanabe, Optical ploperties of Quantum wells with ultrathin-layer superlattices, J. Appl. Phys. 59:2503 (1986)
10. A. Ishibashi, Y. Mori, K. Kaneko, N. Watanabe, A New Connection Rule of Wave Functions at a Heterointerface and Band Discontinuity between GaAs and AlGaAs, J. Appl. Phys. 59:4087 (1986)
11. A. Ishibashi, M. Itabashi, Y. Mori, S. Kawado, K. Kaneko, and N. Watanabe, Raman Scattering from $(AlAs)_m(GaAs)_n$ Ultrathin-layer Superlattices, Phys. Rev. B33:2887 (1986)
12. A. Ishibashi, M. Itabashi, Y. Mori, N. Watanabe, Ratio of LO phonon Intensities in Raman Scattering from $(AlAs)_n(GaAs)_n$ Superlattices, Optoelectronics, devices and Technologies,1:51 (1986)
13. A. Ishibashi, Y. Mori, M. Itabashi, N. Watanabe, A fundamentally New aspect of Electron-Phonon Interaction in $(AlAs)_m(GaAs)_n$ Ultrathin-Layer Superlattices, 18th Int. Conf. Phys. Semicon. vol. 2:1365 (1987)
14. A. Ishibashi, Y. Mori, M. Itabashi, N. Watanabe, Proc. Int. Workshop Future Electron Devices - Superlattice Devices:105 (1987)
15. A. Ishibahsi, K. Funato, and Y. Mori, Ultrathin-Channelled GaAs MESFET with Double-Delta-Doped Layers, Electron. Lett. 24:1034 (1988)
16. A. Ishibashi, K. Funato, and Y. Mori, Heterointerface Field Effect Transistor with 200-A-Long Gate, Jpn. J. Appl. Phys. 27:L2382 (1988)
17. A. N. Broers, W. W. Molzen, J. J. Cuomo, and N. D. Wittels, Electron-Beam fabrication of 80-A Metal Structures, Appl. Phys. Lett. 29:596 (1976)
18. S. Matsui and K. Mori, New Selective Deposition Technology by Electron Beam Induced Surface Reaction, J. Vac. Sci. Technol. B4:299 (1985)
19. B. H. Chin and G. Ehrlich, Formation of Silicon Nitride Structure by Direct Electron Beam Writing, Appl. Phys. Lett. 38:253 (1981)
20. Manasevit, Single Crystal Gallium Arsenide on Insulating Substrate, Appl. Phys. Lett. 12:156 (1968)

21. H. Watanabe, and A. Usui, Atomic Layer Epitaxy, <u>Proc. Int. Conf.</u> <u>GaAs and Related Compounds, Las Vegas</u>:1 (1986)
22. S. D. Hersee, M. Baldy, and P. Assena, The Growth of Quantum Well GaAs/AlGaAs Laser Structures, <u>J. Phys,</u>43:C5-193 (1982)
23. R. Dingle, Confined Carrier Quantum State in Ultrathin Semiconductor Heterostructures, <u>Festkoerperprobleme</u> XV:21 (1975)
24. P. M. Frijlink and J. Maluenda, MOVPE Growth of GaAlAs/GaAs Quantum Well Heterostructures, <u>Jpn. J. Appl. Phys.</u>21:L574 (1982)
25. H. Kawai, K. Kaneko, and N. Watanabe, Photolumicescence of AlGaAs/GaAs Quantum Wells Grown by Metalorganic Chemical Vapor Deposition, <u>J. Appl. Phys.</u>56:463 (1984)
26. N. Watanabe and Y. Mori, Ultrathin GaAs/GaAlAs Layers Grown by MOCVD and their Structural Characterization, <u>Surf. Sci.</u>174:10 (1986)
27. K. Zeeger, "Semiconductor Physics", 3rd edit., Springer Verlag, Berlin (1985)
28. M. Ilgems and G. Pearson, Infrared reflection Spectra of GaAlAs Mixed Crystals, <u>Phys. Rev. B</u>1:1576 (1970)
29. T. Toriyama, N. Kobayashi, and Y. Horikoshi, Lattice Vibration of Thin-layered AlAs-GaAs Superlattices, <u>Jpn. J. Appl. Phys.</u> 25:1895 (1986)
30. C. Colvard, T. A. Gant, M. V. Klein, R. Merlin, P. Fischer, H. Morkoc, and A. C. Gossard, Folded Acoustic and Quantized Optical Phonons in (GaAl)As Superlattices, <u>Phys. Rev. B</u>31:2080 (1985)
31. S. K. Yip and Y. C. Chang, Theory of Phonon Dispersion Relations in Semiconductor Superlattices, <u>Phys. Rev.</u> B30:7037 (1984)
32. See for example, M. Cardona, in "Light scattering in Solids," M. Cardona and G. Guntherodt, ed., Springer Verlag, Berlin, 1975
33. P.Manuel, G.A. Sai-Halasz, L.L. Chang, Chin-An Chang, and L. Esaki, Resonant Raman Scattering in a Semiconductor Superlattice, <u>Phys. Rev. Lett.</u> 37:1701 (1976)
34. J. E. Zucker, A. Pinczuk, D. S. Chemla, A. C. Gossard, and W. Wiegmann, Raman Scattering Resonant with Quasi-Two-Dimensional Excitons in Semiconductor Quantum Wells, <u>Phys. Rev. Lett.</u> 51:1293 (1983)
35. J. E. Zucker, A. Pinczuk, D. S. Chemla, A. C. Gossard, and W. Wiegmann, Delocalized Excitons in Semiconductor Heterostructures, <u>Phys. Rev.</u> B29:7065 (1984)
36. M. Nakayama, K. Kubota, T. Tanaka, H. Kato, S. Chika, and N. Sano, Zone-Folding Effects on Phonons in GaAs-AlAs Superlattices, <u>Jpn. J. Appl. Phys.</u> 24:1331 (1985)
37. R. M. Martin, in "Proc. 2nd Int. Conf. on Light Scatt. in Solids", M. Balkanski, ed., Paris, 1971, p25,
38. Y. Toyozawa, Theory of the Line-Shapes of the Exciton Absorption Band <u>Prog. Theo. Phys.</u> 20:53 (1958)
39. H, Froehlich, Electrons in Lattice Fields, <u>Adv. Phys.</u> 3:325 (1954)
40. G. L. Bir and G. E. Pikus, Theory of the Deformation Potential for Semiconductors with a Complex Band Structure, <u>Sov. Phys. Solid State</u> 2:2039 (1961)
41. B. R. Nag, " Electron Transport in Compound Semiconductors", Springer Verlag, New York (1980)
42. M. Cardona, in "Light scattering in Solids II", M. Cardona and G. Guentherodt, ed., Springer-Verlag, Berlin (1982)
43. W. Poetz and P. Vogl, Theory of Optical Phonon Deformation Potentials in Tetrahedral Semiconductors, <u>Phys. Rev.</u> B24:2025 (1981)
44. C. Kittel, "Introduction to solis state Physics", Wiley, New York (1976)
45. J. N. Schulman and T. C. McGill, Complex Band Structure and Superlattice Electronic States, <u>Phys. Rev.</u>1 B23:4149 (1981)
46. J. N. Schulman and Y. C. Chang, New method for calculationg electronic properties using complex band structures, <u>Phys. Rev.</u> B24:4445 (1981)
47. J. Sanchez-Dehesa and C. Tejedor, Selfconsistent calculation of properties of GaAs-AlAs superlattices with homopolar interfaces, <u>Phys. Rev.</u> B26:5824 (1982)

48. R. C. Miller, D. A. Kleinman, W. A. Nordland, Jr., and, A. C. Gossard, Luminescence studies of optically pumped quantum wells in GaAs-AlGaAs multilayer structures, Phys. Rev. B22:863 (1980)
49. R. Dingle, A. C. Gossard, and W. Wiegmann, Direct observation of superlattice formation in a semiconductor heterostructure, Phys. Rev. Lett, 34:1327 (1975)
50. T. Ishibashi, S. Tarucha, and H. Okamoto, Exciton associated optical absorption spectra of AlAs/GaAs superlattices at 300 K, Inst. Phys. Conf. Ser. No. 63:587 (1981)
51. D. A. B. Miller, D.S. Chemla, d. J. Eilenberger, and P. W. Smith, A. C. Gossard, and W. T. Tsang, Large room-temperature optical nonlinearity in GaAs/GaAlAs multiple quantum well structures, Appl. Phys. Lett. 41:679 (1982)
52. A. C. Gossard, P. M. Petroff, W. Wiegmann, R. Dingle, and A. Savage, Epitaxial structures with alternation-atomic-layer composition modulation, Appl. Phys. Lett. 29:323 (1976)
53. J. P. van der Ziel and A. C. Gossard, Absorption, refractive index, and birefringence of AlAs-GaAs monolayers, J. Appl. Phys. 48:3018 (1977)
54. J. N. Schulman and T. C. McGill, Electronic properties of the (001) interface and superlattice, Phys. Rev. B19:6341 (1979)
55. W. Andreoni, and R. Car, Similarity of (Ga,Al,As) alloys and ultrathin heterostructures: Electronic properties from the empirical pseudopotential method, Phys. Rev. B21:3334 (1980)
56. K. K. Mon, Electronic band structure of (001) GaAs-AlAs superlattices, Solid State Commun. 41:699 (1982)
57. T. Nakayama and H. Kamimura, Band structure of semiconductor superlatices with ultrathin layers (GaAs)n/(AlAs)n with n=1,2,3, and 4, J. Phys. Soc. Jpn, 54:4726 (1985)
58. H. Kamimura and T. Nakayama, Self-consistent band structure calculations of (GaAs)n(AlAs)n superlattices of ultrathin layers with n=1 to 10, Proc. 18th Int Conf. Phys. Semicon.:643 (1986)
69. H. Kamimura and T. Nakayama, Electronic structures and properties of ultrathin layered semiconductor superlattices, Comments Cond. Mat. Phys. 13:143 (1987)
60. M. A. Gell, D. Ninno, M. Jaros, and D. C. Herbert, Zone-folding, morphogenesis, and the role of periodicity in GaAs-AlGaAs (001) superlattices, Phys. Rev. B34:2416 (1986)
61. M. A. Gell, M. Jaros, and D. C. Herbert, Band offsets and zone-folding in GaAs-AlAs (001) superlattices, Superlattice and Microstructures, 3:121 (1987)
62. E. Yamaguchi, Theory of the DX centers in III-V Semiconductors and (001) Superlattices, J. Phys. Soc. Jpn. 56:2853 (1987)
63. N. Hamada, S. Ohnishi, and A. Oshiyama, Energy bands and stable structures of ultrathin-layer semiconductor superlattices, Extended Abstracts 18th Conf. Solid State Devices and Materials,:343 (1986)
64. N. Hamada, S. Ohnishi, Electronic structure calculations of (AlAs)m(GaAs)n superlattices based on full-potential linearized augmented-plane-wave method, Suerlattices and Microstructures, 3:301 (1987)
65. M. Garriga, M. Cardona, N. E. Christensen, F. Lautenschlager, T. Isu, and K. Ploog, Interband transitions in thin-layer GaAs/AlAs superlattices, Phys. Rev. B36:3254 (1987)
66. M. Tanaka, and H. Sakaki, Atomistic models of intergace structures of GaAs-AlGaAs (x=0.2-1) quantum wells grown by interrupted and uninterrupted MBE, J. Cryst. Growth. 81:153 (1987)
67. H. C. Casey, Jr. and M. B. Panish, 'Heterostructure Lasers', Part A, Academic Press, New York (1978)
68. I. Ladny and H. Kressel, Visible CW (AlGa)As heterojunction laser diode, Int. Elec. Dev. Meeting Technical Digest:129 (1976)
69. M. Naganuma, Y. Suzuki, and H. Okamoto, Photoluminescence of GaSb-AlSb superlattices grown by MBE, Inst. Phys. Conf. Ser. 63:125 (1982)
70. G. Griffiths, K. Mohammed, S. Subbana, H. Kroemer, and J. L. Merz,

GaSb/AlSb multiquantum well structures: Molecular beam epitaxial growth and narrow-well photoluminescence, Appl. Phys. Lett. 43:1059 (1983)

71. E. Finkman, M. D. Sturge, and M. C. Tamargo, X-point excitons in AlAs/GaAs superlattices, Appl. Phys. Lett. 49:1299 (1986)

72. E. F. Schubert, A. Fisher, and K. Ploog, The delta-doped field effect transistor (oFET), IEEE Trans. Elec. Dev. ED-33:625 (1986)

73. K. Ploog, M. Hauser, and A. Fisher, Fundamemtal studies and device application of o-doping in GaAs layers and in AlGaAs/GaAs heterostructures, Appl. Phys. A45:233 (1988)

74. T. J. Drummond, H. Morkoc, K. Lee, and M. Shur, Model for modulation doped field effect transistor, IEEE Elec. Dev. Lett. EDL-3:338 (1982)

75. M. B. Das, and M. L. Roszak, Design calculation for submicron gate-length AlGaAs/GaAs modulation-doped FET structures using carrier saturation velocity/charge-control model, Solid State Electron. 28:997 (1985)

76. S. M. Sze, "Physics of Semiconductor Devices," John Willey & Sons, New York (1981)

ASPECTS OF THE GROWTH OF InP/InGaAs MULTI-QUANTUM WELL

STRUCTURES BY GAS SOURCE MOLECULAR BEAM EPITAXY

G J Davies, E G Scott, M H Lyons,
M A Z Rejman-Greene and D A Andrews

British Telecom Research Laboratories
Martlesham Heath
Ipswich IP5 7RE
UK

Abstract

Gas Source MBE including Chemical Beam Epitaxy is shown
to be a promising technique for the growth of
heterostructures involving the incorporation of both
arsenic and phosphorus species.
Planar quantum confined Stark effect modulators/detectors
have been fabricated from InP/InGaAs multi-quantum well
stacks containing 200 wells. The layer sequences have
been analysed by both optical and double crystal X-ray
techniques. The modulator structures have shown excellent
uniformity of the grown layers in both the growth and
lateral dimensions - properties which are essential for
the fabrication of modulator arrays.
4x4 Arrays have been constructed and have shown state of
the art modulation coupled with low leakage currents.

1 Introduction

The use of multi-quantum well layers in novel, low
dimensional structures and devices has been made possible
by recent improvements in semiconductor crystal growth
techniques. The development of Molecular Beam
Epitaxy(MBE) and Metal-Organic Vapour Phase
Epitaxy(MOVPE) has seen the construction and growth of
semiconductor layer structures with control at the atomic
layer limit. This control has been exploited most notably

in the Multi-quantum well laser[1],the MQW avalanche
photodiode[2] and the Quantum confined Stark effect(QCSE)
modulator/detector[3].
The QCS effect is associated with the strong excitonic
absorption resonances that are a feature of multi-quantum
well structures, especially at room temperature. It has
been shown[4] that under the influence of a reverse bias
electric field applied perpendicular to the wells, the
absorption edge undergoes a red shift whilst
simultaneously broadening. This subsequent shift and
broadening can be used to modulate a light signal whose
central frequency is situated close to the absorption
edge at zero field. In this form arrays of the devices
can be used as spatial light modulators/detectors and
have great potential for use in optical interconnects.
    This device imposes very rigorous limits on the growth
process as in order to be effective, large area
uniformity is essential as is the integrity of the well
to barrier ratio. Similarly, the device is in essence a
p-i-n detector relying on absorption in the wells as well
as being able to sustain a constant electric field
perpendicular to the MQW stack. This therefore has the
consequence that a large number of wells are required for
efficient absorption of the radiation but the residual
doping level has, consequently, to be low, to sustain the
electric field. In practice this naturally means a
compromise situation, but 200 wells and barriers with
residual doping levels n~ 10^{15}cm^{-3} are not uncommon.
    This paper will demonstrate the applicability of gas
source MBE to the production of structures suitable for
the fabrication of large area QCSE modulators. It will
examine in depth the uniformity of the grown structures
both in the growth direction and laterally across the
wafer. Analysis of the MQW stacks has been performed
using X-ray multiple crystal diffraction techniques as
well as optical methods. The X-ray technique is shown to
be particularly powerful in this instance especially when
coupled with effective modelling procedures.
    State of the art InP/InGaAs MQW modulators in 4x4 arrays
have been fabricated and shown to act as spatial light
switches in the 1.55um region of the spectrum.

2 Experimental

Gas source MBE has been used as the generic term to
denote the use of any gaseous sources, whether group III
or Group V, in conventional MBE systems. Recently the
various combinations of sources have been classified by
Tsang[5] and this is illustrated in Figure 1.Gas source
MBE is now taken to imply the use of the group V
hydrides, arsine and phosphine, as alternative sources of
arsenic and phosphorus used with conventional elemental
group III sources.An alternative name for this technique
is Hydride source MBE.
The use of the group III sources in the form of metal
alkyls has introduced a further dimension to MBE in that
the potential versatility of source material can be
combined with the ultra sharp interfaces associated with
conventional MBE. The combination of metal alkyl group
III sources with group V hydrides is now known as
Chemical Beam Epitaxy (CBE) whereas the combination of

46

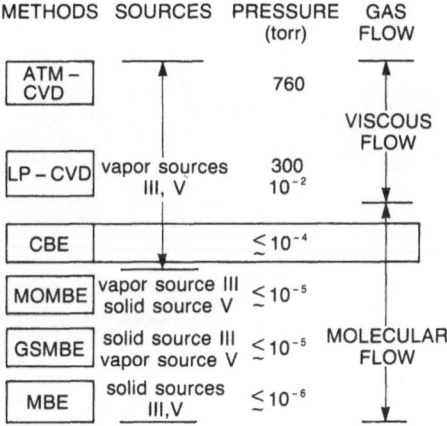

METHODS	SOURCES	PRESSURE (torr)	GAS FLOW

Figure 1 Relationship between various epitaxial
techniques (after ref 5).

group III metal alkyls with conventional elemental group
V sources is now known as Metal-Organic Molecular Beam
Epitaxy (MOMBE).
It is apparent from recent publications[6,7] that there
are significant differences in growth mechanisms between
these nominally similar techniques. It is also apparent
that it is the use of metal alkyl species that effect
these differences.
In conventional and hydride MBE there is no interaction
in the beams and the growth rate is entirely dependent on
the arrival rate of the group III species. These have
unity sticking coefficients on the substrate and so the
growth rate is relatively independent of substrate
temperature. It is only at high temperatures that re-
evaporation takes place. The substrate temperature,
therefore, must be high enough to impart sufficient atom
mobility at the growth interface to ensure that the
crystal grows epitaxially.
In MOMBE and CBE, where alkyl species are present, the
substrate temperature now has two functions. Firstly, it
has to decompose the alkyl then, secondly, impart energy
to ensure sufficient adatom mobility for epitaxy.
The growth rate now is dependent on substrate
temperature. At low temperatures, the growth rate is
dependent on the efficiency of pyrolysis of the alkyl.
There then exists a plateau region where this pyrolysis
occurs at a constant rate. However, the mobile surface
species is likely to be a metal alkyl with at least one
alkyl grouping removed. This has been confirmed from
recent surface studies by Foord at al[8]. This species,
probably a dialkyl would be expected to have a greater
surface mobility for a given substrate temperature than
its elemental counterpart in MBE. This could explain some
of the improved electrical and optical properties

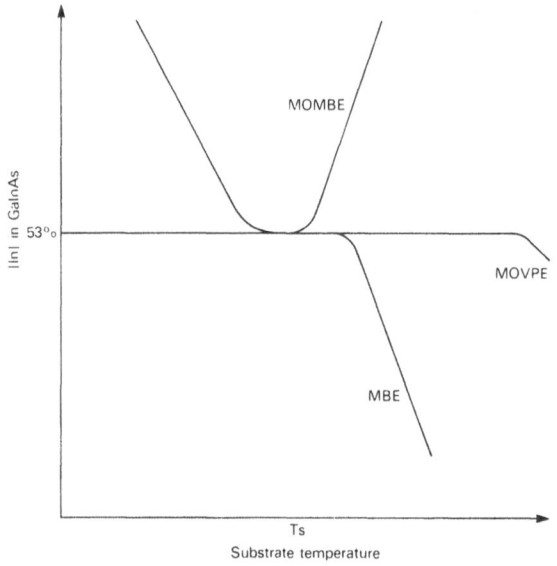

Figure 2   Schematic diagram of the reaction
mechanism differences for nominally
matched InGaAs/InP.

observed for MOMBE and CBE. At higher temperatures a
similar situation occurs to that of the elemental
species.
A further difference in growth mechanism is observed in
the materials system with which this paper is concerned
viz. InGaAs lattice matched to InP. As shown
schematically in Figure 2 we can consider the variation
of In concentration with substrate temperature. In the
case of conventional MBE, the In concentration remains
constant until loss of the least stable binary is
observed, this has been well documented by Scott et al.
[9]. A similar profile is observed for MOVPE but over a
wider temperature range. However, the CBE/MOMBE profile
is noticably different in shape. The low temperature
regime reflects the lower pyrolysation efficiency of the
Ga alkyl compared to the In alkyl, resulting in an
increased In content in the alloy. As the temperature is
increased pyrolysis rates stabilise and lattice matching
becomes more readily achieved, but over a very narrow
range. As the temperature is further increased the In
concentration in the alloy is seen to increase rapidly.
The explanation for this trend, which is the opposite to
that observed with conventional MBE, is not yet fully
understood. However this serves to highlight some of the
differences and potential problems that may occur in the
growth of the layer sequences required for this study.
Using the definitions described above, most of the work
presented in this paper will involve hydride source MBE
but specific reference will also be made to results
obtained by CBE.

The system used at BTRL is similar to that shown
schematically in Figure 3. It is in essence a VG V80H MBE

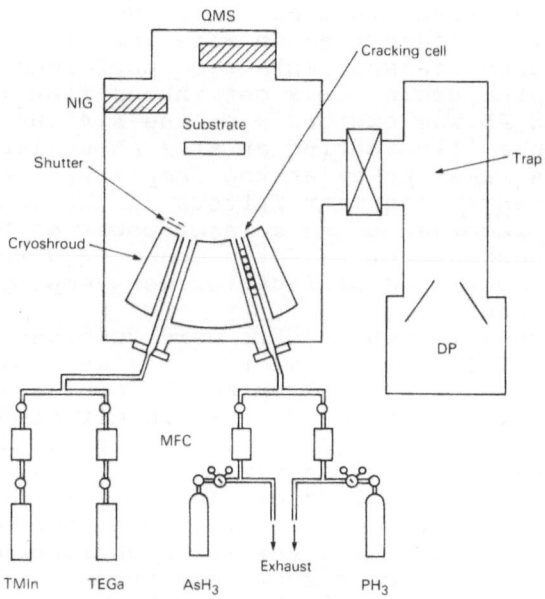

Figure 3 Schematic diagram of gas source MBE
equipment(after ref 7).

system adapted to take gas sources. The group V sources
are supplied as 100% AsH_3 and PH_3 and dried by molecular
sieve. In the pressure regime in which MBE operates the
normal substrate temperatures employed are insufficient
to decompose the hydrides directly and it is therefore
necessary to precrack these species. In our system this
is achieved by thermal pyrolysis of the gases in a
tantalum catalytic cracker as they enter the MBE chamber.
Cracking efficiency is estimated to be in excess of 98%
over the temperature range 900-1000^{0}C.
Control of the flow rates is achieved using mass flow
controllers. Conventional Knudsen effusion cells were
used for the group III sources.
When employed, the metalorganic sources were kept in a
temperature controlled bath and the fluxes set by
precision leak valves. These alkyl species decompose
directly on the substrate surface and so it is only
necessary to ensure that condensation in the introduction
tube does not take place. The substrate temperature was
measured with the sample holder thermocouple, calibrated
with the melting point of InSb(525^{0}C). The pumping system
was a throughput oil diffusion pump to enable the system
to cope with the large volumes of hydrogen that are
evolved. Sulphur doped <100> InP substates were used in
all the growth experiments recorded here. The growth
rates for thick layers of InP and InGaAs were ~1um/h and
~2um/h respectively, at the growth temperature of the
MQW's of 525^{0}C. High quality hetero-interfaces were
achieved by using a group III source temperature ramping
technique to eliminate flux transients due to shutter
operation[10], and by including a growth interruption of

15s at the changeover in group V sources. This latter technique will be described in more detail later. Photoluminescence measurements were performed at 4.2K and 77K with samples cooled in a continuous-flow liquid-He cryostat. The PL was excited with the 514.5nm line of an argon ion laser illuminating an area about 1mm in diameter. The laser power at the sample was controlled by interposing neutral density filters in the beam. Most spectra were recorded using a laser power of $P_0 \sim$ 1mW. A 0.5 metre monochromator, a North Coast Ge detector and conventional lock-in amplification were employed to record the spectra.

Absorbance/transmission spectra were recorded using a Bruker IFS88 FT-IR spectrometer customised for near-IR operation with a tungsten lamp as the source and a Si-on-CaF_2 beamsplitter. The return beam is directed to an externally mounted microscope fitted with a small-area HgCdTe detector(100um on edge).

X-ray rocking curves (plots of reflected intensity as a function of x-ray incidence angle) were recorded using a double diffractometer or a Philips high resolution diffractometer. Rocking-curves were obtained close to the 002 and 004 substrate reflections. The reference crystal in the double crystal diffractometer was an (001)-orientation InP crystal. In the High Resolution Diffractometer, the single reference, or beam-conditioning, crystal used in the conventional double-crystal diffractometer was replaced with a Philips four reflection monochromator which consisted of two (110) grooved Ge monochromators [11]. This can be used with either the 022 or 044 reflection. The 022 Cu Ka1 reflection was chosen for this work because it offered the greater signal to noise ratio. Although the angular divergence is greater for the 022 reflection than for the 044 (approximately double), this is not a limitation when recording well-spaced diffraction features from superlattice structures. The High Resolution diffractometer gave a signal to noise ratio which was approximately two orders of magnitude better than that achieved in a double-crystal diffractometer. This was important when investigating the interface quality in the superlattice structures since much of the information concerning the interface is revealed by the high order satellite peaks. These are very weak and can only be observed if the background noise is $<10^{-5}$ of the intensity of the strongest peaks.

Results

The growth of large area arrays of MQW devices imposes very rigorous limits on the actual growth process. It is the intention in this section to examine the individual priorities that combine to make the final device. In the first instance, the heterointerfaces were examined by the growth of a series of single quantum wells of diminishing thickness from $\sim 100A^0$ to $\sim 10A^0$ and then assessing the layers using low temperature photoluminescence.

It is important to note that the decrease in well
thickness leads to an increasing blue shift in PL
emmision. This enables the device designer to chose the
wavelength of device operation by selecting the
appropriate well thickness. For optical
telecommunications using low loss silica based fibres the
optimum wavelength of operation is 1.55um. This
necessitates an InGaAs well thickness, in a MultiQuantum
Well(MQW) stack of approx. 57A^0; we have chosen a
nominally similar InP barrier thickness. The MQW layers
have been analysed by X-ray rocking curve techniques.
In the case of CBE grown layers, where the gas flows were
initially switched manually, the resulting spectra are
shown in Figure 4. The PL from the reference InGaAs layer
is indicated at ~1.6um whilst the PL from the SQW's shows
the characteristic blue shift in wavelength associated
with quantum size effects. However, the wavelengths of
the QW peaks are all rather longer than would be expected
for the well thickness predicted from the growth times,
the more so for the shorter times. The implication is
that the wells are somewhat thicker than intended. This
may be explained by a residual growth rate of about
10A^{0}min^{-1}, or 4% of normal, which occurs during the
period in which growth is interrupted and the fluxes
equilibrated with the shutters closed.
With this experience, we have repeated the experiment
only this time using a modified vent/run system, where
the TMIn and TEGa gas flows are switched between the
growth chamber and an evacuated vent line. The resulting
SQW PL spectra are shown in Figure 5. We note that the
shortest wavelength PL emission at 954.6nm compares
favourably with the shortest wavelength reported for
InGaAs/InP SQW's by Panish(949nm)[12] and that the

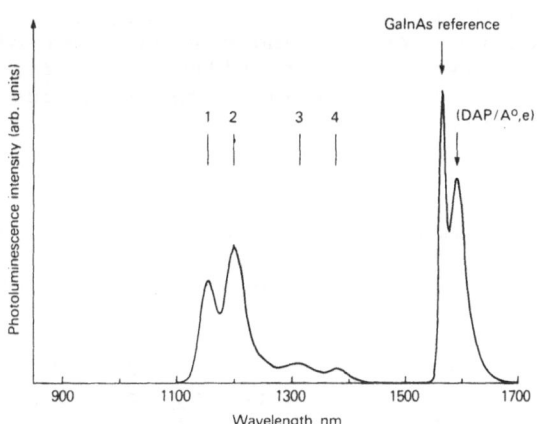

Figure 4 4K PL of manually switched CBE InGaAs/InP
        SQW's. Nominal widths, 1= 22A 31meV, 2=
        27A 36meV, 3= 43A 943meV and 4= 60A 898meV.

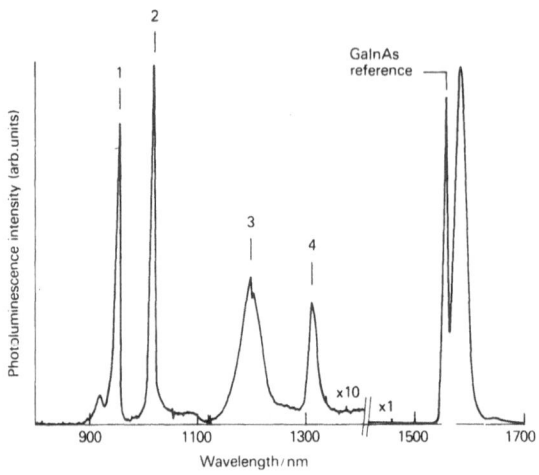

Figure 5 4K PL spectrum of vent/run switched
CBE InGaAs/InP SQW's. 1= 3A 11.5meV,
2= 9A 9.3meV, 3= 18A 39meV and 4=
35A 13.8meV.

linewidths of the two narrowest wells are substantially
less at (11.5meV at 954.6nm and 9.3meV at 1015.8nm) than
either Panish, (40meV at 949 nm) or Claxton (15 meV at
1012 nm)[13]. In so far as the short wavelength and
narrow linewidth of the SQW PL emission is a measure of
layer thickness and interface abruptness, it is clear
that this method of gas switching is at least as
efficient as that for elemental Ga and In in conventional
MBE. Indeed the only narrower reported linewidths are for
longer wavelength emission and presumably broader
wells[14]. We attribute the anomalously larger linewidths
of the two wider wells in our material to free carrier
broadening due to a background doping level of $n=5x10^{15}$-
$1x10^{16} cm^{-3}$. The two narrower wells are nearer the surface
and are likely to be depleted of carriers due to surface
depletion effects.
Skolnick et al[15] have shown that for InGaAs/InP quantum
wells the excitonic emission is an extrinsic process at
liquid He temperatures. The extrinsic nature of the
excitonic emission was attributed to the localisation of
the excitons at potential fluctuations in the quantum
wells. Stolz et al[16] arrived at a similar conclusion
for SQW's in the InGaAs/InAlAs system. This localisation
of the excitons results in an increase in the emission
wavelength and a reduction in peak width. This analysis
therefore produces some uncertainty in the quantification
of interface perfection deduced from 4.2K PL. However as
a qualitative guide it is useful especially if combined
with corresponding data at 77K. This is shown in Figure
6. It is noticable in this instance that the linewidths
and peak positions remain almost unchanged. From the
model proposed by Skolnick, as the excitons become more
mobile they experience more fully the potential
fluctuations in the quantum wells, causing an increase in
their peak emission energy and line width. That this is

indicated there was a large mismatch between the 'GaInAs' and 'InP' layers and implied considerable compositional grading extending through the entire width of the individual layers. Introducing an interrupt time brought a significant improvement in the superlattice peak widths and hence, the compositional uniformity. Rocking curves from samples grown with interrupts all showed a similar pattern of peak intensities with a strong zero-order peak and weak satellite peaks as expected for lattice-matched superlattices. The peak widths (and hence, the uniformity of the superlattices) did not show a steady change. Increasing the interrupt time beyond an optimum value leads to an increase in peak widths.

The group V switching sequence was also investigated. Two types of change-over were studied at each interface: short soak and long soak. In a short soak, the first group V source was closed soon after the metal sources. It was assumed that the residual pressure of this species in the vacuum chamber would be sufficient to protect the sample surface from degradation. The group V source for the next layer would be opened immediately prior to that of the metal sources, with just enough time for the new group V source to stabilize. In contrast, in a long soak, the second group V source is opened soon after the first is closed. This procedure was originally adopted in order to avoid the possibility of surface degradation due to insufficient Group V overpressure. However, the surface is now exposed to the 'wrong' group V species for an extended period, increasing the extent of exchange reactions at the interface. Recently, it has been found that surface degradation does not occur for interrupt times of less than 30s and that consequently, a 'long soak' is unnecessary.
Rocking curves recorded from a number of samples grown with short or long soaks were very similar indicating that under the conditions studied, changing the soak times had little effect on either the uniformity of the superlattices or the interfacial quality.

The quality of interfaces in the GaInAs/InP superlattices grown under optimum conditions, was investigated by a systematic effort to fit experimental and simulated results. A distinctive feature of the experimental curves (fig 8) was the alternating intensity sequence in which odd order satellites were stronger than the preceding even order satellite. Simulations carried out using the dynamical model showed that this characteristic intensity pattern indicated the presence of mismatched layers at both interfaces in the samples, with the mismatch being of opposite sign at the two interfaces.
Figure 10 shows the maximum intensities of the (002) experimental peaks (solid blocks), compared with peak intensities, calculated using the structure factor approach, for three trial structures (open blocks). Assuming perfectly abrupt interfaces, the best fit was found for a structure of 21 monolayers (MLs) of GaInAs and 21MLs of InP (fig 10a). A change to 22 MLs GaInAs and

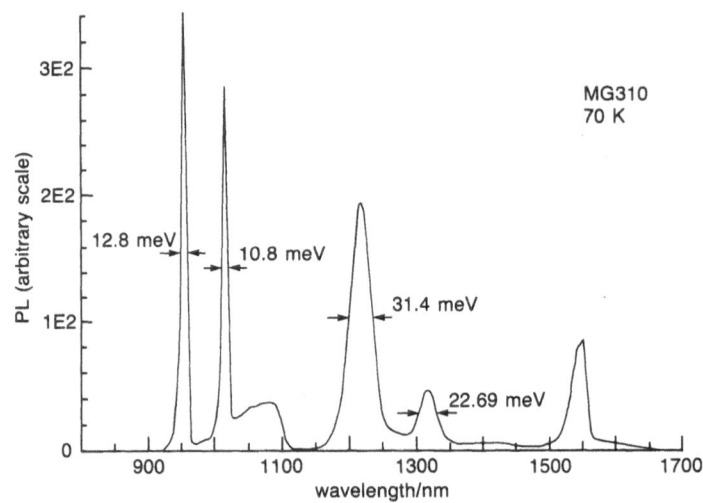

Figure 6 70K PL spectrum of vent/run switched CBE
InGaAs/InP SQW's. The FWHM's are now 1=
12.8meV, 2= 10.8meV, 3= 31.4meV and 4=
22.7meV.

not observed in this instance leads us to believe that,
qualitatively, the interfaces are perfect to near the
monolayer limit.
In the case of the SQW's grown by Hydride/gas source MBE
the SQW's 4K PL spectra were similar to that obtained by
CBE, however, at 70K there was a noticable broadening of
the linewidths. Almost a factor of two in some instances.
From this we would infer that the interfaces in this case
were inferior to those grown by CBE, but the degree to
which the interface broadening has occurred is not
evident from this technique.

In this work we have used almost exclusively the X-ray
rocking curve technique for the critical examination of
the MQW stacks, the layer sequence of which is shown in
Figure 7.
A typical rocking curve from a GaInAs/InP superlattice is
shown in fig 8. The most prominent feature in this curve
consists of two closely spaced peaks: one peak is due to
the substrate reflection, the other is a peak
corresponding to a reflection from a layer with the
average lattice parameter of the superlattice stack. This
is referred to as the zero-order superlattice peak. The
average mismatch of the superlattice may be determined
from the angular separation between the zero-order and
substrate peaks. In addition to the zero-order peak, a
number of weak superlattice peaks are visible; the
angular separation of these peaks is a direct measure of
the period.

More detailed information about the superlattice can be
obtained from consideration of the shapes and intensities

```
    --------------------------------
    InP  4x10^{16}cm^{-3} n-type 1 um

    --------------------------------

    MQW Stack 200 periods
         57A InGaAs Well
         57A InP Barrier

    --------------------------------

    InP  1x10^{18}cm^{-3} n-type 1um

    --------------------------------

    S-Doped  2 Inch InP Substrate
```

Figure 7   Layer sequence of MQW modulator

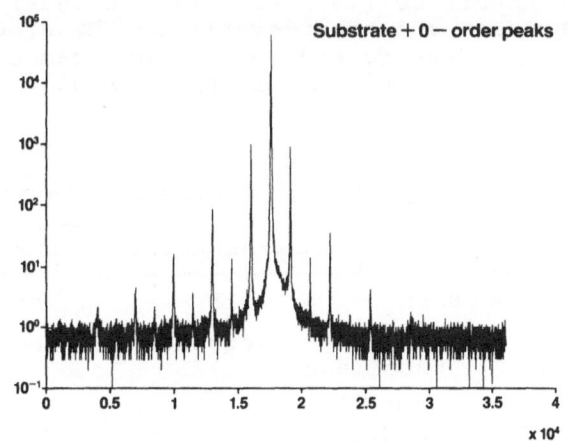

Figure 8   Typical X-ray rocking curve for
           InGaAs/InP MQW stack.

of the superlattice peaks. Thus, variations in the
composition and thicknesses of layers through the
superlattice stack give rise to subsidiary peaks at the
base of the satellite or cause the peak to broaden or
break up into a group of peaks [17]. The intensities of
the superlattice peaks are determined by the way in which
the scattering power and lattice parameter (both
functions of composition) vary within a period. Thus, it
is possible to determine the compositions and thicknesses
of individual layers within a period by analysing the
superlattice peaks intensities. Such analysis can also be
used to investigate interfacial abruptness [18].

In gas-source MBE, switching of the metal sources occurs
very rapidly: growth of a particular layer is controlled
by opening or closing the metal source. The delay between

closing the shutter on one metal source and opening the shutter on the other source is the interrupt time. In contrast, switching of the Group V gas-sources is a slower process with a long pump-down time required to clear the vacuum chamber of the group V alkyl. In order to prevent deterioration of the reaction surface during switching, it is necessary to maintain an overpressure of Gp V species at all times. Switching of the group V species occurs during the interruption of growth, the interrupt time being chosen to ensure that group V pressures have stabilized before growth. Two effects can give rise to compositional grading at the interface: pump-down effects and exchange reactions.

If the interrupt time is too short, then the group V pressures will still be changing as the new layer is growing giving rise to transients. These will give rise to interfacial layers of opposite signs: -ve mismatch at the InP/GaInAs interfaces and +ve mismatch at the GaInAs/InP interfaces,Figure 9. Because As takes longer to pump down than P, the GaInAs/InP interface will tend to be wider. In addition, the greater thermodynamic stability of the As containing solids will also lead to a wider GaInAs interface [19]. However, increasing the interrupt time can add considerably to the growth time of a superlattice structure and in any case cannot be usefully extended beyond the optimum period.

Strained layer(-)   Strained layer(+)

InPInPInP MAsMAsMAsMAsMAsMAs InPInPInP

barrier   well   barrier

Figure 9   Atom layer sequence for InGaAs/InP
          interfaces.

Because there must always be a group V overpressure, there will be a period during the growth interrupt when the wrong group V element will be present (ie. As species over InP or P species over GaInAs). Under these conditions the solid will be thermodynamically unstable with respect to the vapour phase and exchange reactions will occur, giving rise to layers of InAsP (+ve mismatch) at the InP/GaInAs interfaces, and GaInAsP (-ve mismatch) at the GaInAs/InP interfaces. Again, the greater stability of the As containing solids means that interfacial layers, if present, will be more significant at the InP/GaInAs interfaces.

Rocking curves were recorded from a number of samples grown under almost identical conditions. The only change was in the length of the interrupt time. Both the compositional modulation and the uniformity of the superlattice stack were affected by the interrupt time. Continuous growth gave rise to very broad peaks indicating poor composition and thickness uniformity. In addition, the satellite intensity pattern showed strong first order satellites, but a weak zero-order peak. This

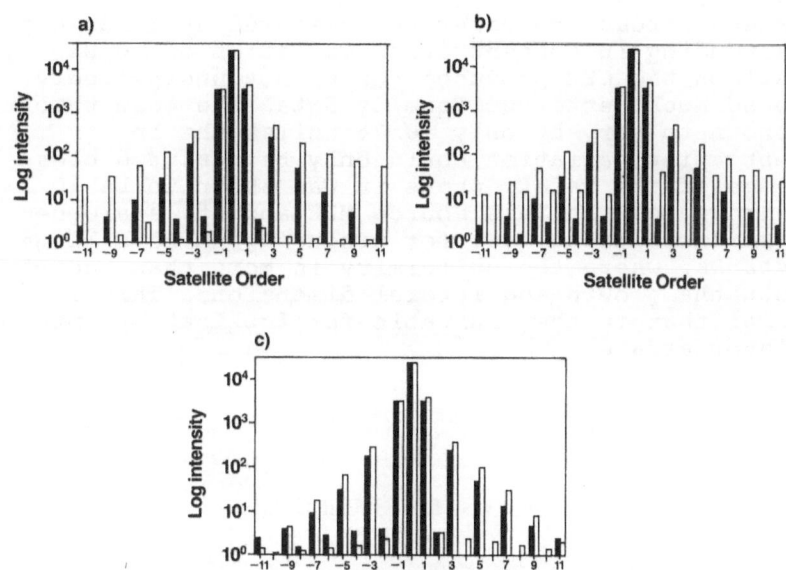

Figure 10   Comparison of experimental(solid blocks) and simulated(open blocks) satellite intensities for periods consisting of a) 21 monolayers InGaAs/21 ML's InP (abrupt Interfaces); b) 22 ML's InGaAs/ 20 ML's InP(abrupt interfaces) and c) 21 ML's InGaAs/ 21 ML's InP(10A interfaces).

20 MLs InP gave rise to a significant increase in the intensities of the even order satellites (fig 10b). Attempts to improve the fit were made by introducing grading at the interfaces. Grading on the group III sub-lattice gave poorer fits than the perfect structure. However, improved agreement between experiment and simulation was obtained by introducing a small amount of grading on either the group V sub-lattice alone, or by assuming equal grading on both sub-lattices. The best agreement between experiment and theory was obtained with a grading ~9-10Å wide on both sub-lattices. The peak intensities assuming a 10Å interface are shown in fig 10c. Wider interfaces resulted in the disappearance of the 11th order satellites, while narrower interfaces predicted more intense high-order peaks than was observed experimentally. If a 9Å interface is assumed, then most of the grading occurs on just two monolayers. Since, as Fewster[20] points out, x-ray measurements do not distinguish between grading and interfacial roughness, this represents a very sharp interface indeed[21].
Having established the degree of perfection of the interfaces produced in the InGaAs/InP MQW stack, the x-ray rocking data can be used to produce area scans measuring both the thickness variation and average alloy composition across the whole 2" wafer. For the MQW stacks produced by gas source MBE the variation in well+barrier

thickness across the wafer was measured as ±4.3% with a
corresponding In concentration variation of ±0.2%. By
comparison the CBE produced MQW's, not unexpectedly,
produced much better uniformity data. The mean thickness
was shown to vary by only ±0.7% whilst the In
concentration variation could only be measured close to
the edge of the wafer, where it was shown to be ±0.1%.
This indicates that gas source MBE and its analogues are
most suitable techniques for producing uniform large area
MQW stacks, where the uniformity is more than acceptable
in both the growth and lateral dimensions. This is
material that is then suitable for fabrication into QCSE
modulator arrays.

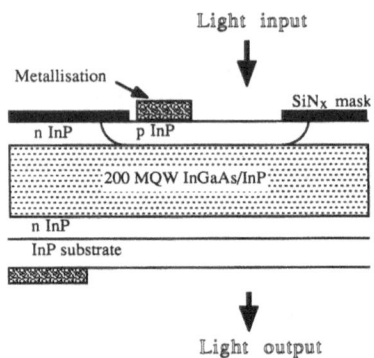

Figure 11   Schematic of QCSE modulator.

A schematic of the QCSE modulator structure is shown in
Figure 11. Planar device processing commences with a
1000A^0 silicon nitride deposition by Plasma Enhanced CVD,
with subsequent definition of windows in the dielectric
by conventional photolithographic techniques. A pn
junction is then formed by zinc diffusion through these
windows to a depth of 1.2um. A reproducible diffusion
depth is obtained by carrying out this stage of the
process in a furnace under a continuous flow of phosphine
and dimethylzinc. Finally, sputtered Ti-Au contacts are
made to individual devices.
A suitable modulator design for use in arrays is an oval
with a total nominal device area of 5900um^2,
incorporating an active area of diameter 45um, together
with an offset bond pad approximately 50um in diameter.
These modulators are set at 125um intervals to enable
easy optical access for subsequent system application,
see Figure 12.
As a consequence of the small device area, and the low
doping of the MQW intrinsic region(gas source MBE), the
capacitance of unbonded modulators is 0.6pF at zero bias
and 0.4pF at an operating voltage of -30V, promising very
fast operation in system use.

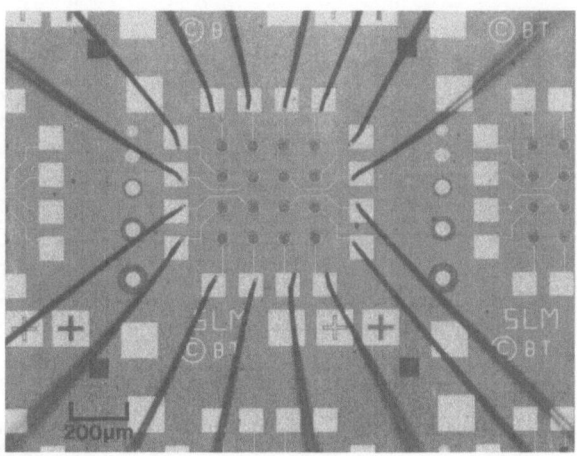

Figure 12   Optical micrograph of a 4x4 modulator
array.

It was mentioned in the introduction that the MQW devices
required a compromise in absorbance(number of layers
grown) and in the field that could be applied across the
device - this is limited by the leakage currents that can
be tolerated and this in turn depends on the residual
doping level in the multilayer stack. The dark currents
of individual devices on a wafer(at appropriate bias
voltages) were measured using an automatic wafer prober.
After selection of appropiate areas of the wafer for
array fabrication, chips from the wafer were cleaved and
then bonded to DIL headers over predrilled apertures,
using a thermally setting conducting epoxy resin. Bonding
to individual modulators was by means of a
thermocompression technique, which enabled the low dark
currents to be maintained through to the completion of
processing of the arrays. Typical reverse bias
characteristics for a 4x4 array are shown in Figure 13.
It can be seen that even at -25V bias the dark current
remains below 100nA for all the devices measured and
there is remarkable uniformity in the data across the
array. This was not true for devices made from the CBE
material at this stage in its development. However, once
the low dark currents have been achieved in CBE grown
material then it is likely to become the preferred
technology.
Absorbance spectra recorded at zero volts bias also
confirm the uniformity of growth and well width already
determined from X-ray measurements. Figure 14 shows
typical absorbance spectra on a device from a 3x3 array
located near the centre of the 2" wafer, with the applied
bias incremented in 5V steps. The difference spectra
shown in Figure 15 demonstrate that 3dB additional
absorption is realised at a bias of -30V and at a
wavelength of 1.525um.

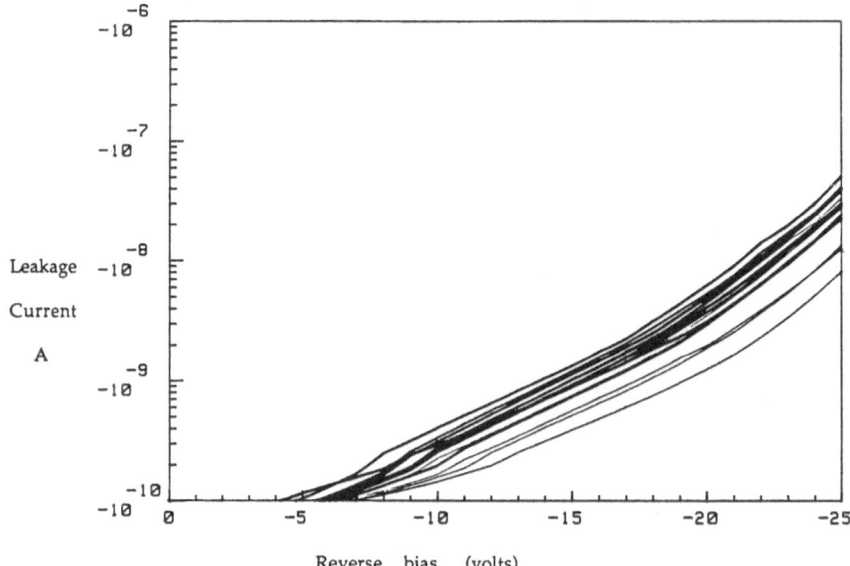

Figure 13   Reverse bias characteristics from each
modulator of the 4x4 array.

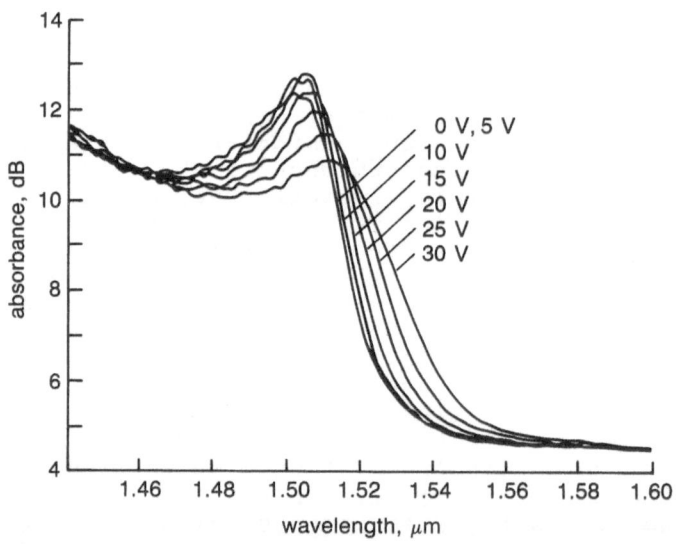

Figure 14   Absorbance spectra of the array
at 5 levels of reverse bias.

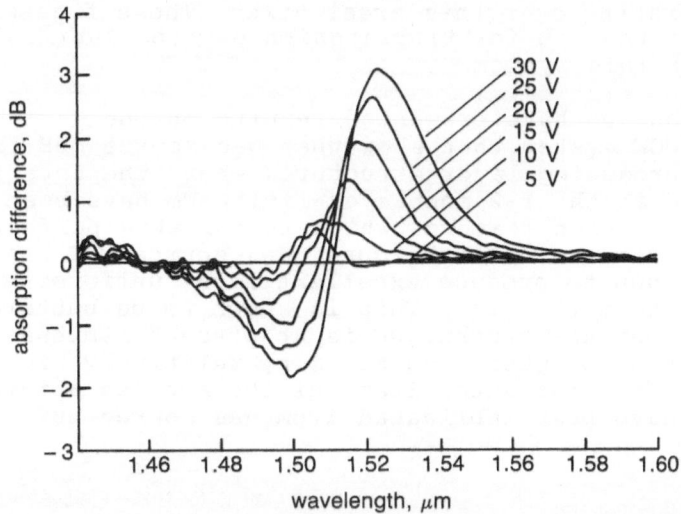

Figure 15 Absorbance difference spectra recorded
at 5 levels of reverse bias.

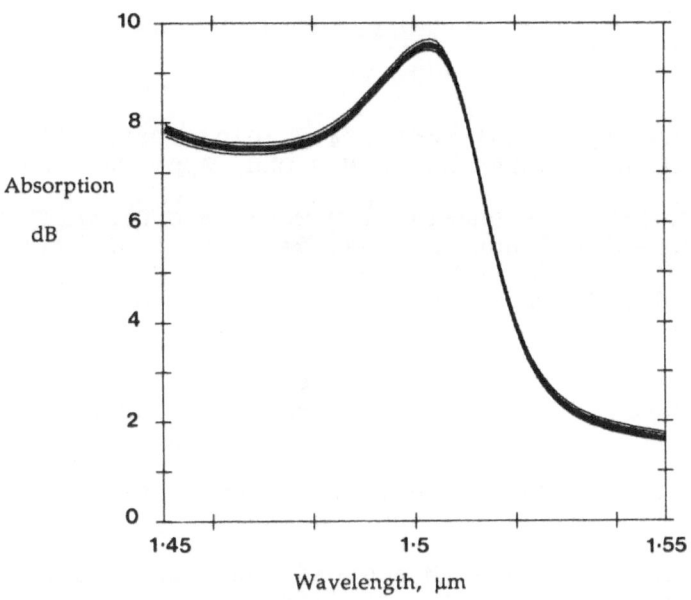

Figure 16 Absorbance spectra from each modulator
in the 4x4 array.

Measurements of the transmission through each device cell of the 4x4 array, Figure 16, show a negligible variation in the absorption edge at zero bias. In operation at the optimum wavelength of Figure 14, this implies a variation of less than 0.1dB in the 3dB figure reported above, for devices operating over this areal array. These figures are state of the art for transmission devices fabricated in this materials system.

In conclusion, we have presented results on the InGaAs/InP MQW system that show that gas source MBE is capable of producing layer structures where the interface integrity is at the 1-2 monolayer limit. We have been able to distinguish features which reveal this perfection by X-ray rocking curve techniques. Gas source MBE has also been shown to produce excellent layer uniformity across the whole 2" wafer. This is shown to be bettered only by CBE but that technique is at present limited in the fabrication of QCSE modulators by relatively high residual leakage currents. State of the art 4x4 arrays of InGaAs/InP have been fabricated from gas source MBE material.

Acknowledgement

We thank the Director of Research and Technology for permission to publish this paper.

REFERENCES

1. W T Tsang, Appl Phys Lett., 49, 1010, 1986.
2. H Temkin, M B Panish and S N G Chu, Appl Phys Lett., 49, 859, 1986.
3. D A B Miller, D S Chemla, T H Wood, A C Gossard, W Wiegmann and C A Burrus, Phys Rev Lett., 53, 2173, 1984
4. D S Chemla, D A B Miller and P W Smith, Opt Eng., 24 556, 1985.
5. W T Tsang, IEEE Circ&Dev., to be published.
6. G J Davies and D A Andrews, Chemtronics, 3, 3, 1988.
7. Y Kawaguchi, H Asahi and H Nagai, Conf. on Solid State
    Devices, Tokyo 1986, 619.
8. J S Foord(Univ Oxford) private communication.
9. E G Scott, D A Andrews and G J Davies, J Crystal Growth, 81, 296, 1987.
10. E G Scott, S T Davey, M A G Halliwell and G J Davies, J Vac Sci Technol., B6, 603, 1988.
11. W J Bartels, J Vac Sci Technol., B1, 338, 1983.
12. M B Panish, J Crystal Growth, 81, 249, 1987.

13. P A Claxton, J S Roberts, J P R David, C M Sotomayor -Torres, M S Skolnick, P R Tapster and K J Nash, J Crystal Growth, 81, 288, 1987.
14. W T Tsang, J Crystal Growth, 81, 261, 1987.
15. M S Skolnick, P R Tapster, S J Bass, A D Pitt, N Apsley and S P Aldred, Semicond Sci Technol, 1, 29, 1986.
16. W Stolz, J Wagner and K Ploog, J Crystal Growth, 81, 79, 1987.
17. M H Lyons and M A G Halliwell, Advanced Materials for Telecommunications eds P A Glasgow, Y I Nissim, J-P Noblanc and J Speight (Paris: Les Editions de Physique), 323, 1986.
18. J M Vandenberg, R A Hamm, A T Macrander, M B Panish and H Temkin, Appl Phys Lett., 48, 1153, 1986.
    J M Vandenberg, M B Panish, H Temkin and R A Hamm, Appl Phys Lett., 53, 1920, 1988.
19. M H Lyons J Crystal Growth 1989 in press.
20. P F Fewster, Philips J Res., 41, 338, 1986.
21. M H Lyons, E G Scott and M A G Halliwell, 'Microscopy of Semicond. Mats.' Inst Phys Conf Ser., 1989, to be published.

QUANTUM WIRES AND RELATED LATERAL SUPERSTRUCTURES

H. Sakaki

Research Center for Advanced Science and Technology (RCAST)
University of Tokyo, Komaba 4-6-1, Meguroku, Tokyo 153

1. QUANTUM WIRES AND OTHER LATERAL SUPERSTRUCTURES (LASS) -CONCEPT AND CLASSIFICATION-

In 1975–6, the first proposal was made to introduce a periodic potential $V(x)$ or $V(x,y)$ of mesoscopic scale within the (x-y) plane of two dimensional electron systems (2DES) (Sakaki, 1975-1976). As shown in Figs 1, this concept of lateral superstructures (LASS) or planar superlattices (PSL) is quite general, since such structures allow the creation of a variety of novel electronic states that cannot be achieved in conventional multilayered quantum structures; they range from isolated quantum wires (QWI) and boxes (QB) to strongly coupled QWI/QB arrays, where the formation of true minibands and minigaps is expected. Although the fabrication of LASS was considered extremely difficult, recent developments of microfabrication technology have blessed the LASS concept and the formation of novel quantum states is demonstrated at least at low temperatures, where the coherence length of electrons gets quite long as compared with the characteristic length of LASS.

Here, we discuss novel properties of LASS and their possible use for the creation of new devices, such as novel FETs and QWI/QB lasers. In particular, we discuss novel features of scattering processes in quantum wires and quantum box arrays to point out the possible suppression of both impurity and optical phonon scattering in such systems. The importance and advantage of adopting quantum wire structures for the realization of quantum interference devices are emphasized . The physical origin for the improved performance of quantum wire and quantum box lasers is discussed.

We emphasize that requirements for the fabrication of these LASS devices are much more stringent than those for the mere observability of quantum state formation. It is because laterally-defined quantum states are formed as long as the following condition (C-1) is satisfied:

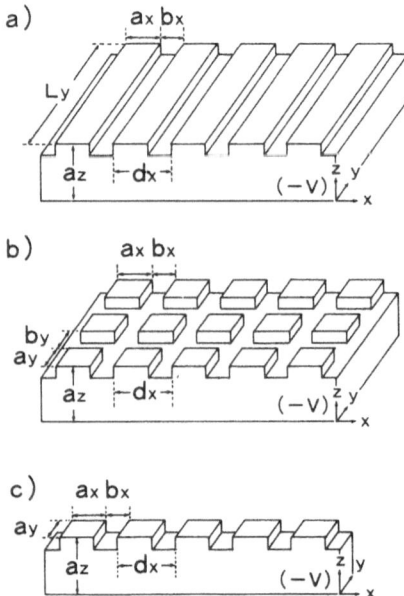

Fig. 1 Various forms of laterally-defined superstructures (LASS). quantum wire (arrays)(a) and quantum box (arrays)(b) and (c). The coupling between adjacent elements leads to the miniband formation.

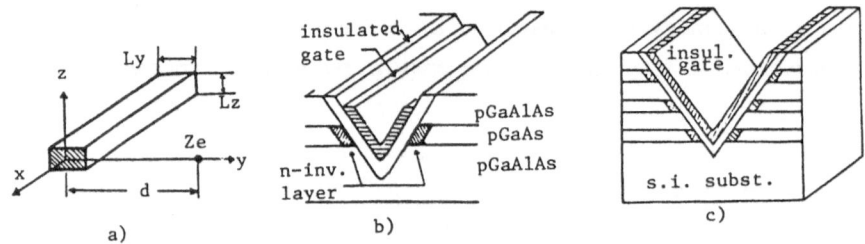

Fig. 2 Various forms of ultrafine semiconductor wires (USW) : a basic structure (a), an insulated-gate FET-type wire (b), which utilizes n-type inversion layers on the edge of layered substrates (INVELS), and the multiple INVELS structures for parallelly-connected operations.

(C-1) the lateral dimension L is small enough that the energy level separation $E(2) - E(1) = 3(\hbar^2/2m)(\pi/L)^2$ is greater than the level broadening ΔE.

In contrast, the most drastic phenomena applicable to novel LASS devices appear only when the following two conditions are additionally fulfilled:

(C-2) the level separation $E(2)-E(1)$ should be greater than the average Fermi energy E_F (or thermal energy kT_e) of carriers so that nearly all carriers are accommodated in the ground level.

(C-3) the quantum level must be sharp with $\Delta E \ll$ carrier energy E_f or kT_e so that the energy distribution of carriers $n(E)$ gets much sharper in QWI/QB systems, while carriers remain unlocalized in QWI systems.

To indicate to which extent these conditions are satisfied by various LASS systems, we call a LASS class-1 system when all the conditions C-1,2,3 are all satisfied. If either C-2 or C-3 is violated, we refer to such systems as class-2 or class-3 LASS systems. Note that it is a class-1 system that is needed in most device applications. For GaAs LASS to be class-1 at room temperature, it is generally necessary to set the lateral dimension L to be less than 200Å. In Sec 6 , recent attempts to prepare LASS with typical lateral dimensions of 100Å are described.

2. SCATTERING PROCESSES IN SINGLE-MODE QUANTUM WIRES BY IONIZED IMPURITIES AND THEIR SIGNIFICANCES

In 1980 it was theoretically predicted (Sakaki 1980) that the ionized impurity scattering of electrons in quantum wires of Fig.2 can be drastically suppressed, if a large number n_e of electrons ($\geq 10^6$ / cm) are accommodated all in the sharply-defined ground subband of a class-1 quantum wire. This is because incident electrons travelling along the wire with the wave number k_i can not be scattered elastically into any states in phase space except a small region in the vicinity of $-k_i$.

Since this backward scattering requires a large exchange of momentum $\Delta k = 2k_i$, its probability is calculated to be quite small. For example, the low temperature mobility limited by bare ionized-impurities is approximately expressed for GaAs wire as 341 cm^2/Vs $\times$ (n_ed) $\exp(2\pi n_e d)$ where n_e is the electron concentration and d the impurity-wire distance. As shown in Fig.3 the mobility reaches as high as 3×10^8cm^2/Vs if n_e is increased to $1 \sim 2\times10^6$ /cm while d is around 200 ~ 100Å.

This suppression of Coulomb scattering process in class-1 quantum wires implies that both the mobility and the mean free path of electrons can be substantially enhanced at low temperatures; the most apparent use of such features is to construct high-speed FETs with this system. Figure 2(b) shows one example of such FETs. One should be aware, however, that advantages of QWI FETs over the conventional modulation-doped FETs appear to be marginal as far as the switching speed is concerned, since the increase of mobility in excess of

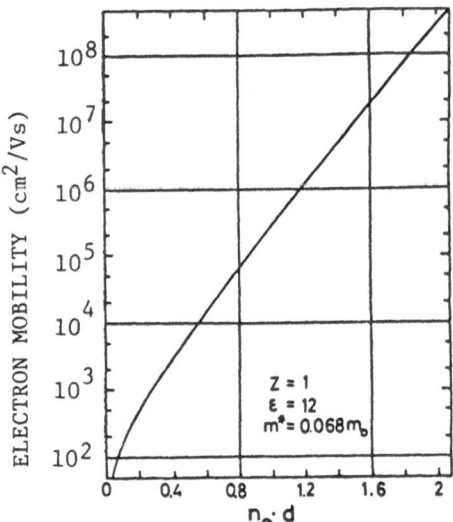

Fig. 3 Impurity dominated mobility of electrons in a class-1 GaAs quantum wire vs electron concentration n_e times d, the impurity-wire distance.

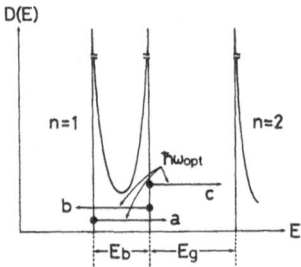

Fig. 4 Schematic illustration of state density $D(E)$ for the ground miniband of a quantum wire superlattice *(QWISL)* as a function of electron energy. Note that neither the absorption (a) nor the emission (b) of optical phonons can take place within the miniband when its width is smaller than the optical phonon energy, which is represented as the length of the arrow. Similarly, the interminiband transition by the absorption of single optical phonon is forbidden when the minigap is wider than the phonon energy (c).

$10^5 cm^2/Vs$ may not much improve the speed limit in usual cases. Hence, the advantage of this system should be explored in other aspects , such as the noise and other characteristics, where the enhancement in coherence of electron-waves and/or the single-mode (SM) feature of QWI systems play greater roles. This will be described in section 4.

3. OPTICAL PHONON SCATTERING IN QUANTUM BOX ARRAYS AND QUANTUM WIRE SUPERLATTICES

It has been recently pointed out by Sakaki(1989) that QWI superlattices or QB arrays of Fig.1(c) can provide a unique method to suppress optical phonon scattering in semiconductors. This unique feature may arise when the miniband structure of such systems satisfies the following three conditions.

(S-1) the first minigap is much wider than the average kinetic energy (E_f or kT_e) of electrons so that electron transport takes place only within the ground miniband.

(S-2) the width of ground miniband is narrower than the optical phonon energy E_{op}(~30meV) so that neither the optical phonon emission nor its absorption is allowed within the miniband.

(S-3) the first minigap is wider than Eop so that no inter-miniband scattering is allowed (except through a multiple phonon process).

Figure 4 shows the density of states of electrons in QWI superlattices for this situation, where no pairs of states can be communicated by the absorption or emission of optical phonons. If the optical phonon scattering is indeed suppressed by this scheme, electrons may exhibit very high mobility($>10^5 cm^2/Vs$) even at high temperatures (Te = 200 ~ 300K) and at high electric fields . The transport in such a miniband will be quite non- linear and may well lead to Bloch-oscillation and other coherent phenomena even at high temperatures. Note that when the miniband width is 20 meV and the period of a superlattice is 100Å, the maximum group velocity of electrons in the QWI superlattice can be as high as $3\times10^7 cm/s$ and can supply reasonably large current or power if a large number of such structures are connected in parallel. A novel scheme to prepare such QWI superlattice will be discussed in section 6.

4. DIFFRACTION AND INTERFERENCE PHENOMENA AND THEIR APPLICATION TO COHERENT TRANSISTORS

As stated in Secs.2 and 3, class 1 quantum wires may serve as a system with small scattering probability and long coherence length. This feature is advantageous in various devices, where the coherent interaction of multiple electron waves play vital roles. The planar superlattice FETs of Fig. 5 first proposed by Sakaki (1975-6) is the first of such examples. This coherent FET makes use of Bragg interaction (diffraction) of electron waves with an artificial periodic potential V(x). This Bragg FET is predicted to exhibit both gate-controlled negative resistance as well as negative transconductances, which were indeed demonstrated recently by Ismail et al(1989). For the same reason, the use of high-mobility single mode (HM-SM) quantum wires is highly desirable in constructing the electrostatic Aharonov-Bohm FETs proposed by Datta et al (1986), because electron wave interference in a ring geometry can be effective only when the electrons maintain their phases and modes. Similarly, the recent discovery of Van Wees et al. (1988) on the resistance quantization of ballistic electrons in ultrashort QWI systems might benefit also from some mobility enhancement since the finiteness of the number of subbands in QWI systems may also contribute to the reduction of scattering rate to some extent.

5. FEATURES OF QUANTUM-WIRE AND QUANTUM-BOX LASERS AND THEIR ORIGINS

In 1982, the use of quantum wires (QWI) and/or boxes (QB) in the active region of laser diodes (LD) was first proposed and analyzed (Arakawa and Sakaki, 1982). Further analysis of such lasers were made subsequently by Asada et al (1986) and Arakawa and Yariv (1986). It has been theoretically found that both the threshold current J_{th} and its temperature variation (dJ_{th}/dT) should dramatically decrease, while the high-frequency response (the maximum modulation frequency) should improve by a factor of $3 \sim 6$. These predicted improvements are due mainly to the sharpening of gain spectrum $g(h\nu)$ or the density-of-states functions of QWI/QB systems (from the quasi-continuum one to δ-function-like one), though some advantages result from the smallness of active layer volume. One should note that the substantial sharpening of the gain spectrum can be achieved only when the two-conditions (C-2) and (C-3) discussed in Section 1 are fulfilled and all the injected carriers are accommodated in the sharply-defined ground level.

6. RECENT ATTEMPS FOR THE FABRICATION OF CLASS-1 LATERAL SUPERSTRUCTRES WITH TYPICAL DIMENSIONS OF 100Å

It is now evident that the most of drastic physical phenomena and their device applications are to be realized only in class-1 LASS system, where the ground quantum level is sharply defined ($\Delta E \ll E_f, kT_e$) and accommodates the major portion of carriers ($E(2) - E(1) > E_f, kT_e$). Since E_f and/or kT_e of typical devices are of the order of 10meV under normal operating conditions, it is important to set a large energy level spacing ($E(2) - E(1) \gg 10meV$) and a small level broadening ($\Delta E < 5meV$) at the same time. This means that one must establish fabrication technology by which one can create various LASS with characteristic length < 250Å with tolerable geometrical inhomogeneities. Despite remarkable progress in lithography, these dimensions are quite difficult to handle. Hence, it appears highly desirable not only to advance the lithographic approach to its limit but also to look for alternative methods to prepare such LASS systems without resort to lithography.

Sakaki proposed to form a QWI structure by using inversion layers on the edge of multilayered quantum wells(1980) which is shown in Fig. 2 (b) . While the preparation of such structures was difficult because of the Fermi level pinning phenomenon at processed GaAs surfaces, the recent progress in the selective facet growth of quantum well structures has

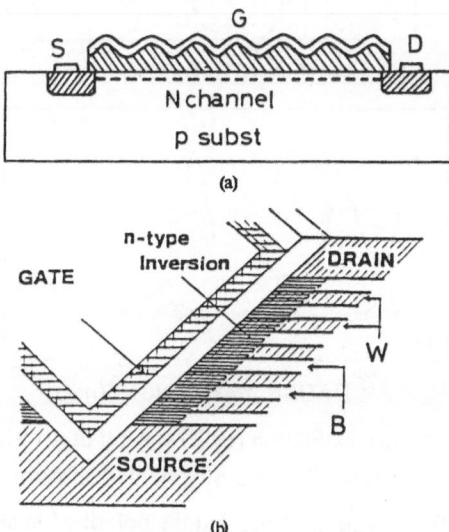

(a)

(b)

Fig. 5 Proposed structures of planar (in-plane) superlattices with gate electrode, in which two-dimensional electrons interact with periodic potential S, D, G, I, and Ch stand for source, drain, gate, insulator and channel, respectively.

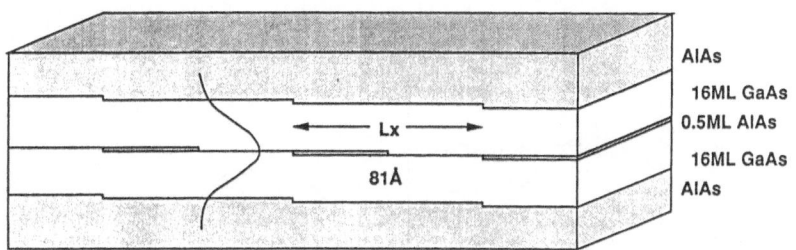

Fig. 6 Schematic illustration of the grid-inserted quantum well (GI-QW) structure, where a periodic array of AlAs bars is inserted in the center of GaAs quantum wells.

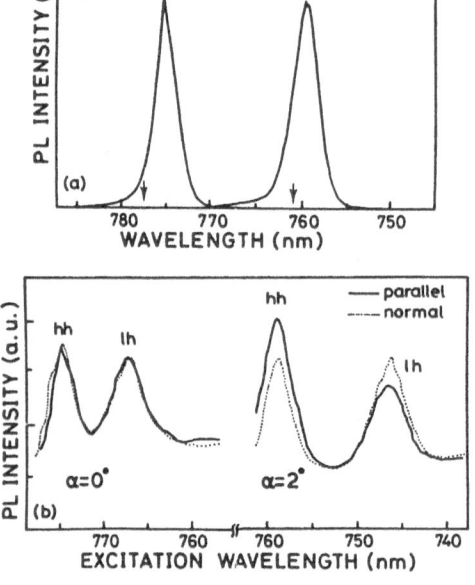

Fig. 7.(a) The photoluminescence excitation (PLE) spectra of grid-inserted quantum wells grown on a flat (100) substrate ($\alpha = 0°$) and a misoriented substrate ($\alpha = 2°$) which is tilted by 2 degrees from the (100) orientation. Note that the polarization angle φ of excitation laser light is set either parallel ($\varphi = -45°$) or normal ($\varphi = +45°$) to the grid. The presence of polarization dependence in the PLE peaks of electron-heavy hole (e-hh) excitons and those of electron-light hole (e-lh) excitons indicates the grid-induced anisotropy in the planar superlattices.

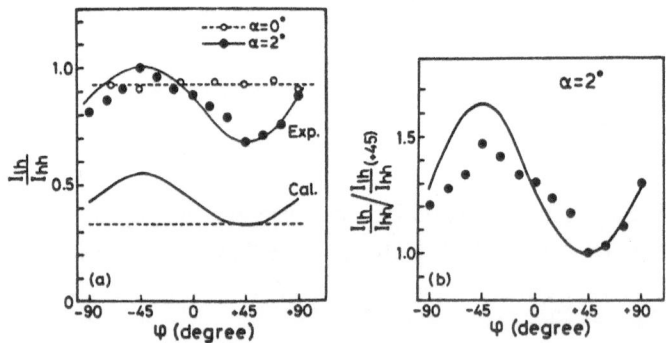

Fig. 7.(b) Ratio I_{lh}/I_{hh} of the peak height of e-hh and e-lh exciton in the PLE spectra of $\alpha = 0°$ and $\alpha = 2°$ is plotted as a function of φ after normalizing it by the ratio at $\varphi = +45°$. The theoretical predictions are also shown.

nearly allowed the fabrication of such systems (Fukui (1989)). As the second alternative approach, Petroff proposed a brilliant method of depositing half-monolayers of GaAs and AlAs alternatingly on vicinal GaAs substrates, where quasi regular steps are expected to be formed (1984). Recent experiments by Fukui et al (1987), Petroff et al. (1988) and Tanaka and Sakaki (1988,1989) have given positive evidence that the formation of laterally modulated structures with short periods (80 ~ 160Å) is indeed possible by the second method, although the composition modulation along the x-axis is not ideally stepwise. Figures 6 shows the structure of a novel planar superlatice, where AlAs grids are inserted in GaAs QWs with typical period of 80 Å (Tanaka, Sakaki 1989). Figure 7 shows the polarization dependent optical absorption of such structures, which indicates clearly the formation of PSL states. Future developments in the material science, electron physics, and devices in LASS systems are likely to open new fields for exciting physics and electronics.

References

For recent review and important references that can not be listed, see,for example, "Physics and Technology of Submicron Structures" ed by H.Heinrich et al(Springer) 1988

Arakawa Y and Sakaki H 1982 Appl. Phys. Lett. **40** 893

Arakawa Y and Yariv A 1986 IEEE J. QE **QE-22** 1887

Asada M, Miyamoto M and Suematsu Y 1986 IEEE J QE **QE-22** 1915

Datta S, Mclloch M. R, Bandyopadhyay S and Lundstrom M. S 1986 Appl. Phys. Lett **48** 487.

Esaki L and Tsu R 1970 IBM J. Res. Developm. **14** 61

Ismail K, Chu W, Yen A, Antoniadis D. A and Smith H. I , 1989 Appl. Phys. Lett. **54** 460

Fukui T and Saito H 1987 Appl. Phys. Lett. **50** 824

Fukui T and Ando S 1989 Electronic Lett. **25** 410

Petroff. M, Gossard A. C and Wiegmann W 1984 Appl.Phys. Lett. **45** 620

Petroff. M, Gaines J, Tsuchiya M, Simes R, Coldren L, Kroemer H, English J and Gossard A. C 1989 J. Cryst. Growth **95** 260

Sakaki H and Sugano T 1975 OYO-BUTSURI (J. of Jpn. Soc. Appl. Phys.) **44** 1131 and Japanese patent.

Sakaki H, Wagatsuma K, Hamasaki J and Saito S 1976a Thin Solid Films **36** 497

Sakaki H, Wagatsuma K, Hamasaki J and Saito S 1976b SEISAN-KENKYU (J.Inst,Industrial Science, University of Tokyo) **28** 34

Sakaki H 1980 Jpn.J.Appl.Phys. **19** L735

Sakaki H 1989 Jpn J. Appl. Phys. (Lett) **28** L314

Tanaka M and Sakaki H 1988 Jpn J.Appl.Phys. **11** L2025

Tanaka M and Sakaki H 1989 Appl. Phys. Lett. **54** 1326

Van Wees B J, Van Houten H, Benakker C W J, Williamson J G, Kouvenhoven L P, van der Marel D, Foxon C T 1988 Phys. Rev. Lett. **60** 848

SPECTROSCOPY ON ONE-DIMENSIONAL ELECTRONIC SYSTEMS

T. Demel, D. Heitmann, and P. Grambow

Max-Planck-Institut für Festkörperforschung
Heisenbergstr. 1, 7000 Stuttgart 80, FRG

Single and multi-layered quantum wire structures have been prepared starting from modulation doped $AlGaAs/GaAs$ heterostructures and multi-
quantum well systems. The energy spectrum of these one-dimensional
electronic systems (1DES) can be characterized by dc magnetotransport
measurements. Quantum wires with 400 to $150nm$ wide electron channels
exhibit typical energy separations of 1 to $3meV$ for the 1D subbands. The
far infrared (FIR) response of these systems is strongly governed by collective effects, which gives the FIR resonances the character of local plasmon
modes. In the multi-layered quantum wire structures optical and acoustical
type of layer-coupled plasmon modes are observed.

INTRODUCTION

One-dimensional electronic systems (1DES) have recently attracted an increasing
interest.[1-17] This interest was initiated in an interactive way by a remarkable improvement of the preparation techniques and by many novel experimental and theoretical
results. 1DES is used here for electronic systems with an energy spectrum that consists
of a set of 1D subbands (i,j= 0,1,2,3,..)

$$E^{ij}(k_y) = \frac{\hbar^2 k_y^2}{2m^*} + E_x^i + E_z^j \quad , \tag{1}$$

where the electrons have a free dispersion only in y-direction (k_y is the electron wave
vector). Such systems are advantageously prepared by starting from originally 2DES[18],
e.g., in layered high-mobility modulation doped $AlGaAs/GaAs$ heterostructures or
quantum well systems or in metal-oxide-semiconductor devices. E_z^j are the quantized
energy levels in the original 2DES with typical separations ranging from 10 to 100
meV, z denotes the direction normal to the layered structures. E_x^i are the quantum
confined energy states due to the additional lateral confinement acting in x-direction.

To achieve for instance a lateral quantization of 2 meV in the $GaAs$ system with
$m^* = 0.065m_o$, one has to confine the electrons into a width w of some 100 nm.
Different arrangements have been used to realize such systems, e.g., split-gate configurations, [1,4,8-13,17] 'shallow'[2] and 'deep'[14-16]-mesa etched structures, or isolation of

75

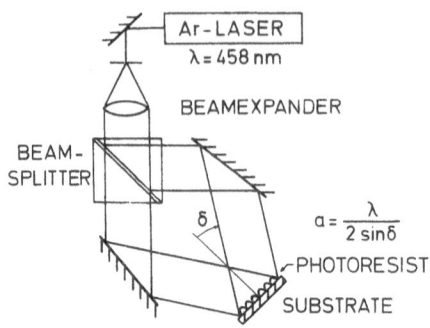

Fig. 1 Schematical set-up for holographic lithography to produce structures with small periodicities.

Ar-LASER
$\lambda = 458\,nm$

BEAMEXPANDER

BEAM-
SPLITTER

δ

$a = \dfrac{\lambda}{2\sin\delta}$

PHOTORESIST
SUBSTRATE

electron channels by ion beam bombardment[5] through narrow masks or with focussed ion beams. These 1DES exhibit unique physical properties, e.g., in the ballistic regime the transport is governed by quantum interference effects.[9,10] In this review we like to discuss some selected dc and in particular far infrared (FIR) experiments on 1DES which highlight some characteristic properties of these systems.

PREPARATION OF 1DES BY HOLOGRAPHIC LITHOGRAPHY AND DRY ETCHING TECHNIQUES

To realize 1DES, lateral structures of typically some 100nm are needed. A very efficient way to produce narrow structures is the method of holographic lithography which is sketched in Fig.1. The superposition of two coherent laser beams results in a sinusoidally modulated intensity pattern which is used to expose a photoresist layer on top of the sample. After the development periodic photoresist stripes are achieved. The periodicity, a, can be controlled via the angle of incidence of the laser beams and the width of the stripes, t, can be varied in a certain range by the exposure and developement time. This holographic method has the advantage that many parallel lines with excellent homogeneity over large areas, both in periodicity and width, can be prepared, which is favourable for many experiments, e.g., optical experiments, FIR and microwave spectroscopy. The photoresist stripes are the starting point for an additional processing to prepare different configurations.

In Fig.2 some of the prepared systems[14-17] , which will be discussed below, are sketched. Figure 2a shows a modulation doped $AlGaAs/GaAs$ heterostructure with photoresist stripes. The whole system is covered with a thin, for FIR radiation semi-transparent film of NiCr which acts as a gate. The photoresist is insulating, thus, if a gate voltage V_g is applied between this gate and the channel, it is possible to deplete the carriers totally in the regime with the narrow distance between gate and channel. This arrangement is similar to a split-gate device. Instead of the photoresist it is also possible to achieve a modulated gate distance by starting with a thicker cap layer (50-300nm) and etch a profile into this cap layer (Fig.2b). In another technique, called 'shallow'-mesa etching, the doped $AlGaAs$ is etched, leaving an 1DES below the remaining doped $AlGaAs$ stripes (Fig.2c). 'Deep'-mesa etched structures are shown in Fig.2d and e. Here, starting from modulation doped $AlGaAs/GaAs$ systems, the etching was performed all the way down into the active $GaAs$ layer. It is not easy to realize the latter structures. The etched surfaces create surface states which trap electrons and deplete the channel laterally. With an optimized $SiCl_4$ process it was possible to overcome this difficulty. We could fabricate structures narrow enough to achieve quantum confinement and nevertheless still had mobile electrons in the system. These methods have been discussed in Ref. 14. The deep-mesa etched structures have some unique advantages. Gated structures normally do not work under illumination with bandgap radiation due to photoinduced leakage currents. Thus deep-mesa etched systems are ideal for photoluminescence and Raman spectroscopy. The deep-mesa etching technique also enables the fabrication of multi-layered quantum wire structures as sketched in Fig.2e by starting from modulation doped multi-quantum well systems. Such systems promise interesting technical applications for the integration of 1DES devices in multi-layered arrangements. These systems also show unique physical properties which are interesting, e.g., to elucidate the dynamic response of 1DES.

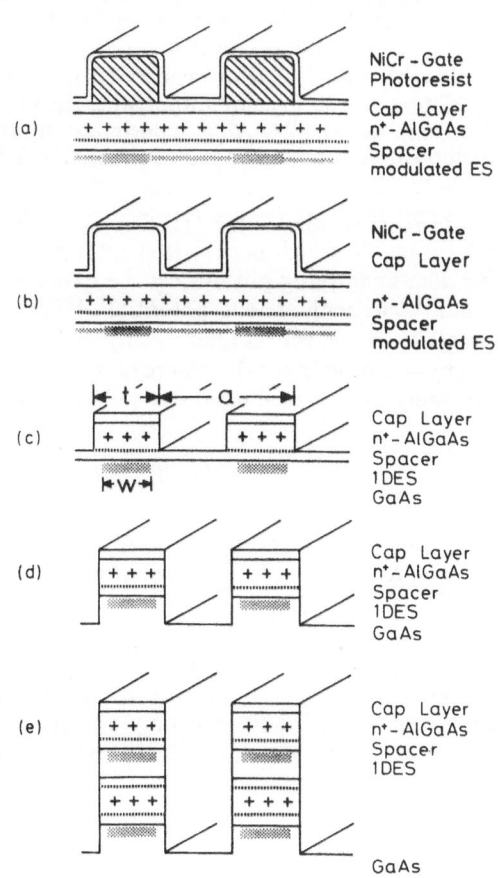

Fig. 2 Some examples of realized periodic arrays of quantum wire structures. (a) and (b) are split-gate configurations where via a gate voltage and a varying distance between gate and channel carriers are depleted leaving isolated quantum wires. In (a) the gate-distance modulation is achieved via a modulated photoresist, in (b) via an etched cap layer. (c)-(e) sketch examples for mesa etched quantum wire structures: (c) shows schematically a 'shallow'-mesa etched structure, (d) a 'deep'-mesa etched single-layer structure, (e) a 'deep'-mesa etched double-layer structure.

Starting material for most of our structures were conventional modulation doped heterostructures. Typically they consisted of a $GaAs$ buffer layer, a $25nm$ $Al_xGa_{1-x}As$ spacer, $50nm$ of n-doped $Al_xGa_{1-x}As$ ($N_d = 1 \cdot 10^{18} cm^{-3}$, $x = 0.3$) and a $10nm$ $GaAs$ cap layer. For structures in Fig.2d we started from electronically decoupled modulation doped MQW systems with two periods where each period consisted of a $25nm$ undoped $Al_xGa_{1-x}As$ spacer, followed by a short-period $AlAs/GaAs$-superlattice (5 periods, each layer $22nm$ thick), a $50nm$ $GaAs$-QW, a $5nm$ $AlAs$ spacer, and a $25nm$ n-doped $Al_xGa_{1-x}As$ layer ($N_d = 1.5 \cdot 10^{18} cm^{-3}$, $x = 0.3$). The cap layer consisted of $25nm$ undoped $Al_xGa_{1-x}As$ ($x = 0.3$) and $2nm$ of $GaAs$. With deep-mesa etching we have prepared multi-layered quantum wire systems with up to five layers.

A first approximation to calculate the energy levels in a 1DES, e.g., in a deep-mesa etched structure as sketched in Fig.2d, is to use an infinite rectangular well model. We find

$$E_x^i = \frac{h^2}{8m^*w^2} \cdot i^2 \quad , \tag{2}$$

where w is the geometrical width. However this is only a very rough approximation. The actual potential depends on charged surface states at the side walls, in other words, on the Fermi level pinning at the surfaces. It depends on the electric field of the remote ionized donors in the $AlGaAs$ stripes as well and on the electrons in the channel itself due to self-consistent screening. The most sophisticated calculations for 1DES so far have been performed for the split-gate configuration (Fig.2b) by Laux et al. [19] It is found that for a small number of electrons in the channel the potential is nearly parabolic. However, with increasing 1D-charge density N_l, very soon self-consistent effects become important, the potential flattens, and the subband separation decreases drastically. Also experimentally it is not easy to characterize the 1DES, i.e., to determine subband spacing, channel width and number of electrons per 1D subband.

In most studies so far 1DES are characterized by the magnetic depopulation of the 1D subbands which occurs if the 1DES is exposed to a perpendicular magnetic field B.[1] The underlying physics can be explained without loss of generality if one assumes a parabolic confinement potential[1] $V(x) = \frac{1}{2}m^*\Omega_0^2 x^2$. In this case the Schrödinger equation can be solved analytically. The magnetic field induces an additional potential $V_B(x) = \frac{1}{2}m^*\omega_c^2(x-x_0)^2$, where $\omega_c = eB/m^*$ is the cyclotron frequency. For this model the energy levels in a magnetic field are given by

$$E^i(k_y, B) = \hbar\Omega(i + \frac{1}{2}) + \frac{\hbar^2 k_y^2}{2m_y^*(B)} \tag{3}$$

with $\Omega^2 = \omega_c^2 + \Omega_0^2$, and $m_y^*(B) = m^*\Omega^2/\Omega_0^2$. This energy spectrum is shown in Fig.3. Since the 1D density of states, $D_{1D}(E, B)$, increases with increasing B the 1D subbands become successively depopulated (Fig.3b), giving rise to oscillations of the Fermi energy. In a transport measurement this leads to Shubnikov-de Haas (SdH) type of oscillations. In a 2DES the number of occupied Landau levels increases with decreasing B, leading, ideally, to an infinite number of SdH oscillations periodic in $1/B$.[18] In a 1DES however, only a finite number of 1D subbands is occupied at B=0, thus only a finite number of SdH oscillations occurs, which are no longer linear in $1/B$.

We would like to demonstrate this behaviour for two types of our samples as shown in Fig. 2c and e. We performed magnetotransport measurements at low temperatures (T=2.2K) in perpendicular magnetic fields B. We defined on some of the one-layered quantum wire structures an active area of $2.5 \times 2.5mm^2$ by chemical mesa etching. Ohmic contacts of a Au/Ge alloy were aligned perpendicularly to the grating in order to measure the dc transport parallel to the stripes. On other samples a quasi-dc response was obtained by measuring the transmission of microwaves (30-40 GHz) through the sample. Since for this measurement no contacts were needed, it was especially useful for samples with multi-quantum well (MQW) wires, because alloyed contacts would have short-circuited the channels of different layers.

In Fig. 4a we show measurements on a shallow-mesa etched sample (Fig.2c). The period of the wires was $a = 500nm$, the remaining width of the n-doped $AlGaAs$ was $t = 250nm$. Via Ohmic contacts the two-terminal resistance was measured as a function of the magnetic field. The dc conductivity shows well pronounced SdH-type oscillations. The important point is that the period of the oscillations is not constant in $1/B$, but exhibits distinct deviations at large values of $1/B$ and correspondingly small B (Fig. 4b). This clearly indicates the 1D character of our structure as was discussed above for the harmonic oscillator model. We calculated the depopulation

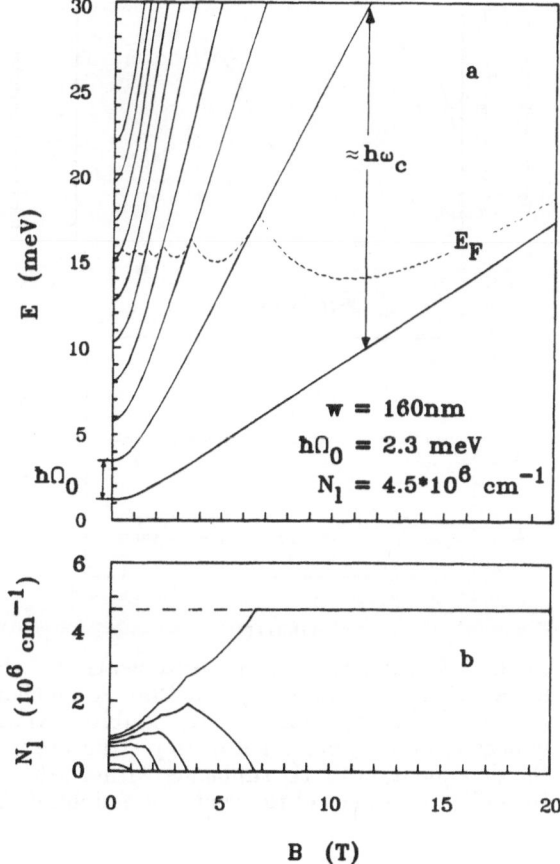

Fig. 3 Energy spectrum for electrons confined in a harmonic oscillator potential and in a magnetic field B. The dotted line shows the oscillations of the Fermi energy for the indicated linear charge density N_l. The lower part shows N_{li} in the different 1D subbands i. The hybrid 1D-subband- Landau-levels become successively depopulated with increasing B.

of the 1D subbands by a magnetic field within this model using Ω_0 and the total 1D carrier density N_l as fitting parameters. The experimental fan chart in Fig. 4b was best described for $\hbar\Omega_0 = 2.3 meV$ and $N_l = 4.5 \cdot 10^6 cm^{-1}$. For these values, six 1D subbands were occupied at B=0. Defining the width w of the electron channel by the amplitude at the Fermi energy: $E_F(B=0) = V(\frac{w}{2}) = \frac{1}{2}m^*\Omega_0^2(\frac{w}{2})^2$, we found $w = 160nm$ which was smaller than the geometrical width $t = 250nm$. Therefore we defined a 'lateral edge depletion region' on either side of length $w_{dl} = \frac{1}{2}(t - w) \approx 45nm$.

'Deep'-mesa etched structures, as shown in Fig. 2d and e, were prepared with different dimensions a and t. Our experience was that for profiles with t about $400nm$ or smaller, no free electrons were left in the channels. This was detected both from zero dc conductivity and also from a missing FIR response. The explanation was that all the carriers were trapped in etching-induced surface states. However, for structures with $t = 550nm$ and periodicities $a = 1100nm$ we found both, dc conductivity and FIR response. As an example we show in Fig. 4c the quasi-dc conductivity of a two-layered quantum wire structure (Fig.2d), measured in microwave transmission as described above. SdH oscillations can be clearly resolved and show again distinct deviations from a linear 1/B behaviour at small B. An analysis within the harmonic oscillator potential model gives for the 1D confinement $\hbar\Omega_0 = 1.5meV$, $N_l = 15 \cdot 10^6 cm^{-1}$ and 16 occupied 1D subbands. From the channel width $w = 320nm$ we deduced that there is a lateral depletion length $w_{dl} = 100nm$ on either side of the wire. Similar values were measured on a number of samples which all had the same doping and were prepared with the same etching process as described above. We expect that w_{dl} should depend particularly on the doping concentration and the etching process. A value of $w_{dl} \approx 500nm$ was reported for chemical mesa etched stripes whose widths ranged from $800nm$ to $2400nm$.[20] Thus with our technology we achieved drastically smaller edge depletion regions which is the crucial point to realize 1DES.

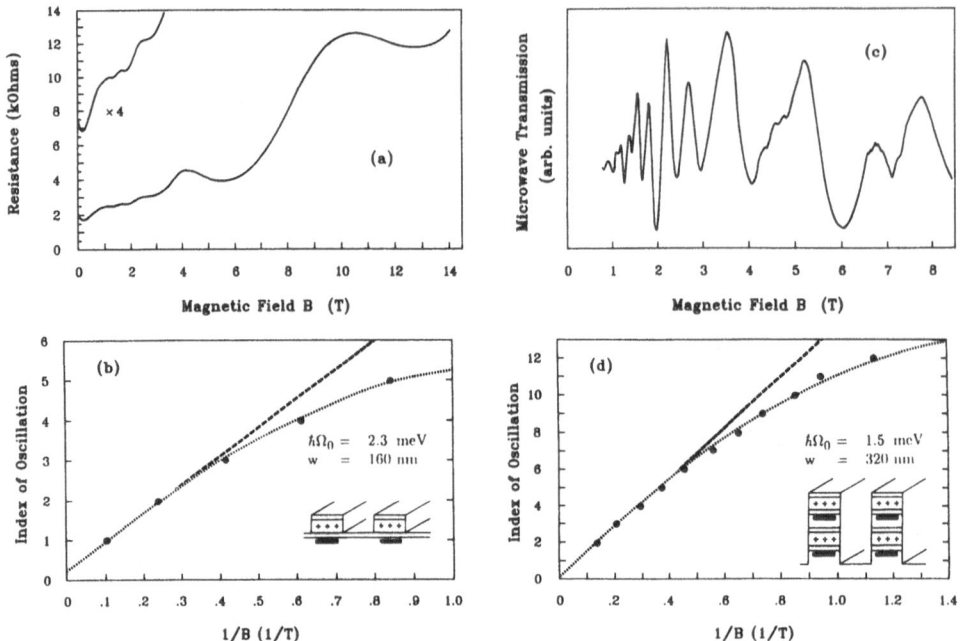

Fig. 4 Magnetotransport measurements on a shallow-mesa etched (a) and on a deep-mesa etched two-layered (c) structure. A fan chart, (b) and (d), for the positions of the maxima in the magnetoresistivity exhibits deviations from a linear 1/B dependence and indicates a one-dimensional energy structure in each sample. The dotted lines show the depopulation of 1D subbands within a harmonic oscillator model, calculated for the indicated values of the confining potentials $\hbar\Omega_0$ and the channel widths w.

FAR INFRARED SPECTROSCOPY ON QUANTUM WIRE STRUCTURES

We have seen above that the $GaAs$ quantum wire structures show typical subband separations of some meV. Thus FIR intersubband resonance spectroscopy appears to be a very promising method to determine the 1D bandstructure. However, it is found that the optical response exhibits a very complex behaviour, where the subband separation can be derived only indirectly. Here in particular the measurements on the multi-layered quantum wire structures are very helpful to understand the nature of the excitations.

The FIR measurements were performed in a superconducting magnet cryostat, which was connected via a waveguide system to a Fourier transform spectrometer. The transmission $T(B)$ through the sample was measured at a temperature of $2.2K$ at fixed magnetic fields B, oriented normally to the surface of the sample. The spectra were normalized to a spectrum $T(B_0)$, where B_0 was chosen in such a manner that the reference spectrum $T(B_0)$ was flat in the frequency region of interest. The resolution of the spectrometer was set to $0.5cm^{-1}$. The FIR measurements were performed in the same set up and under identical conditions as discussed for the characterization of the samples by the dc magnetotransport measurements above.

In Fig.5 we show experimental FIR spectra for the quantum wire structures. At B=0 we observe for the one-layered quantum wire structure (Fig. 2d) a resonance at $\omega_r = 30cm^{-1}$ if the incident electric field is polarized perpendicular to the wires. The resonance shifts with increasing B to higher frequencies. With increasing B a resonance is observed also for parallel polarization with exactly the same resonance frequency as for perpendicular polarization. A very similar behaviour is also found for the multi-layered quantum wire structures. However, the important point is, that in

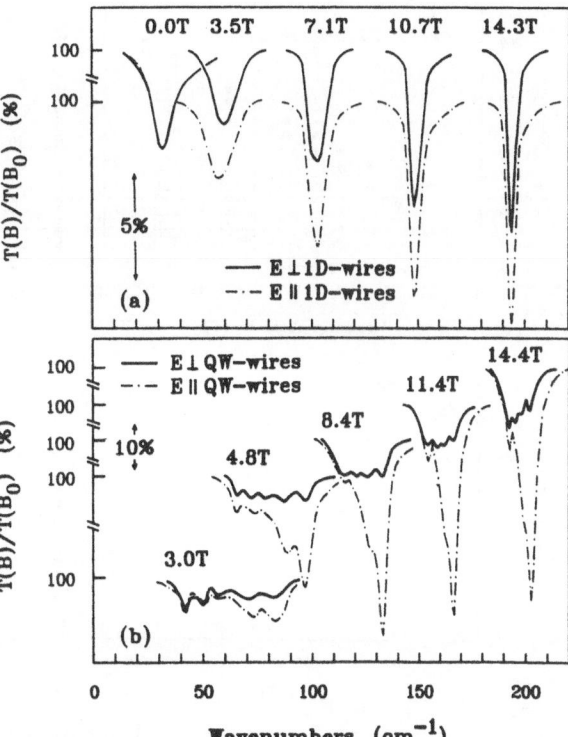

Fig. 5 Experimental FIR spectra, measured on a deep-mesa etched one-layered quantum wire structure (a) and on a five-layered quantum wire structure (b) at indicated magnetic fields B. Full lines and dash-dotted lines denote, respectively, polarization of the incident FIR radiation with the electric field vector perpendicular and parallel to the wires.

a n-layered system up to n resonances are observed, e. g., for the five-layered system in Fig.5b five resonances. The experimental resonance positions for a the two-layered quantum wire system are plotted in Fig.6a on a linear scale, in Fig.6b on a quadratic scale, ω_{ri}^2 versus B^2. From the latter graph we find that the two resonances ω_{r1} and ω_{r2} obey the relation $\omega_{ri}^2(B) = \omega_{ri}^2(B=0) + \omega_c^2$. The same dependence was found also for the single resonance in the one-layered system and for the five resonances in the five-layered structures.

The most surprising result is that these resonance frequencies ω_{ri} in the FIR spectra are significantly higher in energy than one would expect from the 1D subband separation $\hbar\Omega_0$, which was determined from the dc magnetotransport measurements, i.e., at $B = 0$ for the one-layered quantum wire system: $\hbar\Omega_0 = 1meV$, $\hbar\omega_r = 4meV$, for the two-layered system: $\hbar\Omega_0 = 1.5meV$, $\hbar\omega_{r1} = 8meV$, $\hbar\omega_{r2} = 4meV$. However, it is well known that one has to be careful with the interpretation of optical spectra. E.g., for 2DES the FIR intersubband resonance ω_r is shifted with respect to the one-particle transition due to the resonant screening effect of all electrons in the system.[21,18] This collective so-called depolarization effect is characterized by an effective plasma frequency ω_d so that the observed resonance frequency may be written as $\omega_r^2 = \Omega_0^2 + \omega_d^2$. Using this model we find that for the $GaAs$ quantum wires the 1D-intersubband transition is strongly governed by the depolarization effect. E.g., for the one-layered quantum wire structure (Fig.5a) it is $\omega_d^2 = \omega_r^2 - \Omega_0^2 = (4meV)^2 - (1meV)^2 = (3.9meV)^2$. A strongly dominating collective contribution, that was observed here in the mesa etched quantum wire structures, was also found in 1DES of $GaAs$ systems with split-gate configuration (Fig.2a) (see below). After the experimental observation of the FIR resonances in 1DES there have been several theoretical treatments (e.g. Refs. 22-24) where model 1D wave functions are assumed and the optical resonance frequencies are calculated. A general treatment is a very cumbersome task. Thus usually the assumption is made that only one or two 1D subbands contribute to the response. The result of such a calculation in the case that only the lowest subband is occupied in an isolated 1D wire is, e.g., according to Ref. 24,

$$\omega_{res}^2 = \omega_{10}^2 + \omega_{10}e^2 N_l/(\pi\epsilon\epsilon_0) \quad . \tag{4}$$

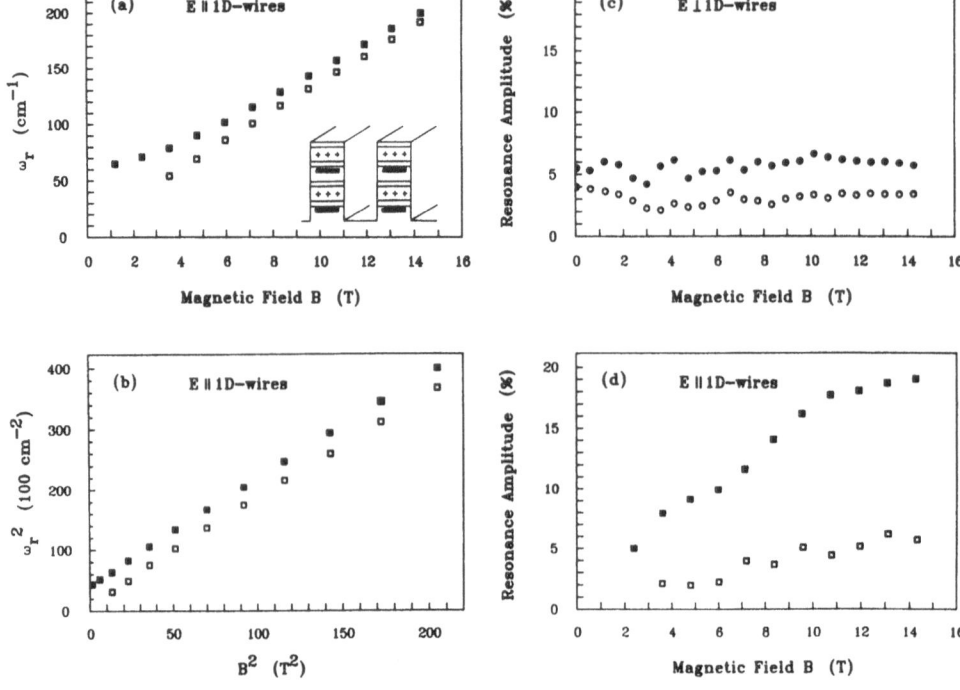

Fig. 6 Experimental resonance positions from FIR spectroscopy on the two-layered quantum wire system are shown on a linear (a) and on a quadratic (b) scale. (c) and (d) depict the resonance amplitude for perpendicular and parallel polarization.

Here ω_{10} is the 1D subband spacing, ϵ the dielectric constant of the material and N_l the linear density. The second term on the right side is the depolarization shift that was introduced above. It describes the resonant screening of the one-particle-excitation due to the collective effects of all other electrons in the system. This effect is well known from 2DES[18,21] and can be directly observed, e.g., in grating coupler induced intersubband resonance experiments on Si-(111) MOS structures.[25] In these 2DES systems the depolarization shift is a small effect (20% frequency increase). If we calculate the depolarization shift for typical conditions in $GaAs$ quantum wires, i.e., assuming a linear density of $N_l = 10^6 cm^{-1}$ and $\omega_{10} = 2meV$, we calculate a polarization shift of $9.5meV$. Thus in accordance with the experiments collective effects are large and important even at very low carrier densities. Actually, the calculation above is an approximation and the calculated value is far beyond the limits of this approach. The question arises if such a model can be applied at all under these conditions. A theoretical treatment of 1DES with many occupied subbands including their interactions and the transition from a 'pure' one-particle resonance to a dominantly collective response is still missing and seems to be a challenging task. The best coincidence between 1D subband separation (measure in dc magnetotransport) and FIR resonance (they differ only by about 30%) was found so far for quantum wires in InSb-MIS systems.[12] The reason for this is a much larger 1D subband spacing of $10meV$ which arises from the very small effective mass ($m^* = 0.014m_0$) of $InSb$. In addition a metal gate close to the 1D system and neighbouring grating stripes decreases the depolarization effect.[23,24]

It has been shown in Ref. 15 that for the conditions here with a strongly dominating collective contribution the FIR response can be much better explained in terms of a 'local' plasmon resonance which we will specify below. Indeed the magnetic field dependence of the FIR resonances and in particular the occurrence of two (n) resonances for two (n)-layered quantum wire structures resembles a plasmon type of excitation. For a two (n)-layered homogeneous 2DES it is known, that the collective excitation spectrum at small wavevectors q consists of two (n) branches.[26–28] Whereas for widely separated electron sheets ($qd \gg 1$, d = separation of the sheets, q is the plasmon wavevector in the plane) the plasmon branches of the individual layers are degenerate, the Coulomb coupling leads for small distances ($qd \approx 1$) to a splitting of the plasmon dispersion. The energetically highest plasmon branch represents an 'in-phase longitudinal oscillation' of all electron layers. The frequencies of the energetically lower plasmon branches are determined by the strength of the coupling between the electron sheets. The lower plasmon branches represent an 'anti-phase' oscillations of the electrons in the different sheets and exhibit a linear q-dispersion at small q. Therefore they are called 'acoustical' plasmons in contrast to the energetically highest 'optical' branch.

However, the resonances observed in our 1DES differ significantly from excitations in a homogeneous system in the following points: (a) Besides the plasmon resonance one would expect to observe a cyclotron resonance in a homogeneous system. This resonance is completely quenched in our deep-mesa etched microstructured samples, all observed resonances ω_{ri} are shifted with respect to ω_c. (b) When we calculate the plasmon frequency of a homogeneous system $\omega_p^2 = \frac{N_s e^2}{2\epsilon\epsilon_0 m^*} q$ (for a recent review on 2D plasmons see, e.g., Ref. 29), using the average dielectric constant $\bar{\epsilon} = 6.9$ for the microstructured region and the average 2D charge density $\bar{N}_{s2D} = N_{s1D}/a$, we find for the one-layered system (Fig.2d): $\omega_p = 23cm^{-1}$ and for the two-layered system (Fig.2e): $\omega_{p1} = 37cm^{-1}$, $\omega_{p2} = 21cm^{-1}$. Thus the experimentally observed resonances ($\omega_r = 30cm^{-1}$ and $\omega_{r1} = 64cm^{-1}$, $\omega_{r2} = 34cm^{-1}$), are significantly higher in energy compared to those of the homogeneous system.

We explain this frequency shift by 'localization' of plasmons in the following sense: Let us assume a single-layer '2DES' which is additionally confined in x-direction to a width w. Then, in a very simple model, we can treat the 2D plasmon mode for the x-direction as a 'plasmon in a box'. The continuous 2D plasmon dispersion $\omega_p^2 = \frac{N_s e^2}{2\epsilon\epsilon_0 m^*} q$ of a homogeneous system[29] with a free wavevector q is now quantized in fixed values of $q = \pi/w_e$ and correspondingly $\omega_{pl}^2 = \frac{N_s e^2}{2\bar{\epsilon}\epsilon_0 m^*} \frac{\pi}{w_e}$. Here $\bar{\epsilon}$ is the average dielectric constant. The effective width w_e is given by $w_e = w(1 + \alpha)$, where α takes account of the phase relation if the plasmon is 'reflected' at the walls of the box. This is of course a very rough model, which totally neglects Coulomb interaction with neighbouring electron stripes and leaves α, so far, undetermined. However, this model explains our experimentally observed upward shift of the resonance frequency with decreasing w. It can also be applied to the two-layer case. It has been demonstrated in Ref. 15 that by projecting the experimentally observed resonance frequencies onto the calculated plasmon dispersion of a homogeneous two-layered system one finds an intersection with the two branches of the plasmon dispersion at nearly the same q-value. If this q-value is converted via $w_e = \pi/q$ into an effective width one finds a value of $w_e \approx 500nm$, which is comparable to the dimensions w and t. The good agreement between the experimentally observed splitting of the two plasmon modes and the calculation confirms the interpretation of layer-coupled plasmon modes and shows that the coupling of the modes is not very different in the microstructures as compared to homogeneous samples.

The observation of layer-coupled type of plasmon modes here is the first at all with FIR spectroscopy. So far these modes have only been detected in laterally homogeneous 2DES by Raman spectroscopy.[26,27] There are several other points noteworthy. (a) The dipole moment for the acoustical branch of a two-layered 2DES with a small separation d in an isotropic surrounding is zero.[30] The fact that we can observe this branch with FIR spectroscopy indicates the strong spatial asymmetry of the exciting fields in the vicinity of the etched profile of the surface grating. This effect is similar to the behaviour of Raman excitation, where, depending on exciting wavelength and angle of incidence only certain plasmon modes give an efficient Raman signal.[27] (b) In Figs. 5 and 6 it is found that the excitation strength of the different plasmon modes depends strongly on the polarization of the exciting fields and on the magnetic field B. In particular the parallel excitation increases much stronger with B. In the five-layered wire structure (Fig.5b), e.g., this excitation strongly dominates the spectrum at B=14.4T. We have so far no qualitative explanation for this effect.

Fig. 7 Experimental FIR spectra, measured on a shallow mesa etched quantum wire structure at B=7.2T. The carrier density in the system has been increased successively (curve (a) to (d)) by bandgap illumination via the persistent photo effect.

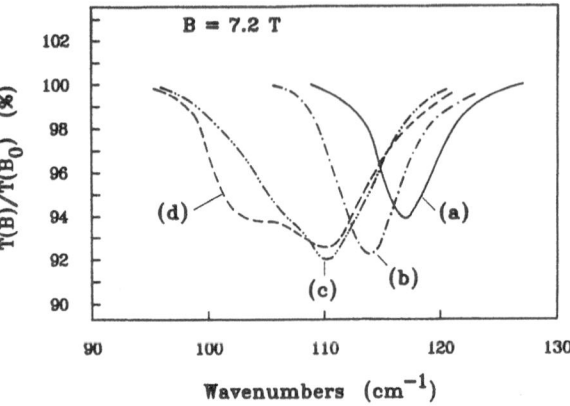

Before we will report additional experiments which elucidate the 'local' plasmon behaviour we like to note that the collective contribution to the excitation in the microstructured systems can be described by various models, such as a depolarization frequency[31] or a geometrical resonance[32,33] . Actually the geometrical resonances considered in slightly wider structures (which have not been tested to be 1DES) in Refs. 32,33 are in many aspects very similar to the response discussed here for the one-layered quantum wire structures and the resonances found in the split-gate structures[13] . Note that for the one-layered system all models give the same dependence $\omega_p^2 = A \cdot N_s / w$. The factor A is slightly different in the various models which reflects the different potential that acts onto the electrons in different configurations. We prefer the local plasmon model, since in this model the layer-coupled plasmon effect of multi-layer quantum wires can easily be described. It also explains experiments for the split-gate configuration, where a continuous transition from an extended plasmon to the local plasmon frequency is found, if via the split-gate the system is tuned from a density modulated system to a 1D system as will be discussed below. For the one-layered system our model of a 'plasmon in a box' is a simplified version of a calculation for a density modulated system by Lai et al.[34] In the latter calculation the coupling between different wires and the 'boundary conditions' for the 'reflection' of the plasmons (i.e., the determination of our α) are treated rigorously.

FIR RESPONSE OF SHALLOW MESA ETCHED STRUCTURES AND SPLIT-GATE DEVICES

Figure 7 shows FIR spectra[17] at B = 7.2T for a shallow-mesa etched sample (Fig. 2e) with, in the dark, a very low electron density of $N_l \approx 10^6 cm^{-1}$. To increase the carrier density via the persistent photoeffect, the sample has been illuminated successively with bandgap radiation. The increase of the carrier density is directly observed by the increased excitation strength. Surprisingly however, the resonance frequency shifts with increasing carrier density to smaller frequencies. This is in contrast to the behaviour of plasmons in a homogeneous system. This decrease can be directly explained within the local plasmon model with $\omega_p^2 \propto N_s / w$. (N_s is here the local 2D density in the wire.) The width of the channel w increases drastically because of the self-consisting screening of the potential from the remote donors. This overcompensates the increase of N_s. Another surprising observation in Fig. 7 is that, after a longer illumination, suddenly two resonances appear (curve (d)), where, with increasing N_s, the frequency of the second resonance approaches the CR resonance frequency ω_c. From dc transport it seems that this second resonance occurs under the condition that there are also electrons induced in the region between the original wires, i.e., the system changes from an isolated wire system to a strongly density modulated system.

We have studied this behaviour also in a split-gate configuration (Fig.2a).[17] In such a system due to the different distances d_k (k=1,2) between channel and gate two different local charge densities $N_{sk} \propto (V_g - V_t)/d_k$ are induced via a gate voltage V_g (V_t

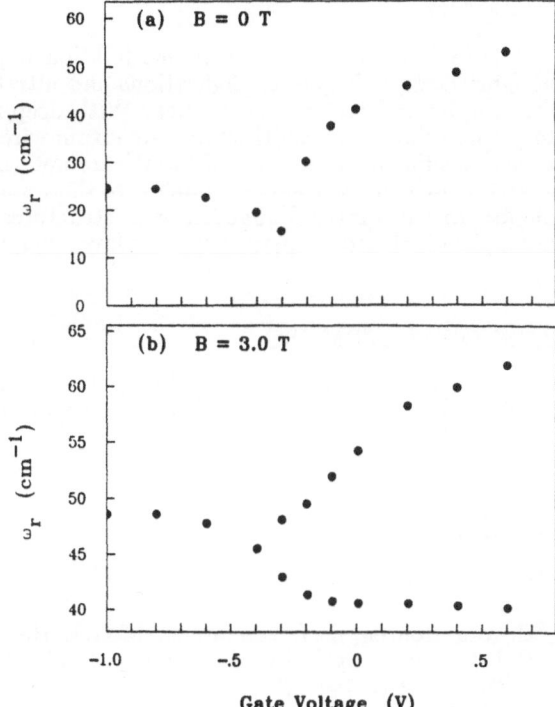

Fig. 8 Experimental FIR resonance positions measured on a split-gate sample (Fig.2a) at B=0 and B=3T. For $V_g \geq$ -0.3V the system is density modulated, below isolated 1D wires are formed.

is the threshold voltage). Figure 8 shows the frequency of the observed FIR resonances in its dependence on the gate voltage V_g. These spectra are very similar to the spectra in Ref. 8. For B=0 and V_g=0 in Fig. 8a the resonance corresponds to the 2D plasmon resonance of the homogeneous system. With decreasing V_g the resonance frequency decreases since the plasmon frequency in a density modulated system is determined by the decreasing averaged charge density.[35-37] At V_g=-0.3V N_{s2} becomes zero and isolated wires are formed. Now we are in the regime of local plasmon resonances. The increase of the resonance frequency here with decreasing V_g shows that the width of the channel decreases faster than N_s. The latter follows also from the self-consistent 1D bandstructure calculations for the split-gate configuration by Laux et al.[19] . The strong increase of the resonance frequency just when the system forms isolated wires indicates the strong rearrangement of the electrons and the sudden decrease of w if the screening of the density modulated system breaks down. For B=3T in Fig.8b the energetically higher resonance is the magnetoplasmon resonance. Both in the density modulated and in the isolated local plasmon regime, as we have seen in Fig. 6b, the resonance frequency obeys the $\omega_r^2(B, V_g) = \omega_r^2(B = 0, V_g) + \omega_c^2$. In the density modulated regime two resonances are observed. The energetically lower resonance is for $V_g \geq$ -0.05V nearly independent of V_g and represents the cyclotron resonance. Very surprisingly, for $V_g \leq$-0.05V, although still in the density modulated regime, the cyclotron frequency increases with decreasing V_g. Also the cyclotron resonance amplitude, which at higher V_g carries the oscillator strength of all electrons, decreases in intensity and transfers oscillator strength to the high energy plasmon type of mode. This latter observation leads us to attribute this upshift of the cyclotron resonance frequency qualitatively to an interacting plasmon-cyclotron excitation. This is the same behaviour that we have also observed for the shallow mesa-etched structure after strong illumination (Fig. 7, curve (d)). Interestingly we find in this part of the density modulated regime with increased 'cyclotron' frequency also deviations from a linear 1/B period of the SdH period, indicating a confinement effect on the electrons without isolated wires. Perhaps it is this confinement that induces the plasmon-cyclotron interaction.

SUMMARY

1DES have been realized in modulation doped $AlGaAs/GaAs$ heterostructures by fabricating split-gate configurations and ultrafine etched structures with optimized lithography and etching techniques. With deep-mesa etching technique it is possible to prepare single and multi-layered quantum wire systems. From dc magnetotransport typical confinement energies of $2meV$ are determined. The FIR response is strongly governed by collective effects which give the resonances the character of local plasmon modes. In multi-layered quantum wire structures a splitting of the plasmon dispersion in longitudinal and acoustical type of layer-coupled local plasmon modes is observed.

ACKNOWLEDGEMENT

We would like to thank K. Ploog for providing us with excellent samples and acknowledge financial support from the Bundesministerium für Forschung und Technologie, Bonn.

REFERENCES

[1] K.-F. Berggren, T.J. Thornton, D.J. Newson, and M. Pepper , Phys. Rev. Lett. **57**, 1769 (1986)

[2] H. van Houten, B.J. van Wees, M.G.J. Heijman, J.P. André, D. Andrews, and G.J. Davies , Appl. Phys. Lett. **49**, 1781 (1986)

[3] J. Cibert, P.M. Petroff, G.J. Dolan, S.J. Pearton, A.C. Gossard, and J.H. English , Appl. Phys. Lett. **49**, 1275 (1986)

[4] T.P. Smith, III., H. Arnot, J.M. Hong, C.M. Knoedler, S.E. Laux, and H. Schmid , Phys. Rev. Lett. **59**, 2802 (1987)

[5] M.L. Roukes, A. Scherer, S.J. Allen, Jr., H.G. Craighead, R.M. Ruthen, E.D. Beebe, and J.P. Harbison, Phys. Rev. Lett. **59**, 3011 (1987)

[6] H. van Houten, B.J. van Wees, J.E. Mooij, G. Roos, and K.-F. Berggren, Superlattices and Microstructures **3**, 497 (1987)

[7] G. Timp, A.M. Chang, P. Mankiewich, R. Behringer, J.E. Cunningham, T.Y. Chang, and R.E. Howard , Phys. Rev. Lett. **59**, 732 (1987)

[8] W. Hansen, M. Horst, J.P. Kotthaus, U. Merkt, Ch. Sikorski, and K. Ploog , Phys. Rev. Lett. **58**, 2586 (1987)

[9] B.J. van Wees, H. van Houten, C.W.J. Beenakker, J.G. Williamson, L.P. Kouwenhoven, D. van der Marel, and C.T. Foxon , Phys. Rev. Lett. **60**, 848 (1988)

[10] D.A. Wharam, T.J. Thornton, R. Newbury, M. Pepper, J.E.F. Frost, D.G. Hasko, D.C. Peacock, D.A. Ritchie, and G.A.C. Jones, J. Phys. **C21**, L209 (1988)

[11] K.-F. Berggren, G. Roos, and H. van Houten , Phys. Rev. **B37**, 10118 (1988)

[12] J. Alsmeier, Ch. Sikorski, and U. Merkt , Phys. Rev. **B37**, 4314 (1988)

[13] F. Brinkop, W. Hansen, J.P. Kotthaus, and K. Ploog , Phys. Rev. **B37**, 6547 (1988)

[14] T. Demel, D. Heitmann, P. Grambow, and K. Ploog , Appl. Phys. Lett. **53**, 2176 (1988)

[15] T. Demel, D. Heitmann, P. Grambow, and K. Ploog , Phys. Rev. **B38**, 12732 (1988)

[16] T. Demel, D. Heitmann, P. Grambow, and K. Ploog, Superlattices and Microstructures **5**,287(1989)

[17] T. Demel, D. Heitmann, P. Grambow, and K. Ploog, to be published

[18] T. Ando, A.B. Fowler, and F. Stern, Rev. Mod. Phys. **54**, 437(1982)

[19] S.E. Laux, D.J. Frank, and F. Stern, Surf. Sci. **196**,101(1988)

[20] K.K. Choi, D.C. Tsui, and K. Alavi , Appl. Phys. Lett. **50**, 110 (1987)

[21] W.P. Chen, Y.J. Chen, and E. Burstein, Surf. Sci. **58**,263(1976)

[22] S. Das Sarma and W.Y. Lai , Phys. Rev. **B32**, 1401 (1985)

[23] W. Que and G. Kirczenow , Phys. Rev. **B37**, 7153 (1988) and , Phys. Rev. **B39**, 5998 (1989)

[24] A.V. Chaplik, submitted to Superlattices and Microstructures

[25] D. Heitmann and U. Mackens Phys. Rev. **B33**, 8269 (1986)

[26]G. Fasol, N. Mestres, H.P. Hughes, A. Fischer, and K. Ploog , Phys. Rev. Lett. **56**, 2517 (1986)

[27]A. Pinczuk, M.G. Lamont, and A.C. Gossard , Phys. Rev. Lett. **56**, 2092 (1986)

[28]J.K. Jain and P.B. Allen , Phys. Rev. Lett. **54**, 2437 (1985)

[29]D. Heitmann, Surf. Sci.**170**,332(1986)

[30]A.V. Chaplik, Surf. Sci. Rept.**5**,289(1985)

[31]W. Hansen, J.P. Kotthaus, A. Chaplik, and K. Ploog, in: High Magnetic Fields in Semiconductor Physics, Proc. of the Int. Conf. Würzburg 1986, Springer Series in Solid-State Sciences 71, ed. by G. Landwehr (Springer, Berlin 1987), p.266

[32]S.J. Allen, Jr., H.L. Störmer, and J.C. Hwang , Phys. Rev. **B28**, 4875 (1983)

[33]S.J. Allen, F. DeRosa, G.J. Dolan, and C.W. Tu, Proc. 17th Int. Conf. Phys. Semicon., San Francisco (1984), edts. J.D. Chadi and W.A. Harrison, p. 313

[34]W.Y. Lai, A. Kobayashi, and S. Das Sarma , Phys. Rev. **B34**, 7380 (1986)

[35]U. Mackens, D. Heitmann, L. Prager, J.P. Kotthaus, and W. Beinvogl , Phys. Rev. Lett. **53**, 1485 (1984)

[36]D. Heitmann and U. Mackens, Superlattices and Microstructures 4,503(1988)

[37]M.V. Krasheninnikov and A.V. Chaplik, Sov. Phys. Semicond. **15**, 19 (1981)

ELECTRONS IN LATERAL MICROSTRUCTURES ON INDIUM ANTIMONIDE

U. Merkt, Ch. Sikorski, and J. Alsmeier

Institut für Angewandte Physik, Universität Hamburg
Jungiusstr. 11, 2000 Hamburg 36, F.R.G.

1. INTRODUCTION

Advances of lithography and etching techniques now make it possible to laterally confine quasi-two-dimensional (2D) electron gases at semiconductor interfaces into wires or dots of widths below 100 nm.[1] Since such widths are less than the mean free path of electrons at low temperatures, one induces quasi one-dimensional (1D) or zero-dimensional (0D) electronic behavior in the wires and dots, respectively, and observes novel phenomena in the transport[2] and optical[3] properties.

We here summarize the present state of our work on these lateral microstructures on InSb. Starting from the 2D electron gas of inversion layers in metal-oxide-semiconductor (MOS) structures,[4,5] we have realized quantum wires and dots that contain 1D and 0D electron systems, respectively. We study resonance transitions between the corresponding discrete levels in the far-infrared (FIR) regime and thus directly prove the lateral quantization.[6,7] The particular advantage of InSb is its small effective electron mass leading to comparatively large energy spacings at a given wire width or dot diameter. Generally, the quantized levels are determined by the external potential that confines the electrons, by the semiconductor band structure, and by the electron-electron interaction. The external potential is not precisely known and we approximate it by a parabolic well. Effects of band structure and electron-electron interaction are particularly interesting for quantum dots because of the analogy between these few-electron systems with real atoms or shallow donors[8] in semiconductors. In contrast to donors, however, we can adjust the size, the shape, and the electron number of our dots.

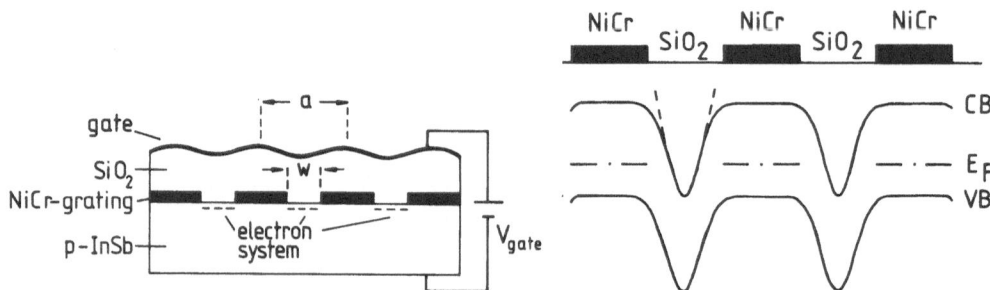

Fig. 1. Schematic cross section of the microstructured field effect device on InSb and its lateral band structure. The figure applies to both quantum wires and quantum dots.

2. EXPERIMENTAL

To perform far-infrared spectroscopy on microstructures, it is necessary to prepare samples with diameters much larger than typical wavelengths of 100 μm that correspond to the lateral quantization energies. Therefore, special preparation techniques are required if one wants to investigate quantization in structures of widths below 1 μm. This problem is overcome with laterally periodic structures in which wires or dots are arranged in arrays of macroscopic areas. We first discuss the idea which underlies both of our structures and then describe the fabrication. The idea is sketched in Fig. 1. NiCr is evaporated onto the substrate in order to pin the Fermi energy E_F above the valence band edge. On the other hand, mobile inversion electrons are induced into the narrow regions between the NiCr by the gate voltage V_g which is applied between the metallic gate and the InSb substrate. Via this field effect wires or dots can be charged without direct contacts to the inversion electrons since the InSb substrate has a finite resistivity R~1 MΩ even at liquid helium temperatures. The resulting lateral potential is sketched in the right hand part of Fig. 1.

For a wire sample of grating constant a≈250 nm and active area ≈3x3 mm², we must fabricate ≈10 000 stripes. Each of them is ≈100 nm wide with little variation along its length of 3 mm. The desired structures are obtained by holographic lithography.[9] After development of the positive photoresist, the resist residues at the exposed areas are removed by dry etching in an oxygen plasma (see Fig. 2). As a result, the InSb surface is covered by periodic photoresist stripes of widths ≤ 100 nm which are used as masks for a NiCr shadowing under an angle of 66°. This way, a NiCr wire grid (d=20 nm) results. Subsequently, the resist stripes are dissolved in acetone (lift–off) and a SiO_2 gate insulator (d≈400 nm) is deposited in a PECVD reactor. Finally, we evaporate a thin homogeneous NiCr film (d≈5 nm) which operates as semi-transparent gate. To enable FIR spectroscopy and DC conductivity measurements simultaneously, 60 nm thick Sn films are evaporated prior to the holographic lithography onto some of our samples. They are alloyed for 2 h at 170 °C in an H_2/Ar atmosphere to provide low-resistance Ohmic contacts to the channels. Via these contacts (see Fig. 2), the quasi-static resistance and its derivative dR/dV_g with respect to the gate voltage are measured in the two directions parallel and perpendicular to the stripes. Schottky-barriers at the sample edges prevent leakage currents between the two contact pairs. A threshold voltage V_t is determined from the conductivity onset along the channels and we use the voltage difference $\Delta V_g = V_g - V_t$ as measure of the density of mobile electrons. There is virtually no quasi-static conductivity perpendicular to the channels.

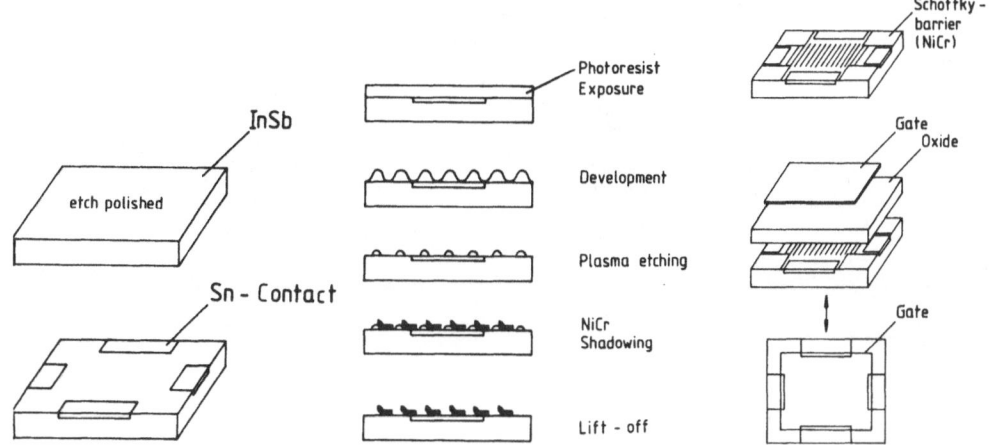

Fig. 2. Preparation of microstructured metal–oxide–InSb devices with four diffused Sn contacts. Chemical polishing, holographic lithography, plasma etching, and plasma-enhanced CVD oxidation are employed.

Quantum dots are prepared with the same techniques in arrays of about 10^8 individual dots again on areas $\approx 3 \times 3$ mm^2. However, after the first exposure the sample is rotated by an angle ⦁ and the photoresist is exposed for a second time. The formula for equal exposure times

$$I(x,y) = I_o \cos^2\left(\frac{\pi y}{a}\right) + I_o \cos^2\left(\frac{\pi x}{a}\sin⦁ + \frac{\pi y}{a}\cos⦁\right) \tag{1}$$

describes the distribution of the light intensity or the resist profile after proper development or etching. Thereby, we assume for simplicity a linear exposure response curve with zero threshold energy of the resist and ignore a phase shift for positive resists. Structures for successively prolonged etching times are modeled in Fig. 3(a–d) for an angle ⦁=90°.

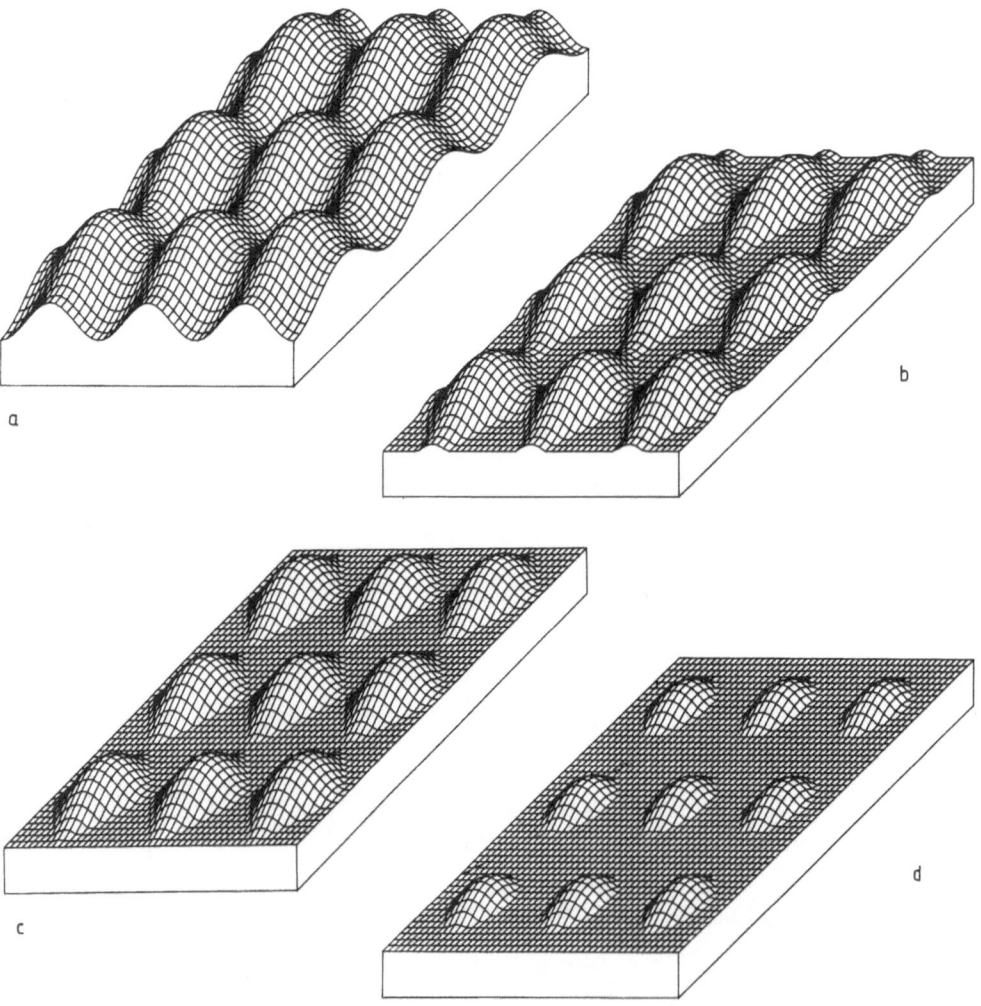

Fig. 3. Simulation of the photoresist profile for prolonged (a–d) developing or etching times. The resist has been exposed twice (⦁=90°).

a   b

Fig. 4. Contour plots of resist profiles after double exposure under angle $\phi=90°$ (a) and $\phi=60°$ (b).

Corresponding contour plots are shown in Fig. 4(a,b) for angles $\phi=90°$ and 60°. These plots demonstrate that, in principle, we can design quantum dots of various shapes like circles, squares, rectangles, and ellipses. Scanning electron micrographs of a finished sample with wires and of a monitor sample with dots are shown in Fig. 5.

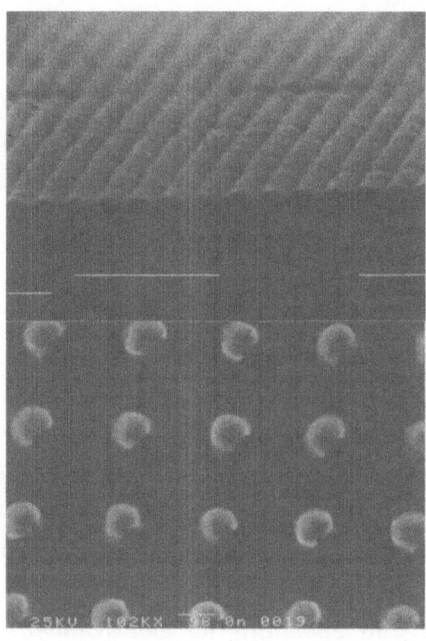

Fig. 5. Micrographs of a finished wire sample (a=250 nm) and a monitor sample with resist dots shadowed with gold for contrast enhancement.

3. THEORETICAL

3.1 Quantum wires

The most simple and convenient model to describe the lateral potential is its approximation by a parabolic well.[10] This has been employed in many experimental as well as theoretical studies.[1,3] In fact, the shape of the conduction band edge near a minimum in Fig. 1 can be approximated by a harmonic oscillator well. Its frequency ω_0 is treated as parameter and no tunneling of electrons between adjacent wells is taken into account. The latter is reasonable, since the height of the barrier between stripes is of order 100 meV and the width of the barrier is approximately given by the grating constant. This model has the advantage that it can be treated analytically in the presence of magnetic fields applied perpendicularly to the original 2D electron layer. Spin-split energies

$$E_{i,n} = E_i + \hbar\omega_h(n+\tfrac{1}{2}) + \frac{\hbar^2 k_y^2}{2m_0^*} \cdot \left(\frac{\omega_0}{\omega_h}\right)^2 \pm \tfrac{1}{2} g_0^* \mu_B B \qquad (2)$$

are given by the subband energies E_i of the original 2D electron system and the hybrid frequency $\omega_h = (\omega_0^2 + \omega_c^2)^{1/2}$ calculated from the cyclotron frequency $\omega_c = eB/m_0^*$. Spin-splitting is described by the effective Landé factor g_0^*. The motion of electrons in x and in z direction is quantized into equally spaced 1D subbands n and into levels i. The electrons are free to move along the inversion channels in the y direction with dispersion that depends on the ratio ω_0/ω_h. In high magnetic fields ($\omega_c \gg \omega_0$), electrons behave 2D-like and observation of Shubnikov-de Haas type oscillations allows us to determine an areal electron density n_c which refers to the channel area. The density of states of the spin-split levels

$$D(E) = \left(\frac{\omega_h}{\omega_0}\right) \left(\frac{m_0^*}{2\pi^2\hbar^2}\right)^{1/2} \sum_{E \geq E_n} (E - E_{i,n})^{-1/2} \qquad (3)$$

yields the Fermi energy E_F, provided the electron density n_ℓ per channel length and the spacing $\hbar\omega_0$ are known. In Eq.(3) the energies $E_{i,n}$ are taken at momentum $\hbar k_y = 0$. In Sec. 4.2 we will determine the parameters n_ℓ and $\hbar\omega_0$ from the experimental magnetoresistance oscillations.

Besides the subband spacing, the electronically active width is most interesting. Commonly, it is distinct from the geometrical one monitored in a scanning electron microscope. The active width is smaller due to depletion regions at the channel edges and must be determined from electrical measurements. We define it in two different ways. The first definition relies on the distortion of the cyclotron motion when the diameter of the cyclotron orbit $2(2n+1)^{1/2}(\hbar/eB)^{1/2}$ becomes comparable to the channel width. A second definition works without the application of a magnetic field. The active width is identified with the spread $w_F = 2\ell_0(2N+1)^{1/2}$ of the wave function at the Fermi level with the characteristic oscillator length $\ell_0 = (\hbar/m_0^*\omega_0)^{1/2}$. We also can define a mean width $\langle w \rangle$ averaged over all populated subbands n=0 to N. When many subbands N+1≫1 are occupied, we obtain the approximate result $\langle w \rangle = \frac{2}{3}w_F = n_\ell/n_c$. The second identity is derived from Eq.(3) and the relation $n_c = m_0^* E_F/\pi\hbar^2$ valid for a 2D electron gas. Note that the density n_c refers to the channel area and that this definition of an areal density is no longer appropriate in the 1D limit N→ 0.

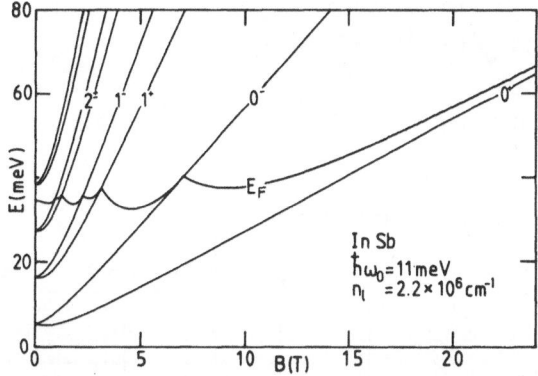

Fig. 6. Calculated 1D subband edges of spin-split levels and Fermi energy vs magnetic field strength. At B=0 we have equally spaced levels of the lateral oscillator potential. The number of occupied subbands at B=0 is equivalent to the number of cusps of the Fermi energy if we ignore spin-splitting for subbands n≥1.

Figure 6 depicts calculated subband edges ($k_y=0$, i=0) and the Fermi energy E_F versus magnetic field strength. The Fermi energy exhibits maxima when it crosses the subband edges and is pinned to the ground level 0^+ in the quantum limit of strong magnetic fields. When it passes through a subband edge, there is a discontinuity in the density of states and in the intersubband scattering. This causes the resistivity oscillations[1] that can be observed in the DC transport.

In the spectroscopic experiments,[11],[12] we measure the transmittance of FIR radiation which is closely related to the high-frequency conductivity of the channels. The real part of the classical conductivity (B=0) for light polarization perpendicular to the channels

$$\sigma_\ell(\omega) = \frac{\sigma_{\ell 0}}{1+\left(\dfrac{\omega^2-\omega_0^2}{\omega}\right)^2 \tau^2} \qquad (4)$$

is obtained with static conductivity $\sigma_{\ell 0}=e^2 n_\ell \tau/m_0^*$ and phenomenological relaxation time τ. For parallel polarization we expect Drude behavior $\sigma_\ell = \sigma_{\ell 0}(1+\omega\tau)^{-1}$.

3.2 Quantum dots

For dots, we consider the harmonic oscillator potential $\frac{1}{2}m_0^*\omega_0^2(x^2+y^2)$ with eigenfrequency ω_0 in a magnetic field along the z direction.[13] The single electron eigenenergies of the lateral motion

$$E_{n,m} = (2n+|m|+1)\hbar \sqrt{\left(\frac{\omega_c}{2}\right)^2 + \omega_0^2} + \frac{\hbar\omega_c}{2} m \qquad (5)$$

depend on the radial $n=0,1,...$ and azimuthal $m=0,\pm 1,...$ quantum numbers.

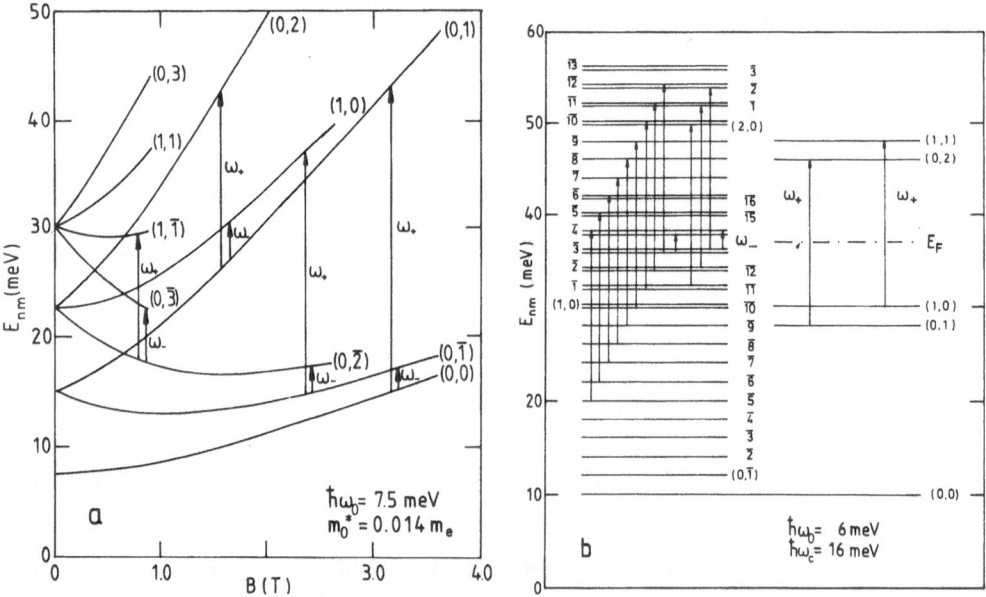

Fig. 7. Eigenenergies of the harmonic oscillator $\frac{1}{2}m_0^*\omega_0^2(x^2+y^2)$ in a magnetic field B∥z. (a) Levels and allowed transitions $\omega_\pm$ for lower quantum numbers (n,m) vs magnetic field strength ($\bar{m}=-m$). (b) Levels and transitions for an electron number $n_0=38$ at a fixed magnetic field strength. From Ref. 7.

Figure 7(a) depicts the energies of the lowest levels versus magnetic field strength and the allowed dipole transitions which have frequencies

$$\omega_{\pm} = \sqrt{(\frac{\omega_c}{2})^2 + \omega_o^2} \pm \frac{\omega_c}{2} \tag{6}$$

and which are excited with circular light polarizations, respectively. At low electron numbers, we can assume that only the lowest 2D subband $E_{i=0}$ is occupied. For B=0 we have equally spaced oscillator levels $(2n+|m|+1)\hbar\omega_o$. In high magnetic fields ($\omega_c \gg \omega_o$), all levels with quantum numbers n=0, m≤0 converge and form a highly degenerate ground state $\frac{1}{2}\hbar\omega_c$. Levels n=1, m≤0 converge to a common excited state $\frac{3}{2}\hbar\omega_c$. Levels with angular momentum m$\hbar$>0 have much higher energies and do not contribute to the signals since they are no longer occupied. In this high magnetic field limit, transitions ω_+ become cyclotron resonances between Landau levels and the electron gas exhibits 2D behavior. Simultaneously, the oscillator strength of transitions ω_- vanishes. These resonances are indeed characteristic of a confined system with a radius comparable or less than the cyclotron radius $l=(\hbar/eB)^{1/2}$. Figure 7(b) shows the eigen-energies for quantization energy $\hbar\omega_o$=6 meV and cyclotron energy $\hbar\omega_c$=16 meV corresponding to a magnetic field B≅2 T. Spin-splitting is not taken into account and an electron number n_0=38 is assumed. For clarity, levels with quantum numbers m≥0 are depicted in the right hand part of Fig. 7(b). Allowed transitions with frequencies ω_+ and ω_- are indicated and the relation $\sum f_-/\sum f_+=\omega_-/\omega_+$ for the sums of oscillator strengths[13] over occupied levels is easily verified with the dipole matrix elements of Table 1.

Table 1. Selection rules, dipole matrix elements, and frequencies $\omega_{\pm}=\hbar^{-1}(E_{n',m'}-E_{n,m})$ of quantum dots in the parabolic well approximation with $L^2=\hbar/m_0^*(\omega_c^2/4+\omega_0^2)^{1/2}$. Resonances $\omega_{\pm}$ are excited by circular polarizations.

(n', |m|')	squared dipole matrix element	frequency m' ≥ 0	frequency m' < 0
n-1, |m|+1	n·L²	$-\omega_-$	$-\omega_+$
n , |m|+1	(n+|m|+1)·L²	$+\omega_+$	$+\omega_-$
n+1, |m|-1	(n+1)·L²	$+\omega_-$	$+\omega_-$
n , |m|-1	(n+|m|)·L²	$-\omega_+$	$-\omega_-$

The classical high-frequency conductivities in a magnetic field

$$\sigma_{\pm}(\omega) = \frac{\sigma_{00}}{1 + (\frac{\omega_0^2}{\omega} - \omega \pm \omega_c)^2 \tau^2} \tag{7}$$

for circular polarizations again are obtained from the classical equation of motion with DC conductivity $\sigma_{00} = en_0\mu$.[14] For linear light polarization, the conductivity is given by the expression $\sigma = \frac{1}{2}(\sigma_+ + \sigma_-)$. At both resonance frequencies $\omega_{\pm}$ the conductivity reaches the same maximum σ_{00}. The ratio of integrated strengths of resonances ω_- and ω_+ is given by the expression ω_-/ω_+ in correspondence with the quantum mechanical result. Hence, the width of the ω_- resonance is smaller than the one of the ω_+ resonance. In other words, the classical theory predicts sharp ω_- resonances as the ratio ω_-/ω_+ becomes small ($\omega_c \gg \omega_0$). This behavior is seen in Fig. 8(a) but even qualitatively is not observed in our experiments with few electrons. Therefore, we also have calculated the Kubo conductivity[15]

$$\sigma_0(\omega) = \frac{\sigma_{00}}{2(\omega_+ + \omega_-)} \left[\omega_+ \cdot \left(\frac{1}{1 + (\omega - \omega_+)^2 \tau^2} + \frac{1}{1 + (\omega + \omega_+)^2 \tau^2} \right) + \right.$$

$$\left. + \omega_- \cdot \left(\frac{1}{1 + (\omega - \omega_-)^2 \tau^2} + \frac{1}{1 + (\omega + \omega_-)^2 \tau^2} \right) \right] \tag{8}$$

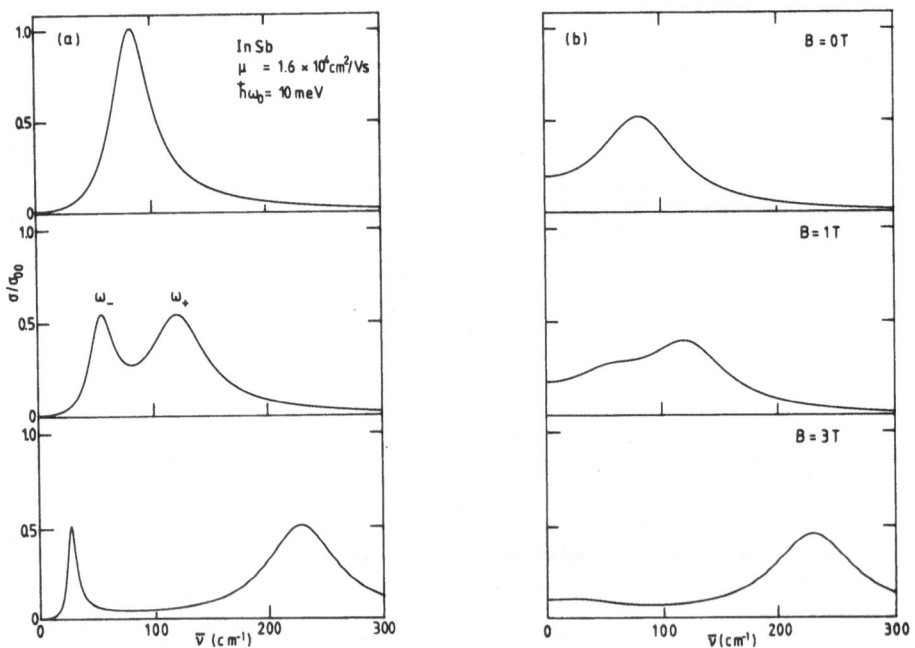

Fig. 8. Classical (a) and Kubo (b) conductivity of quantum dots both with a phenomenological relaxation time τ.

for linear polarization from linear response theory and have plotted it in Fig. 8(b). The ratio of integrated strengths is identical with the classical result, however, the linewidth now is the same for both resonances. This means that the maximum conductivity of the quantum mechanical ω_- resonance is lower than that of the ω_+ resonance which qualitatively agrees with the observation. However, we consider a flaw of Eq.(8) that the conductivity remains finite in the limit $\omega \rightarrow 0$ and also note that for the important situation $\omega_c=0$, $\omega=\omega_0$ the conductivity is only half of the classical value σ_{00} as long as the condition $\omega_0\tau \gg 1$ is realized.

3.3 Influence of the band structure

The energy levels are not only determined by the confining potential but also by the semiconductor band structure and the electron–electron interaction. For narrow-gap materials like InSb the effective mass approximation (EMA) with a constant electron mass m_0^* is not sufficient. The conduction band of InSb is nonparabolic which means that the electron mass m^* increases with momentum or energy above the band edge. Hence, the value m_0^* only stands for the mass at the conduction band edge. In order to obtain a simple and rough description of the influence of nonparabolicity, we ignore the spin–orbit interaction and restrict to terms of order $1/E_g$ within two-level k·p-theory. We start from the effective Schrödinger equation[16] and neglect the so-called Zener terms. The latter is justified for the triangular well which we employ to describe the z direction[17] and for the lateral parabolic well, since in order $1/E_g$ the contributions of the Zener terms are zero and $\frac{1}{4}\hbar^2\omega_0^2/E_g$, respectively. The latter is independent of quantum number. Hence, for a parabolic well the Zener term only causes a shift of the energy scale. We thus obtain the correction

$$\Delta E_{j,\vec{k}}^{k \cdot p} = -\frac{1}{E_g} \cdot \langle j \; \vec{k} \mid [E_{j\vec{k}}^{EMA} - U(\vec{r})]^2 \mid j \; \vec{k} \rangle \qquad (9)$$

of the EMA eigenenergies. In this equation, we have the operator of kinetic energy $T=E-U$, a set of discrete quantum numbers j and the quasi continuous momentum $\hbar\vec{k}$. For a 1D system, we have the discrete quantum numbers $j=(i,n)$ and momentum $\hbar k_y$ along the wire. The effective mass $m_{i,n}^* = \hbar^2 k_y \cdot (\partial E/\partial k_y)^{-1}$ which describes this motion is readily calculated with the well-known matrix elements of the triangular and parabolic well:

$$m_{i,n}^* \cong m_0^* \left(1 + 2 \cdot \frac{\frac{1}{2}\hbar\omega_0(n + \frac{1}{2}) + \frac{1}{3}E_i + \hbar^2 k_y^2/2m_0^*}{E_g}\right) \; . \qquad (10)$$

For a 0D system we have the discrete set j=(i,n,m) and no continuous quantum number since the system is totally quantized. In this case, Eq.(9) gives corrections for discrete energies.

3.4 Electron-electron interaction

It is well-known from experiments in 2D electron systems[4] that the resonance peak does not appear at the frequency ω_0 of the level separation. Instead of this, it is shifted to a value $(\omega_0^2+\omega^2_{dep})^{1/2}$ because each electron feels a dynamically screened electric field which is different from the one of the incident radiation due to the polarization of all the other electrons. This collective effect has been described by the sum of two terms, namely the depolarization and the exciton-like effect. Here we can only discuss the depolarization effect, since the exciton effect has not been treated as yet for 1D or 0D systems. Also, we expect that the exciton effect is small as it is in 2D systems. Generally, the collective contribution $\hbar\omega_{dep}$ can be determined experimentally when the resonance energy is compared with subband spacings obtained from transport measurements that are not affected by depolarization.

The depolarization shift in wires has first been treated classically for a single stripe.[18] In this approach, the depolarization frequency is obtained as the plasma eigenfrequency of a charged ellipsoid of classical width w. The resulting depolarization frequency

$$\omega_{dep} = \left(\frac{\alpha\, e^2 n_\ell}{m_0^* \varepsilon_0 \bar{\varepsilon}\, w^2} \right)^{\frac{1}{2}} \tag{11}$$

contains an effective dielectric constant $\bar{\varepsilon}=(\varepsilon+\varepsilon_{ox})/2$ which averages over the semiconductor and oxide dielectric constant ε and ε_{ox}, respectively. The factor α is of order unity and different values $\alpha=1$ to 2 have been used in the literature.[11] More recently, the collective contributions were treated as intersubband plasmons in the Hartree approximation.[19,20] According to these theories, plasmon excitations exist with arbitrary wave vectors perpendicular to the wires as a consequence of virtual excitations of the 1D subbands. These excitations create charge polarizations of the individual wires in the direction perpendicular to them. Even if the wires are insulated from each other, there is always coupling by the Coulomb force. Analytical expressions were obtained for the two-level approximation in which only the influence of levels N-1 and N is considered for a Fermi energy lying between these levels. This approach is appropriate as long as the conditions $Ry^*/E_F \ll 1$ or $Ry^*/\hbar\omega_0 \ll 1$ with effective Rydberg constant Ry^* are fulfilled for indices $N \gg 1$ and $N \sim 1$, respectively. Note, that the effective Rydberg $Ry^* = 0.7$ meV is comparatively small in InSb. In the absence of tunneling between adjacent channels, one gets the collective contribution[20]

$$\omega^2_{dep} = \frac{e^2 \, \Delta n_\ell}{\pi \varepsilon_0 \, \bar{\varepsilon} m^*_0 \ell^2_0} \cdot J(N) \qquad (12)$$

where Δn_ℓ refers to the difference of electron densities in the adjacent subbands under consideration. The factors

$$J(N) = x \cdot \sum_{n=1}^{\infty} \frac{n}{N} \left[L^1_{N-1} (xn^2) \right]^2 e^{-xn^2} \qquad (13)$$

describe the coupling of depolarization fields of individual wires. In Eq.(13), one has the abbreviation $x = 2\pi^2 \ell^2_0/a^2$ and associated Laguerre polynomials L^1_{N-1}. In case of an isolated stripe we get $J(N) \rightarrow \frac{1}{2}$ independent of index N. Correction factors $(2J)^{1/2}$ by which the depolarization frequency of an isolated stripe should be multiplied in case of a periodic array are depicted in Fig. 9(a) for various transitions $N-1 \rightarrow N$. We restrict to the range of periods $a/\ell_0 > (2N+1)^{1/2}$ since tunneling is excluded in the present theory. The reduction of the depolarization shift is explained by the mutual influence of neighboring channels. The superposition of the polarization fields of all channels becomes more uniform when the period decreases. Then the matrix element of the field between states $N-1$ and N tends to zero due to the orthogonality of the eigenfunctions.[20]

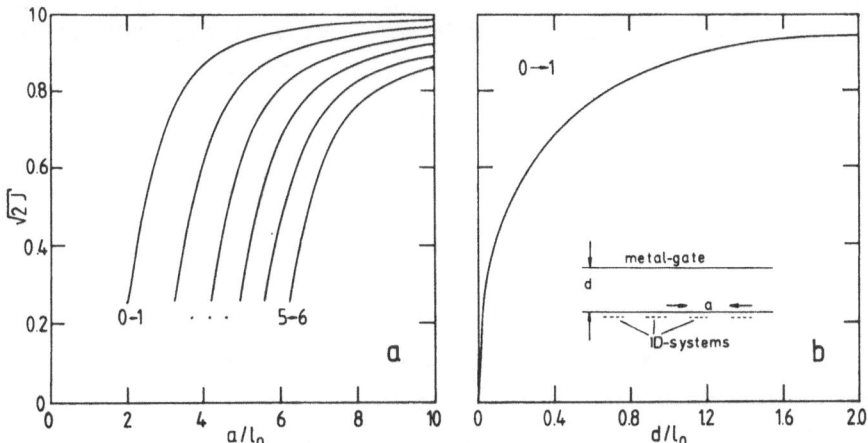

Fig. 9. Correction factors from Ref. 20 for the depolarization shift of a single stripe due to (a) periodicity , (b) screening by a metallic gate. Transitions $N-1 \rightarrow N$ are indicated and periodicity a as well as distance d are normalized with the characteristic oscillator length $\ell_0 = (\hbar/m^*_0\omega_0)^{1/2}$.

The presence of a metallic gate of sheet conductivity $\sigma_\square$ in a distance d above the channels has been considered for a single stripe and the $0\rightarrow 1$ transition.[20] For conductivities $\sigma_\square/\varepsilon_0\ell_0\omega \gg 1$ realized for our Schottky gates with a DC sheet resistivity $R_\square \sim 10 \; \Omega$, screening of the depolarization shift is described by the factor

$$J(\frac{d}{\ell_0}) = \sqrt{\frac{\pi}{2}} \; \frac{d}{\ell_0} \; e^{2(\frac{d}{\ell_0})^2} \left[1 - \mathrm{erf}(\sqrt{2} \; \frac{d}{\ell_0}) \right] \; . \tag{14}$$

The corresponding correction factor for the depolarization frequency is given by the square root $(2J)^{1/2}$ shown in Fig. 9(b). Collective phenomena which lead to plasmon like contributions have also been described for coupled quantum dots.[21]

4. QUANTUM WIRES

4.1 Intersubband spectroscopy

We measure the transmittance of normally incident FIR radiation with a Fourier transform spectrometer.[22] The onset or threshold voltage V_t of the channel conductivity is determined from the onset of absorption of electronic excitations or conductivity as described above. The ratio of transmittances $T(V_g)/T(V_t)$ is plotted in the spectra which allows direct and low noise detection of electronic excitations in tunable electron systems. For small signals, the difference $-\Delta T/T = -[T(V_g)-T(V_t)]/T(V_t)$ is directly proportional to the sheet conductivity σ_s of the inversion electrons. From the Fresnel formulae one obtains the relation[22]

$$-\frac{\Delta T}{T} = \frac{2 \; \mathrm{Real} \; (\sigma_s/Y_0)}{1 + \sqrt{\varepsilon} + \sigma_\square/Y_0} \tag{15}$$

with wave admittance of vacuum $Y_0 = (\varepsilon_0/\mu_0)^{1/2}$ and sheet conductivity $\sigma_\square$ of the metal gate plus stripes. Since the light examines the average areal density n_s, the sheet conductivity σ_s should be related to the conductivity σ_ℓ of a wire via the relation $\sigma_s = \sigma_\ell/a$. This corresponds to the relation $n_s = n_\ell/a$ between average density n_s and density n_ℓ per unit length. To make the compilation of densities complete, we repeat here the above defined density per channel area $n_c = n_\ell/\langle w \rangle = n_s a/\langle w \rangle$.

Normalized spectra[12] for light polarization parallel and perpendicular to the inversion channels are depicted in Fig. 10(a,b) at various gate voltages $\Delta V_g = V_g - V_t$ in the absence of a magnetic field. In parallel polarization we observe Drude type spectra corresponding to the free motion of electrons along the channels. In perpendicular polarization we detect resonances between 1D subbands. The dashed line in Fig. 10(b) is a fit to the spectrum for gate voltage $\Delta V_g = 50$ V according to Eqs.(4) and (15). From the fit we obtain a mobility $\mu = e\tau/m_0^* = 16\ 000$ cm^{2}V^{-1}s^{-1} and a density $n_c = 5.3 \times 10^{11}$ cm^{-2}. Spectra in a magnetic field B=5.4 T are shown in Fig. 10(c,d). For both directions of the light polarization we now observe the same spectral shapes. The resonances are similar to cyclotron resonances of a homogeneous 2D electron system. This is explained by the high magnetic field strength and the related Landau radius $\ell = 11$ nm which is much less than the channel width $w_F \approx 100$ nm. In fact, we observe subband-shifted cyclotron resonances at frequencies

$$\omega = 2\pi c \bar{\nu} = (\omega_c{}^2 + \omega_0{}^2)^{1/2} \ . \tag{16}$$

A small collective contribution to the resonance frequency is ignored in this equation.[20] In strong magnetic fields the cyclotron frequency ω_c clearly exceeds the resonance frequency ω_0 in the absence of magnetic fields. The dashed line in Fig. 10(d) is a cyclotron resonance fit, again for voltage $\Delta V_g = 50$ V, using the 2D Drude magnetoconductivity[5] ($\omega_c \gg \omega_0$) resonance energies is depicted in Fig. 11(b) for two samples of different

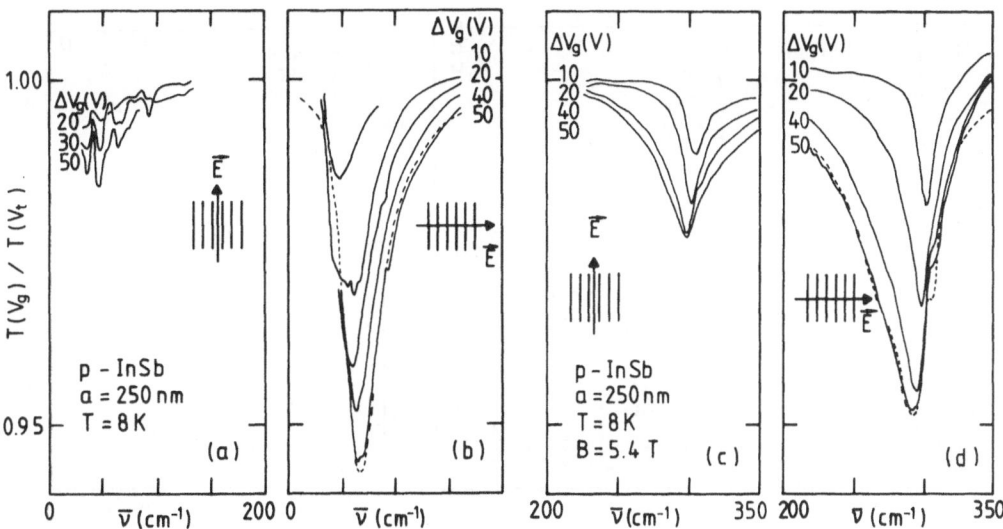

Fig. 10. 1D intersubband resonances (a,b) and subband-shifted cyclotron resonances (c,d) for various gate voltages ΔV_g above threshold. Light polarization is parallel and perpendicular to the wires. The dashed lines represent Drude fits to the $\Delta V_g = 50$ V spectra. From Ref. 12.

and taking into account the occupation of three 2D subbands i=0 to 2. This yields values μ=20 000 cm^{-2}V^{-1}s^{-1} and n$_c$=5.4x10^{11}cm^{-2} for the ground 2D subband i=0 which roughly agree with the ones obtained in zero magnetic fields. Approximately, we find the relation n$_c$/ΔV$_g \approx$1.1x10^{10} cm^{-2}V^{-1} between gate voltage ΔV$_g$ above threshold and average density n$_s$=n$_c$<w>/a for the present sample. Though the spectra are normalized with the transmittance at threshold voltage, the signal measured in parallel polarization in Fig. 10(c) is about a factor of two smaller than the one in perpendicular polarization in Fig. 10(d). This is explained by the polarizing effect of the metal stripes which act as a parallel wire grid and may be described by an effective anisotropic sheet conductivity $\sigma_\square$ in Eq.(15). Since in perpendicular polarization the sheet conductivity is estimated to be much less than the wave admittance Y$_0$, it has been ignored in the denominator of Eq.(15) for the calculation of the line shapes in Figs. 10(b) and 10(d).

Resonance positions versus magnetic field strength[12] are summarized in Fig. 11(a) for three gate voltages ΔV$_g$. The inset once more presents the data in low magnetic fields on an expanded scale. Transmittance spectra could not be taken close to the reststrahlen band $\bar{\nu}$=183–194 cm^{-1} of InSb indicated by the dashed horizontal lines. In the absence of magnetic fields we observe resonance transitions between 1D subbands with resonance frequencies that increase with gate voltage, i.e., number of free electrons induced into the inversion channels. This increase of

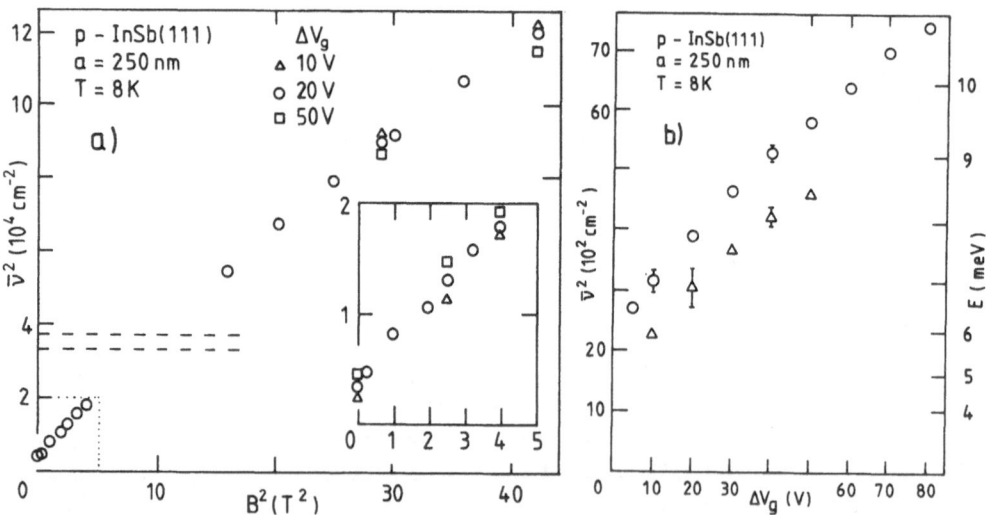

Fig. 11. (a) Resonance positions of 1D subband-shifted cyclotron resonances for three gate voltages ΔV$_g$. The inset shows the data in low magnetic fields on an enlarged scale in the same units. (b) Intersubband resonance positions (B=0) vs gate voltage. From Ref. 12.

resonance energies.[12] In strong magnetic fields above the reststrahlen band, we have subband-shifted cyclotron resonance with frequencies that slightly decrease with increasing gate voltages. This decrease of the cyclotron frequency compares well with the one observed in a homogeneous 2D electron gas on InSb and is a consequence of nonparabolicity.[5]

The data in Fig. 11(b) taken in the absence of magnetic fields allow us to extrapolate the measured resonance energies to zero gate voltage $\Delta V_g = 0$. Thus we determine the subband spacing due to the lateral confinement for vanishing electron density $n \rightarrow 0$. For the two samples, we obtain subband spacings $\hbar\omega_0 = 5.3$ meV and 6.0 meV, respectively. The magnitude of these energies rules out predominance of depolarization in the FIR excitations up to gate voltages $\Delta V_g \approx 50$ V.

4.2 Transport studies

To determine the single electron subband spacings at finite densities, the quasi-static resistance R and its derivative dR/dV_g have been measured[23] in the direction along the stripes via the contacts depicted in Fig. 2. Experimental traces of derivatives dR/dV_g taken at various voltages ΔV_g and a constant current of 1 μA are shown in Fig. 12.

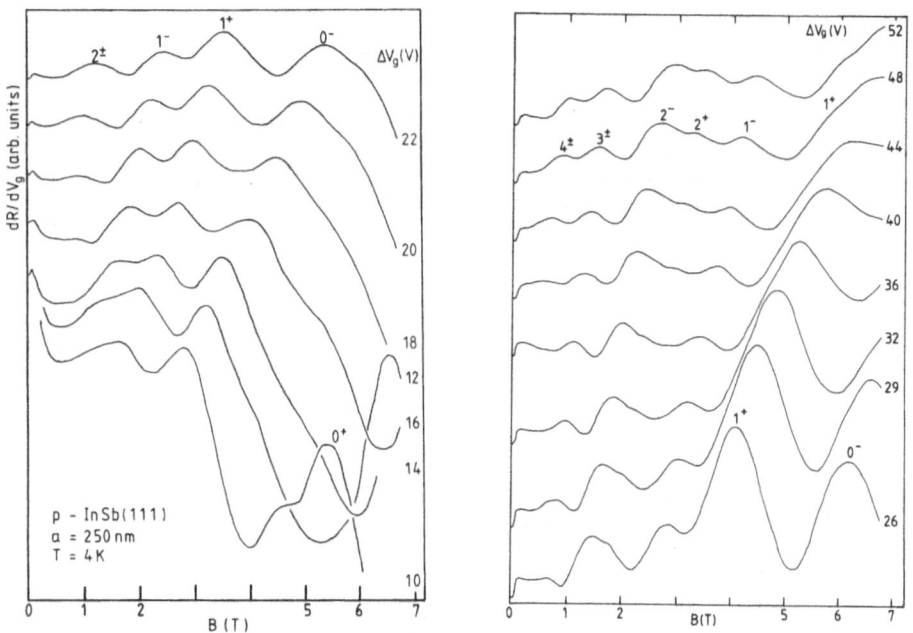

Fig. 12. Derivatives of the magnetoresistance along the inversion channels at various gate voltages ΔV_g above threshold. The oscillation maxima are marked by their subband indices. From Ref. 23.

In this figure we also assign subband indices $n^{\pm}$ to the maxima. There is some weaker structure in the traces that is not due to 1D subband quantization. At magnetic fields B~0.1 T there is a structure whose position does not significantly shift with gate voltage and whose amplitude becomes less pronounced at higher gate voltages. We think that this structure is related to quantum interference causing negative magnetoresistance. The shoulders at magnetic fields B$\approx$4-5 T and voltages ΔV_g=10 to 16 V are due to Shubnikov-de Haas oscillations of the 2D regions between the wire grid and the contacts. An additional oscillation occurs for voltages $\Delta V_g \gtrsim$40 V at magnetic fields B$\approx$5 T and is attributed to an oscillation of the i=1 subband. For various gate voltages ΔV_g, values 1/B of oscillation maxima are given in Fig. 13 versus quantum number $n^{\pm}$. In case of a homogeneous 2D gas, this kind of plot results in straight lines of slopes $2e/hn_s$. For our 1D channels we can draw a limiting straight line only through the points for the highest fields. This is done for gate voltage ΔV_g=52 V in Fig. 13 yielding a channel density n_c=4.2x10^{11} cm^{-2}. In a 2D gas the number of oscillations is principally infinite and the number of observable oscillations is only limited by the attainable mobility ($\mu B \gtrsim$1) and superposition[5] of oscillations due to the occupation of higher subbands. In a 1D system, the number of oscillations is finite due to lateral subband quantization in zero magnetic fields (see Fig. 6). The highest populated hybrid subband can directly be read from Fig. 13. To give an example, subband N=4 is the highest populated one at voltages ΔV_g=44 V and 52 V.

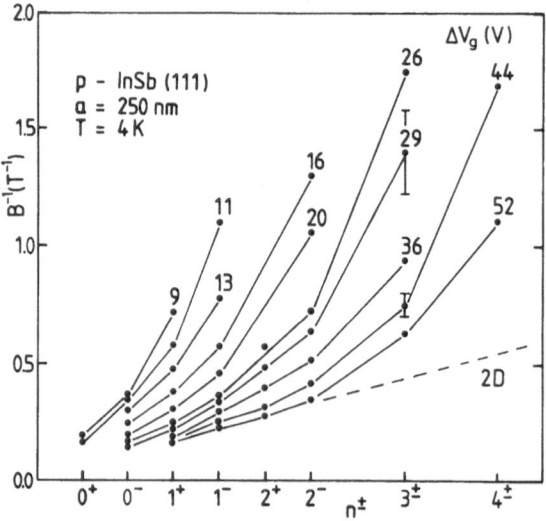

Fig. 13. Fan-chart 1/B vs subband indices of magnetoresistance maxima for various gate voltages ΔV_g. The dashed line for ΔV_g=52 V represents the limiting 2D straight line for high magnetic fields. From its slope, the areal channel density n_c can be deduced.

According to previous studies on low-dimensional systems,[24] maxima in the magnetoresistance R(B) reflect maxima of the Fermi energy. Hence, we rather should take the zeros on the right hand sides of the oscillation maxima dR/dV_g in Fig. 12 for an evaluation according to Eqs.(2) and (3). However, these zeros are difficult to identify in the traces. Since we found by numerical simulation that choosing maxima instead of zeros affects the resulting values n_l and $\hbar\omega_0$ at most by 10%, we adhere to the maxima. In Fig. 14 intersubband resonance energies (circles) are compared with subband spacings (squares) deduced from the magnetoresistance oscillations.[23] The FIR energies exceed the subband spacings and increase more pronounced with gate voltage, i.e., increasing electron density ($n_l/\Delta V_g \simeq 0.075 \times 10^6$ $V^{-1}cm^{-1}$) as a consequence of depolarization. The oscillations have been evaluated in EMA as well as in k·p-approximation. In the limit $n_l \to 0$ both approaches yield the same subband spacing $\hbar\omega_0 \simeq 9$ meV which agrees with the extrapolated FIR resonance energy: Both non-parabolic effects and depolarization vanish in the limit $n_s \to 0$. For the gate voltage $\Delta V_g = 40$ V we deduce a shift of $\hbar\omega_{dep} = 11.6$ meV from Fig. 14 if we compare the FIR data with the k·p values. The classical Eq.(11) gives a value $\hbar\omega_{dep} = 17$ meV ($\alpha=1$) with the mass $m^* = 0.020$ m_e from Eq. (10), the

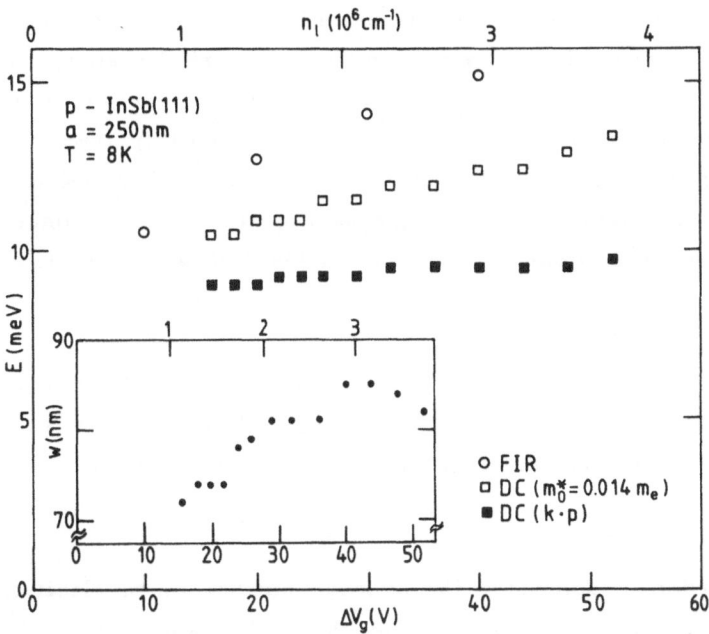

Fig. 14. Subband spacings of 1D inversion channels vs gate voltage ΔV_g or linear density n_l deduced from DC magnetoresistance oscillations using the EMA (open squares) and the k·p-approximation (closed squares). For both approximations the linear densities n_l are almost identical. The inset shows the channel width $\langle w \rangle = n_l/n_c$. Intersubband (FIR) energies (open circles) exceed the subband spacings due to the depolarization shift.

linear density $n_l = 3 \times 10^6$ cm^{-1}, and the width $\langle w \rangle = 85$ nm all obtained from the theoretical description of the magnetoresistance oscillations, as well as from the dielectric constant $\bar{\varepsilon} = 10$. The classical value clearly exceeds the experimental one and a similar result is obtained from the Hartree result of Eq.(12). The theoretical value reduces if one takes into account coupling of wires and screening by the gate according to Eqs.(13) and (14) depicted in Fig. 9(a,b), respectively. The coupling effect yields a reduction of about 10% for our periods a=250 nm and oscillator lengths $l_0 \cong 25$ nm. The screening effect is more difficult to estimate because the Schottky gate is not a constant distance away from the channels. If we put equal the distance to half of the resist height (d≅25 nm), we expect a minor reduction. However, if screening near the channel edges (d~0) is most important, the reduction may be very strong.

5. SPECTROSCOPY OF QUANTUM DOTS

Spectroscopy of dots[7] is carried out with an optically pumped FIR laser with linearly polarized light. The relative change of transmittance is recorded versus the strength B of the magnetic field. Spectra for various laser energies $\hbar\omega$ and gate voltages ΔV_g are shown in Fig. 15. The spectra are almost independent of the polarization direction in the plane as is expected by virtue of sample preparation. Spectra for energy $\hbar\omega = 10.4$ meV resemble cyclotron resonances of a homogeneous 2D gas but the resonance magnetic fields are already shifted considerably ($\Delta B \cong 0.4$ T) to lower field strengths. This directly reflects the spatial quantization in the confining lateral potential. For energy $\hbar\omega = 7.6$ meV, we no longer observe a distinct resonance maximum at finite fields but a monotonic decrease of the relative transmittance when the magnetic field is increased. This is indeed expected from the classical conductivity given in Eq.(7) when the quantization energy of the lateral potential approximately coincides with the laser energy ($\omega \cong \omega_0$). For the energy $\hbar\omega = 3.2$ meV, we again observe distinct but weak resonances at B≅1.5 T. These are the ω_- resonances of Eq.(6) characteristic of a system confined in both lateral directions.

In order to determine the average number n_0 of electrons in a dot and the mobility μ we rely on the classical shapes[5] of cyclotron resonances in strong magnetic fields ($\hbar\omega = 26.6$ meV). Alternatively, they can also be determined from the classical conductivity in Eq.(7) from spectra taken at frequencies $\omega \cong \omega_0$ where we have the relation $t(B=0) \cong 2en_0\mu/\sigma_\square a^2$ for conductivities $\sigma_\square \gg Y_0$. Mobilities $\mu = (B_{1/2})^{-1} \cong 20\ 000$ cm^{2}V^{-1}s^{-1} are

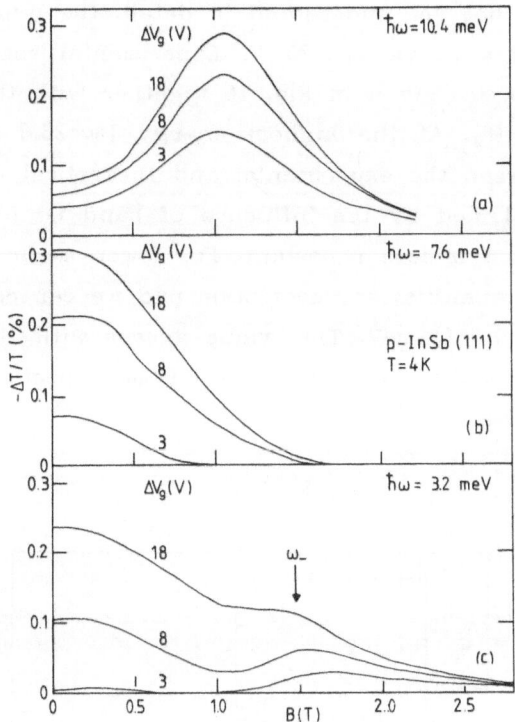

Fig. 15. Far-infrared spectra of quantum dots for three laser frequencies ω and three gate voltages ΔV_g. (a) ω_+ resonance at $B \approx 1$ T for a laser frequency above the quantization frequency ω_0, (b) traces for $\omega \approx \omega_0$, (c) ω_- resonance at $B \approx 1.5$ T for $\omega < \omega_0$. From Ref. 7.

obtained from the fields $B_{1/2}$ where the transmittances have dropped to half of their maximum values at $B \approx 0$. Almost the same electron numbers and mobilities are obtained from both approaches. An evaluation according to the Kubo conductivity of Eq.(8) yields slightly higher electron numbers. Electron numbers n_0, quantum numbers $N = 2n + |m|$ of the highest populated $B=0$ level, and dot radii $r_F = [2\hbar(N+1)/m^*\omega_0]^{1/2}$ at the Fermi energy are summarized in Table 2 for the voltages of Fig. 15. Differences between parameters for frequencies $\omega \approx \omega_0$ (confining oscillator) and $\omega > \omega_0$ (Landau oscillator) are given as experimental uncertainties in Table 2.

Table 2. Electronic dot parameters.

$\Delta V_g(V)$	n_0	N	r_F (nm)
3	3 ± 1	1	54
8	9 ± 1	2	66
18	20 ± 2	3	76

For reasons we do not yet understand in detail, the number of electrons saturates at voltages above $\Delta V_g = 20$ V. Experimental resonance positions for voltage $\Delta V_g = 8$ V are given in Fig. 16 together with theoretical curves calculated from Eq.(6). At the highest energy ($\hbar\omega = 26.6$ meV) there is a shift $\Delta B = 0.8$ T between the experimental and theoretical result. This shift qualitatively is explained by the influence of band nonparabolicity which for the lower branch is less important. For lower magnetic fields, Eq.(6) provides an almost quantitative description and we can estimate a quantization energy $\hbar\omega_0 = (7.5 \pm 1)$ meV. This value agrees with the one which we already deduced from the shape of the $\hbar\omega = 7.6$ meV spectra in Fig. 15.

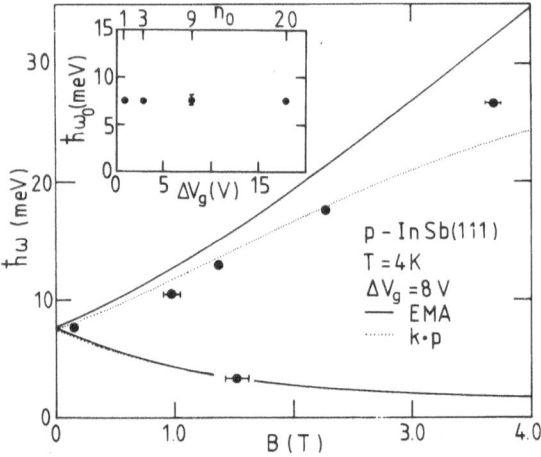

Fig. 16. Zeeman splitting of resonance positions in quantum dots. The inset gives the dependence of the quantization energy $\hbar\omega_0$ on voltage ΔV_g and electron number n_0. The solid lines are calculated from the EMA result in Eq. (6). The dotted lines are obtained from our k·p- approximation for transitions $(0,0) \to (0,1)$ and $(0,0) \to (0,\bar{1})$ ignoring spin splitting.

As is shown in the inset of Fig. 16, the quantization energy does not depend on electron number $n_0 = 3$ to 20 within experimental error. This provides strong evidence that collective depolarization modes[21] which might be expected to become important at higher electron numbers are strongly suppressed in our devices. In fact, macroscopic electric fields are effectively screened by the NiCr Schottky gate since it is evaporated in very close vicinity to the electron system. For the same reason, we do not expect strong electromagnetic coupling of dots. The independence of the excitation energy on electron number is clearly distinct from the sit-

uation in real atoms if one compares, e.g., hydrogen and helium. In our
artificial atoms, however, the electrons are not bound in a Coulomb potential but in a parabolic well as depicted in Fig. 17. This fact and the large
InSb dielectric constant $\varepsilon=17$ strongly modify effects of electron-electron
interaction.[25,26] In particular, no influence is detectable in "helium" since
ground and excited singulet states $S=0$ are shifted by the same energy Δ.
There is no exact cancellation for higher electron numbers and we expect
shifts of fractions of the energy $\Delta=(\pi Ry^* \cdot \hbar\omega_0)^{1/2}$. The smallness of this
parameter may explain our experimental result.

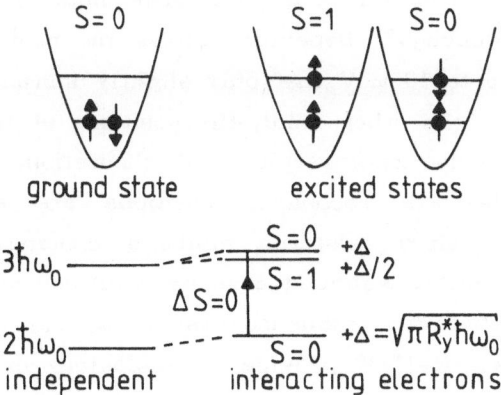

Fig. 17. Electron-electron interaction and resonance energy (see arrow)
for two electrons ("helium") in a dot. Since both ground and excited
singulet states are shifted by the same energy Δ, we expect the same
resonance energy ($\Delta S=0$) as for one electron ("hydrogen"). From Ref. 25.

6. SUMMARY AND CONCLUSIONS

By quasi-static transport[23] and FIR spectroscopy[12] we have studied
quantum wires on InSb exhibiting quantization energies of up to 10 meV
at widths of 100 nm. The magnitude of the quantization energy reflects
the small InSb conduction band mass $m_0^*=0.014\ m_e$ as becomes evident
from a comparison with related wire structures on GaAs heterojunctions[11]
of electron mass $m_0^*=0.066\ m_e$. On these heterojunctions of unrivalled
mobility subband spacings in quantum wires are about 2 meV, i.e., they
are just the same factor smaller as the electron mass is higher.

Spectroscopic studies of electronic excitations in lateral quantum structures are made feasible by the use of arrays of wires or dots prepared on macroscopic areas. By these studies, we gain direct insight into the dynamics of electrons in confined geometries. In particular, at higher electron densities we have to take into account the effect of depolarization in our periodic arrays. The excitations we observe are in fact intersubband-plasmon modes.[19] We have determined the contribution of depolarization to the observed resonance frequency by carrying out transport measurements and FIR spectroscopy on the same wire sample.[23] For the evaluation of Shubnikov-de Haas type oscillations in the magneto-resistance we assume a parabolic well with its characteristic frequency ω_0 as fitting paramter. This should provide a good approximation at least at lower electron densities n_ℓ as we conclude from self-consistent calculations for quantum wires on $GaAs/Ga_{1-x}Al_xAs$ heterojunctions of similar size and electron density.[27] Depending on sample preparation, we obtain subband spacings 5 to 10 meV that only slightly increase with increasing electron density. On the other hand, the positions of the FIR resonances shift to higher energies as a result of depolarization. At lower densities $n_\ell \leqslant 1 \times 10^6$ cm^{-1}, the FIR resonance positions are almost exclusively determined by the single electron subband spacing. At the highest densities $n_\ell \approx 4 \times 10^6$ cm^{-1}, subband spacing and collective depolarization contribute to the observed resonance position by approximately the same amount. Present theories[19,20] provide a satisfactory explanation of these experimental observations, but a really quantitative agreement cannot be expected because of the rather complicated sample geometries and because we do not exactly know the shapes of the confining potentials.

It is not timely to draw conclusions from the very first experiments with quantum dots[7] and we better attempt to give a prospective outlook on these few-electron systems. They have some in common with shallow donors in semiconductors[8] as both are embedded in a medium of dielectric constant $\mathcal{E}$ and as in both the electron motion is characterized by an effective mass m_0^* which is much smaller than the free mass m_e. The size of the dots is of the same order as the effective Bohr radius $a^*=64$ nm, i.e., much larger than the lattice constant. Since the cyclotron energy $\hbar\omega_c$ can readily be adjusted by laboratory magnetic field strengths to exceed the effective InSb Rydberg $Ry^*=0.7$ meV or the binding energy $\hbar\omega_0$ in dots, respectively, both systems are well suited to study the transition from electrically to magnetically bound states in semiconductors. However, there are also significant differences between the two systems: Unlike in shallow donors we can define the size, the shape, and the electron number of the dots by technological means. Also the electrons are not bound in a Coulomb potential but in wells of approximately parabolic shape. So the influence of the electron-electron interaction and of the band structure onto the discrete electronic states is largely different in the two systems and there is a lot of experimental and theoretical work ahead of us to examine and describe quantum dots in semiconductors.

ACKNOWLEDGEMENTS

We thank J. P. Kotthaus and A. V. Chaplik for valuable discussions and the Volkswagenstiftung as well as the Deutsche Forschungsgemein-schaft for financial support.

REFERENCES

1. "Physics and Technology of Submicron Structures", H. Heinrich, G. Bauer, and F. Kuchar, ed., Springer Series in Solid-State Sciences Vol. 83, Springer, Berlin (1988).

2. B. J. van Wees, L. P. Kouwenhoven, H. van Houten, C. W. J. Beenakker, J. E. Mooij, C. T. Foxon, and J. J. Harris, Phys. Rev. B 38:3625 (1988).

3. J. P. Kotthaus, in: "Proceedings of the 19th International Conference on the Physics of Semiconductors", W. Zawadzki, ed., Warschau (1988) Vol. 1, pp. 47-54.

4. T. Ando, A. B. Fowler, and F. Stern, Rev. Mod. Phys. 54:437 (1982).

5. U. Merkt, M. Horst, T. Evelbauer, and J. P. Kotthaus, Phys. Rev. B 34:7234 (1986).

6. W. Hansen, M. Horst, J. P. Kotthaus, U. Merkt, Ch. Sikorski, and K. Ploog, Phys. Rev. Lett. 58:2586 (1987).

7. Ch. Sikorski and U. Merkt, Phys. Rev. Lett., May (1989).

8. Y. Yafet, R. W. Keyes, and E. N. Adams, J. Phys. Chem. Solids 1:137 (1956).

9. T. Demel, D. Heitmann, P. Grambow, and K. Ploog, Appl. Phys. Lett. 53:2176 (1988)

10. D. Childers and P. Pincus, Phys. Rev. 177:1036 (1969).

11. W. Hansen, in: "Festkörperprobleme (Advances in Solid State Physics) Vol. 28", pp.121-140, U. Rössler, ed., Vieweg, Braunschweig (1988).

12. U. Merkt, Ch. Sikorski, and J. P. Kotthaus, Superlattices and Microstructures 3:679 (1987).

13. R. B. Dingle, Proc. R. Soc. London A 211:500 (1952); 212:38 (1952).

14. B. A. Wilson, S. J. Allen, Jr., and D. C. Tsui, Phys. Rev. B 24:5887 (1981).

15. M. Wagner, private communication.

16. W. Zawadzki, J. Phys. C 16:229 (1983).

17. U. Merkt and S. Oelting, Phys. Rev. B 35:2460 (1987).

18. W. Hansen, J. P. Kotthaus, A. Chaplik, and K. Ploog , in: "High Magnetic Fields in Semiconductor Physics", G. Landwehr, ed., Springer Series in Solid State Physics Vol. 71, Springer, Heidelberg (1987), pp. 266-269.

19. W. Que and G. Kirczenov, Phys. Rev. B 37:7153 (1988); 39:5998 (1989).

20. A. V. Chaplik, Superlattices and Microstructures, to be published.

21. W. Que and G. Kirczenov, Phys. Rev. B 38:3614 (1988).

22. E. Batke and D. Heitmann, Infrared Phys. 24:189 (1984).

23. J. Alsmeier, Ch. Sikorski, and U. Merkt, Phys. Rev. B 37:4314 (1988)

24. K. F. Berggren and D. J. Newson, Semicond. Sci. Technol. 1:327 (1986).

25. A. V. Chaplik, private communication.

26. G. W. Bryant, Phys. Rev. Lett. 59:1140 (1987).

27. S. E. Laux, D. J. Frank, and F.Stern, Surf. Sci. 196:101 (1988).

BALLISTIC TRANSPORT IN QUASI-ONE-DIMENSIONAL STRUCTURES

D.A. Wharam, M. Pepper, R. Newbury, D.G. Hasko, H. Ahmed
J.E.F. Frost, D.A. Ritchie, D.C. Peacock§ and G.A.C. Jones

Cavendish Laboratory, University of Cambridge, Madingley Road

Cambridge, CB3 0HE, UK

§Also at GEC Research Centre., Wembley, Middlesex, HA9 7PP, UK

T.J. Thornton

Bellcore, PO Box 7030, Red Bank, New Jersey 07701-7030, USA

U. Ekenberg

Dept. of Physics, Uppsala University, S-75121, Uppsala, Sweden

INTRODUCTION

The split-gate structure illustrated in Plate 1, and modifications thereof, have been fabricated on a variety of high-mobility heterostructures grown by Molecular Beam Epitaxy at the Cavendish Laboratory. The lithographic length of the split-gate channel was defined to be 0.3 μm whilst the defined channel width was 0.5 μm. The significance of the low temperature high-mobility behaviour is to be seen immediately. The eigenstates of momentum are extremely long-lived and give rise to elastic lengths which can be in excess of several microns. Furthermore the inelastic length, which can be extracted from physical phenomena such as universal conductance fluctuations (Thornton 1987) or Aharanov-Bohm oscillations (Ford 1989), is also of the same order of magnitude. Electrons therefore pass ballistically through the narrow constriction, defined by the application of a negative bias to the gate electrodes, the only scattering being specular scattering from the side walls of the confining potential. Furthermore the Fermi wavelength of the two-dimensional electron gas (2DEG), which is given by $\lambda_f = \sqrt{(2\pi/n_s)}$ and is of the order of 50 nm, is comparable to the effective channel width in these devices and hence the quantum nature of the electrons should be significant.

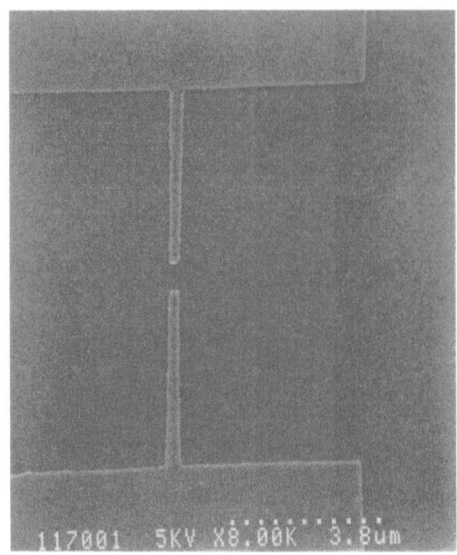

PLATE 1. A Scanning Electron Micrograph of a typical split-gate device. The channel was defined to be 0.5 μm wide and 0.3 μm long. The mean free path in the 2DEG was typically of the order of 5 μm.

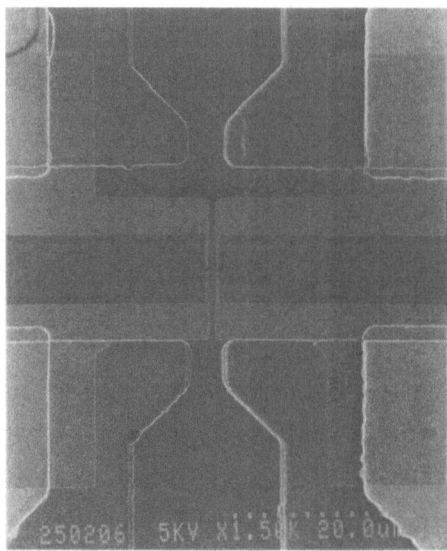

PLATE 2. A Scanning Electron Micrograph of a typical double split-gate structure. Each constriction was lithographically defined to be 0.3 μm long and 0.5 μm wide; the two channels are separated by 1 μm.

1 LOW-TEMPERATURE MEASUREMENTS

In Figure 1 the low-temperature (T ~ 100 mK) resistance of a typical device is plotted as a function of the negative bias applied to the gate electrodes. The resistance was measured between optically defined voltage probes situated at the edge of a 25 μm wide mesa and separated by 100 μm. Hence at zero gate bias the measured resistance was merely that of four squares of 2D material, and was used to determine the low-temperature mean free path of the electrons. The measured mobility of the sample was 83 m^2 V^{-1} s^{-1} which combined with the observed carrier density of 4.2 10^{15} m^{-2} yielded an electronic mean free path of 8.8 μm. At a negative gate bias of -0.5 V there was a step in resistance associated with the depletion of the electrons from the 2DEG beneath the gate and the consequent definition of the narrow channel. As the negative bias was further increased towards pinch-off, the width of the channel narrows and steps between well defined resistance plateaux were observed. The resistance of the channel itself was derived by subtracting the resistance of the surrounding 2DEG and the values of plateaux resistance thereby derived were found to be quantised according to $R = h/2ie^2$ where i was an integer. This quantisation of resistance has been observed in a wide range of fabricated narrow channels of different lengths and different material parameters. However identical devices fabricated on the same wafer could have markedly different gate characteristics suggesting that the particular distribution of scattering centres in the vicinity of the channel played a crucial rôle in the observation of this phenomenon.

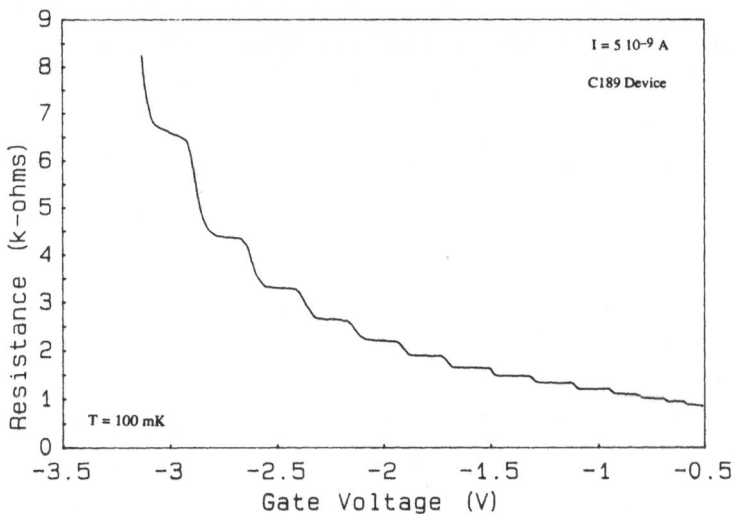

Figure 1. The resistance of a typical split-gate device is plotted as a function of the gate bias applied to the electrodes. The observed quantised steps in resistance are quantised according to $R = h/2ie^2$ (Wharam et al. 1988a).

Semi-classically the resistance of a ballistic narrow channel has been shown to be given by;

$$R = \frac{h}{2e^2} \frac{\pi}{k_f W} \tag{1}$$

where W is the width of the channel (Sharvin 1965, Levinson et al. 1987). This expression is valid providing the elastic length, l, is much larger than the channel width which in turn must be considerably larger than the Fermi wavelength, λ_f; clearly the first constraint was readily satisfied in these structures, however the latter was manifestly not. Typically the Fermi wavelength in the surrounding 2DEG was of the order of 50 nm whilst the channel width varied continuously from 500 nm to zero when the device was pinched-off. Obviously in the régime where the Fermi wavelength was comparable to the channel width quantum size effects were important and significant deviations from the semiclassical formula were expected.

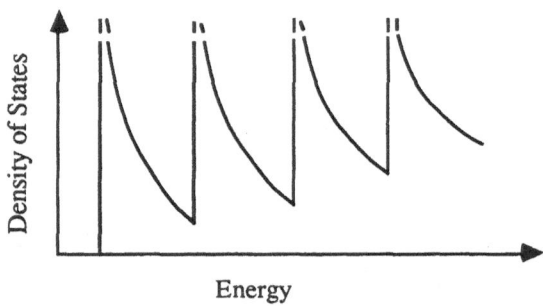

Figure 2. A schematic illustration of the total density of states in a quasi-1D system. Each subband gives rise to a square-root singularity of equal magnitude.

The quantised result is most readily understood within the framework of a one-dimensional subband picture. The contribution to the current from a single subband is given by;

$$I = n(v) \, e^2 \, v_f \, \Delta V \tag{2}$$

where n(v) is the density of states evaluated at the Fermi energy, v_f the Fermi velocity and ΔV the potential drop across the constriction. In one-dimension the product of n(v) and v_f produces an exact cancellation since $n(v) = 1/\pi \, (\partial k/\partial \varepsilon)$ and $v_f = 1/\hbar \, (\partial \varepsilon/\partial k)$ yielding a net current contribution per subband of $e^2/\pi\hbar$. If there are N occupied subbands then the total current is given by $I = (Ne^2/\pi\hbar)\Delta V$ which leads directly to the quantised result. The assumption of one-dimensional statistics used to derive this result is entirely equivalent to the assumption of a quantised transverse momentum as used by van Wees et al. (1988a). This assumption requires, in principle at least, that the channel be considerably longer than it is wide. Close to pinch-off, when the channel became very narrow, this requirement was satisfied, but when the channel was first defined this was clearly not the case and the validity of the above assumptions had to be questioned. In Figure 2 a schematic illustration of the total density of states of a quasi-one-dimensional system is shown. At zero temperature states are occupied up to a well-defined energy, the Fermi energy, such that the required carrier density in the constriction is achieved. As the channel width narrows the energy separation between the one-dimensional subbands increases and eventually an infinity in the density of states associated with the ground state energy of a particular subband passes through the Fermi energy. At this point the subband no longer contributes to the current flowing through the constriction and a transition between two quantised resistance plateaux is observed.

2 TEMPERATURE DEPENDENCE AND LIFETIME BROADENING

In practice the transition between quantised plateaux is never discontinuous, because of both the thermal broadening of the electronic energies at the Fermi energy and the finite lifetime of the momentum eigenstates. The former leads to a thermal spreading in the occupancy of states at the Fermi energy whilst the latter leads to the definition of a new density of states which is no longer truly one-dimensional. At temperatures below $T = \hbar/k_B\tau$, where k_B is Boltzmann's constant and τ the elastic lifetime, the lifetime broadening of the states dominates; in a typical high-mobility heterostructure this corresponds to a temperature of approximately 0.3 K. Above this temperature thermal broadening of the Fermi energy results in a loss of quantum structure. For a typical device the plateaux were not clearly resolved at 4.2 K although there was some evidence of quantum structure, and below 500 mK there was no significant improvement in the quality of the plateaux as the temperature was further decreased. The effects of thermal broadening are shown qualitatively in Figure 3 where the resistance of the $R = h/8e^2$ plateau, corresponding to four occupied subbands, is plotted for a typical device at a number of conveniently accessible temperatures of the dilution refrigerator. Increasing the temperature caused the transition region between plateaux 4 and 5 to move progressively further onto the quantised plateau.

From the data of Figure 1 the effective broadening has been extracted approximately by assuming that the variation of channel width with gate voltage is effectively linear. In addition it was assumed that the ground state energies were given by those of a particle in a box. Thus the effective broadening was determined from the variation in measured gate voltage over a typical transition between plateaux. This yielded energy broadening between plateaux of the order of 0.3 ± 0.1 meV corresponding to thermal broadening of ~ 3 K, or, alternatively, lifetime broadening associated with a mean free path of ~ 0.6 μm. Although

119

this estimate of the broadening cannot be regarded as accurate it is clear that under the experimental conditions of the measurement neither of the above conditions was satisfied and the magnitude of the broadening must be explained by some alternative mechanism.

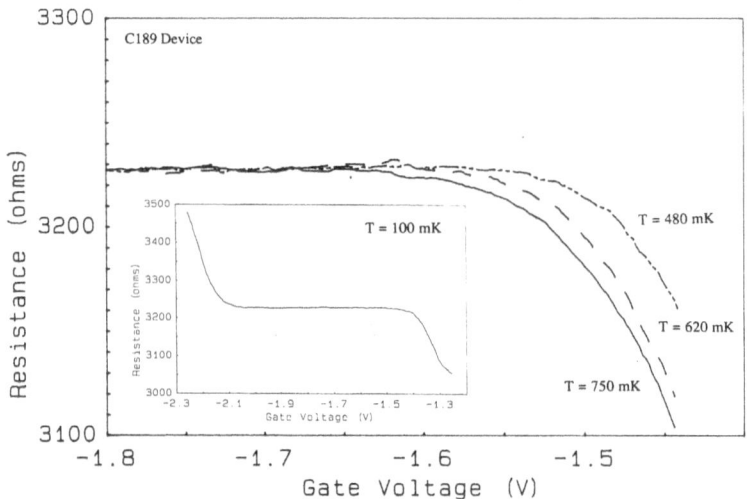

Figure 3. The broadening of the transition region at the edge of the n = 4 quantised plateau is plotted for a range of temperatures in a typical device. For comparison the whole of the n = 4 plateau is plotted as measured at 100 mK.

3 THE APPROACH TO A THEORETICAL DESCRIPTION

Whilst the one-dimensional subband model is capable of explaining the quantisation of resistance as well as the qualitative features of the effects of broadening it is not at all clear why the assumption of one-dimensional statistics is valid for a constriction of small aspect ratio. Furthermore it does not address the effects of quantum coherence between states at the entrance and exit of the constriction. In the subsequent section the somewhat naïve assumption of one-dimensional statistics is justified and alternative descriptions discussed.

3.1 THE ADIABATIC APPROXIMATION

In general the confining potential produced by the application of a negative bias to a split-gate structure is a highly complex function of position whose exact solution is only possible in cases of high symmetry, for example in the case of an infinitely long channel (Laux & Stern 1986, Laux et al. 1988). When the aspect ratio of the constriction is small the width of the confining potential is no longer independent of the position along the channel. There is a smooth variation in the width of the confining potential which is enhanced by the limitations of the fabrication technique; in both plates 1 & 2 a rounding of the split-gate electrodes is clearly visible. This expectation is confirmed by the recent depletion calculations of Davies (1989) and Kumar et al. (1989).

Glazman et al. (1988) have shown that providing the variation of the width, d(x), of the confining potential is smooth at the scale of the Fermi wavelength then an adiabatic separation of the variables in the Schrödinger equation is possible. Thus for a square well potential of variable width the Schrödinger equation becomes;

$$-\frac{\hbar^2}{2m}\frac{d^2\psi}{dx^2} + \varepsilon_n(x)\,\psi = E\,\psi \tag{3}$$

where n is the subband index and the potential $\varepsilon_n(x)$ is given by;

$$\varepsilon_n(x) = \frac{\pi^2 n^2 \hbar^2}{2md^2(x)} \tag{4}$$

and the transverse momentum becomes quantised as has been previously assumed. The semiclassical turning points at the narrowest point of the constriction then determine the number of subbands which contribute to the current. Thus, for $n < [kd_{min}/\pi]$ electron reflection effects are negligible whilst for $n > [kd_{min}/\pi]$ the states decay exponentially along the channel, and the quantised result is again derived. However the finite variation in d(x) about the narrowest point leads to a correction to the conductance of the form;

$$\delta G = \frac{2e^2}{h}[\,1 + \exp(-z\,\pi^2\,\sqrt{2R/d}\,)]^{-1} \tag{5}$$

where R is the radius of curvature at the narrowest point and z is a parameter defining the position on each plateau ($z = k_f d/\pi - n$). This correction, which allows for the finite transmission probabilities of classically forbidden states, leads to a broadening of the transition regions between plateaux which would otherwise be expected to be theta function type discontinuities. This correction is expected to dominate for temperatures below $k_B T = n\hbar^2/m\sqrt{2Rd^3}$. As discussed in section 2 the observed broadening was considerably larger than that expected from either thermal or lifetime broadening. Under the assumption that the additional broadening was given by the conductance correction of equation 5 the data of Figure 1 were used to extract the radius of curvature of the confining potential. A linear variation of k_f with gate voltage was assumed and the derived radius of curvature was found to be 110 ± 10 nm and showed no systematic variation as the width of the channel decreased. The depletion calculations of Davies (1989) are illustrated in Figure 4 and show that, even with perfect lithography, significant rounding of the confining potential is inevitable and that the magnitude of the radius of curvature is of the same order as that extracted from eqn. 5.

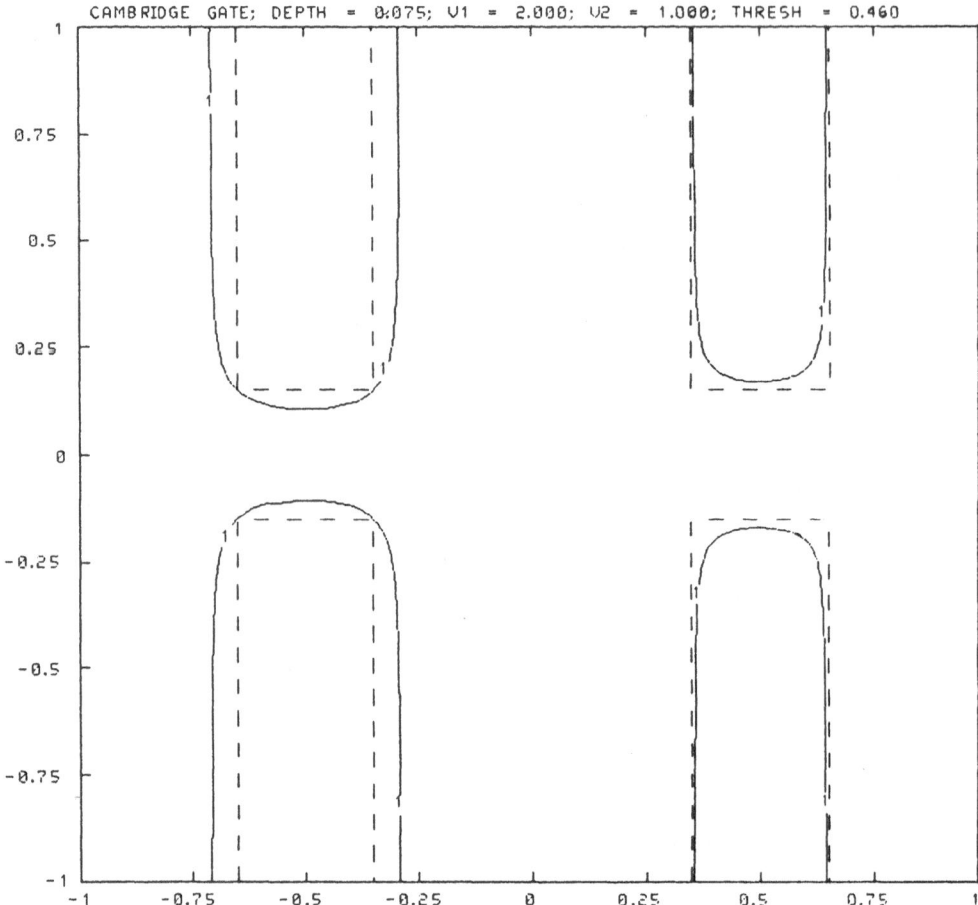

Figure 4. The depletion regions defined by the application of a negative bias to a split-gate structure (Davies 1989).

3.2 THE WIDE-NARROW-WIDE GEOMETRY

The adiabatic model discussed in the previous section assumes that the channel width varies smoothly at the scale of the Fermi wavelength. Whilst this assumption is valid at the centre of the constriction it is not clear that it holds for the constriction entrance and exit where the potential varies rapidly along the device length. A number of papers have addressed the transmission of states through a constriction with abrupt edges using the so-called Wide-Narrow-Wide (WNW) geometry, which is illustrated in Figure 5.

Levinson (1988) has considered the conductivity of such a WNW geometry and included the effects of diffraction at the end of the narrow region and shown that the conductance, G, is given by;

$$G = \frac{e^2}{\pi\hbar} \frac{[T]^4}{4(1 - [T]^2)\sin^2\phi + [T]^4} \qquad (6)$$

where [T] is the transmission coefficient for a state at the Fermi energy and ϕ the phase of the reflected wave. Equation 6 is valid for both quasi-one-dimensional as well as for three dimensional channels. A quantised conductance is obtained only in the absence of diffraction (R = 0) or if there is resonant transmission i.e. sin ϕ = 0. For the 2D case Levinson showed that for d/ λ_f = 1/2 the square of the reflection coefficient is equal to unity but decays extremely rapidly to zero giving rise to a quantised conductance. However equation 6 also predicts a series of resonant peaks in the conductance of the channel before the ($[T]^2 = 1$) limiting value is reached. To date such resonant structure has not been conclusively observed in a single split-gate structure. Levinson suggests that this may be due to thermal broadening of the Fermi surface which would be significant for temperatures $k_B T > h v_f/L$ where L is the channel length; i.e. when thermal broadening is greater then the lifetime broadening associated with the transit time of an electron through the constriction. For a normal device this suggests a temperature in excess of 2 K would be necessary for thermal broadening to smear out the resonances, however the experimental data show no signs of resonances even at temperatures as low as 60 mK.

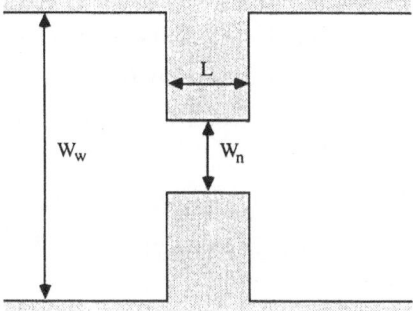

Figure 5. A schematic illustration of the so-called wide-narrow-wide geometry

Szafer & Stone (1989) have also considered a WNW geometry and shown that the quantised result can be readily derived by using both an exact recursive method as well as a mean field approximation (MFA). Furthermore they have shown that the quantisation accuracy, which according to their model can be as good as 0.01%, is perturbed by two effects. Firstly they predict a series of resonant peaks as a consequence of successively constructive and destructive internal reflections within the channel; such resonant structure becomes important for $k_nL > 1$ where k_n is the wavevector of the n-th mode propagating at the Fermi energy. Secondly they predict that evanescent modes within the channel contribute to the conductance when $\kappa_nL < 1$ where κ_n is the magnitude of the imaginary wavevector of the evanescent mode. Thus if the channel length, L, is sufficiently long to damp out any evanescent wave it is also long enough to give rise to resonance peaks. Szafer & Stone have also considered the effect of the constriction tapering upon such resonant structure and have found that when the confining potential is no longer abrupt, as must be the case in any realistic device, there is decreased reflection at the ends of the constriction leading to smaller resonances. However they expect that such resonances should still occur but might be broadened by the temperature of the experiments. They model the temperature dependence of the van Wees (1988a) device and show that resonant structure should persist up to a temperature $T_0 \sim 2.8$ K although the experimental data showed no resonant structure at the temperature of observation, 600 mK.

The effect of constriction tapering has also been considered in a WNW geometry by Escapa & Garcia (1989) who have shown that the introduction of tapered horns to the abrupt junction the resonant structure in the calculated transmission coefficients rapidly disappeared. They conclude that for any practical device the variation of the depletion region in the vicinity of the constriction is sufficiently gradual for transport to be adiabatic.

Kirczenow (1989,1988) has similarly considered a WNW geometry and, by matching wavefunctions at the interface between the broad 2D region and the narrow 1D channel, has demonstrated the existence of conductance plateaux (Figure 6). Within the constriction the electronic wavefunction is expanded in terms of the basis set of eigenfunctions of the Hamiltonian;

$$H_c = -\frac{\hbar^2}{2m}\left(\partial^2/\partial x^2 + \partial^2/\partial y^2\right) + U(y) \tag{7}$$

where $U(y)$ is the confining potential within the constriction. According to the Kirczenow model (1989) the accuracy of the quantisation depends critically upon the form of the chosen potential. In accordance with recent theoretical calculations (Laux & Stern 1986, Laux et al. 1988) Kirczenow considers two different confining potentials; firstly the parabolic potential corresponding to a depleted channel, $U(y) = cy^2 + U_0$, and secondly an infinite square well where $U(y) = U_0$ for $[y] < W/2$, and $U(y) = \infty$ for $[y] > W/2$, which is more appropriate for wider channels. The parabolic well predicts accurate quantisation of only the lowest plateau, whilst the square well without a potential barrier at the interface (i.e. $U_0 = 0$) yields accurately quantised plateaux with the shape of the steps between plateaux being independent of the plateau index. Szafer & Stone (1989) have shown in their MFA that all the steps have the same shape and accuracy as a function of the parameter $\Delta = k_fW/\pi$, which is a consequence of the fact that their transmission coefficients $T_n(\Delta)$ are, in the asymptotic limit, independent of n, the plateau index. The introduction of a potential barrier at the interface between the wide and the narrow regions leads to large deviations away from the quantised values of conductance for high index plateaux. Comparison of these theoretical calculations with experiment is complicated due to a lack of knowledge of the actual confining potential and how this potential changes with applied gate bias. This problem is addressed more fully in section 4. Furthermore, in the theoretical calculations a device of fixed geometry is assumed and the Fermi energy is then swept through the subband structure, whilst in a real device the width of the channel is varied changing the subband energies and in addition changing the Fermi energy as the density of carriers in the channel decreases. Qualitatively however best agreement was expected and was found with the results of the square well when the width of the channel was wide and the depletion regions did not extend completely across the constriction.

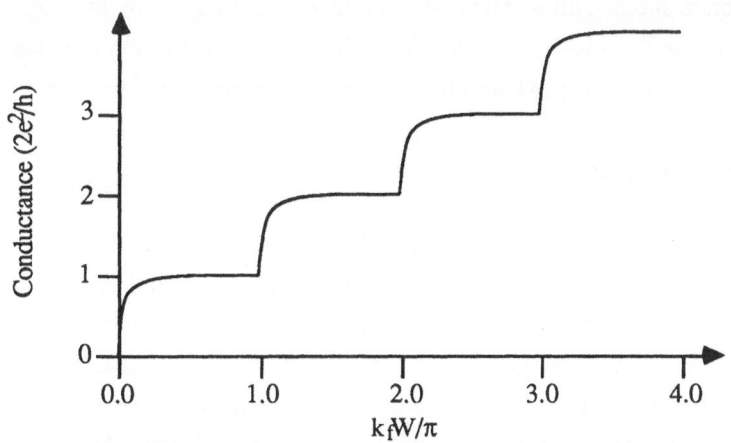

Figure 6. The conductance plateaux as derived by Kirczenow (1989) for a WNW geometry with an infinite square well confining potential in the constriction.

Kirczenow (1988) also addresses the possibility of resonant peaks in the conductance of a ballistic channel. Such resonances are, as discussed by Szafer & Stone (1989), a manifestation of longitudinal resonant states which can be supported by an open-ended tube, and conductance measurements on a ballistic channel are then a spectroscopic method of studying such states. The numerical calculations of Kirczenow (1988) are again based upon matching wavefunctions at the wide-narrow and narrow-wide interfaces. Kirczenow found that the resonant structure was highly dependent upon the aspect ratio of the channel and has suggested that such resonant structure could, in theory at least, be used to determine the length of a ballistic constriction. Qualitatively the resonances persist further into the quantised plateau for higher index plateaux and should lead to significant deviations from the quantised values, even if the resonances are thermally broadened. The temperature dependence of this structure is readily calculated from

$$G_T(\mu) = -\int G_0(\varepsilon)\,\frac{\partial f}{\partial \varepsilon}\,d\varepsilon \qquad (8)$$

which according to Kirczenow's numerical calculations would cause the resonances to disappear for temperatures in excess of $k_B T \sim 0.005\,\varepsilon_f$, corresponding to a temperature of approximately 500 mK in a typical device. However the significance of thermal broadening depends upon the separation of the resonance peaks and hence upon the aspect ratio of the actual device.

Furthermore even for temperatures at which there is significant thermal broadening the resonant structure should still manifest itself in deviations away from the quantised value of conductance for high order plateau and in a finite slope to the plateau region. Such a qualitative trend is, according to Kirczenow, discernible in the data of van Wees et al. (1988a)

3.3 THE EFFECTS OF DISORDER

The fragility of the quantised structure has already been mentioned earlier where it was suggested that the effects of scattering should be minimal for the observation of the quantised resistance. A number of authors have incorporated the effects of scattering into theoretical calculations based upon WNW geometries. Haanappel & van der Marel (1989) have used a tight-binding Hamiltonian and hence calculated the conductance by considering the flux carried by the eigenstates in a nearest neighbour hopping scheme. An impurity was then incorporated in the calculation by the addition of a delta-function potential to the Hamiltonian of a particular lattice site. Such tight-binding schemes have been considered previously in the study of localisation and conductance fluctuation phenomena (MacKinnon 1980, Thouless & Kirkpatrick 1981) and can easily be extended to microstructures of arbitrary geometry. In Figure 7 the numerical results of Haanappel & van der Marel are presented. In the limit of the channel length tending to zero there was still evidence of a steplike conductance even though no plateaux were clearly resolved. As the channel length was increased plateaux developed, however above a certain critical length ($L_{crit} \sim 0.32\,\sqrt{n}\,\lambda_f$) oscillatory structure developed. The effects of impurity scattering are illustrated for an impurity both inside and outside the narrow channel for case b), $L = 0.48\,\lambda_f$. Clearly the presence of an impurity decreased the value of conductance away from its quantised value and the quality of the steps was severely diminished in good qualitative agreement with experimental observations.

Song He and Das Sarma (1989) have similarly used a tight binding Hamiltonian in a WNW geometry and have extended the analysis to include the effects of both elastic and inelastic scattering as well as the effect of a finite temperature. Quantitative comparison with experimental data is precluded by their arbitrary choice of units but their qualitative findings were that both elastic and inelastic scattering led to a loss of the quantised structure; the former gave rise to conductance fluctuations of the same order of magnitude as the plateau steps whilst the latter led to a smoothing of the quantum structure due to a loss of phase coherence within the constriction. Song He & Das Sarma suggested, in agreement with the calculations of Kirczenow (1988), that the quantisation is best at intermediate temperatures where resonant structure on the plateaux was thermally broadened but $k_B T$ remained less than the energy separation between the transversely quantised states.

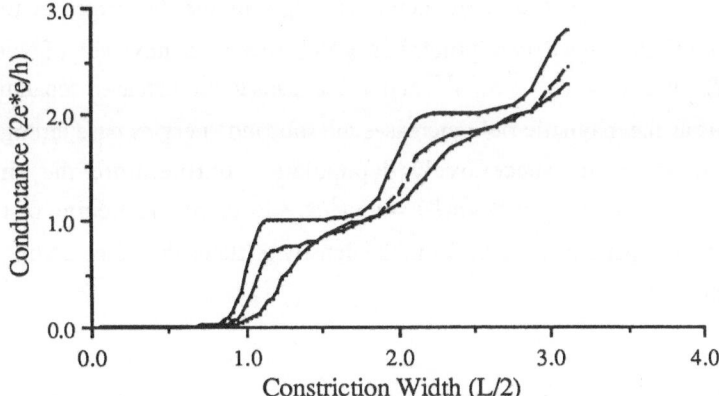

Figure 7. The effect of disorder upon the quantised structure is illustrated as considered in the calculations of Haanappel & van der Marel (1989). The upper full line shows the calculated conductance in the absence of disorder, the dotted shows the effect of introducing an impurity into the channel, and the lower full line the effect of an impurity outside the channel. The width of the channel is measured in units of half Fermi wavelengths.

4 MAGNETIC DEPOPULATION OF ONE-DIMENSIONAL SUBBANDS

The magnetic depopulation of one-dimensional subbands defined using a split-gate geometry was first considered by Berggren et al. (1986). The device structure studied was 15 μm long and transport along the narrow channel was diffusive and hence there was no evidence of resistance quantisation. However as the channel width was narrowed the one-dimensionality of the system manifested itself in the measured magnetoconductance oscillations. Significant deviations away from the 1/B behaviour expected for the conduction maxima were observed and were used to derive the channel width as a function of the applied gate voltage.

The qualitative effects of a magnetic field upon one-dimensional subbands are best understood by considering the behaviour of the 1D subbands of a parabolic confining potential well;

$$V_0(x) = m\omega_0^2 x^2/2 \tag{9}$$

The subband energies are then given by $E_n = \hbar\omega_0(n+1/2)$ and the resulting subband dispersions are parabolic $E_n(k) = E_n + \hbar^2 k^2/2m$. A transverse magnetic field introduces an additional confining potential through the 'magnetic' parabola (Landau and Lifshitz);

$$V_B(x) = m\omega_c^2(x-x_0)^2/2 \tag{10}$$

where $\omega_c = eB/m$ is the cyclotron frequency and $x_0 = \hbar k/eB$ the centre of the magnetic parabola. Combining these two potentials gives rise to a new set of energy levels $E_n = \hbar\omega(n+1/2)$ where $\omega = (\omega_0^2 + \omega_c^2)^{1/2}$ and hence leads to an increased separation between subbands. Thus as the magnetic field increases the subband energies pass through the Fermi energy and the subbands successively depopulate. Furthermore the simultaneous enhancement of the effective mass $m(B) = m\omega^2/\omega_0^2$ leads to a flattening of the subband dispersions and a sharpening of the peaks in the density of states thus accelerating the process of depopulation.

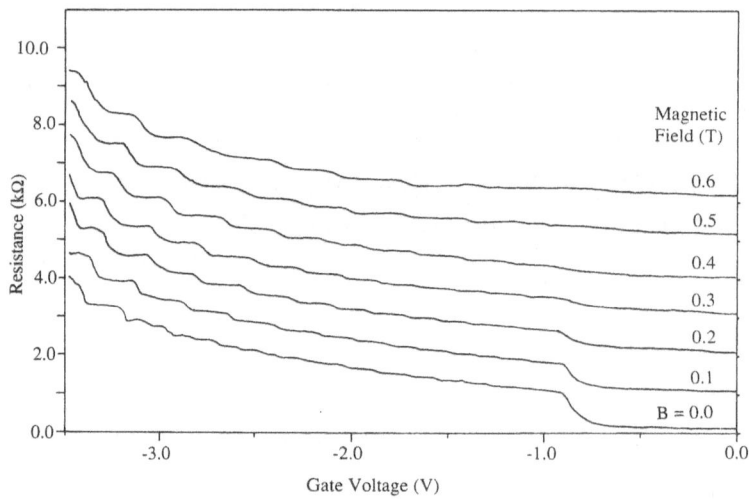

Figure 8. The resistance of a typical device is plotted as a function of gate bias for a number of different magnetic fields. The traces are offset vertically for clarity.

As discussed in section 1 the resistance of a ballistic constriction is determined by the number of occupied 1D subbands. Even in the presence of a magnetic field, when the subbands become hybrid magneto-electric subbands, this quantisation is maintained since the quantisation is independent of the subband dispersion relations. The depopulation of a subband is therefore reflected by a change in the observed value of quantised resistance. Such behaviour is clearly illustrated in the gate characteristics of Figure 8 where the channel resistance is plotted as a function of gate voltage for a number of applied transverse fields. The temperature of the measurements was 300 mK. As the magnetic field is increased the gate voltage required to depopulate a particular subband decreases in qualitative agreement with the model discussed above. However the assumption of a parabolic confining potential is only justified in the limit of a pinched-off channel. Self consistent numerical solutions of the Poisson and Schrödinger equations performed upon both narrow Si MOSFETs (Laux & Stern 1986) as well as split-gate GaAs-AlGaAs heterostructures (Laux et al. 1988) suggest that the confining potential is best given by a flat well with parabolic walls. Accordingly it was decided that the optimum confining potential to describe the channel was;

$$V(x) = m\omega_0^2([x]-t/2)^2/2 \qquad\qquad [x] > t/2$$

$$V(x) = 0 \qquad\qquad [x] < t/2 \qquad\qquad (11)$$

where t is the channel width and ω_0 is an empirical parameter determining the sidewall parabolicity. This potential has been used previously to model MESFET behaviour (Poole et al. 1982), where $\omega_0^2 = e^2N_D/\epsilon m$ as derived from a simple depletion approximation, and more recently heterostructures (Berggren et al. 1986, Berggren & Newson 1986). The above authors all used a semiclassical WKB approximation which works well in the narrow channel limit; however, for wide channels the WKB method predicts energy levels given by;

$$E_n = \hbar^2\pi^2(n+1/2)^2/2mt^2 \qquad\qquad (12)$$

In this limit the potential approaches an infinite square well whose energies are given by the expression above with (n+1/2) replaced by (n+1). Thus the ground state energy is underestimated by a factor of 4 and for a typical channel width considered the ground state energy is approximately three times smaller than expected (Wharam et al. 1989). A novel variational technique was therefore used which gave accurate subband energies in both the wide and narrow channel limits.

As a starting point for the variational calculation the eigenfunctions of a finite square well of given width and height are evaluated. These are then used as a basis set for the evaluation of the matrix elements $\langle\phi_n|H|\phi_m\rangle$, where H is the Hamiltonian derived from equation 11 above. The diagonalisation of this matrix is then equivalent to the determination of the eigenvalues of the Hamiltonian. In theory this calculation is one of infinite dimension, however since only a finite and small number of subbands are occupied the matrix manipulation becomes readily tractable. Typically twenty basis functions are used. Furthermore selecting a finite square well of width slightly larger than the channel width, t, in equation 11 ensures that the corrections to the derived energy levels from higher order terms are negligible. Changing the depth of the square well slightly shifted the calculated eigenenergies by an amount ~ 1 μeV, which is an indication that the basis set gives very accurate eigenvalues. Magnetic fields are incorporated into the calculation via the addition of the magnetic parabola (eqn. 10) to the Hamiltonian, and the subband dispersion calculated by evaluating the eigenvalues for a series of k_y values. The Fermi energy, for given B and t, is then determined by n_l, the number of electrons per unit length, and is evaluated numerically by summing the states along the subband dispersions. This process has been described in detail by Berggren & Newson (1986).

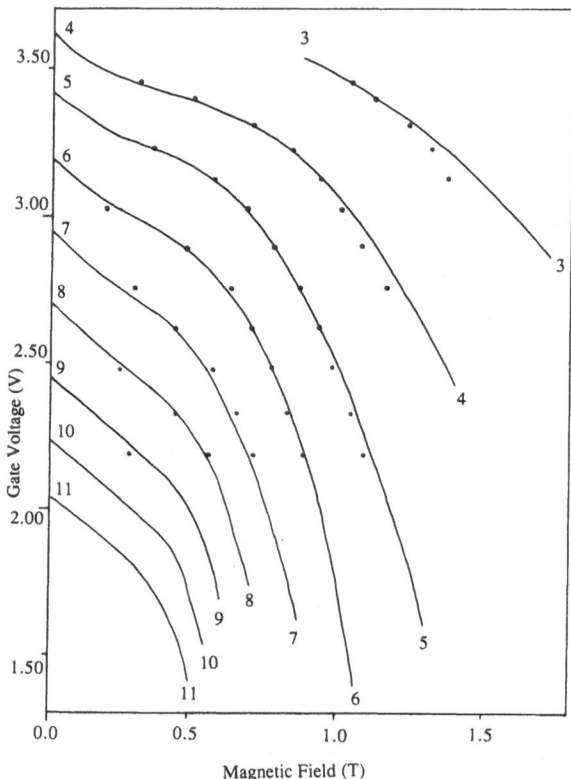

Figure 9. The gate bias required for depopulation is plotted as a function of the applied magnetic field and subband index. The dots define the values calculated from the variational calculation.

Comparison with the experimental data is then facilitated by noting that for a given applied gate bias the channel width and carrier concentration are independent of the transverse magnetic field. Thus it is possible to determine the magnetic field required to depopulate a particular subband as a function of three empirical parameters; the channel width, t, the carrier concentration, n_l, and the side-wall parabolicity, ω_0. In practice the quality of the fit proved to be rather insensitive to the value of ω_0 chosen and, for a given device, ω_0 was assumed to remain constant. The quality of the derived fit is illustrated in Figure 9 where the both the derived (dots) and the experimental (solid lines) gate voltages required to depopulate a given subband are plotted for a number of subbands as a function of the transverse magnetic field. The experimental gate voltages are derived from the point of inflection between two values of quantised resistance. The empirical parameter $\hbar\omega_0$ was taken to be constant and equal to 7.5 meV. This is slightly larger than the value of 5.7 meV calculated for another sample by Laux et al.(1988) and the difference is attributed to the higher doping concentration in the n.AlGaAs region of the device considered here (cf. 10^{18} cm^{-3} as opposed to 6.10^{17} cm^{-3} as used by Laux et al.).

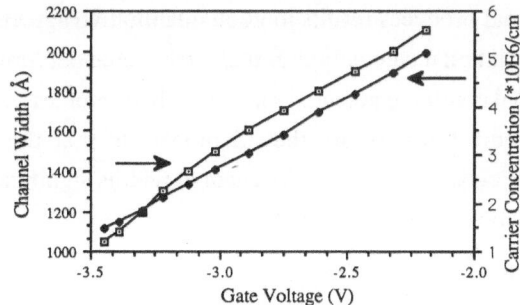

Figure 10. The extracted channel width and carrier concentration are plotted as a function of the applied gate bias.

In Figure 10 the derived channel width is plotted as a function of the applied gate voltage. The channel width varies approximately linearly with gate voltage until pinch-off is approached where the channel narrows more rapidly. Fitting data at small values of applied gate voltage is difficult because of the inherent uncertainty of the point of depopulation when the step in resistance is small. However, assuming the channel variation to be essentially linear with voltage and extrapolating to the voltage for which the channel is first defined ($V_g \doteq -0.8V$) gives a maximum channel width of ~ 3200 Å. This value is somewhat smaller than the lithographically defined channel width of 0.5 μm but this discrepancy is reconciled by assuming that the lateral extent of the depletion regions beneath the defined gates is comparable in magnitude to the vertical depletion length when the channel is first defined. Also illustrated in Figure 10 the derived one-dimensional channel carrier concentration a s a function of the applied gate voltage. The observed linear variation is in qualitative agreement with the numerical calculations of Laux et al. (1988) and furthermore predicts that the channel pinches-off at a gate bias of ~ − 4.0 V in agreement with the experimental data.

The quality of the fit illustrated if Figure 9 is a strong indication that the model potential of equation 11 is indeed an accurate description of the actual confining potential within the channel. This explains why the square well potential considered by Kirczenow (1989) in a WNW geometry gives a quantised conductance in good agreement with the empirical evidence (see Section 3.2).

The quantised conductance of hybrid magneto-electric subbands has similarly been considered by van Wees et al. (1988b). The same qualitative features of magnetic depopulation were observed, and a simple calculation, based upon the Bohr-Sommerfeld quantisation of a square well, was used to derive both the width and the Fermi wavenumber of the narrow constriction as a function of gate voltage. Recently Weisz & Berggren (1989) have undertaken a comparative study of such potential modeling. They have shown that a simple parabolic potential produces results in good quantitative agreement (to within 10 %) with the more exact variational calculation in the narrow channel limit. However for wide channels, when the actual confining potential more closely resembles a square well significant deviations are observed. Furthermore they demonstrate that the semiclassical Bohr-Sommerfeld calculation consistently produces channel widths significantly larger than those derived using the parabolic confining potential.

5 PARALLEL FIELD MEASUREMENTS

A magnetic field affects both the spin state of an electron as well as its orbital motion and it is convenient to investigate such spin effects with the field parallel to the plane of the 2DEG. In this orientation orbital effects are negligible due to the spatial quantisation of 2D systems (Ando et al. 1982), and it is therefore expected that the dominant effect is to lift the spin degeneracy which has been previously assumed. This gives rise to additional plateaux at quantised values of resistance $R = h/2e^2(i+1/2)$. Typical parallel field data are presented for a narrow ballistic constriction in Figure 11; the measurement temperature was 100 mK. The additional plateaux are resolved for fields in excess of 10 Tesla but only for the high index subbands. The high value of field required to observe such spin-splitting is in contrast to the comparatively low fields, typically ~ 2 Tesla, necessary for spin-splitting of the Landau levels in the 2DEG of a GaAs-AlGaAs heterostructure. Such low field splitting is a consequence of the enhancement of the Landé g factor, first proposed by Janak (1969) to explain measurements on Si MOSFETs; g factors as high as 6.2 have been reported by Nicholas et al. (1988) in tilted GaAs-AlGaAs heterojunctions (cf. for bulk GaAs g = − 0.44). However recent differential capacitance measurements performed upon quasi-one-dimensional grating structures have shown a novel anisotropy (Smith,III et al. 1988) with respect to the orientation of the magnetic field. With the magnetic field perpendicular to the plane of the 2DEG spin-splitting of the Landau levels was observed, but with the field parallel to the plane of the 2DEG and perpendicular to the grating no such spin-splitting of the states was resolved even with fields as large as 20 Tesla. The enhancement of the g factor has been explained in 2D systems in terms of an imbalance in the number of spin-up and spin-down electrons (Janak 1969) and requires that the Landau level spacing is much greater then the level broadening. Smith et al. suggest that the same mechanism may be ineffective in the parallel orientation since the 1D subband separations are small, typically ~ a few meV, and that the g factor is close to its bulk value. At 20 T in GaAs this implies a spin-splitting of approximately 0.5 meV and would not be resolvable. Smith et al. also observed an additional anisotropy when the field was parallel to both the plane of the 2DEG and the 1D grating. With the field in this orientation the capacitance oscillations reflecting the 1D subband structure were rapidly damped as the field strength increased.

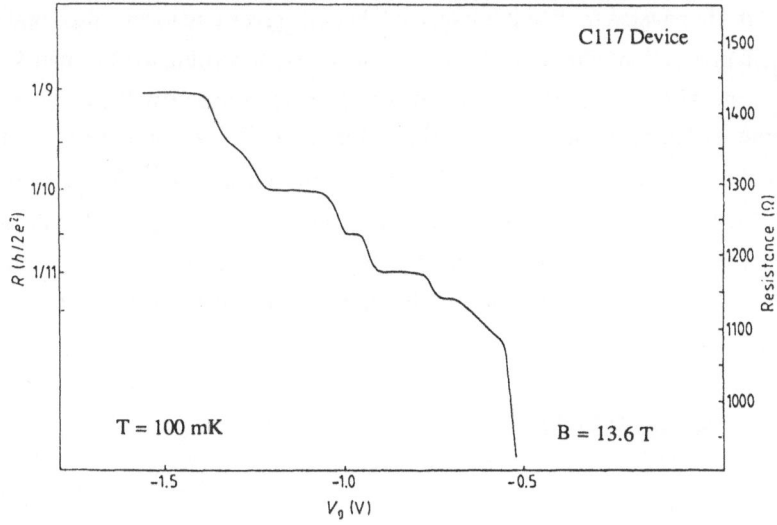

Figure 11. The device resistance is plotted as a function of the applied gate voltage for an applied magnetic field of 13.6 T. The spin-split plateaux at $h/21e^2$ and $h/23e^2$ are clearly resolved.

Several authors have considered the spin-splitting of one-dimensional states in a magnetic field, however, to date, only one paper (Glazman & Khaetskii 1989) permits direct comparison with the experimental results from ballistic channels. Huckestein et al. (1988) have expanded the linear response model of Johnston and Schweitzer (1988) to include the effects of a <u>perpendicular</u> magnetic field. In the weak field limit they predict the emergence of spin-split levels whose 'width' increases with increasing field. For strong magnetic fields Huckestein et al. have computed the energy level shifts as a function of magnetic field and channel width. The conductance is then derived by changing the channel width and calculating the widths for which the spin-split subbands energies pass through a fixed Fermi energy. Qualitatively the results show both the expected magnetic depopulation as well as the emergence of spin-split plateaux which become larger for high index subbands. Direct comparison with the <u>parallel</u> field data is precluded not just because of the absence of magnetic depopulation in this orientation but also because it is impossible to keep the Fermi energy fixed whilst changing the channel width (see Section 4) in direct contradiction to the assumptions of these calculations.

The adiabatic model of quantum transport in a narrow constriction, discussed in section 3.1, has been extended by Glazman & Khaetskii (1989) to incorporate the effects of a magnetic field in both the parallel and perpendicular orientations. With the field in the perpendicular orientation they have shown that the quantisation of resistance is maintained since electronic transport through the constriction remains adiabatic. For a parallel field additional terms are added to the Hamiltonian to include the spin-orbit interaction and the spin-splitting. Since there is a qualitative difference between electronic motion along the channel and the transverse motion it is expected that this gives rise to an anisotropy depending upon the field orientation within the 2D plane. However, according to Glazman & Khaetskii, the energy scale of the spin-orbit interaction remains negligibly small for narrow channels providing the width, d, is much smaller than $\hbar/m\alpha$ where the parameter $\alpha = 10^5$ cm s^{-1} determines the strength of the spin-orbit interaction GaAs. Considering only the spin-splitting of the states Glazman & Khaetskii have shown that the spin-split subband modes switch into the conductance of the channel for different values of their dimensionless parameter $z = k_f d/\pi$ (see Section 3.1). Thus the spin-split levels of the subband of index n are separated by

$$\Delta z_s = n \, g\mu_B B/2\varepsilon_f \tag{13}$$

where μ_B is the Bohr magneton. The separate spin states are then observed when Δz_s becomes greater than δz, the quantum width of the step, i.e. the change in the parameter z required to switch the mode into the conductance. This has been calculated from the earlier adiabatic model (Glazman et al. 1988) and is given by;

$$\delta z = \frac{\sqrt{n\lambda_f/R}}{\pi^2} \tag{14}$$

Combining these two results shows that the subband splitting is only observed for high index plateaux, $n > n_c$, where n_c is given by;

$$n_c = \frac{\lambda_f}{R}\left(\frac{2\varepsilon_f}{\pi^2 g\mu_B B}\right)^2 \qquad (15)$$

For the device considered in the data of Figure 11 the values of Fermi energy and wavelength are $\varepsilon_f = 14$ meV and $\lambda_f = 40$ nm. In addition it is observed empirically that the spin-splitting becomes apparent for subbands with index $n > 9$. Thus the product Rg^2 can be directly evaluated and is equal to 1.1×10^{-7} m. An accurate determination of R, the curvature at the centre of the constriction, is difficult, however the smallest possible value, and hence the largest g factor, would be given by half the lithographically defined channel length. Such an estimate is not unreasonable since the spin-split plateaux were observed very close to device definition when the lateral spread of the confining potential in minimal. Thus assuming R ~ 1500 Å as the smallest possible curvature gives an upper limit for g of 0.86. As an upper limit this value is sufficiently close to the bulk g factor of 0.44 to suggest that in the parallel field orientation there is little or no enhancement of the g factor in such one-dimensional structures.

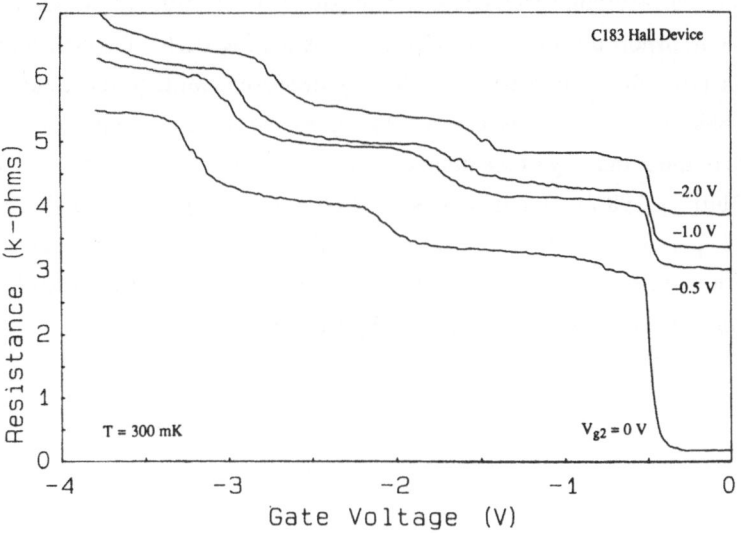

Figure 12. The resistance of the double split-gate structure is plotted as a function of the gate bias applied to one pair of gate electrodes for a number of biases applied to the second pair.

6 DOUBLE SPLIT-GATE STRUCTURES

In Plate 2 a scanning electron micrograph of a typical double split-gate structure is presented. Each split-gate pair defines a channel of lithographic width, W, 0.5 μm and length, L, 0.3 μm, whilst the split-gates are separated by 1.0 μm. In the following section results are presented for a similar device with parameters W = 0.3 μm, L = 0.3 μm and gate separation 0.7 μm. This device was fabricated on high mobility material which had a light induced sheet carrier concentration of 2.34 10^{11} cm^{-2} and a low-temperature mobility of 61 m^2 V^{-1} s^{-1} yielding a mean free path of 4.6 μm which was significantly longer than the total length of the defined channel. Optically defined voltage probes situated at the edge of the mesa 100 μm from the double split-gate structure were used to measure the device resistance whilst voltage probes defined between the split-gates were used to determine the Hall resistance. Both split-gates showed a clear quantisation of resistance although one electrode pair produced five quantised plateaux as the channel width was narrowed whilst the other pair showed six. This discrepancy was attributed to a small difference in the channel width as a result of the fabrication process.

In Figure 12 the total device resistance is plotted as a function of the gate voltage applied to one split-gate pair, V_{g1}, for a number of different voltages, V_{g2}, applied to the other electrode pair (Wharam et al. 1988b). It is immediately apparent that the law of addition of series resistances does not apply for this system. For a gate bias of −1.0 V applied to both pairs separately the channel resistance was given by that corresponding to five occupied subbands (i.e. $R_5 = h/10e^2$), whilst the same bias applied to both gates simultaneously produced a channel resistance of 1.33 R_5 which was considerably less than the factor of 2 expected for ohmic addition. This factor was regarded as a measure of the adiabatic nature of the electronic transport through the device and was used to derive the coefficient for direct transmission through the constriction. The generalised multi-probe Landauer formula (Büttiker 1986a) provides a means for calculating resistances from transmission probabilities and, because of the symmetry of the double split-gate structure, is easily applied to the current problem (Büttiker 1986b, Beenakker & van Houten 1988). For the double split-gate geometry it is possible to define four reservoirs where electronic equilibration occurs. The current, I_α, flowing out of any reservoir is then related to the chemical potentials, μ_β, of the other reservoirs via the formulae first derived by Büttiker;

$$\frac{h}{2e} I_\alpha = \left(N_\alpha - R_\alpha \right) \mu_\alpha - \sum_{\beta \neq \alpha} T_{\alpha\beta} \, \mu_\beta \qquad (16)$$

where N_α is the number of quantum channels in reservoir α, R_α the reflection probability, and $T_{\alpha\beta}$ the transmission probability from reservoir α to reservoir β. Beenakker & van Houten (1988) have hence derived the series conductance for a double split-gate structure;

$$G = \frac{1}{2} \left(\frac{2e^2}{h} N + \frac{2e^2}{h} \left(T_{sd} + \frac{(T_{sr} - T_{sl})^2}{2(N_r - R_r) - T_{sr} - T_{sl}} \right) \right) \qquad (17)$$

where N, the number of occupied subbands defined for a single split-gate, is derived from the normalisation condition $N_s - R_s = T_{sd} + T_{sr} + T_{sl} \sim N$ as observed in the ballistic regime. Note that the above formula is only valid when the full symmetry of the device can be assumed and is thus only applicable in the experimental case when $V_{g1} = V_{g2}$. In the absence of a magnetic field $T_{sr} = T_{sl}$ and the above formula reduces to;

$$G = \frac{1}{2}\left(\frac{2e^2}{h}\right)(N + T_{sd}) \tag{18}$$

similar to the result derived by Büttiker (1986b) in connection with tunneling in double barrier structures. For perfectly adiabatic transmission all electrons pass through the constriction without intermediate equilibration and hence $T_{sd} = N$. Thus the conductance reduces to $2Ne^2/h$ corresponding to the conductance of a single ballistic channel. In the other extreme, where all electrons equilibrate before passing through the second constriction, $T_{sd} = 0$ and the series resistance is ohmic. For the experimental case considered, $V_{g1} = V_{g2} = -1.0$ V, and the resistance of the double split-gate was 1.33 R_S. Thus $T_{sd} = 0.50$ N suggesting that approximately half of the electrons were transmitted adiabatically. Beenakker & van Houten have suggested that the direct transmission probability is enhanced by the collimation of the beam of injected electrons. To determine how this would influence the non-ohmic behaviour of the structure they have derived a classical expression for the conductance assuming a gradual adiabatic tapering of the confining potential. Thus the conductance was found to be given by;

$$G = \frac{1}{2}G_o\left(1 + \frac{W_{max}^2}{2LW_{min}}\right) \tag{19}$$

where G_o is the conductance of a single constriction, W_{max} the width of the channel at its entrance, L the separation of the two split-gates, and W_{min} the minimum constriction width. Beenakker & van Houten have estimated reasonable values for the above parameters when $V_{g1} = V_{g2} = -1.0$ V and derived $G_o/G = 1.3$ in agreement with the experimental data. However similar reasonable estimates for these parameters for the case $V_{g1} = V_{g2} = -2.25$ produce $G_o/G = 1.2$ showing that the collimation improves as the channel narrows whilst the actual data gave $G_o/G = 1.38$ showing no systematic enhancement of the transmission coefficient. Furthermore in the regions where both gate characteristics were situated on quantised plateaux and direct comparison was possible the ratio G_o/G remained approximately constant with values lying between 1.3 and 1.4. Accordingly it has been suggested that transport through the structure reflected rather a conservation of quantum number which, for perfectly adiabatic transport, would imply $G_o/G = 1$. In the device considered the total length of the constriction was, as previously noted, ~ 1.0 µm which was comparable to the electronic mean free path. Thus the fixed ratio of G_0/G reflected the fixed proportion of electrons which were statistically likely to suffer a collision whilst passing through the narrow constriction (i.e. transport was no longer purely ballistic). Further evidence for the conservation of quantum number was found in the persistence of the plateau structure when both narrow constrictions were defined, as is clearly illustrated in the data of Figure 12. For a single split-gate of length 1 μm the quantised resistance has not been observed even for the highest mobility materials considered. Whilst the plateaux were no longer accurately quantised it was still possible to associate unambiguously each plateau with a finite number of occupied subbands, the number of subbands occupied being determined by the narrower of the two constrictions. For an accurately quantised device the step in resistance between plateaux is given by;

$$\Delta R = \frac{h}{2e^2}\left(\frac{1}{i} - \frac{1}{i+1}\right) \tag{20}$$

for the transition between (i+1) and i occupied subbands. The quantisation of this step has been used in the current situation to identify the number of occupied subbands associated with a particular plateau even when the plateau was not accurately quantised. These quantised steps between plateaux are illustrated in Figure 12. Increasing the fixed voltage, V_{g2}, led to the subbands depopulating at smaller values of V_{g1}. The close proximity of the two gate structures inevitably modified the potential well in the neighbouring constriction although this effect was small and not readily computed. Of particular interest was the gate characteristic for non-zero values of fixed voltage. Consider for example the trace defined for $V_{g2} = -0.5$ V when the second subband constriction was first defined. The step in resistance did not correspond to a change in the number of occupied subbands but reflected rather the reduced adiabaticity of the combined constriction.

The principal result illustrated in the data of Figure 12 is that the resistance of two ballistic resistors in series is non-ohmic and, in an ideal device, determined by the number of occupied subbands in the narrower of the two channels. In such an ideal device the resistance is independent of length and reflects a conservation of quantum (subband) number. The device studied was not ideal in as much as the separation of the split-gates was ~ 1 μm whilst the electronic mean free path was 4.6 μm. This conclusion is supported by the recent theoretical calculations of Song He & Das Sarma (1988) who have modeled the conductance for a double Wide-Narrow-Wide structure and shown that, in the absence of disorder, the conductance for the combined structure is that expected for a single narrow channel (see Figure 13).

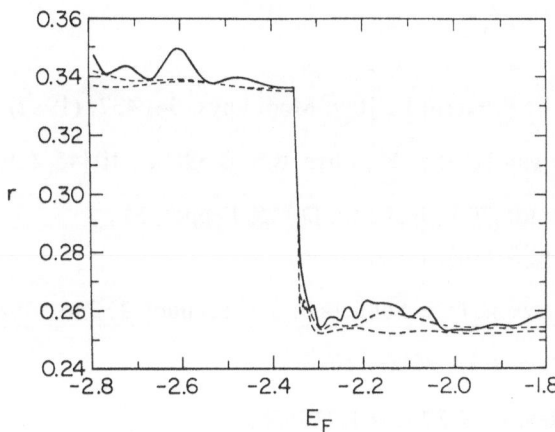

Figure 13. The calculated resistance of a double split-gate device is plotted as a function of Fermi energy. The dashed curves refer to the resistances of the individual gates and the solid shows the combined resistance (Song He & das Sarma 1989).

7 CONCLUSIONS

The resistance of a short, narrow constriction defined in the two-dimensional electron gas (2DEG) of a GaAs-AlGaAs heterostructure has been found to be quantised according to $R = h/2ie^2$ where i has been identified with the number of occupied one-dimensional subbands. The observed quantisation has been explained in terms of the exact cancellation that exists in one dimension between the Fermi velocity and the density of states at the Fermi energy. This result has been justified rigorously by recent theoretical calculations and the two competing models to explain the quantisation have been contrasted.

In the presence of a transverse magnetic field the separation of the subband energies was raised and, as the field was increased, magnetic depopulation of the subband states observed. A variational model for magnetic depopulation has been used to extract useful system parameters, such as the channel width and the 1D carrier concentration. With the magnetic field in the parallel orientation additional spin-split plateaux have been observed but only for fields in excess of ~ 10 Tesla. This has been explained in terms of the adiabatic model for the quantisation and has been used to estimate an upper limit for the Landé g-factor, which shows little or no enhancement relative to the bulk value.

The non-ohmic addition of resistance observed in double split-gate structures has been explained in terms of the adiabatic motion of the ballistic electrons through the channel constriction. This result has been reviewed in the light of recent theoretical modeling of such structures.

REFERENCES

Ando, T., Fowler, A.B. & Stern, F., Rev. Mod. Phys. **54**, 437, (1982).

Beenakker, C.W.J. & van Houten, H., Phys. Rev B **39**(14), 10445, (1989).

Berggren, K.-F.,Thornton, T.J., Newson, D.J. & Pepper, M.,
        Phys. Rev. Lett. **57**(14), 1769, (1986)

Berggren, K.-F. & Newson, D.J., Semicond. Sci. Technol. **1**, 246, (1986).

Büttiker, M., Phys. Rev. B **33**, 3020, (1986a).

Büttiker, M., Phys. Rev. Lett. **57**, 1761, (1986b).

Davies, J.H., Private Communication (1989).

Escapa, L.& Garcia, N., J. Phys. Condens. Matter **1**, 2125, (1989).

Ford, C.J.B., Thornton, T.J., Newbury, R., Pepper, M., Ahmed, H., Peacock, D.C.,
        Ritchie, D.A., Frost, J.E.F. & Jones, G.A.C., Appl Phys. Lett. **54**(1), 21, (1989).

Glazman, L.I., Lesovik, G.B., Khmel'nitskii, D.E.& Shekhter, R.E.,
        JETP Lett. **48**(4), 218, (1988)

Glazman, L.I. & Khaetskii, A.V., Preprint (1989).

Haanappel, E.G. & van der Marel, D., Phys. Rev. B **39**, 5484, (1989).

He, S. & Das Sarma, S., Preprint (1989).

Huckestein, B., Johnston, R. & Schweitzer, L., Preprint (1988) to be published
        "High Magnetic Fields in Semiconductor Physics", Würzburg Conference.

Janak, J.F., Phys.Rev. **178**, 1416, (1969).

Johnston, R. & Schweitzer, L., J. Phys. C **21**, L861, (1988).

Kirczenow, G., Solid State Comm. **68**(8), 715, (1988)

Kirczenow, G., J. Phys. Condens. Matter **1**, 305, (1989).

Kumar, A., Laux, S.E. & Stern, F., Preprint from APS (St. Louis) meeting, March 1989.

Landau, L.D. & Lifshitz, E.M., Quantum Mechanics, 3rd Edn., (Pergamon, Oxford, 1977).

Laux, S.E. & Stern, F., Appl. Phys. Lett. **49**(2), 91, (1986).

Laux, S.E., Frank, D.J. & Stern, F., Surface Science **196**, 101, (1988).

Levinson, I.B., Sukhorukov, E.V. & Khaetskii, A.V., JETP Lett. **45**, 488, (1987).

Levinson, I.B., JETP Lett. **48**, 273, (1988).

MacKinnon, A., J. Phys. C **13**, L1031, (1980).

Mathews, J. & Walker, R.L., Mathematical Methods of Physics, 2nd Edn. (Benjamin, 1970)

Nicholas, R.J., Haug, R.J., von Klitzing, K. & Weimann, G.,
   Phys. Rev. B **37**(3), 1294, (1988).

Poole, D.A., Pepper, M., Berggren, K.-F., Hill, G. & Myron, H.W.,
   J. Phys. C **15**, L21, (1982).

Sharvin, Yu.V., JETP **21**, 655, (1965).

Smith, III, T.P., Brum, J.A., Hong, J.M., Knoedler, C.M., Arnot, H. & Esaki, L.,
   Phys. Rev. Lett. **61**(5), 585, (1988).

Szafer, A. & Stone, A.D., Phys. Rev. Lett. **62**, 300, (1989).

Thornton, T.J., Pepper, M., Ahmed, H., Andrews, D. & Davies, G.,
   Phys. Rev. Lett. **56**, 1198, (1986).

Thornton, T.J., Pepper, M., Ahmed, H., Davies, G. & Andrews, D.,
   Phys. Rev. B **36**, 4514, (1987).

Thouless, D.J. & Kirkpatrick, S., J. Phys. C **14**, 235, (1981).

van der Marel, D. & Haanappel, E.G., Phys. Rev. B **39**(11), 7811, (1989).

van Wees, B.J., van Houten, H., Beenakker, C.W.J., Williamson, J.G.,
   Kouwenhoven, L.P., van der Marel, D. & Foxon, C.T.,
   Phys. Rev. Lett. **60**(9), 848, (1988a).

van Wees, B.J., Kouwenhoven, L.P., van Houten, H., Beenakker, C.W.J., Mooij, J.E.,
   Foxon, C.T. & Harris, J.J., Phys. Rev. B **38**, 3625, (1988b).

Weisz, J.F. & Berggren, K.-F., Preprint (1989).

Wharam, D.A., Thornton, T.J., Newbury, R., Pepper, M., Ahmed, H., Frost, J.E.F.,
   Hasko, D.G., Peacock, D.C., Ritchie, D.A. & Jones, G.A.C.,
   J. Phys. C **21**, L209, (1988a).

Wharam, D.A., Pepper, M., Ahmed, H., Frost, J.E.F., Hasko, D.G., Peacock, D.C.,
   Ritchie, D.A. & Jones, G.A.C., J. Phys. C **21**, L887, (1988b).

Wharam, D.A., Ekenberg, U., Pepper, M., Ahmed, H., Frost, J.E.F., Hasko, D.G.,
   Peacock, D.C., Ritchie, D.A. & Jones, G.A.C.,
   Phys. Rev. B **39**(9), 6283, (1989).

PHONONS IN SUPERLATTICES

Manuel Cardona

Max-Planck-Institut für Festkörperforschung, Heisenbergstrasse 1
D-7000 Stuttgart 80, Fed. Republic of Germany

1. INTRODUCTION

Phonons in semiconductor superlattices have been the object of considerable experimental and theoretical interest during the past few years (for reviews see refs. 1–3 and references therein). The motivation of this work is to obtain a reliable understanding and database of the corresponding eigenvalues and eigenvectors to interpret spectroscopic data on phonons (Raman, IR, EELS, ...) and also to account for electron-phonon interaction effects observed in transport phenomena.

Moreover, the hope has been expressed that optical spectroscopies of superlattices may yield data on phonons of bulk materials somewhat similar to those obtained with neutron spectroscopy. This hope is based on the fact that the artificial periodicity enables optical coupling to modes other than those at $\vec{K} \approx 0$, a fact which may be of interest when neutron data are not available (e.g. for AlAs). This attractive possibility, however, requires a thorough understanding of superlattice phonons and their relationship to those of the corresponding bulk materials. Such understanding has recently been achieved through a combination of Raman spectroscopic data, and macroscopic (continuum models) and microscopic band structure calculations.[1-6]

The simplest idea behind our understanding of superlattice phonons is that the superlattice is nothing but a periodic crystal with a super-unit-cell. Force constants obtainable from neutron data for the bulk constituents can usually be carried over to a lattice dynamical calculation for the superlattice. In the latter, however, large matrices may have to be diagonalized and in this process physical transparency may be lost. Hence a number of macroscopic continuous models have been used. They are based on the idea that if modes exist at a given frequency in both bulk components of the superlattices they will match in some way (boundary conditions) at the interfaces and propagating waves will result. If for a given frequency modes exist only in one of the bulk components the superlattice modes will at that frequency tend to be localized in the pertinent layers decaying rapidly in the medium where no propagation is possible, thus giving rise to the so-called *confined* modes. Of course these modes must still obey Bloch's theorem, i.e., they can be written as modes belonging to one of the irreducible representations of the translation group labelled by the wavevector $\vec{k}$. This is easily done by multiplying the eigenvector of each "confined" layer by the appropriate Bloch factor, summing all of them and normalizing the resulting eigenmode[1]. The resulting mode will have in the case of perfect confinement, no dispersion whatsoever, i.e. will be N-fold degenerate (N = number of layers). This is the situation for optical modes of superlattices made of materials with widely different reduced masses such as GaAs/AlAs and Ge/Si, although in the latter case the Ge-like optical modes overlap with the acoustic continuum and thus are *not strictly* confined. Even in this case, however, because of the very different eigenvectors (optic-acoustic) of the degenerate modes on both sides of the interface, a strong degree of partial confinement results.

The acoustic modes of both media do usually overlap and thus modes propagating with an average $\vec{k}$ (an average speed of sound for small k) result. These modes show splittings at the center and edges of the mini Brillouin zone (MBZ, $\equiv$ reduced BZ) of the superlattice.

2. MECHANICAL AND ELECTROSTATIC BOUNDARY CONDITIONS

The acoustic modes of the bulk give rise in the superlattice to so-called folded (into the mini-BZ) acoustic modes which propagate with an average sound velocity in the long wavelength limit. In this limit the folded modes can be obtained in the continuum approximation where atomic displacements are replaced by strains and forces of atoms by stresses in a layered continuous medium.[7] Acoustic waves in each of the two media are matched at the interfaces with boundary conditions involving the continuity of atomic displacements and stresses.

Because of the simplicity of the continuous models one has tried to develop similar treatments for the optic-like modes in which one atom moves against the other in the bulk Brillouin zone. This can be done in a simple way in cases where the dispersion relations of the optical modes in both media are well isolated from any other modes and thus the optical vibration becomes possible in either one medium or the other (confined modes). This situation can be represented by an *approximate* boundary condition which requires that the optical vibrational amplitude vanishes at the interfaces (neglecting slight penetration into the non-vibrating medium)[8]. This leads to eigenstates for $\vec{u}$ involving in each bulk cell the vibration of one atom against the other, multiplied by a sinusoidal envelope which vanishes at the interfaces. The modes are thus equivalent to standing waves of the bulk modes with wavevector components along the superlattice axis z:

$$k_1 = \frac{\pi}{d_1} m_1, \frac{\pi}{d_1} m_2 \tag{1}$$

where d_1 and d_2 are the thicknesses of layers 1 and 2. Note that in order to fulfil the boundary conditions we must take for the odd (even) modes cosine (sine) standing waves (origin at the center of the layer). In this manner the confined optical modes can be mapped into bulk dispersion relations. In order to do this mapping, however, it has been suggested[8] that $d_1 = a_1 n_1$ and $d_2 = a_2 n_2$ (a_1, a_2 = distance between bulk layers, n_1, n_2 = number of them in the superlattice layer) should be replaced by $d_{1,2} = a_{1,2} (n_{1,2} + \delta)$ with $\delta \approx 1$ in order to take into account penetration into the forbidden medium.

Some confusion into the picture just described appeared for the LO modes of polar materials when the attempt was made to use the macroscopic electrostatic potential as the envelope function. This potential must fulfill Laplace's equation in both media (labelled 1 and 2).

$$\varepsilon_{1,2} \nabla^2 \phi_{1,2} = 0 \tag{2}$$

where ε_1 (ε_2) is the dielectric constant of medium 1(2). There are two ways of fulfilling Eq. (2): either by making $\varepsilon_{1,2} = 0$, in which case confined longitudinal-like modes arise or by making $\nabla^2 \phi_{1,2} = 0$, the latter case leading to so-called interface modes, which will be discussed in Section 3.

The confusion mentioned arises when the attempt is made to impose to ϕ the standard electrostatic boundary conditions, i.e. continuity of E_x and of D_z. For propagation with $k_x \neq 0$ these conditions lead to $\phi = 0$ at the boundaries. Since the mechanical displacement is proportional to the electric polarization, which for $\varepsilon=0$ is equal to $- E/4\pi$, it is easy to see that the $\phi = 0$ boundary condition is incompatible with $u_z = 0$ required for longitudinal standing waves. We have $u_z \sim d\phi/dz$ and it is impossible to have both a sine (cosine) like standing wave and its derivative vanish at the interfaces. Because of this, it has become customary to neglect either the mechanical or the electrostatic boundary conditions, a fact which leads to the same eigenenergies but to envelope functions switching from sine-like to cosine like and vice versa depending on which boundary condition is used. This leads to orthogonal selection rules for electron-phonon interaction and thus for light scattering. Obviously one type of boundary conditions must be more binding than the other. We show below that the mechanical boundary conditions have priority, at least for $k_x \ll \pi/d_{1,2}$

Let us consider a non-polar material. The boundary condition should be that used for u and thus the *envelope of u* should be cosine (sine)-like for an odd (even) index. Let us turn on the polarity slowly. If we take electrostatic boundary conditions the suitable wavefunctions are sin(cosine)-like i.e., odd (even) regardless of how small the polarity may be, e.g. they shift along z by $d_{1,2}/2$ in the case m = 1. This discontinuous nature of the envelope function for incipient ionicity is obviously physically absurd. One would thus expect the mechanical boundary conditions to be dominant at least till the electrostatic restoring forces become larger than the mechanical ones. At least for $k_x \ll k_z$ this never happens and the correct envelope functions are obtained by imposing $\vec{u} \approx 0$ at the interfaces. This fact has been amply confirmed by microscopic lattice dynamical calculations for that superlattice (see Fig. 3 in Sect. 4).

3. INTERFACE MODES: ELECTROSTATIC APPROXIMATION AND LATTICE DYNAMICAL MODEL

The solution of $\nabla^2\phi = 0$ with electrostatic boundary conditions leads to the so-called interface phonons (equivalent to interface or surface plasmons if the dielectric constant is of purely electronic origin). Let us consider the simplest case, that of a planar interface between media 1 and 2 with 1 to the right of (layer z) 2. $\nabla^2\Phi = 0$ can be satisfied by a ϕ which propagates with wave vector k_x along the interface and decays exponentially along z

$$\phi_{1,2}(x,z) \propto e^{ikx} e^{\pm kz} \tag{3}$$

with the +(-) sign in the exponential for medium 2(1). We find the secular equation:

$$\eta(\omega) = \frac{\varepsilon_1(\omega)}{\varepsilon_2(\omega)} = -1 \tag{4}$$

which has a solution in each medium at a frequency somewhere between ω_{LO} and ω_{TO}.

In a superlattice we look for modes of the type of Eq. (3) but Bloch symmetrized, i.e. repeated periodically in each superlattice period with a Bloch factor of wavevector k_z. Note that now both signs in the real exponent of Eq. (3) are possible at each side of an interface. The electrostatic boundary conditions lead then to the secular equation[1]

$$\cos k_z d = \cosh k_x d_1 \cosh k_x d_2 + \frac{\eta(\omega)^2 + 1}{\eta(\omega)} \sinh k_x d_2 \qquad (5)$$

where $\eta(\omega)$ has been defined in Eq. (4). By replacing values of ω and k_x into the r.h.s. of Eq. (5) we obtain the corresponding k_z, i.e., the dispersion relation $\omega\,(k_x, k_z)$ of the interface modes.

It is interesting to note that Eq. (5) is singular for $\vec{k} \to 0$, i.e., the limit $\omega(\vec{k} \to 0)$ is not well defined: it depends on the angle ϕ between $\vec{k}$ and the z-axis along which $\vec{k} = 0$. This is best seen by taking in Eq. (5) the limit $k \to 0$ with $k_x/k_z = \tan \phi$.

We find:

$$<\varepsilon> <\frac{1}{\varepsilon}> \; = \; -\frac{d^2}{d_1^2 + d_2^2} \cot^2\phi \qquad (6)$$

where $<\varepsilon>$ and $<\frac{1}{\varepsilon}>$ are averages of the corresponding quantities of both bulk materials weighted by their thicknesses:

$$<\varepsilon> \; = \; \frac{d_1 \varepsilon_1 + d_2 \varepsilon_2}{d_1 + d_2}, \quad <\frac{1}{\varepsilon}> \; = \; (\frac{d_1}{\varepsilon_1} + \frac{d_2}{\varepsilon_2}) / (d_1 + d_2) \qquad (7)$$

For propagation along $\hat{z}$ $(\vec{k} \parallel \hat{z})$ Eqn (6) becomes:

$$<\varepsilon> <\frac{1}{\varepsilon}> = - \infty \qquad (8)$$

which is fulfilled for $\varepsilon_1 = 0$, $\varepsilon_2 = 0$, $\varepsilon_1 = \infty$ and $\varepsilon_2 = \infty$. We thus recover the standard bulk LO and TO modes of media 1 and 2. Examination of the eigenvectors leads to the conclusion that these modes are equivalent to the $m = 1$ confined modes but shifted along z because of the exclusive use of electrostatic boundary conditions.

For $\phi = \pi/2$ Eq. (6) becomes

$$<\varepsilon> <\frac{1}{\varepsilon}> = - 0 \qquad (9)$$

i.e., either $<\varepsilon> = 0$ or $<\varepsilon>^{-1} = \infty$. The corresponding frequencies can thus be regarded as LO ($<\varepsilon> = 0$) and TO ($<\varepsilon^{-1}>^{-1} = 0$) frequencies for in-plane propagation in an "average" medium with an effective dielectric constant equal to $<\varepsilon>$ for the LO modes and to $<\varepsilon^{-1}>$ for the TO modes. This picture was already proposed by Merlin et al in 1980. We note that if $d_1 = d_2$ these LO and TO frequencies become equal and given by $\varepsilon_1 + \varepsilon_2 = 0$, i.e. we recover the modes of a single interface as given in Eq. 4. A schematic diagram of the dependence and of the $\vec{K} \approx 0$ modes of medium for $d_1 > d_2$, $d_1 = d_2$ and $d_1 < d_2$ is given in Fig. 1.

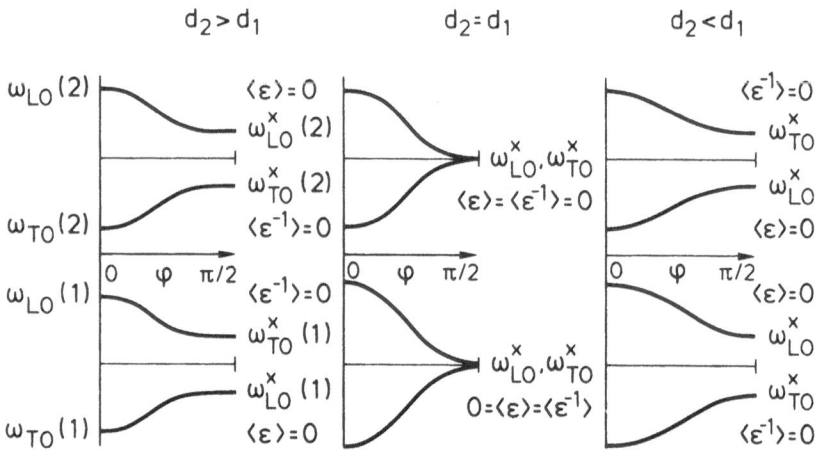

Fig. 1. Schematic diagram of "interface" electrostatic modes of superlattices as obtained with Eq. 6 for $d_2 > d_1$, $d_2 = d_1$ and $d_2 < d_1$. $\omega_{LO}{}^x$ and ω^x_{TO} represent in-plane propagating modes.

We have so far, only compared electrostatic BC's and thus the treatment suffers from the non-fulfilment of the mechanical boundary conditions already pointed out in Section 2. Actually the continuity of E_x at the interface imposes a strong discontinuity in u_x since now $\vec{u} \sim (\varepsilon - 1)\,\vec{E}$ and ε can have opposite signs both sides of the interface. For a single interface, however, u_z is nearly continuous. Hence the eigenmodes obtained in this manner will be incorrect as they were for the confined modes when only electrostatic boundary conditions were imposed. We may, however, expect that the eigenfrequencies obtained with Eq. (5) should bear some resemblance to reality; the condition of continuity of u_x will impose a sharp deviation at the electrostatically calculated eigen vectors near the interfaces. At this point we believe that the usefulness of the result must be tested by means of a lattice dynamical calculation for the superlattice.

A comment concerning the origin of the interface modes is in order. For a non-polar superlattice the confined optical and the interface acoustic modes exhaust all phonon frequencies. (3N for a given $\vec{K}$ where N is the number of atoms per unit cell). Introduction of long range electrostatic interaction should thus not increase the number of modes.

Hence we conclude that the calculated interface modes must be "pulled out" of some of the confined modes, a fact which is particularly evident when we consider that the $m = 1$ confined LO and TO modes are IR active (B_2,E symmetry for the D_{2d} group of the GaAs/AlAs [001] superlattice). Hence these modes should depend strongly on the direction of propagation. An E mode vibrating along x will have the bulk TO frequency for $\vec{k}$ along z ($\phi = 0$) (neglecting bulk dispersion) while it will be shifted up by electrostatic fields when $\vec{k}$ is along the direction of vibration ($\phi = \pi/2$). This is the reason for the shift of this mode from ω_{TO} to the root of $<\varepsilon> = 0$. A similar argument can be made for the first confined LO mode. We thus infer that the $m = 1$ confined modes remain such for $\phi = 0$ with increasing k_z. If we tilt $\vec{k}$, i.e. if we increase ϕ, these modes disperse strongly with ϕ even for $\vec{k} \to 0$ and thus behave as the microscopic counterparts of the electrostatic interface modes. The $m = 1$ TO modes split into two for $\phi \neq 0$: an "interface-like" mode vibrating in the plane defined by $\hat{z}$ and $\vec{k}$ and another non angularly dispersive one vibrating in the direction perpendicular to that plane.

The confined modes for $m = 2$ are not infrared active since the dipole moment of one half of the active layer cancels that of the other half. Hence they should not show any angular dispersion and thus bear no relationship to electrostatic interface modes. The same can be said for any m even confined modes. Other m odd confined modes, however, will have a small infrared activity: $m=3$ for instance has three half waves within the slab. Two of them cancel while one will contribute to the dipole moment thus leading to a weak angular dispersion. Other m odd modes of increasing m will show decreasing but nonvanishing angular dispersion.

The results are illustrated in Fig. 2 for four short period $(AlAs)_m(GaAs)_m$ superlattices ($m = 1, 2, 3, 4$). For the case $m = 4$ the angular dispersion ($\Gamma........\Gamma$) is given. The behaviour discussed above is accurately reproduced for $m = 1, 2, 3$ the interface (I) modes are identified by their discontinuity at Γ.

4. LOW SYMMETRY CASES: GaAs/AlAs[110] SUPERLATTICES

It is of interest to investigate superlattices which can be grown in other, lower symmetry directions such as [012][11] and [110].[5] The $(GaAs)_{n_1} (AlAs)_{n_2}$ [110] superlattices appear in 4 different space groups (2 point groups: D^5_{2d} for $n_1 = n_2 = 1$; C_{2v}[7]

for n_1, n_2 even; C_{2v}^1 for n_1, n_2, odd $\neq$ d and C_{2v}^{20} for m even, n odd or vice-versa). The [100] superlattices, in contrast, appear only in two different space groups (D_{2d}^5 for $m_1 + n_2$ even, D_{2d}^9 for $m_1 + n_2$ odd). The increased number of space and point group possibilities should give rise to a rich phenomenology.

An interesting observation in the [110] case is the fact that the point group C_{2v} includes the two-fold rotation about the z axis while the group of the k-vector of the bulk does not. The longitudinal superlattice and bulk ($\vec{k}$ along [110]) modes mix with the transverse ones vibrating along z while the transverse ones along [1$\bar{1}$0] do not mix. The superlattice modes at k=0 must be odd or even with respect to the twofold rotation about z. Hence m = odd (Eq. 2) longitudinal components must mix with m = even transverse (along z) components and vice-versa. Consequently the modes cannot be mapped exactly on the bulk dispersion relations. The mapping however is perfect for the [1$\bar{1}$0] transverse modes (see Fig. 3). The LO-TO$_z$ mixing offers the possibility of exciting TO$_z$–like modes with Fröhlich electron-phonon interactions through the LO component of TO$_z$. For these processes the m of the LO component should be even. We show in Fig. 4 the envelope function of the displacement vectors of a $(GaAs)_{13}/(AlAs)_{14}$ superlattice as calculated

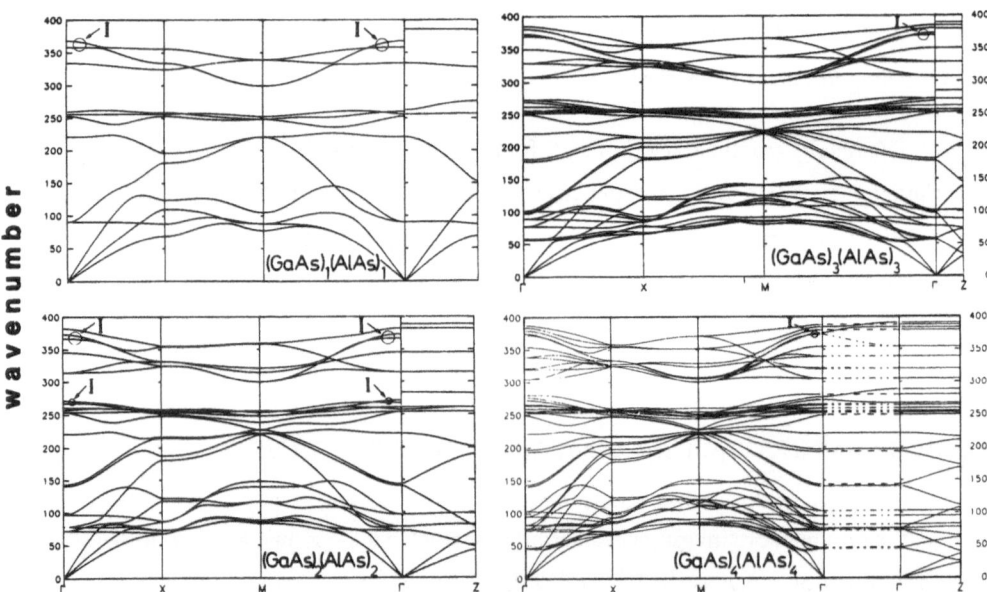

Fig. 2. Phonon dispersion relations of four GaAs/AlAs superlattices with interface modes labelled "I". From Ref. 10.

assuming the same force constants as in bulk GaAs[5, 10] for confined modes (φ=0) and for in-plane propagation (φ = π/2). In all cases it is clear that the displacement vanishes near the interface. With a little good will it is possible to recognize in the in-plane modes (φ = π/2) of dominant m odd character some exponential dependence of the type which corresponds to electrostatic interface modes but this exponential dependence is modified so as to make u vanish close to the boundaries.

In Figure 4 we display the angular dependence of the modes of Fig. 3 for small $\vec{k}$. In the [1$\bar{1}$0] and [001] planes the modes for m odd and small show strong angular dependence as expected for interface modes. For $\vec{k}$ (called q in the figure) in the [001] plane they cross the m even modes which do not have infrared activity in the [001] plane. They do, however, have such activity in the [1$\bar{1}$0] plane, a fact which results in a small angular dispersion and anticrossings with the m odd modes.

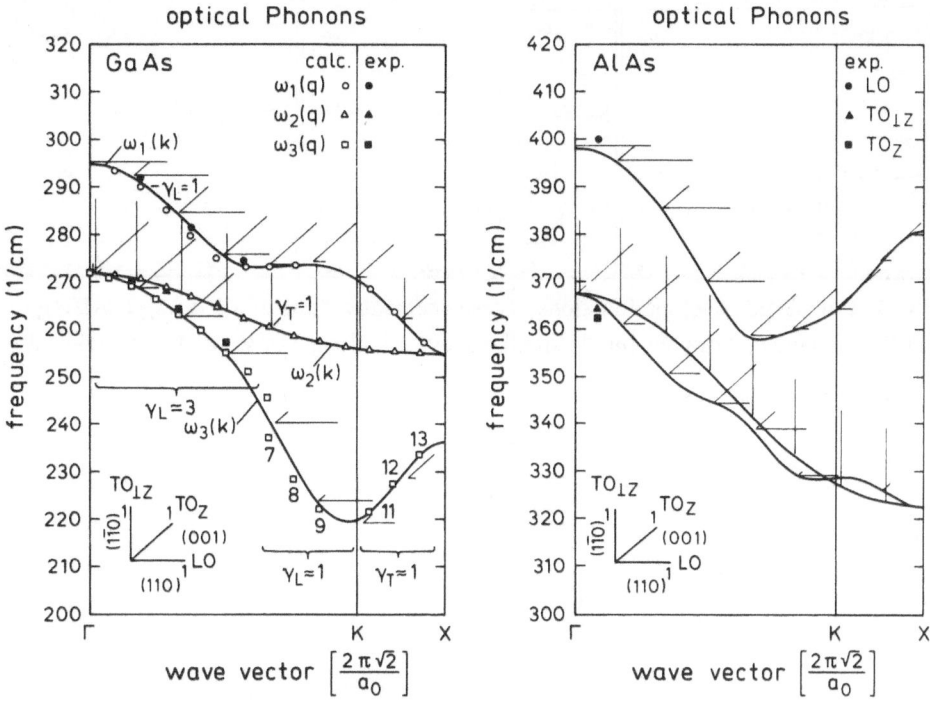

Fig. 3. Experimental (●, ▲, ■) and calculated (○, △, □) confined mode frequencies as a function of confinement wavevector $q_m = [m/(m_i + \gamma_{L,+})] [2\pi \sqrt{2} /a_0]$ for a (GaAs)$_{13}$/(AlAs)$_{14}$ [110] superlattice superimposed on the bulk dispersion curves. The flags represent the composition of the eigenvectors. From Ref. 5.

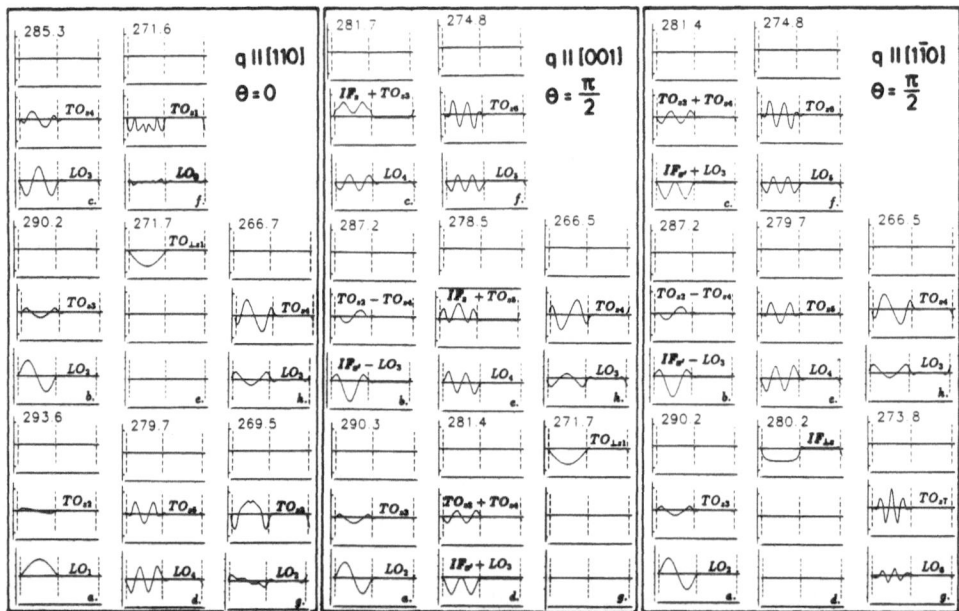

Fig. 4. Calculated envelope functions of displacement vectors of a $(GaAs)_{13}/(AlAs)_{14}$ [110] superlattice. For vanishingly small wavevector along z ($\varphi=o$) and perpendicular to z $\left(\varphi = \dfrac{\pi}{2}\right)$. From Ref. 5.

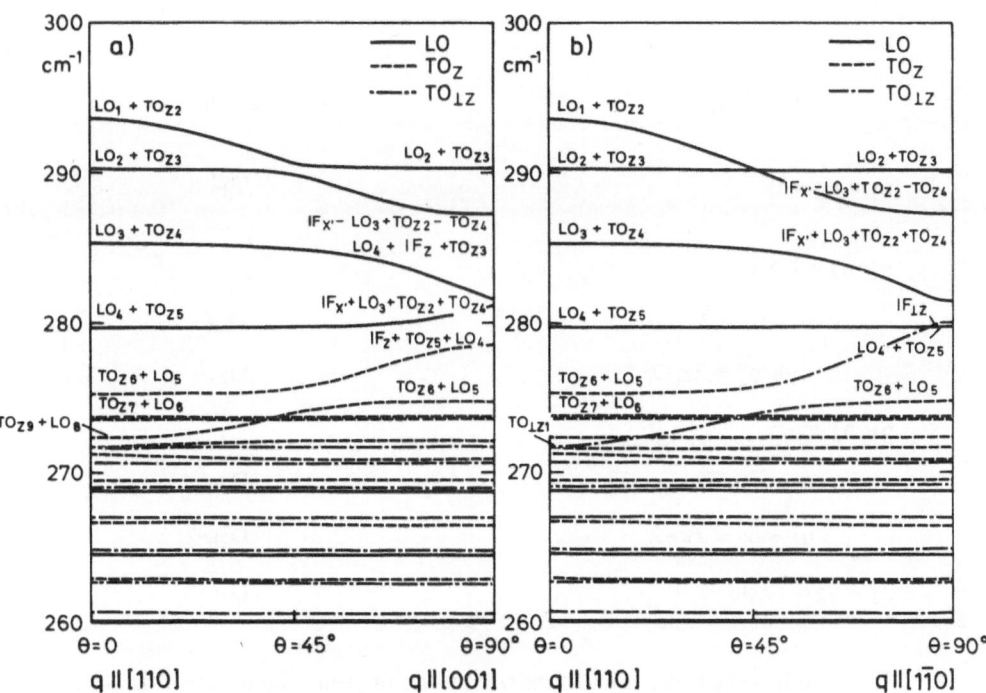

<u>Fig. 5</u>. Calculated angular dispersion of optical modes for vanishingly small wavevector in the superlattice of Fig. 4. (a) in the [1$\bar{1}$0] plane. (b) in the [001] plane. From Ref. 5.

We have just seen that for GaAs/AlAs superlattices along [001] only two space groups obtain. In the Ge_{n_1}/Si_{n_2} case, however, one can have the trivial zincblende (T_d^2) case plus five other space groups (3 point groups). They result from the fact that for $n_1 + n_2 = 4\kappa+2$ the Bravais lattice it is body centered (I) while for $n_1+n_2 = 4\kappa$ (κ is an integer) it is primitive (P), for $n_1 + n_2 =$ odd it is again body centered but with two formula units $[2(Ge_{n_1}/Si_{n_2})]$ per primitive cell n_1(or n_2) even entails a center of inversion while n_1 (or n_2) odd leads to a fourfold rotoinversion axis. The six space groups are[12]

Ge_{n_1}/Si_{n_2}	Space group
$n_1 = n_2 = 1$	T_d^2
$n_1; n_2 \neq 1$ odd	
$\quad n_1+ n_2 = 2\kappa$	D_{2d}^5
$\quad n_1 + n_2 = 2\kappa+2$	D_{2d}^9
$n_1; n_2$ even	
$\quad n_1+ n_2 = 2\kappa$	D_{2h}^5
$\quad n_1 + n_2 = 2\kappa+2$	D_{2h}^{28}
$n_1 + n_2 = 1$ odd	D_{4h}^{19}

It should be easy to distinguish the various point groups by means of i.r. and Raman spectroscopy. The existence of a center of inversion (n_1 or n_2 even) entails mutually exclusive i.r. and Raman active modes. If both n_1 and n_2 are even, the point group is orthorhombic and the transverse modes split into two, polarized along [110] and [1$\bar{1}$0], respectively. This splitting is very large for small n_1 and n_2.[11] We consider this splitting, so far unobserved, to be an excellent fingerprint of the quality of the superlattice. As long as we do not see such effects the quality of the interfaces is to be regarded as poor.

REFERENCES

1.   B. Jusserand and M. Cardona, "Raman Spectroscopy of Vibrations in Superlattices" in "Light Scattering in Solids V", ed. by M. Cardona and G Güntherodt, eds., Springer, Heidelberg 1989

2. M V Klein, IEEE Journal of Quantum Electronics, **33**, 1760 (1986)

3. M. Cardona: "Folded, Confined, Interface, Surface and Slab Vibrational Modes in Superlattices" *Superlattices and Microstructures*, **5**, 27 (1989)

4. T. Tsuchiya, H. Akera and T. Ando, Phys. Rev. B, **39**, 6025 (1989) and H. Akera and T. Ando, Envelope-Function Formalism for Phonons in Heterostructures, Phys. Rev., in press

5. Z. V. Popovic, M Cardona, E. Richter, D. Strauch, L Tapfer, and K. Ploog, Phonon Properties of GaAs/AlAs Superlattice grown in the [110] direction, Phys. Rev. in Press .

6. Articles in this volume by E. Molinari, G. Kanellis, J Menéndez.

7. S. M. Rytov, Soviet Phys. Acoustics, **2**, 68 (1956).

8. A. Fasolino, E. Molinari and K. Kunc, Phys. Rev. Lett. **56**; 1751 (1986); B Jusserand and D. Paquet, Ibid **56**; 1752 (1986)

9. R. Merlin, C. Colvard, M. V. Klein, H. Morkoc, A. C. Gossard, Appl. Phys. Lett. **36**, 43 (1980)

10. E. Richter and D. Strauch, Solid State Commun. **64**, 867 (1987), Also, E. Richter – Diplom Thesis, University of Regensburg 1986.

11. Z V Popovic, M. Cardona, L. Tapfer, K. Ploog, E. Richter and D. Strauch, Appl. Phys. Lett. **54**, 864 (1989), also Phys. Rev. in press

12. M I Alonso, M Cardona and G Kanellis, Solid State Commun. **69**, 479 (1989) and erratum **70**, i, (1989).

RESONANCE RAMAN SCATTERING IN SHORT PERIOD
GaAs-AlAs SUPERLATTICES

J. Menéndez*

Department of Physics, Arizona State University, Tempe, Arizona
USA

A. Pinczuk, J.P. Valladares, L.N. Pfeiffer, and K.W. West

AT&T Bell Laboratories, Murray Hill, New Jersey, USA

A.C. Gossard and J.H. English

Department of Electrical Engineering, University of California, Santa
Barbara, California, USA

INTRODUCTION

New semiconductor growth techniques such as molecular beam epitaxy provide a powerful tool for band structure manipulation.[1] The first optical experiments in the early seventies demonstrated the possibility of fabricating structures whose band gaps differ from those of the bulk parent materials due to quantum confinement effects.[2] Moreover, it soon became apparent that even more dramatic effects could be expected, such as the appearance of new direct optical transitions produced by the folding of the "old" Brillouin zone (BZ) under the superlattice periodicity.[3] This perspective is particularly exciting in the case of indirect gap materials, which might form direct gap superlattices. Recently, Pearsall et al. reported the observation of superlattice-induced optical transitions in Si-Ge structures.[4] These transitions have small oscillator strengths, a carryover of their forbidden nature in the bulk which implies that they can only be observed in superlattices whose periods do not exceed a few atomic layers. This poses a serious material problem, because the superlattice period becomes of the order of the interface roughness. In the specific case of Si-Ge structures, the growth difficulty is compounded by the large lattice mismatch, the different optimal growth temperatures for Si and Ge, and the interchangeability of the two group-IV atoms in the crystal lattice. It is therefore not surprising that efforts in this area have concentrated on GaAs-AlAs structures, whose epitaxial growth is much simpler.

157

Several groups have reported the observation, in very thin GaAs-AlAs superlattices, of electronic structure features qualitatively different from those found in thicker superlattices: non-monotonical thickness-dependence of the E_0-band gap,[5] new direct optical transitions,[6] Γ-X mixing effects,[7] etc. In this paper, we study the problem of the superlattice-induced direct transitions and their oscillator strength using resonance Raman scattering.

Resonance Raman scattering (RRS), i.e., the measurement of the Raman scattering cross section as a function of the laser wavelength, can be viewed as a form of modulation spectroscopy.[8] The Raman cross section shows enhancements when the laser photon energy approaches a critical point in the electronic joint density of states. This capability of Raman spectroscopy was first applied to superlattices by Manuel et al.[9] Subsequent papers by Zucker et al.[10],[11] showed the equivalence of RRS and photoluminescence excitation spectroscopy (PLE) for a determination of optical transition energies in quantum wells, and also demonstrated one of the most important features of RRS: its ability to monitor the spatial extent of the electronic wave functions by comparing the resonance of phonons confined in different layers. More recently, Cardona et al.[12] presented room temperature RRS results for a $(GaAs)_1(AlAs)_1$ superlattice. They observed two resonance peaks, of comparable intensity, at 1.93 eV and 2.15 eV. In random $Al_{0.5}Ga_{0.5}As$ alloys, a single resonance is expected at 2.09 eV,[13] corresponding to the so-called E_0 transition between the p-like valence band states at the Γ-point of the Brillouin zone (BZ) and the s-like conduction band states also at Γ. Cardona et al.[12] assigned the high energy peak in the resonance profile of the $(GaAs)_1(AlAs)_1$ superlattice to the standard E_0 transition. The low energy peak was explained as a new superlattice transition involving the same valence band state as E_0, but a new conduction band state at the center of the superlattice BZ derived from the X point of the zincblende BZ. This transition has reportedly been observed in GaAs-AlAs superlattices in PLE experiments.[6] It is frequently called a "quasidirect" transition because it is indirect in real space, due to the fact that the top of the valence band tends to be localized in GaAs, while the bottom of the conduction band at X tends to be localized in AlAs. To facilitate the discussion, we will use the following convention for the notation of superlattice states:[14] we put a bar over the symbol for a symmetry point in the superlattice and indicate in brackets the corresponding points in the zincblende BZ, using the standard c,v subscripts to denote valence or conduction band states. For example, the two conduction band states involved in the transitions described above are $\bar{\Gamma}(\Gamma_{1c})$ and $\bar{\Gamma}(X_{1c})$. The interpretation of Ref. 12 is not fully confirmed by their own theoretical calculations, which predict a much smaller separation between the two conduction band states and an oscillator strength for the Γ-X transition which is orders of magnitude smaller than observed. To clarify this situation, we performed RRS experiments in $(GaAs)_n(AlAs)_n$ superlattices with n = 1,2,3. We also find, in agreement with Cardona et al.,[12] a two-peak structure, with energy separation in good agreement with theory. Most intriguing, however, we find that the intensity of the Γ-X transition depends very strongly on the superlattice period, and has its maximum for n = 1, for which the two resonance peaks are closest in energy. This suggests that Γ-X mixing is an essential ingredient for the understanding of the optical properties of ultrathin GaAs-AlAs superlattices.

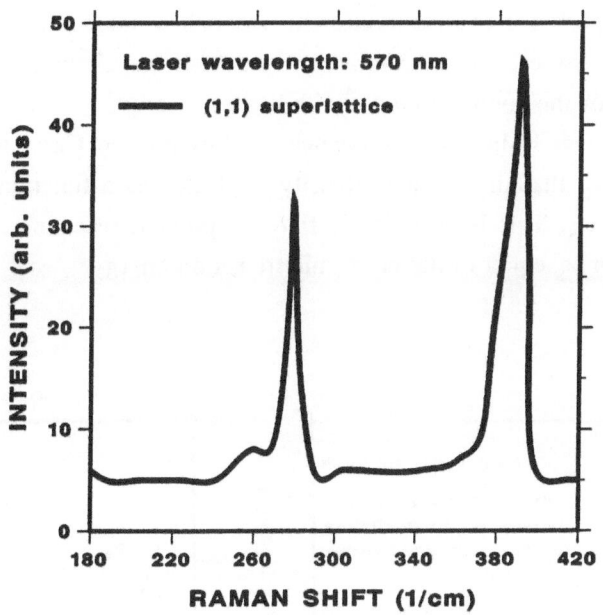

<u>Fig. 1</u>   Raman spectrum of a $(GaAs)_1(AlAs)_1$ superlattice.

EXPERIMENT

Our samples were fabricated by MBE on undoped LEC GaAs substrates with (001) orientation. We used a modified Varian GEN II machine which routinely produces modulation doping structures with mobilities in the 10^7 cm²/V sec range. First a thick GaAs buffer layer was grown, followed by a superlattice structure with a total thickness of several thousand angstrom. The substrates were not rotated during growth, which was interrupted for 2 seconds at all interfaces. We studied three $(GaAs)_n(AlAs)_n$ samples, with n = 1,2,3; which we denote (1,1), (2,2), and (3,3), respectively. The Raman experiments were performed at liquid helium temperatures using several dye lasers pumped with an Ar$^+$ ion laser.

RESULTS

Figure 1 shows a Raman spectrum for the (1,1) sample obtained in the backscattering geometry at the (001) face. For this geometry, only longitudinal optical phonons are Raman allowed. We observe a GaAs-like and an AlAs-like mode. For very thin superlattices, the phonon structure is considerably different from its bulk counterpart. In particular, because of the larger unit cell, more optical modes exist at the Γ-point of the superlattice BZ. These modes are Raman allowed, and are usually refered to as "confined optical modes". The evolution of these modes as a function of the superlattice period and their different resonance intensities will be reported elsewhere. In this paper, we concentrate on the laser wavelength dependence of the Raman intensity for the confined modes with highest energy.

Figures 2 and 3 show the resonance profiles for the GaAs and AlAs modes, respectively. Notice that the Raman intensity is plotted as a function of the scattered photon frequency ω_S. This is because we find the phonon resonances to be outgoing, i.e., they occur for ω_S equal to the electronic transition energy.

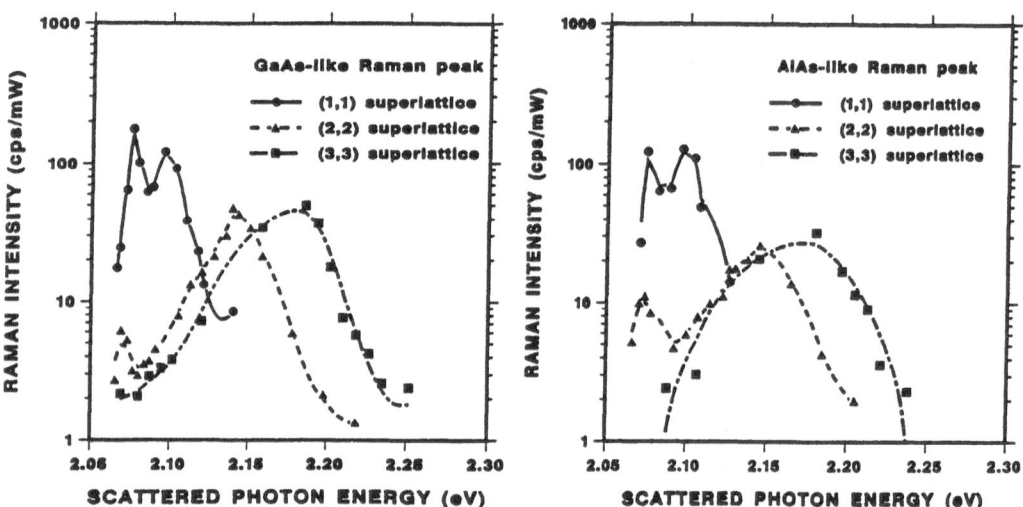

Fig. 2 Resonance Raman profile for the GaAs-like phonons of $(GaAs)_n(AlAs)_n$ superlattices with n = 1,2,3. Solid line: (1,1); dashed line: (2,2); dash-dotted line: (3,3).

Fig. 3 Resonance Raman profile for the AlAs-like phonons of $(GaAs)_n(AlAs)_n$ superlattices with n = 1,2,3. Solid line: (1,1); dashed line: (2,2); dash-dotted line: (3,3).

DISCUSSION

We analyze first the results for the (1,1) sample. We observe, similar to Ref. 12, two resonances, which we also assign to transitions from the top of the valence band to $\Gamma(X_{1c})$ (low energy peak) and $\Gamma(\Gamma_{1c})$ (high energy peak). We also observe these transitions in PLE spectra, and in the case of the quasidirect transition, as a very weak luminescence. The main luminescence lines, apparently of extrinsic nature, appear between 2.00 eV and 2.05 eV.

From Figs. 1 and 2, we obtain for the energies of the $\Gamma(\Gamma_{1c})$ and $\Gamma(X_{1c})$ states 2.095 eV and 2.075 eV, respectively (with respect to the top valence band state). This has to be compared with 2.18 eV and 2.17 eV calculated by Wei and Zunger and 1.95 eV-1.92 eV from Cardona et al.[12] We see that the splitting is in good agreement with theory. On the other hand, as noted above, the quasidirect transition is expected to have an oscillator strength orders of magnitude weaker than the direct transition, in obvious contradiction with the experimental data for our (1,1) sample in Figs. 1 and 2. To better understand the origin of this anomalously large oscillator strength we performed similar experiments in (2,2) and (3,3) superlattices, as shown in Figs. 2 and 3.

For the (2,2) sample, we find that the quasidirect transition, at roughly the same energy as in the (1,1) sample, has a much weaker oscillator strength. At the same time the $\Gamma(\Gamma_{1c})$ state has shifted to higher energy so that the separation between the two conduction band states increases. For the (3,3) sample, the $\Gamma(\Gamma_{1c})$ state is at still higher energy and the quasidirect transition is no longer observed. This systematic trend suggests the following interpretation: The Γ-derived and X-derived states are strongly mixed in the (1,1) sample, so that the quasidirect transition "borrows" intensity from E_0. For the other samples, as the separation between the conduction band states increases, the mixing is reduced until the quasidirect transition becomes too weak to observe in RRS.

The above interpretation, however, is contradicted by theory in one important point. For an odd number of AlAs monolayers, the $\Gamma(\Gamma_{1c})$ and $\Gamma(X_{1c})$ states have opposite parity, so that they do not mix.[15] Mixing occurs for an even number of AlAs monolayers, so that we would expect to see the quasidirect transition for n = 2 only, in disagreement with experiment. Therefore, the interpretation of our results in terms of mixing can only be sustained by assuming the existence of a symmetry braking mechanism, such as interface roughness. If this is the case, the quasidirect transition could be used as an important tool for monitoring the structural quality of the samples. We cannot rule out, however, the possibility that the mixing in the (1,1) sample has an intrinsic origin. The parity selection rule discussed above applies only to states at Γ. For $k_z \neq 0$, the parity of the wave function is no longer well defined, so that mixing can occur.[15,16] In order to sort out the different possibilities, experiments are in progress with samples grown with different interruption times.

The evolution of the E_0 transition with layer thickness is important for an assessment of current theoretical approaches to the band structure of ultrathin superlattices. Effective mass theory, very popular for thick superlattices, is clearly inappropriate in this case, since it cannot account for any mixing phenomenon.[17] Next in the order of complication are tight-binding methods.[18],[19],[20] In particular, a recent tight binding calculation including next nearest neighbors interactions[15] gives a detailed account of the superlattice band structure for n > 3. All these approaches, however, fail to predict a non-monotonical thickness dependence of the E_0 transition. Such a dependence is obtained in multiband calculations,[14,21],[22] albeit with dramatic discrepancies among different authors. According to Wei and Zunger,[14] the interaction of the $\Gamma(\Gamma_c)$ state with higher conduction band states plays an important role in determining the thickness dependence of the E_0 gap. This effect, of course, is missed in one-band calculations.

Our Raman results show an increase in E_0 as n goes from 1 to 3. We find E_0 = 2.095 eV, 2.140 eV and 2.18 eV, respectively. Moreover, considering the fact[23] that optical phonons are expected to be well localized in their respective layers for n ≥ 2, we can conclude from an analysis of the <u>relative</u> resonances of the GaAs and AlAs modes in Figs. 2 and 3 that the states responsible for the E_0 transition are <u>not</u> becoming more localized in GaAs as n increases from 1 to 3. Experiments for n = 4 are under progress. Recent results by Moore et al[24] for a similar sample indicate $E_0 \simeq$ 2.15 eV, suggesting that the largest E_0 gap occurs for n = 3.

Another point of discrepancy between different authors is the relative position of $\Gamma(\Gamma_{1c})$ and $\Gamma(X_{1c})$. We would like to point out here that the interpretation of our results for the (1,1) sample in terms of "intrinsic" mixing necessitates the anticrossing of the conduction bands associated with $\Gamma(\Gamma_{1c})$ and $\Gamma(X_{1c})$, because the intensities of the two resonances is nearly the same. Therefore, the band with the smaller effective mass along the growth direction should be lowest at Γ. Tight-binding calculations by Lu and Sham[15] for somewhat thicker samples show that the effective mass is smaller for the band associated with $\Gamma(\Gamma)$, leading to the conclusion that this state is the lower one. However, pseudopotential calculations[25],[26] show the opposite behavior for n = 1, implying that the $\Gamma(\Gamma_{1c})$ state is above $\Gamma(X_{1c})$.

CONCLUSION

In conclusion, we have investigated systematic trends in the resonant Raman profile of ultrathin GaAs-AlAs superlattices. Our results suggest an explanation in terms of mixing for the anomalously large oscillator strength found for quasidirect transitions in $(GaAs)_1(AlAs)_1$ superlattices. The precise origin of this mixing has not been clarified yet. It could be due to interface roughness, but an intrinsic origin cannot be ruled out. We also find the E_0 transition energy to <u>increase</u> with layer thickness. At the same time, the wave functions remain delocalized. Experiments for still higher values of n and for samples with asymmetric layer thicknesses are in preparation and will contribute to a further clarification of the physics of ultrathin GaAs-AlAs superlattices. .

REFERENCES

* Supported by NSF under Grant No. DMR-8814918

1. A.Y. Cho, in Molecular Beam Epitaxy and Heterostructures, ed. by L.L. Chang and K. Ploog, Nijhoff, Dordretch, 1985, p. 191.

2. R. Dingle, W. Wiegmann, and C.H. Henry, Phys. Rev. Lett. $\underline{33}$, 827 (1974).

3. U. Gnutzmann and K. Clauseker, Appl. Phys. $\underline{3}$, 9 (1974).

4. T.P. Pearsall, J. Bevk, L.C. Feldman, J.M. Bonar, J.P. Mannaerts, and A. Ourmazd, Phys. Rev. Lett. $\underline{58}$, 729 (1987).

5. A. Ishibashi, Y. Mori, M. Itabashi and N. Watanabe, J. Appl. Phys. $\underline{58}$, 2691 (1985).

6. E. Finkman, M.D. Sturge, M.-H. Meynadier, R.E. Nahory, M.C. Tamargo, D.M. Hwang, and C.C. Chang, J. Lumin. $\underline{39}$, 57 (1987).

7. M.-H. Meynadier, R.E. Nahory, J.M. Worlock, M.C. Tamargo, J.L. de Miguel, and M.D. Sturge, Phys. Rev. Lett. $\underline{60}$, 1338 (1988).

8. M. Cardona, Surf. Sci. $\underline{37}$, 100 (1973).

9. P. Manuel, G.A. Sai-Halasz, L.L. Chang, C.-A. Chang, and L. Esaki, Phys. Rev. Lett. $\underline{37}$ 1701 (1976).

10. J.E. Zucker, A. Pinczuk, D.S. Chemla, A.C. Gossard, and W. Wiegmann, Phys. Rev. Lett. $\underline{51}$, 1293 (1983).

11. J.E. Zucker, A. Pinczuk, D.S. Chemla, A.C. Gossard, and W. Wiegmann, Phys. Rev. B $\underline{29}$, 7065 (1984).

12. M. Cardona, T. Suemoto, N.E. Christensen, T. Isu, and K. Ploog, Phys. Rev. B $\underline{36}$, 5906 (1987).

13. D.E. Aspnes, S.M. Kelso, R.A. Logan, and R. Bhat, J. Appl. Phys. $\underline{60}$, 754 (1986).

14. S.-H. Wei and A. Zunger, J. Appl. Phys. $\underline{63}$, 5794 (1988).

15. Y.-T. Lu and L.J. Sham, unpublished.

16. M. Hybertsen, private communication.

17. L.J. Sham, Superlattices and Microstructures $\underline{5}$, 335 (1989).

18. J.N. Schulman and T.C. McGill, Phys. Rev. B $\underline{19}$, 6341 (1979).

19. J.N. Schulman and Y.-C. Chang, Phys. Rev. B $\underline{31}$, 2056 (1985).

20. J. Ihm, Appl. Phys. Lett. $\underline{50}$, 1068 (1987).

21. W. Andreoni and R. Car, Phys. Rev. B $\underline{21}$, 3334 (1980).

22. T. Nakayama and H. Kamimura, J. Phys. Soc. Jpn. $\underline{54}$, 4726 (1986).

23. S.-F. Ren, H. Chu, and Y.-C. Chang, Superlattices and Microstructures $\underline{4}$, 303 (1988).

24. K.J. Moore, G. Duggan, P. Dawson, and C.T. Foxon, Phys. Rev. B $\underline{38}$, 5535 (1988).

25. E. Caruthers and P.J. Lin-Chun, Phys. Rev. B $\underline{17}$, 2705 (1978).

26. J.-B. Xia, Phys. Rev. B $\underline{38}$, 8358 (1988).

PHONONS AND OPTICAL PROPERTIES OF SI/GE SUPERLATTICES

G. Abstreiter, K. Eberl, E. Friess, U. Menczigar
W. Wegscheider, and R. Zachai

Walter Schottky Institut, Technical University Munich
D-8046 Garching, Fed. Rep. of Germany

INTRODUCTION

Short period Si/Ge superlattices are new semiconductor materials whose band structure and consequently whose electrical and optical properties can be changed in a wide range. New device applications are expected on the basis of such layered structures |1|. Recent progress on low temperature molecular beam epitaxial growth |2,3,4,5| allows the realization of high quality Si/Ge superlattices with sharp interfaces and individual layer thicknesses of only a few monolayers. The large lattice mismatch of more than 4 % between the two constituents, however, still causes major problems for the achievement of sufficient total thickness, which is required for the application of new superlattice effects. The concept of strain symmetrization with certain buffer layers |4,6| might be one way to overcome this problem. Various basic properties of such new superlattice materials can be studied, however, also in relatively thin Si/Ge superlattices grown on Si, Ge, and SiGe substrates. In the present article results, obtained mainly in our group are reviewed. The excellent work of various other research groups from all over the world can be found in the literature |7|.

In the next section we describe growth and structural properties which are studied "in situ" by LEED and Auger spectroscopy and "ex situ" by electron microscopy. Phonon properties, as analyzed by Raman spectroscopy, are discussed in the third section. Folded acoustical and confined optical modes lead, for example to information on period length, strain distribution interface sharpness. Phonons are also a sensitive tool for interdiffusion. The paper ends with a short discussion of optical properties of certain strain symmetrized Si/Ge superlattices. Photoluminescence experiments show evidence for new fundamental energy gaps due to band folding, resonant Raman scattering leads to information on higher energy gaps.

GROWTH AND STRUCTURAL PROPERTIES

A specially designed MBE system is used for low temperature

growth of Si/Ge superlattices. The main UHV chamber is equipped with Si and Ge evaporation sources, a quadrupole mass spectrometer for rest gas analysis, a special substrate holder and tools for surface analysis like LEED and AES. The growth conditions are very crucial, especially in the case of ultrashort period superlattices with individual layer thicknesses of only a few monolayers of pure Si and pure Ge. Characteristic growth conditions are for example growth rates of typically one to two atomic layers per minute and vacuum conditions in the low 10^{-10} mbar range during growth. The difference in lattice constant of about 4 % leads to a critical thickness of lattice matched and 2-dimensional growth for Si/Ge heterostructures. Ge can be grown pseudomorphic directly on Si (100), for example, only up to 3 to 4 monolayers at substrate temperatures of about 300°C to 350°C. After 3 atomic layers of Ge we observe already considerable surface roughness by evaluating the energy dependence of LEED spot profile. This indicates the beginning of strain relaxation by formation of misfit dislocations and 3-dimensional growth which is then clearly observed for thicknesses exceeding 6 monolayers of Ge. For the inverse situation of Si on Ge (100) substrate a slightly larger critical thickness is observed, and the crystalline quality is found to be much better with respect to surface flatness.

Fig. 1 shows a cross sectional transmission electron micrograph of a 20 period Si_3Ge_9 superlattice grown on a Ge (100) substrate. No misfit dislocations or defects are observed in the whole sample which was investigated. The growth temperature was as low as 310°C. The lateral lattice constant within the superlattice is equal to that of the Ge substrate. The thickness of the individual Si layers is below the critical value. Consequently the Si layers are extended by 4 % in lateral direction, whereas the Ge layers are unstrained. The total thickness of the superlattice is small enough, that also the second critical thickness, which arises from the extremely asymmetrical strain distribution between the layers, is not reached yet. In an equivalent structure with 120 periods (total thickness ≃ 2000 Å) on the other hand, a lot of misfit defects are observed as shown in |5|.

Fig. 1. Cross-sectional TEM micrograph of a 20 period Si_3Ge_9 superlattice on Ge (100).

To overcome the problem of thickness limitation in these strained layer superlattices, it was proposed to introduce Si_xGe_{1-x} alloy buffer layers, which provide a lateral lattice constant in between those of Si and Ge |3,8|. Recently, we have studied different types of strain adjusting buffer layers |6| which vary in composition and thickness.

Fig. 2 shows a TEM micrograph pf a 30 period Si_6Ge_6 superlattice grown on a Si (100) substrate with a relatively complicated multi layer buffer. It is composed of 30 ML ge, 150 ML Si_3Ge_3 superlattice and 40 ML Ge. The Si_3Ge_3 superlattice was introduced in order to achieve a surface smoothening effect. The spot separation of the diffraction patterns in LEED and TEM are used to determine the difference of the lateral lattice constants between the Si substrate and the Si_6Ge_6 superlattice. We obtain $\delta a/a_{Si} = 2.4 \% \pm 0.1 \%$, which means that the Si layers are extended in lateral direction by 2.4 %, whereas the Ge layers are compressed by about 1.8 %. With this multi-layer buffer we tried to keep the thickness of the region where $a_\parallel$ is changed as small as possible. The clear contrast of the individual layers in Fig. 2 reflects the high quality interfaces whose sharpness has been studied in detail also by Auger electron spectroscopy performed during growth interruptions |3,6|. Fig. 2 demonstrates that short period superlattices with different strain distributions can be achieved on Si, that the quality can be improved by introducing buffer layers, but that there is still a considerable amount of crystalline defects across the whole structure. Further improvement of strain adjusting buffer layers is necessary in order to reduce the defect density within the superlattice and to achieve a crystal quality which is required for improved optical and electrical properties.

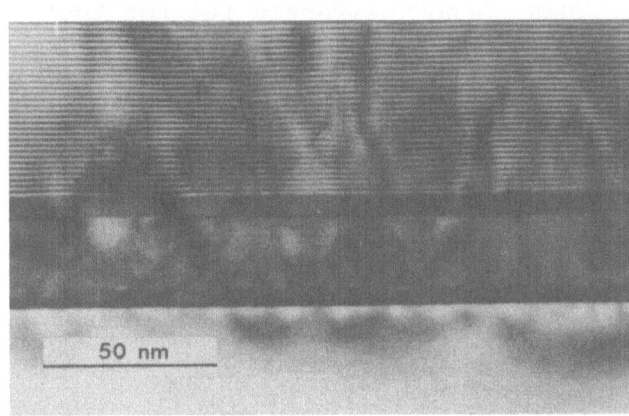

Fig. 2. Cross-sectional TEM micrograph of a 30 period Si_6Ge_6 superlattice grown on Si (100) with a (30 ML Ge/ 150 ML Si_3Ge_3 SL/40 ML Ge) buffer layer

50 nm

PHONON PROPERTIES

First order phonon Raman spectroscopy of Si and Ge reveals one single narrow peak at the energy of the optical phonons close to the Brillouin zone center ($\simeq 520$ cm^{-1} in Si, $\simeq 300$ cm^{-1} in Ge). The main features of SiGe alloys consist of three asymmetrically broadened lines close to the Si and Ge mode and in between at about 400 cm^{-1}. The exact positions and intensities depend on composition and strain |9|. The phonon spectra of Si/Ge superlattices are drastically different from the alloy spectra.

Fig. 3 shows the Raman spectrum of a Si_4Ge_{12} superlattice grown on a Ge (100) substrate. Various additional features are observable which reflect the artificial order of the sample. In the low energy range, below $\simeq 250$ cm^{-1}, several, so called folded acoustical modes appear in the spectrum. In this energy range acoustical phonon branches of Si and Ge overlap, the modes can propagate through the alternating layers of the superlattice. Within the elastic continuum theory these modes can be understood by asuming an average sound velocity whose dispersion shows a backfolding and splitting at the new superlattice Brillouin zone boundary π/d and at the zone center |10,11|. The exact energy of the folded doubletts is very sensitive on the period length d. The simple model, however, is only valid for energies smaller than about 150 cm^{-1} to 200 cm^{-1}. Above these values the acoustic branches deviate from a linear behaviour |12,13,14|. This is important for very small periods which consist of only a few monolayers and for the higher order folded modes shown also in Fig. 3. Already the first folded doublet is in the nonlinear range for period lengths of the order of 10 monolayers or less. Additional information, which can be obtained from folded acoustical modes are for example the fluctuation in the period length which is reflected in a broadening |15| and the abruptness of interfaces, which determine the intensity decay of the higher order modes. The intensities are closely related to the Fourier coefficients of the concentration profile within the sample |10|.

In contrast to the acoustic phonons there is no overlap of the dispersion curve of the optical branches of bulk Si and Ge. Consequently optical modes cannot propagate but are confined within the corresponding slabs and thus provide information on the individual thin layers. This is not perfectly rue for the Ge modes which overlap with the longitudinal acoustical (LA) branch of Si and therefore are expected to have some dispersion in the superlattice direction. The spectrum of Fig. 3 shows a series of closely spaced peaks at about 300 cm^{-1}, the "confined" modes in the 12 monolayer thick Ge layers.

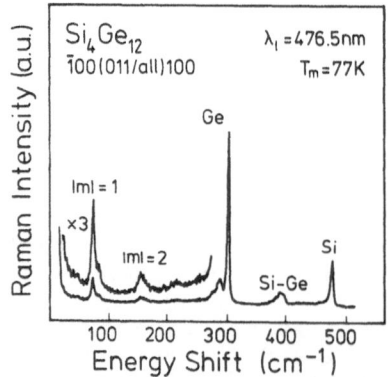

Fig. 3. Raman spectrum of a high quality Si_4Ge_{12} strained layer superlattice on a Ge (100) substrate (from Ref. |5|).

Such confined modes have been observed in Si/Ge superlattices in thin layers of Ge in (110) and (100) direction |16,3| and in Si grown in (100) direction |17|. In a simple model, similar as used for GaAs/Al$_x$Ga$_{1-x}$As short period superlattices |18|, the confined modes can be considered as standing waves which have to fit into the slab. Such an analysis leads to the bulk phonon dispersion of Si and Ge which in this way is accessable to the accurate Raman spectroscopy. One has to be careful, however, especially in the case of Ge where the coupling to the LA branch has to be taken into account |12|. Higher order confined modes are sensitve to interface roughness and are consequently also a good measure of the sample quality.

The sharp phonon mode at about 480 cm^{-1} in Fig. 3 originates from the 4 monolayer thick Si slabs. It is the first order confined optical mode. The major part of the large shift from the bulk value of about 520 cm^{-1} is, however, not due to confinement but due to the large built-in strain. This sample is grown on a Ge substrate. Consequently the lateral lattice constant of the Si layers is increased by about 4 %. This leads to a downward shift of the Si phonon modes of more than 30 cm^{-1}. The energetic position of the Si and Ge optical modes can be used as a sensitive tool to measure the strain distribution in short period Si/Ge superlattices |4,16|.

The weak structure at about 390 cm^{-1} in Fig. 3 is due to Si-Ge vibrations. The strength of this peak is extremely sensitive to interface roughness as can be seen in Fig. 4. A series of spectra of the optical phonon region is shown for a "strain-symmetrized" sample consisting of 12 monolayers of Si and 8 monolayers of Ge. Each spectrum was obtained after a different annealing step. A pronounced increase of the Si-Ge alloy mode, concomitant with a shift of the Ge mode reflects the intermixing at the originally sharp interfaces. From such studies one obtains also information on diffusivities of Si in Ge and Ge in Si, respectively, and how it depends on strain distribution |6,19|.

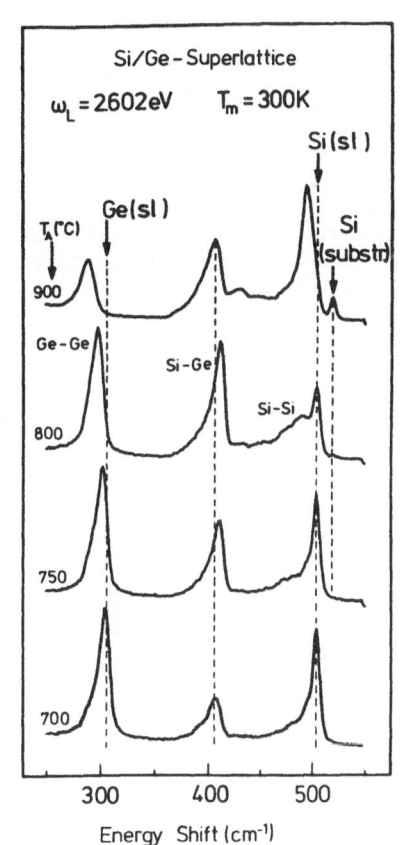

Fig. 4. Raman spectra of a strain-symmetrized Si$_{12}$Ge$_8$ superlattice after annealing steps (from |19|)

Strain, confinement and superlattice effects have a strong influence on the band structure of short period Si/Ge superlattices. Recent experimental results on electro reflectance [20]. photoconductivity [21] and photoluminescence [5,17,22] showed a variety of new optical transitions in strained Si/Ge superlattices in the range from 0,7 eV to 3,5 eV. Band structure calculations [23-27] try to explain these new transitions due to a combination of band folding and strain effects. The assignments, however, are still controversial. Under certain conditions a new quasi-direct fundamental energy gap is expected in the Si/Ge system. This has been proposed already in 1974 by Gnutzmann and Clausecker [28] and more recently by People and Jackson [29] on the basis of zone-folding arguments. Most of the relevant features and the dependence of the energy gap on strain distribution, Ge and Si thickness ratio, and period length can be seen already from a simple picture of band offsets, which takes the splitting of the conduction and valence bands due to strain into account. A quasi direct energy gap requires a folding of the twofold-degenerate Si conduction band minimum along the growth direction ($\Delta_\perp$) back to the zone center. This is best achieved for period lengths of 5, 10 or 20 monolayers, because the Δ minimum is at about 0.8 times the Brillouin zone boundary wave vector in (100) direction. It is however also necessary that the $\Delta_\perp$ minima are lower in energy than the fourfold degenerate in-plane minima ($\Delta_\parallel$). This can be achieved by tensile biaxial strain in the Si layers. The strain has also a large influence on the band alignment. This is shown in Fig. 5 where the band offsets are compared for different lateral lattice constants ($a_\parallel = a_i(Si)$, $a_\parallel = a_i(Si_{0.5}Ge_{0.5})$ and $a_\parallel = a_i(Ge)$). The top of the valence band is always highest in Ge. The band offset for the heavy hole valence band is between 750 meV and 800 meV, nearly independent of strain distribution. The band discontinuity for light holes is decreasing with increasing strain in Si and the average energy is shifting upwards. The conduction band is lowest in Si in all cases (staggered band line-up). The two-fold $\Delta_\perp$ minima shift to low energies with increasing lateral lattice constant, a quasi direct energy gap is expected. The expected fundamental energy gap is decreasing with increasing strain in Si.

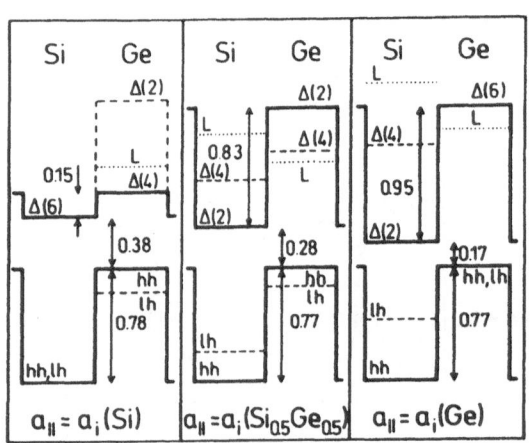

Fig. 5. Band alignment in Si/Ge super-lattice with lateral lattice constants a∥ of Si, $Si_{0.5}Ge_{0.5}$, and Ge in [100] direction (from [17])

First evidence for a quasi-direct energy gap was recently published |5,17,22| for strain symmetrized Si_6Ge_4 superlattices. Strong photoluminescence was observed at about 850 meV from a 2000 Å thick superlattice on a $Si_{0.4}Ge_{0.6}$ buffer layer. The actual strain distribution in the sample was 1.4 % tensile in Si and about 2.7 % compressive in Ge, as determined from Raman spectroscopy. Very recently similar luminescence signals were observed in Si_6Ge_4 superlattices with different strain distribution |22|. Results are shown in Fig. 6. As expected from theory, the luminescence peak energy shifts from 0.84 eV (1.4 % strain in Si) to about 0.77 eV for 2.4 % strain in Si. This strongly supports the ideas of new quasi-direct energy gaps in strained Si/Ge superlattices, however, more work on these structures is necessary to prove unambigiously that the observed radiative recombination is due to intrinsic band structure effects.

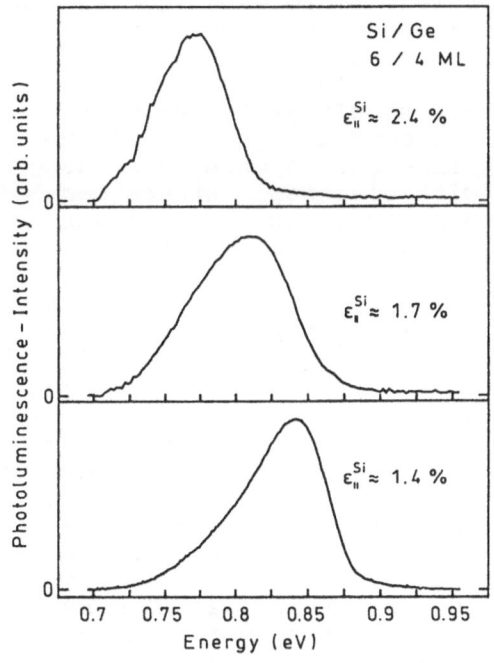

Fig. 6. Photoluminescence of Si_6Ge_4 superlattices with different strain distribution (from |22|)

Electroreflectance has been used as a powerful tool to investigate new energy gaps in Si/Ge superlattices in a wide range |20|. New gaps have been observed both in the near infrared region (direct, indirect) and above 2 eV where the E_1 and $E_1 + \Delta_1$ related gaps are expected. Electroreflectance cannot distinguish between confined and extended electronic states in the superlattice. Recently resonant Raman scattering has been used to get detailed information on the higher energy gaps |30, 31|. Due to the strong confinement of the optical phonons in Si and Ge the resonance enhancement can probe confined electronic states in each layer separately and can distinguish between confined and extended electronic states. The resonance enhancement of the Si and Ge optical phonons for two Si_3Ge_9 superlattices grown on Ge (100) substrates below and above the total critical thickness is shown in Fig. 7. For comparison we show also the resonance curves of Si and Ge like modes of a $Si_{0.25}Ge_{0.75}$ alloy, which was grown on a Ge (100) substrate. The intensities of each Raman peak were measured relative to that of a bulk Si sample mounted next to each superlattice.

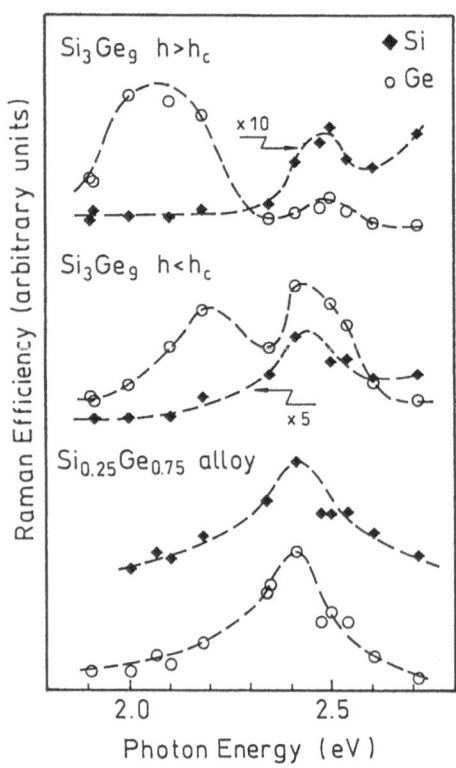

Fig. 7. Resonance enhancement of Si and Ge optical phonons in Si_3Ge_9 superlattices and $Si_{0.25}Ge_{0.75}$ alloy

Two distinct resonance features are observed for the superlattice samples. The Ge optical phonons resonate at about 2.2 eV and 2.45 eV for the asymmetrically strained Si_3Ge_9 superlattice (total thickness 330 Å). The partially relaxed superlattice (total thickness 2200 Å) shows resonance peaks at about 2.1 eV and 2.5 eV. The Si phonon mode resonates only at higher energies. Thus we conclude that the lower resonance corresponds to a energy gap with electronic states confined strongly in the Ge layers. Both modes show a resonance at the higher energy, which is evidence for more extended bands in this energy range. The phonons of the $Si_{0.25}Ge_{0.75}$ alloy show only one resonance peak at about 2.4 eV for all three modes, known as the E_1 and $E_1 + \Delta_1$ resonance. The superlattice resonances evolve from the E_1 gap of the corresponding alloy. A similar splitting has been observed in electroreflectance when the superlattice band structure develops. The Raman resonances of the more extended band gap shifts to higher energies with decreasing period length and/or increasing strain in the Ge layers |31|. The splitting between the two resonances on the other hand is increasing with increasing period length or increasing strain in the Ge layers |31|.

The results reported here are first spectroscopical investigations on details of the new superlattice band structure grown on various substrates. They clearly demonstrate that it is indeed possible to realize new semiconductors by growing short period superlattices, whose bandstructure evolves from the bandstructures of the constituents, but can have properties which are widely different from the corresponding alloys.

ACKNOWLEDGEMENTS

It is our pleasure to thank E. Kasper and his coworkers at the AEG research laboratories in Ulm for the fruitful collaboration especially with respect to the strain symmetrized short period superlattices. The strain symmetrized samples grown by H. Kibbel at AEG were the first one, which showed strong photoluminescence in the infrared spectral region. We had also an excellent collaboration with H. Oppolzer and H. Cerva at the Siemens research laboratories in München-Neu Perlach, where the high quality TEM micrographs were produced. Part of the work was supported financially by the Siemens AG via SFE.

REFERENCES

1. see for example: E. Kasper, SiGe/Si superlattices strain influence and devices, in: "Heterostructures on Silicon: One step further with Silicon," Y. I. Nissim, and E. Rosencher, eds., Nato ASI Series Vol. 160, Dordrecht, (1989).

2. J. C. Bean, L. C. Feldmann, A. T. Fiory, S. Nakahara, and J. K. Robinson, Ge_xSi_{1-x}/Si strained-layer superlattices grown by molecular beam epitaxy, J. Vacuum Sci. Technol. A2: 436 (1984).

3. K. Eberl, W. Wegscheider, E. Friess, and G. Abstreiter, Realization of short period Si/Ge strained layer superlattices, in: ref. 1.

4. E. Kasper, H. Kibbel, H. Jorke, H. Brugger, E. Friess, and G. Abstreiter, Symmetrically strained Si/Ge superlattices on Si substrates, Phys. Rev. B 38: 3599 (1988).

5. G. Abstreiter, K. Eberl, E. Friess, W. Wegscheider, and R. Zachai, Silicon/Germanium strained layer superlattices, Journal of Crystal Growth 95:431 (1989).

6. K. Eberl, E. Friess, W. Wegscheider, U. Menczigar, and G. Abstreiter, Improvement of structural properties of Si/Ge superlattices, E-MRS Meeting, Strasbourg (1989), to be published in: Thin Solid Films.

7. see for example publications in conference proceedings of Refs. 1 and 6.

8. E. Kasper, H. J. Herzog, H. Jorke, and G. Abstreiter, Strain adjustment in Si/Ge superlattices, Mat. Res. Soc. Symp. Proc. Vol. 102:393 (1988).

9. M. A. Renucci, J. B. Renucci, and M, Cardona, Raman scattering in Ge-Si alloys, in: "Light scattering in Solids," M. Balkanski, ed., Flammarion, Paris (1971).

10. C. Colvard, T. A. Gant, M. V. Klein, R. Merlin, R. Fischer, H. Morkoc, and A. C. Gossard, Folded acoustic and quantized optic phonons in (GaAl)As superlattices, Phys. Rev. B31:2080 (1985).

11. H. Brugger, G. Abstreiter, H. Jorke, H. J. Herzog, and E. Kasper, Folded acoustic phonons in $Si-Si_xGe_{1-x}$ superlattices, Phys. Rev. B33:5928 (1986).

12. E. Molinari, A. Fasolino, Calculated phonon spectra of Si/Ge (001) superlattices: Features for interface characterization, Appl. Phys. Lett. 54:1220 (1989).

13. M. I. Alonso, F. Cerdeira, D. Miles, M. Cardona, E. Kasper, and H. Kibbel, Raman spectra of Si_nGe_m superlattices: theory and experiment, preprint.

14. J. White, G. Fasol, R. Ghanbari, C. J. Gibbings, and C. G. Tuppen, Calculation of energies and Raman intensities of confined phonons in Si-Ge strained layer superlattices, in: Ref. 6.

15. M. Ospelt, K. A. Mäder, W. Bacsa, J. Henz, and H. von Känel, Unstrained vs. strained layer epitaxy: Thick Ge-layers and Ge/Si superlattices on Si(100), in: Ref. 1.

16. E. Friess, H. Brugger, K. Eberl, G. Krötz, and G. Abstreiter, Confined optical modes in short period (110) Si/Ge superlattices, Solid State Communications 69:899 (1989).

17. R. Zachai, E. Friess, G. Abstreiter, E. Kasper, and H. Kibbel, Band structure and optical properties of strain symmetrized short period Si/Ge superlattices on Si (100) substrates, in: "19th International Conference on the Physics of Semiconductors," W. Zawadzki, ed., Insitute of Physics, Polish Academy of Sciences, Warsaw (1988).
18. see for example A. K. Sood, J. Menéndez, M. Cardona, and K. Ploog, Resonance Raman scattering by confined LO and TO phonons in GaAs-AlAs superlattices, Phys, Rev. Lett. 54:2111 (1985).
19. H. Brugger, E. Friess, G. Abstreiter, E. Kasper, and H. Kibbel, Annealing effects in short period Si-Ge strained layer superlattices, Semicond. Sci. Technol. 3:1166 (1988)
20. T. P. Pearsall, J. Bevk, L. C. Feldmann, J. M. Bovar, J. P. Mannaerts, and A. Ourmazd, Structurally induced optical transitions in Si-Ge superlattices, Phys. Rev. Lett. 58:729 (1987), T. P. Pearsall, Germanium-Silicon alloys and Heterostructures - Optical and electronic properties, CRC Critical Rev. of Solid State and Materials Sciences, in press.
21. D. V. Lang, R. People, J. C. Bean, and A. M. Sergent, Measurements of the band gap of Ge_xSi_{1-x}/Si strained layer heterostructures, Appl. Phys. Lett. 47:1333 (1985).
22. R. Zachai, K. Eberl, and G. Abstreiter, Photoluminescence in Si/Ge superlattices with different strain distributions, to be published.
23. S. Froyen, D. M. Wood, and A. Zunger, Structural and electronic properties of epitaxial thin-layer Si_nGe_n superlattices, Phys. Rev.B 37:6893 (1988).
24. S. Satpathy, R. M. Martin, and C. G. van de Walle, Theory of electronic properties of the (100) Si/Ge strained-layer superlattices, Phys. Rev. B 38:13237 (1988).
25. M. S. Hybertsen and M. Schlüter, Theory of optical transitions in Si/Ge (001) strained-layer superlattices, Phys. Rev. B 36:9683 (1987).
26. I. Morrison and M. Jaros, Electronic and optical properties of ultrathin Si/Ge (001) superlattices, Phys. Rev. B 37: 916 (1988).
27. M. A. Gell, Effect of buffer-layer composition on new optical transitions in Si/Ge short-period superlattices, Phys. Rev. B 38:7535 (1988).
28. U. Gnutzmann and K. Clausecker, Theory of direct optical transition in an optical indirect semiconductor with a superlattice structure, Appl. Phys. B 36:1310 (1987).
29. R. People and S. A. Jackson, Indirect, quasi-direct, and direct optical transitions in the pseudomorphic (4 x 4)-monolayer Si-Ge strained -layer superlattices on Si (001), Phys. Rev. B 36:1310 (1987).
30. F. Cerdeira, M. I. Alonso, D. Niles, M. Garriga, M. Cardona, E. Kasper,and H. Kibbel, Resonant Raman scattering in short-period Si_nGe_m superlattices, preprint.
31. U. Menczigar, E. Friess, K. Eberl, and G. Abstreiter, Resonant Raman scattering in ultrathin Si/Ge strained-layer superlattices, to be published.

CHARACTERIZATION OF Ge-Si INTERFACES

AND ULTRA-THIN Ge LAYERS BY RAMAN SCATTERING

J. C. Tsang

IBM Research Division
IBM T. J. Watson Research Center
P. O. Box 218
Yorktown Heights, New York 10598

ABSTRACT

The Raman spectrum of an ultra-thin layer of Ge in Si(100) or an ultra-thin layer of Si in Ge(100) can be used to characterize the homogeneity of the layer, the abruptness of the interfaces, the effects of localization on the optical phonons and the spatial extent of critical points which contribute to the resonant Raman scattering. Homogeneous layers with abrupt interfaces require growth at temperatures just above the minimum temperature for epitaxial, crystalline growth. The optical phonons of Ge and Si layers show both local and resonance properties while the E_1 gap in Ge shows localized properties.

INTRODUCTION

The ability to use strain layer epitaxy to grow structures composed of ultra-thin layers of semiconductors such as Ge and Si provides many interesting possibilities for new materials with novel properties. It also raises an interesting challenge with regard to the characterization of these materials. The challenge arises from the fact that the individual layers which make up these materials are generally less than 10 Å thick and often buried under hundreds of angstroms of other materials. The signals from these ultra-thin layers can be small. The presence of both the substrate and overlayers on both sides of these layers will introduce large background signals which complicate the identification of the signals due to the layers.

The interesting properties of the new materials fabricated from these layers provide considerable incentives for meeting this challenge. It has been suggested for example, that short period Si/Ge superlattices when properly strained can show direct absorption edges.[1,2] The quantitative verification of this prediction requires the ability to measure 1) the optical absorption and emission from such a sample, 2) the strain in the 6-10 Å layers of Si and Ge that make up the sample, 3) the composition

175

of the layers, 4) the abruptness of the interfaces between the layers and 5) the electronic structure of the individual layers and the composite system. In addition to the electronic properties of these systems, it has been suggested that the vibrational properties also are of interest.[2,3,4] In contrast to the GaAs-AlAs systems where the Ga and AlAs optical phonons are strictly localized,[5] the Raman active phonons of the Si-Ge system include both localized, and quasi-resonant wavefunctions, and provide a rigorous test of simple one dimensional models of the lattice dynamics of layered systems.[6]

We have used Raman spectroscopy to characterize the properties of individual, ultra-thin, strained layers of Ge grown in Si(100) and Si grown in Ge(100). We have demonstrated the ability of Raman spectroscopy to describe both the thin Ge and Si layers and the still thinner Ge-Si interfaces between these layers for layer thicknesses less than 10 Å. Although the absorption coefficients of opaque semiconductors such as Si and Ge can be $> 10^6$ cm^{-1} for photon energies in the visible,[7] these layers have been non-destructively characterized even when they are covered with more than 50 Å of Si or Ge. The Raman spectra have been used to quantitatively measure the strain in the layers arising from the epitaxial growth on non-lattice matched substrates. The dependence of the abruptness of the Si-Ge interfaces on growth temperature has been studied.[8] Quantitative estimates of the alloy composition of intermixed layers can be obtained from the Raman spectra. The resonant Raman profiles have been used to characterize the spatial extent of the electronic wavefunctions of the host and layers. Beyond the characterization of these materials, a number of interesting questions regarding the properties of the optical phonons in these structures including their local and non-local character have been addressed. The application of Raman spectroscopy to the study of strained layer Si and Ge provides a meaningful test of the use of light scattering from resonant optical phonons as compared to localized phonons in the study of ultra-thin layers and atomic interfaces. Our studies of the vibrational spectra of single, isolated Ge and Si strained layers obviously complement the studies of the properties of strained layer superlattices.[2] The single layer systems are easier to fabricate than the superlattices and show no effects due to interlayer interactions. This is in contrast to the case of the superlattices where interactions between different layers will be significant, especially when dealing with resonant modes.

In this manuscript, we first describe the experimental problems that have limited the use of Raman scattering in the study of individual thin layers of Si and Ge. The capability of modern, imaging photodetectors to solve these problems is shown. The expected perturbations of our thin layer structures on the bulk Raman frequencies, ω_m^o, and the Raman intensities, $I_m^o(\omega_l)$ where m is either Si or Ge and ω_l is the excitation energy, are briefly reviewed. The experimental results are then presented, emphasizing first the description of the Ge and Si layers including the characterization of their homogeneity and the strain in these layers. We then describe experimental results which help to characterize the Si-Ge interfaces and the optical phonons arising from the interfaces. We conclude with a discussion of these results with regard to the characterization of the layers and the interface phonons.

TECHNIQUES

A traditional deterrent to the use of Raman spectroscopy in studying very small samples is the generally small size of the Raman signals produced by such samples.

This problem is illustrated here by considering the intensities of the Raman signals which are generated by 10 Å Ge and Si layers. The scattering efficiency of the first order Raman active, optical phonon at 520 cm^{-1} of bulk silicon is about 6x 10^{-5}Sr^{-1}m^{-1} for an excitation energy of 2.41 eV. The scattering efficiency of the analogous 300 cm^{-1} mode of bulk Ge is about 90 x 10^{-5}Sr^{-1}m^{-1} at 2.41 eV.[9] The scattering efficiencies of the bands which make up the two phonon Raman spectra and are used to describe the phonon density of states of the sample, can be more than two orders of magnitude smaller than the one phonon scattering efficiencies.[10] The scattering efficiency for the first order Raman line from a 10 Å Si layer under 2.41 eV excitation will be about 6x10^{-14} /Sr while the efficiency of Ge will be about 90x10^{-14} /Sr. Excitation at 2.41 eV of the Raman spectrum by the moderate power levels (less than 100 mw) required to guarantee that the samples are not significantly heated during measurement, which could modify the interface properties, means that between 36000 and 540000 photons /Sr-sec will be scattered from a 10 Å layer. Since the indices of refraction of Si and Ge are greater than 3, the effective solid angle inside the sample for the collection of scattered photons outside the sample by a fast f1.2 lens (for example) is small, less than .01 Sr.[11] When all the reflection losses associated with the sample are considered along with the small internal collection solid angle, the finite throughput of the collection optics and monochromator and the finite efficiency of the photon detector, the observed signal levels for first order Raman scattering from a 10 Å layer of Si or Ge will be at the .4 to 10 count/sec level. While this is considerably smaller than the first order scattering from the bulk substrate which supports the epitaxial layer, since the bulk scattering is often spectrally distinct from the layer lines, interference from the bulk scattering usually does not pose a significant problem. This is in fact one of the major advantages of Raman scattering as a probe of these layers since for many other experimental tools such as absorption and reflectivity, the substrate and layer signals overlap completely. The strength of the second order substrate Raman scattering which will often be degenerate with the first order layer scattering can be comparable to the first order Raman scattering from the layer. However, the use of the Raman selection rules for zincblende structure crystals provides a convenient means for discriminating against the second order background scattering in most cases. Since the same selection rules apply for the second order Raman scattering from the substrates and epitaxial layers grown on these substrates, the selection rules cannot be used to discriminate against the substrate second order scattering in studying the second order Raman scattering from our epitaxial layers. Therefore, relatively large substrate generated backgrounds can be expected in studies of the second order layer scattering which will be degenerate with the layer signals. Raman spectroscopy for excitation energies between 1.8 and 3.5 eV in bulk Si and Ge show strong resonance effects which produce significant enhancements of the Raman scattering cross sections. These are in fact responsible for the large value of the Ge scattering efficiency cited above at 2.41 eV.

The experimental requirements for Raman spectroscopy from 10 Å epitaxial crystalline layers of Si or Ge are the ability to detect signals at the .1 to 1 photon per second level and to extract those signals from backgrounds that can be at least an order of magnitude larger. We have shown that while such performance is difficult to obtain from conventional single channel photomultipliers, it can be obtained from the current generation of low noise, imaging photodetectors.[12] We have used a multichannel Raman scattering system with a multistage monochromator and a microchannel plate photomultiplier with a resistive anode to study Raman scattering from individual Ge and Si layers for photon energies between 1.8 and 3.6 eV. Because our detector has photon counting capability and can discriminate against electrical signals due to sources other than the photocathode and because the detector has the ability to spatially resolve regions on a one inch diameter photocathode with an area of about 10^{-4}cm^2, the dark count per resolution element of the system is four to five orders of magnitude smaller than the dark count of a conventional single channel photomultiplier with a one inch diameter photocathode. The photocathode area in a conventional photomultiplier is orders of magnitude larger than the exit slits of the monochromator used to analyze the scattered light. Comparisons of the capabilities of this detector and state of the art charge couple device imagers with more traditional single channel photomultipliers and diode arrays have been prepared by several groups.[13,14]

All of the samples studied in this work were grown by molecular beam epitaxy.(MBE) Two types of samples were prepared. The first was composed of a thin Ge layer on a thick Si(100) buffer layer and covered by an Si cap. The substrate was Si and the surface cleaned and annealed so that growth was always on a 2x1 reconstruction. The growths were monitored by low energy electron diffraction and reflection high energy electron diffraction. The growth temperatures were between 100 C and 600 C. The RHEED data showed a significant dependence of the layer structure on growth temperature, similar to that reported by Sakamoto et al.[15] Well defined RHEED oscillations were only observed at low growth temperatures and for the first 3-5 layers, showing that island growth could occur for thicker layers. The second type of sample consisted of a thin Si layer sandwiched between a Ge(100) buffer layer and a Ge cap layer. The Ge(100) buffer layer was a fully relaxed, 2000 Å thick layer grown by MBE on Si(100). The sample was also capped by a thin, amorphous Si layer to protect the Ge cap against oxidation. In addition to the in situ characterizations and the Raman studies to be described in this paper, these samples have also been studied by medium energy ion scattering, Rutherford Backscattering and ellipsometry.

In figure 1, we show the Raman spectrum of a $\simeq$ 10 Å Si layer grown on Ge(100) and covered by about 50 Å of Ge and a thin layer of amorphous Si for protection against oxidation. This spectrum as obtained under 3.53 eV excitation where the penetration depth of light into Ge is less than 100 Å so that the effective thickness of the Ge substrate for Raman scattering is less than 50 Å. For this reason, and an enhancement of the Si scattering at $E_1 - E_1 + \Delta$ to be discussed latter, the intensities of the Ge substrate scattering at 300 cm^{-1} are comparable to the 10 Å Si layer scattering at 480 cm^{-1}. The sharp line at 480 cm^{-1} is due to Si-Si vibrations analogous to the 520 cm^{-1} mode of bulk Si. The broad structure between 880 and 950 cm^{-1} is the two phonon scattering from the optical modes of the Si layer. Structure at 390 cm^{-1} has been assigned to first order scattering from Si-Ge interface modes.[2] The weaker feature near 770 cm^{-1} is the two phonon scattering associated with the Si-Ge interface. The broad band near 575 cm^{-1} is due to second order scattering from the Ge substrate. Figure 1 clearly shows how one of the great advantages of Raman spectroscopy in the study of thin layers on substrates is its ability to spectroscopically separate different phases and materials.

The Raman spectrum in Figure 1 shows the characteristic features of the Raman scattering from the Si-Ge layer system. These features differ in two significant respects from the well studied GaAs-AlAs system.[5] The most distinctive difference is the appearance of a line at intermediate frequencies between the Si and Ge like modes of the thin layer and the substrate and cap. If we compare the Raman spectrum in Figure 1 to the spectrum obtained from Ge_xSi_{1-x} alloys, we find that the 390 cm^{-1} line corresponds to the Raman scattering from Si-Ge bonds.[16] In the GaAs-AlAs system, the common As sublattice which stretches throughout the sample guarantees there are never any Ga-Al bonds in the sample. Since the first order Raman scattering involves phonons whose eigenvectors describe the motion of the nearest neighbor atomic pairs in the III-V unit cell in opposite directions with equal reduced amplitudes, there is no unique signature of the interface in the first order, non-resonant Raman scattering. There are of course well defined effects due to the finite thicknesses of the layers.[17] In the Si-Ge case, Si and Ge atoms can reside on either of the two sublattices which make up the crystal. Since the Raman active modes are derived from vibrations involving nearest neighbor motions, there are

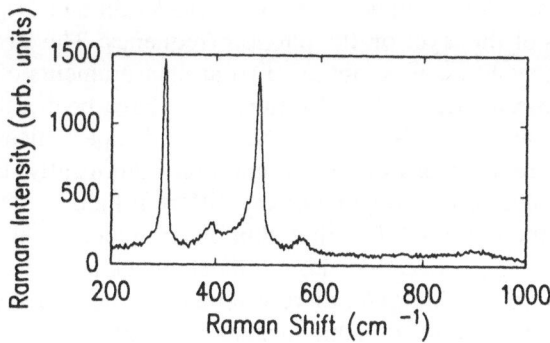

FIGURE 1. The first and second order Raman scattering from $\simeq 10$ Å of Si in Ge(100).

Si-Si, Ge-Ge and Ge-Si like Raman modes. A second, more subtle difference stems from the spatial localization of the different vibrational modes. In the GaAs-AlAs system, the AlAs optical phonons are localized on AlAs sites because they are at higher energies than any of the other modes in the system.[5] The GaAs optical modes are also localized because they lie below the bottom of the AlAs optical phonon branches and above the top of both the longitudinal and transverse acoustic phonons bands of AlAs. In contrast, while the Si modes in the Ge-Si system are localized (since their energies are above the energy of any other vibrations in the system), the longitudinal optical Ge-Si modes are degenerate with the longitudinal acoustic phonons of Si and the Ge-Ge modes are degenerate with both the Si longitudinal and transverse acoustic phonons. This degeneracy means that these modes are not localized modes but are resonances of the Si host system. However, Fasolino et al.[6] and Alonso et al.[18] have shown that for Si-Ge short period superlattices the Ge like modes retain their local characteristics. The vibrational amplitude of the mode is concentrated in the Ge region with only a small part of it in the Si layers. As a result, the Ge modes can be used to describe the Ge layers. The transverse Ge-Si modes of the Ge-Si system fall between the Si optical phonons and the Ge optical and Ge and Si transverse acoustic modes and are localized.[3] Questions have been raised by Molinari et al.[6] about the behavior of the longitudinal optical Ge-Si modes which are degenerate with the LA modes of Si. It has been argued that the longitudinal optical phonons of the Ge-Si interface are delocalized. This is significant since the Raman selections rules for back scattering from an opaque, zincblende structure semiconductor allow only scattering from the longitudinal optical modes. It has been shown experimentally that in Ge_xSi_{1-x} alloys, the Si-Ge longitudinal modes are Raman active.

The Raman active modes of the Si-Ge system can be used to characterize Si and Ge layers and the interfaces between them. The Raman frequency ω_m^l, where m = Si or Ge, of a thin strain layer when the mode is localized in the layer is related to the bulk frequency, ω_m^o, by

$$\omega_m^l = \omega_m^o + \Delta\Omega_s + \Delta\Omega_{loc} \tag{1}$$

where $\Delta\Omega_s$ is the shift in the phonon energy due to strain and $\Delta\Omega_{loc}$ is the effect of the finite thickness of the layer on the phonon frequency. The strain induced shifts in the phonon frequencies can be obtained from measurements of the bulk Raman spectrum under uniaxial stress. Localization shifts have been calculated for finite slab models of various materials[19,20] and short period superlattices.[3,5] If the optical phonons from different regions of the sample mix significantly, then there will be shifts in frequency due to the coupling of the different modes. These shifts will be similar to the shifts observed for the phonon frequencies in one mode alloy systems.[21] The knowledge of ω_m^o depends of course on a knowledge of m. If the layer is an alloy instead of either Si or Ge, there will be alloy induced shifts in the phonon frequencies in addition to the strain and localization shifts.

The relative intensities of the different vibrational modes can also be used to describe the layers. To first order, if the excitation energy is far from any electronic resonances of the sample so that the Raman tensor does not pick out a particular intermediate state, then the relative intensities of the different lines will be determined by the number of bonds of each type in the sample. Changes in the relative intensities of the Si-Si, Si-Ge and Ge-Ge lines in Ge_xSi_{1-x} have been correlated with the number of Si-Si, Si-Ge and Ge-Ge bonds.[16] If I_y (where y is Si, Ge or Si-Ge) is the intensity of the y Raman line, then

$$I_{Ge-Si}/I_{Ge} = 2(1-x)/x \text{ and } I_{Ge-Si}/I_{Si} = (1-x)/2x \tag{2}$$

In a inhomogeneous system under resonant excitation, there will be significant deviations from these relationships. However, the qualitative variation of the relative intensities of the different lines will still reflect changes in the numbers of bonds of different types.

The localized character of optical phonons of the Si-Ge system means that resonant Raman scattering can be used to describe the electronic wavefunctions of these samples. The Raman scattering from group IV semiconductors for excitation energies between 1.7 and 3.5 eV is dominated by resonant contributions from the E_o and

$E_1 - E_1 + \Delta$ gaps. Cardona[22] has shown that the contributions to the Raman tensor, a, for these transitions can be expressed as

$$
\begin{aligned}
a = &- (a_o^2\sqrt{3}\,d_o/16)x \\
&(d\chi^E(\omega)/d\omega_o - (2/\Delta_o)(\chi^E(\omega) - 2(\omega_o/(\omega_o + \Delta_o))^{3/2}\chi^{E+\Delta}(\omega)))/4\pi \\
&+ (a_o^2/4\sqrt{6}\,)x \\
&((-1/2\sqrt{2}\,)d_{1,0}^5 d\chi(\omega)/d\omega + 2d_{3,0}^5(\chi^E - \chi^{(E+\Delta)}/\Delta))/4\pi
\end{aligned}
\tag{3}
$$

where E_o, $E_o + \Delta_o$, E_1 and $E_1 + \Delta$, are the different critical points that can be excited, d_o and $d_{1,0}^5$ are the two band deformation potentials, $d_{3,0}^5$ is the three band deformation potential and $\chi^j(\omega)$ is the electronic susceptibility at ω or the contributions of the E_1 and $E_1 + \Delta$ gaps to the susceptibility. There is a deformation potential connecting each phonon and each electronic state. For the case of a localized phonon and a localized electronic transition, strong resonant enhancement of I_y can occur if both are in the same layer. Since the deformation potential is a short range interaction, there should be no significant enhancement arising from optical phonons localized in one layer and electronic states in a different layer. On the other hand, electronic transitions involving delocalized states can produce a simultaneous enhancement of all the localized phonons.

EXPERIMENTAL RESULTS

In studying the behavior of thin Ge and Si layers grown by MBE, we are interested in both the crystalline quality and purity of the layers and the abruptness of the interfaces between the layers. We first concentrate on how Raman spectroscopy can provide information about the layers and then consider how it can help characterize the interfaces between the layers. In Figure 2, we show the Γ_{25} symmetry Raman spectra of three different thicknesses of thin layers of Ge grown in Si(100) at about 300 C. The Raman scattering from the Si substrate and overlayer was subtracted from all of the curves in Figure 2, leaving behind only the Ge induced scattering. Also shown for reference in Fig. 2 are the Raman spectra of a 100 Å $Ge_{.5}Si_{.5}$ alloy in Si(100) and bulk Ge. The 300 cm^{-1} first order Raman line of bulk Ge is shifted from the analogous line in the 7 and 11 A Ge layers by about 16 cm^{-1}. This is close to the value predicted for the strain induced shift of this phonon frequency due to the pseudomorphic growth of a Ge layer on Si(100).[23] The calculated value is based on uniaxial stress studies of the behavior of these phonons using applied stresses about an order of magnitude smaller than those resulting from the pseudomorphic growth. The Raman spectra of the Ge layers show a clear evolution from alloy like behavior to bulk Ge like behavior. The relative intensities of I_{Ge} and I_{Ge-Si} and Raman shifts ω_{Ge} and ω_{Ge-Si} of the 4 Å Ge layer approach the values obtained for a thick, strained alloy layer. If we scale the intensity of the Raman scattering near 415 cm^{-1} in the alloy to its strength in the 7 and 11 Å Ge layers and then subtract the scaled alloy spectrum from the layer spectra, most of the low energy tails on the Ge lines at 316 cm^{-1} disappear and the resulting difference spectra closely resembles the bulk Ge spectrum except for the shift in frequency which we have explained is close to the theoretically expected value of $\Delta\Omega_s$ for strain layer growth of Ge on Si(100). The linewidths of ω_{Ge} depend on the growth temperature and layer thickness. Linewidths as small as 6-7 cm^{-1} have been observed for 6-11 Å Ge layers. Thicker samples have shown linewidths as large as 25 cm^{-1}. The linewidths of the 4 Å Ge layer resemble the linewidths observed from alloys. The symmetry properties and linewidths of these lines show that they arise from crystalline phases of the samples.

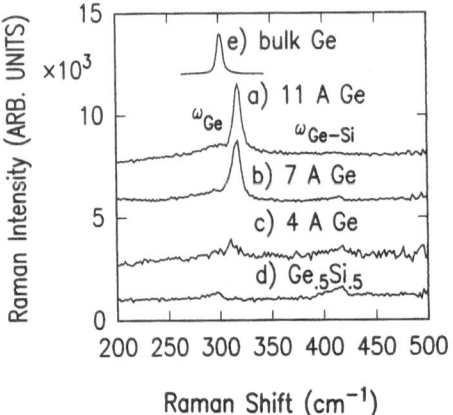

FIGURE 2. The first order, Ge induced Raman spectra of 4, 7 and 11 Å of Ge grown in Si(100) at between 300 and 350 C. The Si substrate scattering has been subtracted. The bulk Ge and Ge$_{.5}$Si$_{.5}$ alloy spectra are for reference.

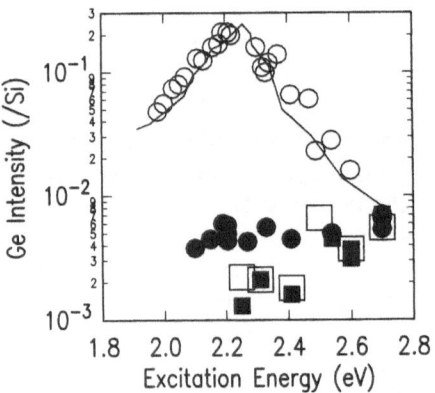

FIGURE 3. The dependence on excitation wavelength of I_{Ge} (open symbols) and I_{Ge-Si} (closed symbols) from a 4 (squares) and 11 Å (circles) thick Ge layer in Si(100). Also shown is the resonant Raman profile of bulk Ge (solid line).

Figure 3 presents the resonant Raman profiles for the 4 (open and filled squares) and 11 (open and filled circles) Å Ge layer samples described in the previous figure. The excitation wavelength dependences of both I_{Ge} and I_{Ge-Si} are given. I_{Ge} and I_{Ge-Si} in Figure 3 are normalized against the intensity of the first order Raman scattering from the Si substrate and overlayer. The Si scattering shows only a small increase in strength for excitation energies between 1.8 and 2.8 eV.[24] Also shown in Figure 3 with a solid line is the resonant Raman profile of bulk Ge. $I_y(\omega)$ for 1.7 ev $< \hbar\omega <$2.7 eV in Figure 3 clearly shows that the Ge vibrations in the 11 Å sample couple strongly to an optical transition near 2.25 eV. This resonance resembles the resonance observed in the bulk. The Ge-Si vibrations in this sample do not couple strongly to the transition enhancing the Ge scattering. The 4 Å Ge layer shows no evidence for an electronic transition near 2.25 eV. It does show evidence for a weak enhancement of I_{Ge} and I_{Ge-Si} at energies above 2.7 eV. The excitation wavelength dependence of the intensities of the Raman scattering from the 4 Å sample is similar to that of bulk Ge_xSi_{1-x} alloys near their $E_1 - E_1 + \Delta$ gaps.[23] The energies of these electronic transitions in the alloys vary continuously with composition between about 2.3 eV for pure Ge and 3.4 eV for bulk Si. $E_1 - E_1 + \Delta$ is about 2.8 eV for $Ge_{.5}Si_{.5}$.

Figure 4 shows the second order Raman spectrum obtained from an 12 Å Ge layer for energies near twice the Raman active, first order phonon energy at 300 cm^{-1}. Also shown in this figure is the second order Raman spectrum of bulk Ge. The various shoulders and peaks in the bulk spectrum have been assigned to different critical points in the two phonon spectrum of Ge.[25] These include 2TO at the X point near 550 cm^{-1}, 2TO at the L point near 570 cm^{-1} and twice the Γ point optic phonon energy at 600 cm^{-1}. Shifts of these energies by amounts calculated from the measured Gruneisen parameters for these modes[25] and the volume change associated with the epitaxial growth of Ge on Si(100) produce values for the pseudomorphically strained Ge layer two phonon critical points that are within the two phonon structure obtained from our 12 Å Ge layer. Figures 2 and 4 show that the Ge-like phonons seen in Raman scattering from 6-12 Å Ge layers grown at low temperatures strongly resemble the Raman spectrum of bulk Ge. The primary perturbation of the epitaxial growth of the thin Ge layer on Si on the phonon spectrum is the strain induced shift of the various phonon frequencies. Our ability to explain the value of ω_{Ge} simply in terms of the effects of strain on the bulk Ge phonon frequency suggests that our Ge layers can be quite pure with relatively little intermixing of Ge and Si.

In figures 5a and b, we show that Raman scattering can be used to similarly describe the properties of thin epitaxial Si layers grown in Ge(100). The Raman spectra under 2.7 eV excitation of three different thicknesses of Si in Ge(100) are plotted in Figure 5 where the thickest layer is about 15 Å and greater than the critical thickness. The Raman spectra are normalized against the strength of the Ge scattering at 300 cm^{-1} from the host. The first order Raman line of bulk Si is at 520 cm^{-1} so the Si Raman active optical phonons in these layers show a large shift to lower energies in contrast to the Ge layers on Si(100) which shifted to higher energies. The predicted shift[23] of the first order bulk Si phonon due to the strain of the pseudomorphic growth on Ge is between 25 and 30 cm^{-1} so that the strained Raman Si phonon should be between 490 and 495 cm^{-1}. The results in Figure 5a resemble those of Figure 2 in our ability to identify Si like Raman scattering and Si-Ge interface scattering. The results differ in that the positions of the Si Raman line can be

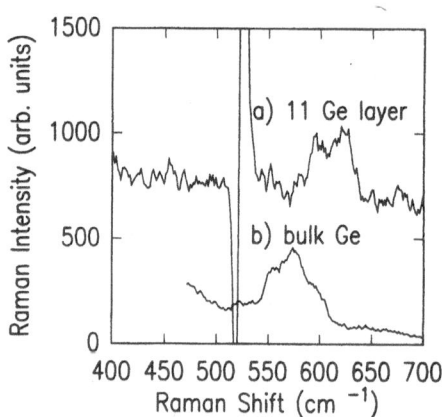

FIGURE 4. The second order Raman scattering for the optical phonons from a 12 Å Ge layer in Si(100). The derivative like structure at 520 cm^{-1} is due to the suppression of the Si substrate scattering. The second order Raman spectrum of bulk Ge is also shown for comparison.

clearly below the expected value assuming only strain as a perturbation. (curves a and b in Fig 5a) As we indicated earlier, the Si mode is rigorously localized and we discuss latter in this paper how localization can explain the observed position of the Si mode in our thin Si layers. In Figure 5b, we show similar spectra excited from these samples at 3.53 eV. Again, the spectra are normalized against the intensity of the Ge host scattering. While the phonon energies are similar in Figures 5a and b, the relative intensities of the Si lines vary tremendously. This change reflects the resonant enhancement of the Si scattering by the $E_1 - E_1 + \Delta$ gap of Si which is near 3.5 eV. Figure 5b shows that this enhancement is much stronger in the thicker Si layers than the thinner layers.

The interface between the Si and Ge layers contains Si-Ge bonds which we showed in Figure 1 produced spectroscopically distinct Raman scattering near 400 cm^{-1}. We have shown previously how growth temperature plays a major role in determining the abruptness of the Si-Ge interface in Si-Ge strain layers. Figure 6[8] presents how the Raman spectrum of a 6 Å Ge layer in Si(100) changes as a function of growth temperature. For growth temperatures below 200 C, there is no epitaxial growth and the layers are amorphous. This is seen through the absence of any sharp Ge induced scattering and the widths of the Ge induced Raman lines.[26] The spectrum in Figure 6a resembles the results of Persans et al.[27] obtained on amorphous Si-Ge superlattices. A 100 C increase in growth temperature results in Raman spectra characteristic of a crystalline phase where $I_{Ge} > I_{Ge-Si}$. Growth at temperatures above 400 C produces Raman spectra that quantitatively resemble the Raman spectra obtained from $Si_{0.5}Ge_{0.5}$ alloys as shown in Figure 6f. We[8] have correlated these results with Medium Energy Ion Scattering studies to show that relatively abrupt interfaces can be obtained in the Si-Ge system only through growth at temperatures below 350 C. Growth at temperatures above 350 C results in the rapid interdiffusion of the Ge and Si layers on the scale of at least 10-20 Å. We have shown that the Raman frequencies of both the Ge and Ge-Si lines for our samples grown at temperatures above 400 C can be completely described by the Raman frequencies of bulk alloys with a frequency shift $\Delta\Omega_s$ due to the strain of pseudomorphic

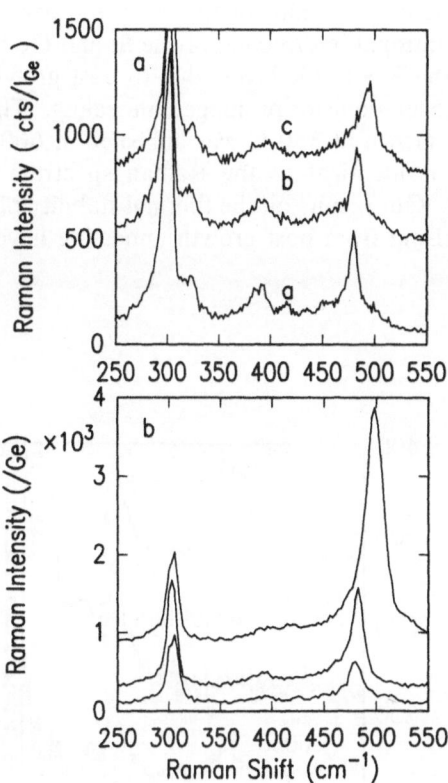

FIGURE 5 a) The Raman spectra of three different thickness of Si grown on Ge(100) by MBE. The thicknesses range between 6 Å and 15 Å and the spectra excited by 2.7 eV light. b) same as above only excited by 3.53 eV light.

growth.[4] On the other hand, the lines obtained from the growths at temperatures between 200 and 350 C are more characteristic of a homogeneous Ge layer with an Ge-Si transition region to the Si host.

The Ge layers whose Raman spectra are shown in Figure 6 were deposited at a growth rate of $\simeq$ 0.3 Å / sec. The time required for the growth of a 6 Å Ge layer was about 20 seconds. The highly non-equilibrium nature of the MBE growth of these Ge strain layers can be inferred from the behavior of these samples on annealing at 600 C. While Figure 6 clearly showed that 20 seconds of growth at 600 C produced a Ge layer whose Raman spectra demonstrated the mixing between the Ge atoms and the Si host and cap layers, we found that the effect of annealing at 600 C an already grown sample for periods of time of the order of 30 minutes did not introduce observable additional intermixing of the Si and Ge layers. Significant additional intermixing of the Si and Ge layers due to post growth annealing of these samples could only be observed after prolonged annealing. This is shown in Figure 6e where a 6 Å sample grown at 350 C was annealed at 600 C for 8 hours. The spectrum in Figure 6e is identical to the Raman spectrum of a dilute alloy of Ge_xSi_{1-x} where x < 0.2. Our results on the thermal stability of the grown interfaces against interdiffusion arising from post growth annealing have been independently

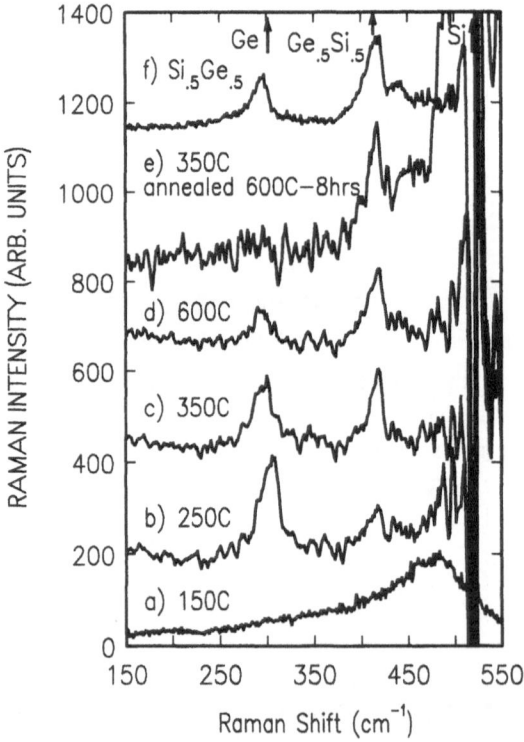

FIGURE 6. . The Raman spectra of 6 Å of Ge grown on Si(100) at 4 different temperatures (a-d) or grown at 350 C and annealed for 8 hours at 600 C (e). (from ref. 8)

verified by Medium Energy Ion Scattering. We[8] have argued that these results imply the presence of the growth surface near the Si-Ge interface can promote the diffusion of the larger Ge atoms towards the Si surface. The results in Figure 6 clearly show the sensitivity of Raman spectroscopy to the amount of interfacial mixing present in an ultra-thin layer.

In Figures 2 and 6, it was shown that the Raman spectra obtained when Ge layers that are < 3 or 4 atomic layers in thickness are grown in Si(100) for growth temperatures below 300 C or for thicker Ge layers when the growth temperature is above 400 C, closely resemble the Raman spectra of strained Ge_xSi_{1-x} alloys. In Figure 7, we demonstrate how Raman scattering from the Si-Ge interface and from a Ge layer in Si(100) can be distinguished spectroscopically. In Figure 7, we plot the values of ω_{Ge} versus ω_{Ge-Si} for 6-12 Å Ge layers. The solid line in Figure 7 shows the expected dependence of ω_{Ge} versus ω_{Ge-Si} for the epitaxial growth of thick Ge_xSi_{1-x} alloys on Si(100). The curve was compiled from the experimental results of Renucci et al.[16] and of Byra[28] for the Raman spectra of bulk alloys and the results of Cerdeira et al.[23] for the strain induced shifts of the phonon frequencies of these alloys when grown on Si(100). We see in Figure 7 that when ω_{Ge} is below 300 cm^{-1}, both ω_{Ge} and ω_{Ge-Si} are consistent with the presence of a Ge-Si alloy. The samples responsible for the points where $\omega_{Ge} < 300$ cm^{-1} were all grown at temperatures above 400 C. However, when $\omega_{Ge} > 310$ cm^{-1}, then our results do not fall on the single phase alloy curve. Since 1) ω_{Ge} for these samples is equal to the bulk value of the Raman active mode of Ge shifted by the effects of the 4% lattice mismatch between Si and Ge as shown by the arrow in Figure 7, 2) the resonant Raman profiles for I_{Ge} of these samples shows a close resemblance to that of bulk Ge with no significant Ge like enhancement of I_{Ge-Si} and 3) I_{Ge-Si} for these samples are the smallest values for these lines which we observe, we conclude our Ge layers are homogeneous and that the 415 cm^{-1} structure in our thin Ge layers on Si(100) is due to Raman scattering from the Ge-Si bonds at the interfaces between the Si and Ge layers. This separation reflects the fact that if the Ge layer is homogeneous and contains no Si atoms, then the Ge-Si modes are excluded from the Ge layer by energetic considerations since they have higher energies than any Ge vibrations. Obviously, if there are Si atoms in the Ge layers, there will be Ge-Si modes in the Ge layers which can mix with the Ge-Si interface vibrations. This coexistence of the Ge-Ge and Ge-Si modes in a layer eventually leads to Ge-Ge and Ge-Si vibrations that can show alloy like behavior. The coupling of the interface modes to the Si layers can be larger given their degeneracy with the Si LA phonon branch but the experimentally observed values of ω_{Ge-Si} and the theoretical results of Alonso et al.[18] suggest that these modes retain considerable local character.

The observation in Figure 3 that the excitation of the $E_1 - E_1 + \Delta$ resonance of Ge produces a strong enhancement of the Ge like Raman scattering in our samples and only a weak effect on I_{Ge-Si} does not apply when the comparable electronic resonance of an Si layer is excited. Figure 8 is the Raman scattering from an epitaxial Si layer in Ge(100) excited at three different photon energies. The Si layer scattering is strongly enhanced relative to the Ge host scattering under 3.5 eV excitation as compared to 2.7 eV excitation. The enhancement of I_{Si} is also seen in I_{Ge-Si} which appears to have grown roughly proportionally in size with I_{Si}. This suggests that the Ge-Si interface modes couple more strongly to the Si layers than the Ge layers since Figure 3 shows no similar enhancement at 2.25 eV.

When the Ge host $E_1 - E_1 + \Delta$ resonance is excited, we observe large changes in the relative strengths of I_{Si} and I_{Ge-Si} in Figure 8. Similar behavior is observed on the part of I_{Ge} with respect to I_{Ge-Si} when the Si host resonance is excited in our 11 Å Ge layer in Si(100).(Figure 9) As suggested by the results in Figure 3, for excitation energies below 2.7 eV, $I_{Ge} \gg I_{Ge-Si}$. However, for higher energy excitation near 3.5 eV where the Si host scattering shows a very strong resonant enhancement due to the Si $E_1 - E_1 + \Delta$ gaps, the Ge induced Raman spectrum is dominated by the Ge-Si mode near 415 cm^{-1} with a significant contribution from the previously very weak scattering near 440 cm^{-1}. We find in Figures 8 and 9 that excitation of the host $E_1 - E_1 + \Delta$ gaps produce Raman spectra which are domi-

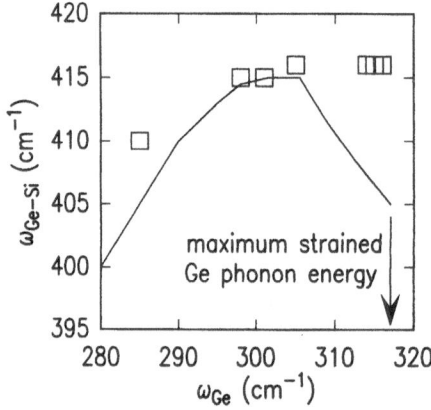

FIGURE 7. The variation of ω_{Ge} versus ω_{Ge-Si} for several different Ge layers in Si(100). The solid line is the expected dependence for strained Ge$_x$Si$_{1-x}$ alloys. The arrow is the position of the pseudomorphically strained Ge first order phonon.

nated by the interface scattering rather than the layer scattering. These figures suggest that the excitation of the host resonance couples more strongly to the interface scattering than the layer scattering from the other side of the interface. It should be noted that the strength of the host Raman scattering, because of the volume of host, makes it difficult to observe any changes in the strength of the host scattering when the layer resonance is excited.

DISCUSSION

We have demonstrated that Raman spectroscopy is a sensitive tool for studying the presence of Ge and Ge-Si bonds in Si hosts and Si and Si-Ge bonds in Ge hosts. Evidence for the observation of vibrational excitations of both the layers and their interfaces has been presented in this paper for layer thicknesses below 3 monolayers

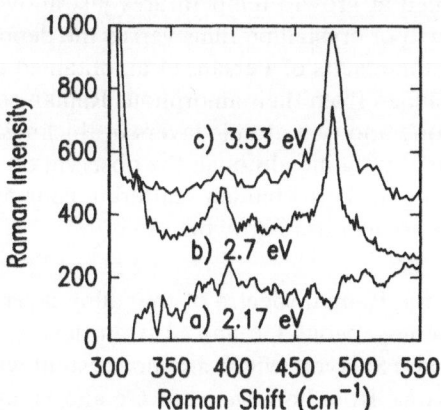

FIGURE 8. The Raman scattering from the Si and the Ge-Si modes from a ultra-thin Si layer in Ge(100) excited at three different photon energies.

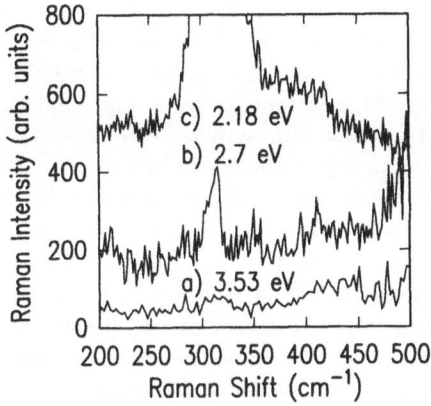

FIGURE 9. The Raman scattering from the Ge and the Ge-Si modes from a 11 Å pseudomorphic, Ge layer in Si (100) excited at three different photon energies.

of Ge and 4 or 5 monolayers of Si. For a given growth rate, the minimum thickness of the Ge-Si interface and the growth temperature will be strongly coupled. The Raman spectra show that we must grow more than 3 layers of Ge on Si(100) to obtain spectra that can be distinguished from that of Ge_xSi_{1-x} alloys. The minimum number of layers required for the observation of Ge like Raman scattering is strongly temperature dependent, increasing as the growth temperature increases. Our low temperature results obtained at growth temperatures just above the minimum temperature for epitaxial growth of crystalline films versus the deposition of amorphous layers are consistent with the results of Persans et al. obtained on amorphous Si-Ge superlattices.[27] They concluded from their amorphous Raman spectra that the Si-Ge interfaces were between one and two atomic layers in thickness. We find that our samples are predominantly crystalline although the observation of weak Ge induced broad line Raman scattering in the forbidden scattering geometry shows that there is some disorder in the low temperature growths.

Our ability to identify the Raman spectra of thin alloy layers means we can also study the properties of homogeneous Ge and Si strain layers. By Occam's razor, Raman spectra of our Ge and Si layers which are inconsistent with Raman scattering from an alloy layer must arise from homogeneous Ge and Si layers. Application of this knowledge to the growth of Si on a Ge substrate means we can also study the properties of thin Si strain layers in Ge(100).

The Raman spectra of the ultra thin Ge layers show strong similarities to the Raman spectrum of bulk germanium both with regard to the frequencies of the Raman active optical phonons and the excitation wavelength dependence of the resonant Raman scattering. The observed values of ω_{Ge}^l differ from ω_{Ge}^0 by an energy that is equal to the strain induced shift of the k=0 optical phonon of Ge as extrapolated from uniaxial stress measurements of this phonon frequency. This fact applies also to the second order Raman scattering which arises from the two phonon critical points of the full optical phonon branch. These conclusions mean that the confinement shifts of these optical phonons are small, probably less than 5 cm^{-1}. This is somewhat surprising given the very narrow widths of these Ge layers in light of past experience based on GaAs-AlAs quantum wells and superlattices.[5] Experimental uncertainties related to this conclusion involve 1) the validity of the linear relationship between strain and the energy of the optical phonons, which is likely to be correct given the good agreement observed between the linear model and results obtained by Cerdeira et al. on thick strained alloy layers,[23] 2) an underestimate of the amount of alloying in our Ge layers which in fact would tend to further reduce the magnitude of any localization effect since the alloying also shifts the mode to lower energies and 3) changes in the strain due to departures from layer by layer growth which is possible for the 11 Å samples but not likely for the 6 Å samples since in situ RHEED showed at least 3-4 oscillations during Ge growth. The small localization shifts we require to explain our results probably reflect the three mode character of the vibrational spectra of the Si-Ge system and the related fact that the Ge modes are not localized modes but resonances of the Si host. As Fasolino et al.[6,29] have shown based on calculations of a one dimensional, interplanar force constant model for Si-Ge superlattices, the non-zero amplitude of the Ge derived resonance modes in the Si host layers coexists with a displacement pattern that remains concentrated in the Ge layer. Localization of the Ge wavefunction would shift the phonon to lower energies given the curvature of the Ge optical phonon band to lower energies for larger wavevectors.[19] Since the long wavelength Si and Ge-Si optical vi-

190

brations with which the Ge layer modes would hybridize are all at higher energies, mixing of the layer modes with the host phonons would produce a shift to higher energies. It is interesting that in the Ge-Si layer system, these two interactions apparently compensate each other to between 10 and 20%.

The resonant Raman profiles of I_{Ge} for our thin Ge layers are quite similar to those of bulk Ge. This is surprising since the gap responsible for the peak at 2.3 eV in bulk Ge is $E_1 - E_1 + \Delta$ and Pollak and Cardona[30] showed that its center of mass should shift strongly to higher energies for the pseudomorphic growth of Ge on Si(100).[23] The position and width of the bulk Raman resonance is due to the three band term in Eq. 3 which is proportional to the difference of the contributions of E_1 and $E_1 + \Delta$ to the susceptibility, $\chi^i(\omega)$. However, Chandrasekhar and Pollak[31] determined that E_1 has a relatively weak dependence on strain compared to $E_1 + \Delta$. If the deformation potentials obtained from uniaxial stress measurements for this gap can be applied to the 4% lattice deformations present in the epitaxial growth of Ge on Si(100), then E_1 will be near 2.08 eV and $E_1 + \Delta$ will be near 2.8 eV for the strained bulk material. The small widths of our layers will shift these energies to higher values due to the confinement of the electron and hole wavefunctions in the Ge layers. The magnitude of the confinement shift for E_1 will depend on the conduction and valence band states that are the involved in the transition. If these states arise from the conduction and valence band states at the L point, the relatively small effective masses of both (the conduction band mass in the (100) direction is about .11 m_o[32]) mean that the E_1 gap for the Ge layer will be well above 2.7 eV, and above the observed peak in Fig. 3. The initial and final states will be delocalized, making it difficult to explain the difference in behavior of I_{Ge} and I_{Ge-Si} where I_{Ge} showed a strong resonant enhancement as compared to the interface mode. However, if E_1 arises from conduction and heavy hole valence band states on the Λ line going from L to Γ[33] the very heavy masses and linear terms in E(k) will result in much smaller localization shifts which would be consistent with the 2.25-2.3 eV resonance energies seen in Figure 3 and the weak coupling of the resonance to the Ge-Si interface modes. Resonant Raman contributions can also arise from the E_o gap as seen in Eq. 2. While band structure calculations on short period superlattices show that there is an E_o like transition near 2.2 eV.[34-37], they also show the electronic states responsible for the transition are delocalized. This would make it difficult to explain the difference between I_{Ge} and I_{Ge-Si} in Figure 3. Assignment of the peak in the resonant Raman scattering curve in Figure 3 to E_1 transitions in our layers does raise questions concerning its similarity to the resonance in the bulk. The bulk scattering arises from the three band terms in Eq. 2 which should be strongly modified in the Ge strain layers because of the strain enhanced splitting of the heavy and light hole bands.

The value of ω_{Si}^l cannot be explained simply by taking ω_{Si}^0 and adding the shift in the frequency of the bulk mode due to strain. The observed extra shifts to lower energies are between 10 and 20 cm^{-1} larger than can be accounted for just by strain. In contrast to the Ge modes, the Si modes are rigorously localized. The localization induced shifts for a free standing slab of Si have been calculated by Kanellis et al.,[18] Campbell et al.[19] and others. The additional shifts which we require are consistent with the theoretical estimates for a 5 to 10 monolayer Si sample. The strong localization of the Si mode and the localization of the electronic states responsible for the E_1 resonance of our Ge layers provide a simple explanation for the failure to observe a resonant enhancement of I_{Si} when the E_1 gap of the Ge host was excited in our Si layer samples. There can be no contribution to the Raman tensor via a deformation potential for phonons and electronic states on different sites.

The resonant Raman profiles in Figure 3 and the Raman spectra shown in Figures 8 and 9 provide information about the spatial extent of the Raman active Si-Ge interface modes. A question has recently been raised concerning this subject by Fasolino et al.[29] who pointed out that the Raman active interface mode is degenerate with the Si longitudinal acoustic phonon branch. The weak coupling of this mode to the Ge layer resonance at 2.25 eV for both the Ge and Si layers shows the mode does not penetrate deeply into Ge. On the other hand, its enhancement under 3.5 eV excitation suggests a stronger coupling to the Si electronic states. Fasolino et al. have argued that the LO Ge-Si mode is completely delocalized in Ge-Si superlattices because of its degeneracy with the Si LA branch. If this was the case, we would expect to see significant differences between the Si-Ge modes of samples consisting of a thin layer of Ge in an Si host and samples consisting of a thin layer of Si in a Ge host. In one case, the interface mode would mix with only 6-10 Å of Si while in the other case, the delocalization would be on a much longer scale. We find that under 2.6 eV excitation, the intensities of the Si-Ge interface modes are roughly comparable for the two cases. ω_{Ge-Si} is about 415 cm^{-1} for the Ge layers and about 380 cm^{-1} for the Si layers. The Ge layer result includes a 15 cm^{-1} shift to higher energies due to strain while the Si layer result includes a 25 cm^{-1} shift to lower energies due to strain. Our results are consistent in both cases with an unstrained interface mode at about 400 cm^{-1}. It suggests that the delocalization of the Raman active Ge-Si interface mode into the Si layer is relatively modest. The enhancements of the intensities of the interface modes with respect to the layer modes at the excitation of the host $E_1 - E_1 + \Delta$ gaps (Figs. 8 and 9) suggest that the coupling of the gaps to the Raman active phonons depends strongly on proximity of the vibrational mode to the electronic state. The interface is necessarily always closer to the host than the layer.

CONCLUSIONS

We have shown how Raman spectroscopy can be used to describe ultra-thin Si and Ge layers which are less than 10 Å thick and covered under 20-50 Å of another material, and have studied the vibrational and electronic states of these layers. This has become possible only in recent years due to substantial improvements in experimental sensitivity associated with the development of new, imaging detectors. The individual channel capabilities of these new detectors greatly exceed those of the previous generation of high efficiency, low noise photodetectors. Since the new detectors also provide the opportunity for multichannel acquisition of very small signals, the increase in sensitivity is very large.

Both the chemistry and physics of the Si-Ge layer system shows interesting behavior when studied by vibrational Raman spectroscopy. The conflict between high quality epitaxial growth and the requirement for abrupt interfaces is severe but can be easily monitored by Raman scattering. The preparation of homogeneous Ge layers less than 10 Å thick with relatively abrupt interfaces requires growth temperatures below 350 C. On the other hand, the growth of crystalline Ge and Si requires temperatures above 200 C and even then, evidence for crystalline disorder has been observed. An advantage of Raman spectroscopy is its ability to non-destructively characterize buried layers. In this case, it has allowed us to show that post growth anneals have a relatively weak effect on the interfacial abruptness of Ge-Si layers raising the possibility that improvements in crystalline quality may be possible through post growth treatments without adverse impact on the interfaces. Though not discussed here, a number of other interesting effects associated with the processing of these material such as the build up of Ge-rich layers at the oxide-alloy interface in the thermal oxidation of the alloys can be studied by Raman scattering.

The lattice dynamics of the Si-Ge system provides an interesting contrast to the GaAs-AlAs system which has been extensively studied in recent years.[5] Although the Ge-Ge and Ge-Si vibrations are resonances of the Si layers, the mixing of the Ge and Ge-Si phonons with the Si vibrations is small. The Ge optical phonons wavefunctions are clearly concentrated at the Ge layers. As a result, they can be used for the characterization of these layers, just as in the case of GaAs where the phonons are rigorously localized. However, the resonant character of these modes should produce some interesting contrasts with the GaAs-AlAs system when Ge-Si superlattices are studied. I_{Ge} should be enhanced at the Si $E_1 - E_1 + \Delta$ gap. Optical phonons in the Ge-Si case should show zone-folding rather than localization shifts associated with the finite size of the Ge wells. It is expected that experimental studies of the TO phonons of the Si-Ge system through the use of either non-(100) oriented samples or of the forward scattering geometry in thinned samples will permit even more detailed characterization of these layers and their interfaces.

The localized character of the Raman active optical phonons of our Ge-Si layer system provides an interesting tool for studying the electronic states of these layers. The resonant Raman scattering profiles from these localized modes can be used to describe the spatial extent of the electronic states responsible for these resonances in the dielectric response. Improvements in the performance of charge coupled device detectors will allow these experiments to be performed at excitation energies as low as 1.2 eV. The state of art in tunable optical sources now provides tunable radiation to energies as high as 4.0 eV. This means that resonant Raman scattering from these localized and quasi-resonant phonons and the truly delocalized, zone folded acoustic modes, of Ge-Si superlattices will be possible through most of the critical points of the dielectric response of these materials.

ACKNOWLEDGMENT The work described in this paper would have been impossible without the collaboration of S. S. Iyer, S. Delage, P. Pukite, M. Copel, and J. Freeouf. Thanks go to B. A. Ek, J. A. Calise and G. Coleman for technical assistance. J. Menendez, E. Burstein, F. H. Pollak, I. P. Ipatova and S. S. Jha provided helpful discussions.

REFERENCES

1. T. P. Pearsall, J. Bevk, L. C. Feldman, J. M. Bonar, J. P. Mannaerts, and A. Ourmazd, Phys. Rev. Lett. 58, 729 (1987).
2. G. Abstreiter, K. Eberl, E. Friess, W. Wegscheider and R. Zacchai, J. Crystal Growth (to be published) (1989).
3. A. Fasolino and E. Molinari, J. Phys.(Paris) Colloq. C5, 569 (1987).
4. J. C. Tsang, S. S. Iyer, P. Pukite and M. Copel, Phys. Rev. B (accepted for publication) (1989).
5. M. V. Klein, IEEE JQE QE22, 1760 (1986).
6. A. Fasolino, E. Molinari and J. C. Maan, Phys. Rev. B39, 3923 (1989).
7. D. E. Aspnes and A. Studna, Phys. Rev. B27, 985 (1983).

8. S. S. Iyer, J. C. Tsang, M. Copel, P. Pukite and R. Tromp, Appl. Phys. Lett. 54, 219 (1989).

9. N. Wada and S. A. Solin, Physica 105B, 353 (1981).

10. P. A. Temple and C. E. Hathaway, Phys. Rev. B7, 3685 (1973).

11. E. Anastassakis and Y. S. Raptis, J. Appl. Phys. 57, 920 (1985).

12. J. C. Tsang, in Dynamics at Surfaces ed. by B. Pullman and J. Jortner, (Reidel, Dordrecht 1984) p. 379.

13. W. P. Aker, B. P. Yip, D. H. Leach and R. K. Chang, J. Appl. Phys. 64, 2263 (1988).

14. J. C. Tsang, in Light Scattering in Solids V ed. by M. Cardona and G. Guntherodt, (Springer, Berlin, 1989) p. 233.

15. K. Sakamoto, T. Sakamoto, S. Nagai, G. Hashiguchi, K. Kuniyoshi and Y. Bando, Japan J. of Appl. Phys. 26, 666 (1987).

16. M. A. Renucci, J. B. Renucci and M. Cardona, in Proc. of 2nd Int. Conf. on Light Scattering in Solids, ed. by M. Balkanski, (Flammarion, Paris, 1971) p. 326.

17. A. K. Sood, J. K. Menendez, M. Cardona and K. Ploog, Phys. Rev. Lett. 54, 2115 (1985).

18. M. I. Alonso, M. Cardona and G. Kannellis, Solid State Commun. 69, 479(1989).

19. I. Campbell and P. Fauchet, Proc. of the 18th Int. Conf. on the Phys. of Semiconductors, ed. by O. Engstrom, (World, Singapore, 1987) p. 1356.

20. G. Kanellis. J. F. Morhange and M. Balkanski, Phys. Rev. B21, 1543 (1980).

21. A. J. Sievers and A. S. Barker, Reviews of Modern Phyics, 47, S2 (1975).

22. M. Cardona in Light Scattering in Solids II, ed. by M. Cardona and G. Guntherodt, (Springer, Berlin, 1982) p. 19.

23. F. Cerdeira, A. Pinczuk, J. C. Bean, B. Batlogg and B. A. Wilson, Appl. Phys. Lett. 45, 1138 (1984).

24. M. A. Renucci, R. Tyte and M. Cardona, Phys. Rev. B11, 3885 (1975).

25. B. A. Weinstein and R. Zallen, in Light Scattering in Solids IV, ed. by M. Cardona and G. Guntherodt, (Springer, Berlin, 1984) p. 463.

26. M. H. Brodsky in Light Scattering in Solids I, ed. M. Cardona, (Springer, Berlin, 1975) p. 205.

27. P. D. Persans, A. F. Ruppert, B. Abeles and T. Tiedge, Phys. Rev. B32, 5558 (1985).

28. W. Byra, Solid State Commun. 12, 253 (1973).

29. E. Molinari and A. Fasolino, Appl. Phys. Lett. 54, 1220 (1989).

30. F. H. Pollak and M. Cardona, Phys. Rev. 172, 816 (1968).

31. M. Chandrasekhar and F. H. Pollak, Phys. Rev. B15, 2127 (1977).

32. M. Cardona, G. Harbeke, O. Madelung and U. Rossler, Physics of Group IV Elecments and III-V Compounds, Vol 17a of Landolt-Bornstein: Numerical Data and Functional Relationships in Science and Technology, ed. by O. Madelung (Springer, New York, 1982) p.91.

33. L. Vina, S. Logothetides and M. Cardona, Phys. Rev. B30, 1979 (1984).

34. M. S. Hybertson and M. Schulter, Phys. Rev. B36, 9683 (1987).

35. S. Froyen, D. M. Wood and A. Zunger, Phys. Rev. B37, 6893 (1988).

36. S. Ciraci and I. P. Batra, Phys. Rev. B38, 1835 (1988).

37. S. Satipathy, R. M. Martin, and C. van de Walle, Phys. Rev. B38, 13237 (1988).

PHONONS OF IDEAL (001) SUPERLATTICES: APPLICATION TO Si/Ge

E. Molinari[a]   and A. Fasolino* [b]

[a] CNR, Istituto "O.M. Corbino", Via Cassia 1216, I-00189 Roma, Italy
[b] CNRS, Service National des Champs Intenses, BP 166, F-38042 Grenoble, France

We present a discussion of the vibrational properties of semiconductor superlattices in terms of matching of bulk waves at interfaces. Qualitative predictions are then compared with the results of actual calculations performed for the physical system Si/Ge. The relevance of our results for structural characterization is discussed.

1.   Introduction

Much interest has been devoted recently to the vibrational properties of semiconductor superlattices (SL's), mostly for structural characterization of the samples. Among the systems other than GaAlAs – which remains the most studied [1,2] – Si/Ge SL's are receiving increasing attention for their potential technological relevance[3]; good quality Si/Ge SL's can indeed be obtained[4-8] in spite of the significant lattice mismatch of bulk Si and Ge, provided that the layer thickness is kept small (a few monolayers) and appropriate substrates are chosen. However, the achievement of atomically abrupt interfaces without intermixing seems still an open problem. Raman experiments have been performed on several SL systems, as well as thin isolated layers of Si (Ge) embedded in Ge (Si), by several groups[5-10].

The combined effect of strain and confinement makes a quantitative description of this system a difficult task. However, much useful information may be derived from model calculations[11-13], which point out qualitative features that can discriminate between SL's with ideal interfaces and SL's with alloying at interfaces.

In Sect. 2 we will also show that a first level of understanding can be obtained from the comparison of the bulk dispersions of the SL components, and that such scheme provides a useful key for the prediction of the qualitative behaviour of a general system. Si/Ge lends itself to being used as a good prototype system in this respect, because – as we will see – it presents a variety of types of superlattice modes. In Sect. 3 we will present calculations for Si/Ge SL's, and in Sect. 4 we will summarize the conclusions relevant to the structural characterization of the SL's.

2.   From the bulk dispersions to the superlattice spectrum

A superlattice can be thought of either as an intrinsically new material with periodicity larger than the bulk components, or as a sequence of "bulk layers" appropriately matched to each other. This second scheme, in principle valid for isolated interfaces or for thick SL's, can hold also for thinner ones in view of the relatively short range interactions involved in lattice dynamics, and we will use it as a guideline through the SL vibrational properties. This amounts to considering the SL spectrum as resulting from the matching of bulk waves at the interface[14].

As the bulk translational invariance is destroyed by the layering, not only bulk Bloch oscillatory waves (bulk solutions with real q_z; q_z being the wavevector component perpendicular to the interfaces), but also evanescent solutions (q_z complex or pure imaginary) must be considered in the matching[14]. The use of complex band dispersions is a standard approach since the early studies of surface electronic or dynamical problems[15].

Let us apply this scheme to (001) grown Si/Ge (Fig. 1). For modes propagating along (001), a great simplification comes from the fact that all the atoms of an atomic layer move together, and longitudinal (L) and transverse (T) vibrations are decoupled.

The dynamics can then be mapped onto that of a linear chain of atomic planes, interacting via different interplanar force constants for different polarizations. These force constants describe the restoring forces felt by an atomic plane – due to the neighbouring atomic planes – when displaced as a whole form the equilibrium positions. However, while a set of interatomic force constants can describe the interactions of atoms for any possible displacement, a set of interplanar force constants can only describe one type of polarization and one direction of propagation. In particular, in the (001) direction a set of symmetric interplanar force constants is needed to describe longitudinal interactions, and a set of asymmetric ones is needed for transverse vibrations[16,17]. In fact the longitudinal displacement of an atomic plane bends and compresses the interatomic bonds on its left as much as it unbends and stretches the ones on its right, which in the harmonic approximation produces the same effect; a transverse displacement, on the contrary, mostly stretches the bonds on one side and mostly bends those on the other side, giving rise to very different force constants[17]. This is the reason why the L dispersion has no gap at $2\pi/a$, while in the T case the asymmetry of the interaction opens a gap at the zone boundary. Being the problem reduced to a one-dimensional (1D) problem, the matching involves only one solution (possibly with complex q_z) in each material. Solutions for the SL will then occur at frequencies where such bulk solutions can be matched to form a Bloch wave with the periodicity of the SL.

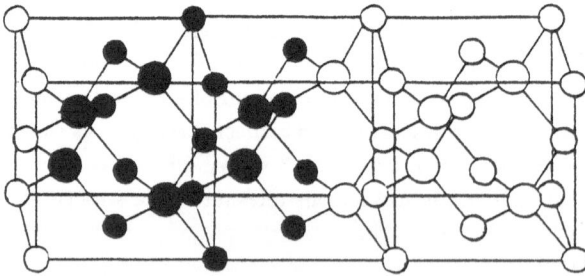

Fig. 1 Geometry of an ideal Si/Ge (001) superlattice: the three dimensional unit cell is shown. Full and empty circles represent Si and Ge atoms respectively. The size of the circles alternates along z, helping to distinguish atoms belonging to successive atomic planes.

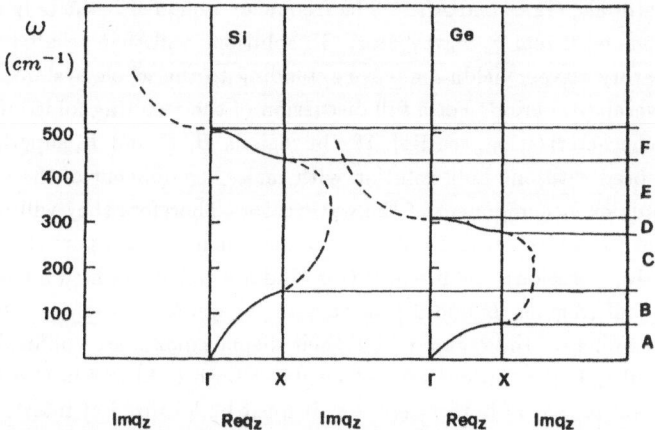

Fig. 2 Sketch of the transverse (001) dispersions of Si and Ge versus complex q_z wavevector. The usual Γ-X dispersion (real q_z) is shown with solid lines. The frequency regions are as follows:

region A:	real q_z	in Si,	real q_z	in Ge;
region B:	real q_z	in Si,	complex q_z	in Ge;
region C:	complex q_z	in Si,	complex q_z	in Ge;
region D:	complex q_z	in Si,	real q_z	in Ge;
region E:	complex q_z	in Si,	imaginary q_z	in Ge;
region F:	real q_z	in Si,	imaginary q_z	in Ge.

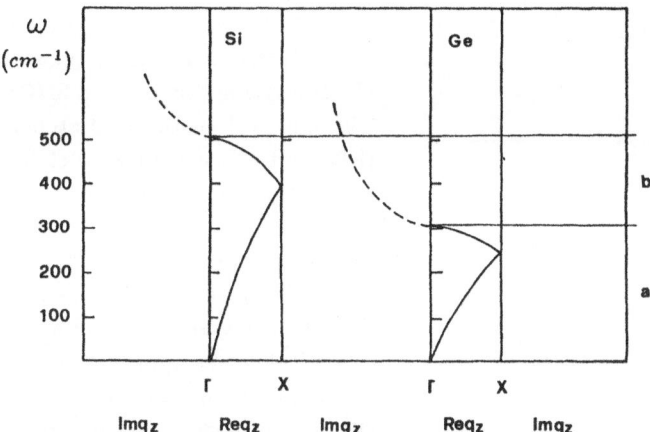

Fig. 3 Sketched longitudinal (001) dispersions of Si and Ge versus complex q_z (same as Fig. 2). The frequency regions are as follows: region a: real q_z in Si, real q_z in Ge; region b: real q_z in Si, imaginary q_z in Ge.

In Figs. 2, 3 we show a sketch of Si/Ge complex bulk dispersions in the T and L polarizations respectively. Although the plots are only schematic, the relative position of the band edges, which is the most relevant aspect for the following discussion, is reproduced.

We start with transverse modes (Fig. 2), which closely resemble the situation in the more studied pair of materials, GaAs and AlAs. The frequency region A is the only one where both bulks have solutions with real q_z wavevector. SL solutions will then result by the matching of two bulk oscillatory waves, yielding a wave extending to the whole system; hence they are sensitive to the overall SL period. For a full discussion of the resulting folded modes and their relevance for SL characterization, see Ref. 18. In regions B, D and F, superlattice solutions can only be composed with one bulk solution with real q_z component of the wavevector, and the other with complex q_z component of the wavevector. Therefore they will be oscillatory in one material and exponentially evanescent in the other (the imaginary part of q_z giving the rate of attenuation of the vibration). Solutions of this kind are called confined modes, of acoustical (region B) or optical (regions D and F) character. They are also referred to as Ge-like or Si-like modes depending on the region where their displacements are confined. In regions C and E, instead, real-q_z bulk solutions are not available both in Si and in Ge: if a SL solution exists, it must be evanescent in both layers, $i.e.$ it must be localized at interfaces; it will have an exponentially damped oscillatory behaviour or a purely exponential decay depending on the complex or purely imaginary value of the wavevector q_z in that layer. In the case of Si/Ge, as will be shown below from the result of actual SL calculations, T interface modes indeed appear in the gap between the two optical branches of Si and Ge (region E). In Fig. 4 we sketch the typical displacements expected for SL modes in the various regions.

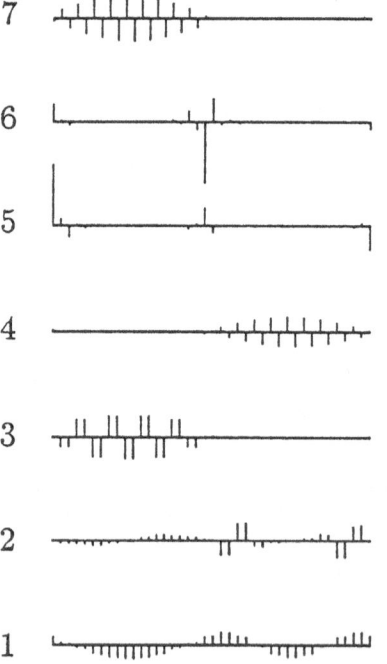

Fig. 4 Sketch of selected Γ-point transverse displacement amplitudes for a Si/Ge superlattice. The unit cell contains 20 atomic planes of Si (left) and 20 atomic planes of Ge (right). The modes are chosen to illustrate the expected behaviour of SL vibrations in the different frequency regions introduced in Fig. 2. *Mode 1* is a typical low-lying acoustical mode (region A); *mode 2* shows the possibility – also in region A – of acoustical modes propagating with very different amplitude in the two materials. *Mode 3* is the highest mode expected in region B (acoustical Si-like confined mode); *mode 4* is the typical topmost Ge-like confined optical mode (region D); then comes the doublet of interface modes (*modes 5 and 6*) falling in region E; finally *mode 7* is the highest Si-like confined optical mode (region F), analogous to mode 4.

Similar arguments can be used for the L spectra (Fig. 3). In this case, as the Ge bulk continuum is entirely contained within the one of Si, the scheme is much simpler: confined modes can only exist above the Ge edge (region b), with Si-like character, while in region a the modes must be propagative in both materials. No gap for real-q_z modes exists, where interface states can localize.

Concerning region a, it should be noted that, at a given frequency, the two real q_z vectors of the modes involved in the matching may be very different, and the displacement of the resulting SL modes may then be strongly reminiscent of the original bulk wavelenghts in each layer. This is already true in the low (linear) part of the dispersion, due to the rather different sound velocities of Si and Ge (see displacement 1 in Fig. 5); in the highest part of the dispersion, close to the LO edge of the Ge continuum, this is even more evident. We anticipate that also the amplitude of vibration in the two layers may be very different (quasi-confined resonances, see below), as exemplified by the displacement 2 of Fig. 5.

The above very simple considerations already yield a great deal of information about the SL phonon spectrum to be expected in Raman experiments. In particular, almost perfect confinement is expected only for T Si-like and Ge-like modes, while it should be less pronounced for the low-lying Si-like L modes which match to a small imaginary q_z component in the Ge layer. Furthermore, we stress that in the L spectrum neither true Ge-like confined modes nor modes localized at the Si-Ge interface and exponentially decaying in both Si and Ge layers can be expected.

However, on the basis of these simple arguments nothing can be said on the possible occurrence of resonant states in the region of overlapping frequency, $i.e.$ states composed of two real oscillatory solutions in the two materials with much bigger amplitude in one of the two layers. Also, the possibility of longitudinal interface states resonant with the continuum of Si cannot be a priori excluded.

We will see in Sect. 3 that from the actual calculation rather sharp resonant quasi confined Ge-like modes appear just below $\omega_{LO}(\Gamma)$ of Ge, close to the edge of overlap, while resonant interface modes with L polarization are not found; on the other hand, interface Si-Ge modes actually appear between the Si and Ge-like confined modes in the T polarization. In conclusion, the simple ideas resulting from the matching of bulk waves can be very useful as a guideline, although they cannot of course replace the actual calculations.

3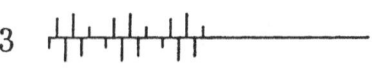

2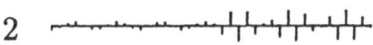

1

Fig. 5 Sketch of selected Γ-point longitudinal displacements for a Si/Ge superlattice (20 atomic planes of Si (left) and 20 atomic planes of Ge (right) in the unit cell). As in Fig. 2, the modes are representative the expected behaviour of SL vibrations in the different frequency regions introduced in Fig. 3. *Mode 1* is typical low-lying acoustical mode (region a). Again, *mode 2* is shown to demonstrate the possible occurrence in region a of acoustical modes propagating with very different amplitude in the two materials. *Mode 3* (region b) is the third of the Si-like confined modes.

3. Calculated superlattice phonons

We calculate the phonon spectra of Si/Ge superlattices along the (001) growth direction by means of a one-dimensional model with interplanar force constants, as done for GaAs/AlAs in Refs. 19, 20. We use the interplanar force constants (extending to third neighbour planes for longitudinal (L) and fifth neighbour planes for transverse (T) modes) calculated[21] for bulk Ge in the local density approximation, to describe both materials, the differences in their bulk spectra being accounted for by their different masses only[16]. In Fig. 6 we compare the resulting bulk Si and Ge phonon spectra along the (001) direction with the experimental data[22,23]. It appears that this "mass approximation" with Ge force constants obviously results in a much better representation of Ge compared to Si. Improving the description of bulk Si is beyond the purpose of this paper, as all the main qualitative features (like the absence of L interface modes) are not affected by this discrepancy, the relative frequency position of the two bulk continua being qualitatively reproduced anyway.

Note also that the effect of strain induced by the lattice mismatch is not included in our model; hence only the sequence of masses in the layering is responsible of the shifts of confined modes.

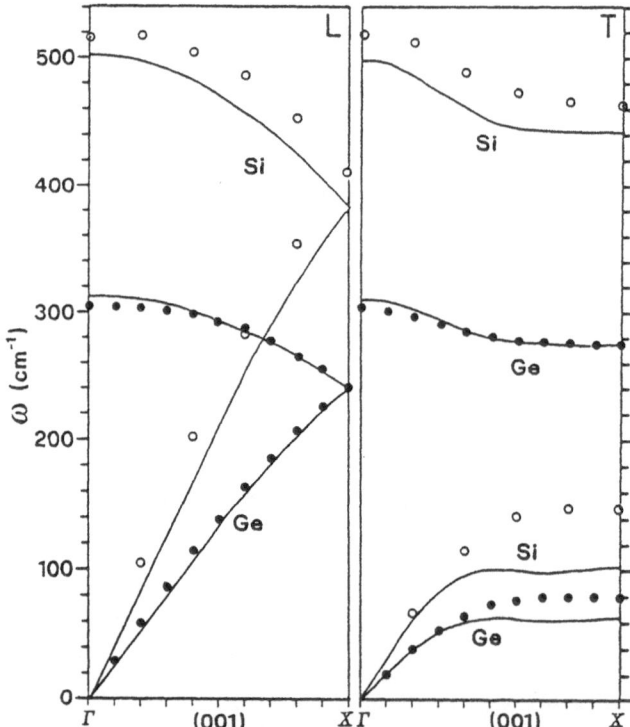

Fig. 6 Longitudinal (L) and transverse (T) bulk phonon dispersions for Si and Ge along the (001) direction. Solid line: calculation with force constants from Ref. 21; full circles: Ge neutron scattering data at 80°K, after Ref. 22; empty circles: Si neutron scattering data at 296°K, after Ref. 23.

In Fig. 7 we show the calculated SL phonon dispersion and Γ-point displacement patterns for a Si_6/Ge_6 SL in the transverse polarization. By inspection of the displacement associated to each mode, its character (confined, propagative...) can be immediately identified. Modes 11, 12 and modes 7, 8 of Fig. 7b are the typical Si-like and Ge-like confined modes respectively. Modes 9 and 10 between the Si-like and Ge-like confine TO modes are the true interface modes, localized on the Si-Ge interface bonds. The occurrence of such modes in the gap is easily understood as the vibration of a different bond (Si-Ge) at the interface, which vibrates at frequencies where neither Si-Si bonds of bulk Si or Ge-Ge bonds of bulk Ge have allowed frequencies of vibration. This situation is analogous to that of InAs/GaSb SL's[24,25,13], but intrinsically different from that of GaAs/AlAs. The interface modes are slightly non degenerate due to the interaction between successive interfaces, as a consequence of the small thickness of the layers. For completeness we also note that confined TA modes in the Si layer (modes 5 and 6) and a quasi-confined resonant acoustical mode (mode 4) arise between the TA(X) of Ge and the TA(X) of Si and at the edge of the Ge TA continuum respectively. We point out that these T modes are forbidden in off-resonance Raman experiments in the standard backscattering configuration from the (001) direction. We will come back to this point in Sect. 4.

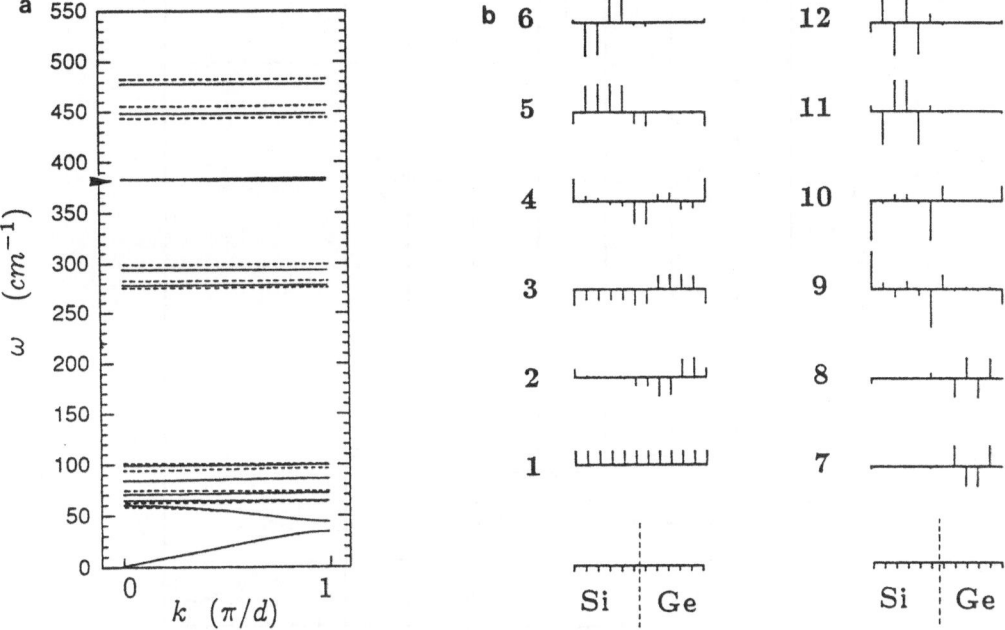

Fig. 7 (a) transverse phonon dispersion of a (001) grown Si_6/Ge_6 superlattice. Solid lines represent modes belonging to one T polarization direction (T1); dashed lines represent the modes of the orthogonal T polarization (T2) which in the SL may be not degenerate with the first ones. The arrow shows the interface modes. (b) amplitude of transverse displacements corresponding to all modes at $k = 0$, in order of increasing frequencies (only displacements with T1 polarization are shown).

Let us turn now to the L polarization. In Fig. 8 we show the higher part of the spectrum of a Si_6/Ge_6 SL and the related Γ point displacement patterns in a more illustrative way, in order to emphasize the analogy of phonon displacements with the wave functions of a particle in a finite well[14]. Note that the analogy is with a particle of negative effective mass, due to the downward dispersion of the LO branches. In this context the upper (lower) edge of the phonon barriers is given by the edge of the Si (Ge) bulk continuum. Within the well, we have the true Si-like confined modes. They are dispersionless and their frequency does not depend on the adjacent layer thickness[11,13] as far as no coupling across the barriers occurs. Note that below the edge of Ge, which represents the edge of the continuum, the displacements in the Ge layer are quite similar to the ones of true confined modes (such as the topmost Si-like modes), but matched to an oscillatory non-evanescent wave in the Si layer. For some of these modes (for example for the highest one) the displacement amplitude in Si is much smaller than in Ge. This is why we call these modes resonant, quasiconfined Ge-like modes. This behavior is more pronounced for thicker Ge layers. For 10 atomic layers of Ge, for example, the first quasiconfined Ge-like mode has a dispersion of less than 2 cm^{-1}, and a frequency which does not depend much on the thickness of the adjacent Si layer. Furthermore, as the displacements of the higher resonant modes are similar to those of true confined modes, the corresponding Raman intensity should be comparable. Notice also that resonances in the layer representing

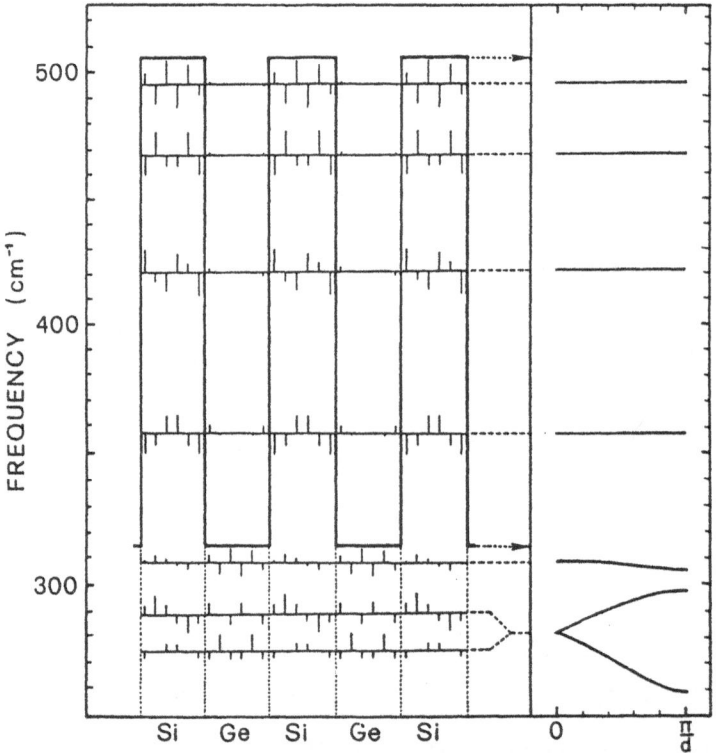

Fig. 8 Confined and resonant longitudinal modes in a Si_6Ge_6 superlattice. *Right*: phonon dispersion along (001) in the superlattice Brillouin zone. *Left*: amplitude of the Γ point longitudinal displacements *vs.* position of the atomic layers along (001). The upper and lower edges of the barriers are given by the edges of the Si and Ge bulk continua respectively, as marked by the arrows.

the well (Si in the present case) may occur. The presence of quasi-confined resonant modes has been predicted also for InAs/GaSb SL's[13], where their sharpness is even more pronounced.

The detailed properties of these modes – like their degree of confinement, their actual number and frequency location – depend strongly on the underlying bulk dispersions and are obviously more sensitive to the adjacent layer and to the details of the interfaces than those of true confined modes. For Si/Ge SL's, we find that in most cases only one or two Ge-like sharp resonances occur.

Clearly no interface modes appear in the L spectrum. By changing the layer thickness and by softening or stiffening the interplanar interface force constant in a wide range of values, we have checked that interface resonances never appear in the L spectrum between the Si-like and the Ge-like modes. It should be noted that for particularly thin superlattices the third Si-like confined L mode, which has a non negligible amplitude of vibration close to the interfaces[12,26], may fall as low as ~ 400 cm^{-1}, and give a contribution to the Raman intensity in that range; such contribution can be however distinguished from that of a true interface mode from its thickness dependence.

Coming back to Ge-like resonant states, it is evident from their displacement (Fig. 8) that they must depend also on the properties of the adjacent Si layer. This fact is evident also from their finite dispersion in the SL Brillouin zone, contrary to true confined modes.

In Fig. 9 we show the full L dispersions of two SL's, having the same Ge thickness as in Fig. 8 but thicker Si-layer, namely $Si_{20}Ge_6$ and $Si_{60}Ge_6$. We notice that the resonant Ge-like state indicated by an arrow remains approximately at the same frequency, with a comparable dispersion. This fact identifies this mode as a property of the Ge layer. However, contrary to Si-like confined modes which would have remained exactly at the same frequency if the adjacent layer was changed, they are sensitive to the variation of Si thickness; for instance, in the case shown the dispersion changes its slope. This is related to the fact that resonant states are not *intrinsically* different from folded propagative modes.

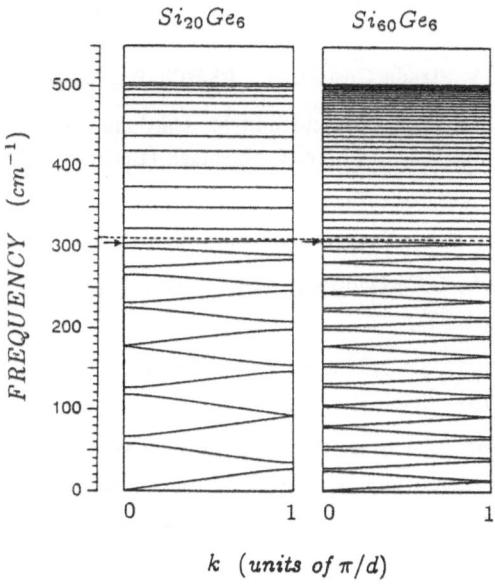

Fig. 9 Dispersion of longitudinal modes along (001) for the $Si_{20}Ge_6$, $Si_{60}Ge_6$ superlattices. The dashed lines represent the edges of the longitudinal continua of bulk Si and Ge; the arrows mark Ge-like resonances (see text).

4. Discussion and conclusions

Several groups have performed Raman experiments in the backscattering geometry from the (001) surface on (001)-grown Si/Ge SL's[5-9] and thin layers[10,27]. By symmetry, only modes with longitudinal polarization can be observed in this configuration if an ideal SL geometry is assumed[28,26].

Besides the observation of acoustical folded doublets, a common feature of these Raman spectra is the presence of three main peaks around 300, 400, 500 cm^{-1}, which are generally attributed to Ge, Si-Ge and Si modes respectively. In particular, the assignment of the peak around 400 cm^{-1} to Si-Ge vibrations is based on the comparison with the $Si_{0.5}Ge_{0.5}$ alloy, which shows a structure at the same energy. The relative intensity of the Si-Ge peak to the others varies significantly from sample to sample; it has been sometimes taken as a measure of the relative numer of Si-Ge bonds and, in turn, of Si-Ge internixing at the interface. By use of the ideas of Sect. 2 and from our results of Sect. 3, it appears that, in the hypothesis of ideally sharp interfaces, modes localized at the interface with wave vector parallel to the growth direction should not appear at all in off-resonance Raman spectra taken in bakscattering geometry. If a thickness-independent peak is instead observed with great intensity between the Si-like and Ge-like peaks, it is most probably due to the Si-Ge mode of the alloy[29] present at the interface. On the contrary, transverse interface modes are predicted and should be detactable in resonant conditions or in different scattering configurations. On the other hand, our calculations support the idea that a contribution to the Ge-Ge peak should come from the resonant Ge-like modes which behave as quasi-confined Raman-active modes.

These conclusions provide simple indications for characterizing interface sharpness of Si-Ge superlattices.

Acknowledgement

One of us (A.F.) would like to thank G. Martinez (SNCI-CNRS, Grenoble) for his hospitality and for many useful discussions. Partial financial support by the italian CNR-CINECA national computing project is acknowledged.

References

*   on leave from: SISSA, Strada Costiera 11, I-34100 Trieste, Italy.

1.   For a recent review see: B. Jusserand and M. Cardona, in *Light scattering in solids 5*, edited by M. Cardona and G. Güntherodt, Springer (Berlin), 1989, p. 49, and references therein.

2.   E. Molinari and A. Fasolino, Superlattices and Microstruct. **4**, 449 (1988).

3.   T.P. Pearsall, J. Bevk, L.C. Feldman, J.M. Bonar, J.P. Mannaerts and A. Ourmazd, Phys. Rev. Lett. **58**, 929 (1987).

4.   J. Bevk, A. Ourmazd, L.C. Feldman, T.P. Pearsall, J.M. Bonar, B.A. Davidson, and J.P. Mannaerts, Appl. Phys. Lett. **50**, 760 (1987).

5.   K. Eberl, G. Grötz, R. Zachai and G. Abstreiter, Journal de Physique C5, 329 (1987); R. Zachai, E. Friess, E. Kasper, and H. Kibbel, in *Proc. 19th Intern. Conf. Phys. Semiconductors, Warsaw, 1988*, edited by W. Zawadzki, Warsaw (1988), p. 487.

6.   J.C. Tsang, S.S. Iyer and S.L. Delage, Appl. Phys. Lett. **51**, 1732 (1987).

7.   D.J. Lockwood, M.W.C. Dharma-wardana, G.C. Aers and J.M. Baribeau, Appl. Phys. Lett. **52**, 2040 (1988).

8.   M. Ospelt, W. Bacsa, J. Henz, K.A. Mäder and H. von Känel, *Proc. IV Internat. Conf. on Superlattices, Microstructures and Microdevices, Trieste, 1988*, Superlattices and Microstructures, in press.

9.   J. Menéndez, A. Pinczuk, J. Bevk and J.P. Mannaerts, J. Vac. Sci. Technol. **B6**, 1306 (1988).

10. S.S. Iyer, J.C. Tsang, M.W. Copel, P.R. Pukite, and R.M. Tromp, Appl. Phys. Lett. **54**, 219 (1989).

11. A. Fasolino and E. Molinari, Journal de Physique **C5**, 569 (1987).

12. E. Molinari and A. Fasolino, Appl. Phys. Lett. **54**, 1220 (1989).

13. A. Fasolino, E. Molinari and J.C. Maan, Phys. Rev. **B39**, 3923 (1989).

14. A. Messiah, *Mécanique quantique*, Dunod, Paris (1962), vol. 1, pp. 81-83.

15. T. E. Feuchtwang, Phys. Rev. **155**, 715 (1967); **155**, 731 (1967).

16. A.S. Barker, Jr., J.L. Merz and A.C. Gossard, Phys. Rev. **B17**, 3181 (1978).

17. K. Kunc and R. M. Martin, Phys. Rev. Lett. **48**, 406 (1982); K. Kunc in *Electronic Structure, Dynamics and Quantum Structural Properties of Condensed Matter*, edited by J. T. Devreese and P.E. van Camp, Plenum Press, New York (1985).

18. B. Jusserand, D. Paquet, F. Mollot, F. Alexandre and G. Le Roux, Phys. Rev. **B35**, 2808 (1987).

19. E. Molinari, A. Fasolino and K. Kunc, in *Proc. of the 18th Intern. Conf. Phys. Semiconductors, Stockholm, 1986*, edited by O. Engström, World Scientific, Singapore (1987), p. 663.

20. E. Molinari, A. Fasolino and K. Kunc, Superlattices and Microstructures **2**, 397 (1986).

21. K. Kunc and P. Gomes Dacosta, Phys. Rev. **B32**, 2010 (1985).

22. G. Nilsson and G. Nelin, Phys. Rev. **B3**, 364 (1971).

23. G. Dolling, in *Symposium on Inelastic Scattering of Neutrons in Solids and Liquids*, IAEA, Wien (1963), vol. II, p.37.

24. A. Fasolino, E. Molinari, and J.C. Maan, Superlattices and Microstructures **3**, 117 (1987).

25. A. Fasolino, E. Molinari, and J.C. Maan, Phys. Rev. **B33**, 8889 (1986).

26. M.I. Alonso, M. Cardona, and G. Kanellis, Solid State Commun. **69**, 479 (1989).

27. G.P. Schwarz, private communication.

28. A. K. Sood, J. Menéndez, M. Cardona and K. Ploog, Phys. Rev. Lett. **54**, 2111 (1985); **54**, 2115 (1985). A thorough discussion of Raman selection rules in Si/Ge SL's is given in Ref. 26.

29. see *e.g.* M.I. Alonso and K. Winer, in *Proc. 19th Intern. Conf. Phys. Semiconductors, Warsaw, 1988*, edited by W. Zawadzki, Warsaw (1988), p. 815; and references therein.

CALCULATION OF PHONONS IN SUPERLATTICES

THE Ge/Si SUPERLATTICES ALONG [0 0 1]

George Kanellis

Physics Department 313-1
Aristotle University of Thessaloniki
54 006 Thessaloniki,Greece

INTRODUCTION

Since the invention of Molecular Beam Epitaxy[1-2], a method which allows for multi-layer heterostructures to be grown with atomically abrupt interfaces and precise thickness control of the constituent compositionally different layers, a considerable and continous interest has been shown for the investigation of the unusual properties of these systems.

Although from the begining the main effort has been directed towards attractive applications[3-4], new phenomena have been discovered[5-7], while new areas have been opened for known phenomena to be studied[8-10].

One particular characteristic that the artificially grown superlattices exhibit is the spatial modulation of the composition, which relaxes strict selection rules for physical processes holding valid for the constituent compounds. Therefore new possibilities appear for probing the corresponding properties of these compounds[8]. More specifically, inelastic light scattering has been proved a powerfull technique in studying numerous properties of bulk semiconductors and surfaces[11]. Superlattices have opened a vast new field for this technique to be used as a tool in studying their electronic and vibrational properties[12] and many very interesting results have already been obtained promising new attractive applications.

Interpretation of these results raised the need to calculate the above properties for a large variety of superlattices which usually are described by large unit cells. There are two major problems to overcome in both types of calculations. The first one, purely computational, is to solve the eigenvalue problem for large and sometimes very large matrices.

The second problem is to find the relation between the eigenstates of the superlattice and those of the constituent compounds. It turns out that some of these states (phonon energies or electronic states) are the same as in one of the compounds, while others result from coupling between states in either compound[13]. It is important that the method of calculation will help to handle these problems.

Calculation of phonons in a simple compound can, in principle, be done either by total energy calculations[14] or by assuming some form for the interaction potential of the atoms (or ions) in the lattice[15]. The first method is computationally "heavy", specially for phonons outside the Brillouin zone center and for complex systems, such as superlattices with large unit cells. Most frequently the second method is used, e.g. some model for the interatomic interactions, is assumed. Depending on the model, a number of parameters has to be evaluated, usually by fitting to experimentally known phonon frequencies in different points of the Brillowin zone and/or to the elastic constants of the compound.

Since dispersion curves of phonons in superlattices have not been measured, direct application of this method is not possible. On the other hand, since the superlattice is composed, in principle, of two compounds (or elements) for which much more information on their dynamical properties is usually available, it is of particular interest to try to relate the phonon properties of the superlattice to the corresponding properties of the constituent compound whenever this is possible. In the case of phonons, this can be done under the assumption that the interactions of a pair of atoms in a compound remain unchanged in the superlattice for the same atoms. The assumption holds valid for short range interactions between atoms well inside a thick layer where the neighborhood of the atoms is the same as in the corresponding compound. For long range interactions the situation is more complicated and it has not been studied up to now.

Moreover, the great variety of possible superlattices made up of the same compounds, with different thicknesses of the successive layers and different orientations, raises the serious problem of reconstructing the appropriate for each case Dynamical matrix using different primitive cells.In order to overcome this problem, a method[16] has been developed which allows for the construction of the appropriate Dynamical matrix, on the basis of some model, provided that the values of the model parameters are known for the constituent compounds (or elements). In the next paragraph we will outline this method.

On the other hand, we have to notice that the acoustic phonon branches for small values of the wavevector,can be adequately described by the elastic continuum model[20]. For optical phonons propagating along the superlattice axis the linear chain model can be used. In this model the entire atomic plane perpendicular to the superlattice axis is mapped onto one "atom" of the linear chain. The interaction forces between atoms in the chain are taken to be the appropriate interplanar interactions in the bulk constituent compounds and can be found either by fitting to the experimentally known phonon dispersion in the compounds or from ab-initio calculations of their total energy[21-23]. Despite the simplicity of the model most of the Raman scattering data have been interpreted using this model. Green's function techniques have also been used in connection to the linear chain model[24].

More detailed Raman spectra[25] on AlAs/GaAs superlattices reveale the need of more elaborate three-dimensional calculations. In a macroscopic approach[26], the influence of the macroscopic field in polar superlattices on the interface modes has been studied. Microscopic three-dimensional calculations are rather limited because of the problems already mentioned and the fact that existing experimental data concern mainly AlAs/GaAs superlattices. The phonon dispersion in GaAs is very well known while in AlAs only a few phonon frequencies have been measured. For this system rigid-ion model[16], bond-charge model[27] and shell model[28] calculations have been reported, based on the assumption that the interatomic interactions in AlAs are the same as in GaAs.

DESCRIPTION OF THE METHOD

A crystalline structure can be described in terms of a primitive cell which is assumed to contain s atoms, labeled by the index κ,having position vectors $\mathbf{x}(\kappa)$ with respect to the origin of the cell. The choise of this cell is not unique. Usually the primitive cell is defined by the smallest three non-coplanar translation vectors $\mathbf{a}_1$, $\mathbf{a}_2$, $\mathbf{a}_3$ and is labeled by the triple index $\mathbf{l} = (l_1, l_2, l_3)$ such that the position vector of its origin is given by,

$$\mathbf{x}(\mathbf{l}) = l_1\mathbf{a}_1 + l_2\mathbf{a}_2 + l_3\mathbf{a}_3. \tag{1}$$

and the position vector of the κ atom in the $\mathbf{l}$ primitive cell is,

$$x(I,\kappa) = x(I) + x(\kappa).$$ (2)

We will refer to this description of the structure as the *original description*. With the above primitive cell a reciprocal lattice is associated whose primitive translations are defined by the relation,

$$a_i \cdot b_j = 2\pi\delta_{ij} , \qquad 1 < i,j < 3$$ (3)

and the Wigner-Seits cell of this reciprocal lattice is the *first Brillouin zone* (BZ) of the structure, to which we will refer as the first *original* BZ (OBZ). Assuming cyclic boundary conditions and within the adiabatic and harmonic approximations[15,17] for the potential energy of the interatomic interactions, we can construct a $3s \times 3s$ Dynamical matrix, $\underline{D}^o$, (first DM) whose eigenvalues and eigenvectors give the (squares of the) phonon frequencies and the normal modes of vibrations respectively, for each value of the wavevector lying in the first OBZ.

It is now possible to choose a multiple cell, a *supercell* N_o times larger than the primitive one, defining new non-coplanar translation vectors by the relation,

$$a_i' = \Sigma_j N_{ij} a_j ,$$ (4)

where the integers N_{ij} constitute the matrix $\underline{N}$ and N_o equals its determinant. The supercell contains $N_o \cdot s$ atoms which belong to the primitive cells whose origins are in turn contained in the supercell. The supercells of the structure are labeled by the triple index $L = (L_1, L_2, L_3)$ such that their origin has the position vector,

$$x'(L) = L_1 a'_1 + L_2 a'_2 + L_3 a'_3 ,$$ (5)

and the primitive cells within each supercell are labeled by the triple index $n = I_o$, the index of the primitive cell in the original description contained in the $L = 0$ supercell. Following this procedure to describe the structure, we can label the I primitive cell of the structure by the pair indices (L,n) which satisfy the relation,

$$I = L\underline{N} + n.$$ (6)

Accordingly, the (I,κ) atom in the original description is now labeled as (L,n,κ) and has position vector,

$$x'(L,n,\kappa) = x'(L) + x'(n,\kappa) ,$$ (7)

where the vector $x'(n,\kappa)$ is the position vector of the same atom with respect to the origin of the Lth supercell.

We will refer to this latter way of describing the structure, as the *superstructure description* (SD). From the above relations it is obvious that the position vector of the same atom in the two notations is,

$$x(I,\kappa) = x'(L,n,\kappa).$$ (8)

Describing the structure in terms of the supercell, a new reciprocal lattice can be defined, with primitive translations given by,

$$a_i' \cdot b_j' = 2\pi\delta_{ij} . \qquad 1 < i,j < 3$$ (9)

The first BZ defined on the new reciprocal lattice has a volume N_o times smaller than the first OBZ and it will be called, first *superstructure* BZ (SBZ).

For a wavevector k lying in the first OBZ we can set,

$$k = m \cdot b' + k' ,$$ (10)

where **k'** is a wavevector lying in the first SBZ and the triple index $\mathbf{m} = (m_1, m_2, m_3)$ labels the N_0 small first SBZ's inside the larger first OBZ.

Apparently, the two descriptions of the structure are equivalent and can be used to calculate the properties of a perfect infinite crystal. The common choise is the OD, as it has already been discussed.

Assume now that we choose a supercell to describe the structure and in order to calculate the phonon energies and eigenvectors, we construct a $3N_0 \cdot s \times 3N_0 \cdot s$ Dynamical matrix, $\underline{D}^s(\mathbf{k'})$, (second DM) in the frame of the same above mentioned approximations with respect to the interaction potential. Solving this second DM, for some value **k'** of the wavevector in the first SBZ, we will find the same solutions as if we had solved the first DM, $\underline{D}^o(\mathbf{k})$, for the N_0 values of the wavevector **k** (in the first OBZ) given by the relation (10) for the N_0 values of the triple index **m**.

It is now possible to construct a block-diagonal matrix $\underline{D}^b(\mathbf{k'})$, (third DM) in the $3N_0 \cdot s$-dimensional space, whose the N_0 $3s \times 3s$ diagonal blocks are the $\underline{D}^o(\mathbf{k})$ matrices (first DM) for the N_0 values of the wavevector **k** given by the relation (10), e.g.

$$\underline{D}^b(\mathbf{m}, \mathbf{m'} \mid \mathbf{k'}) = \underline{D}^o(\mathbf{mb'} + \mathbf{k'})\delta_{mm'}. \tag{11}$$

The two matrices $\underline{D}^s$ and $\underline{D}^b$ having exactly the same eigenvalues are related by a similarity transformation,

$$\underline{D}^b(\mathbf{k'}) = \underline{G}^{-1}\underline{D}^s(\mathbf{k'})\underline{G} \ , \tag{12}$$

which depend only on the structure and the choise of the supercell. The elements of the matrix $\underline{G}$ are given by[16],

$$g(\alpha, \beta; \kappa, \kappa'; \mathbf{n}, \mathbf{m}) = (N_0)^{-1/2} \exp[i\mathbf{mb'} \cdot \mathbf{x'}(\mathbf{n}, \kappa)]\delta_{\alpha\beta}\delta_{\kappa\kappa'} \tag{13}$$

where α, β denote cartesian coordinates, κ and κ' are atom labels in the primitive cell and δ is the Kronecker delta.

From eq.(12) it is obvious that the matrix $\underline{D}^s$ can be immediatly constructed for any choise of the supercell, provided that we know the matrix $\underline{N}$, which defines the new translation vectors for the SD and the matrix $\underline{D}^o$, which is the usual Dynamical matrix for the compound.

In order to show how the above results can be applied to the problem of calculating the phonons in a superlattice, let us assume that the superlattice is composed of two compounds (or elements) A and B with matched lattice constants, whose the dynamical properties are known. The procedure will be the same in cases where the cations or the anions are the same or no ions are common to the compounds, or the superlattice is composed of two elements. We choose the matrix $\underline{N}$ which defines the appropriate supercell for the superlattice. This will provide the indices **n** and **m** for the primitive cells contained in the supercell and the first SBZ's contained in the OBZ. In some of the primitive cells in the supercell, say the first N_1, the ions belong to compound A while in the rest N_2 ($= N_0 - N_1$) they belong to compound B. From the matrix $\underline{D}^o(\mathbf{k})$ for the compound A we can construct a block-diagonal matrix $\underline{D}^b(\mathbf{k'})$ for the N_0 values of the wavevector and using eqs.(12) in the inverse sense we obtain the matrix $\underline{D}^{sA}(\mathbf{k'})$ appropriate to describe the phonons in the compound A using the supercell as "primitive" cell. In exactly the same way we obtain the matrix $\underline{D}^{sB}(\mathbf{k'})$ appropriate to describe the phonons in the compound B using the same supercell as primitive cell. From these two matrices it is now possible to obtain another matrix $\underline{D}^{SL}(\mathbf{k'})$ appropriate to describe the phonons in the superlattice in a first approximation in the following way: The upper-left $3N_1 s \times 3N_1 s$ diagonal block of the matrix $\underline{D}^{SL}(\mathbf{k'})$ describes interactions between ions (sublattices) belonging to compound A and hence can be taken to be the corresponding block of the matrix $\underline{D}^{sA}(\mathbf{k'})$. The lower-right $3N_2 s \times 3N_2 s$ diagonal block of the matrix $\underline{D}^{SL}(\mathbf{k'})$ describes interactions between ions (sublattices) belonging to compound B and hence can be taken to be the corresponding block of the matrix $\underline{D}^{sB}(\mathbf{k}')$. The remaining two off-diagonal blocks of the matrix $\underline{D}^{SL}(\mathbf{k}')$ describe interactions between ions (sublattices) belonging to both compounds A and B and in a first approximation can be assumed to be equal to the average of the corresponding interactions in the two compounds. Finally the self-terms of the matrix $\underline{D}^{SL}(\mathbf{k}')$ have to be corrected to preserve translational invariance.

The matrix constructed this way will give, within the above mentioned approximations, the phonon frequencies and eigenvectors for the superlattice composed from N_1 layers of the compound A and N_2 layers of the compound B. No restriction has been imposed to the values of the wavevector. The whole procedure can be done on a computer and can be easly applied for differend supercells and layer thicknesses. There are two main points to be checked with respect to the assumptions, in comparison to the experiment: a) if the approximation of the unknown interactions between pairs of ions, which do not occur in either of the compounds, as average interactions of the corresponting ions in both compounds, is satisfactory and b) if the interactions of a pair of ions which occur in one of the compounds, remains the same in the environment of the superlattice.

In order to illustrate the method and its flexibility with respect to applications to the various superlattices which can be constructed from the same compounds or elements, we will calculate the phonon dispersion in several Ge/Si superlattices grown along the [0 0 1] direction.

Ge/Si SUPERLATTICES ALONG [0 0 1]

Ge and Si crystallize with the diamond structure (space group O_h^7 or Fd3m) with two atoms per primitive cell. Although there is a lattice mismatch, it is possible to grow thin strained superlattices along the [0 0 1] direction[18].

It has been reported[19] that depending on the number of monatomic layers of each element superimposed along the growth direction, superlattices that belong to different space groups can be obtained. Apart from the trivial case, where monatomic layers of Si and Ge are alternatively superimposed along [0 0 1], yielding the zincblende (ZB) structure (space group T_d^2 or F43m), five more space groups can be found. For some of these superlattices the dispersion of phonons propagating along the growth axis has been calculated on the basis of the Keating model[29] and the phonon symmetries at $\mathbf{k}=0$ have been discussed[19].

Here we will present some details of the calculation and new results on the phonon dispersion for wavevectors along and/or perpendicular to the superlattice axis. The model parameters are the same as in ref.19, i.e. one bond stretching force constant α and one bond bending force constant β. The values of these parameters for Ge ($\alpha_{Ge}=38.67$ N/m, $\beta_{Ge}=11.35$ N/m) ware taken from ref.30. The values of the same parameters for Si ($\alpha_{Si}=43.74$ N/m, $\beta_{Si}=11.54$ N/m) ware obtained by fitting only the experimental[31] optic phonon branches along the directions Δ, Σ and Λ in the Brillouin zone. Coulomb interactions and strain effects are neglected while the lattice constant for the superlattices was taken as the mean compositionally weighted lattice constant.

All Ge/Si superlattices (except the ZB case) belong to one of the following two categories with respect to the type of lattice they exhibit: P-type (simple) lattice, if the number of primitive cells N_0 of the diamond structure contained in the primitive cell of the superlattice (the supercell) is even, and I-type (body-centered) lattice if the number of primitive cells N_0 of the diamond structure contained in the supercell is odd.

The P-type lattices occur in two different space groups depending on the number of monatomic layers each element contributes to the superlattice: space group D_{2h}^5 (Pmma-orthorombic) if each element contributes an even number of monolayers and space group D_{2d}^5 (P4m2-tetragonal) if each element contributes an odd number of monolayers. Assuming for the primitive translations of the diamond structure the usual form,

$$\begin{pmatrix} \mathbf{a}_1 \\ \mathbf{a}_2 \\ \mathbf{a}_3 \end{pmatrix} = (a/2) \begin{pmatrix} 0 & 1 & 1 \\ 1 & 0 & 1 \\ 1 & 1 & 0 \end{pmatrix} \cdot \begin{pmatrix} \mathbf{x}_0 \\ \mathbf{y}_0 \\ \mathbf{z}_0 \end{pmatrix}, \qquad (14)$$

we can obtain the primitive translations for the superlattices of this type from eq.(4) where the transformation matrix is given by,

$$\underline{N} = \begin{pmatrix} 0 & 0 & 1 \\ 1 & -1 & 0 \\ N_0/2 & N_0/2 & -N_0/2 \end{pmatrix} \qquad . \qquad (15)$$

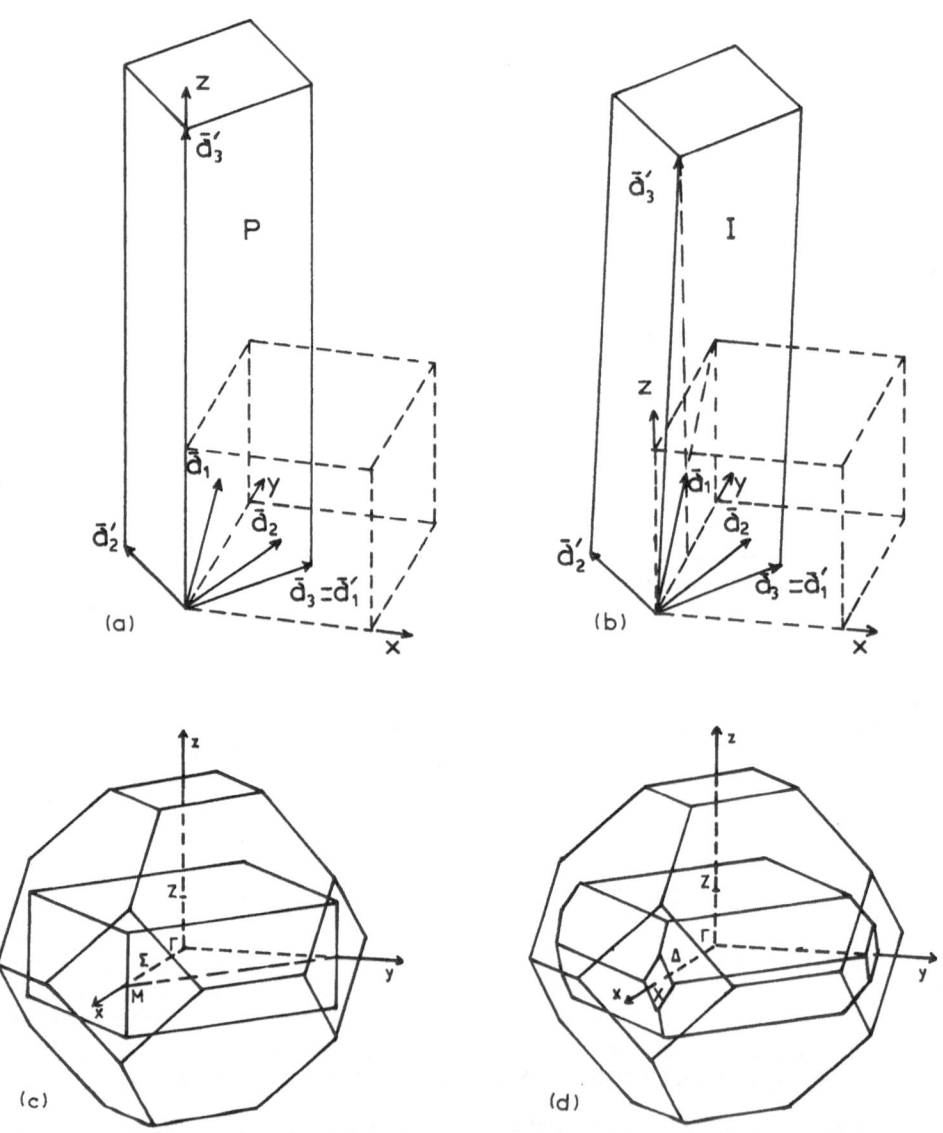

Fig. 1. Primitive cells for Ge/Si superlattices and corresponding Brillouin zones: (a) for P-type, (b) for I-type lattices. The dashed cube is the diamond crystallographic cell. Primitive translations as described in the text. Brillouin zones for P-type (c) and for I-type (d) lattices are shown inside the Brillouin zone of the diamond structure.

The so obtained supercells are illustrated in Fig.1(a) where both sets of the primitive translations are shown. In Fig.1(c) the coresponding first SBZ is shown inside the first OBZ (appropriate for the diamond structure).

The I-type lattices occur in three different space groups: D_{2h}^{28} (Imma-orthorombic) if each element contributes an even number of monolayers to the superlattice, D_{2d}^{9} (I4m2-tetragonal) if each element contributes an odd number of monolayers and D_{4h}^{9} (I4$_1$/amd-tetragonal) if the one element contributes an even number and the other an odd number of monolayers. In this latter case the supercell contains the double number of monolayers. The primitive translations for the superlattice can be obtained from eq.(4), where the transformation matrix is given now by,

$$\underline{N} = \begin{pmatrix} 0 & 0 & 1 \\ 1 & -1 & 0 \\ (N_0+1)/2 & (N_0-1)/2 & (1-N_0)/2 \end{pmatrix} \tag{16}$$

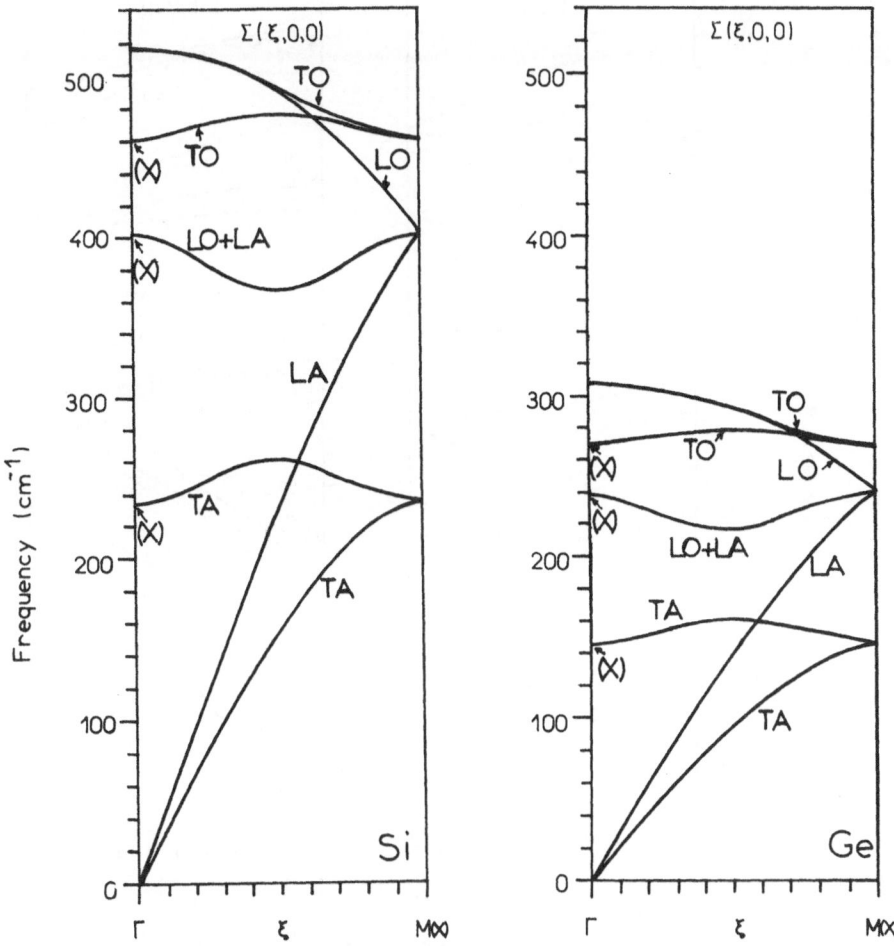

Fig. 2.   Calculated phonon dispersion along [1 0 0] for Si and Ge, as obtained when the structure is described in terms of a supercell containing two primitive cells along [0 0 1]. Notation of the Brillouin zone as for the P tetragonal lattice. The corresponding points of the diamond Brillouin zone are given in parentheses. ξ is the reduced wavevector coordinate.

The supercell for I-type lattices is shown in Fig.1(b) together with the two sets of primitive translations. In Fig.1(d) the corresponding first SBZ is shown inside the first OBZ (for the diamond structure).

Phonons in P-type Ge/Si superlattices

In order to illustrate the effect of the zone folding on the phonon branches and for the sake of comparison with forthcoming results for a Ge_2Si_2 superlattice, we present in Fig.2 the phonon dispersion ,for pure Si and pure Ge, for wavevectors along the [1 0 0] direction. For the calculation of these curves the structure is described in terms of a supercell along [0 0 1], two times larger ($N_O=2$) than the usual primitive cell defined by the translations of eq.(14). In this figure we use the notation of the folded zone (first SBZ).

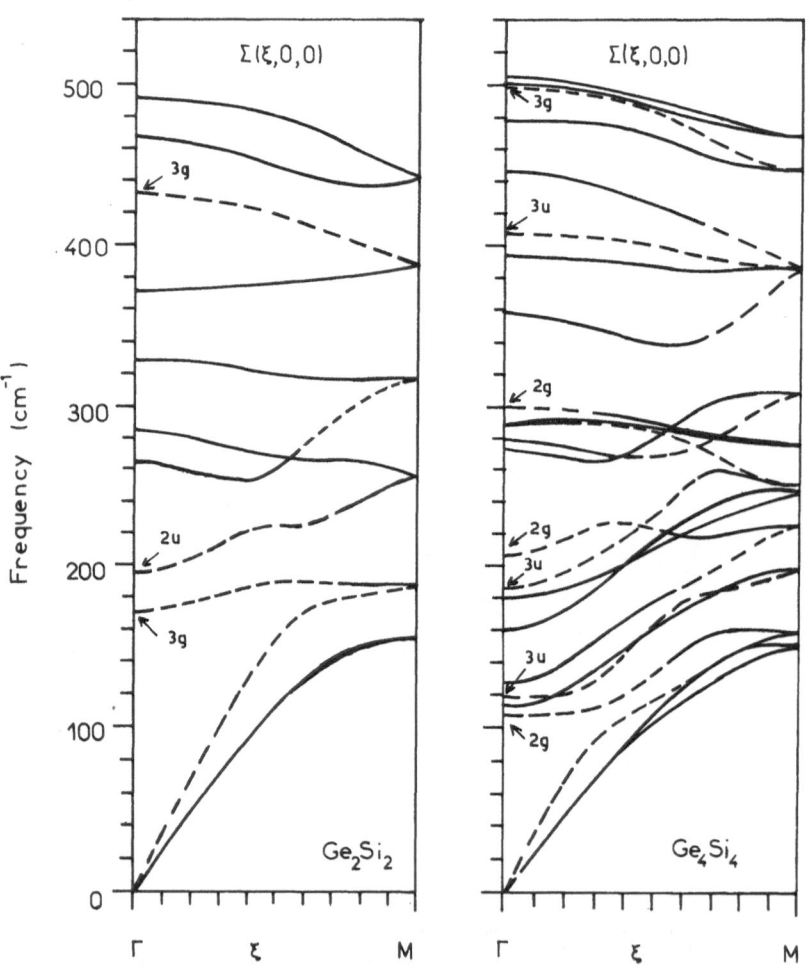

Fig. 3.   Calculated phonon dispersion along [1 0 0] for Ge_2Si_2 and Ge_4Si_4 superlattices (D_{2h}^5 space group). Solid and dashed lines represent transverse and longitudinal modes, respectively (main character). The labels give the phonon symmetries at Γ point(label g stands for A_g while the rest reffer to B symmetries). The symmetry of the remaining modes at Γ point is given in ref.19. ξ is the reduced wavevector coordinate.

In Fig.3 the phonon dispersion along [1 0 0] is shown for a Ge_2Si_2 and a Ge_4Si_4 superlattices (space group D_{2h}^5-orthorombic) for both of which the dispersion along [0 0 1] has been reported and the phonon symmetries at $\mathbf{k}=0$ have been discussed in ref.19, where the labels of the unlabeled phonons can be found (Figs.1 and 2 of ref.19). The labels 2g, 3u etc stand for B_{2g}, B_{3u} and so on. The calculation for the Ge_2Si_2 superlattice uses a combination of the two matrices which give the results of Fig.2 as it has been explained in the preceding section. By comparison with the results of Fig.2 the anticrossing of modes of mainly longitudinal character can be clearly seen. The modes of vibration in this direction have mixed polarization and their characterization as longitudinal (dashed lines) or transverse (solid lines) is indicative of their main character. Similar arguments apply for the dispersion along [1 0 0] in the Ge_4Si_4 superlattice.

Superlattices whose the primitive cell (the supercell) contains an even number of diamond primitive cells and each element contributes an odd number of monolayers have the symmetry of the D_{2d}^5 space group (tetragonal). The thinner superlattices in this category are the Ge_1Si_3 and Ge_3Si_1. Hence, in the primitive cell of the superlattice

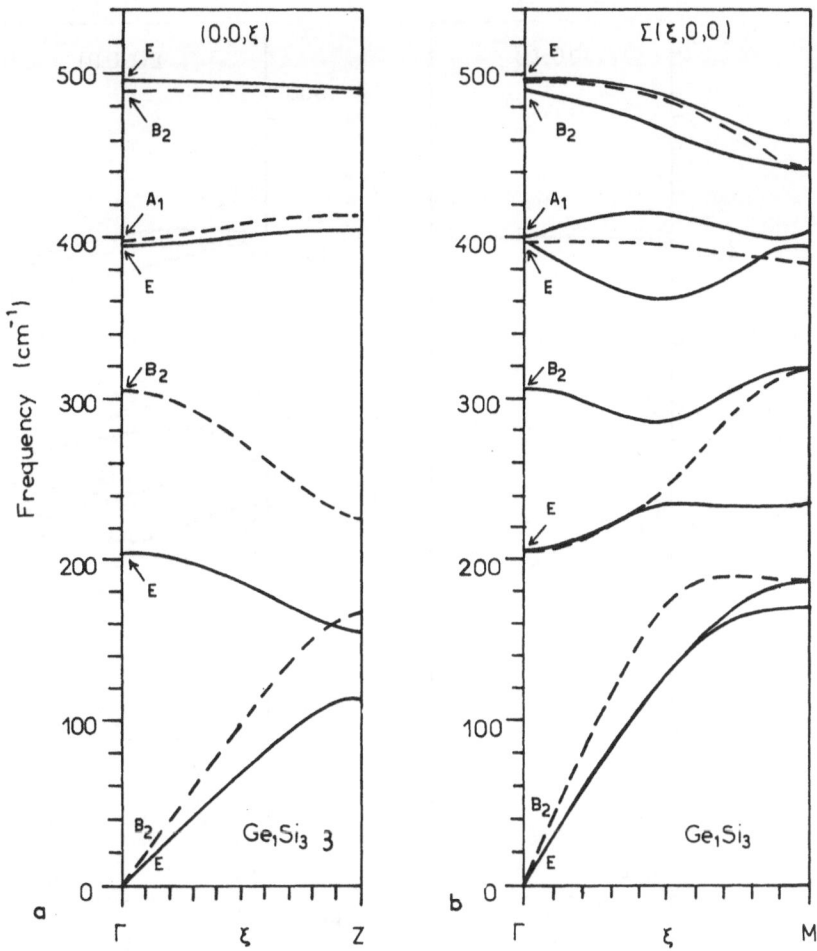

Fig. 4.   Calculated phonon dispersion (a) along [0 0 1] and (b) along [1 0 0] for Ge_1Si_3 (D_{2d}^5 space group) superlattice. Solid and dashed lines represent transverse and longitudinal modes, respectively. (In (b) the phonon polarization is mixed). The labels give the phonon symmetries at Γ point. ξ is the reduced wavevector coordinate.

there are 1 Ge and 3 Si atoms (or 1 Si and 3 Ge atoms for the latter superlattice). The Ge atom is at *d* Wyckoff site (point symmetry 4m2) and from the 3 Si atoms, one is at *b* (4m2) and the other 2 at *g* Wyckoff sites[32] (point symmetry 2mm). Using the nuclear site group analysis method[33] we find that the reducible representation at **k**=0 splits as:

$$\Gamma = A_1(\alpha_{xx} + \alpha_{yy}, \alpha_{zz}) + 3B_2(\alpha_{xy}) + 4E(\alpha_{xz}, \alpha_{yz}),$$

where the *d* and *b* sites contribute 1 B_2 and 1 E mode each. B_2 modes are longitudinal while the E modes are transverse. Both are also infrared active. In Figs.4 and 5 the phonon dispersion curves along [0 0 1] and [1 0 0] are given for Ge_1Si_3 and Ge_3Si_1 superlattices, respectively. In both of these cases the single type atoms (Ge in the former and Si in the latter) lie at the interfaces. Modes involving mainly motion of these atoms have an interface character. A discussion on the confinement of phonons in Ge/Si superlattices can be found in ref.34.

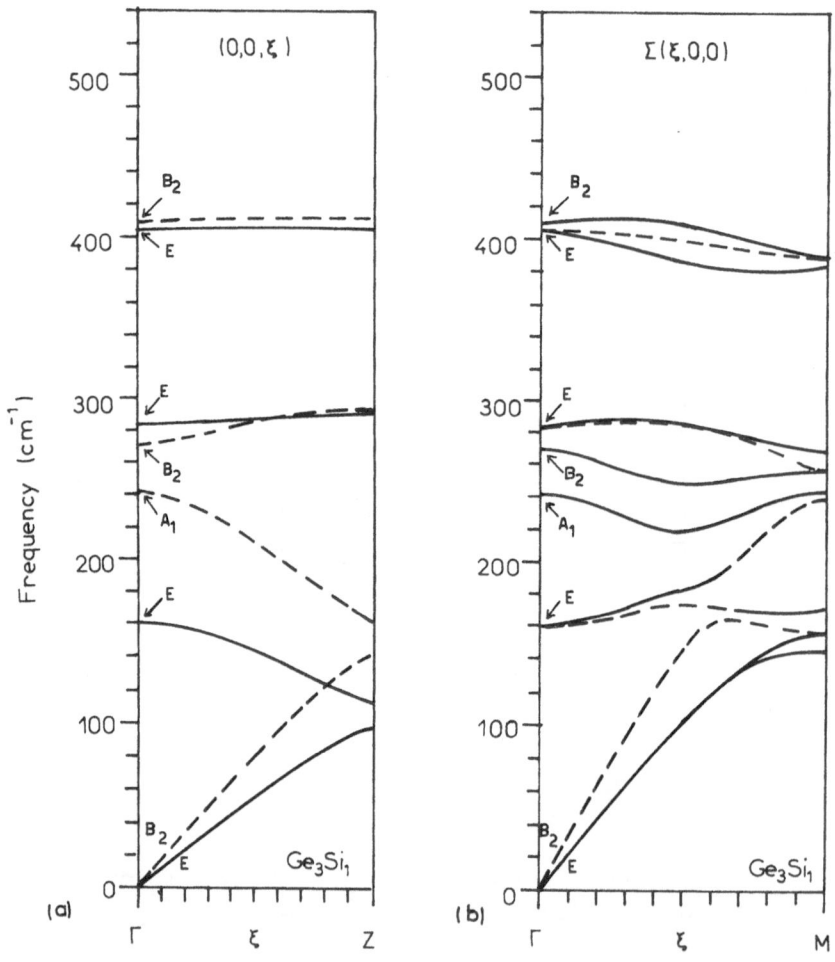

Fig. 5.   Calculated phonon dispersion, (a) along [0 0 1] and (b) along [1 0 0] for Ge_3Si_1 (D_{2d}^5 space group) superlattice. Solid and dashed lines represent transverse and longitudinal modes, respectively. (In (b) the phonon polarization is mixed). The labels give the phonon symmetries at Γ point. ξ is the reduced wavevector coordinate.

Phonons in I-type Ge/Si superlattices

In this category belong the Ge_3Si_3 (space group $D_{2d}{}^9$) and the Ge_1Si_2 (space group $D_{4h}{}^{19}$) superlattices, for both of which the phonon dispersion along [0 0 1] and the symmetries of the $\mathbf{k} = 0$ modes have been discussed in ref.19, where the symmetry labels are given for all of the modes (see Figs.3 and 4 of ref.19). In Fig.6 we present the dispersion curves for phonons propagating along [1 0 0]. All of the modes are of mixed polarization and the dashed and solid lines denote their main character, longitudinal and transverse respectively.

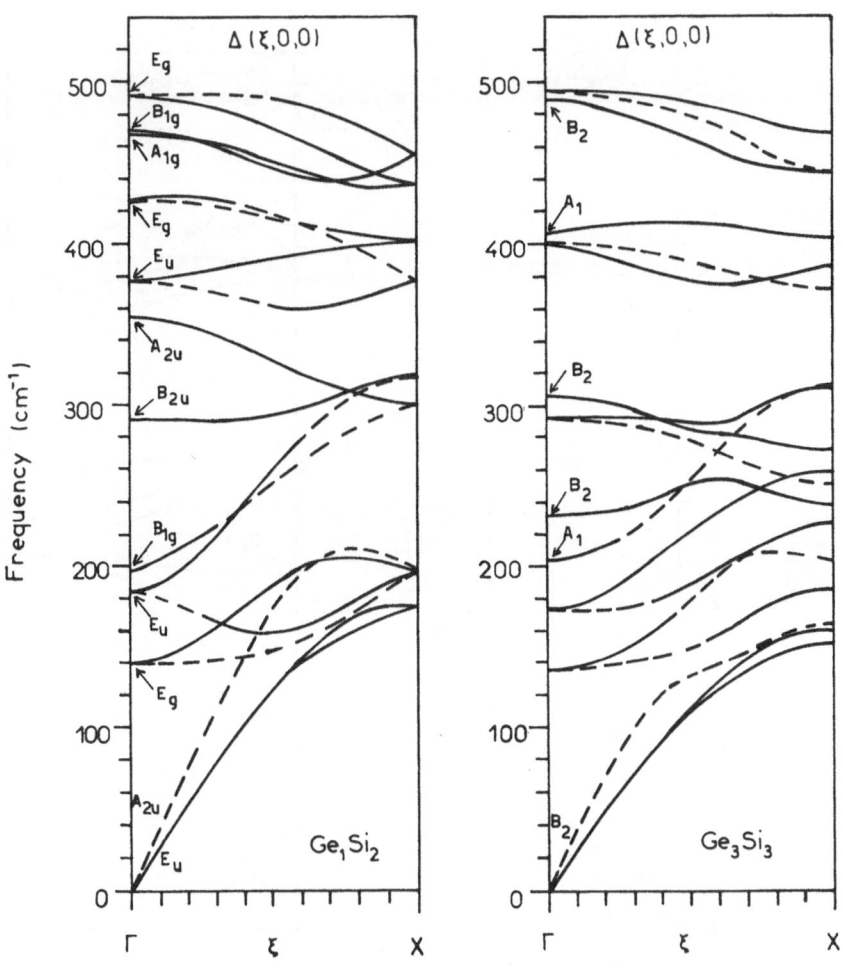

Fig. 6. Calculated phonon dispersion along [1 0 0] for Ge_1Si_2 ($D_{4h}{}^{19}$ space group) and Ge_3Si_3 ($D_{2d}{}^9$ space group) superlattices. Solid and dashed lines represent transverse and longitudinal modes, respectively (main charac- ter). The labels give the phonon symmetries at Γ point. The symmetry of the remaining modes at Γ is given in ref.19. ξ is the reduced wavevector coordinate.

In Fig.7 the phonon dispersion curves along [0 0 1] and [1 0 0] are given for the Ge$_2$Si$_1$ superlattice (space group D_{4h}^{19}) which obviously displays the same phonon symmetries as the Ge$_1$Si$_2$ superlattice already discussed but for different frequencies. From the displacement of a mode of given symmetry along the frequency axis (compare with Fig.4 of ref.19) the relative participation of Ge and Si atoms in the mode can be deduced. Modes of the same symmetry occuring in about the same frequency in the two superlattices are almost pure Ge-Si modes.

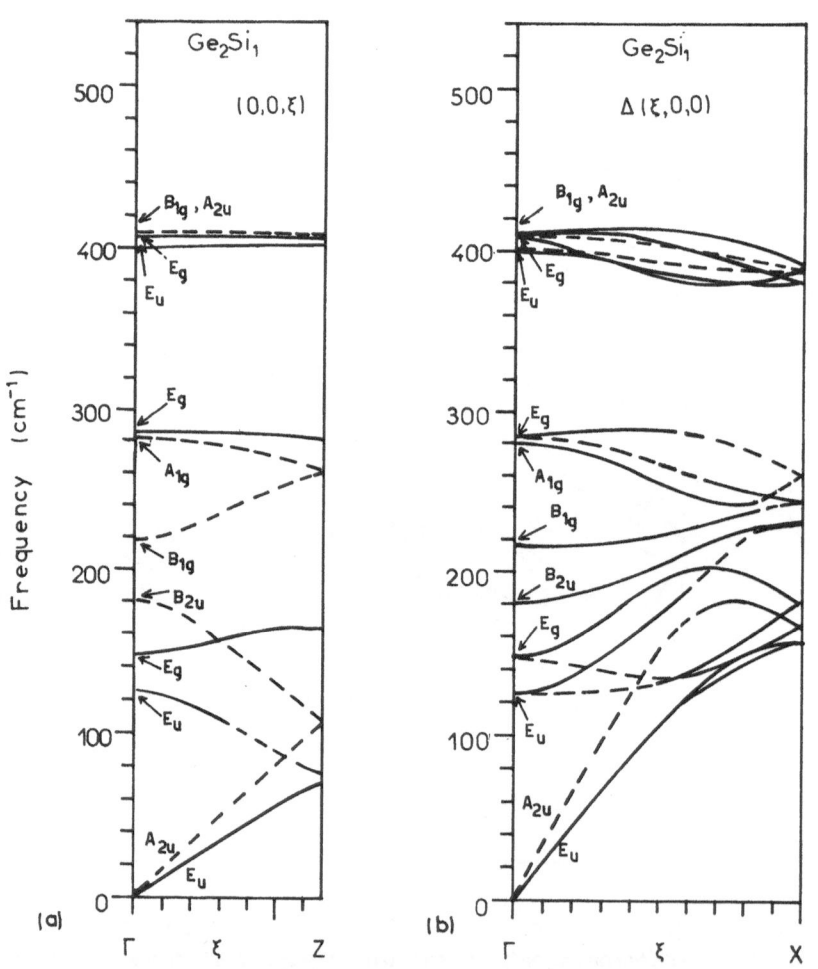

Fig. 7.   Calculated phonon dispersion along (a) [0 0 1] and (b) along [1 0 0] , for Ge$_2$Si$_1$ (D_{4h}^{19} space group) superlattice. Solid and dashed lines represent transverse and longitudinal modes, respectively. (In (b) the phonon polarization is mixed). The labels give the phonon symmetries at Γ point. ξ is the reduced wavevector coordinate.

One more case which has not been considered previously is the Ge_2Si_4 and Ge_4Si_2 superlattices belonging to the space group D_{2h}^{28} (orthorombic). In this case all of the atoms occupy, in pairs, e Wyckoff sites. Each such site contributes one of A_g, B_{2g}, B_{3g}, B_{1u}, B_{2u} and B_{3u} irreducible representation and hence three modes belong to each one of these symmetries at $\mathbf{k}=0$. The phonon dispersion along [0 0 1] and [1 0 0] for the two, above mentioned, superlattices are given in Figs.8 and 9 respectively, where the modes are labeled by the indices of the representations. Dashed and solid lines represent longitudinal and transverse modes respectively for the dispersion along [0 0 1]. For the dispersion along [1 0 0] where the polarization is mixed the lines denote the main character of the modes. More discussion on the phonon properties in all of the above cases can be found in ref.19.

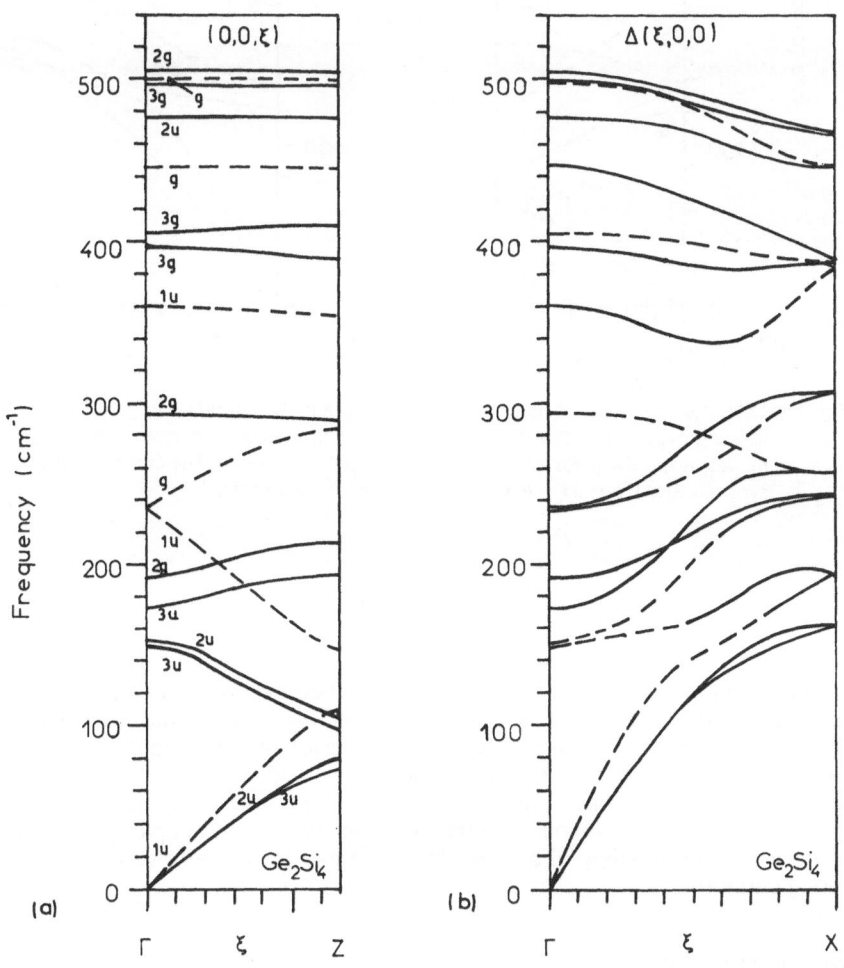

Fig. 8.   Calculated phonon dispersion, (a) along [0 0 1] and (b) along [1 0 0] for Ge_2Si_4 (D_{2h}^{28} space group) superlattice. Solid and dashed lines represent transverse and longitudinal modes, respectively. (In (b) the phonon polarization is mixed). The labels give the phonon symmetries at Γ point. (ξ is the reduced wavevector coordinate).

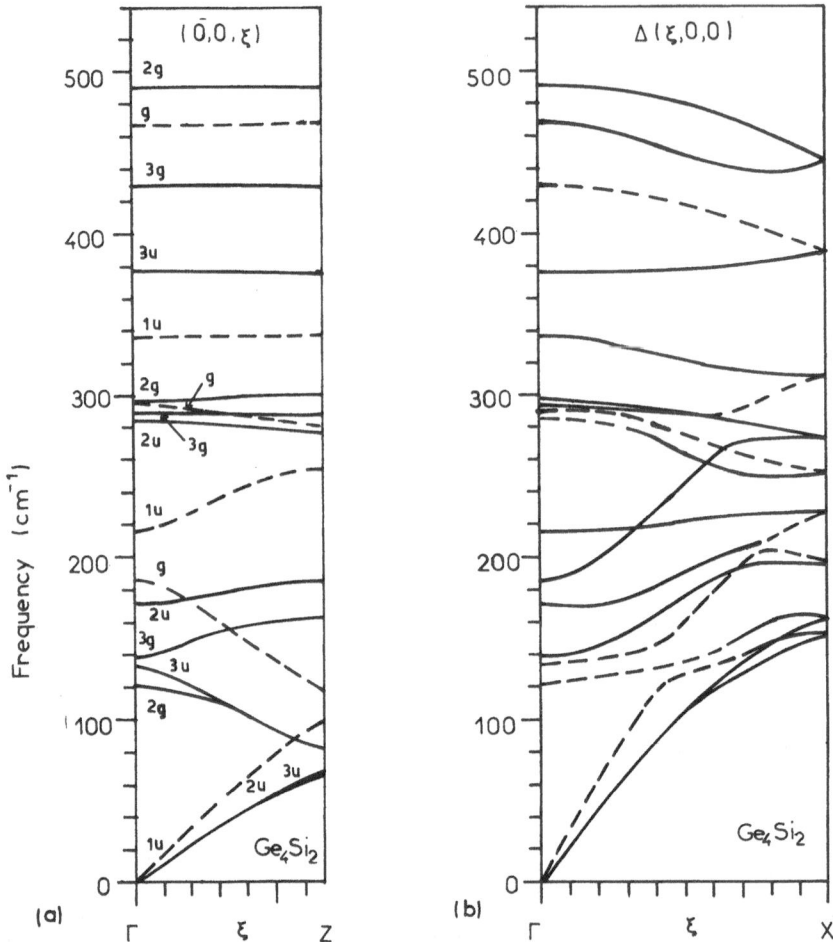

Fig. 9. Calculated phonon dispersion (a) along [0 0 1] and (b) along [1 0 0], for Ge_4Si_2 (D_{2h}^{28} space group) superlattice. Solid and dashed lines represent transverse and longitudinal modes, respectively. (In (b) the phonon polarization is mixed). The labels give the phonon symmetries at Γ point. ξ is the reduced wavevector coordinate.

In conclusion, calculations of the phonon dispersion curves in two directions of high symmetry in the BZ have been presented for several Ge/Si superlattices, which occur in different space groups, depending on the number of monolayers that each element contributes to the superlattice. The calculations are based on a Keating model and the values of the parameters have been chosen to describe experimentally known phonon frequencies of the bulk elements. The method of calculation is simple and well suited to account for the large variety of superlattices that two (or more) given compounds (or elements) can form.

ACKNOWLEDGEMENT

This work has been partially suported by the Greek Ministry of Research and Technology.

REFERENCES

1.   A.Y.Cho and J.R.Arthur,in: "Progress in Solid State Chemistry", J.O.McCaldin and G.Somorjai, eds., Pergamon Press, New York,(1975),10:157.
2.   M.B.Panish, Science, 208:916 (1980)
3.   L.Esaki and R.Tsu, IBM J.Res.Develop.,14:61 (1970)
4.   J.P.van der Ziel,R.Dingle,R.C.Miller,W.Wiegman and W.A.Nordland Jr. App. Phys. Lett.,26:463 (1975)
5.   L.Esaki and L.L.Chang, Phys.Rev.Lett.,33:495 (1974)
6.   K.von Klitzing,G.Doreda and M.Pepper, Phys.Rev.Lett., 45:494 (1980)
7.   D.C.Tsui,H.L.Stormer and A.C.Gossard, Phys.Rev.Lett., 48:1559 (1982)
8.   C.Colvard,R.Merlin,M.V.Klein and A.C.Gossard, Phys.Rev.Lett.,45:298 (1980)
9.   E.E.Mendez,G.Bastard,L.L.Chang and L.Esaki, Phys.Rev.B26:7101 (1982)
10.   M.V.Klein, IEEE J.Quantum Electron.,QE-22:1760 (1987)
11.   See for instance: "Light Scattering in Solids III", M.Cardona and G.Guntherodt, eds., Springer, Heidelberg, (1982) and references therein.
12.   B.Jusserand and M.Cardona,in "Light Scattering in Solids V", Springer, Heidelberg,in press.
13.   H.M.Polatoglou,G.Kanellis and G.Theodorou,in "NATO Advanced Research Workshop on Band Structure Engineering in Semiconductor Microstructures", R.A. Abram, ed., Plenum Press,New York,(1989); Phys.Rev.B,in press.
14.   "Ab Initio Calculation of Phonon Spectra",J.T.Devreese, V.E.Van Doren and P.E.Van Camp, eds.,Plenum Press,New York,(1983).
15.   M.Born, K.Huang, "Dynamical Theory of Crystal Lattices", Oxford University Press, London,(1954).
16.   G.Kanellis, Phys.Rev.B35:746 (1987);Solid St.Commun. 58:93 (1986)
17.   A.A.Maradudin,E.W.Montroll,G.H.Weiss and I.P.Ipatova,in "Solid State Physics", Suppl. 3, H.Ehrenreich,F.Seitz and D.Turnbull,eds.,Academic Press,New York, (1971)
18.   H.Brugger and G.Abstreiter, in Proc.of the 3rd Int.Conf. on Modulated Semiconductor Structures,Montpellier,1987 A.Raymond and P.Voison,eds.,published in J.de Physique Coll.C5,48:321 (1987)
19.   M.I.Alonso,M.Cardona and G.Kanellis, Solid St.Commun. 69:479 (1989); Corrigentum,ibid,70:   (1989)
20.   S.M.Rytov, Akoust.Zh.2:71 (1956) [Sov.Phys.Acoust.2:68 (1958)]
21.   K.Kunc and R.Martin, Phys.Rev.Lett.48:406 (1982)
22.   E.Molinari,A.Fasolino and K.Kunc, Superl.Microstr.2:397 (1986)
23.   A.Fasolino,E.Molinari and J.E.Maan, Phys.Rev.B33:8889 (1986)
24.   B.Djafari-Rouhani,J.Sapriel and F.Bonnouvrier, Superl.Microstr. 1:29 (1985)
25.   R.Merlin,C.Colvard,M.V.Klein,H.Morkoc,A.Y.Cho and A.C.Gossard, Appl. Phys. Lett. 36:43 (1980)
26.   R.E.Camley and D.L.Mills, Phys.Rev.B29:1695 (1984)
27.   S.Yip and Y.C.Chang, Phys.Rev.B30:7037 (1984)
28.   E.Richter and D.Strauch, Solid St.Commun.64:867 (1987)
29.   P.N.Keating, Phys.Rev.145:637 (1966)
30.   R.M.Martin, Phys.Rev.B1:4005 (1970)
31.   G.Dolling,in"Inelastic Scattering of Neutrons in Solids" Vol.II,IAEA,Vienna,(1963) p.37; G.Nilsson and G.Nelin, Phys.Rev.B6:3777 (1972)
32.   "International Tables for Crystallography",T.Hahn,ed. D.Reidel Publishing Company,Dordrecht,Holland (1983)
33.   D.L.Rousseau,R.P.Bauman and S.P.S.Porto, J.Raman Spectroscopy 10:253 (1981)
34.   A.Fasolino and E.Molinari, J.Phys.(Paris) C5:569 (1987)

STRUCTURAL AND OPTICAL PROPERTIES

OF PERIODIC FIBONACCI SUPERLATTICES

D. Paquet, M.C. Joncour, B. Jusserand[*]
F. Laruelle, F. Mollot, and B. Etienne[**]

[*] C.N.E.T. Laboratoire de Bagneux, [**] C.N.R.S.-L2M
196 avenue Henri Ravéra. 92220 Bagneux. France

I) INTRODUCTION

After the discovery of quasicrystals exhibiting fivefold symmetry and thus a lack of periodic long range order[1], a good deal of work has been performed to understand the crystallography of such structures and the nature of their elementary excitations. The main results are that the Fourier transform of the quasilattice (which generalizes the notion of periodic reciprocal space) is made of a dense but countable (d.c.) set of delta functions[2,3], and that the allowed energies for excitations span a Cantor set[4,5]. The simplest models were developped for the paradigmic one dimensional (1D) quasicrystal: the Fibonacci superlattice.

The first realization of an artificial Fibonacci superlattice is due to Merlin and coworkers[6], who grew and studied GaAs/AlAs structures. X-ray[7] and Raman[8,9] scattering experiments gave strong evidence of such d.c. spectra. To our knowledge, only a few studies are available on their electronic and optical properties[10,11].

Our aim in this paper is to understand the building up of such unusual properties by considering a true Fibonacci superlattice as the limit of a series of periodic structures, called Periodic Fibonacci Superlattices (PFS), with unit cells of increasing size and complexity. We shall present and discuss X-ray, Raman scattering and excitation of the luminescence experiments performed on different series of PFS grown, in the GaAs/AlAs system, by molecular beam epitaxy. We shall show that most of these results can be easily explained with the help of an analytical expression of the PFS structure factor.

II) WHAT IS A PERIODIC FIBONACCI SUPERLATTICE?

1) Real space

Let us recall that a Fibonacci sequence[12] S_j of order j is obtained by j successive applications of the transformation rule $A \rightarrow AB; B \rightarrow A$. Starting from $S_{-1}=B$, one obtains successively $S_0=A$, $S_1=AB$, $S_2=ABA$, $S_3=ABAAB$ and so on. One easily sees that $S_j=S_{j-1}S_{j-2}$. The sequence S_j contains P_j elements A and P_{j-1} elements B, P_j beeing the jth Fibonacci number given by the recurrence law $P_j=P_{j-1}+P_{j-2}$ with the seeds $P_{-2}=1$ $P_{-1}=0$. As j increases, the ratio P_j/P_{j-1} converges towards the golden mean $\tau=(1+\sqrt{5})/2$. A Fibonacci superlattice is obtained by piling, along the growth axis, two types of layers A and B, according to the limiting sequence S_∞. Such a superlattice is clearly non periodic.

On the contrary, the jth Periodic Fibonacci Superlattice PFS_j is a standard superlattice with its unit cell constructed according to the sequence S_j. If one denotes d_A and d_B the respective thicknesses of the building layers A and B, the period of PFS_j is $d_j=P_j d_A+P_{j-1}d_B$. The X-ray diffraction pattern for wavevector k along the growth axis is expected to exhibit superlattices reflections at $k_{s,j}=s.(2\pi)/d_j$, where the satellite index s is any positive or negative integer. In the limit where the sequence index j goes to infinity, one recovers a d.c. spectrum.

A very useful trick to visualize these Fibonacci stackings is the so-called projection method[3]. Let us consider a two dimensional (2D) rectangular periodic lattice with primitive translations $\vec{a_j}$ and $\vec{b_j}$ with $a_j=(d_A d_j/P_j)^{1/2}$ and $b_j=(d_B d_j/P_{j-1})^{1/2}$. Draw the straight line D joining the lattice points $(0,0)$ and (P_j,P_{j-1}) and two other lines $D^\uparrow$ and $D^\downarrow$ parallel to D and containing respectively the points $(a+\epsilon,0)$ and $(0,b-\epsilon)$ (see Fig 1a) where ϵ is a vanishing small quantity so that these two lines avoid all the lattice points. The stripe define by these two latter lines contains a unique broken line joining nearest neighbour lattice points, whose projection on D defines a periodic 1D tiling that identifies with PFS_j. The A blocks are the projections of the horizontal links with origins in the upper stripe $S^\uparrow$, and conversely the B blocks are the projections of the vertical ones (origin in the lower stripe $S^\downarrow$) . As j increases, the primitive translations lengths hardly change, but the slope of line D, written in primitive translation units, converges towards $1/\tau$. This is the old geometrical trick deviced by Diophantes[13] to exhibit irrational numbers! In this limit, one recovers after projecting on D the quasiperiodic Fibonacci superlattice.

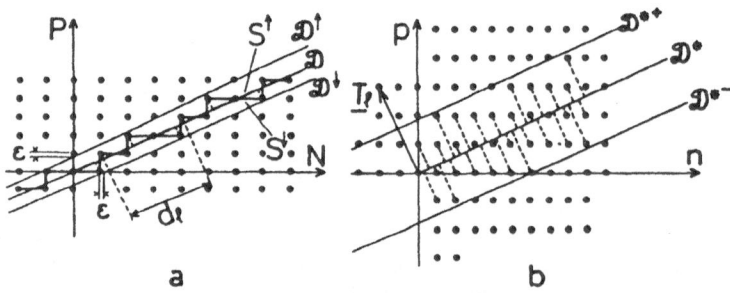

FIG. 1. (a) Geometric construction of PFS_j real lattice from the two-dimensional Bravais lattice. (b) Geometric construction of PFS_j reciprocal space from the two-dimensional reciprocal lattice.

2) Reciprocal space

The construction of the PFS_j reciprocal space is somewhat more subtle: let us first draw the reciprocal lattice of the previous 2D real one (see fig 1b). It is again a rectangular one with primitive vectors $\vec{a_j^*}$, $\vec{b_j^*}$ with $a_j^*=(2\pi)/a_j$ and $b_j^*=(2\pi)/b_j$. The PFS reciprocal space is the line D^*, reciprocal of its real space (line D), perpendicular to the 2D reciprocal vector $\vec{T_j}=(-P_{j-1},P_j)$ and passing through the origin. As shown by Duneau and Katz[3] in the quasiperiodic case, the allowed k vectors are the orthogonal projections of the 2D reciprocal lattice one, labeled by their integer coordinates (n,p), onto the reciprocal line D^*. This defines an absolute labelling of the allowed wave vectors

$$k_{n,p}=\frac{2\pi}{d_j}(nP_j+pP_{j-1})$$

which is associated to the satellite index s through

$$s_{n,p;j}=nP_j+pP_{j-1} \qquad (1)$$

Note that $k_{n,p}$ hardly changes with j, while $s_{n,p,j}$ diverges. Indeed, the projection direction $\overrightarrow{T_j}$ being rational, we get an infinite set of 2D vectors (n,p) with the same projection $k_{n,p}$. If one defines a new stripe S' in reciprocal space delimited by the two lines D'± obtained from D' by the translations $\pm\overrightarrow{T_j}$, one gets a one to one correspondence between the reciprocal points (n,p) lying within S' and the satellite index $s_{n,p,j}$. As j increases, the width of S' increases, thus leading to a proliferation of allowed k vectors (see fig 2). When j goes to infinity, the satellite index does not make anymore sense, and the k vectors become

$$k_{n,p}^{\infty} = \frac{2\pi}{\tau d_A + d_B}(n\tau + p)$$

showing the quasiperiodic character of the true Fibonacci superlattice. Furthermore this provides a hierarchical scheme to classify the k-vectors: let us associate to each 2D reciprocal lattice point (n,p) the lowest Fibonacci sequence index $j_{n,p}$=j1 for which (n,p) is a representative point, i.e. lies within the stripe S_{j1}'. For sequence j1-1, (n,p) is not a representative point but defines a unique 1D wave-vector $k_{n',p'}$ with (n',p') belonging to S_{j1-1}'. With this filiation pattern, the allowed wave-vector $k_{n,p}$ in PFS_{j1} appears as issued from the splitting of line $k_{n',p'}$ in PFS_{j1-1}.

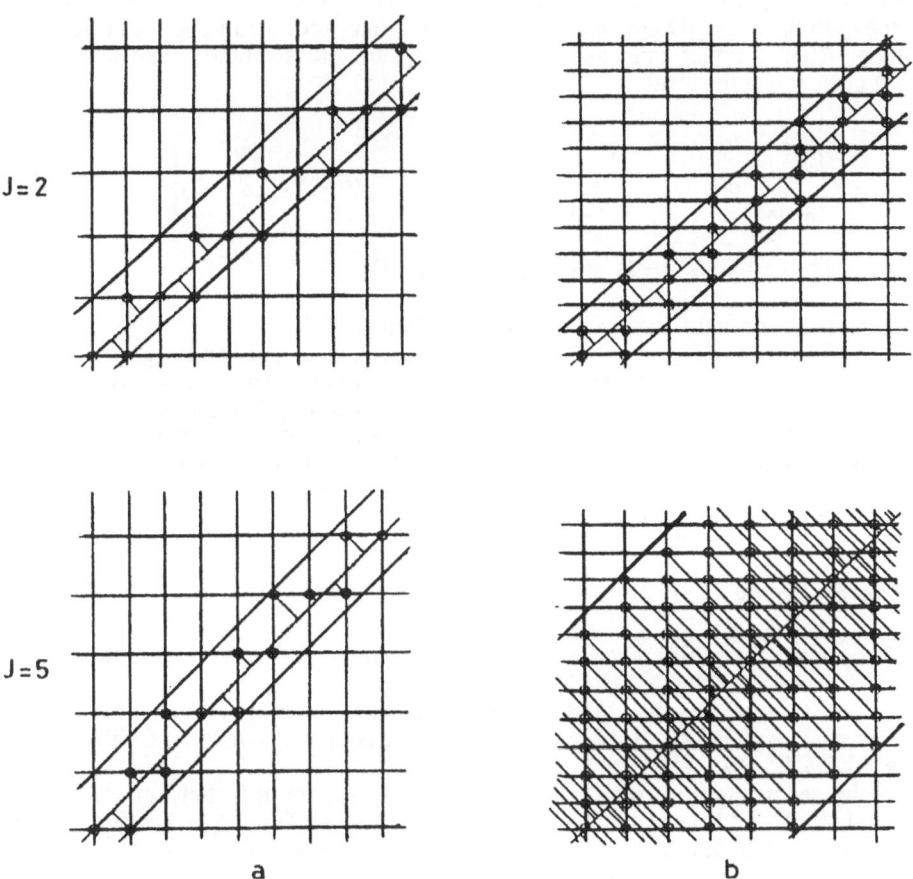

J = 2

J = 5

a   b

FIG. 2. Graphical illustration of the projection method in both (a) real and (b) reciprocal spaces for generations j=2 and j=5. Beside the slight deformation of the two-dimensional lattices and the slight change in the projection directions with increasing j, note the widening of the stripe in the reciprocal space, which leads to a proliferation of allowed one-dimensional wave-vectors.

III) STRUCTURE FACTOR

Let us now compute the Fourier transform $\phi(k)$ of any quantity $f(x)$ exhibiting Fibonacci invariance[14]. $f(x)$ is defined
- by the geometry of PFS_j, that is d_A, d_B and j
- by its value within each tiling layer $f_A(y)$ or $f_B(y)$, where y vanishes at the origin of the corresponding layer.

$f(x)$ could be the electron density, a relevant photoelastic tensor component, or the effective electron potential in an effective mass approximation... Denoting N and P the integer coordinates of the real 2D lattice and $L_{N,P}$ the distance between the coordinates origin and the projection on D of the point (N,P), we get

$$f(x) = \sum_{N,P} \{1_{(N,P) \in S} f_A(x - L_{N,P}) + 1_{(N,P) \in S} f_B(x - L_{N,P})\}$$

The Fourier transform takes the form

$$\phi(k) = \sum_{N,P} \{1_{(N,P) \in S} \phi_A(k) + 1_{(N,P) \in S} \phi_B(k)\} e^{ikL_{N,P}}$$

where ϕ_A and ϕ_B are the Fourier transform of f_A and f_B. Using the standard projection method[3], this can be transformed into a sum over the whole 2D reciprocal lattice. Furthermore, for a finite j, the contributions of all the equivalent vertices (n,p), which are projected onto the same 1D allowed wave-vector $k_{n,p}$ interfere to build up the corresponding amplitude. After a rather long and tedious calculation, these interference terms can be summed up[14], to get the result

$$\phi(k) = \left\{ \sum_{(n,p) \in S_j^*} \delta(k - k_{n,p}) \frac{-i\pi}{d_j \sin((-1)^j k_{n,p} d_{j-1}/2)} \exp(i(-1)^j k_{n,p} d_{j-1}/2) \right\} \times$$

$$\{\phi_A(k)(\exp(ik_B) - 1) - \phi_B(k)(\exp(ikd_A) - 1)\} \qquad (2)$$

This structure factors splits into two multiplicative terms:
- The first one, called sampling function, which depends only on the geometry but not on the physical content of the tiling blocks, selects the allowed wave-vectors $k_{n,p}$ with a given weight roughly inverse proportional to the distance $q_{n,p}$ of the representative vertex (n,p) to line D^*. It generalizes the notion of standard crystals reciprocal lattice.
- The second one, called form factor, does not depend on the generation index j, but only on the characteristics of the two tiling blocks, and generalizes the notion of unit cell structure factor.

IV) EXPERIMENTS

We studied three different series of PFS, grown by molecular beam epitaxy, which differ by the content of the building blocks A and B.

In the first type[15] (series 1), designed for Raman scattering, A and B are made of a single material: $A=(GaAs)_8$; $B=(AlAs)_{12}$; total thickness of the sample $L_t \approx 6300$Å; sequence indices $j=1,2,3,4,5,6,9$.

In the second type[16] (series 2α and 2β), designed for the study of the electronic structure, both A and B are composed of a GaAs well and a $Ga_{1-x}Al_xAs$ barriers. The wells have the same width, but the barriers differ either by their width (case α) or by their height -i.e., Al composition (case β). Their characteristics are:

Series 2α: $A=(GaAs)_{10}(Ga_{0.77}Al_{0.23}As)_{11}$; $B=(GaAs)_{10}(Ga_{0.77}Al_{0.23}As)_{24}$; $L_t \approx 6500$Å; $j=1,2,3,4,5,9$.

Series 2β: $A=(GaAs)_{10}Ga_{0.77}Al_{0.23}As)_{11}$; $B=(GaAs)_{10}(Ga_{0.67}Al_{0.33}As)_{11}$; $L_t \approx 6500$Å; $j=1,2,3,4,7,9$.

1) X-ray diffraction

Although each sample was carefully characterized by X-ray diffraction, we shall just discuss single diffraction spectra on series 1 and 2α. Typical spectra performed around the (0,0,2) GaAs reflection are displayed on fig 3.

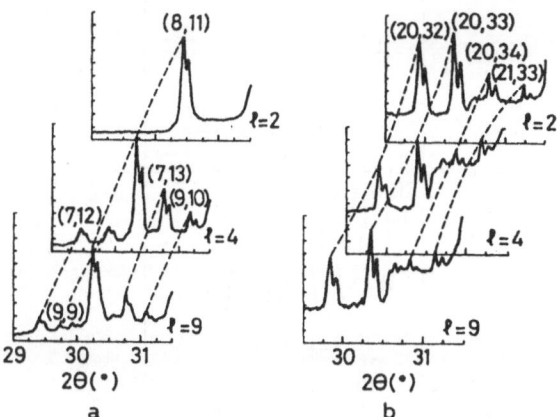

FIG. 3. X-ray single diffraction patterns of Periodic Fibonacci Superlattices l=2,4 and 9, performed around the (002) GaAs reflection. The lines are labelled by the absolute indices (n,p) Each line appears twice, due to the Cu Kα₁,Kα₂ doublet. (a) is series 1. (b) is series 2α.

Three important facts must be underlined[14]:

1) For a given j, the hierarchy in the line amplitudes differ between the two series, as one would expect from the different contents of the building blocks. This comes, in the expression of the structure factor (eqn 2), from the difference in the form factors.

2) Within a given series, the spectra are very similar, with an increasing complexity as j increases. The strongest lines appear already in the early generations, and are rapidly stabilized in strength and position. The new lines appearing in PFS_{j+1}, when compared with PFS_j, are relatively small. The form factor is here the same, but is sampled by sampling functions converging towards the quasiperiodic superlattice one. The increase of complexity corresponds to the enlarging of stripe S*, which selects new vertices in the 2D reciprocal space. These corresponding new lines are associated with vertices whose distance to D* is larger, and thus to smaller sampling amplitudes. The lines can be labelled by the absolute biindices (n,p), and the filiation pattern discussed in section II can be followed (see fig 4).

3) Diffraction patterns around GaAs reflections (0,0,0) and (0,0,2) are very similar (see fig 5). Indeed, if one neglects altogether the small lattice mismatch between GaAs and AlAs and the slight dispersion of the atomic structure factors, it is easy to show from equ 2 that both the sampling function and the form factor become periodic with period $k_0=2\pi/d_0$, where d_0 is the common thickness of GaAs and GaAlAs monolayers. The exact computation of the structure factor of PFS_6 series 1, using equ 2 and including both the strain due to lattice mismatch and the dispersion, fits rather well with the experiments (see fig 5), except for some lines like s=50, which, although forbidden, appear experimentally. This breaking of the Fibonacci selection rules can be shown to be a fingerprint of interface imperfections[14,17].

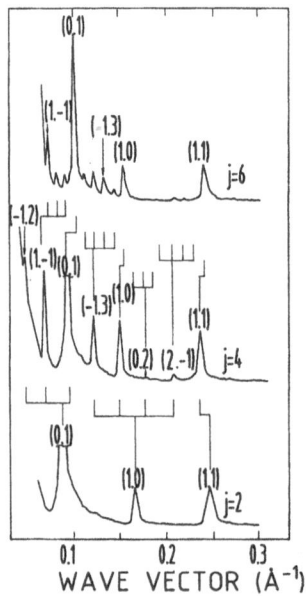

FIG. 4. X-ray diffraction patterns of Periodic Fibonacci Superlattices (j= 2,4 and 6) series 1, performed around the GaAs (0,0,0) reflection. The lines are labelled by the absolute indices (n,p), and their filiation is sketched.

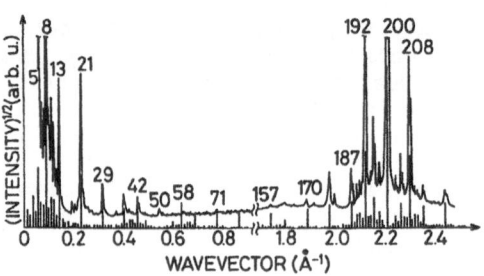

FIG. 5. Experimental X-ray diffraction pattern of PFS_6 series 1 compared with the theoretical prediction deduced from Equ 2. Satellite number 200 nearly corresponds to the (002) GaAs substrate reflection.

2) Raman scattering

Raman scattering experiments[15,18] were performed on samples of serie 1, to study the folded acoustic modes (see Fig 6). Here again the spectra, although somewhat similar, become more bushy as j increases, the more intense lines being already stabilized in frequency and intensity for j=4.

To understand this result, let us first recall the basic physics of light scattering on acoustic waves in superlattices. The longitudinal eigendisplacement field u(z), propagating along the superlattice growth axis z at a given angular frequency Ω, fullfills

$$\frac{\partial}{\partial z}\left[C(z)\frac{\partial u}{\partial z}\right] = -\Omega^2\rho(z)u(z)$$

where C(z) and $\rho(z)$ are respectively the relevant elastic constant and mass density.

On the other hand, the scattering activity of such sound waves is mainly due to the photoelastic process[19] (Brillouin effect and not the conventional Raman one)[20], and the Raman intensity for a Stockes shift Ω and a transferred wave vector q is given by

$$I_q(\Omega) \propto \frac{n(\Omega)+1}{\Omega} \left| \int_{-\infty}^{+\infty} P(z)\frac{\partial u}{\partial z}e^{-iqz}dz \right|^2$$

where P(z) is the relevant photoelastic constant profile, and n(Ω) the thermal Bose factor. This intensity thus reflects the interference of two contributions:
- that of the probed excitation: the displacement field
- that of the response function: the photoelastic profile.

At first approximation, the acoustic impedance modulation of GaAs/GaAlAs superlattices is small. If one neglects it, the eigendisplacements in PFS_j are plane waves propagating at the sound velocity v_j, which can be obtained by simply adding, within a period, the transit time through each layer[21]:

$$\frac{d_j}{v_j} = P_j\frac{d_A}{v_A} + P_{j-1}\frac{d_B}{v_B}$$

Denoting K=Ω/v_j the scattering phonon wave vector, the Raman intensity takes the form

$$I_q(\Omega) \propto \frac{n(\Omega)+1}{\Omega}K^2 \left| \int_{-\infty}^{+\infty} P(z)e^{i(q+K)z} \right|^2$$

and is thus proportional to the squared photoelastic structure factor, which can be computed according to equ 2.

As a consequence, the Raman lines appear as doublets of equal intensity $I_{n,p}$ with Stockes shifts $\Omega_{\pm}=v_j|k_{m,p}-q|$. This provides again an absolute labelling of the lines and a filiation pattern as j increases. This is exemplified in fig 6 where each line of a doublet $\Omega_{\pm}$ is put in correspondence with a vertex of a new 2D reciprocal space derived from the one of section II-2 by the following transformations:
- a rotation has been performed to get an horizontal D*.
- the origin of D* has been shifted by the backscattering transfer wave-vector q.
- the D* coordinate has been multiplied by the sound velocity v_j, in order to get a frequency scale.

One verifies again that the line intensities decrease as the distance of the associated vertex to D* increases, as predicted by the behaviour of the sampling amplitude in equ 2. We thus get a good semiquantitative description of the Raman spectra. A quantitative one could be obtained while taking account of the acoustic impedance modulation which accounts for the asymmetry of the doublet intensities[22].

229

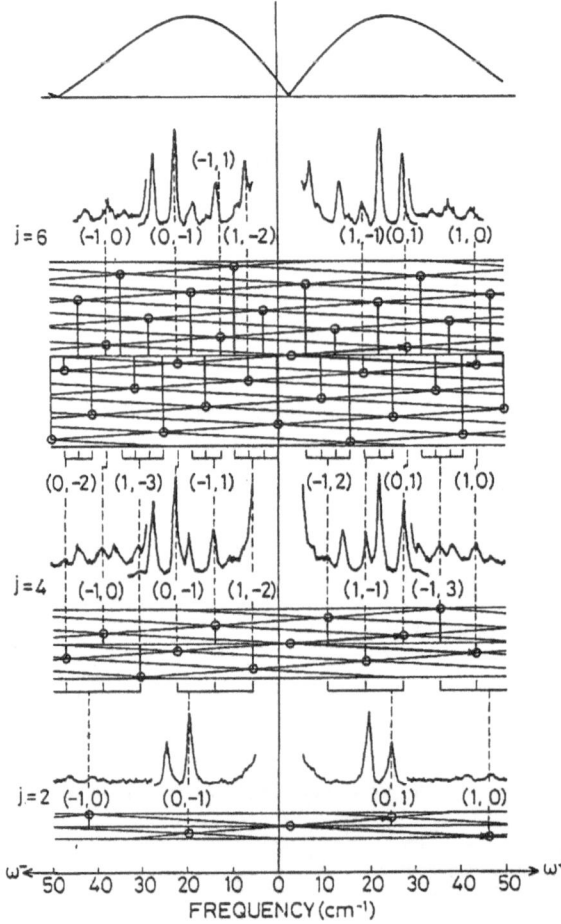

FIG. 6. Graphical comparison between the acoustic Raman spectra of PFS$_j$ for j=2,4,6 and the prediction of formula 2. The Stockes spectra are represented twice on the axis ω^+ and ω^-, which correspond to positive and negative wave-vectors in the photoelastic profile Fourier transform. The lines in each spectra are connected to their corresponding vertices in the two-dimensional reciprocal space, and are labelled by the absolute indices (n,p). The filiation of the lines when j increases is exhibited. The photoelastic form factor is shown at the top of the figure.

3) Excitation of the luminescence

Before presenting experimental results, let us first discuss the electronic structure of PFS series 2, specially the conduction electron one. Both tiling blocks A and B consist of the same GaAs quantum well denoted w and of different GaAlAs barriers b_A and b_B, b_B being always more difficult to tunnel through than b_A. Furthermore the structure has been deviced in order that, whatever its surrounding, the quantum well w binds only one conduction state.

In a very primitive approximation, the PFS 1D conduction hamiltonian takes the form

$$H = -\frac{h^2}{2m^*}\frac{d^2}{dz^2} + V(z)$$

where m^* is the common effective mass in GaAs and AlAs, and $V(z)$ the conduction band minimum profile along the growth axis. Using the plane wave basis $|k>$, we are led to diagonalise the hamiltonian matrix

$$<k_1 | H | k_2> = \frac{h^2 k_1^2}{2m^*}\delta(k_1 - k_2) + V(k_2 - k_1)$$

where $V(k)$ is the Fibonacci Fourier transform of $V(z)$, which can easily be computed using equ. 2. A first estimate of the gaps can be obtained using degenerate first order pertubation theory: degenerate plane waves with wave vectors $\pm k_{n,p}/2$ are coupled by the potential components $V(\pm k_{n,p})$. This leads to a gap value $G_{n,p}=2V_{n,p}$, the 2D vertex belonging to the stripe S_j^*. Generalizing our knowledge on X-ray and Raman scattering, we get the following semiquantitative picture: as the Fibonacci index 1 grows, a hierarchical structure appears for the gaps. They can be labelled by the potential biindices (n,p), the largest gaps appearing already in the earlier generations. Furthermore, as the bands of interest are built with $P_l + P_{l-1}$ quantum wells by unit cell, the relevant k-vectors correspond to satellite indices $s = 1, 2, ... P_l + P_{l-1}$, and thus range between $k_{0,0}$ and $k_{1,1}$ (see equ. 1). When j goes to infinity, the electron k-values at which a gap opens become dense, and the allowed energies form a Cantor set.

Another qualitative picture of the band structure can be obtained from a local point of view[23]. In PFS_1 (unit cell AB), all the wells occur in the sequence $b_B w b_A w b_B$. From these two quantum wells, one can construct a binding state $|2b>$ and an antibinding one $|2a>$. Each of this type of states gives rise, by hybridization through b_B, to a miniband. We thus get two bands separated by the gap $G_{0,1}$. The corresponding allowed energy ranges are labelled 2 and 4 on fig 7.

In PFS_2 (unit cell ABA), one gets the repetition of the three-wells sequence $b_B w b_A w b_A w b_B$ which leads to three bands, separated by the two gaps $G_{0,1}$ and $G_{1,0}$, originating from the three-wells binding $|3b>$, not binding $|3nb>$, and antibinding states $|3a>$ (energy ranges 1, 3 and 5 in fig 7). The unit cell of any PFS_j, for j larger than two, can be constructed with these two new units (the two-wells block and the three-wells one). In case α, none of the energy ranges labelled from one to five overlap, and the whole hierarchy of gaps is issued from binding, not binding or antibinding between similar states centered at increasing distances from each other. In case β, the energy ranges 4 and 5 overlap: this means that one can hybridize states $|2a>$ and $|3a>$ centered on neighbouring blocks. Therefore the gaps around these energies must be rather small. Indeed, a careful calculation, using the transfer matrix method and the scaling properties of the bandwidths as j increases[11], evidence the existence of a single energy E_{deloc} (indicated by a dotted line in fig 7) for which an eigenstate exists for any j. This state is delocalized over the whole structure. It can be shown[24] that the existence of such an extended state is not specific of the Fibonacci piling, but is only due to the commutation, for the specific energy E_{deloc}, of the two tranfer matrices which describe the tunneling through barriers b_A and b_B. An extended state would therefore exist, at energy E_{deloc} in any superlattice built by piling randomly the two blocks A and B. The square of wave functions $|3b>$, $|2b>$ and of the delocalized one are drawn in fig 8, for PFS_6 series 2β.

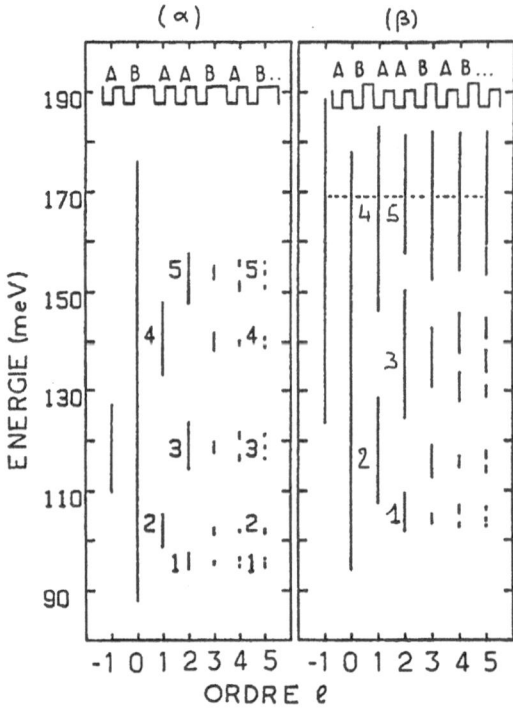

FIG. 7. Conduction allowed energies of PFS$_l$ as a function of the generation index l, for series 2α (left) and 2β (right). The labels 1 to 5 define groups of bands in energy relation with those of PFS$_1$ and PFS$_2$. The dotted line in case β indicates the energy E_{deloc} for which a delocalized states exists in the limiting PFS$_\infty$.

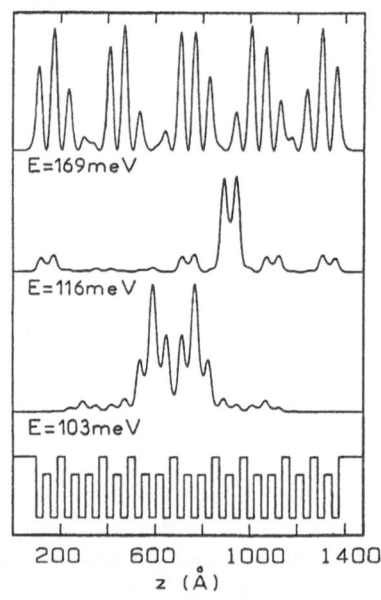

FIG. 8. Square modulus of electronic wavefunctions for three different allowed energies of the finite j=6 Fibonacci superlattice (series 2β). The insert shows the electronic potential profile. Bottom: ground state associated with the energy range labelled 1 in fig 7. It is a binding combination of two adjacent three-wells binding states. Middle: state in the energy range labelled 2, mainly formed of a two-wells binding state. Top: extended state corresponding to the energy E_{deloc}.

We are now in a position to present and discuss the excitation spectra obtained with a dye laser at low excitation (typical power density 100 mW/cm²), which are drawn on fig 9. The structures labelled EHn (respectively ELn) are attributed, according to their energy, to transitions between heavy (respectively light) holes and electrons whose common hybridization in the two or three wells clusters is labelled from 1 to 5 according to the previous discussion (see fig 7).

As expected, the transitions observed in case α for orders l=3,4 and beyond (not shown in the figure) are explained by the transitions for orders 1 and 2. The spectra l=3 and l=4 mainly differ by the onset, for l=4, of a small bump between lines EH3 and EH4 which could be attributed to the upper electron band in energy range 3 obtained by antibinding two non-binding states of two three-wells adjacent clusters. If one agrees with this interpretation, we get an experimental evidence that the electron wave functions remain coherent at least on a six-wells cluster, that is on a distance greater than 320Å. Nothing changes at higher order because the new gaps are small and the oscillation strength of the new transitions are certainly weak.

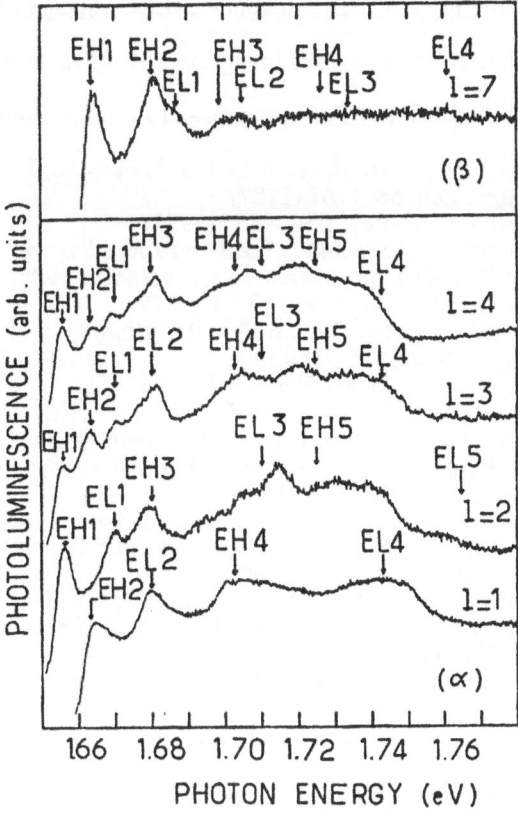

FIG. 9. Photoluminescence excitation spectra of PFS, l=1,2,3,4 series 2α (bottom) and of PFS, series 2ß (top) at T=2K. Arrows indicate calculated electron-heavy-hole (EH) and electron-light-hole (EL) transition energies for l=1 and l=2. Labels refer to fig 7.

In case ß (displayed here only for l=7), there is less structure on the high energy side of the excitation spectrum, in agreement with weaker band gap opening and weaker selection rules due to wavefunction delocalization. Unfortunately we do not get any convincing experimental proof for the existence of the delocalized high energy state.

V) CONCLUSION

Periodic Fibonacci Superlattices appear to be the ideal structures to understand the building up of dense but countable spectra in ideal quasiperiodic Fibonacci superlattices. They form, in the sense of approximation of the golden mean by rational numbers, the best series of periodic approximants to the quasiperiodic superlattice. Due to their periodic nature, the whole set of standard concepts like Brillouin zone and Bloch states remains useful, but notions specific to quasicrystals as the periodic reciprocal superspace are also relevant. In a way, they pave the bridge between standard periodic and quasiperiodic superlattices. We have shown that X-ray and Raman spectra are quantitatively and geometrically understood when using a new analytical expression of the structure factor, which explains the hierarchy of the lines. Excitation of the luminescence experiments revealed a part of the complicated band structure, and allowed to give a lower bound to the electron coherence length.

References

1) D. Shechtman, I. Blecht, D. Gratias and J.W. Cahn. Phys. Rev. Lett. **53**, 1961 (1984)
2) D. Levine and P.J. Steinhardt. Phys. Rev. Lett. **53**, 2477 (1984)
3) M. Duneau and A. Katz. Phys. Rev. Lett. **56**,1168 (1986)
4) M. Khomoto, L.P. Kadanoff and C. Tang. Phys. Rev. Lett. **50**, 1870 (1983)
5) S. Ostlund, R. Pandit, D. Rand, H.J. Shellnhuber E.D. Siggia. Phys. Rev. Lett. **50**, 1873 (1983)
6) R. Merlin, K. Bajema, R. Clarke, F.Y. Juang and P.K. Bhattacharya. Phys. Rev. Lett. **55**, 1768 (1985)
7) T. Todd, R. Merlin, R. Clarke, K.M. Mohanly and J.D. Axe. Phys. Rev. Lett. **57**, 1157 (1986)
8) M.W.C Dharma-warnada, A.H. MacDonald, D.J. Lockwood, J.M. Baribeau and D.C. Houghton. Phys. Rev. Lett. **58**, 1761 (1987)
9) K. Bajema, R. Merlin. Phys. Rev. **B36**, 4555 (1987)
10) R. Merlin, K. Bajema, R. Clarke and J. Todd, in *Proceedings of the 18th International Conference on the Physics of Semiconductors, Stockholm, 1986*, edited by O. Engström (Word Scientific, Singapore, 1987), p 675
11) F. Laruelle and B. Etienne. Phys. Rev. **B37**, 4816 (1988)
12) L. Fibonacci, in *Liber Abaci* (Pisa, 1202)
13) Diophantes, in *Arithmetica* (Alexandria, ≈ 350)
14) D. Paquet, M.C. Joncour, F. Mollot and B. Etienne. Phys. Rev. **B**, in press (15 May 1989). A preliminary version can be found in *Proceedings of the 19th International Conference on the Physics of Semiconductors, Warsaw 1988*, edited by W. Zawadski (Polish Academy of Science 1988) p 357
15) B. Jusserand, F.Mollot, M.C. Joncour, B. Etienne, in *Proceedings of the 18th International Conference on Modulated Semiconductors Structures, Montpellier 1987*, J. de Phys. (Paris). Colloq **48** C5-577 (1987)
16) F. Laruelle, V. Thierry-Mieg, M.C. Joncour and B. Etienne, in *Proceedings of the 18th International Conference on Modulated Semiconductors Structures* Montpellier 1987, J. de Phys. (Paris). Colloq **48** C5-529 (1987)
17) M.C. Joncour, D. Paquet, F. Laruelle, F. Mollot, B. Etienne. Zeitschrift für Kristallographie **185**, 447 (1988)
18) B. Jusserand, D. Paquet, F. Mollot, M.C. Joncour and B. Etienne. Phys. Rev. **B39**, 3683 (1989)
19) C. Colvard, T.A. Gant, M.V. Klein, R. Merlin, R. Fisher, H. Morkoc and A.C. Gossard. Phys. Rev. **B31**, 2080 (1985)
20) B. Jusserand and D. Paquet. Phys. Rev. Lett. in press (May 1989)
21) B. Jusserand and D. Paquet, in *Heterojunctions and Semiconductor Superlattices*, edited by G. Allan, G. Bastard, N. Boccara, M. Lanoo and M. Voos (Springer 1986), p 108
22) B. Jusserand, D. Paquet, F. Mollot, F. Alexandre and G. Leroux. Phys. Rev. **B35**, 2808 (1987)
23) F. Laruelle. Phd Thesis. Université Paris Sud (Orsay) (1988) unpublished.
24) F. Laruelle, D. Paquet and B. Etienne, to be published in *Proceedings of the 3d International Conference on Superlattices, Microstructures and Microdevices (Trieste 1988)*

RAMAN SCATTERING FROM PHONONS IN QUASIPERIODIC SUPERLATTICES

BASED ON GENERALIZATIONS OF THE FIBONACCI SEQUENCE

T.A. Gant[a], D.J. Lockwood[a], J.-M. Baribeau[a],
and A.H. MacDonald[b]

[a]National Research Council, Ottawa, Canada
[b]Physics Department, Indiana University, Bloomington IN

INTRODUCTION

Recently there has been a great deal of interest in the structural, vibrational, and electronic properties of nonperiodic superlattices.[1] This work has been stimulated by the discovery of quasicrystals[2] and the realization that 1-D analogs of quasicrystals could be created artificially in multilayer systems.[3] By far the majority of the work in these systems has concentrated on quasiperiodic Fibonacci superlattices.[4] The Fibonacci structure is a particular case of a class of quasiperiodic structures defined by the recursion relation[5]

$$S_j = (S_{j-1})^n S_{j-2}.$$ (1)

By defining the basic building blocks S_1 and S_2 in terms of layers of different materials and thicknesses we have attached a basis to the quasiperiodic lattice. Table 1 illustrates how the recursion relation (1) is used to build up the first 5 generations in terms of S_1 and S_2.

The structures defined by Eq. (1) are characterized by an irrational number, τ, which may be defined as the ratio of the number of S_2 blocks to S_1 blocks in S_j in the limit $j \to \infty$. For n=1, Eq. (1)

Table 1. The first five generations of the quasiperiodic structures defined by Eq. (1), in terms of the basic building blocks S_1 and S_2.

n=1 (Gold)	n=2 (Silver)	n=3 (Bronze)
S_1	S_1	S_1
S_2	S_2	S_2
$S_2 S_1$	$S_2^2 S_1$	$S_2^2 S_2 S_1$
$S_2 S_1 S_2$	$(S_2^2 S_1)^2 S_2$	$(S_2^3 S_1)^3 S_2$
$S_2 S_1 S_2 S_2 S_1$	$((S_2^2 S_1)^2 S_2)^2 S_2^2 S_1$	$((S_2^3 S_1)^3 S_2)^3 S_2^3 S_1$

defines the Fibonacci sequence with $\tau = (1+\sqrt{5})/2$ (the golden mean). For n=2, $\tau = 1+\sqrt{2}$ (the silver mean), and for n=3, $\tau = (3+\sqrt{13})/2$ (the bronze mean), and in general

$$\tau = (n+(n^2+4)^{1/2})/2. \tag{2}$$

We have carried out Raman scattering measurements on a variety of quasiperiodic Si-Si$_{1-x}$Ge$_x$ superlattices for n=1,2, and 3. It is of interest to study the vibrational spectrum, as it has been shown for periodic superlatices that Raman scattering can provide information on the structure factor which complements that obtained from X-ray diffraction.[6] Raman scattering also allows the investigation of unusual features in the acoustic wave propagation in quasiperiodic systems. The results obtained for systems close to and far from self-similarity are analyzed in terms of a Rytov model for layered systems, and values for the ratio of photoelastic coefficients are deduced.

THEORY

The photoelastic model[6] has been successfully used to model Raman scattering by acoustic phonons in periodic superlattices and in Fibonacci superlattices. The Raman cross section is proportional to

$$\sum_j \frac{1}{\omega_j} \left| \int dz\, e^{-iqz} P(z) \frac{\partial U_j}{\partial z} \right|^2 \delta(\omega-\omega_j)(n(\omega)+1) \tag{3}$$

where P(z) is the photoelastic constant, q is the scattering wavevector, ω is the Stokes shift, and U_j and ω_j are the j'th eigenfunction and eigenfrequency calculated from a Rytov (layered elastic continuum) model. This expression greatly simplifies if we assume that the eigenfunction are plane waves. In this model, the intensity is proportional to

$$\sum_k |P(k)|^2 \omega(n(\omega)+1)\delta(\omega-v|k+q|) \tag{4}$$

where P(k) is a Fourier component of the photoelastic constant and v is the sound velocity of the superlattice. This simpler expression gives equal intensities to the two members ($\pm k$) of a doublet, which is its main drawback. The asymmetry of a doublet will be enhanced when one of its members is near a gap in the density of states (in a periodic superlatice, this means near the zone center or zone edge). In this case, the more complete expression in Eq. (3) is necessary.

In order to calculate P(k) we need to know the structure factor of the quasiperiodic lattice. This can be calculated by expressing the quasiperiodic lattice as a projection from two dimensions.[7,8] We use the projections given by Holzer[9] to obtain the Gold, Silver, and Bronze structures considered in this paper.   We will need to calculate partial structure factors for the two sublattices.   For the Fibonacci case, this can be done by using the inflation symmetry: the separate sublattices are themselves complete Fibonacci lattices.[10]   This is only true for Fibonacci lattices; we must use another method for the Silver and Bronze cases.   This can be accomplished by changing the limits of integration in the projection method, as described below.

The projection method is illustrated in Fig. 1.   The 2-D lattice is defined by

$$L(x,y) = \sum_{m,n} \delta(x-an)\delta(y-bm). \tag{5}$$

Now we define a rotated coordinate system ξ,η:

$$\begin{pmatrix} \xi \\ \eta \end{pmatrix} = \begin{pmatrix} \cos\theta & \sin\theta \\ -\sin\theta & \cos\theta \end{pmatrix} \begin{pmatrix} x \\ y \end{pmatrix}. \tag{6}$$

We take the points of L with $x \geq 0$, $y \geq 0$ in the region $-b\cos\theta < \eta < a\sin\theta$ and project them onto the line $\eta = 0$.   The angle θ is such that $\tau = (a/b)\tan\theta$.   By choosing $\tau = (n+(n^2+4)^{1/2})/2$ we obtain the Gold, Silver, or Bronze structures for n = 1, 2, or 3.   Other values of τ give other quasicrystalline classes; see Holzer.   The blocks S_1 and S_2 have thicknesses $d_1 = a\cos\theta$ and $d_2 = b\sin\theta$, respectively.   If we choose a=b then $d_2/d_1 = \tau$ and the structure is self-similar.   Now consider a line $\eta = a\sin\theta - b\cos\theta$.   This line divides the two sublattices.   The points in the region $a\sin\theta - b\cos\theta < \eta < a\sin\theta$ contribute to the S_1 sublattice;   those in the region $-b\cos\theta < \eta < a\sin\theta - b\cos\theta$ contribute to the S_2 sublattice.

The S_1 and S_2 sublattices are given by the expression

$$Q_i(\xi) = \int_{-\infty}^{\infty} d\eta \, R_i(\eta) \, L'(\xi,\eta) \tag{7}$$

for i=1,2, respectively, where $L'(\xi,\eta)$ is simply L(x,y) with the restriction $m,n \geq 0$ and with x,y expressed in terms of the coordinates ξ,η.   The functions $R_i(\eta)$ in Eq. (7) are defined by

$$R_1(\eta) = \begin{cases} 1, & a\sin\theta - b\cos\theta < \eta < a\sin\theta \\ 0 & \text{otherwise} \end{cases} \tag{8a}$$

and

$$R_2(\eta) = \begin{cases} 1, & -b\cos\theta < \eta < a\sin\theta - b\cos\theta \\ 0 & \text{otherwise} \end{cases} \tag{8b}$$

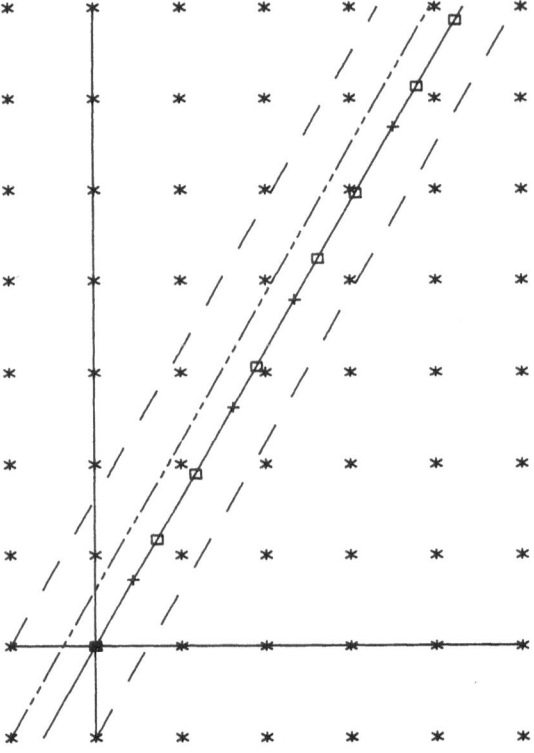

Fig. 1. An illustration of the projection method for the Fibonacci case. The 2-D lattice $L(x,y)$ is denoted by asterisks (*). Points lying between the two dashed lines are projected onto the solid line through the origin. The double dashed line divides the points which form the S_1 and S_2 sublattices. The S_1 sublattice points are denoted by +, and the S_2 sublattice by squares.

The structure factors $S_1(k)$ and $S_2(k)$ are the Fourier transforms of $Q_1(\xi)$ and $Q_2(\xi)$, respectively. Using the method of Zia and Dallas,[8] the Fourier transform of $Q(\xi)$ can be expressed as a convolution integral:

$$S(k) = \int_{-\infty}^{\infty} dp \, R(-p) \, L'(k,p) \tag{9}$$

where the appropriate subscript (1 or 2) should be added to S and R. $R(p)$ is the Fourier transform of $R(\eta)$ and $L'(k,p)$ is the 2-D Fourier transform of $L'(\xi,\eta)$. The results for the structure factors are

$$S_1(k) = \frac{2}{dd'} \sum_{l,m} \frac{1}{p_{lm}} e^{ip_{lm}(\tau-1/2)d'} \sin(p_{lm}d'/2)\delta(k-k_{lm}) \tag{10a}$$

$$S_2(k) = \frac{2}{dd'} \sum_{l,m} \frac{1}{p_{lm}} e^{ip_{lm}(\tau/2-1)d'} \sin(p_{lm}\tau\, d'/2)\delta(k-k_{lm}) \tag{10b}$$

where $d \equiv d_1+\tau d_2$, $d' \equiv (d_1 d_2/\tau)$, $k_{lm} \equiv 2\pi(l+m\tau)/d$, and $p_{lm} \equiv 2\pi(md_1-ld_2)/(dd')$. The wavevectors k_{lm} and p_{lm} are components along ξ and η of a reciprocal lattice vector associated with the original 2-D lattice.

We can now use the structure factor to obtain the Fourier components of the photoelastic constant. We assume that $P(z)$ takes on a functional form $P_1(z)$ in an S_1 layer and $P_2(z)$ in an S_2 layer. Then $P(k)$ takes the form

$$P(k)=P_1(k)S_1(k)+P_2(k)S_2(k) \tag{11}$$

where the form factors $P_1(k)$ and $P_2(k)$ are the Fourier transforms of $P_1(z)$ and $P_2(z)$. Now we need to consider the specific form of S_1 and S_2. We assume that S_1 and S_2 are made up of layers of two different constituents, A and B, as follows: $S_1= \{A,B_1\}$, $S_2= \{A,B_2\}$, where layer A has thickness d_A, layer B_1 has thickness d_{B1}, and layer B_2 has thickness d_{B2}. Then $d_1=d_A+d_{B1}$ and $d_2=d_A+d_{B2}$. If we assume that $P(z)=P_A$ in an A layer and $P(z)=P_B$ in a B layer, then

$$P_1(k) = \frac{1}{\pi k} \left[(P_A-P_B) e^{-ikd_A/2} \sin(kd_A/2)+P_B\, e^{-ikd_1/2} \sin(kd_1/2)\right] \tag{12a}$$

and

$$P_2(k) = \frac{1}{\pi k} \left[(P_A-P_B) e^{-ikd_A/2} \sin(kd_A/2)+P_B\, e^{-ikd_2/2} \sin(kd_2/2)\right]. \tag{12b}$$

Substituting Eqs. (10) and (12) into Eq. (11) gives

$$P(k) = \frac{2}{dd'} \sum_{l,m} e^{-ip_{lm}(\tau+1)d'/2}\, \frac{\sin(p_{lm}(\tau+1)\, d'/2)}{p_{lm}}\, \frac{(P_A-P_B)}{\pi}\, e^{-ikd_A/2}\, \frac{\sin(kd_A/2)}{k}\, \delta(k-k_{lm}) \tag{13a}$$

for $k \neq 0$, and

$$P(k) = \frac{2}{dd'}\, \frac{d'}{4\pi}\, [P_A(1+\tau)d_A+P_B(d_{B1}+\tau d_{B2})]\, \delta(k) \tag{13b}$$

for $k = 0$. Equation (13a) has the form (total structure factor) $\times$ (form factor). This occurs because the A layers in S_1 and S_2 are the same thickness. Note that for $k \neq 0$, $P(k)$ is proportional to P_A-P_B, but for $k = 0$ (i.e. the Brillouin peak) $P(k)$ is proportional to the average value of the photoelastic constant. This allows us to extract information about the ratio of the two photoelastic constants P_A/P_B, provided we obtain good data on the intensity of the Brillouin peak.

To investigate the effect of self-similarity we let $d_2 = \tau d_1$. Then $d' = d_1$ and $p_{lm} = 2\pi(m - l\tau)/d$. We can see in Eq. (13a) that the structure factor is largest when $p_{lm} = 0$. If we let l and m become adjacent numbers in the generalized Fibonacci sequence, then $m/l \approx \tau$ and p_{lm} is small. We define these numbers as follows: $F_j = nF_{j-1} + F_{j-2}$, $F_0 = 0$, $F_1 = 1$. We can establish the following identities:[11] $F_j + \tau F_{j+1} = \tau^{j+1}$ and $F_{j+1} - \tau F_j = (-1)^j \tau^{-j}$. If $l = F_j$ and $m = F_{j+1}$ then $k_{lm} = 2\pi\tau^{j+1}/d$ and $p_{lm} = 2\pi(-1)^j \tau^{-j}/d$. Peaks corresponding to a multiple of a power of τ are also seen; for example $l = 2F_j$, $m = 2F_{j+1}$, $k_{lm} = 4\pi\tau^{j+1}/d$, $p_{lm} = 4\pi(-1)^j \tau^{-j}/d$. For increasing j the structure factor goes to a constant; the Raman intensity is then determined by the form factor $\sin(kd_A/2)/k$ which goes to zero at $k = 2\pi/d_A$.

EXPERIMENT

The samples investigated were all synthesized in a Vacuum Generators V80 MBE system. All samples consist of Si-$Si_{1-x}Ge_x$ on (100) Si substrates. The growth process has been described previously.[12] The prescription given beneath Eq. (11) was used in growing the samples: $S_1 = \{A, B_1\}$, $S_2 = \{A, B_2\}$, where A and B represent Si or $Si_{1-x}Ge_x$ layers. The $Si_{1-x}Ge_x$ alloy layers are typically 20%-30% Ge. For many of the samples there was no B_1 layer, in which case the prescription is $S_1 = \{A\}$, $S_2 = \{A, B\}$. This latter prescription has been used previously for Si-SiGe Fibonacci superlattices.[10] The work on GaAs-AlAs Fibonacci superlattices has tended to use a more complicated prescription where A and B each contain both GaAs and AlAs layers.[3,13] All of the equations above continue to hold with $d_{B1} = 0$.

The structural properties of the various superlattices were studied by double-crystal X-ray diffraction. Rocking curves were recorded in a (+,-) geometry using a (100) Si first crystal and Cu K_α radiation.[14] Rocking curves about the (400) reflection were measured to determine the strain perpendicular to the growth direction. In this geometry the satellite reflections are observed at angles θ_{lm} given by

$$\sin\theta_{lm} = \frac{2\lambda}{\langle a \rangle} + \frac{\lambda}{2d}(l + m\tau) \tag{14}$$

where $\langle a \rangle$ is the average lattice constant.

The Raman spectra were taken at room temperature and were excited with various Ar+ and Kr+ lines. A pseudo-Brewster angle backscattering geometry was used which is close to exact backscattering inside the sample due to the large index of refraction. The incident light was polarized in the plane of scattering. The polarization of the scattered light was not analyzed. The spectrum was analyzed with a Spex 14018 double grating spectrometer. Most of the spectra were taken with a 50 μm slit width to achieve a high resolution. The high quality of the samples allowed us to observe the Brillouin peak under these conditions for most samples.

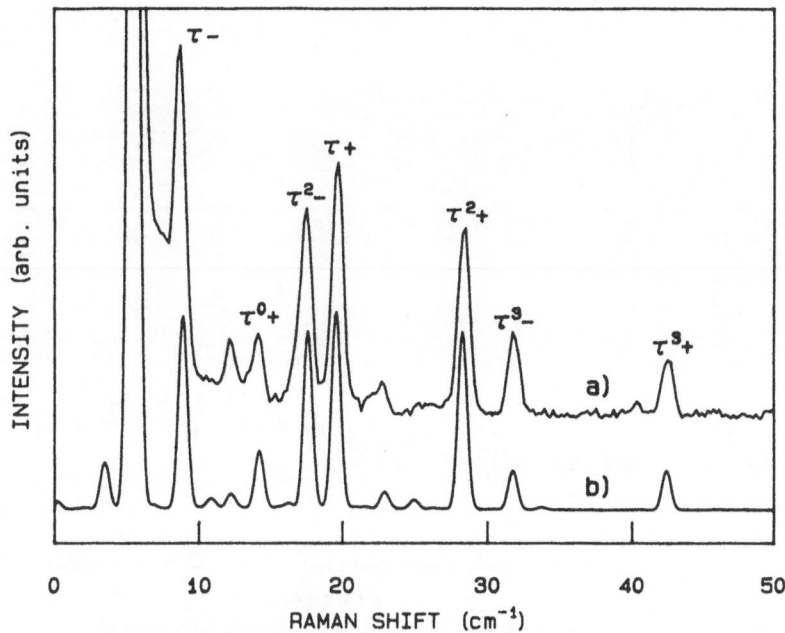

Fig. 2. The (a) experimental and (b) calculated Raman
spectrum of a self-similar Fibonacci superlattice
(sample 352) excited with the 458 nm Ar+ laser line.
The peak labels (τ^p) correspond to $k_{lm} \equiv 2\pi\tau^p/d$.

EFFECTS OF SELF-SIMILARITY ON THE SPECTRUM

<u>Gold</u>

We have examined two Fibonacci superlattices, one close to self-
similarity and one far from self-similarity. These samples were both
grown according to the scheme $S_1 = \{A\}$, $S_2 = \{A,B\}$. Figure 2 shows
measured and calculated Raman spectra on a self-similar Fibonacci
superlattice (sample 352) with A={82 Å Si}, B={52 Å $Si_{0.7}Ge_{0.3}$}. The
layer thicknesses are as measured by transmission electron
microscopy and the Ge concentration is based on growth parameters.
The peaks are labelled as powers of τ rather than with the (l,m)
notation to point out the self-similarity. Peaks corresponding to four
powers of τ ($\tau^0 - \tau^3$) are labelled in the figure. There is a peak at 11
cm^{-1} which is present in the calculated curve but is lost in the noise
in the experimental curve. This peak corresponds to τ^{-1}. Some of the
weaker peaks (not labelled) correspond to multiples of powers of τ;
the doublet at 12.5 cm^{-1} and 23 cm^{-1} corresponds to $2\tau^{-1}$.

The degree of agreement between the calculated and experimental
curves in Fig. 2 is remarkable; the biggest discrepancy is the 11 cm^{-1}
peak. This demonstrates how accurate the simple plane wave model
can be. A possible explanation for the good fit is that the calculated

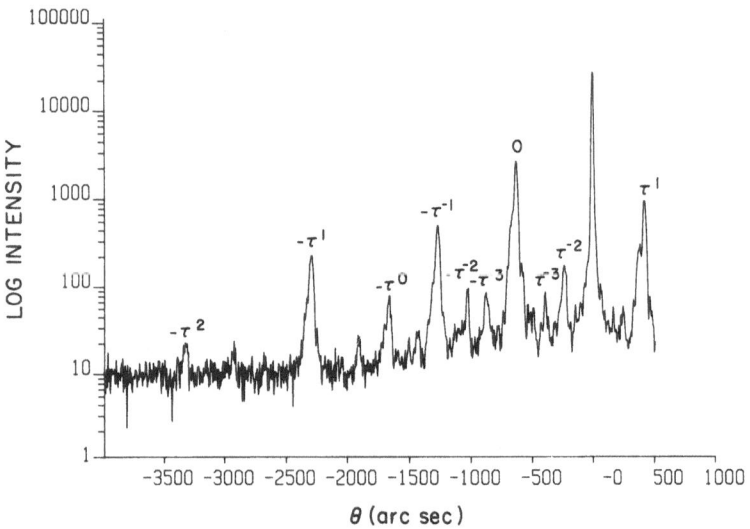

Fig. 3. (400) X-ray rocking curve on a self-similar Fibonacci superlattice (sample 352). The peak labels should be multiplied by τ when comparing with the Raman case (Fig. 2).

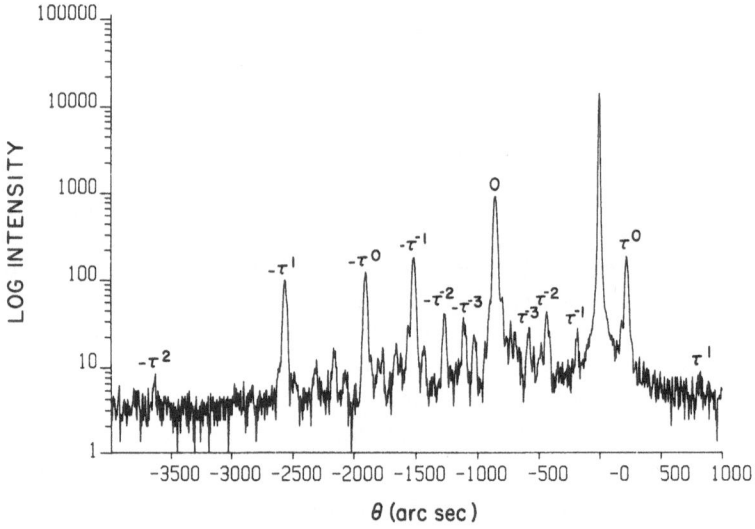

Fig. 4. (400) X-ray rocking curve on a Fibonacci superlattice far from self-similarity (sample 797). The peak labels should be multiplied by τ when comparing with the Raman case (Fig. 5).

spectrum is rather insensitive to small variations in the parameters. The growth parameters $d_A = 84$ Å, $d_B = 48$ Å, and $x = 0.3$ were used as the starting point for the fit. These were treated as adjustable parameters to obtain the best fit. The ratio d_A/d_B was adjusted to match the relative intensities of the peaks (except the Brillouin peak). We obtained several different sets of experimental data on this sample and the parameters d_A, d_B, and x obtained from fitting. Each set varied from 83.5 Å, 55.8 Å, and 0.29 to 81.9 Å, 62.5 Å, and 0.16, respectively. In each case, the quality of the fit was very good.

Figures 3 and 4 show X-ray rocking curves on sample 352 and on a sample far from self-similarity (sample 797). For sample 797, A = {50 Å Si}, B = {80 Å $Si_{.8}Ge_{.2}$} (based on growth parameters). The strong peak (unlabelled) in each curve is due to the substrate. It is striking that these curves look so similar. This provides a demonstration of the fact that the structure factor consists of many densely spaced sharp peaks irregardless of the degree of self-similarity.

Raman data on sample 797 is shown in Fig. 5. The Raman spectrum does not exhibit the rich structure seen for the self-similar sample (Fig. 2). The three strongest peaks in the Raman spectrum — (1,0), (0,1), and (1,1) (corresponding to τ^0, τ^1, and τ^2 respectively) — are also the three strongest peaks in the X-ray data. The fall-off in the Raman intensity to higher frequencies is clearly not due to the

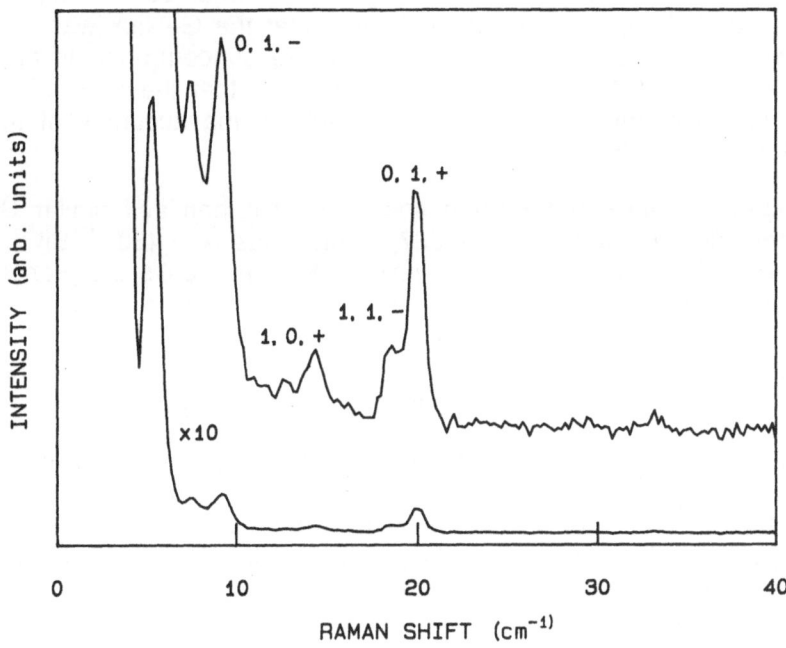

Fig. 5. Raman spectrum of sample 797 recorded with the 458 nm Ar+ laser line. The same curve is plotted on two different scales to show the Brillouin peak at 5 cm^{-1}.

form factor. Since d_A is smaller for this sample (50 Å instead of 80 Å) the zero of the form factor $\sin(kd_A/2)/k$ is pushed out to higher frequency than in sample 352.

Comparing Fig. 5 with Fig. 2, we see that the peaks present in Fig. 5 (except for the peak at 7.5 cm^{-1} which we could not assign) correspond with the strong peaks in Fig. 5. The difference is that the τ^2+ and the $\tau^3\pm$ are missing in the in Fig. 5; otherwise the spectra are very similar. These same peaks are weaker in the X-ray data for sample 797 than for sample 352, which indicates that the drop-off in intensity is due to the effect of the lack of self-similarity on the structure factor.

Silver

Figure 6 shows the Raman spectrum of a self-similar Silver quasiperiodic superlattice, sample 694, for which A = $Si_{0.75}Ge_{0.25}$, d_A = 28.8 Å; B = Si, d_{B1} = 28.8 Å, d_{B2} = 115.2 Å (based on growth parameters). Two powers of τ are exhibited in the spectrum: τ = (0,1) and τ^2 = (1,2). In addition, we see peaks corresponding to (1,1) and (1,3) which also represent a factor of τ. The spectrum is more complicated here; there seem to be two different incommensurate series of peaks showing the powers of τ.

The calculated curve in Fig. 6 agrees very well with the experimental curve. Even the weak features at 14, 38, and 51 cm^{-1} agree well with the theory. The best fit was obtained with d_A = 32.7 Å, d_{B1} = 28.8 Å, d_{B2} = 115.1 Å, indicating that the Ge flux was somewhat higher than intended; thus the Ge concentration is higher than the x = 0.25 that was intended. The fact that the fit is so good is an indication of the sample quality as well as the accuracy of the model used for fitting.

Figure 7 shows the Raman spectrum of a non self-similar Silver quasiperiodic structure (sample 687), which has A = {60 Å Si}, B2 = {44 Å $Si_{0.68}Ge_{0.32}$} (no B1 layer). The same peaks are present as for sample 694. These spectra exibit the same phenomenon as we saw for the Fibonacci superlattices. The same peaks are present for both the self-similar and non self-similar samples, but the spectrum for the self-similar sample extends to higher frequency shifts.

The calculated and experimental curves in Fig. 7 do not agree quite as well as those in Fig. 6. Theory and experiment are in poor agreement for the intensity of the weak features. The (0,1) peaks in the experimental curve show a large asymmetry which cannot be reproduced by the plane wave model.

Bronze

Figure 8 shows a Raman spectrum for a self-similar Bronze quasiperiodic superlattice (sample 690) for which A = $Si_{0.75}Ge_{0.25}$, B = Si, d_A = 28.8 Å, d_{B1} = 28.8 Å, d_{B2} = 168 Å. This spectrum shows two

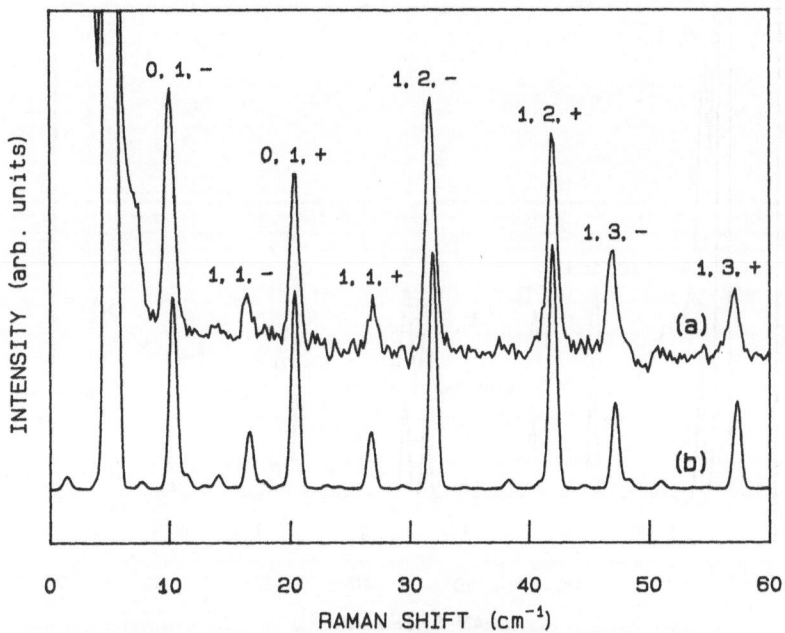

Fig. 6. The (a) experimental and (b) calculated Raman spectrum of a self-similar Silver quasiperiodic superlattice (sample 694) recorded using the 468 nm Kr+ laser line.

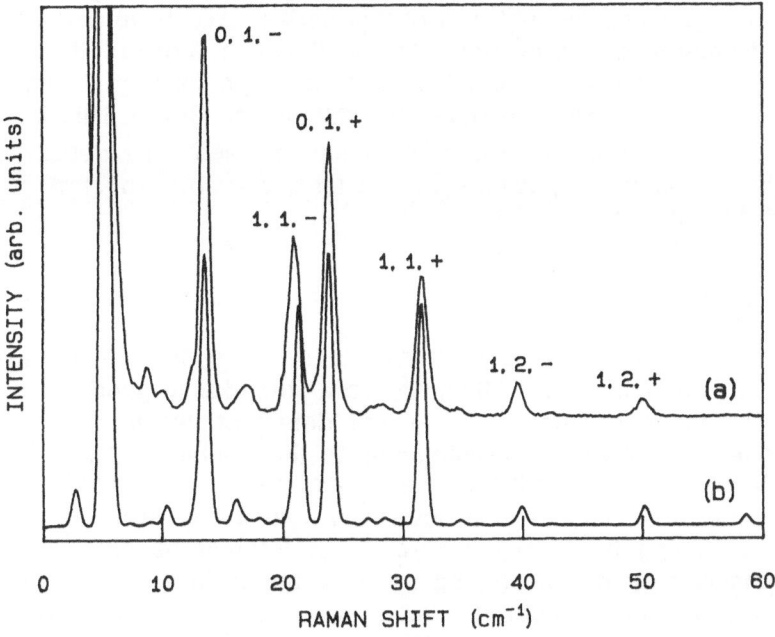

Fig. 7. The (a) experimental and (b) calculated Raman spectrum of a non self-similar Silver quasiperiodic superlattice (sample 687) recorded using the 458 nm Ar+ laser line.

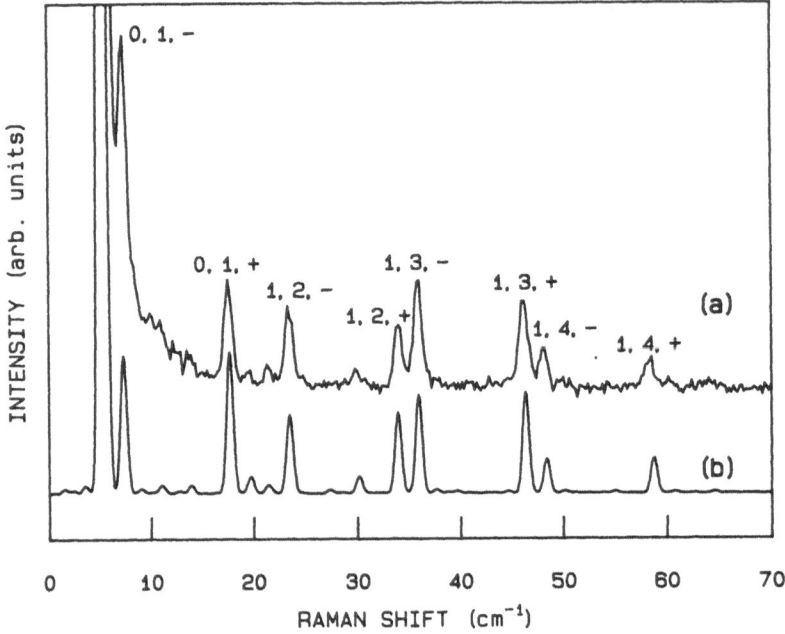

Fig. 8. The (a) experimental and (b) calculated Raman spectrum of a self-similar Bronze quasiperiodic superlattice (sample 690) recorded using the 468 nm Kr+ laser line.

factors of τ: $\tau = (0,1)$ and $\tau^2 = (1,3)$. The peaks at $(1,2)$ and $(1,4)$ are of comparable strength and are not part of this series. We have seen a general trend emerge as we go from Gold to Silver to Bronze: the self-similar Silver and Bronze samples have more peaks that do not follow the power series τ^p than does the self-similar Gold. This is probably due to the larger value of τ for these samples: the density of peaks which do follow the power series is less, so more spectral weight is shifted to other peaks.

WAVELENGTH DEPENDENCE

We have demonstrated that, in many cases, the simple plane wave model we have been using for fitting shows a remarkably good agreement with the experiment. The major defect of this model is the fact that it does not give any asymmetries to the doublets. By analogy with periodic superlattices, we expect these asymmetries to be greatest when one member of a doublet is near a gap in the density of states. Examining these asymmetries is of interest because it provides a possible means of probing the stop bands in the dispersion curve, which were the original motivation for studying quasiperiodic superlattices.[3] We chose sample 687 (the non self-similar Silver) for this study because it exhibited a strong asymmetry for the strongest doublet (0,1). Figure 9 shows Raman spectra on sample 687 at four different laser wavelengths. The top curve, at 458 nm, shows the greatest asymmetry. As the wavelength is increased, the (0,1,+) peak increases in intensity relative to the (0,1,-) peak. Note also the increase in intensity of the (0,1,+) peak relative to the (1,1,-) peak. In the bottom curve, at 514.5 nm, these peaks are on the verge of crossing over, and the (0,1,+) is strongest relative to the (1,1,-) at this wavelength. This is very similar to what happens when approaching the zone center in a periodic superlattice. At the point where two branches cross, there is an anti-crossing behavior; near this point the phonons are standing waves rather than plane waves. The two peaks which are crossing show a strong asymmetry in their intensities due to one peak being Raman active and the other Raman forbidden. In Fig. 9, we do not actually see the peaks cross over because we did not have any laser lines available at the appropriate wavelengths. If the interpretation above is correct then after the peaks cross over the (0,0,+) will be weaker than the (1,1,-); i.e. the structure at 22-23 cm^{-1} will look the same as in the bottom curve, but the two peaks will have switched places. This effect is difficult to see in periodic superlattices because the zone center and zone edge are usually not easily accessible in a Raman experiment.

PHOTOELASTIC CONSTANTS

The fits to the relative intensities of the Brillouin line and the major peaks in the acoustic phonon spectrum provide estimates for

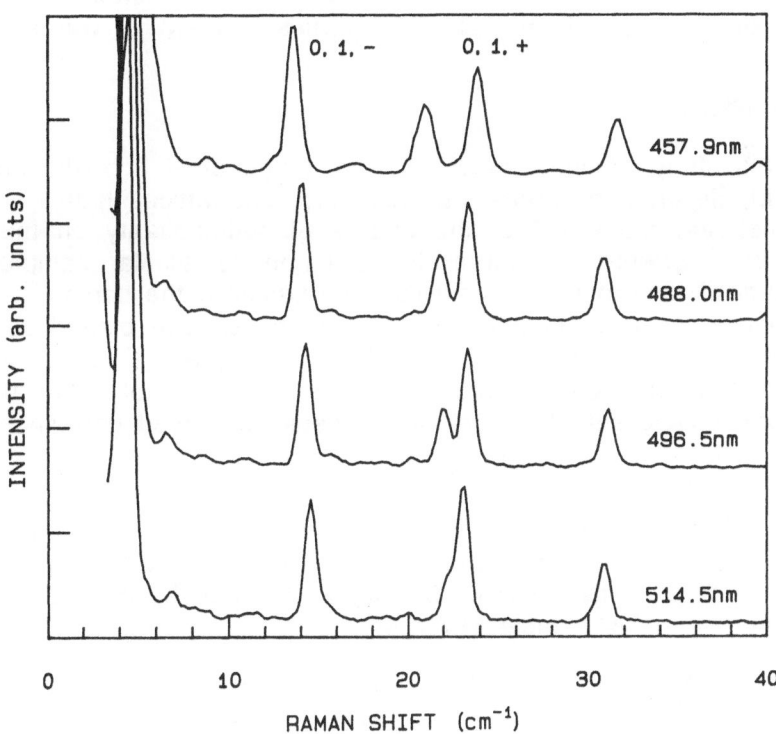

Fig. 9. Raman spectra at four different Ar+ laser
wavelengths on sample 687.

the ratio $P = P_{SiGe}/P_{Si}$ of the photoelastic coefficient of the alloy to that of Silicon for each superlattice. As mentioned previously, the layer thicknesses, the alloy concentration, and P were allowed to vary in the fits. This resulted in greater uncertainty in the fitted parameters, particularly the Ge concentration x, than if some of the parameters were known accurately and thus could be fixed in the fit. We have also included a wide range of peak frequencies in the fit with equal weighting. This will also introduce some error in P, as the lines at higher frequency could have lower intensity than expected due to imperfect superlattice interfaces.[15] From the line intensity analysis of a number of samples with 0.2 < x < 0.45 with exciting wavelengths in the range 457.9—514.5 nm we found that generally speaking the values for P lay in the range 3—9. Such values are consistent with a recent determination[16] of P for one periodic superlattice with x = 0.5.

The wavelength dependence of P for two samples is shown in Table 2. It should be emphasized that the value of P determined from the fits is insensitive to the other parameters. The wavelength dependence on a particular sample thus provides reliable data on the wavelength dependence of P for a particular (but poorly known) value of x.

It can be seen from Table 2 that the ratio of photoelastic coefficients decreases with increasing wavelength. A very steep decrease was found previously for x=0.5.[16] Our results show that as x becomes smaller the wavelength dependence becomes less pronounced and is very slight for $x \approx 0.2$. This behavior is consistent with a strong dependence of the optical properties of $Si_{1-x}Ge_x$ with x.[17]

CONCLUSIONS

This study of light scattering from longitudinal acoustic phonons in Gold, Silver, and Bronze quasiperiodic superlattices has demonstrated the effect of the degree of self-similarity on the spectrum. Spectra of self-similar and non self-similar samples with the same d tend to be very similar, with peaks at the same frequencies. The main difference is that the spectra extend out to higher frequency shifts for the self-similar samples. This is due to the effect of the self-similarity on the structure factor. For the self-similar Gold sample, most of the peaks observed correspond to

Table 2. Wavelength dependence of the ratio P of the photoelastic constant of $Si_{1-x}Ge_x$ to that of Si. The first row is sample 352 and the second row is sample 687.

x	P at laser wavelength (nm)					
	457.9	468.0	476.5	488.0	496.5	514.5
0.2	5.0±0.3		4.2	4.5		
0.3		8.0		8.0	7.4	7.5

powers of τ. As we go from Gold to Silver to Bronze we observe more and more peaks which do not correspond to powers of τ.

In general, the observed spectra are well represented by a photoelastic model within the plane wave approximation. However, when the observed peaks are close gaps in the phonon spectrum the plane wave approximation is no longer appropriate. In this case, a full Rytov model must be used to calculate the phonon eigenfunctions.

The ratio of $Si_{1-x}Ge_x$ to Si photoelastic coefficients was found to lie in the range 3—9 for $0.2 < x < 0.45$ using exciting wavelengths in the range 457.9—514.5 nm. The ratio decreases with increasing wavelength.

REFERENCES

1. See, for example, the review by R. Merlin, IEEE J. Quantum Electr. 24, 1791 (1988).
2. D. Schechtman, I. Bloch, D. Gratias and J. W. Cahn, Phys. Rev. Lett. 53, 1951 (1984).
3. R. Merlin, K. Bajema, R. Clarke, F.-Y. Juang and P. K. Bhattacharya, Phys. Rev. Lett. 55, 1768 (1985).
4. See the review by A. H. MacDonald, in "Interfaces, Quantum Wells, and Superlattices", C. R. Leavens and R. Taylor, eds., Plenum, New York, 1987, p. 347.
5. G. Gumbs and M. K. Ali, Phys. Rev. Lett. 60, 1081 (1988).
6. C. Colvard, T. A. Gant, M. V. Klein, R. Merlin, R. Fischer, H. Morkoç, and A. C. Gossard, Phys. Rev. B 31, 2080 (1985).
7. V. Elser, Phys. Rev. B 32, 4892 (1985).
8. R. K. Zia and W. J. Dallas, J. Phys. A 18, L341 (1985).
9. M. Holzer, Phys. Rev. B 38, 1709 (1988).
10. M. W. C. Dharma-wardana, A. H. MacDonald, D. J. Lockwood, J.-M. Baribeau and D. C. Houghton, Phys. Rev. Lett. 58, 1761 (1987).
11. These identities, given in Ref. 4 for the Fibonacci case, easily generalize to $n \neq 1$.
12. J.-M. Baribeau, T. E. Jackman, P. Maigné, D. C. Houghton, and M. W. Denhoff, J. Vac. Sci. Tech. A 5, 1898 (1987).
13. D. J. Lockwood, A. H. MacDonald, G. C. Aers, M. W. C. Dharma-wardana, R. L. S. Devine, and W. T. Moore, Phys. Rev. B 36, 9286 (1987).
14. J. M. Baribeau, Appl. Phys. Lett. 52, 105 (1987).
15. D. J. Lockwood, J.-M. Baribeau and P. Y. Timbrell, J. Appl. Phys. 65, 3049 (1989) and references therein.
16. J. He, J. Sapriel and H. Brugger, Phys. Rev. B 39, 5919 (1989).
17. J. Humlicek, M. Garriga, M. I. Alonso and M. Cardona, J. Appl. Phys. 65, 2827 (1989).

CAN WE TUNE THE BAND OFFSET
AT SEMICONDUCTOR HETEROJUNCTIONS?

Stefano Baroni[a], Raffaele Resta[a,b],
Alfonso Baldereschi[b,c], and Maria Peressi[a]

[a]*Scuola Internazionale Superiore di Studi Avanzati (SISSA)*
Strada Costiera 11, I-34014 Trieste - Italy

[b]*Institut Romand de Recherche Numérique en Physique des Materiaux (IRRMA)*
Ecole Polytechnique Fédérale de Lausanne
PHB-Ecublens, CH-1015 Lausanne - Switzerland

[c]*Dipartimento di Fisica Teorica dell'Università di Trieste*
Strada Costiera 11, I-34014 Trieste - Italy

The long-standing problem of determining which interface-specific properties affect the band offset at semiconductor heterojunctions is readdressed using a newly developed theoretical approach. The actual interface is considered as a perturbation with respect to a reference periodic system (virtual crystal). By comparison with state-of-the-art self-consistent calculations, we show that linear-response theory provides a very accurate description of the electronic structure of the actual interface in a variety of cases, and sheds light on the mechanisms responsible for the band offset. Results are presented for a number of lattice-matched junctions, both isovalent and heterovalent. It is shown that—within linear response theory—band offsets are genuine bulk properties for isovalent interfaces, whereas they do depend on the atomic structure of the junction for polar interfaces between heterovalent semiconductors. In the latter case, however, the interface-dependent contribution to the offset can be calculated—once the microscopic geometry of the junction is known—from such simple quantities as the lattice parameters and dielectric constants of the constituents. Perspectives for extending the theory to non-lattice-matched systems are also briefly discussed.

I. Introduction

Semiconductor superlattices have attracted a steadily growing attention since they were introduced by Esaki and Tsu in the early seventies[1]. Much of their enormous technological interest stems from their unique transport properties along the growth direction. The key parameter which governs such properties is the valence-band discontinuity across semiconductor heterojunctions[2].

In spite of more than a decade of concerted theoretical and experimental efforts, only very recently have the basic mechanisms responsible for the band offsets begun to be

revealed. Experimentally, spectroscopic and transport properties are often not accurate enough
to discriminate between competing theoretical models, and even give—in some occasions—
conflicting results. From the theoretical point of view, the difficulty basically originates from the
long-range Coulomb interaction which makes the average electrostatic potential of an infinite
system ill-defined[3], and the potential lineup between two semi-infinite systems depending in
principle on the detailed structure of the interface. The question naturally arises whether band
offsets are determined solely by the bulk properties of the two constituents, or if rather there
exist some interface-specific phenomena which affect them and could therefore be used to *tune*
them.

Much of the recent theoretical work in this field is concerned with such vague concepts as
interfacial "bare dipoles", "screening dipoles", "charge transfers", and so on. The purpose
of the present paper is twofold: On one side, we show that a quantitative description of
phenomena occurring at semiconductor interfaces can be achieved—using ideas from macroscopic
electrostatics—without making use of such ill-defined concepts; On the other hand, we develop
a new theoretical tool to get insight into the mechanisms governing those phenomena.

Our approach consists in viewing the actual interface as a perturbation with respect to a
suitably chosen reference periodic system (virtual crystal), and to treat this by linear-response
theory[4]. The present formulation of our theory—as well as the calculations performed so far—
is limited to lattice-matched interfaces; the extension to the lattice-mismatched case is under
way. This approach—besides being quantitatively very accurate for a number of cases where
it has been checked against state-of-the-art self-consistent calculations—sheds new light on the
mechanisms governing potential lineups. It turns out that—within linear-response theory and
in the lattice-matched case—band offsets are genuine bulk properties of the two constituents for
interfaces between isovalent semiconductors and for non-polar interfaces between heterovalent
semiconductors. For polar interfaces between heterovalent semiconductors—on the contrary—
they do depend on the detailed atomic structure of the interface. However, once this structure
is known—either experimentally or by theoretical total-energy minimization—the structure-
dependent contribution to the lineup can be rigorously calculated by elementary electrostatics.

In Sec. II we introduce the basic concepts and terminology about the band-offset problem.
In Sec. III we discuss in detail the concept of "macroscopic average" already introduced by some
of us in order to deal with electrostatic effects in semi-infinite systems[5]. In Sec. IV we present
our linear-response approach to the problem, together with applications to a few paradigmatic
cases. Section V finally contains our conclusions.

II. Band offsets vs. potential lineups

From a theoretical point of view, the band offset across the junction between two semiinfinite
crystals A/B is defined in terms of the *local density of states*, $n(\epsilon, \mathbf{r})$. Far from the interface on
the two sides of the junction, $n(\epsilon, \mathbf{r})$ is a periodic function of $\mathbf{r}$. Averaging with respect to $\mathbf{r}$*
yields the bulk densities of states of the two materials, $g_{A,B}(\epsilon)$. The band offset is *by definition*
the difference between the valence-band edges of the two densities of states, as defined above.
The reason why one has in principle to make use of the local density of states to define the band
offset—and one is not allowed to work directly in terms of integrated bulk densities of states—
is the long-rangeness of the Coulomb interaction. Because of this, the density of states of an
infinite system is defined only up to an arbitrary constant in the energy. This situation is often

---

* The precise meaning of such an average will be revealed in Sec. III.

described associating single-particle energies with "removal energies". As in an infinite system there is no "elsewhere" to remove an electron, it would follow that single-particle eigenvalues are ill-defined quantities. This argument is incomplete, if not wrong. In fact, were the interactions finite-range, the removal energy from a macroscopic but *finite* sample would *not* depend on shape or surface effects and would have therefore a well defined thermodynamic limit. It is

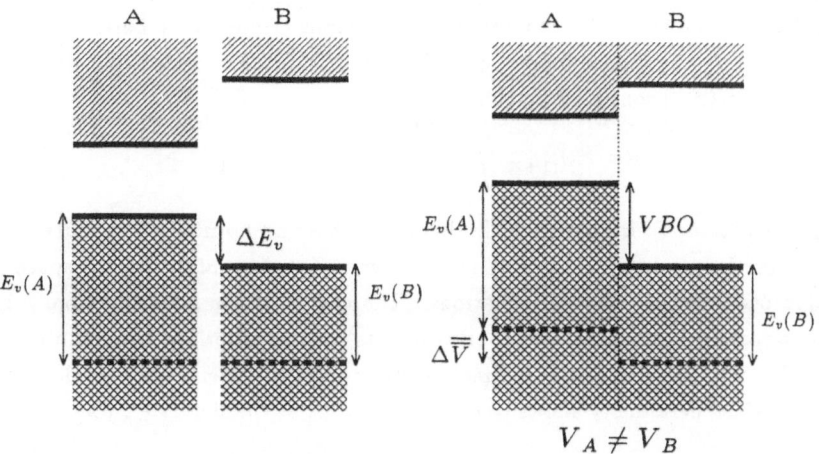

Figure 1. Valence-band offset, VBO, vs. electrostatic potential lineup, ΔV. ΔE_v is the difference between the bulk valence edges, referred to their own average electrostatic potentials.

precisely the long-rangeness of the Coulomb interaction which makes removal energies depend on the detailed structure of the surface[3] (i.e. on the value of the *surface dipole***) and hence ill-defined in the thermodynamic limit. Because of this, it is convenient to split the valence-band offset, VBO, into two contributions: the first one, ΔE_v, is characteristic of the two individual bulks; the second, ΔV, depends in principle on the actual structure of the interface:

$$\text{VBO} = \Delta E_v + \Delta V, \tag{1}$$

where ΔE_v is the difference between the two band edges, when the single-particle eigenvalues are measured with respect to the (arbitrary) average of the electrostatic potential in each bulk material, while ΔV is the difference between the average electrostatic potentials on the two sides of the interface. The lineup ΔV, which is a macroscopic electrostatic quantity, contain all of the interface effects; according to Eq. (1), the remaining microscopic quantum effects are all embedded in ΔE_v. The latter quantity can be obtained from two distinct bulk calculations; in particular, many-body effects on the quasiparticle spectra only affect ΔE_v[6,7]. As we are interested here into interface-specific phenomena, we will pay little attention to such effects.

** In Sec. III we will give an unambiguous meaning to the concept of interface dipole.

The relevant information from the interface local density of states is contained in a single figure: the lineup of classical electrostatic potentials, ΔV, on the two sides of the junction. The splitting given in Eq. (1) is to some extent arbitrary, as short-range contributions to the average potential can be included in ΔE_v as well as in ΔV. In our applications, we choose to include exchange-correlation and short-range electrostatic contributions to the offset in ΔE_v, whereas ΔV is defined as the lineup of the electrostatic potential generated by the electron charge distribution, plus a neutralizing uniform background.

III. The macroscopic average

IIIa. Basic theory. The concept of macroscopic average is a basic one in classical electromagnetism. Any macroscopic quantity $f^{(macro)}(\mathbf{r})$ is related to its microscopic counterpart $f^{(micro)}(\mathbf{r})$ through a convolution:

$$f^{(macro)}(\mathbf{r}) = \int w(\mathbf{r} - \mathbf{r}') f^{(micro)}(\mathbf{r}') \, d\mathbf{r}', \qquad (2)$$

where $w(\mathbf{r})$ is real, nonzero in some neighborhood of $\mathbf{r} = 0$ and normalized to unity over all the space[8]. It is a trivial matter to show that such an averaging commutes with the space and time differentiation occurring in Maxwell equations. The choice of the filter function w is largely arbitrary; however each macroscopic problem has its own appropriate lower limit of relevant lengths and this sets the size of the w function to be used[9]. When studying electrostatics in crystalline materials, periodicity allows to take length scale as small as $\Omega^{1/3}$, where Ω is the unit cell volume.

In order to illustrate this, let us consider the textbook example of dielectric polarization in a constant field. The induced charge and field show microscopic (or local-field) oscillations in the bulk region; these however are lattice periodic and can be easily washed out to recover the macroscopic picture, where the bulk polarization is uniform. The filter function doing the job is the characteristic function of the unit cell: $w(\mathbf{r}) = 1/\Omega$ inside the unit cell and vanishing outside. Maximum symmetry can be imposed choosing a Wigner-Seitz cell. After such macroscopic filtering, the induced charge in a uniformly polarized crystalline sample can be nonzero only in the neighborhood of the surface, while decaying to zero in the bulk. The decay length is the longest amongst the following two: (a) the length scale of the filter, $\sim \Omega^{1/3}$; (b) the typical screening length appearing in the microscopic dielectric response of the given material. Incidentally we notice that (a) and (b) happen to be of the same order in typical covalent materials[10].

Let us now come back to the semiconductor heterojunction problem which is our main concern here. Matters are simple for interfaces between two lattice-matched materials, where the above defined characteristic function w is material-independent. Macroscopically averaged quantities (charges, fields, potentials) show no microscopic oscillations on either side of the interface and recover the macroscopic limit in the two bulks. Conversely, deviations from the macroscopic value define the interface region.

Before the concept of macroscopic average was introduced into the heterojunction problem by some of us[5], there has been a long-standing confusion about the nature of the interfacial dipolar charge distribution which gives rise to the potential lineup. A "screening" dipole was defined as originating from "charge transfer" across an idealized interface[11]. This idealized "bare" configuration was then defined as the simple juxtaposition of slabs of bulk materials. Of course, there is no clearcut way of defining a boundary between two materials (not even a "bare" one), the only physically meaningful things being the chemical nature and positions of the ions. As a matter of fact, different geometries (such as the shape of the idealized "bare" interface, or even its crystallographic orientation) give rise to different screening dipoles, whereas the potential lineup turns out to be quite independent of them. We conclude that such dusky concepts as "screening dipoles" or "charge transfers" are unnecessary, and even misguiding since they lead to attribute a physical meaning to the ideal reference interface which instead has none. The macroscopic average, on the contrary, allows to define an interface dipole and to blow up any genuine interface features using the interface charge as the only ingredient, with no subtraction and no choice for boundary. The interface charge profile so obtained is exactly related, via the Poisson equation, to the potential lineup.

IIIb. **Macroscopic averages from slab-adapted unit cells.** We show in Fig. 2 contour plots of the electronic valence charge of $(GaAs)_3(AlAs)_3$ (001) [12]. This supercell has two equivalent interfaces: one at the figure center and one at the figure borders. At this scale, the presence of interfaces is hardly detectable.

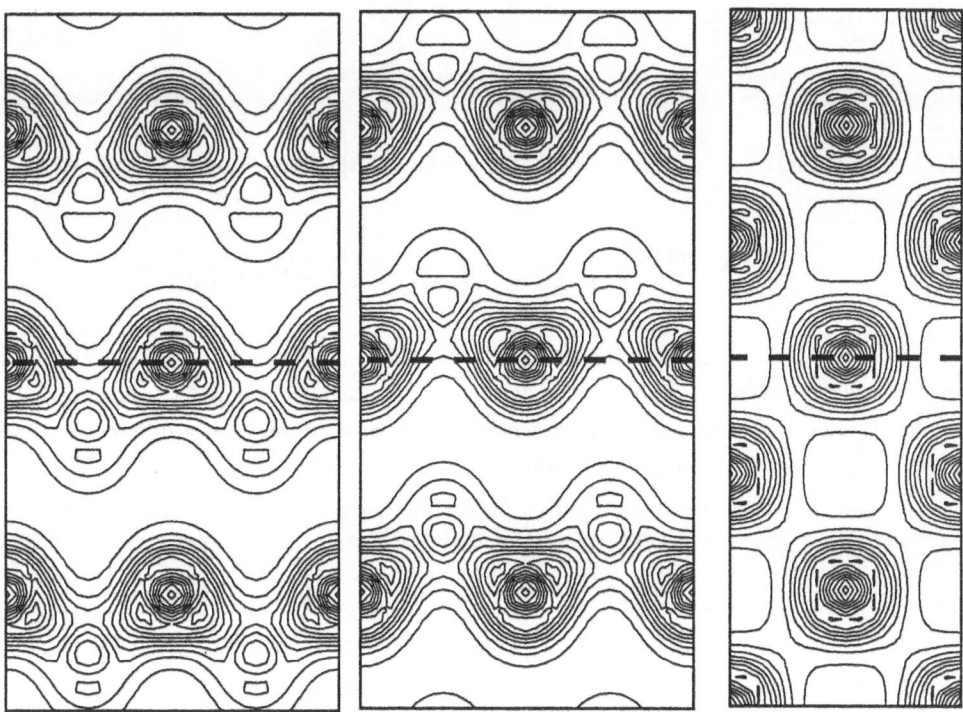

Figure 2. Charge density of $(GaAs)_3(AlAs)_3$ (001). The plots are centered at an interface anion; $GaAs$ down; $AlAs$ up. Left: (110) plane; middle: $(\bar{1}10)$ plane; right: (010) plane.

Since this geometry is periodic in the (x, y) planes, the first obvious simplification is to consider planar averages as functions of the z-coordinate only: $\bar{f}(z) = \frac{1}{S} \int_S f(x, y, z) dx dy$. From the three-dimensional electronic density we get the one-dimensional density $\bar{n}(z)$ and potential $\bar{V}(z)$ shown in Fig. 3a. This shows two different—though closely similar—periodic functions on the two sides of the interface, which join smoothly across it. Since this is a lattice-matched case, there is a unique linear period a. Any interface effect is due to the *difference* between these periodic functions; at this scale this is barely visible, being almost masked by bulk oscillations.

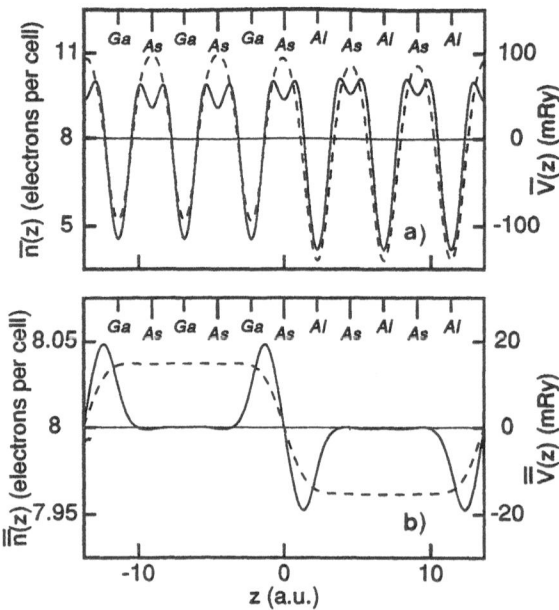

Figure 3. Planar (a) and macroscopic (b) averages of the charge density (solid lines) and the electrostatic potential (dashed lines) of $(GaAs)_3(AlAs)_3$ (001).

The next step is to pass the function $\bar{n}(z)$ (or $\bar{V}$) through the (one-dimensional) filter defined by the convolution:

$$\bar{\bar{n}}(z) = \frac{1}{a}\int_{z-a/2}^{z+a/2} \bar{n}(z')\ dz' = \frac{1}{a}\int \Theta\left(\frac{a}{2} - |z - z'|\right)\bar{n}(z')\ dz', \tag{3}$$

where Θ is the step function. The results are shown in Fig. 3b. As for the charge, the figure shows, at a magnified scale, a dipole profile, localized in the neighborhood of the interface. The electrostatic potential varies across the interface from a constant value to a different one, the lineup Δ being related to the dipole moment of the charge profile:

$$\Delta V = \frac{e^2}{4\pi}\int z\big(\bar{\bar{n}}(z) - n_o\big)\ dz, \tag{4}$$

where n_o is the average electronic density of the two bulks (eight electrons per cell, in the case of elemental or binary semiconductors). We note that Eq. (4) allows to define in an unambiguous way the concept of *interface dipole* for any interface, and also, in particular, for a surface.

IIIc. Three-dimensional macroscopic averages. The macroscopic averages shown so far have been built in two steps: first the (x,y) planar average and then the one-dimensional convolution. It is easy to realize that this construction is equivalent to the three-dimensional convolution of Eq. (2), where the filter function $w(\mathbf{r})$ has been taken as the characteristic function of a suitably shaped (slab-adapted[3]) unit cell. Other choices are possible, all giving the same macroscopic values but different resolution and different shapes of the interface charge. In the previous section we have obtained the planarly averaged interface profile; suppose instead we want to investigate on which *sites* this dipole charge is located. To this aim, we need (x,y) resolution: the natural and unbiased choice for w is to use the characteristic function of the bulk Wigner-Seitz cell as a filter function.

Starting again from the three-dimensional charge of Fig. 1, we obtain the macroscopically averaged charge shown in Fig. 4, using a magnified spacing between contours. This kind of macroscopic average goes—as the previous one—to the constant value n_o in the two bulk regions; it deviates from n_o in the neighborhood of the interface, but this deviation is no longer (x, y) independent. Instead it is periodic in the (x, y) planes and shows very clearly that the dipoles responsible for the lineup are localized at the interface anions (and nearby bonds) for this (001) geometry. Such a localization can be defined only to a resolution of $\Omega^{1/3}$, intrinsic to the macroscopic average itself.

Figure 4. Three-dimensional macroscopic average of the electron density in $(GaAs)_3/(AlAs)_3$ (001), projected onto the (110), ($1\bar{1}0$), and (010) planes. A characteristic function of the bulk Wigner-Seitz cell is used in this case as a filter function. The shaded area indicates regions where the density is lower than eight electrons per cell. $GaAs$ down, $AlAs$ up.

IIId. Extension to the lattice-mismatched case. We consider here the more general case of an interface between two non lattice-matched materials, A and B. Keeping the same attitude as before, we look for a filter function w which is as localized as possible, although still able to recover macroscopic electrostatics in the bulk regions of both materials. The solution consists in filtering two times, using the functions w_A and w_B appropriate to each material in turn. This double filtering can be recast in terms of the single filter function:

$$w(\mathbf{r}) = \int w_A(\mathbf{r} - \mathbf{r}')w_B(\mathbf{r}') \, d\mathbf{r}', \tag{5}$$

which explicitly shows commutativity. We notice at this point that w_A is localized within $\sim \Omega_A^{1/3}$ and w_B within $\sim \Omega_B^{1/3}$, while w is localized within $\sim \Omega_A^{1/3} + \Omega_B^{1/3}$. Since usually $\Omega_A^{1/3} \sim \Omega_B^{1/3}$, we loose resolution by about a factor of two with respect to the lattice-matched geometry. This is however the best one can do in this case. For the special case A=B, the use of w is *not* equivalent to the use of w_A. Both give of course the same macroscopic physics, but w_A gives better resolution (*i.e.* faster convergence to bulklike values) in the interface region.

IV. Linear-response theory of the potential lineup

IVa. **General considerations.** The ill-defined nature of the average electrostatic potential of an infinite solid was fully acknowledged only quite recently. The reason has to be searched in a widespread conceptual attitude: before self-consistent calculations were made possible by modern computers and computational techniques, it was customary to think of infinite systems as made up by *rigid* atomic-like charge distributions:

$$n_{cryst}(\mathbf{r}) = \sum_{\mathbf{R}} n_{loc}(\mathbf{r} - \mathbf{R}). \tag{6}$$

It is easy to recognize that—under the hypothesis of Eq. (6)— the average potential inside the crystal is indeed a well defined constant, provided the localized charge distributions n_{loc} are neutral, carrying no dipole nor quadrupole. The precise meaning of such an affirmation is that the potential drop across the surface of any semiinfinite sample would not depend on the details of the surface structure. Intuitively, this is so because the prescription on the localized charge distribution implicitly fixes all the relevant details of the surface density profile (i.e. of the "surface dipole"). From a more formal point of view, we observe that the average potential of any finite system can only be defined through a limiting procedure:

$$\langle V \rangle = \lim_{\mathbf{q} \to 0} \frac{4\pi e^2}{q^2} \tilde{n}(\mathbf{q}). \tag{7}$$

Of course—in a periodic solid—Eq. (7) is meaningless because the lattice structure factor makes the Fourier coefficients of the crystal density to be defined only at *discrete* wavevectors. However, any *prescription* to interpolate between the discrete physical values of $\tilde{n}_{cryst}(\mathbf{q})$ with a continuous function allows to calculate the limit, provided it exists[13]. A recipe to decompose the crystal density into a sum of localized densities is in fact equivalent to a prescription to interpolate between its Fourier coefficients, and allows to give a precise meaning to Eq. (7). Such a limit exists provided the long-wavelength behavior of the localized charge distribution is $\tilde{n}_{loc}(\mathbf{q}) \sim \alpha q^2 + \mathcal{O}(q^3)$. The absence of constant, linear, and anisotropic quadratic terms in the long-wavelength behavior of n_{loc} is summarized by the property that it is neutral, bearing no dipole nor quadrupole***. Of course, a recipe to calculate Eq. (7) would also solve the lineup problem: the potential lineup across a semiconductor interface would simply be the difference between the average potentials of the two infinite solids, as calculated from Eq. (7)[13]. It is important to notice that such a procedure provides a value of the potential lineup and of the band offset which is independent of the interface orientation, of its detailed structure (i.e. its abruptness), and which also display transitivity (i.e. the offset between two materials AC is the sum of the AB and BC offset).

Two models have been recently proposed for the band-offset problem which exploit the idea of Eq. (6). In the first one, the basic building blocks of the infinite solid are assumed to be neutral atomic charge distributions[14]; in the second one—put forward by some of us for common-anion (or common-cation) heterostructures[5]—the elementary bricks are postulated to be cation-centered (or anion-centered) Wigner-Seitz bulk cells. The first approach proved to be

---

*** For further details, see the discussion of Sec. IVb.

successful in a great number of cases, albeit its accuracy (~ 0.1 eV) is limited by the crudeness of the assumptions it relies on. The second one was found to be much more accurate in the case of $GaAs/AlAs$ interfaces (~ 0.01 eV), but it is of less general applicability.

We showed above how a suitable decomposition of the bulk charge densities into localized distributions makes the band lineup independent of interfacial details. The key observation leading to formulate the interface problem in terms of linear-response theory is that—in order to have such an independence— it is not necessary to decompose the full bulk densities according to Eq. (6). It is instead sufficient that such a decomposition holds for the *difference* between the actual bulk densities and a suitable reference crystal (which may also be one of the two bulks themselves). In order to minimize such a difference, we choose the reference crystal to be the *virtual crystal* whose ionic pseudopotentials are the averages between the ionic pseudopotentials of the two bulks:

$$V_{virt}^{o}(\mathbf{r}) = \sum_{\mathbf{R}} \left(\langle v_C(\mathbf{r} - \mathbf{R}) \rangle + \langle v_A(\mathbf{r} - \mathbf{R} - \boldsymbol{\delta}) \rangle \right), \qquad (8)$$

where $\langle v_A \rangle = \frac{1}{2}(v_{A_1} + v_{A_2})$ is the average of the anionic potentials, and analogously $\langle v_C \rangle$ is the average cationic potential; $\boldsymbol{\delta}$ is the distance between the anionic and cationic sublattices. The bare perturbation leading from the virtual crystal to the real ones is:

$$\Delta V_{1,2}^{o}(\mathbf{r}) = \pm \sum_{\mathbf{R}} \left(\Delta v_A(\mathbf{r} - \mathbf{R}) + \Delta v_C(\mathbf{r} - \mathbf{R} - \boldsymbol{\delta}) \right), \qquad (9)$$

where $\Delta v_{A,C} = \frac{1}{2}(v_{A_1,C_1} - v_{A_2,C_2})$, and the upper (lower) sign refers to crystal 1 (2).

Figure 5. Electron-density maps of $(GaAs)_3(AlAs)_3$ (001) projected onto the (010) plane (left panel), together with the corresponding virtual-crystal density (mid panel), and their difference (right panel). The latter has been plotted with a magnified level spacing ($\times 40$). *GaAs* down; *AlAs* up.

In Fig. 5 we compare the electron-density map of $(GaAs)_3(AlAs)_3$ (001) projected onto the (010) plane, with the corresponding virtual-crystal density. The smallness of the difference between the density of the superlattice and that of the virtual crystal suggests that the superlattice can be accurately described using low-order perturbation theory starting from the virtual-crystal solution. The charge densities of crystals 1 and 2 are

$$n_{1,2}(\mathbf{r}) = \langle n(\mathbf{r}) \rangle + \int \chi(\mathbf{r}, \mathbf{r}') \Delta V_{1,2}^{o}(\mathbf{r}') d\mathbf{r}' + \mathcal{O}(\Delta V^2), \tag{10}$$

where $\langle n \rangle$ and χ are the electron density and density-response function of the virtual crystal. According to Eqs. (9) and (10), the differences between the charge density of the two bulks and that of the virtual crystal is indeed—to linear order in the difference between the bare potentials—the sum of localized distributions:

$$\Delta n(\mathbf{r}) \approx \sum_{\mathbf{R}} (\Delta n_C(\mathbf{r} - \mathbf{R}) + \Delta n_A(\mathbf{r} - \mathbf{R} - \boldsymbol{\delta})) \tag{11a}$$

$$\Delta n_{A,C}(\mathbf{r}) = \int \bar{\chi}(\mathbf{r}, \mathbf{r}') \Delta v_{A,C}(\mathbf{r}') d\mathbf{r}'. \tag{11b}$$

In the following, we examine in detail the consequences of Eqs. (11), in order of increasing complexity: we consider first the case of a common-anion (or common-cation) interface (such as $GaAs/AlAs$); then we study the case where both cations and anions differ on the two sides of the junction, but they still have the same valence charge (general isovalent heterojunctions such as $InP/GaInAs$); finally, we deal with the case of heterovalent heterojunctions such as $Ge/GaAs$, where qualitatively new effects may occur.

IVb. Common-anion heterojunctions. Let us start examining the case of a lattice-matched common-anion heterojunction. For the sake of definiteness, we consider the case of $GaAs/AlAs$, but our conclusions will also apply to other common-anion (or common-cation) systems. The appropriate reference crystal in this case is the virtual crystal $\langle Ga_{0.5}Al_{0.5} \rangle As$. This is an artificial zincblende-structure crystal whose anion is As and whose cation is represented by a pseudopotential which is the arithmetic average of the Ga and Al potentials. We then consider the linear response of the reference crystal to a *single* isovalent substitution, where a mixed cation, $\langle Ga_{0.5}Al_{0.5} \rangle$ is replaced by a Ga (or Al) ion. The Fourier transform of the variation of electrostatic potential induced by the perturbation is:

$$\Delta \tilde{V}_{Ga}(\mathbf{q}) = \frac{4\pi e^2}{q^2} \Delta \tilde{n}_{Ga}(\mathbf{q}), \tag{12}$$

where $\Delta n_{Ga} = -\Delta n_{Al}$ is the electron density induced by the perturbation, which has the full point symmetry of the site (T_d), being the perturbing potential spherically symmetric. The long-wavelength behavior of $\Delta \tilde{n}_{Ga}(\mathbf{q})$ is:

$$\Delta \tilde{n}_{Ga}(\mathbf{q}) = Q - i\mathbf{d} \cdot \mathbf{q} - \frac{1}{2} \mathbf{q} \cdot \overleftrightarrow{\mathbf{D}} \cdot \mathbf{q} + Aq^2 + O(q^3), \tag{13}$$

where $Q = \int \Delta n_{Ga}(\mathbf{r}) d\mathbf{r}$ is the net displaced charge; $\mathbf{d} = \int \mathbf{r} \Delta n_{Ga}(\mathbf{r}) d\mathbf{r}$ is the induced dipole, $D_{\alpha\beta} = \int (x_\alpha x_\beta - \frac{1}{3} r^2 \delta_{\alpha\beta}) \Delta n_{Ga}(\mathbf{r}) d\mathbf{r}$ is the corresponding quadrupole moment, and $A = \frac{1}{6} \int r^2 \Delta n_{Ga}(\mathbf{r}) d\mathbf{r}$ is the second spherical moment of the induced charge. Because of symmetry, Δn does not carry dipole ($\mathbf{d} = 0$) nor quadrupole ($\overleftrightarrow{\mathbf{D}} = 0$) moments, whereas— within LRT—isovalent substitutional impurities carry no net charge ($Q = 0$).

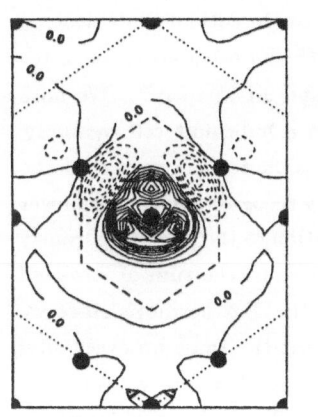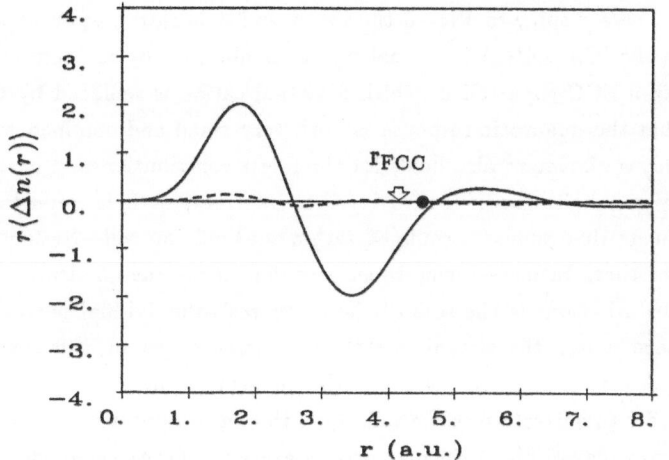

Figure 6. Electron density response of the $\langle Ga_{0.5}Al_{0.5}\rangle As$ virtual crystal to a single $\langle Ga_{0.5}Al_{0.5}\rangle \to Ga$ cationic substitution. a: Contour plots of the linear term $\Delta n^{(1)}(\mathbf{r})$ in the (110) plane (dashed lines indicate a negative density response); the intersection of the plane with cation-centered 2-atoms and 16-atoms FCC Wigner-Seitz cells are indicated with dashed and dotted lines. b: Radial spherical averages of the linear (solid line) and quadratic (dashed line) density responses. The average radii of the Wigner-Seitz cell and neighbor distances are also indicated.

The bare perturbation leading from the virtual crystal to the most general physical system (bulk, superlattice, or alloy) is:

$$\Delta V^{\circ}(\mathbf{r}) = \sum_{\mathbf{R}} \sigma_{\mathbf{R}} \Delta v_{Ga}(\mathbf{r} - \mathbf{R}), \tag{14}$$

where $\sigma_{\mathbf{R}}$ is an Ising-like variable whose value is $+1$ if the $\mathbf{R}$-site cation is Ga, or -1 if it is Al. According to Eqs. (11), the corresponding variation in the charge density is:

$$\Delta n(\mathbf{r}) \approx \sum_{\mathbf{R}} \sigma_{\mathbf{R}} \Delta n_{Ga}(\mathbf{r} - \mathbf{R}). \tag{15}$$

The Fourier transform of the Coulomb potential generated by this charge distribution is:

$$\Delta \tilde{V}(\mathbf{q}) = \frac{1}{N\Omega} \sum_{\mathbf{R}} \sigma_{\mathbf{R}} \, e^{i\mathbf{q}\cdot\mathbf{R}} \, \frac{4\pi e^2}{q^2} \, \Delta \tilde{n}_{Ga}(\mathbf{q}) = 4\pi e^2 A \sum_{\mathbf{R}} \sigma_{\mathbf{R}} \, e^{i\mathbf{q}\cdot\mathbf{R}} + O(q), \tag{16}$$

where N is the number of cations in the crystal and Ω is the volume of the unit cell. From Eq. (16) it follows that the potential drop across the interface between two semiinfinite regions characterized by different concentrations x and y of Ga atoms ($Ga_xAl_{1-x}As/Ga_yAl_{1-y}As$) is given by:

$$\Delta V = \frac{4\pi e^2}{\Omega} A \times 2(x - y), \tag{17}$$

independent on interface orientation, and even abruptness[15]. In particular, the lineup across the $GaAs/AlAs$ interface is $\Delta V = \frac{8\pi e^2}{\Omega} A$.

261

We display in Fig. 6 the linear and quadratic responses to a single cationic substitution in the $\langle Ga_{0.5}Al_{0.5}\rangle As$ virtual crystal as obtained by self-consistent (SCF) calculations on a 16-atom FCC supercell in which a virtual cation is replaced by Ga and Al in turn[12]. We note that the quadratic response is both very small and confined within a bulk unit cell (actually, this confinement also holds for the linear contribution and is responsible for the success of the Wigner-Seitz model for $GaAs/AlAs$ interfaces[5]). It is important to recognize that—whenever quadratic response is confined within one bulk unit cell—its contribution to the lineup rigorously vanishes. In fact—owing to such confinement—the quadratic response to the sum of localized perturbations is the sum of the responses to individual perturbations. Furthermore, since we have chosen the virtual crystal as a reference system, this response is the same on every site, independently of the sign σ_R: the quadratic contribution to the lineup therefore vanishes. In order to better assess the validity of the linear approximation for describing an actual interface, we display in Fig. 7 the macroscopic averages of the electron densities and electrostatic potentials of $(GaAs)_3(AlAs)_3$ grown along the (001), (110), and (111) directions. Continuous lines indicate results of full Self-Consistent (SCF) supercell calculations. Results obtained within LRT are plotted with a dashed line which is barely distinguishable from the previous one, thus indicating the excellent quality of the linear approximation in this case. The value of the potential lineup so obtained is 0.41 eV, independent on interface orientation. In order to obtain the band offset, we have to add the difference of the bulk band edges $\Delta E_v = 0.04$ eV (see sec. II), to obtain VBO ≈ 0.45 eV. This value is obtained neglecting many-body[7] and relativistic effects which amount to ≈ 0.1 and ≈ 0.03 eV respectively. The resulting value of the VBO would be ≈ 0.58 eV Experimental values range from 0.45 to 0.55 eV[16]. We stress that many-body and relativistic effects can be obtained by bulk calculations on the two components and—whatever their values may be—their presence does not affect our conclusions on the potential lineups which contain all of the relevant interface-specific information.

IVc. General isovalent heterojunctions. We consider now the more general case of a III-V/III-V interface in which both cations and anions differ on the two sides of the junction. As a prototypical case, we consider the $InP/Ga_{0.47}In_{0.53}As$ system, where gallium-indium arsenide is treated in the virtual-crystal approximation, considering an X cation which is represented by the appropriate average of Ga and In pseudopotentials ($X = \langle Ga_{0.47}In_{0.53}\rangle$). It is important to distinguish this kind of virtual crystal from further averaging between X and In (or P and As) potentials which will be used as a starting point of our linear-response treatment. To this end, we define an average cation $C = \langle In_{0.5}X_{0.5}\rangle \equiv \langle In_{0.715}As_{0.285}\rangle$, and an average anion $A = \langle P_{0.5}As_{0.5}\rangle$, and the corresponding reference crystal as the CA zincblende-structure virtual crystal. The actual interface is then recovered substituting C with In on the left and X on the right of the interface, and analogously the anion A with P and As.

Using LRT, the final interface density profile simply is:

$$n(\mathbf{r}) \approx \langle n(\mathbf{r})\rangle + \Delta n_A(\mathbf{r}) + \Delta n_C(\mathbf{r}), \tag{18}$$

where Δn_A and Δn_C are the density responses to anion and cation replacements respectively. In Fig. 8 we display the macroscopic average of the electron-density distributions in $(CP)_3/(CAs)_3$, $(InA)_3/(XA)_3$, and $(InP)_3/(XAs)_3$, (001) superlattices. In the latter case, we also report the density distribution obtained superimposing the two previous ones (note that the macroscopic average of the full perturbed charge coincides with the average of the response charge, within a constant which is the average of the virtual-crystal charge). The fact that the two curves are hardly distinguishable signals again that LRT is an excellent approximation also in this case.

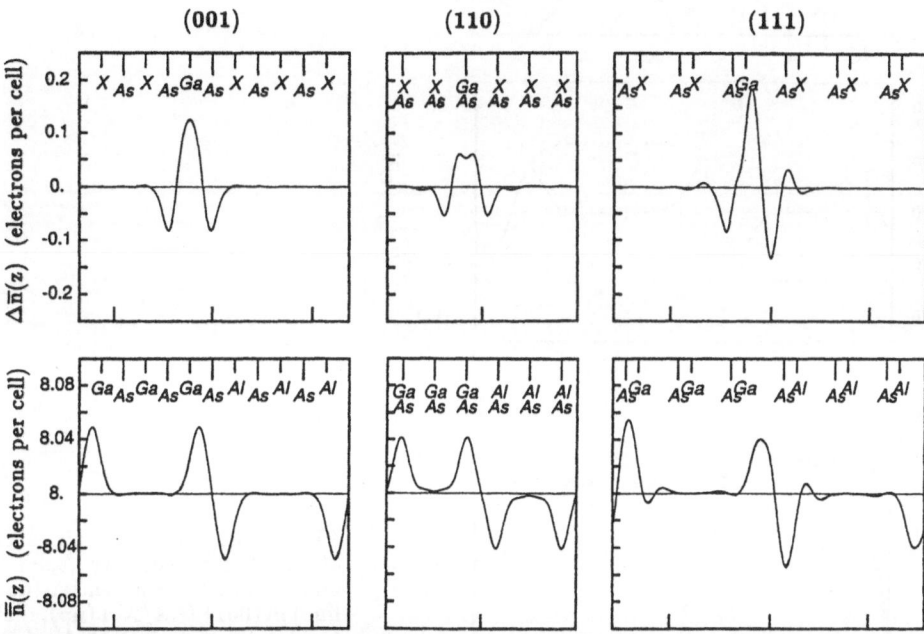

Figure 7. Top panels: charge-density response of the virtual crystal $\langle Ga_{0.5}Al_{0.5}As\rangle$ to a single planar cationic substitution. Bottom panel: macroscopic averages of the electron charge densities of $(GaAs)_3(AlAs)_3$. Continuous lines indicate results from full self-consistent calculations, whereas dashed lines indicate results from LRT. Left panels: (001); center: (110); right: (111).

In Fig. 9 we report the superlattice electron densities and potentials for the (001), (110), and (111) orientations. Despite the fact that the density profiles look very different in the three cases, the resulting potential lineups practically coincide with each other, as predicted by LRT.

In Fig. 10 we show a contour plot of the linear electron density response of the virtual crystal to single $C \rightarrow In$ and $A \rightarrow P$ substitutions. The geometry used to perform this calculation is the same as described in the previous section in the case of the $GaAlAs$ system. The spherical average of the linear and quadratic density responses are also reported. As in the case of $GaAlAs$, the rearrangement of the electron density occurs almost entirely within a bulk FCC Wigner-Seitz cell centered at the substitutional site. However, in this case, there are also some nonnegligible tails spilling out, as evidentiated more clearly by the spherical-average plots. From the data reported in Fig. 10 we can calculate the anion contribution to the potential lineup, following Eq. 17. Analogous calculations for a cation substitution would yield the corresponding contribution to the lineup. Our results are: $\Delta V_{cation} = +0.34$ eV, $\Delta V_{anion} = -0.58$ eV, $\Delta V_{tot} = -0.24$ eV independent on orientation, and in excellent agreement with the values obtained by SCF calculations on superlattices (-0.25 eV). Adding the bulk band edges ($\Delta E_v = 0.55$eV) we obtain a value for the total band offset of 0.31 eV, in good agreement with recent measurements (0.35 eV[17]). We note that—analogously to the $GaAlAs$ case, this figure should be corrected by many-body effects. Moreover, disorder in the $GaInAs$ region is also responsible for further self-energy effects which should also shift the corresponding band edge. We stress however that—analogously to many-body effects—the disorder-induced self-energy is a pure bulk effect, independent of the electronic structure of the interface.

IVd. Heterovalent interfaces. Consider now the interface between two heterovalent semiconductors, such as $Ge/GaAs$. In this case, the appropriate virtual crystal is a zincblende

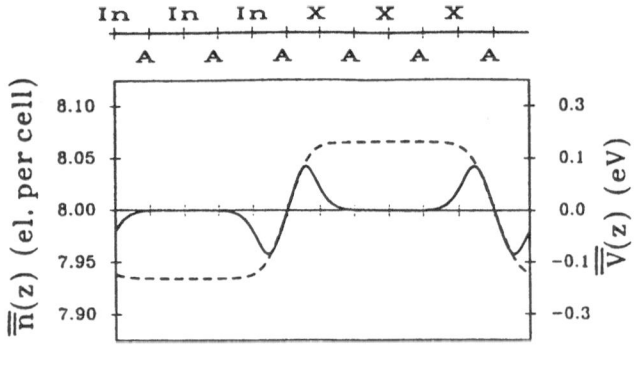

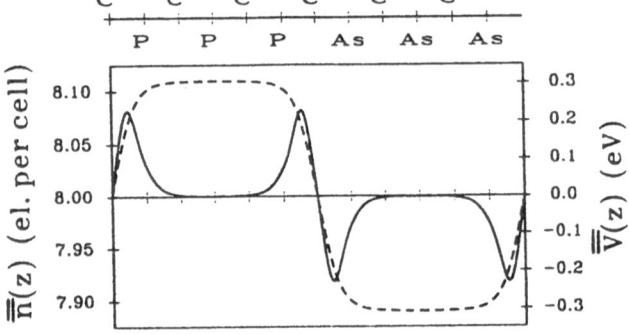

Figure 8. Macroscopic average of the electron density (solid line) and electrostatic potential (dashed line) in (001) InA/XA (up), CP/As (middle), and InP/XAs (down) superlattices.

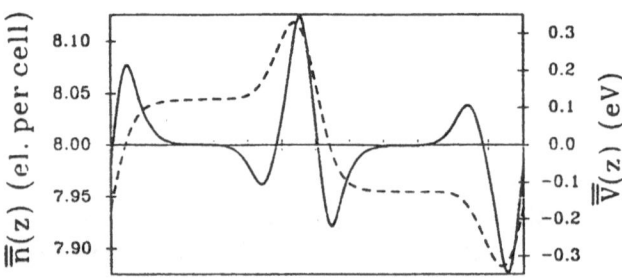

$\langle Ge_{0.5}Ga_{0.5}\rangle/\langle Ge_{0.5}As_{0.5}\rangle$, whose cation has valence charge 3.5, while the anion has valence charge 4.5. The relevant (bare) localized perturbations which transform a virtual ion in a real one are:

$$\Delta v_C = \pm \frac{1}{2}\left(v_{Ga} - v_{Ge}\right); \quad \Delta v_A = \pm \frac{1}{2}\left(v_{As} - v_{Ge}\right), \tag{19}$$

where the upper sign transforms a virtual cation (anion) into a gallium (arsenic) ion, whereas the lower one transforms both of them into germanium ions. The bare perturbations of Eq. (19) carry a net charge ± 0.5. therefore—to first order in the perturbation—the total charge (bare + screening) induced by them is $\pm \frac{1}{2\epsilon_\infty}$, where ϵ_∞ is the static electronic dielectric constant of the virtual crystal (in this paper we will assume that—for lattice matched heterojunctions— atoms maintain their ideal positions even in the proximity of the interface). We stress that this result is *exact* within LRT and fully takes into accounts self-consistency and local-field effects.

According to the previous discussions, the Fourier transform of the total charge induced by

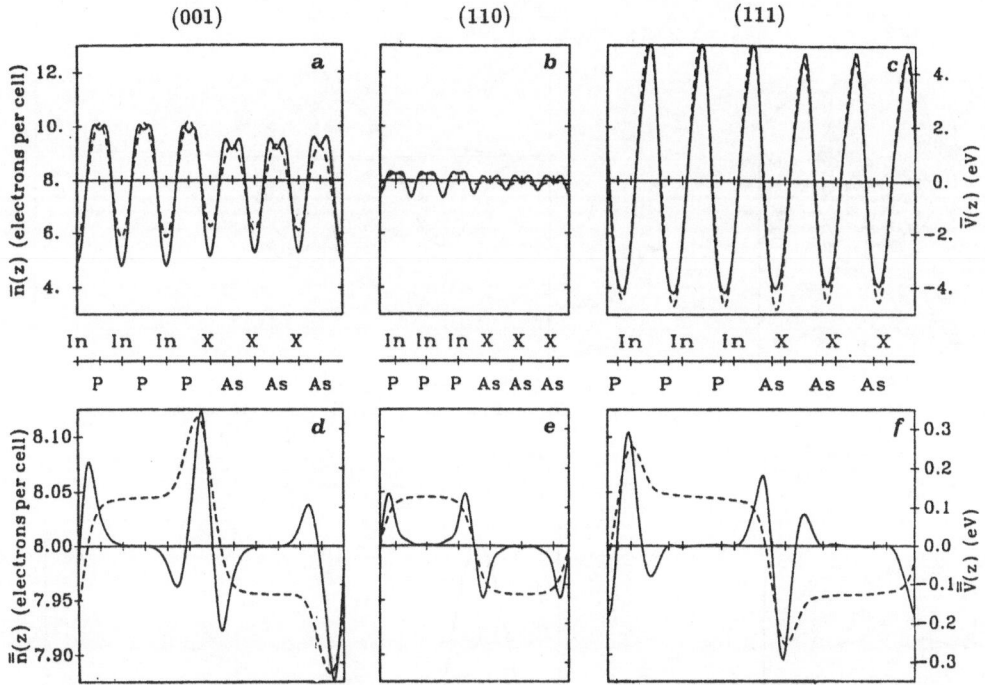

Figure 9. Planar averages of the electron density (solid line) and potential (dashed line) of $(InP)_3/(Ga_{0.47}In_{0.53}As)_3$ superlattices. a: (001); b: (110); c: (111); d,e,f: corresponding macroscopic averages.

one localized perturbation is:

$$\Delta n_{A,C}(\mathbf{q}) = \pm \left(\frac{1}{2\epsilon_\infty} + A_{A,C}q^2 \right) + \mathcal{O}(q^3). \tag{20}$$

In the spirit of LRT, we split the potential lineup into two contributions:

$$\Delta V = \Delta V^{(0)} + \Delta V^{(2)}, \tag{21}$$

due to the constant and quadratic terms of Eq. (20). The latter is by construction the lineup due to neutral localized perturbations. According to the discussion of the preceding section, $\Delta V^{(0)}$ is independent of the orientation and abruptness of the interface. The former is formally equivalent to the lineup generated by an assembly of *point charges* of absolute value $\frac{1}{2\epsilon_\infty}$. $\Delta V^{(0)}$ does depend on the atomic structure of the interface (orientation, abruptness, relaxation, and so on). However—once this structure is known, either experimentally or by independent theoretical calculation—$\Delta V^{(0)}$ can be calculated from elementary electrostatics.

In order to illustrate the above concepts, we examine in some details the cases of $Ge/GaAs$ (001) and (110) interfaces. In the (001) direction, the virtual crystal is made of an alternating stacking of cationic and anionic planes, carrying a surface charge density of 3.5 and 4.5 electrons per unit surface cell. This is schematically shown in Fig. 11. In the (001) direction, there exist two inequivalent interfaces, according to whether the $GaAs$ region is Ga- or Al-terminated. The point-charge contribution to the potential lineup, $\Delta V^{(0)}$, is due to the planar charge distributions sketched in Fig. 11. Inspection of the corresponding macroscopic averages shows that these interfaces are *charged* and therefore thermodynamically unstable. In fact, a net

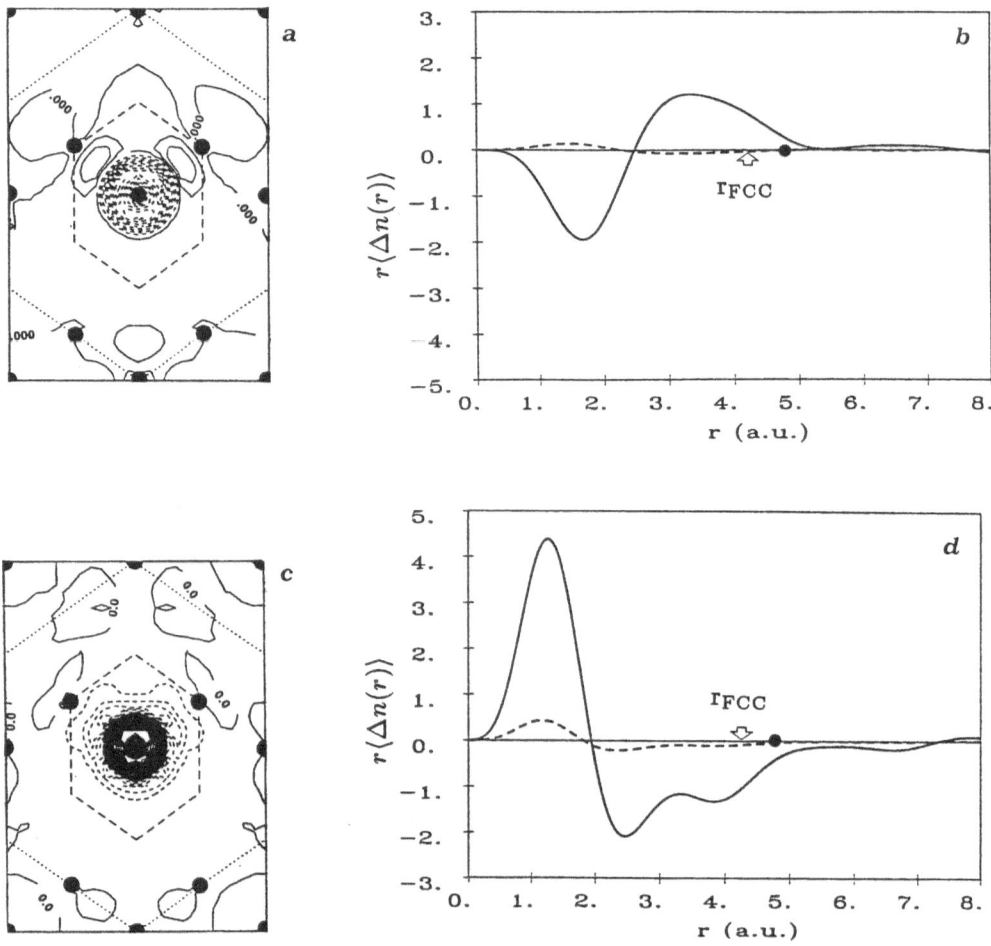

Figure 10. Electron density response of the $\langle In_{0.5}X_{0.5}\rangle\langle As_{0.5}P_{0.5}\rangle$ virtual crystal to a single cation (a,b: $\langle In_{0.5}X_{0.5}\rangle \to In$) or anion (c,d: $\langle As_{0.5}P_{0.5}\rangle \to P$) substitution. a,c: Contour plots of the linear term $\Delta n^{(1)}(r)$ in the (110) plane (dashed lines indicate a negative density response); the intersections of the plane with cation-centered 2-atoms and 16-atoms FCC Wigner-Seitz cells are indicated with dashed and dotted lines. b,d: Radial spherical averages of the linear (solid line) and quadratic (dashed line) density responses. The average radii of the Wigner-Seitz cell and neighbor distances are also indicated.

interface charge would give rise to macroscopic electric fields which would alter the *bulk* energy of the system. The simplest geometries which give rise to a neutral interface are those where there is a mixed Ga/Ge or As/Ge plane. They are schematically illustrated in Fig. 12. The point-charge contribution to the potential lineup is obtained from the macroscopic average of the distributions of Fig. 12 (divided by ϵ_∞ to account for the electronic response) through Eq. (4). Their value is $\Delta V^{(0)} = \pm\frac{\pi e^2}{2a_0\epsilon_\infty}$, where a_0 is the lattice parameter of the virtual crystal. We note that the *average* between the potential lineups of the two inequivalent interfaces equals just the contribution due to the *neutralized* perturbations, $\Delta V^{(2)}$. In the (110) direction, the virtual crystal is made of atomic planes with two ions (one cation and one anion) per unit surface cell. The planar perturbations leading to the physical interface are therefore neutral. We conclude that $\Delta V^{(0)}$ vanishes and that the lineup coincides in this case with the average lineup of the two inequivalent (001) interfaces considered above. This was observed empirically by Kunc and

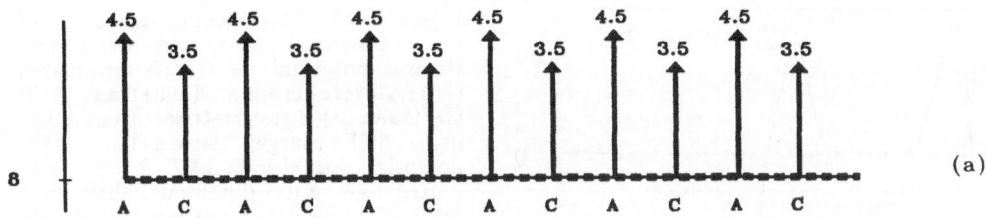

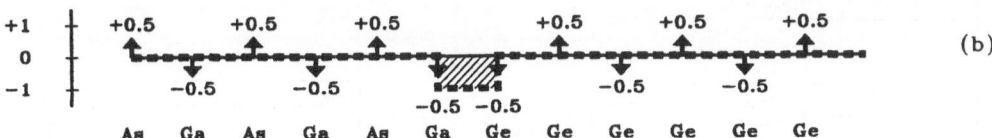

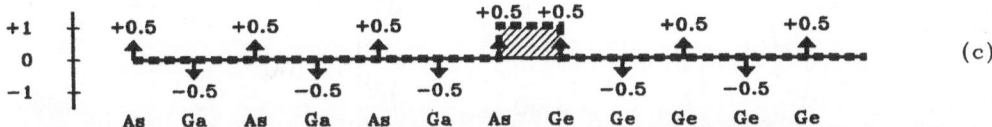

Figure 11. a: Distribution of planar charges for the succession of cationic planes of the CA virtual crystal in the (001) direction. b: Charge distribution for the perturbation leading to a Ga-terminated interface. c: Charge distribution for the perturbation leading to an As-terminated interface. Arrows indicate delta functions. Dashed lines indicate the macroscopic average of the point-like charge distributions: the corresponding scale (electrons per cell) is displayed on the y axis.

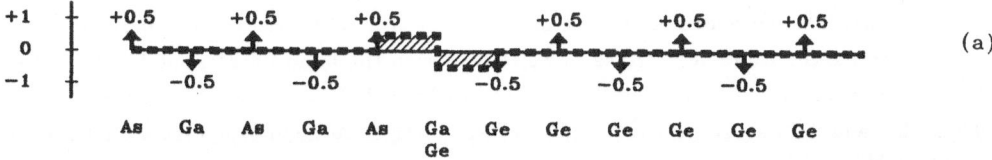

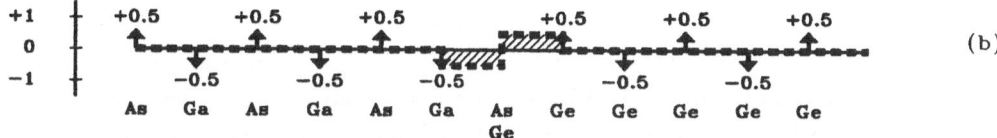

Figure 12. Point-like charge distributions leading form a CA virtual crystal to Ga/Ge- or As/Ge-terminated $Ge/GaAs$ (001) interfaces.

Martin a few years ago[18]: LRT provides a natural and simple explanation of this fact.

In Fig. 13 we report the macroscopic average of the electron charge density and electrostatic potential of $Ge/GaAs$ (001) and (110). Also in this case the full SCF calculation is very close to the predictions of LRT. One may notice that the absolute value of the potential lineup is much larger in the (001) cases that in the (110) one. In fact, this is the *electronic* contribution

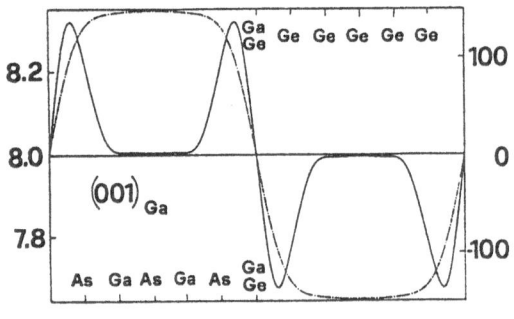

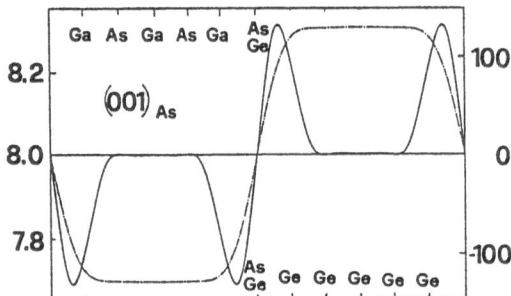

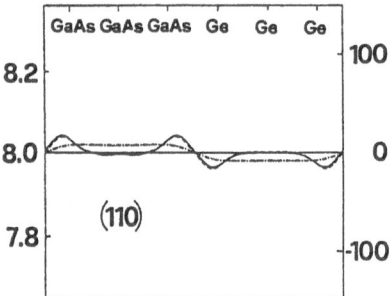

Figure 13 Macroscopic averages of the z-projected charge densities and Hartree potentials of Ga/Ge-terminated (001), As/Ge-terminated (001), and (110) Ge/GaAs $3 + 3$ superlattices. Continuous line: SCF charge; dashed line: SCF potential; dotted line: LRT charge; dash-dotted line: LRT potentials. Left scale (charge) is electrons per zincblende cell; right scale (potential) is mRy.

to the lineup. To this one must add the bare point-charge contribution which is of opposite sign and comparable magnitude. The total lineup is -0.22, $+0.06$, and -0.54 eV for the (110), $(001)_{Ga}$, and $(001)_{Al}$ interfaces. LRT and SCF calculations give identical results (to the figure reported here). This data show that the small nonlinear effects have virtually no effect on integrated quantities such as the potential lineups. The fact that the (110) lineup is not exactly the average between the (001) ones even within LRT is a consequence of the finite thickness of our superlattice. The effect due to the size of our supercell is estimated to be ~ 0.01 eV, while LRT can describe the isolated interface as well. To obtain the band offsets, one has to add to the potential lineups the difference in the bulk band edges, $\Delta E_v = 0.76$ eV: this gives an offset of 0.54, 0.82, and 0.22 eV respectively. The corresponding experimental figures are 0.56, 0.55, and 0.60 eV[19].

The agreement between theory and experiment is very good for the (110) case and poor for the (001) geometries. We don't think that the quality of the calculations is different in the two cases, nor that the latter deserve more sophisticated theoretical tools. We believe instead that the disagreement between theory and experiment in the (001) cases is a direct consequence of atomic intermixing occurring at the interface. We have seen how sensitive is the lineup to the actual atomic arrangement at polar interfaces: it is a simple exercise to find different geometries having very different lineups. It is still an open question to determine the mechanisms responsible for the interface atomic arrangement in epitaxial structures: in our belief equilibrium thermodynamics is not sufficient to predict them and kinetic effects may play a fundamental role. Whatever such mechanisms may be, our results show that—for polar interfaces—the band offset is a very sensitive quantity to characterize the quality of the interface.

We conclude mentioning some potential applications of the peculiarities of polar interfaces to device design. The lineup between inequivalent interfaces for a same crystallographic orientation could be exploited to induce a potential drop within different regions of a same material. Consider for instance Ge with just a double atomic intralayer of $GaAs$ oriented along (001). Along the lines exposed above, it is simple to show that the energy necessary to bring an electron

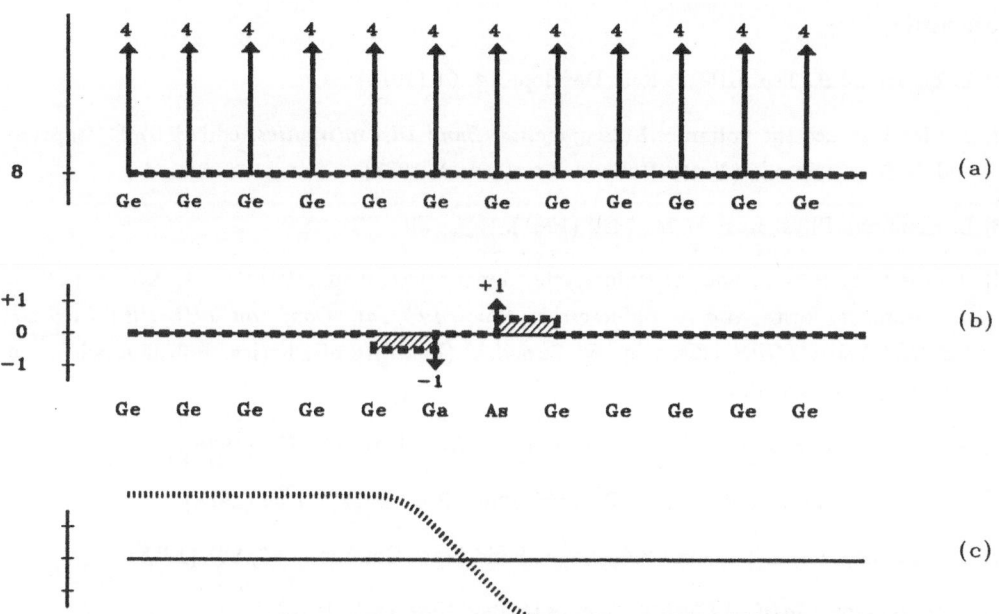

Figure 14. a: Distribution of planar charges for *Ge* in the (001) direction. b: Charge distribution for the perturbation corresponding to a *GaAs* double layer. c: Electrostatic potential generated by the charge distribution (b).

at the top of the valence band from the left to the right of the double layer is $\Delta V = \frac{\pi e^2}{a_0 \epsilon_\infty}$. the situation is schematically illustrated in Fig. 14. The possibility of obtaining such an effect depends very much on the sharpness of the double layer. It is yet unknown whether a sufficient quality can be achieved with present growth techniques, but we believe that, if this were the case, this would open new and very interesting perspectives for device design.

V. Conclusions

We summarize the main results exposed in this work. First of all, we have developed the concept of *macroscopic average* which allows to extract relevant interface phenomena from first-principles calculations on interfaces and superlattices in a direct and physically meaningful way. Second, we have illustrated the basic concepts of our linear-response approach to mixed semiconductor systems. Starting from this, we have then shown that—for lattice matched heterojunctions—the band offset at isovalent interfaces is a bulk quantity (i.e. independent of any interface detail) for isovalent interfaces. For heterovalent interfaces, the offset may depend on the details of the atomic arrangement at the interface. However, once this arrangement is known, the interface-dependent contribution to the offset can be calculated "on the back of an envelope" from such simple (and experimentally accessible) quantities as the bulk lattice constant and dielectric constants of the constituents. We suggest that this dependence can be usefully employed to characterize the crystallographic quality of polar interface and also, perhaps, to design new devices with very interesting properties. We have neglected on purpose the very important case of mismatched heterojunctions which has close relations with the problem of *absolute deformation potentials*[20,21] and which is currently under study. For the time being, we content ourselves to just mention that—for group-IV semiconductors (i.e. for systems with vanishing ionic effective charges)—the band offset continue to be essentially a bulk property (in some generalized sense)[21], while it may be not for other systems.

References

[1] L. Esaki and R. Tsu, IBM J. Res. Develop. **14**, 61 (1970).

[2] See for instance the volume: *Heterojunction Band Discontinuities*, edited by F. Capasso and G. Margaritondo (North-Holland, Amsterdam, 1987).

[3] L. Kleinman Phys. Rev. B **24**, 7412 (1981).

[4] R. Resta, S. Baroni, and A. Baldereschi, Superlattices and Microstr. **5**, XXXX (1989); S. Baroni, R. Resta, and A. Baldereschi. *Proc. 19th Int. Conf. on THE PHYSICS OF SEMICONDUCTORS*, edited by W. Zawadzki (Institute of Physics, Polish Academy of Sciences, Wroclaw, 1988), p. 525.

[5] A. Baldereschi, S. Baroni, and R. Resta, Phys. Rev. Lett. **61**, 734 (1988).

[6] R.W. Godby, M. Schlüter and L.J. Sham, Phys. Rev. B **37**, 10159 (1988).

[7] S.B. Zhang, D.Tománek, and S.G. Louie, Solid State Commun. **66**, 585 (1988).

[8] J. D. Jackson *Classical Electrodynamics* (Wiley, New York, 1975).

[9] F. N. H. Robinson, *Macroscopic Electromagnetism* (Pergamon, Oxford, 1973).

[10] R. Resta, in: *Festkörperprobleme*, vol. **XXV**, edited by P. Grosse (Vieweg, Braunschweig, 1985), p. 183.

[11] D.M. Bylander and L. Kleinman, Phys. Rev. B **34**, 5280 (1986); *ibid.* **36**, 3229 (1987); Phys. Rev. Lett. **59**, 2091 (1987).

[12] All the SCF results presented in this paper have been obtained within the local-density approximation (LDA), using norm conserving pseudopotentials and plane-wave basis sets. The technical ingredients are the same as in Refs. 4,5 .

[13] Of course, the value of the limit depends on the actual prescription chosen. A "good" choice is one which provides average values such that differences of them are good approximations to the interface potential lineups.

[14] C. Van de Walle and R.M. Martin, Phys. Rev. B **35**, 8154 (1987).

[15] Intuitively, this is simply related to the difference between the "left" and "right" $q \to 0$ Fourier transform of σ_R, i.e. to their averages in two macroscopic regions far from and on opposite sides of the interface.

[16] W.I. Wang and F. Stern J. Vac. Sci. Techn. B **3**, 1280 (1985); D. Arnold, A. Ketterson, T. Henderson, J. Klem, and H. Morkoc, J. Appl. Phys. **57**, 2880 (1985); G. Danan, B. Etienne, F. Mollot, R. Planal, A.M. Jean-Luois, F. Alexandre, B. Jusserand, G. Le Roux, J.Y. Marzin, H. Savary, and B. Sermage, Phys. Rev. B **35**, 6207 (1987).

[17] D.V. Lang, M.B. Panish, F. Capasso, J. Allam, R.A. Hamm, A.M. Sergent, and W.T. Tsang, Appl. Phys. Lett. **50**, 736 (1987).

[18] K. Kunc and R.M. Martin, Phys. Rev. B **35**, 8154 (1987).

[19] R.W. Grant, J.R. Waldrop, and E.A. Kraut, Phys. Rev. Lett. **40**, 656 (1978); J.R. Waldrop, E.A. Kraut, S.P. Kowalczyk, and R.W. Grant, Surf. Sci. **132**, 513 (1983).

[20] C.G. Van de Walle and R.M. Martin, Phys. Rev. Lett. **62**, 2028 (1989).

[21] R. Resta and S. Baroni, Bull. Am. Phys. Soc. **34**(3), 832 (1989).

OPTICAL PROPERTIES AND BAND ALIGNMENTS OF

III-V HETEROSTRUCTURES

Karen J. Moore

Philips Research Laboratories
Cross Oak Lane, Redhill, Surrey, U.K.

ABSTRACT

Photoluminescence (PL) and photoluminescence excitation (PLE) studies have been made of both short period, $m=n \leqslant 8$, $(GaAs)_m$-$(AlAs)_n$ super-lattices (SL) and pseudomorphic InGaAs-GaAs structures. Systematic investigations of the electronic properties, band alignments and super-lattice effects highlight important differences between these systems arising from the nature of the lowest conduction band (CB) state and the resulting subband dispersion in thin samples. In $(GaAs)_m$-$(AlAs)_n$ SLs PL and PLE measurements reveal that for $m=n \geqslant 4$ the lowest CB state is a folded X_z minimum. PLE spectroscopy also provides information about the higher energy direct Γ-Γ transitions. The observed energy gaps are compared with a simple Kronig-Penney description of the electronic states and the limits of the model as the layer thicknesses are decreased are explored. Optical data is also reported on both a 25Å $In_xGa_{1-x}As$-400Å GaAs quantum well (QW) structure and on a 25Å $In_xGa_{1-x}As$-100Å GaAs SL structure ($x=0.12$). Low temperature PL spectra exhibit sharp peaks and in the PLE spectrum of the QW sample the $n=1$ heavy-hole exciton peak is clearly resolved from the onset of the excited states and the continuum edge. These observations provide the first direct measurement of the exciton binding energy in a strained sample and for a thickness at which we expect the binding energy (as a function of QW width) to have gone through a maximum. In the 5 well SL sample, there is significant overlap of the electron and hole wavefunctions of adjacent wells, leading to a mixing and a splitting of the ground state into 5 discrete levels. The predicted splittings are observed in the PLE spectrum of this sample.

1. INTRODUCTION

The range of III-V materials which can be combined to produce a heterostructure provides a wealth of opportunity for physics investig-ations and a new flexibility in the design of novel device structures. The driving force behind the large scale investigations of these systems depends at least in part on the maturing of growth techniques such as molecular beam epitaxy (MBE) and metal-organic chemical vapour phase deposition (MOCVD). Whilst the greatest effort has certainly been centred on the near lattice-matched GaAs-AlAs heterojunction, the ability to grow epitaxial layers which incorporate a finite strain, is the real key to extending material flexibility and design parameters. This is reflected in the range of strained layer systems which have increasingly been studied in the last few years. In this paper, the results are limited to the lattice-matched GaAs-AlAs system and the pseudomorphic $In_xGa_{1-x}As$-GaAs system.

2. $(GaAs)_m$-$(AlAs)_n$ SHORT PERIOD SUPERLATTICES

Extensive optical investigations of confinement effects in GaAs-$Al_xGa_{1-x}As$ multiple quantum well (MQW) structures have revealed a wealth of information about their subband structure and electronic properties. Broadly speaking, for the case of isolated MQWs, in which the lowest energy states of one QW are electronically decoupled from the equivalent states in adjacent wells, the optical data are well described by a number of theoretical models of varying degrees of sophistication. However as we start to introduce coupling into the problem by reducing the individual layer thicknesses we progress from an essentially 2D picture to the quasi-3D dispersion of a superlattice (SL). A number of fundamental questions regarding the electronic properties of $(GaAs)_m$-$(AlAs)_n$ short period SLs must then be addressed: How closely does the system approximate to that of the corresponding AlGaAs alloy? From a spectroscopic viewpoint, is the lowest energy transition direct, indirect, or pseudo-direct and does it change as a function of layer thickness? What is the extent of the dispersion in the z-direction i.e. normal to the plane of the samples, of the lowest CB state and does this give rise to new SL phenomena in the optical spectra? How complete is our theoretical understanding of these structures?

From a theoretical point of view the question of the electronic structure of $(GaAs)_m$-$(AlAs)_n$ superlattices (SLs) has been attacked by a variety of methods from the simplest Kronig-Penney calculation,[1], through envelope function type calculations[2] to the sophisticated microscopic methods based on tight-binding,[3,4] empirical and self-consistent pseudo-potentials,[5-7] and local-density[8] or augmented spherical wave calculations.[9] It is fair to say that there is no concensus amongst these calculations (sometimes even using the same method) as to the nature of the electronic structure of the (m+n) SL as the number of monolayers m or n is varied. For example, Nakayama and Kamimura[7] have used a self-consistent pseudo-potential scheme to predict that the (1+1) GaAs-AlAs SL has its lowest conduction-band state at the R point of the SL Brillouin zone, making the smallest gap indirect, while for m=n=2,3 and 4, the lowest state is predominantly Γ-like in character[5] and the SLs have a direct energy gap. In contrast, Andreoni and Car, using a non-self-consistent, supercell, empirical pseudo-potential scheme found that the band gap of the m=n=1,2,3 and 4 SLs were "pseudo-direct" with the lowest state having predominantly the character of the folded X-point state. Andreoni and Car found a small shrinkage of the lowest gap between the (1+1) SL and the (4+4) SL of only 16 meV. Qualitatively these latter results are in accord with the effective mass calculations presented here. In this simplest of the calculational schemes the lowest state for the m=n SL, when m=n<10, is also predicted to be the folded AlAs X state. Although the scale of the change of the pseudo-direct gap is found to be about an order of magnitude larger, shrinking by $\approx$ 120 meV between the (1+1) SL and the corresponding (4+4) system compared to the 16 meV variation found by Andreoni and Car.[5]

In this section of the paper results are reported of the low temperature (<8K) photoluminescence (PL) and photoluminescence excitation (PLE) measurements made on short period, nominally m=n, $(GaAs)_m$-$(AlAs)_n$

274

SLs grown by MBE. Values of m ranged between 2 and 8. The measurements are contrasted with other low temperature PL observations on similar GaAs-AlAs short period structures grown either by MBE[10-14] or MOCVD.[15,16] Some of this work claims to present evidence supporting a change from a direct to indirect band gap with decreasing period, while other workers[15,16] studying nominally the same structures conclude that their observations are in accord with the PL expected from an $Al_{0.5}Ga_{0.5}As$ random alloy. The collected experimental results are also compared to our simple envelope function[17] and effective mass type calculations of the subband edges of the Γ, folded X, and non-folded X minima as the SL period (m+n) is varied.

2.1 Growth Details and Structural Analysis

The GaAs-AlAs samples studied here were all grown in a Varian Gen-II MBE system. Growth conditions were as close as reproducibly possible to those used previously.[18] All the layers were deposited on (001)-oriented semi-insulating GaAs substrates held at a nominal substrate temperature of 630°C. The substrate was rotated during growth which was continuous. The growth sequence was as follows: (a) a 1 μm GaAs buffer, (b) ≈ 1 μm of alternating layers of GaAs and AlAs (the SL region), and finally (c) a 0.1 μm GaAs capping layer. The layers were undoped, although there is a background acceptor concentration of about 2×10^{14} cm^{-3}. In the SL region of the sample the GaAs and AlAs layer thicknesses were nominally equal, i.e. m=n. Values of m ranged between 2 and 8 for the samples to be discussed here.

All the samples have been examined using X-ray diffraction techniques. A diffractometer scan along (001) gave satellite peaks in the vicinity of (000) and (00L) Bragg reflections. The satellite peak positions and intensities allow one to comment on the periodicity of the SL and on the interface sharpness, respectively. The high resolution diffractometer profile close to the (115) reflection was compared with simulated profiles using dynamical diffraction theory to yield the average mole fraction, x, of the SL.[19] Measurements[20] on the specimens studied here revealed that all the samples had an average Al fraction close to 0.5, with average periods within 0.5 monolayers (MLs) of their nominal values. However, preliminary investigations of the number and strength of satellite peaks for these samples shows some evidence for a gradual variation in the Al fraction across the interfaces. This effect is most pronounced for the (2+2) sample. That such an effective grading exists is not at all surprising when one considers that the samples have been grown without any interrupts.

2.2 Photoluminescence: Experiments and Results

Photoluminescence and photoluminescence excitation spectra were recorded with the samples mounted on the cold finger of a variable temperature (4-300K), continuous flow cryostat. For the PL measurements the samples were excited with either the 5173Å line of an Ar^+ laser or at 5713Å by an Ar^+ pumped dye (Rhodamine-6G) laser. The same dye laser arrangement provided the tunable source for the PLE measurements. The luminescence was collected and analysed by a double grating monochromator and detected with a cooled GaAs photomultiplier and associated photon-counting electronics. Results on $(GaAs)_m-(AlAs)_n$ SLs where m=n=2,4,5,6 and 8 monolayers are discussed here. Some of these results have also been reported elsewhere.[21]

Figure 1 shows the 5K PL and PLE spectra from the (6+6) sample. For convenience the spectral regions are labelled (a), (b) and (c). Regions (a) and (b) illustrate the PL and PLE associated with the pseudo-direct X-Γ transitions, while region (c) shows the PLE spectrum in the vicinity of the direct Γ-Γ transition. The PL spectrum in fig.1(a) was recorded using an excitation wavelength of 5145Å. Theoretically, we anticipate that the lowest conduction band state of this sample would be at X_z, in AlAs, and that the thickness of the GaAs is sufficient to allow only weak coupling of the states in adjacent AlAs layers. The sample is thus akin to the "type-II" samples previously studied by ourselves[22,23] and by Finkman and co-workers[14,24] and we continue to assign the emission, at 1.8990 eV, to the recombination of localized excitons built from electron states at X_z in AlAs and holes at Γ in

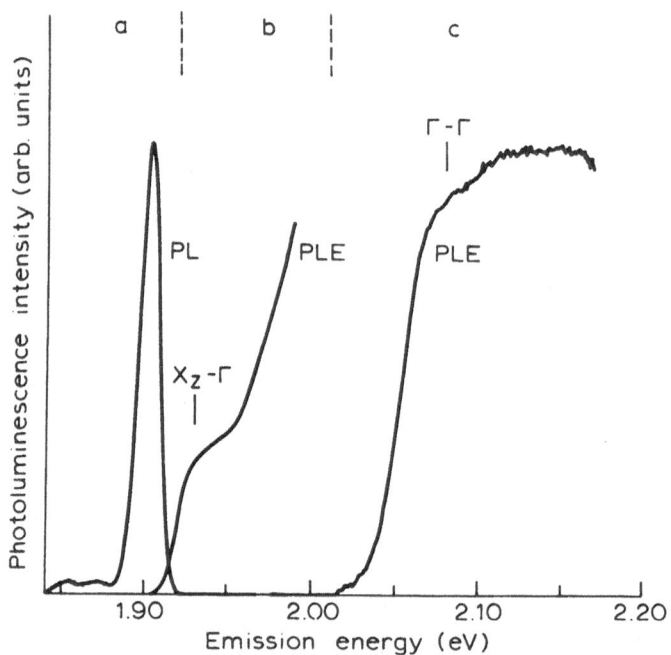

FIG. 1. Low temperature (5K) photoluminescence (PL), region (a), and photoluminescence excitation (PLE) spectra, regions (b) and (c) of the (6+6) GaAs-AlAs superlattice. Note that there is an increase in the gain of the detection system between regions (c) and (b).

GaAs. Setting the detection energy at 1.902 eV and performing PLE over the energy range spanned by (b) reveals a distinct feature at ≈ 1.926 eV. At this energy excitons are being created via states at the X_z minima. As in the previous study[23] of the nature of the pseudo-direct transition, its localized character is inferred from the temperature dependence of the PL and PLE peaks. The localization energy of the exciton, deduced from the shift between PL and PLE peaks, is about 26 meV for this sample. Continuing to scan the excitation source over region (c) reveals a strong rising edge, interpreted as the direct Γ-Γ absorption, whose mid-point is close to 2.06 eV. No clear excitonic, $\Gamma-\Gamma$, peaks are discernible for samples of this dimension and smaller, in accord with measurements reported earlier by Nagle and co-workers.[13]

Consider now the (5+5) and (4+4) SL samples. For illustrative purposes, the (4+4) sample is discussed, but almost all the remarks could be made equally well about the (5+5). The 5K PL and PLE spectra from the (4+4) sample are displayed in figs.2(a)-2(c). The PL spectrum from this sample is similar to the wider (6+6) and (8+8) samples. The emission is dominated by a peak of 1.9927 eV which is intrinsic in nature, again involving the $X_z-\Gamma$ transition, whilst the lower energy shoulder is most likely extrinsic in origin. The evidence supporting these assignments

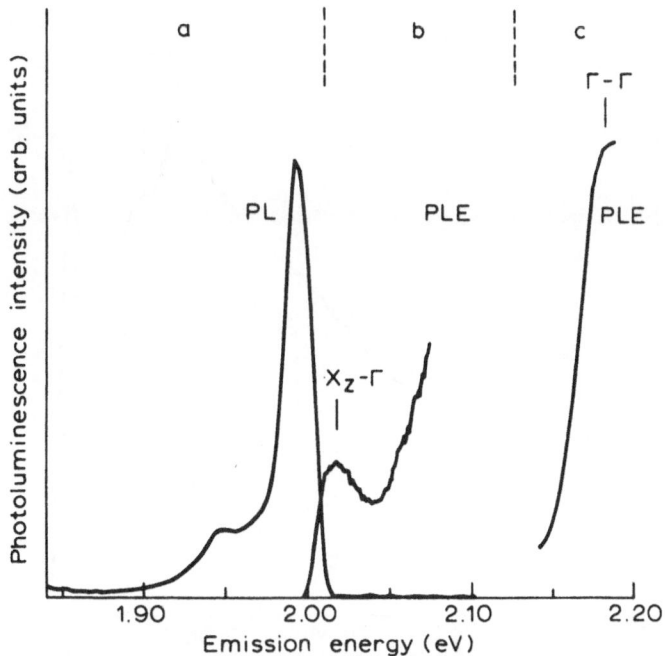

FIG. 2. Low temperature (5K) photoluminescence (PL), region (a), and photoluminescence excitation (PLE) spectra, regions (b) and (c) of the (4+4) GaAs-AlAs superlattice. Note that there is an increase in the gain of the detection system between regions (c) and (b).

comes from a number of observations such as the excitation power dependence of the spectrum. The intensity of the higher energy line increases in a linear fashion with increasing excitation power, while the intensity of the low energy feature saturates at around $8W\ cm^{-2}$. The intrinsic PL peak from this sample is somewhat broader and slightly asymmetric when compared with the (6+6) spectrum, indicating that there is perhaps more than one contribution to the lineshape. Recording the PL spectrum with very low excitation power, $< 0.5W\ cm^{-2}$, reveals two lines split by about 14 meV. These are possibly due to parts of the sample with different periods, for example (4+4) in one region and (4+3) in another. Setting the detection energy at 1.980 eV to record a PLE spectrum over the energy range of region (b) reveals a distinct peak at 2.017 eV which is assigned to the creation of $X_z-\Gamma$, free excitons. Continuing the PLE scan to higher energy shows a sharply rising edge whose onset is close to 2.15 eV with the top of the step below 2.18 eV. The strong increase in PLE signal is once more attributed to the direct $\Gamma-\Gamma$ absorption in the SL.

The 5K PL spectrum for the (2+2) sample, which is shown in fig.3, is similar to that reported by Nagle et al.[13] Three distinct peaks riding on a large, probably impurity-related background can be identified. The highest energy peak is most likely excitonic in origin while the two peaks at lower energy are consistent with being bulk phonon replicas of the excitonic feature.[16] This spectrum most closely resembles that of an

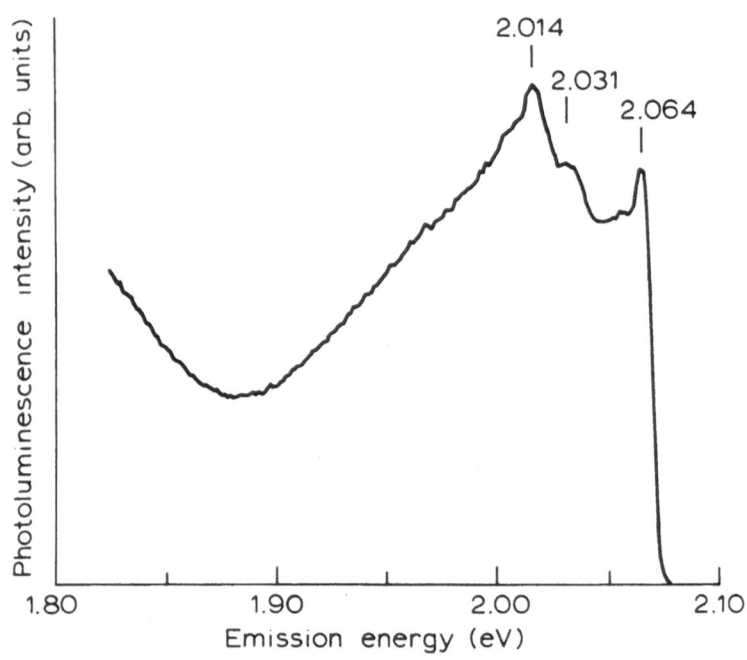

FIG. 3. Low temperature (5K) photoluminescence spectrum of the (2+2) GaAs–AlAs superlattice.

$Al_{0.5}Ga_{0.5}As$ alloy. That this should be the case can once more be understood by considering the growth conditions. Remember that growth is continuous for these samples so that it is extremely unlikely that the interfacial regions can be considered to be extended flat areas. Most likely the in-plane extent of the interfacial islands is smaller than the scale of the excitonic probe and if we consider that the minimum step height is 1 ML then the now quasi-three-dimensional exciton is essentially sampling an alloy-like region of the crystal. This effect, of course, becomes more pronounced as the layer dimensions become smaller, manifesting itself also in the X-ray analysis as a grading of the Al fraction at the interfaces.[20]

2.3 Calculations and Discussion

Our calculations of the subband structure of GaAs-AlAs, (001)-oriented multiple quantum wells and SLs have been made within the envelope function approximation. We specifically concerned ourselves with the subband structure associated with the Γ point and X points. Coupling between the states belonging to different symmetry points in the Brillouin zone of the bulk was not considered. Any effects of elastic strain, due to the small difference in lattice constant between the binary compounds, were also neglected. To proceed with the calculations we adopted a conduction to valence band offset ratio of 67:33, in line with the most accurate optical determination of this quantity.[18,25] Neglecting any confinement effects and assuming this offset ratio of 67:33 makes the X point in AlAs the lowest X state at the GaAs-AlAs heterojunction. Within our simple model this means that (i) electrons at Γ see a Kronig-Penney potential whose size is determined by the Γ-Γ separation of the conduction band extrema between GaAs and AlAs, while (ii) electrons at one of the X points experience a Kronig-Penney potential determined by the separation between the X point of AlAs and the X point of GaAs.

The lowest subband for the electron at the Γ point in the GaAs was calculated using our version of Bastard s implementation of the Kane model[2] which naturally includes non-parabolicity of the Γ electron band. Subband minima for the various X points were calculated by ensuring continuity of the envelope function and the product of (1/m*) and the derivative of the envelope function at each of the heterointerfaces. We neglected any complications associated with the "camel s back" structure of both GaAs and AlAs and assumed parabolic dispersion for an electron state precisely at the X point. The effective masses used in the calculations are shown in Table 1. Note that the anisotropic nature of the X minima means that those X minima which have a component of momentum parallel to the growth direction, (001) and labelled X_z, have a much larger confinement mass than those whose momenta lie in the layer planes, i.e. (100) and (010) which we label X_x and X_y. Often we will refer to the latter pair of states as X_{xy}. Some uncertainty exists in the literature as to the precise values of the X point effective masses either in GaAs or in AlAs, however the important point to remember is that the longitudinal mass at the X point is large, whereas the transverse mass is small.

For the calculation we make here, the factor of almost 6 between the confinement mass of the X_z mininum versus that of the X_{xy} minima (1.1 m_0 compared to 0.19 m_0) means that the folded X_z state is always at a lower confinement energy than the unfolded X_{xy} states in any GaAs-AlAs multiple quantum well or SL system. This conclusion would be modified somewhat if the elastic strain between the AlAs and GaAs layers were important as has recently been suggested by Drummond and co-workers.[26] This question was addressed more closely in our earlier

TABLE 1. Effective masses in units of m_0 used in the model calculations

	m_e^{Γ} (001)	m_{hh}^{Γ} (001)	m_{1h}^{Γ} (001)	m_e^{z} (001)	m_e^{x} (100)
GaAs	0.0665	0.34	0.094	1.3	0.19
AlAs	0.15	0.752	0.16	1.1	0.19

publication,[21] but since none of the observations made below about GaAs-AlAs SLs depend explicitly on the system being strained it is left out of the calculation.

In fig.4 the calculated variation of the Γ-Γ, X_z-Γ, and X_{xy}-Γ subband energy gaps as a function of the $(GaAs)_m$-$(AlAs)_m$ SL period (2m) are shown. Also given for reference are the direct and indirect energy gaps of an $Al_{0.5}Ga_{0.5}As$ bulk sample. Γ refers to the lowest conduction band state in the GaAs, while the X states refer to the minima in the AlAs layers. Also shown in this figure is a collection of the available spectroscopic data on m=n GaAs-AlAs short period SLs. Comments on the comparison of this data either with the theory, with our own data, or between sets of data are given below. As noted earlier, we find the X_z state is always below the unmixed X_{xy} state and for periods smaller than 22 ML it becomes the lowest conduction band state of the whole system[17] with the valence to lowest conduction band line-up taking on a staggered or type-II arrangement. Unlike the situation of fixed GaAs thickness and decreasing AlAs layer thickness[23] the m=n SL can never revert back to a type-I or straddled arrangement for the valence to lowest conduction band lineup. From these calculations we can say that when the period of m=n, $(GaAs)_m$-$(AlAs)_n$ SLs is smaller than 20 ML we would expect to observe the allowed, no-phonon recombination of an electron of predominantly AlAs X character, i.e. folded X_z with a hole at the Γ point localized mainly in the GaAs layers and not an electron at the X_{xy} minima as suggested by Finkman et al.[14,24]

To some extent this picture is borne out in the ability to described theoretically the spectroscopic data on type-II GaAs-AlAs quantum wells in terms of the folded X_z state being the lowest in the system[23] using the same calculation. Furthermore, recent optically detected magnetic resonance measurements on type-II GaAs-AlAs QWs,[27] as a function of orientation, have revealed that for QWs with AlAs layers thinner than $\approx 55 \text{Å}$ the X_z conduction band valley has the lowest energy.

2.4 Comparison with Experiment

The energy positions of the direct Γ-Γ related transitions and the X-related transitions are compared with our decoupled, X and Γ, Kronig-Penney type calculations in fig.4. Also plotted is a large selection of other available data and marked are the band edge positions of the indirect, X and direct gaps for an $Al_{0.5}Ga_{0.5}As$ alloy. All the Γ related observations plotted refer to positions determined from PLE and are all most likely excitonic in origin. To compare with the calculated band-band transitions an additional binding energy, of the order of 10 meV,[28] needs to be added to the observations. The X related data are all assumed to be intrinsic PL peaks and therefore need to be corrected by a similar amount for the exciton binding energy,[29] and, in some cases, by an additional amount to account for the localised nature of the emission.[18]

It is clear that the energy position of the Γ-Γ transitions is extremely well reproduced down to the (4+4) SL. Below (4+4) the experimental data is scarce, however the one reported measurement by Nagle et al.[13] lies well below both our calculations and similar ones made by these authors. It is extremely surprising that such a simple model works so well for layers of these dimensions. The three-band Kane model on which it is based is really only applicable over a limited energy range close to k=0 yet for the (4+4) sample the bottom of the electron subband is ≈ 500 meV shifted from the GaAs conduction band edge. Assuming that the structural integrity of the SL is maintained for the (2+2) sample examined in ref. 13 then clearly by (2+2) the model has failed.

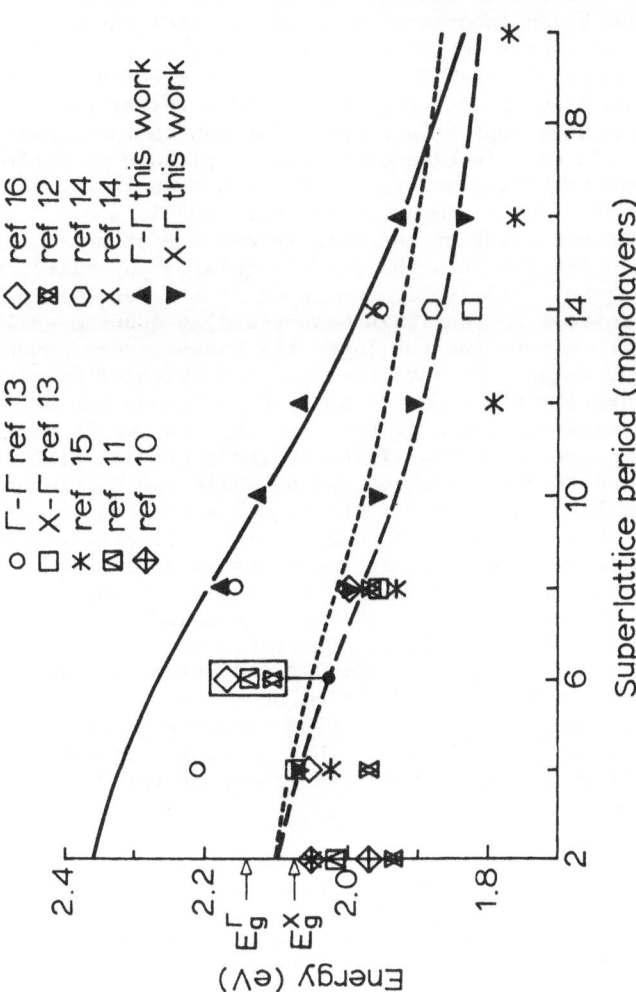

FIG. 4. Comparison between our model calculations of the principal subband–subband energy gaps of $(GaAs)_m$–$(AlAs)_m$ superlattices and a variety of available experimental data. The solid curve is the calculated Γ–Γ gap, the dotted curve is the X_{xy}–Γ gap, and the dashed curve corresponds to the X_z–Γ gap. Also marked on the plot are the direct and indirect gaps of an $Al_{0.5}Ga_{0.5}As$ bulk sample.

Drawing conclusions about the applicability of the simple decoupled model to the description of X related transitions is a little more difficult, in part contributed to by the quite wide spread in experimental observations on samples which are nominally identical. This problem becomes particularly acute for SL dimensions of (4+4) and below. The combination of our X_z-Γ observations and the parameters we have chosen to use in the calculation persuades us that presently the limit of these calculations is at (6+6) monolayers. However, this statement is not intended to be categorical and a number of factors should be borne in mind before considering this conclusion as absolute.

There are a number of obvious limitations to our simple model. One source of discrepancy could be an incorrect value of the X-electron wave vector in the GaAs. We have chosen to calculate its k vector using the band edge value of the effective mass. The detailed band structure in bulk GaAs in the vicinity of the X points is complicated and a better model description should use some more appropriate approximation to the complex, non-parabolic dispersion. On an effective mass sort of picture an improved dispersion would correspond to having a larger effective mass in the GaAs rather than the band edge value of $1.3\ m_0$ that we have used in our rudimentary model. One further thing to consider is our use of an idealised square well potential for both the Γ and X related superlattices. In a recent publication Nelson, Miller, Tu, and Sputz[30] have determined the binding energies of excitons in very thin GaAs-(AlGa)As quantum wells, where interruption of growth meant that the layer thicknesses were known to be in discrete monolayer steps. An envelope function calculation was employed to determine accurately the lowest subband eigenvalues and a correction made to the eigenvalues based on the fact that the wells were not precisely square. Evidence for a transition of the potential profile over a monolayer was claimed from the work of van de Walle and Martin on the ideal, (001) GaAs-AlAs heterojunction.[31] Were such a grading of the potential profile present then it would, of course, alter the eigenvalues of the system we are considering, becoming an increasingly important effect as the superlattice dimensions become smaller. The net effect would be to increase the value of all the eigenvalues[32] of the system and qualitatively move the calculations toward our experimental observations. As a general point, the spread in data for samples where (m+n)⩽8 indicates the clear need for some improved structural characterisation of similar samples, perhaps using Raman spectroscopy.[33] If the samples really have the dimensions that the authors suggest they have, then something is clearly questionable in the identification by each of the groups of their spectral features as being intrinsic.

The optical data described above has revealed that for $(GaAs)_m$-$(AlAs)_n$ SLs with m=n⩾4 the lowest lying conduction band state is a folded X_z minimum. This has important consequences when we consider the extent to which the SLs are electronically coupled. Since the confinement mass of the X_z minimum is large ($1.1\ m_0$) there is relatively little dispersion in the z direction of the lowest conduction band state of the system. We calculate a miniband width for the X_z electron state of a (6+6) SL ≈ 1.2 meV. The higher lying Γ electron state is of course a truly extended SL state with a calculated miniband width ≈ 76 meV for a (6+6) structure. The features observed in the PLE spectrum in the X_z-Γ transition region of these structures are therefore well described by a 2D picture. In order to observe real SL effects in the optical data of GaAs-AlAs structures, samples must be carefully designed to ensure that the lowest conduction band minimum is the direct Γ electron state with a large dispersion in the z direction. Examples are the (9+3) and (9+2) SLs investigated by Moore and co-workers.[23] Alternatively one can look to different heterostructures combining two direct semiconductors such as GaAs-Al_xGa_{1-x}As (x < 0.45) or InGaAs-GaAs. The latter combination is discussed below.

3.   InGaAs-GaAs SUPERLATTICES

The ability to grow strained layer heterostructures without the generation of misfit dislocations not only widens the choice of material combinations, it also greatly increases the ability to control the electronic and optical properties of such structures. By carefully selecting the appropriate heterostructure and substrate it is possible to design systems in which the QW is under either compression or tension. For the case of $In_xGa_{1-x}As$-GaAs QWs the InGaAs is under compression in the layer plane, while the lattice constant is extended in the direction normal to the interfaces. Elastic tetragonal distortion of the zinc blende lattice does not destroy the crystalline quality of the epitaxial layer provided the InGaAs thickness is kept below a strain dependent critical value.[34,35] The effect of the lattice mismatch in this system is two-fold: The hydrostatic pressure component increases the energy gap of the InGaAs whilst the uniaxial stress lifts the degeneracy of the valence band edge at the centre of the Brillouin zone so that the heavy hole band moves up and the light hole band moves down relative to the unstressed band. Considerable attention has been devoted to the growth of pseudomorphic InGaAs-GaAs QWs and high optical quality material has been demonstrated, grown both by MOCVD[36] and by MBE.[37,38] This success has been accompanied by optical studies of the QW electronic properties and subband structure. It has been clearly demonstrated that data from a range of techniques, including photoluminescence, absorption and photo-reflectance, when combined with the appropriate theoretical model, can provide an excellent means of addressing fundamental questions about this system.

Below are reported the results of low temperature PL and PLE measurements made on InGaAs-GaAs quantum wells and superlattices grown by MBE. The samples studied comprised InGaAs layers of thickness 25Å with a nominal In fraction of 0.12, and GaAs layers of either 400Å or 100Å. The optical quality of the material is demonstrated by PL linewidths which are comparable with the best reported in the literature. Furthermore, differences in the PLE spectra of isolated QW samples and coupled SLs are evidenced by the observation of new features in the continuum region of the first heavy-hole exciton peak.

3.1   Growth Details

The layers were grown in a Varian modular GEN II molecular beam epitaxy system. The substrate temperature during growth was estimated to be $580^{0}C$ which is in the temperature range where significant re-evaporation of In occurs. This loss was compensated for by using a growth rate about 1.5 times higher than that nominally required for the desired In composition. This method was found to give quite reproducible results for several sets of layers grown at this substrate temperature. Growth rates for both GaAs and InGaAs (and thus InAs) were measured using the RHEED oscillation technique on a GaAs monitor slice prior to growth. The GaAs growth rate was always set at 1 μm/hr for these layers and a flux of As_4 was supplied which was just sufficient to ensure an arsenic stabilised surface for both GaAs and InGaAs. The GaAs (001) undoped substrates were mounted in In free holders and rotated at 20 rpm during growth to ensure lateral uniformity. The growth sequence was as follows: (a) a 1 μm GaAs buffer, (b) 5 periods of $In_xGa_{1-x}As$ and GaAs layers and (c) a GaAs capping layer. The InGaAs layers were 25Å thick while the GaAs layers were either 100Å or 400Å thick. The samples were grown with a fixed nominal composition of $x = 0.12$ and none of the layers was intentionally doped. The total thickness of the GaAs capping layer plus superlattice region was fixed at 0.27 μm for both structures. The InGaAs layers were deliberately kept below the reported critical layer thickness for this composition.[34,35] Furthermore, X-ray diffraction analyses[39] indicated that indeed the lattice mismatch is elastically accommodated entirely within the InGaAs layers.

3.2 Theoretical Background

In considering the description of InGaAs-GaAs QWs grown on a (001) GaAs substrate (and buffer) then the InGaAs layers will be under biaxial compression in the layer planes and subjected to a shear, uniaxial stress in the growth direction. The GaAs substrate and buffer regions are far thicker than the epitaxially deposited InGaAs layers and provided we do not exceed the critical layer thickness then we are justified in assuming that it is only the InGaAs layers that are elastically strained.

The biaxial strain, ε, is given by

$$\varepsilon = (a_2 - a_1)/a_1$$

where a_1 and a_2 are the lattice constants of InGaAs and GaAs respectively. The lattice constant of the alloy is calculated by assuming a linear variation between the binary end-members.

As noted above, the InGaAs is under compression in the layer planes ($\varepsilon < 0$) and the net effect of the hydrostatic and uniaxial components of the strain is to increase the band gap relative to the unstrained value and remove the degeneracy of the light and heavy hole valence bands. If we denote the unstrained band gap of the $In_xGa_{1-x}As$ by $E_o(x)$ then the strained conduction to heavy-hole (C-HH) and strained conduction to light-hole (C-LH) band gaps are given by[40]

$$E^{C-HH} = E_o(x) + \delta E_H - \delta E_s$$

$$E^{C-LH} = E_o(x) + \delta E_H + \delta E_s - (\delta E_s)^2/2\Delta_o$$

where Δ_o is the spin-orbit splitting in the InGaAs and δE_H and δE_s are the hydrostatic shift and shear splitting components of the strain respectively. These last two quantities are expressed in terms of the elastic stiffness constants (C_{ij}) and appropriate deformation potentials as[40]

$$\delta E_H = 2a\left[(C_{11} - C_{12})/C_{11}\right]\varepsilon$$

$$\delta E_s = b\left[(C_{11} + 2C_{12})/C_{11}\right]\varepsilon$$

The parameters a and b are the interband hydrostatic and uniaxial deformation potentials respectively. Once again, values of a and b for the ternary are determind by a linear interpolation between the values for the binary end-members. The values used in our calculations are gathered together in Table 2.

Of course, in the systems we are concerned with we not only have to account correctly for strain induced effects but also for quantum confinement effects. We have calculated the position and width of the appropriate electron and hole subbands within the envelope function approximation; matching conditions used ensured continuity of the particle wavefunction and "current" at each interface. Necessary input parameters into the calculations are the variations in effective masses as a function of In fraction, the variation of the unstrained gap $E_o(x)$, the change in Δ_o with In fraction and, of course, the appropriate band offsets. The variations in the physical parameters with In fraction are given in Table 3.

TABLE 2. Deformation potentials and elastic stiffness constants used in the calculation of the strain InGaAs band gaps. Values for the ternary are found from linear interpolation.

Material	a(eV)	b(ev)	C_{11} $(10^{11} dyn/cm^2)$	C_{12} $(10^{11} dyn/cm^2)$
GaAs	-9.8	-1.7	11.88	5.38
InAs	-5.9	-1.8	8.33	4.53

In all our comparisons we have used a band offset, assumed to be independent of the In fraction and equal to a value of Q_c of 0.67. Our definition of Q_c is such that this quantity corresponds to that fraction of the energy gap difference between the GaAs barrier and the strained conduction to heavy-hole band gap of InGaAs that appears in the conduction band. This particular value means that for GaAs barriers we always have a mixed situation for the heavy- and light-holes; the heavy-holes being confined in the InGaAs wells whilst the light-holes reside predominantly in the GaAs. Clearly the values of some of the parameters will be critical in comparing the calculated and measured transition energies, and it is fair to say that many of the parameters remain uncertain. This seems to be particularly true of the deformation potential, a. We will not attempt to justify our use of the offset fraction here but note that given our information with regard to well width and In fraction, we find that this offset ratio adequately describes our assigned transitions and splitting between the light- and heavy-holes.

At the growth temperature we have used In desorption from the surface is occurring. This is a difficult regime to work in and the amount of In in the layers will depend sensitively on the precise temperature of the substrate. Rather than rely on growth information to calibrate the In fraction in the layers we have determined the In fraction by calculating the energy of the lowest allowed transition as a function of In fraction until agreement is obtained. In this procedure we assume that the well width is close to the nominal value. Our calculations show that in fact the lowest levels are rather insensitive to both the layer thickness and to the band offset.

TABLE 3. Variation of materials parameters as a function of In fraction x. All masses are in units of m_0 and all energies in eV.

Quantity	Variation with In fraction, x
Unstrained gap (4K)[37]	$E_0(x) = 1.519 - 1.5387x + 0.475x^2$
Electron mass	$m^*_e = 0.0665 - 0.0435x$
Heavy-hole mass	$m^*_{hh} = 0.34$
Light-hole mass	$m^*_{lh} = 0.094 - 0.062x$
Spin-orbit splitting	$\Delta_0 = 0.341 - 0.09x + 0.14x^2$

3.3 Superlattice Effects

The PL and PLE measurements were recorded using the same experimental arrangement described earlier. In this case however, the excitation source was an Ar^+ pumped dye (styrl-9) laser which extended the tunable wavelength range out to 9000Å.

Figure 5 shows the 4K PL and PLE spectra from a sample with 5 periods of 25Å $In_xGa_{1-x}As$ and 100Å GaAs. The PL spectrum was recorded using an excitation wavelength of 8060Å. The emission peak is measured at 1.4717 eV and has a FWHM of 2.2 meV, a good indication of the quality of the sample. The PLE spectrum was recorded with the detection energy set at 1.469 eV. The three most prominent peaks are the 1s states of the n=1 heavy- and light-hole excitons, at 1.4715 eV and 1.4926 eV respectively, and the GaAs free exciton peak at 1.5150 eV, originating from the unstrained GaAs buffer layer and substrate. The coincidence of the fundamental heavy-hole exciton peak, observed in PLE, with the PL peak identifies the emission as free exciton recombination. The best fit to this data, assuming the nominal layer thicknesses, is obtained by taking an In composition of 14½%, allowing a few meV for the exciton binding energies. Notice that the GaAs exciton is extremely well resolved from the continuum states, giving a direct confirmation of the bulk GaAs exciton binding energy of about 4 meV. In addition to the fundamental exciton peaks there are 3 distinct features in the PLE spectrum between the heavy- and light-hole peaks, labelled a, b and c, and a further feature above the light-hole labelled as d. The appearance of these peaks in the spectrum is a consequence of the electronic coupling of adjacent wells. Discussion as to the origin of this structure follows below.

One of the most significant differences between the GaAs-AlAs SLs already reported in this paper and the $In_{0.14}Ga_{0.86}As$-GaAs system is the importance of electronic coupling in samples with relatively thick barrier layers. The combined effect of a smaller electron effective mass and much lower barrier height in the pseudomorphic structures results in considerable subband dispersion in 25Å structures, even with 100Å barrier layers. The transition from an isolated QW to a SL is characterised by a change in the density of states, or the absorption curve, from a stepwise function in the purely 2D picture, to a quasi 3D model in which the continuum edges associated with each subband are smeared out by an energy which describes the miniband width of that state. For an infinite SL structure of the dimensions described above, the heavy-hole states remain essentially localised in the InGaAs layers, whilst the electron and light-hole states are significantly coupled. We calculate miniband widths of 20 meV, 1 meV and 26 meV for the electron, heavy-hole and light-hole respectively. However real minibands are only formed in SL structures with a large number of repeats of well and barrier layers. For the case of a small, finite number of periods a system intermediate between the isolated QW case and the true SL case is created. When the barriers are sufficiently thin to couple the wavefunctions in adjacent wells, the degenerate single well states mix, and split into multiple, discrete but delocalised states. Each of the resulting excited states being made up of a linear combination of the exact envelope functions of the isolated wells. Hence the total number of states derived from the original ground state corresponds to the number of periods in the sample. In the GaAs-AlGaAs system, considerable attention has been devoted to understanding the properties of coupled QW structures. The predicted splitting of the discrete well states into symmetric and anti-symmetric components has been observed in photo-luminescence[41] and in excitation spectroscopy[42,43] for the case of coupled double QWs, and in absorption measurements[44] for both double and triple QWs. However, to our knowledge, similar studies in strained layer materials have not yet been reported.

The sample studied in figure 5 is a coupled five well structure, hence we anticipate 4 excited electron states in addition to the ground state already identified. Assuming the nominal layer thicknesses, we expect an energy difference of 21 meV between the ground state exciton and the fifth electron to heavy-hole state. In the PLE spectrum of figure 5 we cannot unambiguously identify the SL structure without further experimental evidence. However, we note that the energy splitting between the (e1-hh1) and (e1-ℓh1) exciton peaks is exactly 21 meV. Therefore we ascribe the features a, b and c to the 2nd, 3rd and 4th excited electron

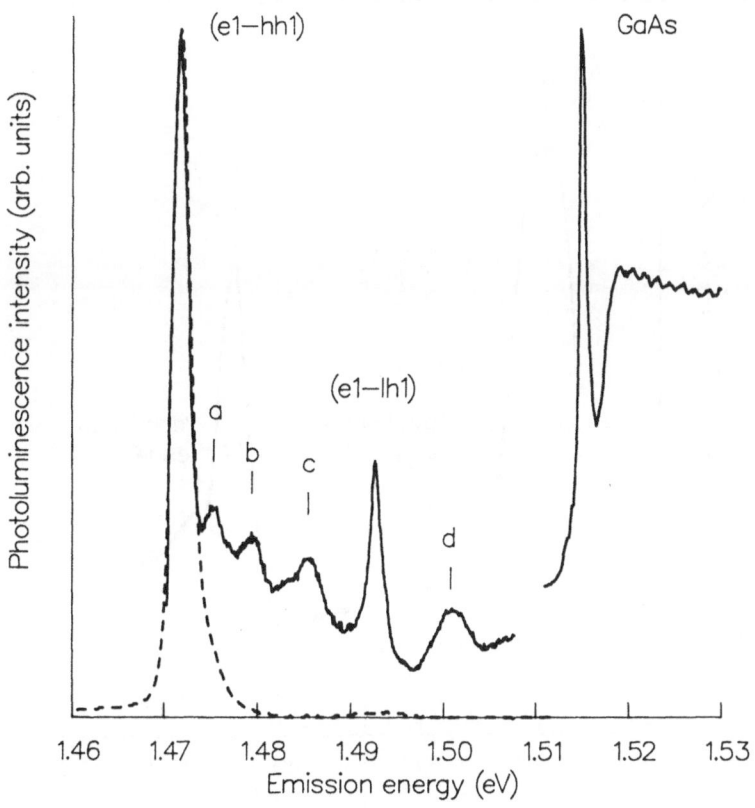

FIG. 5. Low temperature (4K) photoluminescence (dashed line) and photoluminescence excitation (solid line) spectra of the 25Å 100Å $In_xGa_{1-x}As$-GaAs (x=0.145) superlattice. Note that there is an increase in gain of the detection system below 1.510 eV.

to heavy-hole exciton peaks and suggest that the 5th state is in fact degenerate with the fundamental light-hole exciton. Similarly, peak d is attributed to the second excited electron to second excited light-hole exciton. In this case, the subbands of both carriers are coupled, and we have to add the non-degeneracy of the light-hole to the splitting of the electron state already described. Hence the splitting of peak d from the fundamental (e1-ℓh1) peak is approximately twice the corresponding heavy-hole splitting. Strong evidence to support these interpretations is obtained by contrasting the result on this sample with data on an isolated QW sample, discussed in section 3.4.

3.4 Measurement of the Exciton Binding Energy

Figure 6 shows the 4K PLE spectrum from a sample with 25Å
$In_xGa_{1-x}As$ and 400Å GaAs. The spectrum was recorded in two parts.
The position of the 1s state of the heavy-hole exciton (e1-hh1) was
measured by setting the detection energy to the low energy tail of the
emission line. The remainder of the spectrum was then recorded with the
detection energy set to the peak of the PL line at 1.469 eV. For this
sample the PL and the fundamental exciton peak measured in the PLE
spectrum were coincident within 0.5 meV and again this identifies the
emission from this sample as free exciton recombination. The second
dominant feature in fig.6 is the ground state light-hole exciton (e1-lh1)
measured at 1.495 eV. The best fit to this data, assuming the nominal

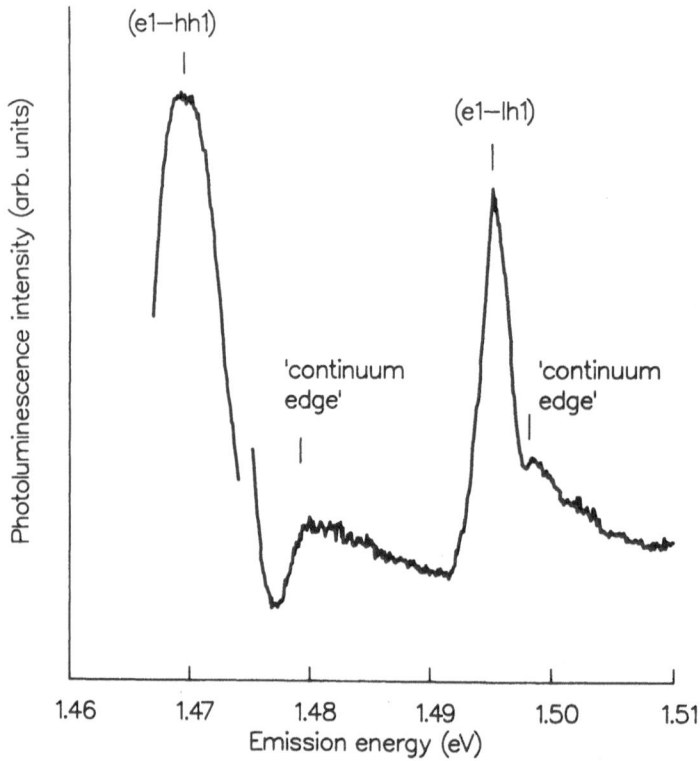

FIG. 6. Low temperature (4K) photoluminescence excitation spectrum of
the 25Å-400Å $In_xGa_{1-x}As$-GaAs (x=0.15) quantum well. Note
that there is a change in the detection energy below 1.474 eV.

layer thicknesses, is obtained by taking an In composition of 15%. The
thickness of the GaAs barrier layers in this structure was deliberately
chosen to ensure that adjacent InGaAs layers were effectively
electronically isolated. Therefore we anticipate that all the features
observed in the excitation spectrum of this sample should be adequately
described by a simple 2D model in which the carriers are confined to the
layer planes. This is borne out when we study the shape of the excitation
spectrum. On the high energy side of the (e1-hh1) peak a shoulder is
resolved which we interpret as representing the onset of the excited
states of the heavy-hole exciton which are not resolved from the
continuum. Similar structure, although less well resolved, is observed
on the high energy side of the (e1-lh1) peak, representing the light-hole
exciton continuum edge.

To date such measurements have been almost entirely limited to GaAs-AlGaAs QW structures. In this system a number of groups[46,47] have reported PLE spectra showing the n=1 1s heavy-hole exciton feature clearly resolved from its continuum states. Furthermore, in high optical quality materials, the excited 2s state of the heavy- and light-hole excitons have also been observed,[48] again in GaAs-AlGaAs QWs. The appearance of the 2s peaks in the spectra reported by ourselves,[48] allowed us to precisely determine the difference in binding energy between the 1s and 2s states. Combining this measurement with the calculated 2s exciton binding energy produced an accurate determination of the binding energy of the 1s state. In GaAs-AlAs QWs we have measured exciton binding energies as high as 15 meV in a sample with 42Å wells,[49] while D.F. Nelson et al.[30] have deduced an exciton binding energy of 25 meV for a GaAs-$Al_{0.4}Ga_{0.6}As$ structure with a well width of 17Å. Theoretically, the exciton binding energy is expected to reach a maximum as the QW width is reduced, determined by an interplay of increasing two-dimensional confinement and increasing penetration of the barriers by the exciton wave function. As the well width approaches zero, the binding energy should then decrease and approach the exciton binding energy of the barrier material. Chomette, and co-workers[28] have used PLE and PL measurements as a function of temperature to determine exciton binding energies in small period GaAs-$Al_{0.3}Ga_{0.7}As$ SLs. They reported a decrease in the heavy-hole binding energy with decreasing SL period. In the all-binary GaAs-AlAs system we note a drop in the heavy-hole exciton binding energy by about one half between measurements on QW[48,49] and SL[50] structures. These observations are consistent with an increase in the 3D character of the exciton as the subband states become extended SL minibands. However to our knowledge the predicted "turn-over" of the exciton binding energy for isolated QWs has never been observed experimentally.

In the PLE spectrum of the 25Å InGaAs QW, illustrated in fig.6, we are not able to resolve the excited 2s exciton from the continuum states. However, a good estimate of the heavy-hole exciton binding energy can be obtained from the measured splitting between the 1s exciton state and its continuum edge. From this data we estimate a heavy-hole exciton binding energy of 9.0 meV. This we believe is the first direct measurement of the exciton binding energy in strained samples of significant (>5%) In fraction. Furthermore, calculations[51] for the $In_xGa_{1-x}As$-GaAs (x=0.14) system predict that the exciton binding energy reaches a maximum at approximately 50Å InGaAs. Hence this result is particularly significant because the thickness of the InGaAs layer in the sample is less than the thickness anticipated for the "turn-over" of the exciton binding energy. We have made similar measurements on additional samples as a function of well width which confirm this result, including an exciton binding energy of 6.5 meV for an isolated QW sample with 14Å InGaAs wells. This data, together with our calculations of the 1s binding energy in this system, are the subject of another publication.[52]

It should be noted that there are significant differences between the data shown in figures 5 and 6. Recall that the two samples studied are nominally identical except for the thickness of the GaAs layers. The first sample (fig.5) is a coupled SL structure with 100Å barriers, while in the second sample (fig.6) the thickness was increased to 400Å, producing an isolated QW structure. The features a, b, c and d in figure 5, attributed to the excited states of the coupled SL and should be absent from the PLE spectrum of the QW sample. This is clearly our observation, and is substantial evidence in support of our interpretation. Hence these additional peaks are a direct result of the overlap, between adjacent wells, of the electron and hole wavefunctions, leading to a mixing and split levels in the conduction and valence bands.

4. CONCLUSIONS

The optical properties of both short period, $m=n\leqslant 8$ $(GaAs)_m-(AlAs)_n$ SLs and pseudomorphic InGaAs–GaAs structures have been investigated. For $(GaAs)_m-(AlAs)_n$ SLs careful PL and PLE measurements have revealed that for $m=n\geqslant 4$ the lowest lying CB state is a folded X_z minimum and the SLs can be considered a "pseudo-direct" materials. This has important consequences when considering the extent to which the SLs are electronically coupled. The large confinement mass of the X_z minimum means that there is relatively little dispersion in the z-direction of the lowest CB state in this system. All the features observed in the PLE spectra are therefore well described by a two-dimensional model of the subband states. For these samples the smallest energy gap does not approximate to that of the $Al_{0.5}Ga_{0.5}As$ alloy until the period is reduced to only 4 ML. A simple decoupled, Kronig-Penney type model of the principal energy gaps in this system shows surprisingly good agreement with the observed direct, $\Gamma-\Gamma$ transitions for samples with $m=n\geqslant 4$ ML. For these samples, favourable comparison of this simple model with measured $X_z-\Gamma$ transitions presently breaks down below $m=n=6$. This is most likely due to an underestimate of the energy of the AlAs, X_z-subband minima and the use of an idealized square-well profile. Finally, cross comparison of many of the published spectroscopic data reveals the need for much more detailed microscopic information on the lateral structure of the sample on the scale of the exciting probe, i.e. exciton diameter, as compared to the "average" quantities yielded by the current X-ray analysis.

Optical measurements on InGaAs–GaAs structures have highlighted the relative importance of superlattice effects in this system. In a structure with 25Å InGaAs and 100Å GaAs significant electronic coupling is anticipated and new features are observed in the PLE spectrum due to the additional dispersion along k_z. These peaks are attributed to the excited states of the coupled 5 well SL system. Such features are readily observed in this system because of the extended nature of the lowest direct CB state. Evidence to support these assignments has been obtained by studying a similar sample with much thicker barrier layers. This ensured that the subbands were electroncally isolated and accordingly the ground states are degenerate in this structure. The PLE spectrum of the 25Å InGaAs–400Å GaAs QW reveals a n=1 1s heavy-hole exciton peak clearly resolved from the onset of the excited states and the continuum edge. This observation provides an estimate for the exciton binding energy of 9.0 meV in this sample. This result is particularly significant because the well width in this sample is less than the thickness at which the exciton binding energy is predicted to have gone through a maximum, for the isolated QW case.

ACKNOWLEDGEMENTS

I am indebted to many of my colleagues at Philips for their contributions to parts of this work. In particular I am most grateful to Geoff Duggan for his valuable and continual support in the theoretical aspects of these investigations. My thanks are also extended to Karl Woodbridge, Christine Roberts, Tom Foxon and David Hilton, all of whom were involved in the growth of the samples, and to John Orton for his tireless support and encouragement.

REFERENCES

1. R. L. de Kronig and W. J. Penney, Proc. R. Soc. London, Ser. A130, 499 (1930).
2. G. Bastard, Phys. Rev. B 25, 7584 (1982).
3. J. N. Schulman and Y. C. Chang, Phys. Rev. B 24, 2445 (1981).
4. J. N. Schulman and T. C. McGill, Phys. Rev. 19, 6341 (1979).
5. W. Andreoni and R. Car, Phys. Rev. B 21, 3334 (1980).
6. M. A. Gell, D. Ninno, M. Jaros and D. C. Herbert, Phys. Rev. B 34, 2416 (1986).
7. T. Nakayama and H. Kamimura, J. Phys. Soc. Jpn. 54, 4726 (1985).
8. D. M. Bylander and L. Kleinman, Phys. Rev. B 36, 3229 (1987).
9. R. Eppenga and M. F. H. Schuurmans, Phys. Rev. B 38, 3541 (1988).
10. A. C. Gossard, P. M. Petroff, W. Weigmann, R. Dingle and A. Savage, Appl. Phys. Lett. 29, 323 (1976).
11. J. P. van der Ziel and A. C. Gossard, J. Appl. Phys. 48, 3018 (1977).
12. T. Isu, De-Sheng Jiang and K. Ploog, Appl. Phys. A 43, 75 (1987).
13. J. Nagle, M. Garriga, W. Stolz, T. Isu and K. Ploog, J. Phys. (Paris) Colloq. 29, C5-495 (1987).
14. E. Finkman, M. D. Sturge and M. C. Tamargo, Appl. Phys. Lett. 49, 1299 (1986).
15. A. Ishibashi, Y. Mori, M. Itabashi and N. Watanabe, J. Appl. Phys. 58, 2691 (1985).
16. N. Kobayashi and Y. Horikoshi, Appl. Phys. Lett. 50, 909 (1987).
17. G. Duggan and H. I. Ralph, Proc. SPIE 792, 147 (1987).
18. P. Dawson, K. J. Moore and C. T. Foxon, Proc. SPIE 792, 208 (1987).
19. P. F. Fewster and C. J. Curling, J. Appl. Phys. 62, 4145 (1987).
20. P. F. Fewster, (unpublished).
21. K. J. Moore, G. Duggan, P. Dawson and C. T. Foxon, Phys. Rev. B 38, 5535 (1988).
22. K. J. Moore, P. Dawson and C. T. Foxon, J. Phys. (Paris) Colloq. 29, C5-525 (1987).
23. K. J. Moore, P. Dawson and C. T. Foxon, Phys. Rev. B 38, 3368 (1988).
24. E. Finkman, M. D. Sturge, M-H. Meynadier, R. E. Nahory, M. C. Tamargo, D. M. Hwang and C. C. Chang, J. Lumin. 39, 57 (1987).
25. G. Danan, B. Etienne, F. Mollot, R. Planel, A. M. Jean-Louis, F. Alexandre, B. Jusserand, G. Leroux, J. Y. Marzin, H. Savary and B. Sermage, Phys. Rev. B 35, 6207 (1987).
26. T. J. Drummond, E. D. Jones, H. P. Hjalmarson and B. L. Doyle, Proc. SPIE 796, 2 (1987).
27. H. W. van Kesteren, E. C. Cosman, P. Dawson, K. J. Moore and C. T. Foxon, Phys. Rev. B, (to be published).
28. A. Chomette, B. Lambert, B. Deveaud, F. Clerot, A. Regreny and G. Bastard, Europhys. Lett. 4, 461 (1987).
29. G. Duggan and H. I. Ralph, Phys. Rev. B 35, 4152 (1987).
30. D. F. Nelson, R. C. Miller, C. W. Tu and S. K. Sputz, Phys. Rev. B 36, 8063 (1987).
31. C. G. van de Walle and R. M. Martin, (unpublished).
32. J. N. Schulman, J. Vac. Sci. Technol. B 1, 644 (1983).
33. M. Cardona, T. Suemoto, N. E. Christensen, T. Isu and K. Ploog, Phys. Rev. B 36, 5906 (1987).
34. I. J. Fritz, S. T. Picraux, L. R. Dawson, T. J. Drummond, W. D. Laidig and N. G. Anderson, Appl. Phys. Lett. 46, 967 (1985).
35. M. Gal, P. C. Taylor, B. F. Usher and P. J. Orders, J. Appl. Phys. 62, 3898 (1987).
36. D. C. Bertolet, J-K. Hsu, S. H. Jones and K. M. Lau, Appl. Phys. Lett. 52, 293 (1988).
37. G. Ji, D. Huang, U. K. Reddy, T. S. Henderson, R. Houdre and H. Morkoç, J. Appl. Phys. 62, 3366 (1987).
38. L. Goldstein, M. N. Charasse, A. M. Jean-Louis, G. Leroux, M. Allovon and J. Y. Marzin, J. Vac. Sci. Technol. B 3, 947 (1985).

39. P. F. Fewster (unpublished).
40. F. H. Pollak, Surf. Sci. 37, 863 (1973).
41. H. Kawai, J. Kaneko and N. Watanabe, J. Appl. Phys. 58, 1263 (1985).
42. E. S. Koteles, Y. J. Chen and B. S. Elman, Proceedings of the
     International Meeting on Excitons in Confined Systems, Roma (1987).
43. C. Delalande, U. O. Ziemelis, G. Bastard, M. Voos, A. C. Gossard and
     W. Wiegmann, Surf. Sci. 142, 498 (1984).
44. R. Dingle, A. C. Gossard and W. Wiegmann, Phys. Rev. Lett. 34, 1327
     (1975).
45. K. J. Moore, G. Duggan, K. Woodbridge and C. Roberts, (unpublished).
46. R. C. Miller, D. A. Kleinman, W. T. Tsang and A. C. Gossard, Phys.
     Rev. B 24, 1134 (1981).
47. M. H. Meynadier, C. Delalande, G. Bastard, M. Voos, F. Alexandre and
     J. L. Lievin, Phys. Rev. B 31, 5539 (1985).
48. P. Dawson, K. J. Moore, G. Duggan, H. I. Ralph and C. T. Foxon, Phys.
     Rev. B 34, 6007 (1986).
49. K. J. Moore and P. Dawson, (unpublished).
50. K. J. Moore, G. Duggan, P. Dawson, C. T. Foxon, N. J. Pulsford and
     R. J. Nicholas, Phys. Rev. B 39, 1219 (1989).
51. G. Duggan, (unpublished).
52. G. Duggan, K. J. Moore, K. Woodbridge and C. Roberts, (unpublished).

INDIRECT GaAs/AlAs SUPERLATTICES

Marie-Helene Meynadier

ATT Bell Laboratories
600 Mountain Avenue
Murray Hill, NJ 07974, USA

ABSTRACT

The optical properties of indirect GaAs/AlAs superlattices reflect a band structure in which the admixture of states issued from various points of the Brillouin zone plays an important role. We decribe here several experiments in which the degree of mixing between X and Γ states is monitored by tuning the energy spacing between the corresponding minima. Cw and time resolved photoluminescence experiments, in and out of the presence of an electric field, allow the determination of the strength of the potential responsible for the mixing. Under higher optical excitation the character of the photoluminescence changes drastically from a mixed indirect exciton into a very efficient electron-hole plasma recombination. The bandgap energy of this plasma increases with excitation density as a result of the separation of electron and holes along the growth axis.

INTRODUCTION

AlAs and the GaAlAs ternary alloys with an Aluminum content of more than 43% have their conduction band minimum at or close to the X point of the Brillouin zone. For pure AlAs, and given the recently corrected valence band offset for the GaAs/AlAs heterojunction [1], this X minimum lies about 200 meV above the Γ minimum in GaAs (see Figure 1). This relatively small energy separation can be further reduced by the effect of confinement, e.g. if the GaAs layer shows quantum size effects. The energy of the ground bound state of a GaAs quantum well of thickness ≈ 35 A is indeed expected to line up with the AlAs X minimum. For even thinner quantum wells and short period superlattices the conduction minimum has been shown to be at the X minimum, in AlAs [2,3]. Because the hole bound state remains, however, of Γ character and mostly localized in GaAs for all thicknesses, such superlattices are indirect in reciprocal and real (Type II) spaces.

Confinement effects at the X point in AlAs as well as a slight lattice mismatch between GaAs and AlAs further complicate this picture. Simple effective mass arguments predict the lifting of the 6-fold X valley degeneracy into an X_z doublet, with momentum along the growth axis, and a $X_{x,y}$ quadruplet, with momentum in the plane. Taking into account the contrary effects of strain and confinement one expects X_z to be lowest for AlAs layer thicknesses below 80 A, and highest for those above 80 A. Several experiments have recently confirmed this ordering for small AlAs thicknesses [4-6].

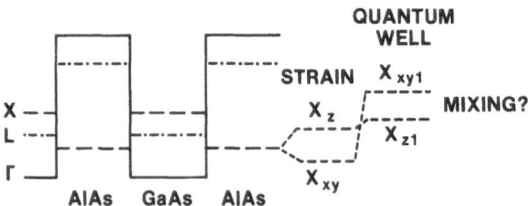

Figure 1. Profiles of the Γ, X and L conduction extrema in a GaAs/AlAs heterostructure.

A more detailed analysis of the band structure of indirect GaAs/AlAs superlattices [7-9] reveals, besides, the existence of superlattice mixing between the different points of the Brillouin zone. The superperiodicity is along <001> and therefore mixes the <000> (Γ) and <001> (X_z) states together, as well as the <010> and <100> (X_y and X_x). Such effects of zone folding and mixing are very similar to those encountered in Si/Ge superlattices [10,11] and depend critically on parity effects, i.e. on wether the GaAs and AlAs layers have odd or even numbers of atomic planes.

Optical experiments have been widely used to understand the band structure of these superlattices [2,3,12,13]. This paper will describe several optical experiments performed on indirect superlattices of relatively large periods (40 to 120 A). Section I reviews general absorption and emission characteristics, including a brief description of the phonons assisting the indirect transition. Section II describes more specifically time-resolved photoluminescence measurements; in Section III is presented the results of an electric field induced switching between direct and indirect transitions. The two latter experiments yield an estimate of the strength of the Γ-X mixing whose origins are discussed in section IV. Finally photoluminescence measurements taken under high optical excitation, which show the formation of an electron-hole plasma with new characteristics, will be presented in section V.

I- GENERAL OPTICAL PROPERTIES

Unlike Si/Ge superlattices, GaAs/AlAs indirect superlattices show strong photoluminescence at low temperature. For the samples that will be discussed here the main photoluminescence line has been assigned to the recombination of an indirect exciton made of a hole of Γ, GaAs character and an electron of X, AlAs character. As will be emphasized in Section II, recombination of this exciton is made allowed by mixing of the X conduction state with zone center conduction states lying at higher energy. At high temperature, only Γ recombination (of higher energy) is observed, probably the result of non-radiative effects taking over at the X minimum and of the small thermal distribution of electron in the fast decaying Γ subband.

Phonon replicas of the indirect exciton are are also observed in the low temperature photoluminescence spectra such as that of a 28/60 A superlattice displayed in Figure 2. The strength of these transitions with respect to the no-phonon line has been found to vary with the superlattice structure [14]. The number and frequency of the modes involved can be obtained with relatively good precision from Figure 2 which

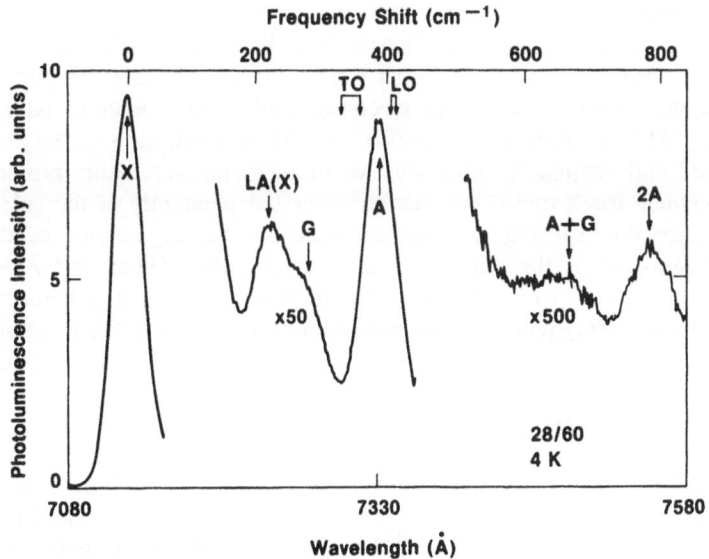

Figure 2. Low temperature photoluminescence spectrum of a 20/60 A superlattice. X is the zero-phonon line of the indirect exciton. LA(X) is a zone edge acoustical mode. G and A are a GaAs and AlAs mode respectively and appear in combinations in the two-phonon replicas.

displays one phonon as well as two-phonon replicas. The lowest energy mode is found in ≈ 220 cm^{-1} and is likely to be a longitudinal acoustical mode at the X point. The two other are a GaAs and an AlAs optical mode in 285 and 395 cm^{-1} respectively. In another series of samples with thinner AlAs layers (20/20 A) the GaAs mode is identical but the AlAs mode is found in 385 cm^{-1}. None of these frequencies can be straightforwardly identified with bulk optical frequencies at the X and Γ points. In particular, the AlAs mode involved is at substantially lower frequency than the one participating in bulk indirect transitions [15], although these are known with a relatively poor accuracy. At any rate the observed AlAs mode frequency rules out a zone center (confined or not) mode. It is not either that of a confined X mode: the frequency increases (385 to 395 cm^{-1}) with increasing AlAs layer thickness (20 to 60 A), contrary to what is expected for the positive dispersion of the LO branch at the X point.

On the other hand, both the GaAs and the AlAs frequencies are in remarquable agreement with those measured in the Raman spectra of similar superlattices in resonance with the direct transition, and attributed to interface modes [16,17]. This agreement is further corroborated by (i) the increase of the AlAs mode frequency upon increasing the AlAs thickness and (ii) the strength of the electron-multiphonon interaction which is evidenced here by the high order replicas, all features observed in the former Raman experiment. It is therefore very likely that the same phonons are involved in the two processes. The interface nature of such modes makes them good candidates to assist an electron-hole recombination which takes place essentially at the interfaces, between evanescent tails. However, if these modes effectively result from long range Coulomb interaction [17] their vibrationnal properties will not fit the momentum requirements of the indirect transition. The observed replicas would therefore be allowed by Γ-X mixing just as the zero phonon line. If this interpretation is correct, we have here evidence for an extremely strong coupling, maybe due to the interface, "localized" nature of the recombination process.

Excitation spectra of the zero-phonon or replicas of the indirect exciton show an extremely weak absorption for the indirect transition, and a strong turn-on at the energy of the direct transition between Γ electrons and holes. Recent photoconductivity measurements [18] have shown well resolved excitonic absorption peaks associated with the fundamental and excited X subbands. All of these measurements provide values for the spacing between the X and Γ conduction states independently of the hole confinement energies. Because this spacing depends linearly on the alignment of the bandgaps, accurate measurements of the valence band offset between GaAs and AlAs have been obtained [3,19] and found to be between 530 and 550 meV (corresponding to a 67 to 66 % conduction band offset) with an error bar given by the mass at the X point.

II TIME RESOLVED PHOTOLUMINESCENCE

As mentioned above the recombination of the indirect exciton is forbidden in the absence of scattering mechanisms to the Γ valley. The strength of the interband optical transition is therefore directly proportionnal to the degree of mixing between the Γ and X wavefunctions. Former experiments have shown the radiative decay of the exciton to be non-exponential, suggesting that the mechanism for scattering is random disorder at the superlattice interfaces [2,20]. However experiments done at higher excitation intensity show exponential decays which could indicate that the lowest excitonic states are bound on impurities and defects while at higher energy the excitons are more delocalized [21].

Photoluminescence time decay experiments have been done in that higher density regime on a series of samples with different layer thicknesses and resulting energy spacings between the X and Γ subbands. The results, summarized in Table 1, show lifetimes ranging from 0.22 to 5 μs for period ranging from 40 to 80 A. Extremely long lifetimes (several ms) have been obtained by another group in GaAs/GaAlAs/AlAs superlattices where the spatial overlap between electron and hole wavefunctions have been further reduced [22].

The radiative lifetime is found to scale inversely with the energy spacing E(X)-E(Γ) between the direct and indirect transitions. A perturbation approach, valid for those spacings much larger than the potential responsible for the mixing V_{mix}, gives the ratio of the indirect to the direct radiative rates are

$$\frac{\omega(X)}{\omega(\Gamma)} = \left\{ \frac{V_{mix}}{E(\Gamma) - E(X)} \right\}^2$$

Following such an approach we find the values for V_{mix} listed in Table I, ranging from 1.2 to 3 meV (the direct radiative rate has been measured in similar, but direct superlattices to be of the order of $5 \ 10^9 \ s^{-1}$ [19]). This strength as well as the origin of this potential will be further discussed in Section IV.

TABLE I. *Radiative rates, corresponding amount of mixing and deduced strength of the mixing potential as measured from the time decay of the photoluminescence.*

layer thickness A	$E_\Gamma - E_X$ meV	ω_X $10^6 \ s^{-1}$	mixing -	V_{mix} meV
19/60	202	.26	$0.5 \ 10^{-4}$	1.45
22/60	180	.3	$0.6 \ 10^{-4}$	1.39
25/60	127	.46	$0.9 \ 10^{-4}$	1.21
28/60	104	.75	$1.5 \ 10^{-4}$	1.27
22/20	168	1	$2 \ 10^{-4}$	2.37
19/19	100	4.5	$9 \ 10^{-4}$	3

III ELECTRIC FIELD INDUCED ANTICROSSING OF X AND Γ STATES

Adjusting the energy spacing between X and Γ can be obtained, rather than by changing sample structures, by a variety of external parameters; magnetic field, pressure [6,23] and electric field effects have different effects on the X and Γ valleys. We have inserted an indirect superlattice (35/80 A) into the i region of a p-i-n structure with GaAs contact layers. Because of the larger effective bandgap in the i region, forward current is limited to tunneling at low temperature. The structure could therefore be biased either forward (up to 6V) or reverse. The effects of the electric field (intensity and shifts of the photoluminescence lines) could be monitored under both biases and were found to be symmetric with respect to a bias of + 1.75 V, which we therefore take to be the flat band condition. Fields at other applied voltages were estimated from that value and the assumption that the electric field is uniform along the 0.5 μm intrinsic region.

Under no eletric field the photoluminescence signal is a weak peak in 1720 meV. Upon applying a field its intensity increases strongly, while another peak appears in 1735 meV. This confirms that the former peak is associated with the indirect transition, as for Type II superlattices one expects an increased optical element when the electron and hole wave function are brought closer by an electric field [24]. The peak in 1735 meV, about 15 meV higher in energy, has the characteristic behavior of a direct transition and is assigned to recombination involving the Γ electronic state. The indirect peak is also found to move towards higher energy with increasing field, as shown in Figure 3. This is also the signature of electron-hole separation. In quantum wells the bound states experience a red shift under application of electric field [25]. In a Type I superlattice confined state energy shifts ΔE and ΔH for the electron and the hole add up, resulting in an overall red shift of the interband transition. In a Type II, however, the centers of symmetry of the electron and hole wavefunctions are separated by a half period d/2. For a given hole state there will therefore be one electron state at a eFd/2 higher voltage and another one at an eFd/2 lower one. It is easy to see that the overlap of the hole

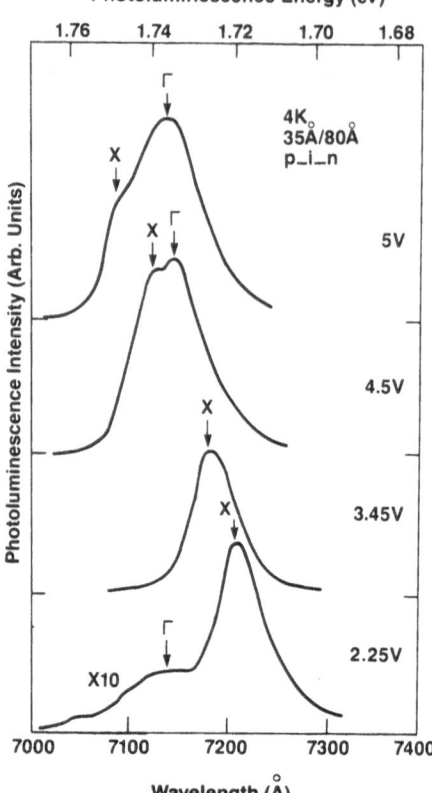

Figure 3. Low temperature photoluminescence spectra of a 35/80 Å superlattice at various applied forward voltages. The peaks labeled X and Γ are identified as due to the recombination of electrons in the X and Γ valleys, respectively, with holes at Γ.

wavefunction will be best with that of the electron at highest energy, resulting in a shift of the transition energy with field equal to

$$eFd/2 + \Delta E + \Delta H$$

There is therefore an overall blue shift of the photoluminescence which had ben precedently observed [23]. At large electric fields (here around 10^5 V/cm and above) the red Stark shifts tend to saturate and we indeed observe a linear dependence $+eFd/2$ of the photoluminescence energy. At lower fields, on the other hand, one can extract the linear term from the overall shift to obtain the sum of the Stark shifts of the electron and hole. This has been plotted in Figure 4 against the square of the electric field. The expected linear dependence is well reproduced. Since the hole Stark shift, because of the much smaller hole quantum well, is negligible with respect to the electron one, the slope on Figure 4 measures the electron mass along the growth axis. A simple perturbation analysis within an infinite well model gives here a mass of $1.2m_o$. This is in very close agreement with the value quoted in the litterature for the longitudinal mass at the X point in AlAs [27]. It therefore identifies the X orbital involved as X_z, since the confinement mass for the other two orbitals would be substantially lighter (about $0.2\ m_o$). This assignement has been since then corroborated by other experiments [5,6].

The energy shift of the indirect and direct transitions are plotted in Figure 5 as a function of the applied field. No red Stark shift can be observed for the direct transition due to the narrowness of the GaAs quantum well (35 Å). However, due to the large blue shift of the indirect transition there is an avoided crossing of the two energies at a field of about 4.5 10^4 V/cm. At field above that value the lowest optical transition is a direct one and the indirect transition rapidly fades as the X subband empties. The splitting between the two transition energies at crossover evidences again the degree of admixture between the two conduction states. The strength of the mixing potential appears directly as the half of this splitting i.e. about 1 meV, in consistent agreement with our former estimate.

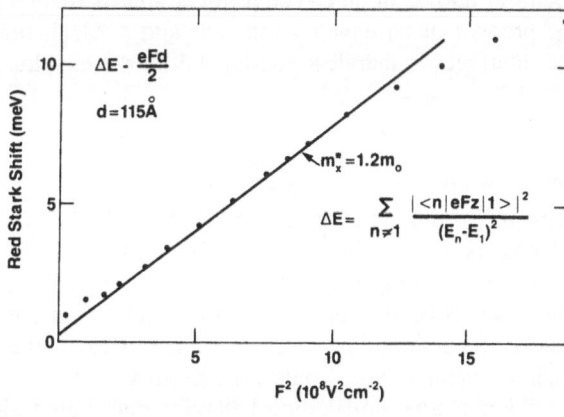

Figure 4. Quantum Stark shift (linear component substracted) versus the square of the electric field, at low fields. The solid line is calculated for a mass of $1.2m_o$.

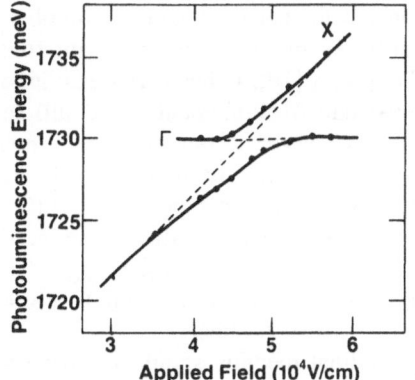

Figure 5. Energy of the direct and indirect transitions in the region of the anticrossing. The dotted line indicates the $+eFd/2$ variation of the type II transition. The Stark shift is saturated.

IV STRENGTH AND ORIGIN OF THE MIXING POTENTIAL

X-Γ mixing has been measured in GaAlAs alloys of composition close to the direct-indirect crossover, where it originates from random fluctuations of the atomic potentials [27]. The strength of the matrix element involving the X and Γ Bloch wavefunctions has been measured to be of the order of 200 meV.

There are in the superlattices studied here two possible mechanisms for wavefunction mixing. On mirrors the alloy situation, and results from the presence of alloy disorder at the interface. This scattering is difficult to model since it depends quite strongly on the local structure of the interface fluctuations. A signature for that mechanism would be the presence of mixing in the $X_{x,y}$ valleys since there should be no wavevector selection. At this point it is unclear whether this is the case. However, the fact that the authors in Ref. [18] observe well defined peaks in their absorption spectra as opposed to steps demonstrates the presence, at least in their samples, of non-random mixing for the X_z orbital. Random interface fluctuations will include all wavevectors and result in a step-like absorption [19].

Non-random mixing, on the other hand, is brought in by the discontinuity of the atomic potential at the interfaces even when ideal. This appears even for single interfaces, for example in the transmission characteristics of a single AlAs barrier embedded in GaAs for perpendicular transport [28]. Recent calculations of superlattice structures have all predicted this type of mixing [7-10]. A potential of strength comparable to what has been deduced from the experiments described above has been found by the authors of [9]. One can use a more simplistic approach to reconcile the strength of this potential to that found in alloys. In an effective mass approximation both the Γ and X states are written as the product of an envelop function and a Bloch function. Bloch wavefunction mixing by the interfaces is therefore modulated by the envelop function overlap and should write

$$V_{mix} = <\psi_X|\psi_\Gamma>.<u_X|V|u_\Gamma>$$

where the first part is the overlap of the wavefunctions and the second one the matrix element for mixing as it appears in the bulk. For the samples studied the envelop function overlap is in the 10^{-1} range. The experimental measure of V_{mix} leads to a "bulk" potential of about 10 to 20 meV which should be compared to the 200 meV measured in the alloy. Note that the strength of such an interface scattering mechanism should drop quickly for non-abrupt interfaces. Also, because the interfaces are perpendicular to <001>, such a potential should only mix X_z to Γ.

For a true superlattice behavior one would also expect this non-random mixing to cancel for certain parities. The <001> Fourier component of such a superpotential is non nul only if the period is an even number of layers. This is equivalent to the better known condition for mixing associated with zone folding, which has been heavily debated for the Si/Ge system [10]. Other parity restrictions have been predicted for the individual layers of GaAs and AlAs although these still seem to be controversial [8,9]. No effect of parity has been observed to our knowledge neither in the Si/Ge system or in GaAs/AlAs. This might be the result of monolayer fluctuations at the interface which would smear out parity effects between superlattices different by only one monolayer per period. Note also that any selection rule valid for reduced Brillouin zone center states of the superlattice will be reversed at the zone edge. Therefore, for those structures with very narrow minibands (i.e. very little well to well coupling, which applies here), one would not expect to see such effects.

A final mention should be made to the strength of the mixing as it appears in the "semi-direct" optical matrix element, which is of the order of 10^{-3}-10^{-4} for X-Γ spacing of the order of 100 meV (see Table 1). This value, corresponding to a wavefunction mixing in the percent range, can be probably further increased by about one order of magnitude by "delocalizing" the wavefunctions and increasing their overlap. The strongest increase of mixing, however, can be obtained by reducing the energy separation between the two minima, a task easier to realize in the GaAs/AlAs system than in the Si/Ge one.

V ELECTRON-HOLE PLASMA RECOMBINATION

All the results discussed above have been obtained in that excitation density regime where excitonic effects are prevalent, i.e. for in plane carrier densities below 10^{11} cm^{-1} in each layer. Owing to the long carrier lifetimes, however, it is relatively easy to excite carrier densities well in excess of that value. We observe then a transformation of the recombination process at the lowest X minimum from an exciton to that of a plasma. More striking is the apparition at a certain excitation threshold of another emission band of higher energy with a strongly non-linear intensity dependence on the excitation level.

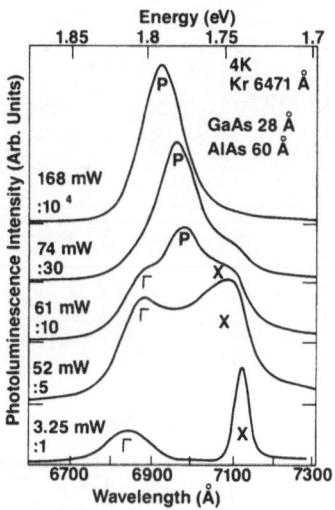

Figure 6. Low (10 K) temperature photoluminescence spectra of a 28/60 A superlattice under different excitation densities. The threshold for the apparition of the high energy plasma, noted P_t, corresponds to a density of about 3 kW.cm^{-2}. The high and low energy emission lines in the low excitation spectrum are due to the direct and indirect excitons.

This can be seen in Figure 6, which displays the photoluminescence spectra of a 28/60 A superlattice taken with excitation ranging from 3.25 to 168 mW. At low pump power the spectrum consists of the X exciton in 1.740 eV and a broader emission line from the direct transition, noted Γ, in 1.82 eV. As the pump power increases the X band transforms into a plasma while at about 50 mW of incoming power (about 3 kW/cm^2) a new band labeled P appears in 1.78 eV. This line shifts to higher energy with increasing pump power. Its intensity is a strongly superlinear function of the incoming intensity, as can be seen from Figure 7 where the overall integrated photoluminescence intensity has been plotted.

Time resolved measurements show that while the X exciton is long lived (about 1 µs) even in the plasma phase, this new emission decays in a few ns time scale. A typical photoluminescence decay is displayed in Figure 8 for an energy of 1.81 eV. Contributions from the Γ band can be seen as a shoulder originating from a short lived peak resolution-limited by the equipment. In turn, the P line decays with a sligthly slower rate. It also appears with a delay with respect to the incident pulse This delay has been found to vary with pump power from 2.5 ns at threshold to 0.5 ns at a power four times higher. Such times are quite too long to reflect any intervalley transfer phenomenon. They suggest, rather, the existence of separate phases or regions within the excited spot (about 50 µm in diameter).

A lineshape analysis of the P line in terms of the recombination of an indirect electron hole plasma gives good results except for the low energy part which is probably broadened by disorder [29]. Values for the bandgap, carrier densities and temperature can be obtained from such a fit. It is found that the bandgap increases roughly linearly with increasing pump power, by up to 10 meV over the range of excitation used. The

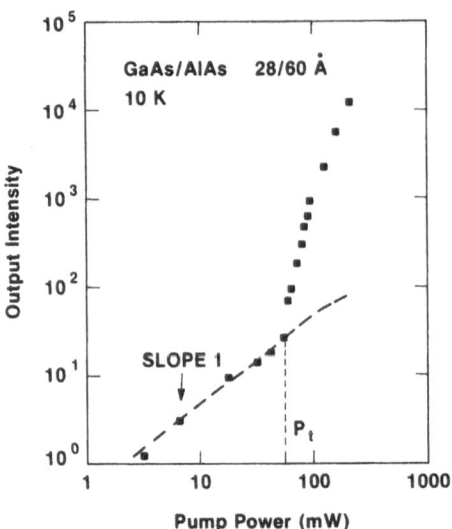

Figure 7. Integrated photoluminescence intensity versus incident power. The P line appears above 3 kW.cm^{-2}. At lower pump power the X exciton shows a linear pump power dependence.

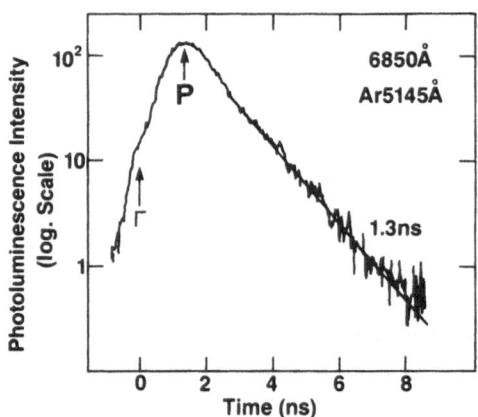

Figure 8. Photoluminescence time decay taken at 6850 A. The shoulder at short times is a contribution from Γ. The P emission is delayed by about 2 ns.

temperature ranges from 50 to 65 K depending on the pump power. The density of charges, however, is not found to change over the range of excitation density investigated. We find it to be in excess of 10^{12} carriers.cm^{-2} for all possible configurations of X in-plane mass, i.e X_z or $X_{x,y}$.

It is difficult to reconcile the bandgap increase with the constance of the carrier density. This again suggests two phases, one of a constant density (liquid ?) plasma and a gaseous phase with a density increasing with pump power. Two possible mechanisms can then contribute to the bandgap increase. As the pump power increases the plasma phase increases in size and occupies regions of higher energy due to interface fluctuations; also, if the plasma regions are small in size the bandgap will be influenced by the surrounding gas. This gas should have an increasing bandgap with increasing density. In Type II superlattices the separation of charges produces band bending at the GaAs/AlAs interfaces which results in an increased bandgap energy. A full description of this requires a self consistent calculation and to take into account the contrary effects of bandgap renormalization [30]. Rough estimates can however be obtained from a simplistic plane capacitor model and are of the same order of magnitude as the observed shift in the 10^{12} carrier.cm^{-2} range.

It should be noted that the energy at which this plasma radiates (1.760 eV at threshold) is 20 meV above that of the X_z exciton, i.e more than 10 meV above that of the unrenormalized bandgap for a reasonable exciton binding energy. It is possible that this plasma lies in the $X_{x,y}$ valley, which is in that energy region. This valley also has a larger in-plane mass and degeneracy which make it more likely to host a dense or liquid plasma. We have no clue, however on why would this phase, and in these valleys, have such a high radiative efficiency (about 10^3 higher than the X_z exciton).

CONCLUSION

We have reviewed several optical experiments performed on indirect, Type II GaAs/AlAs superlattices of relatively large periods. The excitonic recombination is beginning to be reasonably well understood as resulting from the admixture of Γ and X states. This in turn has allowed us to experimentally measure the strength of the mixing potential (about 1 meV in the samples investigated), a parameter of importance to be compared with the results of the numerous theories being developed for this system. Two other exciting aspects of the optical properties of such superlattices have yet to be further investigated. One is the electron-phonon coupling mechanism which we find to be very efficient for what we think are interface modes. The other is the high-density optical response; our data suggests the presence of two phases including one of density constant with increasing pump density. Whether this is a liquid or an expanding gas, it displays several characters which are specific to the peculiar band structure of this system.

Acknowledgements

I wish to thank my collaborators J.L. de Miguel, E. Finkman, R.E. Nahory, M.D. Sturge, M.C. Tamargo and J.M. Worlock for their participation in this work. I am also grateful to G. Bastard for several fruitful discussions and to R. Martin for expert technical assistance.

REFERENCES

[1] R.C Miller, A.C. Gossard, D.A. Kleinmann and O. Munteanu, Phys. Rev. **29**, 3740 (1984).
[2] E. Finkman, M.D.Sturge and M.C. Tamargo, Appl. Phys. Lett. **49**, 1299 (1986).

[3] G. Danan, B. Etienne, F. Mollot, R. Planel, A.M. Jean-Louis, F. Alexandre, B. Jusserand, G. Le Roux, J.Y. Marzin, H. Savary and B. Sermage, Phys. Rev. B 35, 6207 (1987).

[4] M.H. Meynadier, R.E. Nahory, J.M. Worlock, M.C. Tamargo, J.L. de Miguel and M.D. Sturge, Phys. Rev. Lett. 60, 1338 (1988).

[5] H.W. van Kesteren, E.C. Cosman, F.J.A.M. Greidanus, P. Dawson, K.J. Moore and C.T. Foxon, Phys. Rev. Lett. 61, 129 (1988).

[6] B. Gil, P. Lefebvre, H. Mathieu, F. Mollot and R. Planel, Proc. 19th Conf. Phys. Semiconductors, Warsaw 1988, to be published.

[7] J. Ihm, Appl. Phys. Lett. 50, 1068 (1987).

[8] D. Z.-Y Ting and Y.C. Chang, Phys. Rev. B. 37, 4359 (1987).

[9] Y.T. Lu and L.J. Sham, to appear in Phys. Rev. B.

[10] M.S. Hybertsen and M. Schluter, Mat. Res. Soc. Symp. Proc., Vol 102 edited by R.T. Tung, L.R. Dawson and R.L. Gunshor (1987).

[11] G. Abstreiter, H. Brugger, K. Eberl and R. Zachai, Surf. Sci. 174, 640 (1986).

[12] B.A. Wilson, IEEE J. Quantum Electron. 24, 1763 (1988).

[13] K.J. Moore, P. Dawson and C.T. Foxon, Phys. Rev. B 38, 3368 (1988).

[14] C. Bonner and R. Spitzer, private communication.

[15] Semiconductors, edited by O. Madelung, M. Schulz and H. Weiss, Landolt-Bornstein, Group 3, Vol.17 (Springer verlag, Berlin, 1983).

[16] M.H. Meynadier, E. Finkman, M.D. Sturge, J.M. Worlock and M.C. Tamargo, Phys. Rev. B 35, 2517 (1987).

[17] A.K. Sood, J. Menendez, M. Cardona and K. Ploog, Phys. Rev. Lett. 54, 2115 (1985).

[18] J. Barreau, K. Khirouni, T. Amand, J.C. Brabant, F. Mollot and R. Planel, to appear in J. Appl. Phys.

[19] E. Finkman, M.D. Sturge, M.H. Meynadier, R.E. Nahory, M.C. Tamargo, D.M. Hwang and C.C. Chang, J. Lumin. 39, 57 (1987).

[20] F. Minami, K. Hirata, K. Era, T. Yao and Y. Masamuto, Phys. Rev. B 35, 2875 (1987).

[21] M.D. Sturge and M.H. Meynadier, unpublished results.

[22] R.C. Spitzer, B.A. Wilson, C.E. Bonner, L.N. Pfeiffer and A.M. Glass, to be published.

[23] G. Danan, F.R. Ladan, F. Mollot and R. Planel, Appl. Phys. Lett. 51, 1605 (1987).

[24] E.E. Mendez, G. Bastard, L.L. Chang, L. Esaki, H. Morkoc and R. Fischer, Phys. Rev. B 26, 7101 (1984). Note than in a superlattice with a finite miniband there is rather a blue shift due to Wannier-Stark localization (see next reference), but this effect should be negligible for the superlattices investigated here.

[25] J. Bleuse, G. Bastard and P. Voisin, Phys. Rev. Lett. 60, 220 (1988).

[26] S. Adachi, J. Appl. Phys. 58, R1 (1985).

[27] M.D. Sturge, E. Cohen and R.A. Logan, Phys. Rev. B 27, 2362 (1983).

[28] A.C Marsh, Semicond. Sci. Technol. 1, 320 (1986).

[29] Details of this calculation will be published elsewhere.

[30] P. Hawrylak, to appear in Phys. Rev. B.

OPTICAL TESTING OF PROBABILITY DENSITIES

IN QUANTUM WELL EIGENSTATES

Jean-Yves Marzin and Jean Michel Gérard[*]

Laboratoire de Bagneux[**]
Centre National d'Etudes des Télécommunications
196 Av. Henri Ravera 92220 Bagneux, France

INTRODUCTION

Probing on a microscopic scale the electronic or structural properties of solids has always constituted a challenge. In this field, the insertion and study of local probes inside the material is a fruitful approach. With this respect, the case of isoelectronic impurities in the study of semiconductors gave nice illustrations of which kind of informations can be obtained. The study of the excitons trapped on the isolated impurity N in $GaAs_xP_{1-x}$ yielded an insight on the local arrangement of atoms around the impurity and thus on the disorder in these materials[1]. The choice of the isoelectronic impurities to study bulk three-dimensional semiconductors is however restricted. In particular, in optical studies, they yield poor information unless they cause additional optical transitions situated in the band gap. In three dimension materials, they must thus create strong attractive potential either for electron or holes. The impurities have thus to be far in the periodic classification from the element they will be substituted to. Their size will be very different and so the localized character of the potential they create is also lowered by the complicated deformation they induce in the host matrix.

When planes of isoelectronic impurities can be inserted in a structure, the situation is far simpler. On one hand, any attractive potential will result in a bound state, and on the other, the host matrix will remain unstrained. When the plane position is varied in the structure, one can obtain information on the spatial variations of a given quantity along the direction perpendicular to the probe plane. Such planes allowed to obtain the penetration depth of a surface state on Ag (111), using Au as an impurity[2].

In bi-dimensional systems such as quantum wells or superlattices, the growth axis z constitutes a preferential direction. Most of their near band gap electronic properties arise from the modulation of the potential along this axis and are adequately described by effective mass calculations[3]. In this frame, the $k_x=k_y=0$ eigenfunctions are products of a slowly varying envelope function (on the scale of the constitutive bulk material unit cell) and a rapidly varying function having the periodicity of the bulk material. In the crudest approximation, the envelope functions are eigenfunctions of one-dimensional hamiltonians, one for each of the bands of interest near the band gap, namely the electrons heavy and light holes, of the form:

$$H = \frac{P_z^2}{2m_e^*} + V_e(z)$$

for the electrons, and equivalent expressions for the heavy and light holes. $V_e(z)$ describes the variations of the conduction band minima in the constitutive materials. The eigenenergies of H are the $k_x=k_y=0$ eigenenergies for the quantum well or superlattice. The squared envelope function for a given state is the probability density of finding the particle in this state in a slab parallel to the (x,y) plane, averaged in the z direction on the bulk material unit cell which we will refer to as envelope probability density.

Of course, these quantities are important, because they contain information on the spatial distribution in the structure of electrons and holes. Whereas the validity of the effective mass approach is well established in numerous systems, the ingredients of the effective mass potential (the band discontinuities) are often not known so that it appears interesting to try to obtain the probability densities experimentally.

Doping with charged impurities is one possible way and was elegantly used in $Si/Si_{0.5}Ge_{0.5}$ strained layer superlattices[4]. A series of samples were grown with a donor doping plane which position was moved from sample to sample. A steep increase of the in-plane electron mobility was observed when the doping plane went from Si to $Si_{0.5}Ge_{0.5}$. This observation proved that the spatial separation of the impurity ions and carriers was obtained, thus showing that the lowest energy electron level was mainly localized in Si.

In another experiment[5], the envelope function for the higher energy hole state was experimentally determined, again in a doped quantum well. The structure was a GaAs/GaAlAs multi quantum well (with wide wells, up to 745 Å) with an asymmetric p doping in the barrier, so that a sheet of holes, located at one interface, was created. Raman scattering resonances occurred at energies corresponding to transitions between the uppermost (occupied) hole state and the conduction band states. The intensity of the different resonance peaks were explained in terms of the overlaps between the electrons and the hole wave functions. Assuming the electron wavefunctions corresponded to sine functions, these overlaps were thus the Fourier Transform components of the hole wave function, which could then be determined.

In these experiments however, little information is gained on the wave functions shapes in the undoped system, due to the extended change in the potential profile due to charge transfer mechanisms.

In this paper, we show on the test system constituted by GaAs/GaAlAs quantum wells, that planar isoelectronic probes can be used successfully to extract the envelope probability densities in the few first electron eigenstates. This kind of study is made possible by the great improvement in the control of the deposition of ultrathin layers by Molecular Beam Epitaxy. The planar probe consisted of ultrathin layers of InAs or AlAs, inserted in the structures. The paper is organized as follows: in a first part, we will describe some properties of delta perturbations of a quantum well and justify thus our analysis method, and discuss in a second part our experiments and results.

DELTA PERTURBATION OF AN IDEAL QUANTUM WELL

In this first part, the principle of the method is described. First, the effects of a delta perturbation on a one-dimensional Hamiltonian are discussed. We show in a first section that it is possible, in principle, to extract the probability densities in all states from the knowledge of the eigenenergies of the perturbed and unperturbed problems. In a second section, we treat the practical case of a limited number of discrete levels.

1- Unlimited set of levels

Consider the one-dimensional Hamiltonian:

$$H_0 = \frac{P_z^2}{2m} + V(z) \tag{1}$$

having a discrete set of eigenstates $|i>_0$, of energies $E_{i,0}$ and wavefunctions $\Psi_{i,0}(z)$. Consider now the perturbed problem:

$$H_a \;=\; \frac{P_z^2}{2m} + V(z) + W\delta(z-a) \tag{2}$$

Let us assume the eigenstates of H_a are $|j>_a$, with energies $E_{j,a}$ and wavefunctions $\Psi_{j,a}(z)$. If $|j>_a$ is developed on the basis formed by the $|i>_0$ under the form:

$$\Psi_{j,a}(z) \;=\; \sum_i C_i^j \Psi_{i,0}(z) \tag{3}$$

then

$$E_{j,a}\Psi_{j,a}(z) \;=\; \sum_i C_i^j \{E_{i,0} + W\delta(z-a)\}\Psi_{i,0}(z) \tag{4}$$

projecting on $|l>_0$ yields:

$$E_{j,a}C_l^j \;=\; C_l^j E_{l,0} + \sum_i C_i^j W\Psi_{l,0}^*(a)\Psi_{i,0}(a) \tag{5}$$

$$=\; C_l^j E_{l,0} + W\Psi_{l,0}^*(a)\Psi_{j,a}(a)$$

extracting C_i^j from (5) and injecting in (3) yields:

$$\Psi_{j,a}(z) \;=\; \Psi_{j,a}(a)\sum_l \frac{W\Psi_{l,0}^*(a)\Psi_{l,0}(z)}{E_{j,a}-E_{l,0}} \quad, \tag{6}$$

if $E_{j,a}$ differs from all the $E_{l,0}$ or

$$\Psi_{j,a}(z) \;=\; \Psi_{l,0}(z) \tag{7}$$

if $E_{j,a}=E_{l,0}$ (in this case $\Psi_{l,0}(a)=\Psi_{j,a}(a)=0$).

If one knows the $|\Psi_{l,0}(a)|^2$, the eigenenergies of H are: all the E_l. where $\Psi_{l,0} = 0$, and the roots of

$$G_a(E) \;=\; \sum_{\substack{l \\ \Psi_{l,0}(a)\neq 0}} \frac{|\Psi_{l,0}(a)|^2}{E-E_{l,0}} \tag{8}$$

The shape of function $G_a(E)$, which is is the projection of the Green function $1/(z-H_0)$ on the state of wave function $\delta(z-a)$, is indicated in Figure 1. It is a strictly decreasing function between two adjacent values of $E_{l,0}$ corresponding to states with $\Psi_{l,0}(a)\neq 0$. The nth solution of H will thus be in the segment $[E_{n,0},E_{n+1,0}]$, where $E_{n,0}$ are the eigenenergies of H_0 numbered in sequence of increasing energies if W is positive. If W is negative, there is of course one solution which energy $E_{0,a}$ is lower than the fundamental $E_{0,0}$ of H_0, and $E_{n,a}$ for $n \neq 0$ is in the segment $[E_{n+1,0},E_{n+2,0}]$.

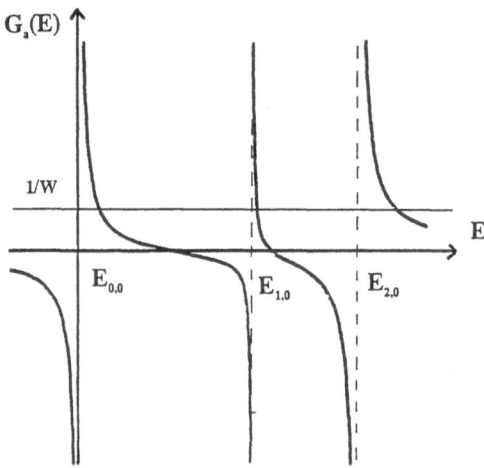

Fig 1. Schematic plot of the function $G_a(E)$ defined in equation (8). $G_a(E)=1/W$ gives eigenenergies of the perturbed Hamiltonian H_a.

For a small perturbation , the energy change in level n is just given by:

$$E_{n,a} - E_{n,0} = W|\Psi_{n,0}(a)|^2 \qquad (9)$$

so that the probability densities in the eigenstates of H_0 are readily obtained from the $E_{n,a}$. For large W, the eigenstates of H_0 are admixed and Equations (6) and (7) still allow to determine the $|\Psi_{l,0}(a)|^2$ from the knowledge of the eigenenergies of the perturbed and unperturbed problems from:

$$\sum_{\substack{l \\ E_{l,0} \neq E_{j,a}}} \frac{|\Psi_{l,0}(a)|^2}{E_{j,a} - E_{l,0}} = \frac{1}{W}$$

$$(10)$$

$$\Psi_{l,0}(a) = 0 \qquad if \qquad E_{l,0} = E_{j,a}$$

If the localized perturbation $W\delta(z-a)$ is scanned by varying a, the $|\Psi_{l,0}(a)|^2$ can be determined as a function of a.

Of course, only a finite set of eigenenergies $E_{j,a}$ will be determined experimentally, and the perturbation will not be scanned continuously, so that further approximations have to be made to extract the probability densities for a real system.

2- Extraction of the P.D.E's from a limited set of levels

To simplify the analysis, we will further assume in this part that all measured $E_{j,a}$ differ from the $E_{n,0}$ (assumption which does not change the generality of what follows). Let us assume now that only the N lower eigenenergies are known for H_a and H_0. The first Equation of (10) can be rewritten under the form:

$$\sum_{l=0}^{N-1} \frac{|\Psi_{l,0}(a)|^2}{E_{j,a} - E_{l,0}} = \frac{1}{W} - \sum_{l > N-1} \frac{|\Psi_{l,0}(a)|^2}{E_{j,a} - E_{l,0}} = \frac{1}{W_j(a)} \qquad (11)$$

The simplest approximation is to neglect the terms coming from the remote bands: this corresponds to expand the $|i>_a$ on the basis of the $|n>_0$ with n=0 to N-1. One can also partly account for the effect of the remote bands by assuming that $W_j(a)$ is independent of a, and equal to W_j. W_j depends of j due to the contributions of the remote levels of index n>N. System (11) can then be written under a matricial form:

$$M^a . \rho(a) \quad = \quad \bar{U} \tag{12}$$

where M is an NxN matrix, $\rho(a)$ and $\bar{U}$ N dimension vectors with:

$$M_{ij}^a \quad = \quad \frac{1}{E_{i,a} - E_{j,0}} \tag{13}$$

and

$$\bar{U}_j \quad = \quad \frac{1}{W_j}$$

$\rho_j(a)$ is the approximate value of the probability density for unperturbed state j at point a $|\Psi_{j,0}(a)|^2$. To determine these quantities, one needs the values of $\bar{U}$, which can be obtained by normalizing the N functions $\rho_j(a)$. One finally obtains:

$$\rho(a) \quad = \quad (M^a)^{-1} (\overline{M^{-1}})^{-1} u \tag{14}$$

with:

$$(\overline{M^{-1}}) \quad = \quad \int_{z_a}^{z_b} (M^a)^{-1} da \tag{15}$$

and where u is the vector with all components equal to 1.

In (15) $[z_a, z_b]$ is either the extension of the eigenfunctions or covers one period of the structure if it is periodic.

Finally, in the experiments, a is not scanned continuously. The integral in (15) is then replaced by a discrete sum over the sampled values. We will now examine the consequences of these different approximations.

3- Limitations of the method

The limitations of the method are twofold.
On the experimental point of view, a large number of samples have to be realized. In the case of a GaAs/GaAlAs multi quantum well structure, the thicknesses of the wells and barriers, the alloy composition and the probe have to be very similar in the whole series of samples. In practice, this limits the number of samples to about 10, which can be grown in the same run and the same day. In our case, the size of the tested well (160 Å) was chosen so that the step in the position of the probe is 15 Å, and the first half of the well was tested. This leads to a cut-off in spatial frequency for the measured probability densities. The second experimental limitation comes from the following compromise: the potential created by the probe has to be large enough to entail measurable shifts in the eigenenergies, but not too large so that the effects of the restriction on the number of observed levels are not too severe. For the examined wave functions, the probe has to be planar and localized enough. This last condition is well satisfied in most systems where an effective-mass type model is adequate to describe the eigenstates. The potential created by one plane of isoelectronic atoms (in our case the cations) will lead to a localized perturbation for the envelope wavefunctions. If it is not too strong, in the case of semiconductors, the probe potential will not couple the electron and hole systems. For electrons (resp. holes), W will be roughly equal to the conduction (resp. valence) band discontinuity between the host and probe semiconductors multiplied by the unit cell size.

Finally, in the case of the multi quantum well system, one has to extract the electron or holes eigenenergies from experiments, which can be done as we will see in next part.

On the calculation point of view, the main limitation is due to the limited number of observed levels. It is a priori system dependent; the precision is better for the lowest energy levels and also improves when the perturbation is decreased.

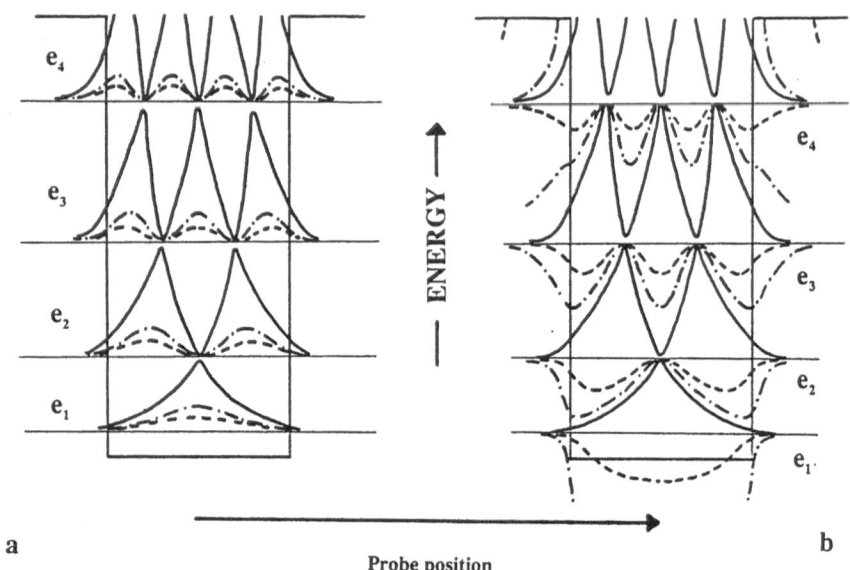

Fig 2. Plot of the calculated bound electron levels energies in a GaAs well of thickness d=200 Å with AlGaAs barriers, perturbed by a delta potential, as a function of the location of this probe. In a) the potential is repulsive and in b) attractive. The results of the calculation are given for 3 different strengths of the potential (in units of $\hbar \pi^2/(m^*d)$). The dashed curves correspond to 0.2, the dashed-point to 0.4 and the full line ones to 20.

4- GaAs/Al0.3Ga0.7As Quantum Well

In the case of a GaAs quantum well of width d with $Al_{0.3}Ga_{0.7}As$ barriers, where the electron (holes) are treated in an effective mass model, the Hamiltonian has the form (1), where V(z) is the conduction (resp. valence) band minimum profile. The eigenenergies of H are then shown in Fig. 2 as a function of the perturbation location inside the quantum well, for various attractive (Fig 2a) and repulsive (Fig 2b) perturbative potentials. The ordering of the perturbed energies with respect to the unperturbed ones clearly satisfy what is stated above. The limit for which a zero order perturbation should hold in this case is $(W/d) \ll \hbar^2\pi^2/(md^2))$. Figure 3 shows the electron envelope wave functions for the unperturbed and perturbed problems, using an InAs (attractive) monolayer as a perturbation. Finally, the probability densities corresponding to the 3 first electron levels can be deduced from the calculated 3 first perturbed and unperturbed eigenenergies, following Equations (14) and (15). The effect of using a finite set of levels (and a non delta potential) would lead to slightly distorted curves as compared to effective mass probability densities. The high frequency components to build up the perturbed wave-functions come from the remote states, so that the unperturbed wave functions on which the probability densities are obtained using (14) and (15) are also averaged. The typical averaging length corresponds roughly to the k value at the lowest energy remote level (here of the order of $d/(4*\pi)$=12 Å).

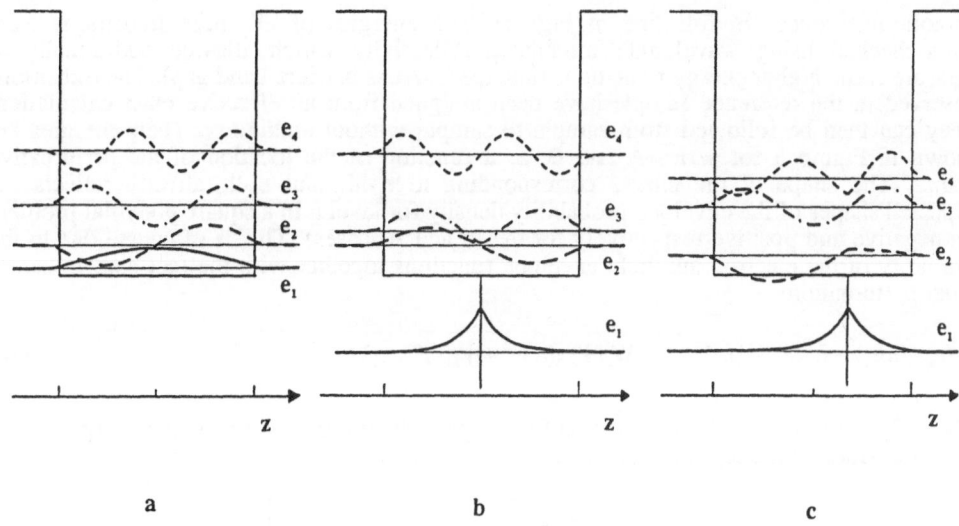

a   b   c

Fig 3. Calculated wave functions for the first electron levels in a 200 Å wide GaAs quantum well, unperturbed in a) and with an attractive probe in b) and c). Here the strength of the probe potential (in reduced units) is -0.4, corresponding to 1 InAs monolayer.

EXPERIMENTAL REALIZATION

This second part is devoted to the experimental results obtained on the test GaAlAs/GaAs system.

1- Samples growth and characterization

The basic structure consists in 6 GaAs wells, 160Å wide, separated by 85 Å $Al_{0.3}Ga_{0.7}As$ barriers, grown by Molecular Beam Epitaxy on a GaAs buffer layer deposited on a GaAs (100) substrate. Two series of samples were grown. In the first series (A), one In atomic plane was inserted at the same location in each quantum well or barrier. In eight samples, the position of the In was varied, probing the first grown half of the (symmetric) quantum well. One additional sample was grown with the In plane in the second half of the well, and finally, a reference sample without In was realized. All ten samples were grown in a single run to ensure a better reproducibility of the sample parameters and in particular in the In content per well. In the second series (B), the probes were Al planes and six samples (including a reference sample) were grown.

For series A), the position of the zeroth order satellite observed using X-Ray double diffraction profiles allows to obtain the In content per well: for the whole series, it is equal to 1.05 ± 0.05 monolayers. For series B, the composition of the barriers were obtained from optical measurement of their band gap, and an Al content per well of 0.9 monolayers was then deduced from the growth conditions.

2- Optical measurements

It is a priori difficult to extract the eigenenergies experimentally. However, with reasonable assumptions, they can be obtained from the measured optical transitions energies.

Figure 4 shows typical optical spectra obtained on a reference sample (b), and on samples with the In (a) or Al (c) located at the center of the wells. The different lowest energy transitions were obtained by photoluminescence and excitation of the photoluminescence (in full line in Fig. 4). The energies of all these transitions were cross-checked using wavelength modulated reflectivity which allowed additionally to measure some higher energy transitions (and the GaAlAs barriers band gap). The transitions observed in the reference sample have been assigned from an effective mass calculation. They can then be followed from sample to sample without ambiguity. Their energies are shown in Figure 5 for series A and B, as a function of the location of the perturbative planes. The shape of the curves corresponding to e_n-hh_n and e_n-lh_n already reflects the expected shapes of the envelope probability density for level n in a square potential problem (in negative and positive respectively for the In and Al cases). This is of course due to the similarity of the electron and hole envelope functions together with the fact that, in zeroth order perturbation:

$$E_a(e_n - hh_n) - E^0(e_n - hh_n) \;=\; W_e |\Psi^e_{n,0}(a)|^2 + W_{hh}|\Psi^{hh}_{n,0}(a)|^2 \qquad (16)$$

where E_a is the energy of the transition for the perturbation in a and E_0 is the energy for the same transition in the reference sample.

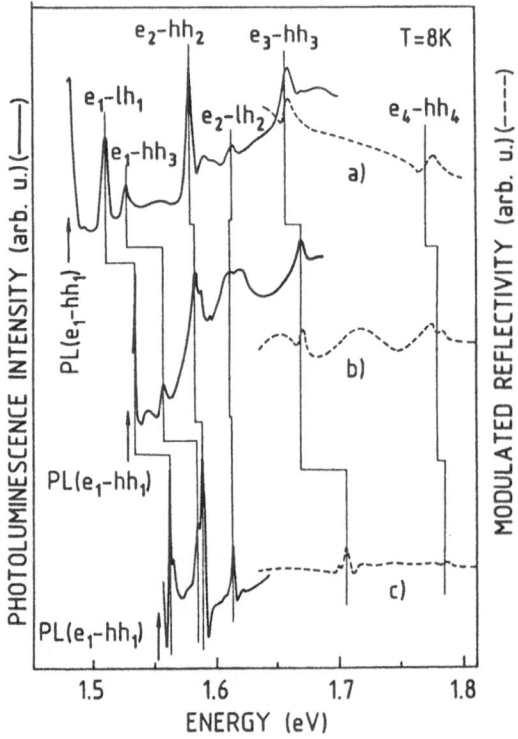

Fig 4. Experimental Photoluminescence excitation (full line) and modulated reflectivity (dashed lines) spectra obtained at 8 K on: a) the sample of series A with the In plane at the middle of the quantum well, b) the reference sample, c) the sample of series B with the Al plane at the middle of the quantum well. The arrows show the energies where the luminescence is observed.

One can note however that the shape of the fundamental transition for series A is not as smooth as the expected envelope function for a square potential. This is due to the strength of the potential created by the In plane. As illustrated in Fig. 2, the wave function of the state which is linked on the delta attractive potential may have an extent which is much smaller than the well width. In this case, this state will be insensitive to the presence of the quantum well barriers, unless the perturbation plane is close enough to the barrier. In other words, the perturbation potential is two strong to use zeroth order perturbation theory and obtain accurate probability densities.

This is supported by the estimation of W_e for the electrons, according to the conduction band discontinuities[6] between GaAs and (strained) InAs (or AlAs). The perturbation strength should be of the order of -2.1 eV.Å (resp. 2.9 eV.Å) when InAs (resp. AlAs) is used. The GaAs quantum well width is 160 Å, so that from part 1, it has to be compared to $\hbar^2\pi^2/(m_e d)$=5.6 eV.Å .We thus expect a significant mixing of the quantum well states induced by the perturbation in both cases. The somewhat more complicated analysis described in the first part of the paper has thus to be used.

First, the eigenenergies have to be extracted from the optical data, and further approximations have to be made. The first one concerns the light hole states. In series A, the perturbation consists in one InAs monolayer, strained on GaAs. Due to the strain effect, the effective potentials resulting from the In plane are different for the heavy and light holes. In previous experiments[7], we had noticed that a one monolayer ultrathin InAs quantum well imbedded in a GaAs matrix binds an electron and heavy hole state, but hardly binds a light hole level. In other words, we have $W_{lh} \ll W_{hh}$ or W_e. A reasonable approximation is thus to assume that the In perturbation does not affect the light hole states, which energies are taken to be equal in the whole series A to those in the reference sample.

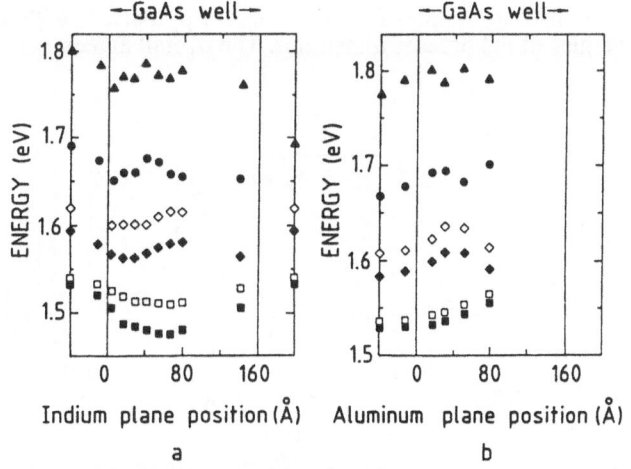

Fig 5. Plot of the different transitions energies for series A in a) and series B in b), as a function of the indium (a) or aluminum (b) probe position inside each period of the structure. ■:e_1-hh_1, □:e_1-lh_1, ◆:e_2-hh_2, ◇:e_2-lh_2, ●:e_3-hh_3, ▲:e_4-hh_4. The transitions energies plotted at 40 Å and 200 Å are those of the reference samples.

In the case of the Al perturbation this approximation is more questionable. Nevertheless, in the case of a repulsive perturbation, another argument can be used. If we examine the two lower energy transitions for series B in Fig. 5, we note that the energy difference between the two is small in all samples. We have seen in the first part that for a repulsive perturbation, the first state energy in the perturbed problem stays bounded by the first two states energies in the unperturbed problems. For the heavy hole system, due to the large effective mass, hh_1 and hh_2 are very close in energy, hh_1 (and hh_2 to a smaller extent) perturbed level energy will be merely affected by the perturbation. The parallelism between curves corresponding to hh_1-e_1 and lh_1-e_1 (and between curves hh_2-e_2 and lh_2-e_2) show that lh_1 energy (and lh_2) does not strongly vary either: the energy variations are dominated by the electron levels energy variations. The assumption on the light hole levels was thus made for both series.

The second assumption concerns the exciton binding energies. For hh_n-e_n (lh_n-e_n), they were taken to be equal in all samples to those measured in Ref.8 (6.5meV for hh_1-e_1 (resp. 9 meV for lh_1-e_1)) in a sample similar to the reference samples. For hh_n-e_m or lh_n-e_m with n different of m, they were taken to be zero.

The fact we consider these quantities (eigenenergies for the light hole states, excitons binding energies) constant for all samples is more important than their exact values, if one wants to extract the electron envelope probability densities. This is because their value disappear from most of the energy differences in matrix M^a of Equ (14). The analysis of the data then relies mainly on the correct assignment of the experimental transitions.

From differences between the energies of the observed transitions, the eigenenergies for the 3 first electron and heavy hole states are extracted (for both series) for the different perturbation locations. The symmetry of the well is then used when writing the probability densities normalization conditions (15).

3- Results

The probability densities extracted for the three first electron levels are shown in Fig. 6 where the results obtained for both series (independently) have been gathered. The effective mass calculated curves are shown for comparison. The indicated error bars (shown here for 3 standard deviations) include the experimental errors on the transition energies, and on the values of the light hole and exciton binding energies, but not on the possible fluctuations of the parameters from sample to sample. The error bars for ρ_n increase with n, due to a poorer precision on the energies of the broader transitions. The overall agreement is satisfying.

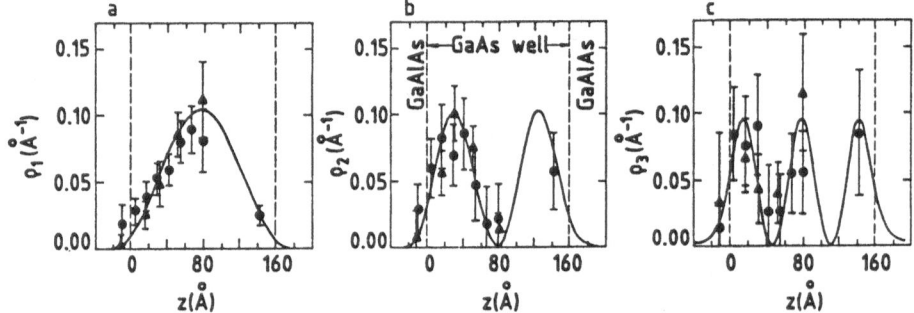

Fig 6. Experimental probability densities envelopes for e_1 (a), e_2 (b), and e_3 (c) electron levels. Full circles and triangles correspond to the values deduced from series A and B, respectively.

In principle, similar curves can be obtained for the heavy hole states, but the energy variation of these levels is found, in agreement with our analysis, to be much smaller than those for the electron levels (for all levels in series B, and all but the first one in series A). As a consequence the uncertainty on the hole probability densities is very large.

The normalization conditions (15) allow to obtain the W_j. The correction due to the remote levels is larger for the excited levels, and one expects that W_1 should be closer to W. For series A and B, we obtain:

$W_e = -2.4 \pm 0.3$ eV.Å for the In plane and

$W_e = 2.5 \pm 1.5$ eV.Å for the Al plane.

The uncertainty on W is larger for the Al case mainly because the energy variations of the first electron level is smaller than in the In case and also because the number of samples is lower.

These values are quite in agreement with those obtained when considering the conduction band discontinuities between GaAs, AlAs and strained InAs.

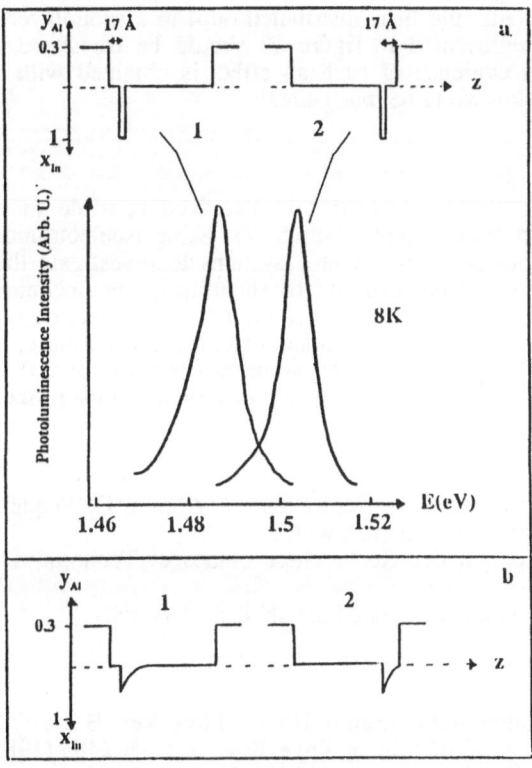

Fig 7. a) Photoluminescence spectra of 2 samples containing an In probe at different positions, designed to be symmetric with respect to the center of the well. Intended composition profiles are also shown for each sample.
b) Schematic In composition profile for both samples, explaining the difference in their emission energy .

4- Informations on the probe itself

We had the idea of developing this technique while we were studying what constitutes one of our probes, namely ultrathin InAs layers. We were concerned in establishing the In concentration profile when one monolayer of InAs is deposited by M.B.E. on GaAs and covered with GaAs. Several indications (from surface studies[9] or microscopy[10]) show that the profile should be smoothed due to In segregation when GaAs is grown on top of InAs. One possible method to test this asymmetry is to include one InAs monolayer in a quantum well. If there is an In segregation, the position of the probe is displaced in the growth direction. Two samples with a symmetric position of the probe with respect to the center of the well should exhibit the same optical transitions if there were no segregation. Figure 7a shows the low temperature photoluminescence spectra for such a pair of samples. Sample 1 is designed with the In plane located inside the well at 17 Å from the first grown interface, whereas for sample 2, it is at the same distance from the second interface (see inset of Fig. 7a). The emission energy for sample 1 is 17 meV lower than that of sample 2. A similar effect might result from the asymmetry of the GaAs/GaAlAs interfaces, but would produce smaller energy shifts, so that this phenomenon is neglected here. Let us assume that the In segregation moves the average position of the indium plane of a distance d_s in the growth direction (Fig. 7b). The typical slope S for the energy of e_1-hh_1 transition is around 1.7 meV.Å^{-1} in this range of In plane location, and the difference in energy E(2)-E(1) for this transition in the 2 samples is given by $2d_sS$. We thus deduce from the measured shift in emission that d_s is about 5 Å. This value has to be compared with that deduced from the

315

profile determined in ref. 9, characterized by an exponential decrease of the form $\exp(-z/d_s)$, with $d_s=13$ Å. It is in better agreement with the result obtained from high resolution microscopy, indicating that the In is distributed on 4 to 5 monolayers[10]. The experimental probability densities represented in figure 12 should be translated of d_s, which merely affects the results. No evidence of such an effect is obtained with the Al probe, where segregation effects are known to be much smaller.

CONCLUSION

Our experiments clearly show that it is indeed possible to extract the envelope probability densities in semiconductor structures, using isoelectronic planar probes. This kind of experiments can be useful in new systems to investigate the electron (or holes) confined states. In particular, one can directly show, using that technique, whether a system is of type I or of type II by following the evolution of the optical transition energies when a probe is inserted either in one or the other material. Furthermore, we have shown that repulsive or attractive localized potentials can be successfully used, though attractive probes lead to a better precision. This should allow a broad choice in the probe material.

ACKNOWLEDGMENT

The authors acknowledge M. Bensoussan, I. Abram, D. Paquet, G. Bastard and M. Voos for their fruitful comments on this work.
* Member of the Direction des Recherches, Etudes et Techniques (French Ministry of Defense)
** Laboratoire de Bagneux is associated to C.N.R.S. (UA 250)

REFERENCES

1 H. Mariette, J. Chevallier and P. Leroux-Hugon, Phys. Rev. B 21, 5706 (1980)
2 T.C. Hsieh, T. Miller and T.C. Chiang Phys. Rev. Lett. 55, 2483 (1985)
3 see e.g. G. Bastard, Phys. Rev. B 24, 5693 (1981)
4 G. Abstreiter, H Brugger, T Wolf, H. Jorke and J.H. Herzog, Surf. Science 174, 640 (1986)
5 T. Suemoto, G. Fasol and K. Ploog, Phys. Rev. B37, 6397 (1988)
6 Priester C., Allan G. and Lannoo M., Phys. Rev. B 38, 9870 (1988)
7 J.M. Gerard and J.Y. Marzin, Appl. Phys. Lett. 53, 568 (1988)
8 R.T. Collins, L. Vina, W.I. Chang, L. Esaki, K.v. Klitzing and K. Ploog, Phys. Rev. B 36, 1531 (1987)
9 C. Guille, F. Houzay, J.M. Moison and F. Barthes, Surf. Science 189-190, 1041 (1987)
10 C. d'Anterroches,

PHOTOEMISSION FROM AlGaAs/GaAs HETEROJUNCTIONS AND QUANTUM WELLS UNDER NEGATIVE ELECTRON AFFINITY CONDITIONS

Franco   Ciccacci

Istituto di Fisica, Politecnico di Milano
piazza L. Da Vinci 32, I-20133 Milano, Italy

1. INTRODUCTION

The possibility of lowering the work function Φ in several materials by adsorption of alkali metals has been extensively used in the pioneer works on photoemission, making it possible to perform experiments with conventional light sources. This procedure has proved to be very effective in semiconductors, leading to the development of Negative Electron Affinity (NEA) photocathodes [1]. Under such conditions, the bottom of the bulk conduction band lies above the potential barrier at the surface, so that electrons excited to the conduction band can escape into vacuum. In practice NEA condition is achieved in heavily p-doped semiconductors (to encourage downward band bending at the surface) by adding in Ultra High Vacuum (UHV) environment a thin film (one to some atomic layers) of cesium rich oxide on the clean semiconductor surface. The activation procedure, developed for technological purposes, is well known, at least for gallium arsenide, and NEA GaAs photocathodes are widely used in photomultiplier tubes.

On the other hand, in GaAs absorption of circularly polarized light of energy slightly greater than the band gap gives rise to a spin-polarized electron population in the conduction band [2]. By combining these two effects very efficient spin-polarized electron sources based on NEA GaAs photocathodes excited by circularly polarized light have been realized [3], contributing to the renewed scientific interest in this field.

Photoemission studies of bulk GaAs under NEA conditions have been performed by analyzing both the energy distribution of the emitted electrons with high resolution spectrometers and their spin-polarization with the use of a Mott polarimeter. Valuable information on the band structure and the conduction electron dynamics have been obtained [4,5]. Similar studies have also been applied to semiconductors grown *in situ* by Molecular Beam Epitaxy (MBE) [6,7].

We review the results obtained on such materials as well as on expecially designed semiconductor structures, namely AlGaAs/GaAs heterojunctions (HJ), quantum wells (QW) and superlattices (SL) grown *ex situ* by MBE or Metal-Organic-Chemical-Vapor-Deposition (MOCVD).

In section 2, the photoemission efficiency as a function of the photon energy $h\nu$ is discussed. Next section 3 deals with the analysis of the kinetic energy of the photoemitted electrons, while the analysis of their spin-polarization is the subject of section 4. In section 5 the issue of spin-polarized electron sources is addressed.

2. QUANTUM YIELD

Here we discuss the photoemission quantum yield (Y), defined as the number of photoemitted electrons per incident photon, as a function of hv. In the case of a clean semiconductor surface, photoemission is obtained at energies larger than the difference between the vacuum level and the valence band (i.e. the sum of the electron affinity and the band gap E_g), which means some 5.5 eV in p-type GaAs. However, activation of the surface by coadsorption of cesium and oxygen, decresing the electron affinity, leads to a drastic reduction of about 4 eV of the threshold, so that for a NEA cathode the threshold is set by the semiconductor band gap. Photoemission experiments can thus be performed at relatively low photon energies: to measure quantum yield spectra monochromatized quartz halogen W-filament lamps are commonly used as light sources.

The yield of a NEA cathode is much larger than the one of conventional cathodes even at higher photon energies. This fact can be understood on the basis of the three steps model for photoemission [1]. Such a model is based on the assumption that photoemission is a bulk rather than purely surface event. Even though it represents a rough approximation in many cases when only few atomic layers are involved in the process, it mantains its validity in the opposite situation and describes correctly photoemission from NEA cathodes. In this model the photoemission process is broken down into three consecutive and indipendent events: i) optical absorption, ii) transport of the excited electrons to the surface (the electron beam may suffer several scattering processes losing energy and relaxing its spin-polarization), iii) escape ionto vacuum across the surface. The second process is drastically different in the case of NEA condition with respect to the clean surface. In fact, it is controlled by the electron mean free path for electrons excited into a final state above the vacuum level. This corresponds to the conduction electron diffusion length in the NEA case, where the vacuum level at surface lies below the bottom of conduction band, whereas is much smaller (few to tens of Å) for the clean surface, owing to the higher electron energies needed to overcome the surface barrier. Therefore in the NEA case electrons excited much deeper in the crystal can be photoemitted, resulting in an essentially bulk process. Carrying out the calculations in the three steps model [1], it turns out that the photon energy dependence of the quantum yield Y essentially reflects the behaviour of the optical absorption coefficient $\alpha(hv)$, with threshold at the band gap.

Figure 1 shows the quantum yield curve for bulk GaAs grown by different techniques and activated to NEA with the same procedure (full and dashed lines, right-hand axis) [6]. Apart from minor differences discussed in ref. 6, the two curves exhibit the same behaviour with a sharp increasing of the yield corresponding to the energy gap: as soon as the photon energy exceeds the GaAs gap, optical absorption takes place and electrons are photoemitted.

In Figure 2 the results obtained in the case of $Al_xGa_{1-x}As$ alloys grown by MBE and activated to NEA *in situ*, are shown for different values of x, the Al-content [7]. The spectra are similar (and similar to the GaAs one), but shifted according to the shift in the energy gap. From these curves the energy gap of the alloy E_G (direct for x < 0.45, indirect for x > 0.45) has ben determined and found in agreement with previously reported values. At photon energies corresponding to E_G a sharp threshold with large yield is seen for all the alloys with direct gap. When x increases beyond 0.45 the alloy becomes an indirect gap semiconductor and a drastic reduction of the yield values is observed. The possibility of making all the operations (growth, activation and photoemission measurements) *in situ* was particularly important for this study due to the high reactivity of the Al-containing surface. In such a way, surface contamination occurring during exposure to air was avoided. The contaminants, in fact, can not be completely removed from such surfaces by heating the sample in UHV as it is the case for the GaAs surface.

A similar problem is not present in the structure shown in Figure 3a. It represents a heterojunction consisting of an $Al_{0.3}Ga_{0.7}As$ alloy covered by a GaAs overlayer of thickness d, thinner than both the photon absorption depth (0.3 to 1 μm) and the conduction electron diffusion length (~ 1 μm). Therefore the photoemission current is due to electrons originating from either the GaAs overlayer or the AlGaAs substrate, thus carrying information on the energy levels and vertical transport mechanism in the HJ.

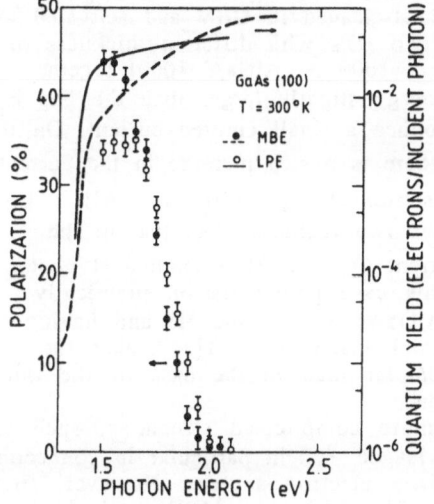

Fig.1. Photoelectron quantum yield (full and dashed lines) and spin polarization of NEA MBE and LPE GaAs

Fig. 2. Quantum yield of NEA $Al_xGa_{1-x}As$ labeled with the value of x.

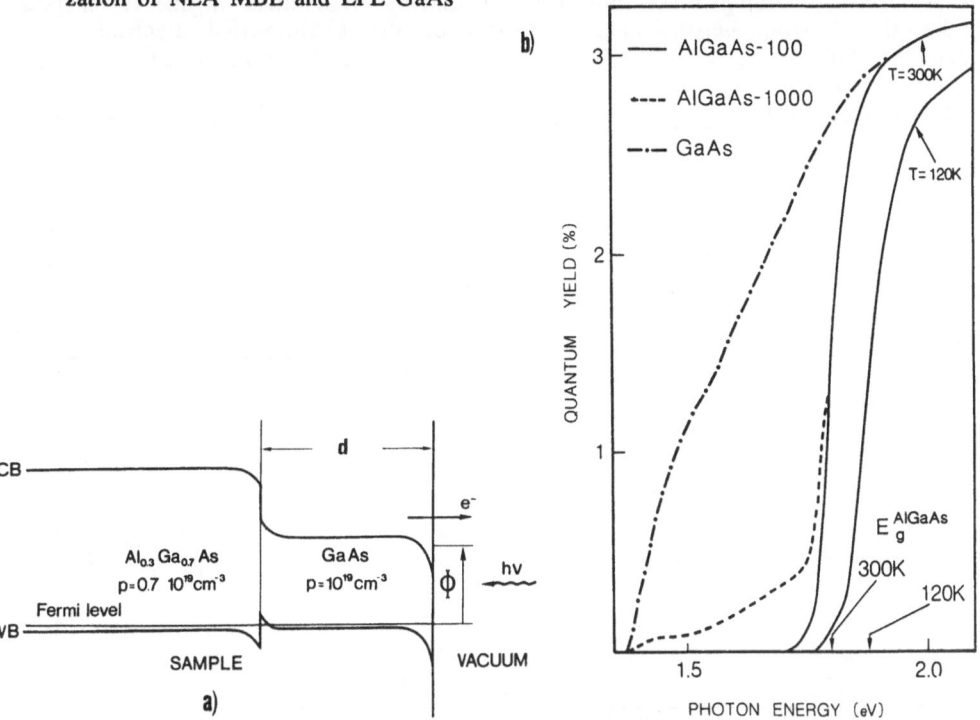

Fig. 3. a) Principle of NEA photoemission and schematized band profile of the AlGaAs/GaAs heterojunctions. $d = 100$ Å: AlGaAs-100. $d = 1000$ Å: AlGaAs-1000. b) Photoemission yield at 120 K and 300 K from AlGaAs-100 (solid lines). Normalized yields at 300 K from AlGaAs-1000 (dotted line) and from bulk GaA (dashed-dotted line) are also shown. The position of the AlGaAs gap are indicatedby arrows.

The sample is uniformely heavily p-doped (high p-doping level at surface is necessary for achieving NEA). Because of this high acceptor concentration and the Fermi level alignement, the tops of the valence band are lined up in the two parts of the HJ, whereas the separation between the bottom of the conduction bands is equal to the difference of the gaps ($E_G - E_g$ = 380 meV, at room temperature). Since the surface layer consist of GaAs, samples may be grown *ex situ* (e.g. by MBE or MOCVD, in a separate chamber) and then heat-cleaned in UHV and activated to NEA. Figure 3b shows the quantum yield for two HJ's with different thickness of the GaAs overlayer (d =100 Å: AlGaAs-100 and d =1000 Å: AlGaAs-1000) together with the curve for bulk GaAs [8]. Photons of energy slightly larger than the gap E_g of GaAs are absorbed in the overlayer and produce a small emitted current. On the other hand, when $h\nu > E_G$, the AlGaAs gap, there is a step increase in the current, corresponding to absorption in the AlGaAs substrate. The gap difference $\Delta E_g = E_G - E_g$ can be easily measured from the onset of the two regimes. Note that in the AlGaAs-100 sample a very small amount of current ($Y < 10^{-5}$) is emitted at photon energies smaller than E_G, showing that in this case photoemission practically originates only from electrons excited in the AlGaAs part of the HJ and having reached the surface after crossing the 100 Å thick GaAs layer. This makes the analysis of the photoemission process much simpler than in the case of the thicker overlayer, as will be shown in the next section.

Application of photoemission · studies to more complicated structures, such as GaAs/AlGaAs superlattices, has been also envisaged [9], in particular in connection with the hope of realizing very highly polarized electron sources. However, first attempts in this direction with heavily Be-doped SL's grown by MBE *in situ* , did not reveal any difference with respect to the bulk AlGaAs [10]. More recently experiments on quasi-intrinsic SL's covered by a p-doped GaAs overlayer have been performed, yielding clear indication of discrete transitions in the Y spectra and directly showing perpendicular transport of the photoexcited electrons through the AlGaAs barriers [11]. Figure 4a shows the band profile of an $Al_{0.32}Ga_{0.68}As$/GaAs SL covered by a 1200 Å thick p-doped GaAs overlayer activated to NEA. The corresponding quantum yield is shown in Fig. 4b [12]. The two curves

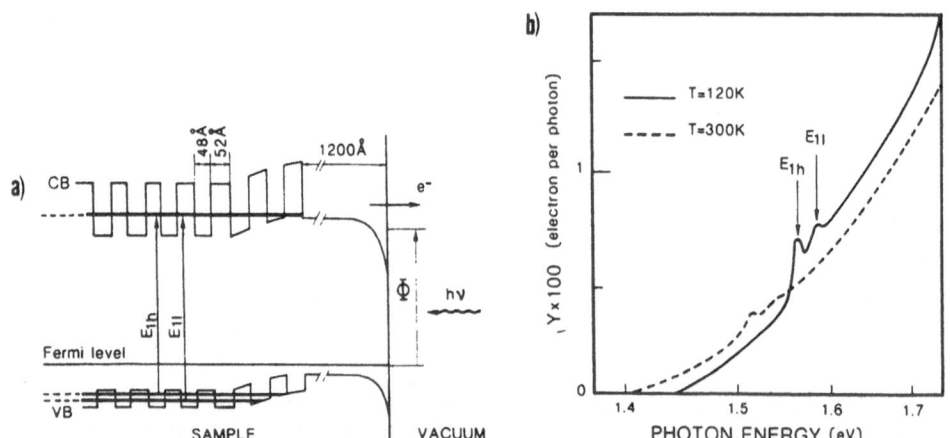

Fig. 4.   a) Band profile of a NEA AlGaAs/GaAs superlattice covered by a heavily p-doped 1200 A thick GaAs overlayer. b) Quantum yield at 120 K and at room temperaturefrom the same sample; the transitions E_{1h} and E_{1l} are indicated.

refer to spectra taken at room temperature and at 120 K, the shift of the threshold being related to the gap change with temperature. The peaks appearing in the curves (more evident at low temperature) are due to the optical transitions E_{1h} and E_{1l} from the quantum levels of the heavy-hole (hh) and light-hole (lh) valence bands to the quantum levels of the conduction bands. The electrons photoexcited in the SL are able to tunnel through the barriers and diffuse to the surface where they are emitted: a quantum mechanical description of the SL state in contrast to incoherent tunneling from well to well is necessary to interpret the data [13].

Figure 5a shows the band profile of a sample containing ten quantum wells with a thinner GaAs overlayer. In this case it is possible to observe several features in the yield spectrum [14], as shown in Fig. 5b. After the low energy onset corresponding to band gap excitation in the overalyer, several stuctures, indicated by arrows in Fig. 5b, are seen. They are related to transition between localized states in the QW's, while the last one occurring at 1.84 eV is due to band gap excitation in the $Al_{0.27}Ga_{0.73}As$ substrate. The relative amplitude of the QW's related structures to GaAs band gap absorption ($h\nu < 1.56$ eV) indicates that all the 10 wells contribute to the photoemission current, thus confirming the large efficiency of electron tunneling through the barriers towards the surface. From the energy of the transitions it is possible to estimate how the AlGaAs-GaAs gap discontinuity ΔE_g is distributed between valence and conduction band. A value for the conduction band offset ranging from 0.6 ΔE_g to 0.7 ΔE_g is obtained [14,15].

3. ELECTRON ENERGY DISTRIBUTION

As discussed above, the quantum yield measurement is spectroscopically equivalent to an optical absorption spectrum. However, in a photoemission experiment additional information is obtained by analyzing the number of photoemitted electrons versus their kinetic energy as measured by an electron spectrometrer. The so obtained energy distribution curves (EDC's) reflect the electronic distribution in the crystal modified by energy losses during transport to the surface and affinity cutoff.

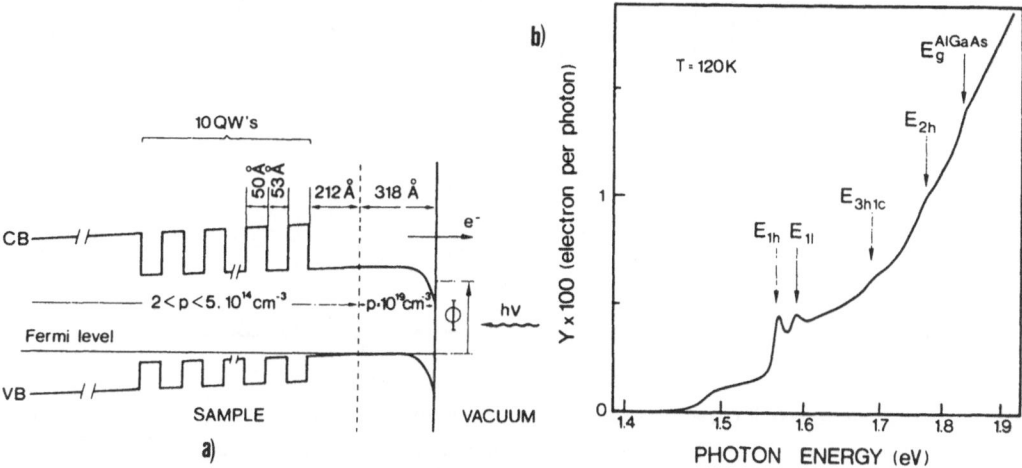

Fig. 5. a) Band profile in real space of the $Al_{0.27}Ga_{0.73}As$ 10 QW's sample (Figure not inscale for clarity). b) Photoemission yield: several transitions between the discretelevels in the wells are indicated.

321

For the sake of clarity, we shall briefly discuss the case of bulk GaAs (a full discussion can be found in ref. 4). Since the diffusion length and the light absorption depth are both much larger than the mean free path between collision with optical phonons, most of the electrons are thermalized before reaching the band bending region. There they may lose more energy and be emitted into vacuum, if their energy is above the vacuum level. It results that most of the photoemitted electrons appear in the low energy side of the EDC (the so called Γ peak) originating from electrons first thermalized to the bottom of the conduction band. On the other hand, non-thermalized, ballistic electrons appear on the high energy side of the EDC. The situation can be analyzed referring to Fig. 6, where a scheme of the GaAs band structure is shown. It is seen that the high energy threshold of the distribution must correspond to electrons excited from the hh valence band (transition 1) and ballistically emitted into vacuum. Each final state of the various vertical transitions, induced by absorption of photons with energy $h\nu$, is a source for subsequent thermalization cascade or for photoemission of ballistic electrons. The same holds for the electrons accumulated in the side minima (L and X) before further thermalization in the lowest minimum Γ or emission. A structure is expected in the EDC for each source of photoemitted electrons. Since at low temperature the electrons can only lose energy, in this case the position of the source is given by the high energy threshold of the structure. When this position depends on the excitation energy $h\nu$ it is due to ballistic electrons photoexcited from one of the valence bands (hh, lh, and spin-orbit-split band: Γ_{8h}, Γ_{8l} and Γ_7, respectively). In fact, in the parabolic band approximation, the kinetic energy ε_c of an electron photoexcited to the conduction band depends on the energy being $\varepsilon_c = (h\nu - E_g)(1 + m_e^*/m_h^*)^{-1}$, where m_e^* (m_h^*) is the electron

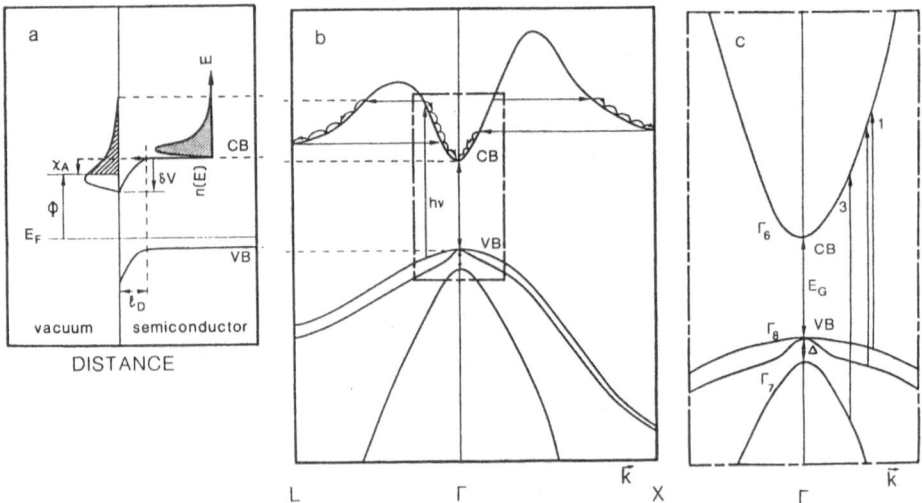

Fig. 6.　Conduction (CB) and valence (VB) bands of bulk NEA GaAs; a) in real space showing the band bending δV; b) in k-space throughout the Brillouin zone; c) near the Γ point with the vertical transitions induced by photons of energy $h\nu$, from the heavy hole, the light hole and the spin-orbit-split valence bands (1, 2 and 3 respectively).

(hole) effective mass [16]. A structure with a position indipendent of hν, instead, arises from electrons first accumulated in subsidiary minima of the conduction band before being ballistically emitted. Therefore from a set of EDC's obtained for different hν both the location of the L and X minima and the dispersion of the valence and conduction band can be determined. An accurate study has been indeed reported for bulk GaAs [4].

Figure 7 shows the EDC and its derivative versus the kinetic energy of the photoelectrons, as measured with a high resolution (20 meV) electron analyzer [17], from the SL of Fig. 4a at low tmperature (120 K), under excitation with the 1.83 eV line of a Kr^+ laser. The structures are better revealed in the derivative curve. A set of EDC derivatives obtained with the monochromatized light from a W-filament lamp is presented in Fig. 8 [12]. The result are identical to the ones obtained in bulk GaAs. In particular features indipendent of photon energy are seen, from whose onset the position of the Γ, X and L minima are determined (for the extrapolation procedure see ref. 4). Other structure are also present, whose energy position is consistent with high energy ballistic electrons all originating in the GaAs overlayer. No new structure is apparent, neither at the position of the SL miniband nor from the SL continuum. We may then conclude that no ballistic electrons from the SL reach the surface through the 1200 A thick GaAs overlayer, even though the yield curve (Fig. 4b) evidences the contribution of the electrons excited in the SL to the total current.

The EDC and its derivative from the QW sample of Fig. 5a excited by a He-Ne laser (hν =1.96 eV) is shown in Fig. 9. Also in this case no structure related to

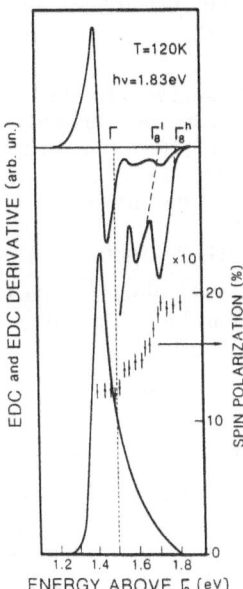

Fig. 7.

EDC (full line, lower panel) and EDC derivative (upper panel) of the SL for hν = 1.83 eV at 120 K. Energy is referred to the top of the valenceband Γ_8. The spin polarization vs electron kinetic energy is also shown in the lower panel.

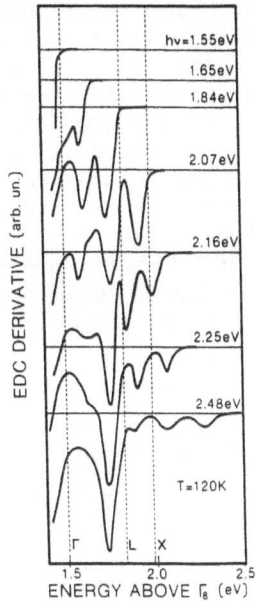

Fig. 8.

EDC derivatives from the SL sample at 120 K taken at different photon energies. The bulk GaAs Γ, L and X positions are indicated by dotted lines.

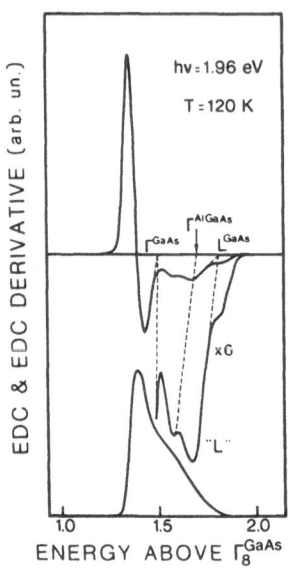

Fig. 9. EDC (lower panel) and EDC derivative
(upper panel) of the QW's sample at
120 K, for hν = 1.96 eV.

electrons excited into discrete levels of the wells (very well visible in the Y curve, see Fig. 5b) can be detected. An extra peak, however, appears in the EDC derivative as compared to bulk GaAs . Such a peak is present at the same kinetic energy in EDC's taken at other photon energies larger than 1.84 eV, i.e. the $Al_{0.27}Ga_{0.73}As$ gap. It is therefore attributed to ballistic electrons injected in the GaAs overlayer from the bottom of the $Al_{0.27}Ga_{0.73}As$ conduction band. Using the above mentioned extrapolation procedure [4], we find the energy position of $Al_{0.27}Ga_{0.73}As$ conduction band relative to the GaAs one. The conduction band offset is then directly measured and found in very good agreement with the value obtained from the energy position of optical transition as measured in the Y curve [14]. This determination, however, is somewhat unprecise because of the presence of the broad L peak, due to electrons thermalized in the GaAs L minimum.

Figure 10 shows the EDC and its derivative for the AlGAAs-1000 HJ. The dominant contribution to the current appears at energy lower than Γ_{AlGaAs} corresponding to the bottom of the $Al_{0.3}Ga_{0.7}As$ conduction band. It arises from electrons firstly thermalized in the $Al_{0.3}Ga_{0.7}As$ Γ minimum and then injected into the GaAs overlayer which is too thin to allow complete thermalization. This can be seen by comparing the shape of this EDC with that of a bulk GaAs sample shown in the inset of Fig. 10. For hν < E_G all electrons are iussed from the GaAs overlayer. For hν > E_G all the photons which are not absorbed in the overlayer excite electrons in the thick $Al_{0.3}Ga_{0.7}As$ bottom layer. From the known band structures [16] and the Fermi level location in these two regions, the ballistic electrons originating from $Al_{0.3}Ga_{0.7}As$ should appear at a kinetic energy very close to that of the electrons from the GaAs overlayer. In fact, in the EDC's the usual GaAs structures are observed and, in addition, a new peak appears at a fixed energy, Γ_{AlGaAs} , when hν > E_G . It is due to electrons first thermalized at the bottom of the AlGaAs conduction band, injected in GaAs, and photoemitted before complete thermalization in the overlayer. At variance with the case of the QW sample, however, in this case the separation between the bottom of the GaAs and alloy conduction bands, due to the Fermi level alignment, is equal to the gap difference. Therefore the measurements do not yield the conduction band offset but the gap difference ΔE_g . The position of the peak is indeed in agreement with what found in the yield curve (Fig. 3b).

In the case of the AlGaAs-100, as noted above, the situation is simpler since electrons photoexcited in the thin GaAs layer do not contribute sizably to the emitted current. Here, the photoemission process may be analyzed as follows. Electrons are mainly photoexcited in the AlGaAs part of the sample over the distance of the light absorption depth (1μm at $h\nu = 1.96$ eV). Practically all these electrons are rapidally thermalized to the bottom of the AlGaAs conduction band and subsequentely diffuse towards the interface. There they are injected into the GaAs layer at a kinetic energy $\varepsilon_c ' = E_G - E_g$ corresponding to a velocity $v = 10^6$ m/s, then ballistically reach the surface after a time $t = d/v = 10^{-14}$ s, and either appear into vacuum or are reflected   back into the sample.

The EDC's (see Fig. 11) clearly show that again the photoelectrons do not thermalize in the GaAs overlayer. In this case the hot-electron effect is more pronounced than in the AlGaAs-1000 HJ, where the top layer is thicker (cf. Fig. 10), leading to an almost rectangular shape of the EDC. The fact that the distribution is not sharply peaked at E_G evidences, however, an efficient energy relaxation in the 100 Å GaAs layer. From the EDC of Fig. 11 and others taken at various excitation energies $h\nu$ (He-Ne laser and the infrared and visible lines of a Kr^+ laser) information on the band structure of the alloy have been obtained [8]. The photon energy depentent structures are found almost at the same position as in GaAs, since the electron to hole effective mass ratio $m_e{}^*/m_h{}^*$ is practically indipendent of the Al-content of the alloy [16]. A new experimental determination of the position of the X-subsidiary minimum, resulting 140 meV above the Γ minimum, in $Al_{0.3}Ga_{0.7}As$ is also achieved.

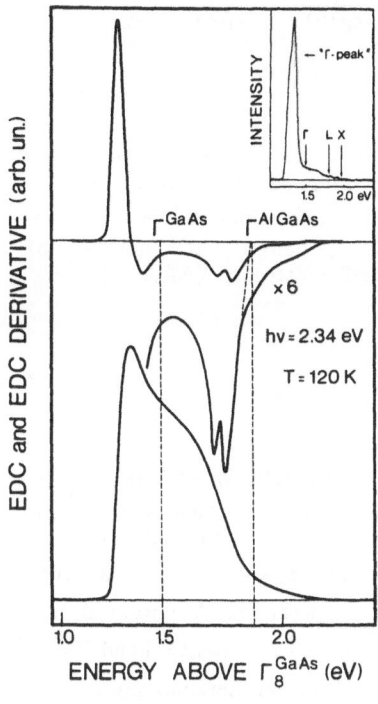

Fig. 10.

EDC and EDC derivative of the HJ AlGaAs-1000, for $h\nu = 2.34$ eV The inset shows the EDC from bulk GaAs in the same conditions.

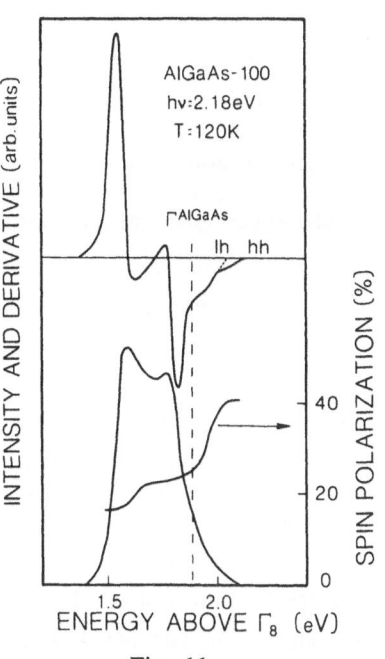

Fig. 11.

EDC and EDC derivate of the HJ AlGaAs-100, for $h\nu = 2.18$ eV. The location of the AlGaAs conduction band minimum is indicated.

4. SPIN-POLARIZATION

Absorption of circularly polarized light of energy slightly larger than the gap in a zincblend semiconductor gives rise to spin-polarized conduction electrons [2]. This is the so called optical pumping of conduction electrons [18]. The polarization of electrons in the conduction band is determined by the dipole selection rules for light absorption, resulting in a maximum degree of polarization $P = (n_+ - n_-)/(n_+ + n_-) = 50\%$ (here n_+ and n_- are the number of excited electrons with spin parallel or antiparallel to the to the light wavevector). As shown in Fig. 12, the value of $P = 50\%$ arises from the fact that at Γ the transitions to the conduction band minimum ($m_j = \pm 1/2$) involve the degenerate hh and lh bands ($m_j = \pm 3/2$ and $m_j = \pm 1/2$

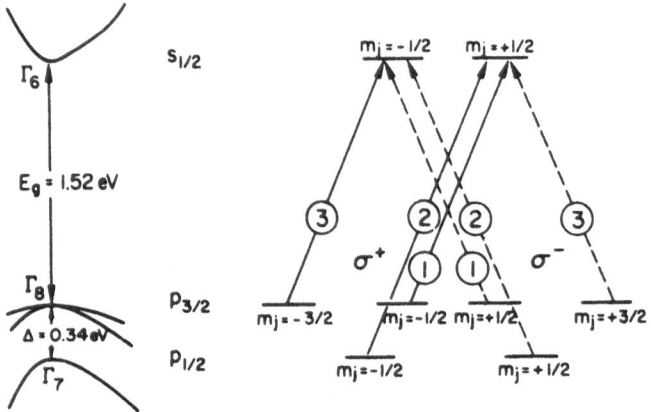

Fig. 12. Left. scheme of the band structure of GaAs near the Γ point. Right. atomic-like picture for the optical transitions. The allowed transitions for circularly polarized light σ^+ (σ^-) are shown as full (dashed) lines. The circled numbers give the relative intensities. Changing the helicity of the light revers the spin direction of the excited electrons. (After 3).

rispectively), and from the $\Delta m_j = +1$ ($\Delta m_j = -1$) selection rule for σ^+ (σ^-) circularly polarized light. The net polarization is given by the relative probabilities (3:1) of the transition from the hh band with respect to that from the lh band. For higher photon energies, P is expected to decrease very slowly for $E_g < h\nu < E_g + \Delta$, where Δ is the spin-orbit splitting of the valence band at the Γ point [19]. A strong decrease is instead expected at the onset of transitions from the split-off valence band ($h\nu > E_g + \Delta$) with angular momentum $J = 1/2$ (see Fig. 12). Sevenreal depolarization mechanisms, whose efficiency increases with temperature, may contribute to the reduction of the spin-polarization during the electron lifetime in the conduction band [19,20].

If the surface is activated to NEA, electrons excited to the conduction band can escape into vacuum and spin-polarized photoelectrons are emitted from the cathode [21]. The polarization of the emitted electrons can be measured by the so called Mott detector, in which the electron beam is accelerated to high voltage (typically some 100 keV) and then scattered by a thin gold foil [22]. Two different experimental procedures are possible. In both cases circularly polarized light, produced by means of either a Pockel's cell or a Babinet-Soleil compensator, is sent onto the sample. In the first set-up, the total current of photoelectrons is sent to the detector, and the polarization of the beam is measured as a function of $h\nu$. In the second one, the photon energy is kept fixed, and P is measured as a function of the electron kinetic energy.

An example of the first type is shown in Fig.1 (full and empty dots, left-hand axis), in the case of GaAs at room temperature. The maximum value of P (45%) is very close to the theoretical one for the MBE-grown sample, indicating that the depolarization mechanisms are much less effective in this case as compared with the bulk GaAs, where a much smaller value is obtained (36%) [6]. Furthermore, in both cases a drastic reduction of P is seen for photon energies larger than $E_g + \Delta$ (Δ = 0.34 eV in GaAs), as expected.

Similar results have also be obtained for the direct gap (x < 0.45) MBE-grown alloys [7]: large values of P near the threshold and strong decrease for $h\nu > E_G + \Delta$. As for the quantum yield, the spectra for different Al-content x are shifted in energy according to the shift of the gap. From the dropoff in the polarization curve the spin-orbit splitting of the valence band Δ in the alloy has been determined as a function of x. Moreover spectra taken on indirect gap samples (x > 0.45) show that indirect transition from the top of the valence band at Γ to the X conduction band minimum do not give rise to optically pumped elctrons with significant polarization. Note that for indirect transitions the dipole selection rules, which allow the optical orientation of the electron spin in direct transitions, do not hold. A careful analysis of the maximum value of P vs x indicates also that the depolarization mechanisms become more effective with increasing Al concentration in the alloy [7].

On the other hand, by analyzing the polarization versus energy distribution curves (PEDC's) obtained for various excitation energies, the conduction electron dynamics as well as the depolarization mechanisms which affect the polarization of the emitted electrons, can be investigated. Absorption of circularly polarized light with energy $h\nu$ promotes electrons from the valence band (hh, lh, and spin-orbit-split band) into the conduction band at kinetic energies and with spin-polarizations depending on the photon energy and the particular transition involved. A detailed analysis beyond the naive atomic-like picture given above (cf. Fig. 12), which is meaningless for electron k vector different from zero (Γ point), is indeed necessary [5]. Moreover, due to the cubic symmetry, the so obtained theoretical values of P have to be averaged over all the electron moment directions. For instance, the maximum theoretical value of P is equal to 2/3, in spite of a 100% polarization of the electrons excited from the hh band with k parallel to the direction of light propagation (surface normal). This results from angular averaging and could not be increased by angular selection in vacuum since momentum anisotropy is relaxed by the Cs-oxide layer [5].

Without entering a detailed discussion, we note that the PEDC's obtained from the SL are very similar to those for bulk GaAs. An example for $h\nu$ = 1.83 eV is shown in Fig. 7 (dots, right-hand axis). The values of spin-polarization are actually reduced, indicating that the spin relaxation mechanisms in the Sl are more effective than in GaAs. The various features in the PEDC reflect the final state of optical transition or the effects of spin relaxation during the thermalization cascade. As can be seen from Fig. 7, the most energetic electrons carry the largest polarization. These electrons are excited from the highest valence band, i.e. the Γ_{8h} hh-band and emitted without any energy loss; they retain the largest polarization P_{max} since they escape very rapidly from the solid. In fact, each electron promoted into the conduction band with a spin $S(k)$ suffers coherent precession in the internal magnetic field $w(k)$ due to the lack of inversion symmetry in III-V compounds, the intensity of which increases with the kinetic energy ε_c in the conduction band [19]. The total spin momentum along the surface normal is decreased by this effect. In the case of bulk GaAs, the ballistic electron mean free path is deduced from the experimental P_{max} [5].

The main contribution to the current emitted by a NEA photocathode arises from the Γ peak region at low energy (left-hand side of the PEDC in Fig. 7). Its polarization determines the overall polarization of the electron beam in vacuum, when no energy selection is performed. The thermalized electrons undergo several spin relaxation mechanism during their lifetime in the crystal [20]. These processes are phenomenogically described by means of a spin relaxation time for electrons at the bottom of the conduction band.

A set of PEDC's measured at different photon energies in the HJ AlGaAs-1000 shown in Fig. 3a, is presented in Fig.13. In Fig. 14 comparison is made between the PEDC's measured in bulk GaAs and in the HJ, obtained with a He-Ne laser at 120 K [15]. The high energy sides of the curves correspond to ballistic electrons which have the same polarization P_{max} in both samples, confirming that in the HJ these electrons originate only from the GaAs part of it. The thermalized electron polarization reflects the absorption ratios in the two parts of the sample and is predominally due to electrons first excited in $Al_{0.3}Ga_{0.7}As$ since $\alpha d < 1$. For a given $h\nu > E_G$ this polarization is larger than the values in bulk GaAs, because the electrons excited in the alloy with a smaller kinetic energy $\varepsilon_c{}'$ suffer less depolarization. Before emission into vacuum they are injected in the overlayer and

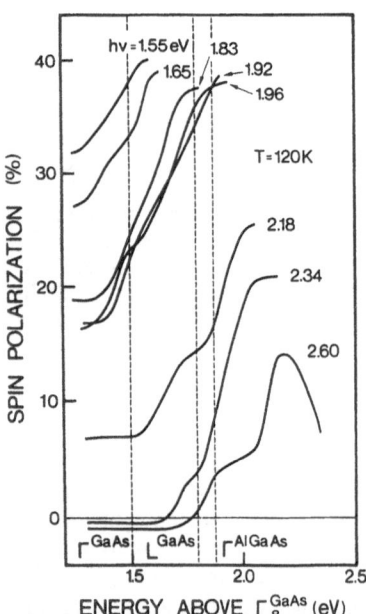

Fig.13. PEDC's taken on the HJ AlGaAs-1000 for different photon energies. Electron energies are referred to the top of GaAs valence band $\Gamma_8{}^{GaAs}$.

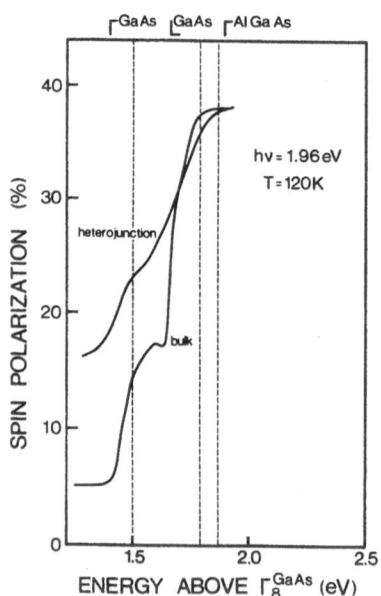

Fig. 14. PEDC's taken at 120 K on the HJ AlGAAs-1000 and on bulk GaAs at 120 K for $h\nu = 1.96$ eV.

undergo some additional depolarization so that they escape into vacuum with still a significant polarization (16% at hν = 1.96 eV, rather than 5% as in GaAs, see Fig. 14). Note also that for hν = 2.34 eV, corresponding to a maximum creation energy ε_c' of 370 meV in $Al_{0.3}Ga_{0.7}As$, the thermalized electron polarization is zero. In bulk GaAs, it goes to zero for an energy ε_c of 500 meV [5]. This suggest again that the spin relaxation mechanism due to the precession in the w(k) field, may be more effective in the alloy, which is more ionic than GaAs.

In finishing this section it is worthwhile to remark that more work is surely necesary in order to better understand the electron spin dynamics in SL's and QW's. It should also be noted that high quality samples are needed, since it has been experimentally found that the electron spin-polarization, via the spin depolarization time, is a very sample-dependent parameter, much more than the others measured in photoemiossion.

5. SPIN-POLARIZED ELECTRON SOURCES

The interest for solid state polarized electron sources has been expanding considerably in atomic- [22], surface- [23], and high-energy-physics [24], since the first experiments on spin-polarized photoemission from NEA GaAs [21]. Spin-polarized electron sources, based on photoemission from NEA GaAs excited by circularly polarized light have beeen developed [3]. It should be noted here that, as discussed in the previous section, high polarization of the photoelectrons is achieved only for photon energies slightly larger than the gap: a matched infrared laser diode is then commonly used. One of the most important features of such sources is the possibility of reversing the polarization of the beam (obtained by simply revesing the polarization of the light, e.g. from σ^+ to σ^-), as required from most experiments. Moreover a very high figure of merit P^2I, where P is the polarization and I is the beam current, can be achieved. Since in the following we shall be concerned only with souces based on NEA photocathodes, we normalize for the light intensity, defining a new figure of merit $F = P^2Y$, where Y is the quantum yield.

In order to improve considerably the performances of the source, one has to look for a material where a large polarization of the conduction electrons is achievable and a high Y is present. From the analysis presented in the previous section, it appears that removal of the cubic symmetry would be necesary. The reduction of symmetry will define a preferred axis in the crystal thus avoiding averages over k. It will also lift the degeneracy between the hh and lh bands at Γ, giving rise to a 100% theoretical value for P_{max}.

Several proposals have been put forward in this direction. The most immediate approach is to introduce an uniaxial stress in GaAs in order to lower the symmetry. This requires, however, a relevant experimental effort in combining UHV and NEA surface preparation technologies with high uniaxial stress: such a goal has not yet been accomplished [25]. Photoluminescence studies have suggested the use of AlGaAs/GaAs SL's [9], but as we have seen above [10,12], no high values of P have been obtained up to now: the subject surely still needs careful investigations. In any case quantum yields considerably lower than in GaAs or in AlGaAs have to be expected, because the highly polarized electrons are confined in the GaAs QW's and must escape through a tunnel process (compare the yields of Fig.'s 1 and 2 to that of Fig. 4b). Other materials with lower symmetry than the zincblend, such as chalcopyrites or hexagonal CdSe, have been proposed. Also in this case, however, first attempts [26] have not been successful, very likely because of the lack of single-crystal samples with high p-doping level, as necessary for NEA.

Finally, we note that the conditions necessary to obtain high Y are expected to be fullfilled in the artificial semiconductor $AlGaAs_2$, obtained from the alternate deposition of GaAs and AlAs layers on (001) oriented substrates. Moreover, the presence of two distinct cations sublattices must remove the cubic symmetry and cause a tetragonal splitting of the degenerate hh and lh bands. A quantitative evaluation of this splitting, as well as of the orbital composition of the states at the valence and conduction band edges, are needed in order to estimate the degree of polarization achievable for optically pumped conduction electrons. This has been done on the basis of a self-consistent *ab initio* relativistic band sturcture calculation for this system [27], and the use of an AlGaAs$_2$ photocathde as highly polarized electron source has been recently suggested [28]. It should be mentioned, however, that the growth of such materials requires an extreme control of the epitaxial techniques; the experimental operation of this source has not yet been investigated.

An alternative approach to improve the source, is to make its use simpler. This can be done, for instance, by allowing the use of a visible photon energy for optical pumping in a region where convenient light sources are available and still obtain high spin-polarization with good intensity. Experiments have been succesfully performed on III-V direct-gap semiconductors with energy gap larger than the GaAs one, namely $Al_xGa_{1-x}As$ [7] and GaP_yAs_{1-y} [29] alloys, matched to the photon energy of a He-Ne laser (hν = 1.96 eV), obtaining comparable values of F around 1.5 - 2 x 10^{-3}.

The PEDC shown in Fig. 14 for the AlGaAs/GaAs HJ, suggests that similar results should also be obtained in heterojunctions with a thin GaAs overlayer. We can combine, in fact, the advantage of matching the gap of the semiconductor to the energy of a convenient light source, together with that of an easy and stable activation of the GaAs surface. This very favorably compares with the above mentioned attempts of gap matching: i) GaPAs, in which the surface composition is not well controlled, since both Phosphorus and Arsenic evaporate very easily from the surface, and ii) AlGaAs which, as noted above, must be grown and used *in situ* to avoid heavy contamination.

In Fig. 15 we compare the EDC's and PEDC's from bulk GaAs and two HJ's with different thickness of the overlayer, whose yields have been reported in Fig. 3b [8]. The curves have been obtained under He-Ne laser excitation at 120 K. The mean

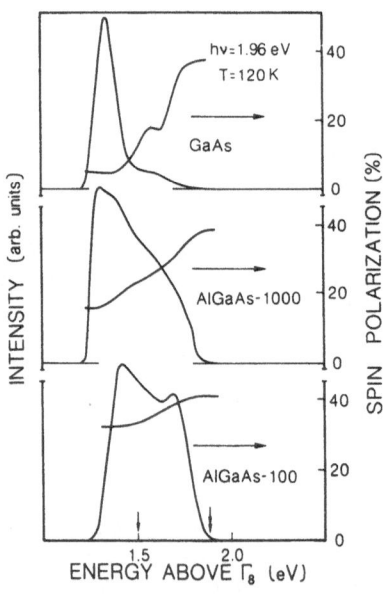

Fig. 15. EDC's and PEDC's taken at 120 K using a
He-Ne laser as light source (hν = 1.96 eV)
or bulk GaAs (top), AlGaAs-1000 (middle)
and AlGaAs-100 (bottom). The locations of
the bottom of the conduction band in AlGaAs
and in GaAs are indicated by arrows.

value of P for the AlGaAs-100 sample is 36% at 120 K and 30% at room temperature, corresponding to values of F equal to 3.9 x 10^{-3} and 2.7 x 10^{-3}, respectively (to be compared to the values of F = 4.2 x 10^{-3} and 3 x 10^{-3} in bulk GaAs [3]). It is also seen in Fig. 15 (lower panel) that P does not vary much with the kinetic energy of the emitted electrons (32.5% < P < 42.5%), and the EDC is not sharply peaked at low energies. These results are very important. In fact, in normal photocathodes, due to Cs release from the surface and/or contaminants accumulation on it, the vacuum level rises with time, so that there is a variation of the electron affinity, from negative to zero to positive, which drastically reduces the emitted current. This mechanism is described by the photocathode lifetime, resulting of the order of ten hours for GaAs in UHV conditions [3,6]. The electron affinity increasing is reflected by a low energy cut-off moving towards higher energies in the EDC's. It is then clear that small variations of the electron affinity of the GaAs surface, affect   the emitted electron intensity and the polarization much less in the case of the HJ than in bulk GaAs (compare top and bottom panels in Fig. 15). The AlGaAs-100 source is expected to be very stable, and we indeed observe decay time costants as long as a week in UHV conditions.

At present, work is in progress in our laboratory in Milan to use such heterojunction as a spin-polarized electron source in spin resolved inverse photoemission experiments.

ACKNOWLEDGEMENTS

It is a pleasure to thank S.F. Alvarado who introduced me into this field. The more recent experiments (on HJ's, QW's, and SL's) have been done at the Laboratoire de Physique de la Matière Condensée, Ecole Polytechnique Palaiseau, in collaboration with the group of G. Lampel, to whom I am deeply indebted. Critical reading of the manuscript by M. Sancrotti is also acknowledged.

REFERENCES

1. W.E. Spicer, Appl. Phys. 12, 115 (1977), and references therein.
2. G. Lampel, in *Proceeding of the 12th International Conference on the Physics of Semiconductors, Stuttgart 1974* , ed. by M.H. Pilkuhn (Teubner, Stuttgart 1974) p.743.
3. D.T. Pierce, R.J. Celotta, G.C. Wang, W.N. Unertl, A. Galejs, C.E. Kuyatt, and S.R. Mielczarek, Rev. Sci. Instrum. 51, 478 (1980).
4. H.J. Drouhin, C. Hermann, and G. Lampel, Phys. Rev. B 31, 3859 (1985).
5. H.J. Drouhin, C. Hermann, and G. Lampel, Phys. Rev. B 31, 3872 (1985).
6. S.F. Alvarado, F. Ciccacci, S. Valeri, M. Campagna, R. Feder, and H. Pleyer, Z. Phys. B 44, 259 (1981).
7. F. Ciccacci, S.F. Alvarado, and S. Valeri, J. Appl. Phys. 53, 4395 (1982).
8. F. Ciccacci, H.J. Drouhin, C. Hermann, R. Houdré, and G. Lampel, Appl. Phys. Lett. 54, 632 (1989).
9. R.C. Miller, D.A. Kleinmann, and A.C. Gossard, in *Proceedings of the 16th International Conference on the Physics of Semiconductors, Edingburgh 1978* , ed. by B.L.H. Wilson (Institute of Physics, London 1979) p. 1043.
10. S.F. Alvarado, F. Ciccacci, and M. Campagna, Appl. Phys. Lett. 39, 615 (1981).
11. R. Houdré, C. Hermann, G. Lampel, P.M. Frijlink, and A.C. Gossard, Phys. Rev. Lett. 55, 734 (1985).
12. F. Ciccacci, H.J. Drouhin, C. Hermann, R. Houdré, and G. Lampel, Vuoto XVI, 185 (1986).
13. R. Houdré, C. Hermann, G. Lampel, and A.C. Gossard, Physica Scripta T 13, 241 (1986).
14. F. Ciccacci, H.J. Drouhin, C. Hermann, R. Houdré, G. Lampel, and F. Alexandre, in *Excitons in Confined Systems* , Springer Proceedings in Physics 25, ed. by R. Del Sole, A. D'Andrea, and A. Lapiccirella (Springer-Verlag, Berlin 1988) p. 185.

15. F. Ciccacci, H.J. Drouhin, C. Hermann, R. Houdré, G. Lampel, and F. Alexandre, Solid State Electron. 31, 489 (1988).

16. For the band parameters of the AlGaAs alloy, see S. Adachi, J. Appl. Phys. 58, R1 (1985).

17. H.J. Drouhin, and M.Eminyan, Rev. Sci. Instrum. 57, 1052 (1986).

18. F. Meier and B.P. Zakharchenya (ed.'s), *Optical Orientation* , Series Modern Problems in Solid State Sciences, Vol. 8 (Elsevier, Amsterdam 1984).

19. M.I. D'yakonov and V.I. Perel', Sov. Phys. JEPT 33, 1053 (1971) [Zh. Eksp. Teor. Fiz. 60, 1954 (1971)].

20. G. Fishman and G. Lampel, Phys. Rev. B 16, 820 (1977).

21. D.T. Pierce, F. Meier, and P. Zuercher, Appl. Phys. Lett. 26, 670 (1975).

22. J. Kessler, *Polarized Electrons* , Springer Series on Atoms and Plasma 1 (Springer Verlag, Berlin 1985).

23. R. Feder (ed.), *Polarized Electrons in Surface Physics* , Advances Series in Surface Science (World Scientific, Singapore 1985).

24. C. Sinclair, in *High Energy Physics with Polarized Beams and Polarized Targets* , ed. by C. Joseph and J. Soffer (Birkhauser Verlag, Lausanne 1981) p.27.

25. P. Zorabedian, SLAC Report 248 (Standford University, 1982).

26. S.F. Alvarado, F. Ciccacci, H. Riechert, and M. Campagna, unpublished.

27. N.E. Christensen, E. Molinari, and G. Bachelet, Solid State Commun. 56,125 (1985).

28. F. Ciccacci, E. Molinari, and N.E. Christensen, Solid State Commun. 62, 1 (1987).

29. D. Conrath, T. Heindroeff, A. Hermanni, N. Ludwig, and E. Reichert, Appl. Phys. 20, 1955 (1979).

WANNIER-STARK QUANTIZATION AND BLOCH OSCILLATOR IN BIASED SEMICONDUCTOR SUPERLATTICES

G. Bastard and R. Ferreira

Groupe de Physique des Solides de l'ENS
24 rue Lhomond F-75005 Paris, France

ABSTRACT

We review salient features exhibited by electronic states in semiconductor superlattices subjected to a static electric field applied parallel to the growth axis. When the field is large enough the band structure is destroyed and replaced by evenly spaced levels : the Wannier-Stark ladder. The Wannier-Stark ladders are calculated to survive the broadening due to static scatterers (ionized impurities, interface defects). We briefly compare the average spatial localization of the Wannier-Stark states to their semiclassical counterpart derived from the Bloch oscillator formalism.We show that the field-induced quantization leads to an effective blue shift of the band-to-band absorption edge as well as to the existence of oscillatory structures which are periodic in $1/_F$.

WANNIER-STARK STATES

About thirty years ago Wannier[1] proved that the eigenstates of a hamiltonian whose potential energy is the sum of a periodic component V(z) and of linearly varying term eFz are discrete and evenly spaced in eFd, where d is the period of the underlying lattice (Wannier-Stark ladders). The corresponding eigenfunctions are spatially localized. These results contrast with the zero field (F = 0) case where the periodic potential leads to a spectrum which is quasi-continuous and organised into allowed and forbidden bands. Wannier's results were derived under the assumption of a one-band analysis, i.e. by neglecting any field-induced couplings between the different eigenstates. A number of papers[2-6] have examined the validity of these assumptions and it is now believed[6] that Wannier's findings retain their validity with the restrictions that i) the Wannier-Stark bound states are actually virtually bound (due to the field-induced escape to regions where eFz is large and negative) and ii) that resonant tunnel effects delocalize the carrier when a state of a given one-band

Wannier-Stark ladder coincides in energy with another state belonging to a different ladder.

The key parameter which controls the spatial extension of a member of a Wannier-Stark ladder is $f = eFd/I\lambda I$, where $I\lambda I$ is the nearest neighbour transfer integral (the zero field bandwidth Δ is equal to $4 I\lambda I$). This means that $f \approx 10^{-3} - 10^{-2}$ in usual solids (where $d \approx 6$ Å and $I\lambda I \approx 1$ eV) but may exceed unity in semiconductor superlatices (where $d \approx 60$ Å and $I\lambda I \approx 15$ meV). We show in Fig. (1) the site dependence of the probability of finding the carrier in the pth site when it is in the vth Wannier-Stark state i.e. when its energy is

$$\varepsilon_v = E_1 + veFd \qquad (1)$$

corresponding to the eigenfunction:

$$\phi_v(z) = \sum_n J_{v-n}(-2/f)\varphi_{loc}(z-nd) \qquad (2)$$

where $v, n \in Z$ and E_1 is the confinement energy of an unbiased, isolated quantum well (wavefunction φ_{loc}). In this simplest tight-binding scheme the site probability is equal to $J^2_{v-n}(2/f)$ where J_k is the Bessel function of order k. Therefore, it exhibits a more than exponential localization around the vth site when f increases. The localization is almost complete when $f = 4$, i.e. when the potential energy drop over a superlattice period is equal to the zero field bandwidth.

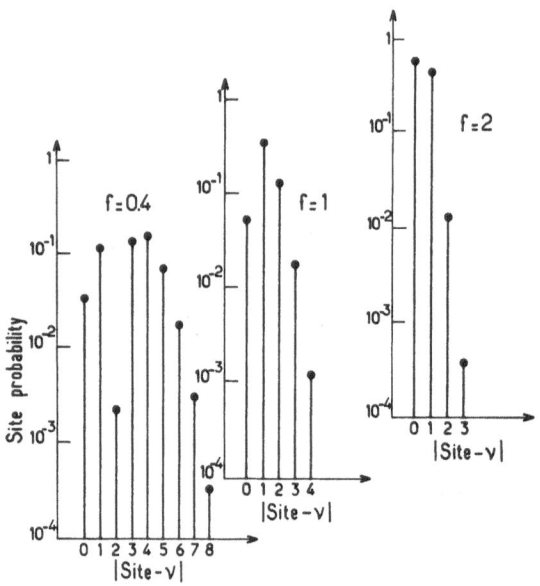

Fig. 1. Site dependence of the site probability for the vth Wannier-Stark state calculated in the one-band nearest neighbour approximation.

Another insight on the localization properties of the Wannier-Stark states is obtained by calculating the time dependent probability $P_j(t)$ of finding the carrier in the atomic state centered on the jth period at time t if it was in that centered on the 0th period at time $t = 0$.

After some calculations one obtains:

$$P_j(t) = J_j^2 \left(4/f \sin(\tau/2) \right) \tag{3}$$

where:

$$\tau = eFdt/\hbar \tag{4}$$

$P_j(t)$ thus oscillates with time with an angular frequency which is equal to the Bloch frequency:

$$\omega_B = eFd/\hbar \tag{5}$$

It is tempting to connect the quantum $P_n(t)$ to a corpuscular description of an electron leaving the site 0 at $t = 0$ and travelling against the field $(n>0)$ or along the field $(n<0)$. We remark that $P_n(t)$, $n>0$, is very small for a certain time period, then raises to reach a maximum at t_n, decays with oscillations, and then grows again $(t>\pi/\omega_B)$ to repeat the previous sequence backward and vanishes at $t = 2\pi/\omega_B$, where a new cycle will start. The first maximum exists only if $t_n<\pi/\omega_B$. We loosely interpret t_n as the time needed by the electron to travel from the site 0 to the site n. A plot of t_n versus n for different electric field strengths reveals several features:

1) an almost linear increase of t_n with n in the limit of small fields (such that $t_n<<\pi/\omega_B$). The slope is close from that of a free particle travelling with constant velocity $2I\lambda Id/\hbar$, which is that of an electron in an unbiased superlattice with an energy corresponding to the center of the miniband (E_1).

2) a steep rise of t_n with n when t_n approaches π/ω_B, to the extent that once $n>n_{max}$ (where $t_{nmax} = \pi/\omega_B$) t_n should be considered as infinite.

3) a decrease of n_{max} when F increases.

4) an increase of t_n with F for a fixed n.

5) identical results for P_n and P_{-n}.

Point 1) is roughly in agreement with a corpuscular interpretation. The latter is however invalidated by the other points. In particular, the electric field acts in the opposite way of our intuition since it decelerates (point (4)) the electron in both the upfield and downfield motions. In addition, $P_n(t_n)$ decreases with n, on account of the normalization condition: upon increasing t_n there is more and more of the probability of finding the electron at sites $m<n$ (in the extra oscillations of $P_{m<n}$) and thus less and less available for the maximum of P_n. The corpuscular interpretation is thus of little relevance for that particular aspect of the Wannier Stark ladder. A diffusive pattern may be more correct: the quantum uncertainty on the electron position measured from the 0th period at time t is $\Delta(t) = (<z^2>_t - <z>_t^2)^{1/2}$. Since $<z>_t$ vanishes, we are left with:

$$\Delta(t) = (\Delta^2(0) + 8d^2/f^2\sin^2(\tau/2))^{1/2} \tag{6}$$

Eq.(6) nicely illustrates the oscillatory spreading of the electron wavefunction with time. We further notice from (6) that letting t going to infinity after having let F to vanish (i.e. for the flat band superlattice or for $\omega_B t \ll 1$ if $F \neq 0$) leads to a recovery of the diffusive motion for an electron with the superlattice effective mass at the subband edge.

The bound state nature of the Wannier-Stark state holds for an arbitrary weak electric field, a result of our one band assumption together with our lack of accounting for broadening effects. The latter can be included in the Born approximation for the scattering time. Let us focuse our attention on static scatterers and compute the lifetime of the edge of the νth Wannier-Stark subband (since in actual superlattices the carrier motion in the layer plane is free and gives rise to two dimensional subbands attached to each member of the Wannier-Stark ladder). As static scatterers we have considered either gaussian interface defects with an in-plane characteristic extension a, depth b and areal concentration N_{def} or coulombic impurities located on the interfaces (areal concentration: N_{imp}). Owing to the translational invariance of the structure along the growth axis (interface defects and impurities are assumed to sit on all the "inverted" interfaces of the assumed infinite superlattice) the lifetime τ_ν is independent of ν. Some approximations should be made in order to render the algebra tractable . They essentially amount to retaining in the three centers integrals of the type $< \varphi_{loc} (z-md)I f(z-pd)I \varphi_{loc}(z-m'd) >$ only the terms which correspond to $m = m' = p$ if f is a function peaked around the pth cell. In which case one obtains after some algebra:

$$1/\tau(k= 0) = (2\pi e^2/\kappa)^2 N_{imp}/\hbar e F d \sum_j 1/j \; I_{00}^2(0, z_{imp}, 2m^*/\hbar^2 jeFd) \; x$$
$$\sum_p J_p^2 J_{p-j}^2 \tag{7}$$

where:

$$I_{mm'} (n, z_{imp}, Q) = < \varphi_{loc}(z-md) I \exp (- Q I z-z_{imp}-nd I) I \varphi_{loc}(z-m'd) > \tag{8}$$

where z_{imp} is the impurity position in the nth cell. Similarly, for gaussian interface defects we have found:

$$1/\tau(k= 0) = (\pi a^2 V_b P_{int})^2 m^* N_{def}/\hbar^3 \sum_j \exp(- m^* a^2 jeFd/\hbar^2) \sum_p J_p^2 J_{p-j}^2 \tag{9}$$

where P_{int} is the integrated probability of finding the carrier between $-L/2$ and $-L/2+b$, V_b is the barrier height, m^* the carrier effective mass (for the bulk material and assumed to be position-independent) and L the quantum well thickness. Fig.(2) shows a plot of $\omega_B \tau$ versus 2/f. One notices the strong increase of the scattering time upon increasing F (decreasing 2/f). It is clearly associated with the wavefunction localization of the νth Wannier-Stark state around the νth period when F increases. In fact, under many experimental situations the Wannier-Stark state is little broadened ($\omega_B\tau >$ or $>> 1$) or, equivalently, the Bloch oscillator executes many oscillations before undergoing a scattering event. The acoustical phonon scattering and, more so, the optical phonon scattering should be more severe, still exhibiting the same qualitative $\omega_B\tau$ improvement at high field due to the field induced localization.

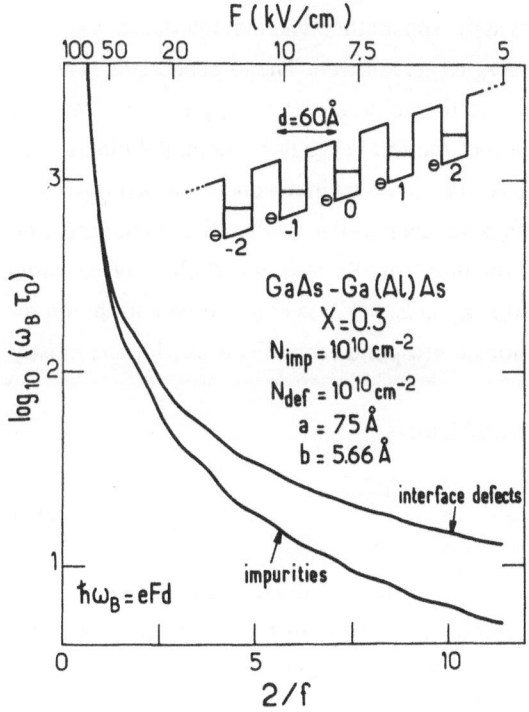

Fig. 2. The decimal logarithm of $\omega_B\tau_0$, where ω_B is the frequency of the Bloch oscillator and τ_0 the relaxation time is plotted against 2/f (lower scale) or against F (upper scale) for a 30 Å - 30 Å $GaAs-Ga_{0.7}Al_{0.3}As$ superlattice. Interface defects and ionized impurity scatterings have been considered.

WANNIER-STARK STATES VERSUS BLOCH OSCILLATOR : THE SPATIAL LOCALIZATION

It may be worth comparing the one-band tight binding predictions of the average extension of the νth Wannier-Stark state with those derived from a semi-classical description of the carrier motion : the Bloch oscillator model[12]. In this semi-classical description the band index n is a constant and one uses the Newton law

$$\hbar dq/dt = -eF \tag{10}$$

together with the relationship between the carrier velocity and energy

$$v_z = 2|\lambda|d/\hbar \sin(qd) \tag{11}$$

to obtain :

$$z(t) = z_0 + 2d/f \cos\{ d(q_0 - eFt/\hbar)\} \tag{12}$$

: the Bloch oscillator centered at z_0 undergoes periodic oscillations of amplitude $2d/f$ with the frequency $\omega_B = eFd/\hbar$. The oscillation amplitude diverges when $f \to 0$ to eventually become comparable with the crystal size. In the latter case Eq. (10) is no more valid and should be replaced by one which accounts for boundaries at the superlattice-vacuum interfaces.

As usual in semiclassical mechanics all energies are allowed. The total energy ε is at any time the sum of the kinetic energy $\varepsilon_n [q(t)]$ and of the potential energy $eFz(t)$. Classical oscillators are characterised by an equipartition of the total energy between the kinetic and potential contributions. The Bloch oscillator instead displays a kinetic energy which fluctuates around the band center (E_1) and, at any time, ε reduces to $E_1 + eFz_0$. To enable a better comparison between the semiclassical and quantum predictions we shall only examine situations where $\varepsilon = E_1 + veFd$. This means that z_0 in Eq. (12) is equal to vd and therefore that both quantum and semiclassical descriptions correspond to an electron which is mostly localised near the vth period.

The average spatial extension of the Bloch oscillator is

$$< (z(t) - vd)^2 >_{class}^{1/2} = d\sqrt{2}/f \tag{13}$$

where the classical average is taken over one period $(2\pi/\omega_B)$ of the oscillation. On the other hand, one can show by using the nearest neighbour tight-binding description of the vth Wannier-Stark state that the quantum average of $(z - vd)^2$ is equal to

$$< (z - vd)^2 >_{quant} = < \varphi_{loc}|z^2|\varphi_{loc}> + 2\sum_p p^2 J_p^2(2/f) \tag{14}$$

The first term is the quantum uncertainty on the electron position in the unbiased, isolated quantum well state E_1, while the second term accounts for the spreading of the vth state over periods adjacent to the vth one.

Performing the summation in Eq. (14) we have found that the value of $[<(z - vd)^2>_{quant} - < \varphi_{loc}|z^2|\varphi_{loc}>]^{1/2}$ exactly coincides with the semi-classical result $d\sqrt{2}/f$. That the Bloch oscillator result coincides with the Wannier-Stark predictions in the weak field limit $(f \ll 1)$ is not unexpected. On the other hand, the agreement at strong field (apart from the $< \varphi_{loc}|z^2|\varphi_{loc}>$ term) is perhaps more striking since the semi-classical formalism can be expected to break down when the external perturbation varies appreciably over a period as compared to the bandwidth, i.e. if $f \gg 1$.

The good agreement between the semiclassical and quantum descriptions is also witnessed by comparing the site dependences of the integrated probability distributions P_j of finding the particle around a given site j (i.e. between (j - 1/2)d and (j + 1/2)d) while its energy is ε_ν. We may approximate P_j in the quantum case by $< \nu I \varphi_{loc} (z - jd) >^2$, i.e. by $J^2_{\nu-j} (2/f)$. The semi-classical P_j is obtained as

$$P_j^{class} = 1/\pi \int_{(j-\frac{1}{2})d}^{(j+\frac{1}{2})d} \omega_B dz / Iv_z I \qquad (15)$$

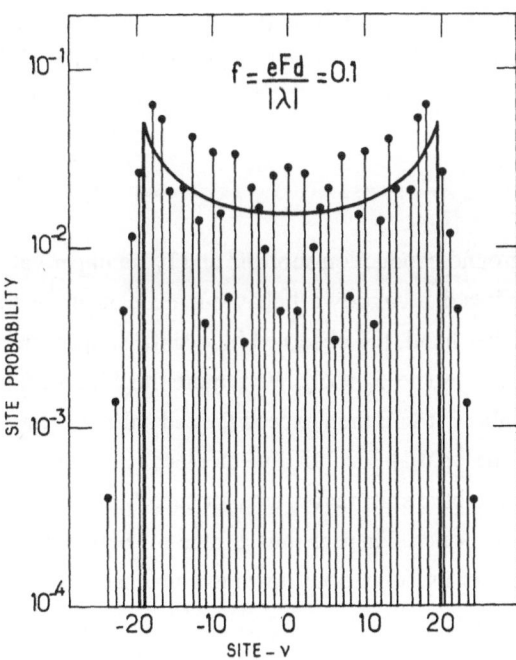

Fig. 3. The integrated probability distribution of finding the particle around a given site while being in the νth Wannier-Stark state is plotted against the difference between ν and the site number. Full circles : Wannier-Stark description. Thick solid line : Bloch oscillator model. The line is actually drawn through the calculated points for each site to help the eye.

The comparison between the two quantities is shown in Fig. (3) for a small value of the reduced electric field (f = 0.1). The semi-classical P_j's have been joined together to facilitate the comparison with the quantum results. It is seen that the semi-classical description accounts fairly well for the overall shape of the quantum result except for nodes which are absent in the former description but present in the latter one. Moreover, as might have been anticipated from the previous discussion, the good agreement between both descriptions persists in the strong field limit (f > 1).

We have recently examined[7] the consequences of the Wannier-Stark localization on the optical properties of semiconductor superlattices and shown that the absorption coefficient associated with interband transitions should display an effective <u>blue</u> shift due to the field-induced insulation of a given well from its neighbours. The latter is linked to the quenching of the tunnel coupling between wells whose eigenstates become misaligned by eFd. This blue shift, as well as oscillatory structures which are periodic in 1/F and arise from the evenly spaced nature of the Wannier-Stark spectrum, have recently been evidenced[8-10]. This has provided a clear cut experimental answer to the question of the very existence of the Wannier-Stark quantization which, up to now, was a tantalizing concept. Some of the properties of the Wannier-Stark states have recently been reviewed by us[11] and are summarized in Appendix A. Let us rather discuss the superlattice parameters which have to be designed in order to observe the field-induced blue shift of the band-to-band absorption edge at reasonable electric field strengths. For definiteness we shall only consider superlattices built out of two host materials of equal layer thicknesses. We search for periods d such that there is an almost complete localization for electron states (the holes due to their much smaller $|\lambda|$'s will then be automatically localized) i.e.

$$eF_c = \Delta_c \qquad (16)$$

where Δ_c is the zero field bandwidth of the ground conduction subband and F_c an upper value of the electric field beyond which field-induced breakdown behaviour may occur in the sample. We take $F_c = 100$ kV/cm. On the other hand, the observed blue shift is, apart from excitonic effects, equal to $1/2 (\Delta_c + \Delta_v) \approx 1/2 \Delta_c$ since $\Delta_c \gg \Delta_v$. We require this blue shift to be larger than 10 meV in order to be observable : it will then exceed the red shift associated with the reinforcement of excitons when the spatial localisation of electrons and holes increases with increasing field strength. A large blue shift favors smaller superlattice periods while the localisation criterion favors the larger ones. There is thus a trade off between both requirements which actually narrows the range of useful periods down to a few nanometers.

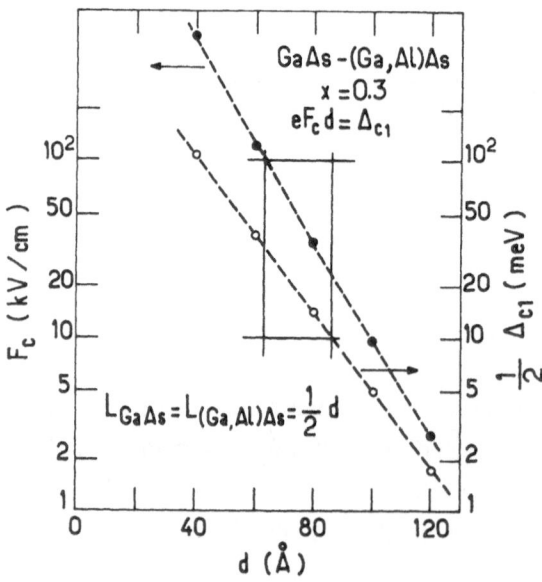

Fig. 4. Calculated period (d) dependences of the ground subband halfwidth ($1/2 \Delta_c$) and of the critical field F_c in GaAs-Ga(Al)As.

Figs. (4-6) show the period dependence of 1/2 Δ_c and F_c for three superlattice systems : GaAs-Ga$_{0.7}$Al$_{0.3}$As (Fig.4) ; Ga$_{0.47}$In$_{0.53}$As-InP (Fig.5) and Ga$_{0.47}$In$_{0.53}$As-Al$_{0.48}$In$_{0.52}$As (Fig. 6). The rectangles define the range of superlattice periods where $F_c \leq 10^5$V/cm and 1/2 $\Delta_c > 10$ meV. It is seen that periods of ≈ 70 Å are favorable to observe the blue shift in the three systems and, eventually, to use it for optical modulators.

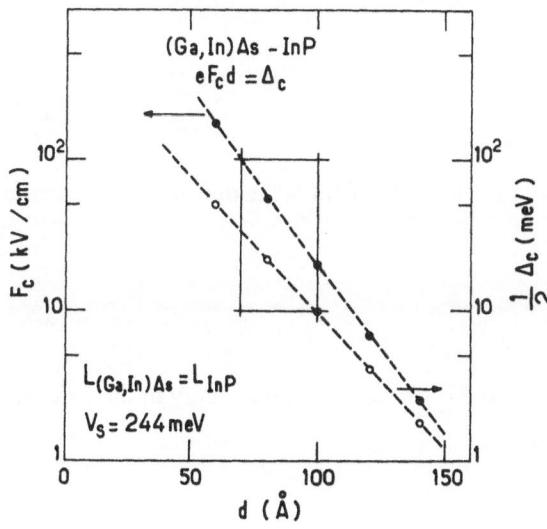

Fig. 5. Calculated period (d) dependences of the ground subband halfwidth (1/2 Δ_c) and of the critical field F_c in InP-Ga$_{0.47}$In$_{0.53}$As.

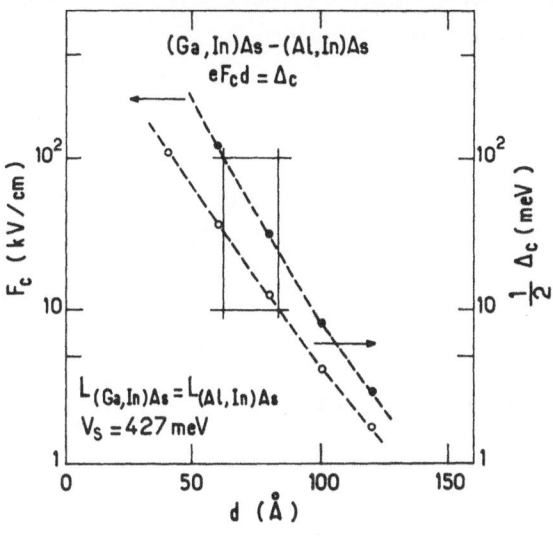

Fig. 6. Same as in Fig. 5 but for Ga$_{0.47}$In$_{0.53}$As-Al$_{0.48}$In$_{0.52}$As.

APPENDIX A: SOME ALGEBRA OF THE WANNIER STARK STATES

In the simplest tight binding scheme the biased superlattice wavefunctions are expanded on the isolated quantum well wavefunctions taken as orthonormalized:

$$<\varphi_{loc}(z-nd)I\varphi_{loc}(z-md)> = \delta_{nm} \tag{17}$$

$$\phi_\nu(z) = \sum_n c_n \varphi_{loc}(z-nd) \tag{18}$$

and the c_n's fulfil the eigenvalue equation:

$$c_{n+1} + c_{n-1} = f(n-\nu)c_n \tag{19}$$

which is the recursion relation for the Bessel functions. In (19) f is the dimensionless electric field strength:

$$f = eFd/I\lambda I \tag{20}$$

and the eigenenergies ε_ν are related to ν and to E_1, the confinement energy in the isolated well, by:

$$\varepsilon_\nu = E_1 + \nu eFd \tag{21}$$

In (21) ν is a relative integer to comply with the translational invariance of the hamiltonian which implies that if $\phi_\nu(z)$ is an eigenstate with energy ε_ν, $\phi_\nu(z+d)$ is also an eigenstate with energy ε_ν-eFd and is identical to $\phi_{\nu-1}$ whose energy is $\varepsilon_{\nu-1}$.

One of the independent solution of (19) diverges when n becomes very large. Thus, it should be discarded if the superlattice is infinite and we are left with:

$$c_n = J_{\nu-n}(-2/f) \tag{22}$$

where J_n is the Bessel function of integral order n. It is easily checked that the Wannier Stark wavefunctions are normalized. Moreover the νth state peaks at the νth period since:

$$<\nu I z I \nu> = \nu d \tag{23}$$

while the off diagonal matrix elements of the z operator are equal to:

$$<\nu I z I\mu> = d/f (\delta_{\mu,\nu+1} + \delta_{\mu,\nu-1}) \tag{24}$$

A measure of the spatial spreading of ϕ_ν is provided by calculating the average of $(z-\nu d)^2$ over ϕ_ν. We have found:

$$<\nu \mid (z-\nu d)^2 \mid \nu> = <\phi_{loc} \mid z^2 \mid \phi_{loc}> + 2d^2/f^2 \qquad (25)$$

where the first term accounts for the spatial spreading in the isolated quantum well bound state while the second term arises from the spatial spreading of the Wannier Stark state over several superlattice periods. As expected, the spreading is ν independent as, in fact are any averages over ϕ_ν of functions centered on the νth period. Although the Wannier Stark and Landau levels ($(n+1/2)\hbar\omega_c$) spectra are evenly spaced, they profoundly differ in that the former extends to infinity in both directions while the latter admits a lower eigenvalue, corresponding to n=0. The usual rule on the nodes of the eigenfunctions applies to the Landau level spectrum, namely the ground state has no node, the first excited state has one node etc., while it does not apply to the Wannier Stark state since all the ϕ_ν have the same number of nodes (a countable infinity), a property which is evident if one notices that two Wannier Stark states ϕ_ν, ϕ_μ are the same wavefunctions, but with their argument displaced from one another by a translation of $(\nu-\mu)d$. These differences in the nodal behaviour of the eigenfunctions of the two kinds of ladders leads to the ν independence of the averages over $\mid\nu>$ of operators which are centered on the νth site, while the same sort of averages are n dependent for the Landau level ladders.

In some accasions, it proves useful to calculate $<\nu \mid e^{-iqz} \mid \mu>$. We have found:

$$<\nu \mid e^{-iqz} \mid \mu> = e^{i(\nu-\mu)\pi 2}\, e^{-iqd(\nu+\mu)/2} < \phi_{loc} \mid e^{-iqz} \mid \phi_{loc}> J_{\mu-\nu}\,(\,-4/f\,\sin(qd/2)) \qquad (26)$$

APPENDIX B: PERTURBATION OF THE WANNIER STARK LADDER BY A WEAK IN PLANE MAGNETIC FIELD.

The quantum motion of an electron moving in a biased superlattice (**F**//**z**) and subjected to an in plane magnetic field may be described by the hamiltonian

$$H = \{\, p_x^2+(p_y - eBz/c)^2 + p_z^2\}/2m^* + eFz + V(z) \qquad (27)$$

where $V(z)$ is the periodic potential and where spin effects are neglected. The transverse gauge $\mathbf{A} = (0, -Bz, 0)$ has been used to connect the vector potential **A** to the magnetic field **B** (**B**// **x**). The eigenstates of (27) still factorize:

$$\Phi(\mathbf{r}) = 1/\sqrt{S}\, \exp i(k_x x + k_y y)\, \varphi(z) \qquad (28)$$

and $\varphi(z)$ is the eigenstate of:

$$H = \hbar^2 (k_x^2 + k_y^2)/2m^* + p_z^2/2m^* + V(z) + z(eF - \hbar\omega_c k_y) + 1/2m^*\omega_c^2 z^2 \qquad (29)$$

We are interested in the weak B regime where the spatial localization is provided by the electric field term while the term quadratic in B is a perturbation of the Wannier Stark ladder:

$$\varepsilon_v(k_x, k_y) = vd(eF - \hbar\omega_c k_y) + \hbar^2(k_x^2 + k_y^2)/2m^* \qquad (30)$$

where ω_c is the cyclotron frequency and the effective electric field is the sum of the external (F) and Hall ($-Bv_y/c$) fields. The projection of the eigenstates on the localized wavefunction φ_{loc} centered at $z = nd$ is equal to $J_{v-n}(-2/f_{eff})$ where:

$$f_{eff} = f(1 - Bv_y/Fc) \qquad (31)$$

Thus, the carrier velocity should be small enough to allow f_{eff} to be close enough from f. This means that the perturbative treatment is valid only near the subband edges.

The average of the magnetic perturbation over an unperturbed state $|v\rangle$ is equal to:

$$\Delta\varepsilon_v = 1/2m^*\omega_c^2 \{v^2 d^2 + 2d^2/f_{eff}^2 + \langle\varphi_{loc}| z^2 |\varphi_{loc}\rangle \} \qquad (32)$$

where the last term can be dropped since it is both v and k_y independent. The next term in the perturbation expansion vanishes. Thus, the eigenvalues of (29) can be approximated by:

$$\eta_v = eFvd + \hbar^2 k_x^2/2m^* + \hbar^2\{k_y - vd/l^2\}^2/2m^* + m^*\omega_c^2/f_{eff}^2 \qquad (33)$$

where l is the magnetic length. The last term in (33) can be expanded in power of F_{Hall}/F since this parameter is assumed to be small. Dropping again terms which are both v and k_y independent η_v can be re-arranged in the form:

$$\eta_v = eFvd + \hbar^2/2m^*\{k_x^2 + \{k_y - vd/l^2 + 2m^{*2}\lambda^2/\hbar (\omega_c/eF)^3\}^2\} + $$
$$v(\hbar\omega_c/eFd)^3 (d/l)^2 2m^* d^2\lambda^2/\hbar^2 \qquad (34)$$

The shift in the k_y origin is immaterial for the density of states which, in our perturbative approach will retain its regular staircase shape. The in-plane magnetic field will however modify the spacing of the Wannier Stark ladder by a term proportionnal to B^4. In interband optical absorption, it should be best measured between the ± 1 and ± 2 "oblique" transitions (to avoid excitonic effects as much as possible). For B = 10T, F = 5×10^4 V/cm, d = 6nm, λ = 17 meV, m^* = 0.07m_0 (parameters adapted to GaAs-Ga$_{0.7}$Al$_{0.3}$As superlattices) the shift is only 1.8 meV, a very small effect. A more easily visible effect may be associated with the shifts of the electron and hole subband extrema in opposite directions. It may significantly blur the onset of the subband- to subband transitions. Detailed calculations are however needed to fully analyze this complex behaviour.

ACKNOWLEDGEMENTS

One of us (R. F.) wishes to thank the CAPES (Brazil) for financial support. The Groupe de Physique des Solides de l'ENS is UA (17) at the CNRS. The work has been partly supported by CNET contract (number 888B041).

REFERENCES

1) Wannier, G. H., Rev. Mod. Phys. 34, 645 (1962).

2) Zak, J., Solid State Physics vol. 27 (New-York : Academic Press).

3) Saitoh, M., J. Phys. C 5, 914 (1972).

4) Luban, M. and Luscombe, J. H., Phys. Rev. B 34, 3674 (1986).

5) Emin, D. and Hart, C. F., Phys. Rev. B 36, 7353 (1987).

6) Bentolosa, F., Grecchi, V. and Zironi, F., J. Phys. C 15, 7119 (1982).

7) Bleuse, J., Bastard, G. and Voisin, P., Phys. Rev. Lett. 60, 220 (1988).

8) Mendez, E. E., Argullo-Rueda, F. and Hong, J. M., Phys. Rev. Lett. 60, 2426 (1988).

9) Voisin, P., Bleuse, J., Bouche, C., Gaillard, S., Alibert, C. and Regreny, A., Phys. Rev. Lett. 61, 1639 (1988).

10) Bleuse, J., Voisin, P., Allovon, M. and Quillec, M., Appl. Phys. Lett. 53, 2632 (1988)

11) Bastard, G., Bleuse, J., Ferreira, R. and Voisin, P., (1988) to appear in Superlattices and Microstructures.

12) See e.g. Ashcroft, N. W. and Mermin, N. D., "Solid State Physics", Holt, Rinehart and Winston, New-York (1976).

ELECTRONIC RAMAN SCATTERING IN PHOTOEXCITED QUANTUM WELLS: FIELD EFFECTS AND CHARGE-DENSITY DOMAINS

R. Merlin

The Harrison M. Randall Laboratory of Physics
The University of Michigan
Ann Arbor, MI 48109-1120, U. S. A.

INTRODUCTION

Raman scattering has been shown to be a powerful tool for the study of the electronic properties of space-charge layers and quantum-well structures (QWS), particularly those based on the GaAs-Al$_x$Ga$_{1-x}$As system.[1,2] In this work, the focus is on *photoexcited* QWS.[3-7] Compared with modulation doping, photogenerated electron-hole systems present the disadvantages of the non-uniformity of the plasma and the non-equilibrium nature of the photoexcitation process. Advantages include the ease of the tuning of the carrier density and plasma neutrality. In particular, band discontinuities can be easily inferred from Raman measurements of intersubband separations in photoexcited QWS.[4,5] This is unlike modulation-doped samples where the intersubband energies strongly depend on the (poorly-known) electrostatic potential due to the charge transfer.

The outline of this article is as follows. After a brief discussion of (electronic) scattering mechanisms and selection rules, a review of early results on the (intersubband) Stark effect[6] and on coupled cyclotron-intersubband excitations[7] is presented. The final section considers a preliminary report on *charge-density* domains, a phenomenon resulting from nonlinearities associated with the photoexcitation process and the response of a two-component QWS plasma.

First-order Raman processes deal with the creation (or annihilation) of a single elementary excitation such that its energy and momentum balances the difference between the energies and momenta of the incident and scattered photons. Typically, spectra in the energy range 1–100 meV and momentum range 10^4–10^6 cm^{-1} can be probed using the Raman technique. In the quasi-two-dimensional systems of interest, these ranges cover those of intersubband and intrasubband excitations, transitions involving Landau and impurity levels, plasmons and other collective modes.[1,2] Here, one should note that resonant tuning of the photon energy is required in most cases for the observation of electronic scattering.[1,2] Reflecting the relatively low carrier densities accessible to doping or photoexcitation, the corresponding *non-resonant* scattering cross sections are vanishingly small.[1,2] For GaAs–Al$_x$Ga$_{1-x}$As heterostructures, experiments are usually performed at resonances associated with the E_0-gap and $E_0+\Delta_0$-gap of GaAs.[1,2] Under favorable conditions, signals from systems with densities as low as $\sigma \approx 5\times10^9$cm^{-2} can be observed.[6]

Figure 1 shows examples of diagrams contributing to intersubband scattering (there is a total of two and six diagrams for two- and three-step processes, respectively; those shown give the strongest resonant enhancement). In the two-step process,[8] the incident photon creates an electron-hole pair in the c_n, h_m subbands. This is followed by the recombination of the electron in the c_0 subband with the h_m-hole to give

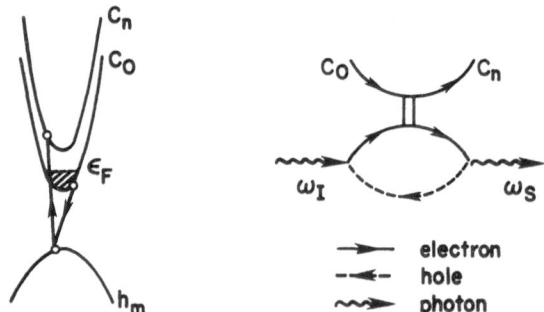

FIG. 1   Diagrams showing two (*left*) and three (*right*) step processes leading to electronic scattering. Electron and hole subbands are labelled c_i and h_i, respectively; ε_F is the Fermi energy. Coulomb interaction is represented by the double bars.

the scattered photon. The net result is a $c_o \to c_n$ excitation. The two-step resonance involves only the outgoing channel and, unlike electron scattering, it does not allow hole scattering at the $E_0 + \Delta_0$-gap. The three-step mechanism[9] does not discriminate between electrons and holes. Because of its higher order, it predicts stronger (and narrower) resonances. In the diagram, the electron created by the incident photon of frequency ω_I interacts with a second electron which undergoes a $c_o \to c_n$ transition. It then recombines with the hole producing the scattered photon of frequency ω_s. Early results in the literature support the two-step interpretation of intersubband scattering.[1,2] However, more recent work[9] and some of the data discussed here favor the three-step mechanism. On general grounds, it is expected that both processes should contribute to the scattering. Unfortunately, there have not been as yet any reports on hole-scattering in the vicinity of the $E_0 + \Delta_0$-gap so as to determine the relative importance of the two contributions.

States at the bottom of the conduction band of GaAs show mainly s-character leading to uncoupled spin and orbital degrees of freedom. In GaAs-Al$_x$Ga$_{1-x}$As structures, Raman scattering provides information on the spectra of spin- and charge-density fluctuations of electrons.[1,2] The Raman tensors associated with spin and charge fluctuations are, respectively, antisymmetric and symmetric.[1,2] Therefore, the two contributions can be separated by the appropriate choice of scattering configurations. Spin excitations are commonly probed in the $z(x',y')\bar{z}$ geometry, where z is normal to the layers and x',y' are along the [110] and [1$\bar{1}$0] directions. The configuration $z(x',x')\bar{z}$ allows charge, but not spin excitations. In the range of carrier densities found in experiments, final state (many-body) effects are relatively small and the spectrum of spin-density fluctuations approximate that of a single electron.[10] Accordingly, the positions of intersubband spin-density peaks give, to a very good approximation, the *bare* intersubband energies.[1,2] Charge-density excitations are collective modes exhibiting positive depolarization shifts with respect to the bare excitations, due to their associated (macroscopic) electric field. In addition, they mix with longitudinal-optical (LO) phonons giving rise to hybrid modes. These effects have been extensively considered in the literature, particularly for electrons in modulation-doped structures.[1,2] For *holes*, the experiments[11,12] have so far failed to reveal evidence of depolarization shifts. Although not pursued here, an example of hole scattering from a photoexcited system is shown in Fig. 2. The selection

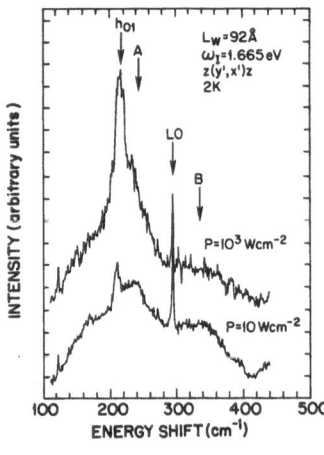

FIG. 2 Raman spectra of a quantum-well of width L_W=92Å showing scattering by intersubband hole excitations (h_{01}). A and B denote acceptor transitions and LO is a confined longitudinal optical mode of the GaAs slab. P is the power density. The photon energy is resonant with the HH2 exciton.

rules are consistent with the fact that hole states transform like Γ_6 giving $\Gamma_6 \otimes \Gamma_6 \equiv \Gamma_1 + \Gamma_2$ for the symmetries of the Raman tensor.[12]

MAGNETIC FIELDS: COUPLED CYCLOTRON-INTERSUBBAND EXCITATIONS

In quasi-two-dimensional electron systems, the motions in the confinement plane (x,y) and perpendicular to it are coupled for magnetic fields $\vec{B}$ at angles $\theta \neq 0$ with respect to z.[13-15] This leads to excitations of mixed character such as combined intersubband-cyclotron resonances which exhibit avoiding-crossing behavior near degeneracies.[13-15] Subband-Landau level coupling has been extensively studied using infrared[16-20] and, to a lesser extent, Raman scattering[7] techniques. Figure 3 shows Raman spectra of a QWS with well-width L_W=700Å at various tilt angles and fields. The

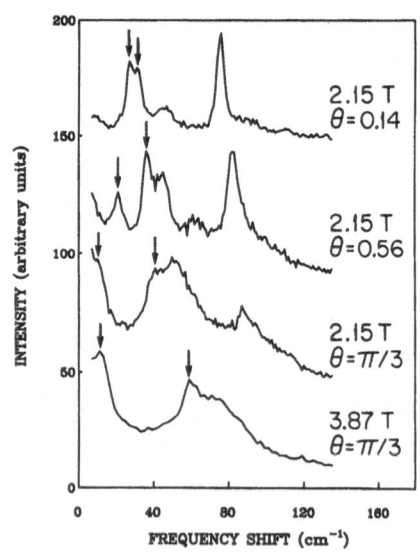

FIG. 3 Raman spectra of a 700-Å-thick quantum well for different magnetic fields and tilt angles (θ). Arrows indicate coupled cyclotron-intersubband excitations. Additional features are higher-lying electronic transitions. The data were obtained at the $E_0 + \Delta_0$ resonance.

350

arrows label the hybrid excitations derived mainly from the cyclotron mode and the $c_0 \to c_1$ intersubband transition; their splitting increases with tilt angle and field. Raman data on coupled modes for a sample with $L_w=460\text{Å}$ are plotted in Figs. 4 and 5. The experiments are in very good agreement with

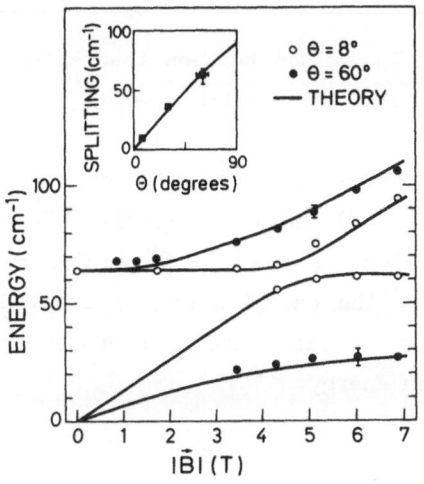

FIG. 4   Energy of the coupled intersubband-Landau level excitations as a function of magnetic field. θ is the tilt angle. The quantum well width is $L_w=460\text{Å}$. Theoretical curves were obtained from the expressions for parabolic wells. The inset shows coupled-mode splitting versus θ at a field of $B=4.8$ T, which corresponds to $\hbar\omega_c=E_{01}$.

the predictions of the parabolic-well model,[15] which is exactly solvable. In particular, we note that the lower branch in Fig. 4 approaches $E_{01}\cos\theta$ at large fields, as predicted by the model. The reason why the lowest-lying eigenstates of a square well in a magnetic field can be described by

FIG. 5   Energy of the coupled modes as a function of $\cos\theta$ (constant field). The theoretical results correspond to parabolic wells. $L_w=460\text{Å}$.

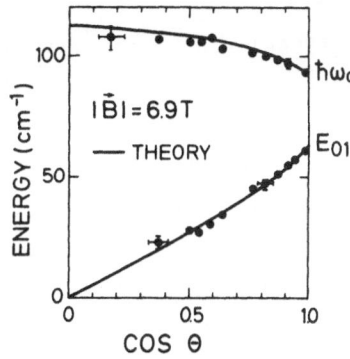

the parabolic model is simply the fact that the corresponding ground and first excited subband eigenfunctions are very similar (the comparison is valid for infinite wells). The matrix elements involved in the coupled-mode problem for the two cases differ by less than 2%.[7] The eigenenergies of the parabolic model can be combined and scaled in such a way to give expressions which are independent of the well-width and θ. This is shown in Fig. 6. The excellent agreement with Raman data on samples of various widths and at different tilt angles supports the validity of the parabolic approach.

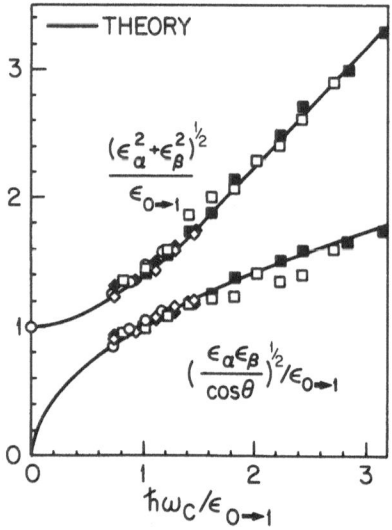

FIG. 6   Comparison between theoretical results for parabolic wells and experimental data on samples with L_w=700Å (*squares*), 460Å (*diamonds*) and 380Å (*circles*) showing scaling. Open (closed) symbols denote θ=8° (60°). ε_α and ε_β are the energies of the coupled modes. $\varepsilon_{0\to1}$ is the intersubband transition energy.

INTERSUBBAND STARK EFFECT

The effect of electric fields on intersubband excitations has been considered in many publications.[6,21,22] Such studies bear on possible device applications including high speed infrared modulators and photodetectors. Results of Raman experiments on a photoexcited QWS (L_w=264Å) are reproduced in Figs. 7 and 8. As expected,[21] the intersubband

FIG. 7   Raman spectra of a 264-Å-thick QWS showing the $c_0 \to c_1$ intersubband transition at different external voltages. The inset shows a schematic energy diagram (not to scale). The laser energy is in the vicinity of the HH5 resonance (three step scattering process).

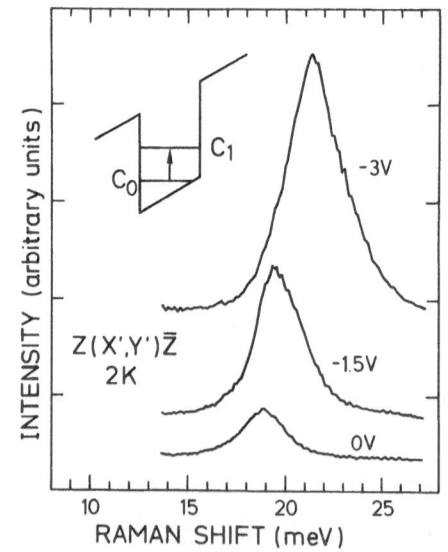

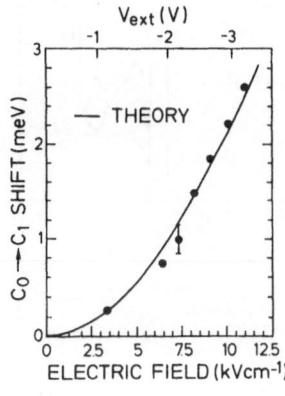

FIG. 8 Comparison between measured (solid circles) and calculated $c_o \rightarrow c_1$ Stark shifts.

peak shifts to higher energies with increasing field. An interesting feature of the spectra is the field-induced increase in the scattering intensity due to parity mixing.[6] In addition, the peak width *decreases* with field.[6] The latter effect likely reflects contributions due to surface roughness which become less important at higher fields as the intersubband separation depends more on the field and less on the width of the well.[6]

The experiments discussed in this (and the previous) paragraph involve photogenerated densities that are sufficiently low so that the electrostatic (Hartree) contribution of the plasma to the total potential can be ignored. Electric field effects at large concentration of electrons and holes are considered below.

CHARGE-DENSITY DOMAINS

For infinitely deep wells and in the absence of fields, the charges associated with the photoexcited electrons and holes completely cancel. This section deals with charge imbalance due to the presence of electric fields. Recent work[23] has shown that the consideration of the plasma potential leads to many interesting phenomena such as bistability, negative differential capacitance and nonlinear screening. Theoretical results for the electrostatic potential as a function of carrier density (σ) are shown in Fig. 9.[23] The constant external field is due to uniformly distributed charges $\pm Q$ of a parallel-plate capacitor. The data correspond to Hartree calculations for a neutral plasma in a well with infinitely large barriers (the mass of the electron, m, and the hole-mass are assumed to be equal). V is *half* the potential drop across the well in units of λ/e, where $\lambda = \hbar^2/(2mL_w^2)$. The carrier density, σ, and the external charge, Q, are given in units of $\lambda/(2\pi e^2 L_w)$ and $\lambda/(4\pi e L_w)$, respectively. $U(z)$ is the total potential and $\Psi_o(z)$ is the self-consistent electron or hole eigenfunction.

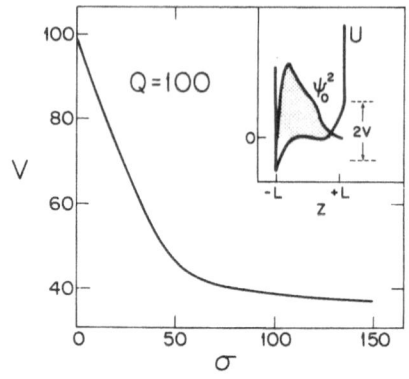

FIG. 9 V as a function of σ (constant Q). Inset: diagram of the system showing ψ_o^2 (z) and $U(z)$ for $Q=\sigma=100$.

At low carrier concentrations, V varies linearly with σ. With increasing σ, V crosses over to a regime showing a much weaker dependence on the density. Using WKB methods, it can be proved that $V \propto \sigma^{-1/4}$ for $\sigma \to \infty$.[23] The combination of screening nonlinearities and those of the equations determining the photoexcited plasma density can lead to bistable behavior, i.e., to the existence of two stable σ-solutions (*domains*) for a given set of parameters.[23] The results in Fig. 10 reveal evidence of domain formation in the spectra at -5.2 V and -6.0 V, showing two sets of coupled intersubband-LO modes.[24] The data were obtained on a sample consisting of 40 periods 212 Å GaAs/160 Å $Al_{0.7}Ga_{0.3}As$ at the $E_0+\Delta_0$-resonance. The densities (gained from the positions of the coupled modes) are in the range 8.0×10^{10}-1.5×10^{11} cm^{-2}. Experiments at slightly lower power densities,

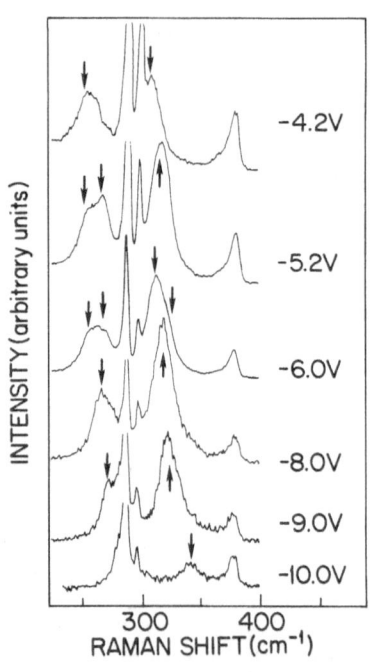

FIG. 10 Raman spectra of a 40-period multiple QWS (L_w=212Å; barriers are 160-Å-thick $Al_{0.7}Ga_{0.3}As$). Arrows denote coupled intersubband-LO modes. The scattering geometry is $z(x',x')\bar{z}$ and $\hbar\omega_I$= 1.8899 eV.

i.e., lower carrier concentrations, and at the E_0-resonance show a single domain in the whole voltage-range. This indicates that the domains relate to the photoexcitation process and, furthermore, that the instability is not due to *sequential resonant tunneling*[25] (the measured positions of the intersubband peaks and the photoinduced *I-V* response do not support the latter possibility either).[24] However, a clear picture of the origin of the bistability is yet to be uncovered. Resonant photoexcitation is known to lead to bistable behavior,[23] but the observed dependence of the concentration on the voltage does not completely agree with this interpretation.

Acknowledgments - The author has benefitted from collaborations with K. Bajema, R. Borroff, D. Gammon, D. A. Kessler, N. Mestres and C. Tejedor. This work was supported by the U. S. Army Research Office under Contract No. DAAL-03-89-K-0047 and the URI program Contract No. DAAL-03-87-K-0007.

REFERENCES

1. G. Abstreiter, R. Merlin and A. Pinczuk, Inelastic Light Scattering by Electronic Excitations in Semiconductor Heterostructures, IEEE J. Quantum Electron. **QE-22**: 1771 (1986).

2. A. Pinczuk and G. Abstreiter, Spectroscopy of Free Carrier Excitations in Semiconductor Quantum Wells, in *Light Scattering in Solids V*, M. Cardona and G. Güntherodt, eds., Topics in Applied Physics **66**, Springer, Berlin (1989), Ch. 4.

3. A. Pinczuk, J. Shah, A. C. Gossard and W. Wiegmann, Light Scattering by Photoexcited Two-Dimensional Electron Plasma in GaAs-(AlGa)As Heterostructures, Phys. Rev. Lett. **46**: 1341 (1981).

4. J. Menéndez, A. Pinczuk, D. J. Werder, A. C. Gossard and J. H. English, Light Scattering· Determination of Band Offsets in GaAs-AlGaAs Quantum Wells, Phys. Rev. B **33**: 8863 (1986).

5. J. Menéndez, A. Pinczuk, D. J. Werder, S. K. Sputz, R. C. Miller, D. L. Sivco and A. Y. Cho, Large Valence-Band Offset in Strained-Layer InGaAs-GaAs Quantum Wells, Phys. Rev. B **36**: 8165 (1987).

6. K. Bajema, R. Merlin, F. -Y. Juang, S. -C. Hong, J. Singh and P. K. Bhattacharya, Stark Effect in GaAs-AlGaAs Quantum Wells: Light Scattering by Intersubband Excitations, Phys. Rev. B **36**: 1300 (1987).

7. R. Borroff, R. Merlin, R. L. Greene and J. Comas, Observation of Coupled Quasi-Two-Dimensional Electronic Excitations in Tilted Magnetic Fields, Surface Sci. **196**: 626 (1988).

8. E. Burstein, A. Pinczuk and D. L. Mills, Inelastic Light Scattering by Charge Carrier Excitations in Two-Dimensional Plasmas: Theoretical Considerations, Surface Sci. **98**: 451 (1980).

9. G. Danan, A. Pinczuk, J. P. Valladares, L. N. Pfeiffer, K. W. West and C. W. Tu, Coupling of Excitons with Free Electrons in Light Scattering from GaAs Quantum Wells, Phys. Rev. B **39**: 5512 (1989).

10. S. Katayama and T. Ando, Light Scattering by Electronic Excitations in n-Type GaAs-AlGaAs Superlattices, J. Phys. Soc. Jpn. **54**:1615 (1985).

11. A. Pinczuk, H. L. Störmer, A. C. Gossard and W. Wiegmann, Energy Levels of Quasi two Dimensional Holes in GaAs-(AlGa)AS Quantum Well Heterostructures, in *Proceedings of the 17th International Conference on the Physics of Semiconductors*, J. D. Chadi and W. A. Harrison, eds.,Springer, Berlin (1984), p. 329.

12. D. Gammon, R. Merlin, W. T. Masselink and H. Morkoç, Raman Spectra of Shallow Acceptors in Quantum-Well Structures, Phys. Rev. B **33**: 2919 (1986).

13. T. Ando, Theory of Intersubband-Cyclotron Combined Resonances in the Silicon Space-Charge Layer, Phys. Rev. B **19**: 2106 (1979)

14. See, e.g., T. Ando, A. B. Fowler and F. Stern, Electronic Properties of Two-Dimensional Systems, Rev. Mod. Phys. **54**: 437 (1982), and references therein.

15. J. C. Maan, Combined Electric and Magnetic Field Effects in Semiconductor Heterostructures, in *Two-Dimensional Systems, Heterostructures, and Superlattices,* G. Bauer, F. Kuchar and H. Heinrich, eds., Springer Series in Solid State Science 53, Springer, Berlin (1984), p. 183; R. Merlin, Subband-Landau-Level Coupling in Tilted Magnetic Fields: Exact Results for Parabolic Wells, Solid State Commun. **64**: 99 (1987).

16. W. Beinvogl and J. F. Koch, Intersubband-Cyclotron Combined Resonance in a Surface-Charge Layer, Phys. Rev. Lett. **40**: 1736 (1978).

17. Z. Schlesinger, J. C. M. Hwang and S. J. Allen Jr., Subband-Landau-Level Coupling in a Two-Dimensional Electron Gas, Phys. Rev. Lett. **50**: 2098 (1983).

18. G. L. J. A. Rikken, H. Sigg, G. J. G. M. Langerak, H. W. Myron and J. A. A. J. Perenboom, Subband-Landau-Level Spectroscopy in GaAs-AlGaAs Heterojunctions, Phys. Rev. B **34**: 5590 (1986).

19. A. D. Wieck, J. C. Maan, U. Merkt, J. P. Kotthaus, K. Ploog and G. Weimann, Intersubband Energies in GaAs-GaAlAs Heterojunctions, Phys. Rev. B **35**: 4145 (1987).

20. S. Huant, M. Grynberg, G. Martinez and B. Etienne, Observation of a Gap in the Coupled Intersubband Cyclotron Resonance Excitations in a Quasi Two-Dimensional Electron Gas, Solid State Commun. **65**: 457 (1988).

21. D. Ahn and S. L. Chuang, Intersubband Optical Absorption in a Quantum Well with an Applied Electric Field, Phys. Rev. B **35**: 4149 (1987).

22. A. Harwit and J. S. Harris Jr., Observation of Stark Shifts in Quantum Well Intersubband Transitions, Appl. Phys. Lett. **50**: 685 (1987).

23. R. Merlin and D. A. Kessler, Photoexcited Quantum Wells: Nonlinear Screening, Bistability and Negative Differential Capacitance, unpublished.

24. N. Mestres, A. McKiernan, R. Merlin, J. Oh and P. K. Bhattacharya, Observation of Charge-Density Domains in Photoexcited Quantum-Well Structures, to be published in Surface Sci.: *Proceedings of the 4th International Conference on Modulated Semiconductor Structures*, L. L. Chang, R. Merlin and D. C. Tsui, eds.

25. L. Esaki and L. L. Chang, New Transport Phenomenon in a Semiconductor Superlattice, Phys. Rev. Lett. **33**: 495 (1974).

LINEAR AND NONLINEAR OPTICS OF CONFINED EXCITONS

I. Balslev

Fysisk Institut, Odense Universitet
DK 5230 Odense M, Denmark

1. INTRODUCTION

The optical properties of direct gap excitons in semiconductors are understood in great detail in the regime of linear response and little influence of surfaces and interfaces. In this case the quantum mechanics of the electron-hole motion and the propagation of electromagnetic waves represent phenomena which conceptually and computationally are well separated: stationary states of the quantum mechanics are mapped into poles in the complex linear susceptibility which is to be inserted into Maxwell's equations.

Deviations from an extended homogeneous system occur when surfaces become important or in case of micro-structures such as quantum wells. In both cases the translational symmetry is lost, and the relative and the center-of-mass motion of the electron-hole pair can not be separated. This situation calls for a theoretical basis without the usual truncation of the configuration space when going from the electron-hole dynamics to the electrodynamics.

Beyond the regime of linear response there is often a pronounced need for a truely dynamic approach, i. e. a theory without expansions in terms of steady-state solutions. This is the case because there is often no direct connection between the response to strong short external light pulses and the steady state in the presence of spectrally narrow sources.

In the present paper we shall discuss the linear and nonlinear optics of confined excitons in the frame work of a two-band density matrix theory. Following the ideas of Stahl[1,2] we employ a real-space, real-time representation of this approach. This is convenient for describing the highly nontrivial surface effects of the linear optics of excitons[2,3] as well as the short pulse dynamics in the nonlinear optical regime[4].

We shall here primarily be concerned with the nonlinear absorptive response near an excitonic resonance, i. e. the dynamical Stark effect, particularly in case of near-resonance pumping explored by Myzyrowicz et al.[5] and Von Lehmen et. al.[6]. We present results from real-time numerical integration relevant for pump-and-probe experiments. The calculational procedure allows us to study the influence of pump pulses with a duration comparable to the dephasing times of the system. We study the magnitude of the dynamical Stark effect as function of the well thickness and the difference between spectra with spectral filtering of the probe before and after the sample.

357

In Sec. 2 we present the basic equations of motion of the relevant density matrices and relate the interband density with the macroscopically observable excitonic polarization. Section 3 brings a short review of the relevance of the approach to the linear exciton optics of the half space and the quantum well geometry. In Sec. 4 we go to the nonlinear optical regime and discuss the approximation and simplifications necessary for a reasonably simple analytical or numerical treatment. Sec. 5 contains a presentation and a discussion of the calculated results while Sec. 6 brings a conclusion and an outlook.

2. THE TWO-BAND DENSITY MATRIX APPROACH

We assume here a direct gap semiconductor idealized to have two parabolic, isotropic bands. The density matrix in a real-space representation is written as

$$\rho(\mathbf{r}_1,\mathbf{r}_2,t) = \left\{ \begin{array}{ll} C(\mathbf{r}_1,\mathbf{r}_2,t) & Y^*(\mathbf{r}_1,\mathbf{r}_2,t) \\ Y(\mathbf{r}_1,\mathbf{r}_2,t) & \delta_B(\mathbf{r}_1-\mathbf{r}_2)-D(\mathbf{r}_1,\mathbf{r}_2,t) \end{array} \right\} \tag{1}$$

C is the conduction band submatrix, D is the valence band submatrix in the hole representation, Y is the interband density matrix, and δ_B is a suitably broadened delta-like function. In relation to the optical properties near the band gap the relevant electromagnetic quantity derived from ρ is the interband polarization

$$P(\mathbf{r}) = M_o (Y(\mathbf{r},\mathbf{r})-Y^*(\mathbf{r},\mathbf{r})), \tag{2}$$

M_o is the dipole moment of the allowed transition. The diagonal components $C(\mathbf{r},\mathbf{r})$ and $D(\mathbf{r},\mathbf{r})$ are equal to the density at $\mathbf{r}$ of electrons and holes, respectively.

Applying the Hartree-Fock decoupling of the equations of motion one finds the following closed set of equations[7] for the submatrices Y, C, and D:

$$\dot{Y}_{12} + i\Omega_{eh}Y_{12} + X^Y_{12} = \frac{iM_o}{\hbar}(E_1\delta_{12} - E_1C_{12} - E_2D_{21}) \tag{3}$$

$$\dot{C}_{12} + i\Omega_{ee}C_{12} + X^C_{12} = -\frac{iM_o}{\hbar}(E_1Y_{12} - E_2Y^*_{21}) \tag{4}$$

$$\dot{D}_{12} + i\Omega_{hh}D_{12} + X^D_{12} = -\frac{iM_o}{\hbar}(Y_{12}E_{21} - Y^*_{21}E_1) \tag{5}$$

The numerical subscripts are abbreviations for the coordinates in the sense $Y_{12} \equiv Y(\mathbf{r}_1,\mathbf{r}_2)$ etc. The Ω operators describing the propagation in the $\mathbf{r}_1,\mathbf{r}_2$ configuration space are given by

$$\Omega_{eh} = \omega_g - \frac{V_{12}}{\hbar} - \frac{\hbar}{2m_e}\nabla^2_1 - \frac{\hbar}{2m_h}\nabla^2_2 + \frac{e}{\hbar}(\phi^h_1-\phi^e_2) \tag{6}$$

$$\Omega_{ee} = -\frac{\hbar}{2m_e}(\nabla^2_2-\nabla^2_1) + \frac{e}{\hbar}(\phi^e_1-\phi^e_2) \tag{7}$$

$$\Omega_{hh} = -\frac{\hbar}{2m_h}(\nabla^2_2-\nabla^2_1) - \frac{e}{\hbar}(\phi^h_1-\phi^h_2) \tag{8}$$

where

$$V_{12} = \frac{e^2}{4\pi\varepsilon |r_1 - r_2|} \tag{9}$$

is the direct Coulomb interaction dielectrically screened by the valence electrons (ε), m_e, m_h are effective masses and ω_g is the gap frequency. The potentials ϕ^e, ϕ^h are composed of two contributions: the selfconsistent electromagnetic potential and the pseudopotentials from a surface or from isoelectronic spatial structures. We neglect here the influence on the propagation of the transverse part of the electric field, usually included via the vector potential A[7]. Therefore the propagators (6-8) are unable to describe the Stark effect in case of pumping far below the band gap as done by Fröhlich et al.[8-9]. Besides the term V_{12} in (6) the Coulomb interaction also gives rise to Fock-type exchange terms X^Y, X^C, X^D given by[7]

$$X_{12}^Y = \frac{i}{\hbar}\int (V_{13}-V_{23})(Y_{11}C_{32} - D_{13}Y_{32})d^3r_3 \tag{10}$$

$$X_{12}^C = \frac{i}{\hbar}\int (V_{13}-V_{23})(Y_{31}^* Y_{32} + C_{13}C_{32})d^3r_3 \tag{11}$$

$$X_{12}^D = \frac{i}{\hbar}\int (V_{13}-V_{23})(Y_{31}^* Y_{32} + D_{13}D_{32})d^3r_3 \tag{12}$$

3. LINEAR RESPONSE

Let us briefly consider the case of linear response. Then, in the equation of motion for Y we neglect X^Y and terms containing C and D. Y is driven by the electric field from a source at the origin of relative space, i.e. at regions in space where $r_1 \approx r_2$:

$$-i\hbar\dot{Y}_{12} + \left[\hbar\omega_g - V_{12} - \frac{\hbar^2}{2m_e}\nabla_1^2 - \frac{\hbar^2}{2m_h}\nabla_2^2 + e(\phi_1^h - \phi_2^e)\right]Y_{12} = M_0 E_1 \delta_{12} \tag{13}$$

A very appealing rigorous theory for the famous halfspace problem of exciton polaritons[2,10,11] is based on (13) with $e\phi^e$ and $e\phi^h$ containing infinite repulsive barriers at the crystal surface[1,3]. The corresponding boundary conditions for Y then provide the 'additional boundary condition' (ABC), searched for since the mid 1960's. Furthermore, solving the coupled set of equations involving (13) and Maxwell's equation:

$$c^2\nabla^2 E(r) = \frac{\partial^2}{\partial t^2}\left[E(r) + \frac{M_0}{\varepsilon_0}(Y(r,r) + Y^*(r,r))\right] \tag{14}$$

($\tilde{\varepsilon}$ is the residual dielectric constant[3]) one can account for the well known dead-layer problem[10-12,13] of the coupling between external photons and bulk polaritons. The computational problems of treating (13) with an arbitrary mass ratio are not yet fully overcome.

Of similar rigorousity is the approach of D'Andrea and Del Sole[14]. Their theory is based on an expansion in terms of evanescent waves which are solutions to (13) with $\dot{Y}=-i\omega Y$ and $E=0$ (ω is the optical frequency). The expansion coefficients are to be chosen to minimize the amplitude of the total wave function at the surface. So far the large orbit excitons and continuum states has been neglected in this expansion. Little is known about the consequences of such an approximation.

We next consider a quantum well with the width W and oriented so that the z axis is the growth direction. Here the potentials ϕ^e, ϕ^h contain the band offsets forming confining barriers for electrons and holes. Computational problems call for simplifications and approximations:

a) If the light propagation direction is parallel to the z-axis then there is full translational symmetry in the plane of the quantum well, and so Y then depends only on z_1, z_2, the projection $\bar{\rho}$ of r_1-r_2 onto the xy-plane and the time t.

b) An expansion of Y in terms of products containing standing waves in the z direction can often be truncated to include only one pair of sublevels. The condition is that the exciton binding energy is small compared to the spectral distance to the sublevel pairs, and that the wave length is long compared to the well width W.

In this restricted case we may approximate Y_{12} as follows:

$$Y_{12} = u_h(z_1)\ u_e(z_2)\ Y(\bar{\rho}) \tag{15}$$

where u_h (u_e) is a confinement wave function of the hole (electron). The equation of motion for $Y(\bar{\rho})$ can be obtained by averaging over the confinement wave functions. The result is:

$$(-i\hbar\frac{\partial}{\partial t} + \hbar\omega_g' - \frac{\hbar^2}{2\mu}\nabla^2_\rho - v(\bar{\rho}))\ Y(\bar{\rho}) = m_0 E\ \delta(\bar{\rho}) \tag{16}$$

where

$$v(\rho) = \frac{e^2}{4\pi\varepsilon}\int dz \int dz' \frac{1}{\sqrt{\rho^2+(z-z')^2}} u_e^2(z)\ u_h^2(z'), \tag{17}$$

$$m_0 = M_0 \int dz\ u_e(z)\ u_h(z), \tag{18}$$

and ω_g' is the gap frequency associated with the sublevel in question. $v(\rho)$ is Coulomb-like ($v(\rho) \propto 1/\rho$) for $\rho \gg W$ and logarithmic for $\rho \ll W$. The resonances of quantum well excitons are found by solving the homogeneous version of (16). As shown in numerous papers the ground state binding energy is 4 times the bulk exciton Rydberg in the two dimensional limit (W = 0) and drops monotonously towards the bulk value as W increases. If $W = a_B$ ($\equiv$ the bulk exciton Bohr radius) and the confining barriers are assumed infinitely high then the binding energy associated with the lowest sublevel pair is 2.3 times the bulk Rydberg.

4. NONLINEAR OPTICS OF EXCITONS: THE DYNAMICAL STARK EFFECT

Let us still consider a quantum well. When going beyond the linear regime we must also expand C and D in terms of confinement wave function. Retaining only one term as in (15) we get:

$$C_{12} = u_e(z_1)\ u_e(z_2)\ C(\bar{\rho}) \tag{18}$$

$$D_{12} = u_h(z_1)\ u_h(z_2)\ D(\bar{\rho}) \tag{19}$$

Inserting (14, 18-19) into (3-5) and averaging over the confinement wave functions we get the following coupled equations:

$$(-i\hbar\frac{\partial}{\partial t} + \hbar\omega_g' - \frac{\hbar^2}{2\mu}\nabla^2 - v(\bar{\rho}))\ Y(\bar{\rho}) +$$

$$+ 2\int d^2\bar{\rho}'\left[v(\bar{\rho}')Y(\bar{\rho}')C(\bar{\rho}'-\bar{\rho})-\tilde{v}(\bar{\rho}')C(\bar{\rho}')Y(\bar{\rho}'-\bar{\rho})\right] = m_0 E\ (\delta(\bar{\rho})-2C(\bar{\rho})) \tag{20}$$

$$-i\hbar\frac{\partial}{\partial t}C(\bar{\rho}) +$$

$$+ \int d^2\bar{\rho}'\ v(\bar{\rho}')\left[Y^*(\bar{\rho}')Y(\bar{\rho}'-\bar{\rho})-Y(\bar{\rho}')Y^*(\bar{\rho}'-\bar{\rho})\right] = -m_0 E\ (Y(\bar{\rho})-Y^*(\bar{\rho})) \tag{21}$$

where $\tilde{v}(\rho)$ is an interaction potential which includes a Hartree type contribution[4]. We have used that $C(\bar{\rho}) = D(-\bar{\rho})$, and so the D equation is omitted. A further simplification is to put $\tilde{v}(\rho) = v(\rho)$ (corresponding to the absence of induced charges as obtained for infinite barriers and equal sublevel indices of the two bands[4]). Eq. (21) implies that $C(\bar{\rho})$ is real.

Eq. (20-21) can be brought into agreement with the theory developed by Schmitt-Rink, Haug and others[15-20] by means of appropriate Fourier transformations and special analysis of the Hartree-terms[4,16]. Unlike the works in Refs. 15-20 we emphasize here the real time representation because we want to study the short pulse behaviour.

A common application of (20-21) is based on two superimposed electric fields: E_{pm} (pump) and E_{pr} (probe) $\ll E_{pm}$. We first solve (20-21) in the presence of E_{pm} alone thereby obtaining Y_{pm} and C_{pm}. Then we insert for E the sum $E_{pr} + E_{pm}$ driving the density matrices $Y_{pm} + \delta Y$ and $C_{pm} + \delta C$. The second set of equations can be linearized so that it becomes linear in E_{pr}, δY and δC.

In the simulated experiment we shall assume that suitable spatial filtering allows the detection of $\delta P \propto \delta Y(0)$ without admixture of $P_{pm} \propto Y_{pm}(0)$ from the pump. We shall concentrate on the relative absorption $-\Delta I/I$ of the probe light. If the transmitted light is filtered through a monochromator the detected quantity is

$$- \left.\frac{\Delta I(\omega)}{I(\omega)}\right|_{filtered} \propto \frac{Im(E_{pr}^{*\omega} \delta Y^{\omega}(0))}{|E_{pr}^{\omega}|^2} \qquad (22)$$

where superscript ω refers to a Fourier transformation. Without a monochromator after the sample, the relevant quantity is

$$- \left.\frac{\Delta I}{I}\right|_{unfiltered} \propto \frac{\int dt \; Im(E_{pr}^{*} \delta Y(0))}{\int dt |E_{pr}|^2} \qquad (23)$$

In realistic applications of (20-21) one must include irreversible dephasing processes. Simplest is a T_2-like relaxation obtained by replacing $\partial Y/\partial t$ by $\partial Y/\partial t + \Gamma Y$. As the actual absorption near exciton lines is not well described by Lorentzian line shapes a realistic modelling of a pump and probe experiment is to use two different dephasing rates, one for the dephasing of the pump excitation Γ_{pm}, one for the dephasing of the probe excitation Γ_{pr} to be inserted into the above mentioned second (linearized) set of equations.

5. Calculated results

We have calculated the dynamical Stark effects assuming both pump and probe pulses to be Gaussian having peak fields E_{pm}^o, E_{pr}^o, center frequencies ω_{pm}, ω_{pr}, and full-width-half-maximum durations (for the intensity) t_{pm}, t_{pr} of pump and probe, respectively. The detuning, i. e. the difference between the pump frequency and the resonance is denoted δ. We apply the rotating wave approximation and the above fields are meant as the rotating part only. We study 1s-excitons and use a value for the bulk exciton Rydberg equal to 4 meV relevant for GaAs. We use a dephasing rate for the probe excitation of $\Gamma_{pr} = 1.5$ meV.

We perform numerical real-time integrations using 4 equations of motion, namely (20-21) for C_{pm} and Y_{pm} (complex), and the corresponding linearized equations for δC and δY (complex). We calculate two types of spectra: one based on filtering the response in case of a very short pulse: t_{pr} = 20 fs, and one based on a long (spectrally narrow) probe pulse: t_{pr} = 0.9 ps without filtering after the sample. Details of the numerical integration is given in Ref. 4.

It turns out that the two situations give similar spectra in the case that the pump pulse is long (spectrally narrow) enough. In this limit (t_{pm} ≥ 6 ps) we use Γ_{pm} = 0 and obtain spectra shown in Fig. 1. For Stark shifts less that 6 meV we find satisfactory agreement between the numerical integration and a perturbative treatment[4]. (As the latter is valid only to second order in E_{pm}^o there is a poor agreement for case c in

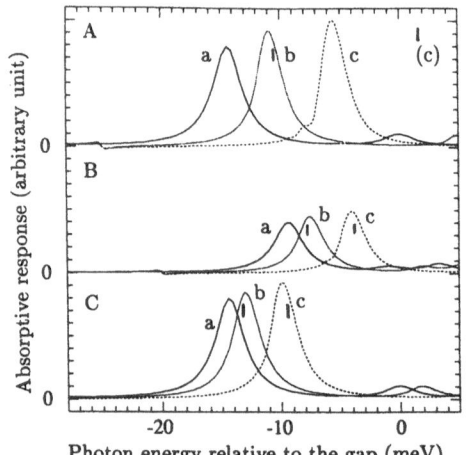

Fig. 1   Absorptive response for various values of detuning δ, peak pump fields E_{pm}^o and width W of an infinite barrier quantum well. The curves marked (a) are the unperturbed spectra while curves (b) and (c) have $M_o E_{pm}$ = 1.4 and 3.2 meV, respectively. Fig. 1A and 1C shows the spectra for the limit W = 0 with δ = -10 meV and 20 meV, respectively. In Fig. 1B, W = a_B and δ = -10 meV. Other important parameters are Γ_{pm} = 0 and Γ_{pr}= 1.5 meV. The vertical bars shows the shifts derived from a perturbative treatment[4].

Fig. 1A.). The results in Fig. 1 deviate considerably from those of a two-level system[21] obtained by neglecting the exchange terms (10-12). Thus the Author must correct his previous position concerning reliability of the approximation in Ref. 21.

Note that the Stark shift is much weaker for the quantum well with W = a_B (Fig. 1B) than for the (hypothetical) two-dimensional limit (W = 0, Fig. 1A). Surprisingly, the Stark shift for W = a_B is also weaker than the pure three-dimensional limit[4]. We have no simple explanation for this anomalous lack of monotonous behaviour between the limits W = 0 and W = ∞. As a consequence any interpolation scheme between these limits must be abandonned.

Next we consider a shorter duration of the pump pulse. Fig. 2 shows the spectra for t_{pm} of the order of a dephasing time T_2 (=Γ_{pr}^{-1}). We still neglect the dephasing of the pump and consider zero delay between the pulses. With spectral filtering after the sample (Fig. 2A) the absorption peak becomes distorted and an oscillatory behaviour developes on the red side of the absorption line. This is the remainder of the coherent oscillations at negative delay discussed by Koch et al.[18]. The effect is

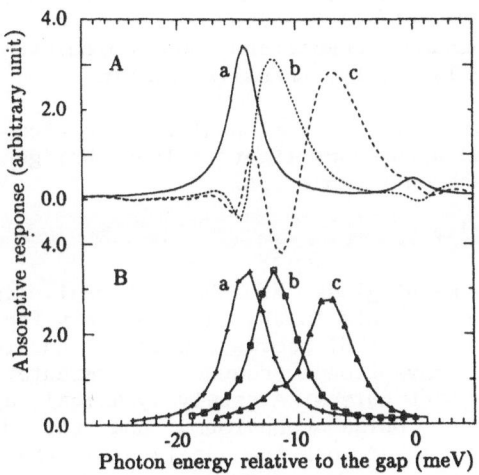

Fig. 2   Absorptive response for a finite duration (t_{pm} = 0.9 ps) of the pump pulse. Fig. 2A shows are filtered response to a short probe pulse (t_{pr} = 20 fs), δ = -10 meV, W = 0 and pump fields given by $M_0 E_{pr}^0$ = 0 (a), 1.6 meV (b), and 3.2 meV (c). For comparison we show in Fig. 2B the corresponding unfiltered response to a spectrally narrow probe pulse (t_{pr} = 0.9 ps) centered at plotted frequency. The parameters for the filtered and unfiltered spectra are otherwise identical. Other important parameters are Γ_{pm} = 0, Γ_{pr} = 1.5 meV, and zero delay between the pulses.

absent if the filtering of the probe pulse is before the sample (see Fig. 2B). In any case the absorption line is less shifted and significantly broadened when compared to the steady-state result (Fig. 1). This difference is due to the fact that within a probe dephasing time T_2 the pump pulse does not have a well defined amplitude, and the average of this amplitude is less than the peak value. It is important to note that the experiments show a pronounced broadening at high pump fields[5-6]. This is probably due to the finite duration of the pump pulse used.

6. Conclusion and outlook

We have applied the two-band density matrix theory to two types of complications in exciton optics: 1) the break down of translational symmetry near surfaces and interfaces, and 2) the optical nonlinearities associated with very short light pulses.

A conventional theory of bulk linear response leads to a susceptibility $\chi(k,\omega)$ where the k-dependence accounts for the nonlocal response. Near surfaces and interfaces one often generalizes the susceptibility and introduces the real-space response function $\chi(r_1, r_2, \omega)$:

$$P^\omega(r_2) = \int \chi(r_1, r_2, \omega) \, E^\omega(r_1) dr_1^3 \qquad (24)$$

However, with such a complicated structure of χ it is easier to abandon the susceptibility concept all together. Instead one can solve (13-14) with appropriate boundary conditions and calculate directly the observables such as reflectivity, transmission etc.

Turning to the main topic, the nonlinear optics of excitons, the properties associated with short light pulses brings us in a similar situation. The conventional expansion:

$$P = \chi^{(1)} E + \chi^{(3)} E^3 + \ldots \qquad (25)$$

becomes increasingly meaningless as the external light pulses become shorter and more intense. This is the case not only for so strong fields that higher order terms in (25) become comparable to lower order terms, but also because the conventional frequency representation of $\chi^{(3)}, \chi^{(5)} \ldots$ is not relevant for short pulses. A truely dynamical approach is needed for exploring the short pulse behaviour, and the directly observable quantities must be extracted form the model rather than susceptibilities.

The gross features of the dynamical Stark effect of excitons are well understood on basis of previous steady-state theories[15-20] and the present results for short external pulses. However, a very important aspect of the dynamical Stark effect is essentially unexplored in experiments and in theory, namely the polarization properties[22-23]. For example, how does the Stark shift depend on the mutual polarization directions of pump and probe. Furthermore, how strong is the excitonic Kerr effect, i. e. the transmission for crossed polarizers in the probe beam (before and after the sample), and pump polarization at 45° to the probe polarizers.

Acknowledgements

The author is have benefited greatly from cooperation and discussions with A. Stahl, R. Zimmermann and J. Hvam.

References

1. A. Stahl, phys. stat. sol. (b) **94**, 211 (1979), **106**, 575 (1981)
2. A. Stahl, I. Balslev, *'Electrodynamics of the Semiconductor Band Edge'*, Springer Tracts in Modern Physics Vol. **110** (Springer Verlag, Berlin, 1987)
3. I. Balslev, A. Stahl, phys. stat. sol. (b) **111**, 531 (1982), I. Balslev, Comm. Mod. Phys. B (Cond. Mat. Phys) **13**, 21 (1987)
4. I. Balslev, R. Zimmermann, A. Stahl, submitted to Phys. Rev. B (1989)
5. A. Mysyrowicz, D. Hulin, A. Antonetti, A. Migus, W.T. Maaslink, H. Morkoc, Phys. Rev. Lett. **56**, 2748 (1986)

6.   A. Von Lehmen, D.S. Chemla, J.E. Zucker, J.P.Heritage, Optics Lett. **11**, 609 (1986)

7.   A. Stahl, Z. Phys. **B 72**, 3271 (1988)

8.   D. Fröhlich, A. Nothe, K Reimann, Phys. Rev. Lett. **55**, 1335 (1985)

9.   D. Fröhlich, R. Wille, W. Schlapp, G. Weimann, Phys. Rev. Lett. **59**, 1748 (1987)

10. J.J. Hopfield, D.G. Thomas, Phys. Rev. **132**, 563 (1963)

11. V.M. Agranowich, in *Surface Excitations*, ed. by V.M Agranowich and R. Loudon (Elsevier, Science Publishers B.V., Amsterdam, 1984) p. 513

12. F. Evangelisti, A. Frova, F. Patella, Phys. Rev. **B9**, 4253 (1974)

13. I. Balslev, phys. stat. sol. (b) **88**, 155 (1978).

14. A. D'Andrea, R. Del Sole, Phys. Rev. **B38** , 1197 (1988)

15. S. Schmitt-Rink, D.S Chemla, Phys. Rev. Lett. **57**, 2752 (1986)

16. S. Schmitt-Rink, D.S. Chemla, H. Haug, Phys. Rev. **B37**, 941 (1988)

17. R. Zimmermann, phys. stat. sol. (b) **146**, 545 (1988)

18. S.W. Koch, N. Peyghambarian, M. Lindberg, J. Phys. C: Solid State **21**, 5229 (1988)

19. R. Zimmermann, M. Hartmann, phys. stat. sol. (b) **150**, 379 (1988)

20. C. Ell, J.F. Müller, K. El Sayed, L. Banyai, H.Haug, phys. stat. sol.(b) **150**, 393 (1988)

21. I. Balslev, A. Stahl, Solid State Commun. **67**, 85 (1988), phys. stat. sol. (b) **150**, 412 (1989)

22. A. Stahl, private communication.

23. M. Joffre, D. Hulin, A. Migus, to be published.

MAGNETO-EXCITONS IN GAAS/GAALAS QUANTUM WELLS

L. Viña

Instituto de Ciencia de Materiales-C.S.I.C., and Departamento de Física Aplicada. Universidad Autónoma, E-28049 Madrid, Spain

Abstract.- We present photoluminescence excitation spectra of a p^+- i- n^+ GaAs/ $Ga_{0.65}Al_{0.35}As$ heterostructure, with the intrinsic region consisting of five isolated wells (160A). Two different sets of experiments will be presented. In the former experiment, a small magnetic field of 0.5T is applied perpendicularly to the wells in order to enhance the oscillator strength of the excited states of the heavy-hole excitons $[h_1(x)]$ and to remove the Kramers degeneracy. These states are tuned by means of an electric field, parallel to the magnetic field. In the latter experiment, the pseudo-absorption of the quantum well is studied as a function of magnetic field. Due to the high quality of the sample, the n–s series of the heavy-hole and light-hole excitons are resolved at fields as low as 1T. The features in the spectra are identified as excited states of excitons by comparison with calculations that take into account exciton mixing in the presence of electric and magnetic fields.

INTRODUCTION

Two-dimensional semiconductors are an ideal system to study the properties and energy spectrum of excitons. The absorption spectra of undoped quantum wells (QW's) are dominated by the coulomb interaction between electrons and holes. The spectra exhibit sharp peaks, corresponding to excitons associated with electrons and holes of the different confined subbands in the conduction and valence bands,[1] rising on a step-like background characteristic of two-dimensional systems. A conspicuous property of these excitons is their existence at conditions where they are absent in bulk materials: excitons in GaAs/$Ga_{1-x}Al_xAs$ QWs are observed, even at room temperature, in the presence of high electric fields,[2] and the presence of high density, light-induced, plasma seems not to destroy the excitonic character of interband transitions.[3] Their binding energy is enhanced with respect to bulk crystals, and the first excited state (2s), although its oscillator strength is ~25 times smaller than that of the 1s-state, may be observed in high quality samples.[4-6]

We have used photoluminescence excitation spectroscopy (PLE) to study the pseudo-absorption of a p^+- i- n^+ GaAs/ $Ga_{0.65}Al_{0.35}As$ heterostructure, with the intrinsic region comprising five isolated wells (160A) and barriers of 250A. PLE presents a number of advantages compared with other optical techniques. it yields information not only on ground states, as in the case of photoluminescence, but also on higher excited states of excitons. The experiments can be carried out in as-grown samples avoiding the thinning procedures necessary to perform direct absorption measurements, which add an unknown amount of strain, and, finally, it does not involve complicated lineshapes due to admixtures of ϵ_1 and ϵ_2, as it is the case in modulation techniques.[7] The main drawback of PLE measurements lies in the determination of oscillator strengths, since they can be strongly affected by non-radiative channels in the samples. However, a good estimation can be obtained in good quality samples.[8]

External electric fields, applied perpendicularly to the layers in semiconductor QW's, are a very useful tool to tune the energy of excitons. The Stark shifts of confined excitons depend strongly on their effective masses,[9-11] and therefore, with the choice of an appropriate well width, coupling between different excitons can be achieved. This tuning yields to level crossing, a method widely used in atomic spectroscopy to reveal fine and hyperfine structure in the spectra.[12] We will use this approach, together with a small magnetic field, as a perturbation to increase the oscillator strength of the excitons, in order to resolve high angular momentum states by coupling with optically active s-state excitons.

The magnetoabsorption of semiconductors has received great attention in the past because it provides very useful information about band structure parameters.[13-20] In the case of undoped QW's, the experimental studies have been used to determine the exciton effective Rydberg by extrapolation of high- and intermediate-field data to zero magnetic field.[15-18] In our studies we have used a very high quality sample, which allows us the direct observation of intense, resolved, sharp-line structure at fields as low as 1T. We identify the lines in the spectra as ground and excited states of confined excitons by comparison with a hydrogenic-like theory which takes into account the complexity of the valence band structure and the effects of exciton mixing in the presence of electric and magnetic fields.[20]

OBSERVATION OF HIGH-ANGULAR MOMENTUM EXCITONS BY ELECTRIC FIELD INDUCED COUPLING

The low-temperature absorption of bulk GaAs near the direct band edge is also dominated by the coulombic interaction between electrons and holes. In spite of the numerous studies dedicated to excitons in semiconductors, fine structure in bulk materials has been resolved and identified only up to the second excited state (3s).[21,22] Figure 1 shows the best up-to-date absorption spectrum of an ultra-pure GaAs sample.[21] The ground state and the first two excited states of the free exciton are clearly resolved, together with other sharp peaks corresponding to donor-bound excitons. The application of an external magnetic field has led to the observation of fine structure in the excited states, but the different lines have not been successfully identified.[23]

With the improvement in sample quality, fine structure in the excitonic spectrum of GaAs/Ga$_{1-x}$Al$_x$As QW's has become resolvable.[4,5,11,24] The quasi-two-dimensional character of these systems plays an important role in the richness of their optical spectra: for strictly two- or three-dimensional semiconductors, hydrogenic levels with the same principal quantum number (n) are degenerate with respect to the orbital angular momentum quantum number (l). However, the reduction of symmetry from T_d (in bulk) to D_{2d} (in QW's) lifts this degeneracy. For a 160A QW, calculations have shown that the binding energy of $h_1(2p)$ is 0.4meV larger than that of $h_1(2s)$ (Ref. 25). Although $h_1(2p)$ is not optically active, it can be observed if a small electric field is applied to the QW parallel to the growth direction, and the measured $h_1(2p)$ – $h_1(2s)$ splitting amounts to 0.45meV,[11] in very good agreement with the theoretical predictions. This observation has led us to study in more detail the excitonic energy spectrum of QW's.

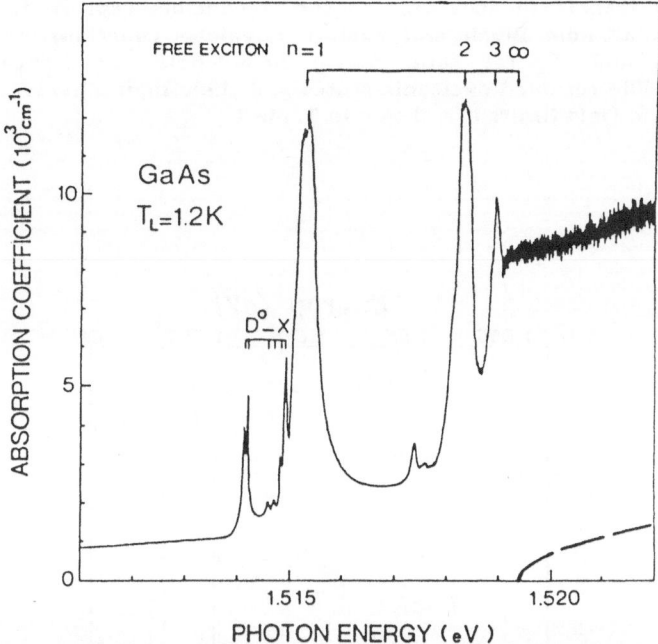

FREE EXCITON n=1   2 3∞

GaAs
T_L=1.2K

D^0-X

Fig. 1. Absorption spectrum at 1.2K of ultra-pure GaAs near the band gap edge. Donor-bound excitons (D^0-X) and the n=1,2,3 free-exciton peaks are shown, together with the band gap determined by extrapolation to n=∞. The dashed line depicts the expected square-root dependence in the absence of electron-hole interaction. (After R.G. Ulbrich and C. Weisbuch, unpublished.)

Figure 2 shows a PLE spectrum of our sample at 5K in the presence of an electric field of ~2kV/cm. Besides the ground state of the heavy-hole and light-hole excitons, $h_1(1s)$ and $l_1(1s)$, respectively, their first excited states are also clearly seen as sharp peaks, $h_1(2s)$ and $l_1(2s)$. The rest of the structures correspond to a weak-forbidden transition between the first electron subband and the third heavy-hole subband, $h_{13}(1s)$; a 2p excited state of h_{12}, and a peak labeled h_{12a}, which appears as a consequence of electric-field induced mixing between excited states of l_1 and the h_{12} ground state.[26-28] The two-dimensional character of the QW is seen in the step-like shape of the density of states. A small shoulder on the low-energy side of $h_1(2s)$ has been identified as the 2p excited state of the heavy-hole exciton.[11,29] Since the heavy states $h_1(x)$, where x stands for excited, are at slightly higher energies than the light exciton $l_1(1s)$, their energy separation will decrease with increasing electric field, and an interaction between excitonic states belonging to the same symmetry will occur.

Excitation spectra in the region of $l_1(1s)$ and $h_1(x)$ are shown in Fig.3 for several electric fields and σ^+ and σ^- polarizations. A small magnetic field of ~0.5T, applied perpendicularly to the layers, breaks the Kramers degeneracy and enhances the oscillator strength of weakly bound states. The former effect enables the resolution of very close lines by using circularly polarized light, which excites states associated to holes with spin-up and spin-down components selectively. In the presence of an electric field the different excitons interact and they cannot unambiguously be classified as heavy or light anymore.

369

However, they transform according to the irreducible representations of the direct product of hole Bloch and exciton envelope functions,[20] and can be labeled as Γ_7 and Γ_6 (at zone center heavy-hole and light-hole states, respectively). The resolved excitonic states and their main character in the low and high electric field limits are shown in Table I.

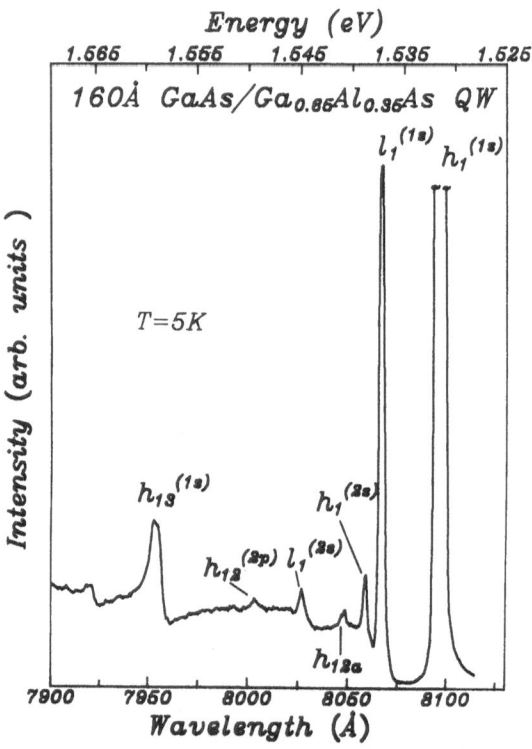

Fig. 2. PLE spectrum of a 160A GaAs/Ga$_{1-x}$Al$_x$As quantum well in the presence of an electric field of ~2kV/cm. See text for the labeling of the states. Note the high quality of the sample in the sharpness of the lines.

With increasing electric field the energy separation between $l_1(1s)$ and $h_1(x)$ decreases, therefore increasing the interaction between them. $l_1(1s)$ shares its oscillator strength with the excited states of the heavy-hole exciton and they become observable. $h_1(2p-)$, the state with the largest binding energy of $h_1(x)$, is the first to suffer the coupling and is clearly resolved as a peak at a field of 10.4kV/cm in σ^+ configuration, whereas it can be seen only as a shoulder, $h_1(2p+)$ in σ^-, due to its closeness to $l_1(1s)$. At this point let us recall that in a two-dimensional system the third component of the angular momentum can take only two values $\pm m$. At 19.8kV/cm a new line (x) in σ^+ appears in the spectral range. This peak cannot unambiguously be labeled as $h_1(3p-)$ or $h_1(3d+)$ because of the difficulty to resolve these states, which from the calculations should be only ~0.4meV apart. As we shall see below, the theory seems to favor the d character, but this result is very sensitive to the choice of band structure parameters. In σ^- two new structures are resolved at fields of

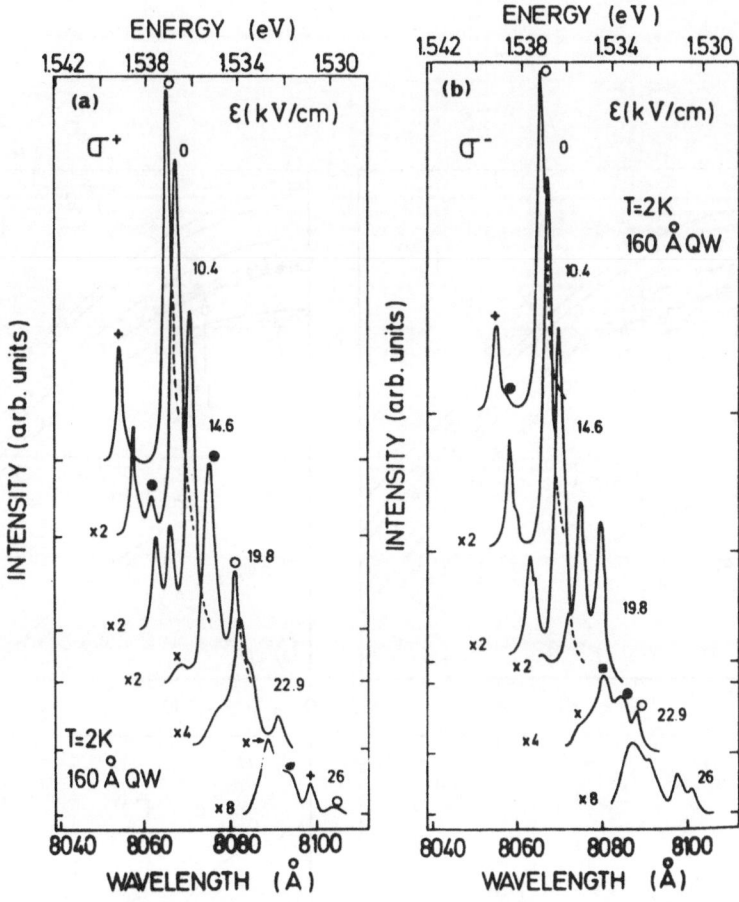

Fig. 3. PLE spectrum of the 160A QW at several electric fields normal to the well and a magnetic field of 0.5T parallel to the electric field. a) and b) are excited with σ^+ and σ^- polarized light, respectively.

Table I. Labeling of excitonic states in the low and high electric field limits indicating their main character. The symbols correspond to those used in Fig. 3.

	σ^+			σ^-		
	low field	high field		low field	high field	
$\Gamma_7{}^{(1)}$	$h_1{}^{(1s)}$ (Δ)	$h_1{}^{(1s)}$ (Δ)	$\Gamma_7{}^{(1)}$	$h_1{}^{(1s)}$ (Δ)	$h_1{}^{(1s)}$ (Δ)	
$\Gamma_6{}^{(1)}$	$l_1{}^{(1s)}$ (o)	$h_1{}^{(2p-)}$ (o)	$\Gamma_6{}^{(1)}$	$l_1{}^{(1s)}$ (o)	$h_1{}^{(2p+)}$ (o)	
$\Gamma_7{}^{(2)}$	$h_1{}^{(2s)}$ $(+)$	$h_1{}^{(2s)}$ $(+)$	$\Gamma_7{}^{(2)}$	$h_1{}^{(2s)}$ $(+)$	$h_1{}^{(2s)}$	
$\Gamma_6{}^{(2)}$	$h_1{}^{(2p-)}$ $(\bullet)$	$h_1{}^{(3p-)}$	$\Gamma_6{}^{(2)}$	$h_1{}^{(2p+)}$ $(\bullet)$	$h_1{}^{(3d-)}$	$(\bullet)$
$\Gamma_6{}^{(3)}$	$h_1{}^{(3p-)}$	$h_1{}^{(3d+)}$ $\}(\bullet)$	$\Gamma_6{}^{(3)}$	$h_1{}^{(3d-)}$ $(\blacksquare)$	$h_1{}^{(3p+)}$	
$\Gamma_6{}^{(4)}$	$h_1{}^{(3d+)}$ $\}(x)$	$l_1{}^{(1s)}$ (x)	$\Gamma_6{}^{(4)}$	$h_1{}^{(3p+)}$	$h_1{}^{(4d-)}$ $\}(\blacksquare)$	
			$\Gamma_6{}^{(5)}$	$h_1{}^{(4d-)}$ $\}(x)$	$l_1{}^{(1s)}$ (x)	

371

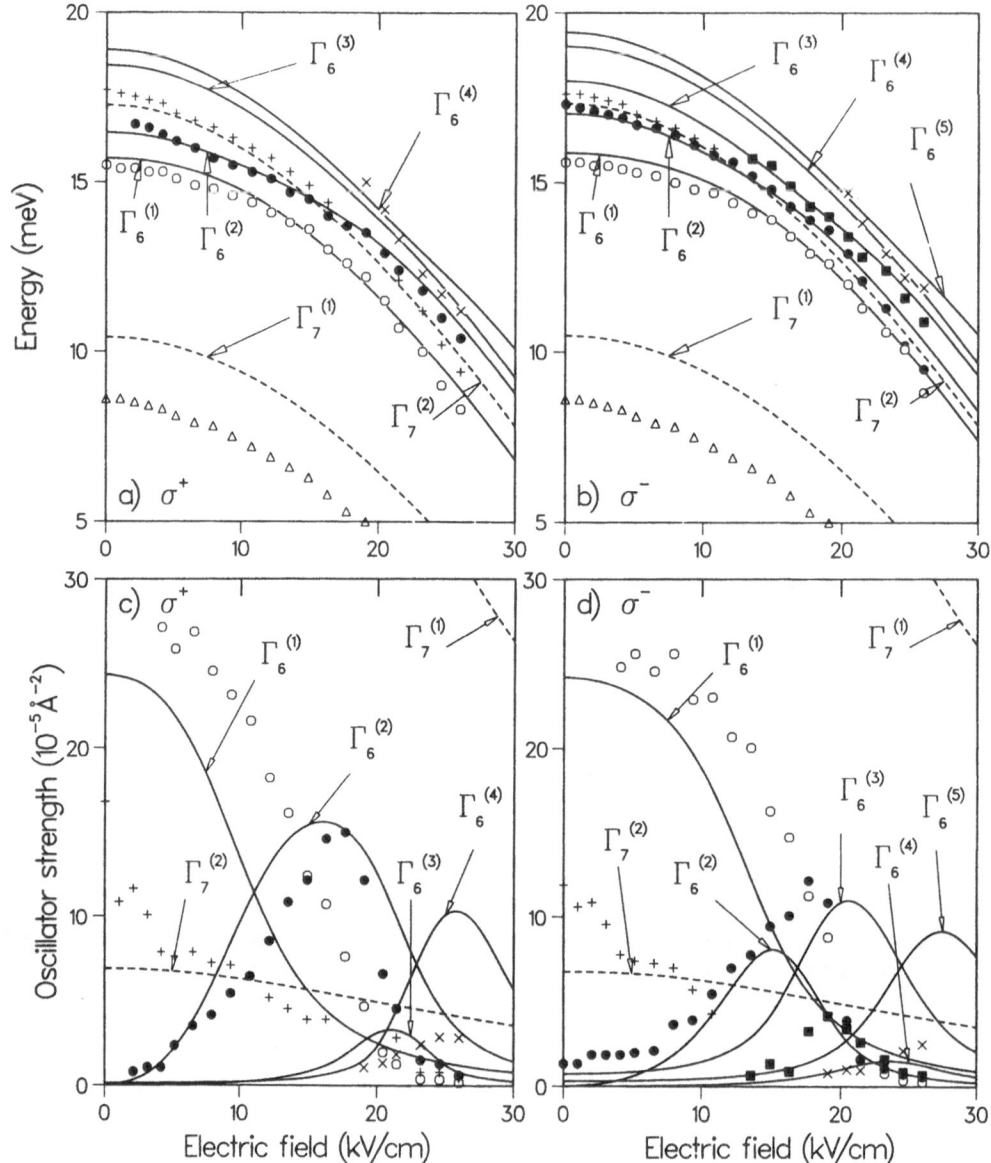

Fig. 4. Stark shifts, a) and b), of the ground states of the light–hole (l_1) and heavy hole (h_1) excitons and the excited states of h_1 for σ^+ and σ^- polarizations, respectively. The electric field and a magnetic field of 0.5 T are applied normal to the layers. c) and d) are the oscillator strengths of the levels shown in a) and b), respectively. The lines correspond to the results of the calculations. The states are labeled according to Table I.

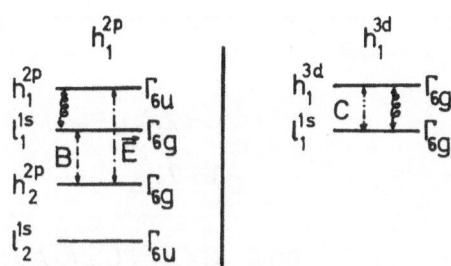

Fig. 5. Scheme of one of the possibles channels for the coupling of the excited
states of the heavy-hole exciton, $h_1(2p)$ and $h_1(3d)$, with $l_1(1s)$, the
ground state of the light-hole exciton.

~20kV/cm. One of them, $\Gamma_6{}^{(3)}$ (■), is assigned to $h_1(3d-)$, while the second may
have a *3p+* or *4d-* character; a definitive classification is not possible for the
same reasons mentioned above. Note that the spectra are fairly different for
the two polarizations, and that the resolution of all these states would have
been very toilsome without the use of the magnetic field.

The oscillator strengths and energy shifts of the excitons are plotted in Fig.
4 for both light polarizations. The solid lines correspond to the results of the
calculations for Γ_6 states, while Γ_7 states are depicted by dashed lines. The
points show the experimental results obtained from the PLE spectra, except
those for $h_1(1s)$ ($\Gamma_7{}^{(1)}$) taken from photoluminescence spectra. The *2s* heavy
state belongs to a different symmetry (Γ_7) than the rest of the high-lying
states, and therefore is not expected to interact with them. This is borne out
by the experiments, particularly in the σ^+ configuration, where $h_1(2s)$ crosses
the $\Gamma_6{}^{(2)}$ level without interaction. There is, however, a disagreement between
the calculated and the measured dependence of its oscillator strength with
electric field. The larger observed quenching of $h_1(2s)$ originates probably on
the presence of non-radiative channels in the sample and in the fact that PLE
is measuring the emission of $h_1(1s)$, which is also decreasing with field.

The agreement between theory and experiment is remarkably good in spite
of the uncertainties in the Luttinger parameters used in the calculations. The
oscillator strengths are very sensitive to the choice of these parameters,
especially to the value of K, which determines the effective g factor of the
holes.[24] We have chosen the most commonly used band structure parameters
(see Table III in Ref.20), without trying to adjust them to the experimental
findings. However, the experimental Zeeman splittings of $l_1(1s)\downarrow$ – $l_1(1s)\uparrow$ and
$h_1(2p+)\downarrow$ – $h_1(2p-)\uparrow$ of 0.15meV and 0.5meV, respectively, are very well
reproduced by the theory which obtains 0.18meV and 0.6meV, respectively. Also
the splittings of the *s* states of the heavy-hole excitons are negligible compared
to those of light-hole excitons, a fact that can be only explained if
non-diagonal terms in the Hamiltonian are taken into account.

The theory predicts that with increasing electric field the oscillator
strength of $l_1(1s)$ [$\Gamma_6{}^{(1)}$] is subsequently shared with $\Gamma_6{}^{(k>1)}$ states. For the same
principal quantum number, *n*, the coupling is expected to be stronger for *d*
than for *p* states. The coupling between excitons takes place through
non-diagonal terms in the exciton Hamiltonian (see Eq. 5 in Ref.20). One of the
mixing paths between the excited states of h_1 and l_1 is schematically
represented in Fig. 5, where, for the sake of clarity, we neglect the spin
directions, since only states observed in a given polarization may interact with
each other. The electric field couples $h_1(2p)$ with $h_2(2p)$, which on his hand
mixes with $l_1(1s)$ through the lowering operator *k-* contained in the

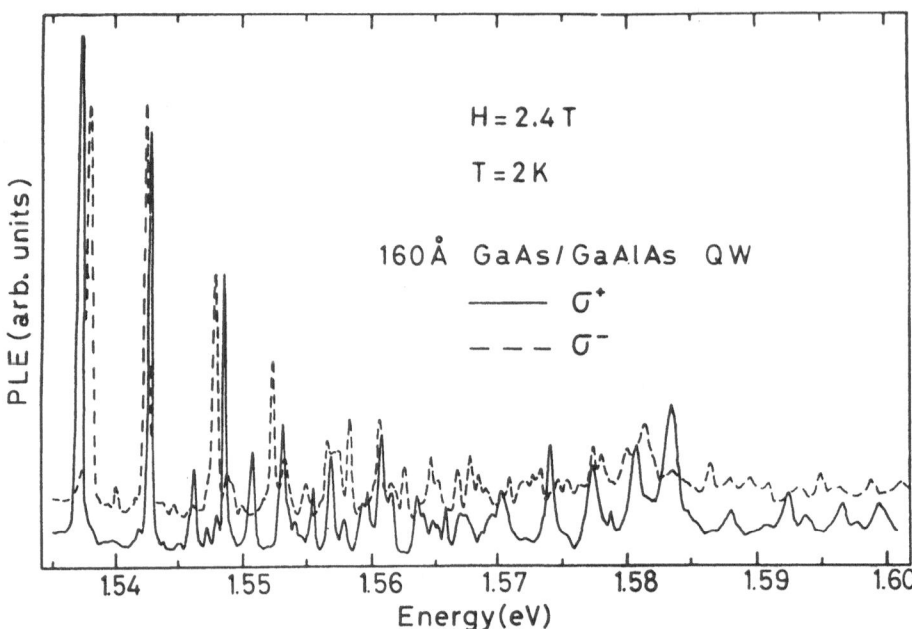

Fig. 6. Excitation spectra in the presence of a magnetic field of 2.4T normal to the layers of the quantum well for right- (solid line) and left-handed (dashed line) circularly polarized light. A small, residual, electric field of ~5kV/cm is present in the p-i-n structure.

non-diagonal B term.[20] An alternative path is the coupling of $l_1(1s)$ with $l_2(2s)$, which then mixes with $h_1(2p)$ by k^-. The electric field enhances the interaction by decreasing the energy splitting between heavy-hole and light-hole states and by increasing the coupling between adjacent, confined subbands. d states may already couple with $l_1(1s)$ in the absence of an electric field,[29] since they belong to the same irreducible representation, Γ_{6g}, of D_{4h}: the coupling arises from the non-diagonal C term (Eq. 5 in Ref. 20), which contains square powers of k^- and of the rising operator k^+. In this case the electric field mainly enhances the coupling through the tuning of the states and the mechanism is more efficient than in the case of p states since it is a one-step process.

HIGHLY RESOLVED MAGNETO-EXCITONS

It has been pointed out that the observable transitions in QW's in a finite magnetic field are all excitons and that none can be due to unbound electron-hole pairs.[19] Lines with a weak magnetic field dependence have been attributed to ground states of excitons, however, most of the lines in the experiments have been interpreted in terms of Landau-level transitions which are a posteriori corrected for excitonic effects.[15-18,30] The high quality of our sample allows us to resolve very sharp peaks in the spectra already at low magnetic fields, when the magnetic length $(\hbar c/eH)^{1/2}$ is larger than the exciton radius, and the states have to be treated as truly bound electron-hole pairs. This field range correspond also to the regime of validity of the theory that we use to interpret our experiments.[20]

Figure 6 depicts PLE spectra of the 160A GaAs QW at 2.4T, for right (solid line) and left (dashed line) circularly polarized light. The spectrometer was set at the peak of the heavy-hole exciton ground state (1.532eV, not shown in the figure), which at this field did not shown any measurable Zeeman splitting. The sample was excited with the light from an LD700-dye laser pumped by a Kr⁺-ion laser. The spectra were recorded at 2K with magnetic fields, from a polyhelix resistive magnet, applied normal to the layers in the Faraday configuration. The heavy-hole exciton emission was normalized by measuring the laser intensity, through a beam splitter, with a silicon photodiode. It is worthwhile to emphasize the richness of the spectra in this energy range of only ~65meV. The σ- spectrum appears more complicated than the corresponding σ⁺ counterpart, which is a general fact for all fields in the range of magnetic field studied.[31] By comparison with Fig. 1, we observe significant changes in the density of states, which is modified from a step-like shape (0T) to an almost vanishing density in the regions between the excitonic transitions, and the spectra become fully discrete.

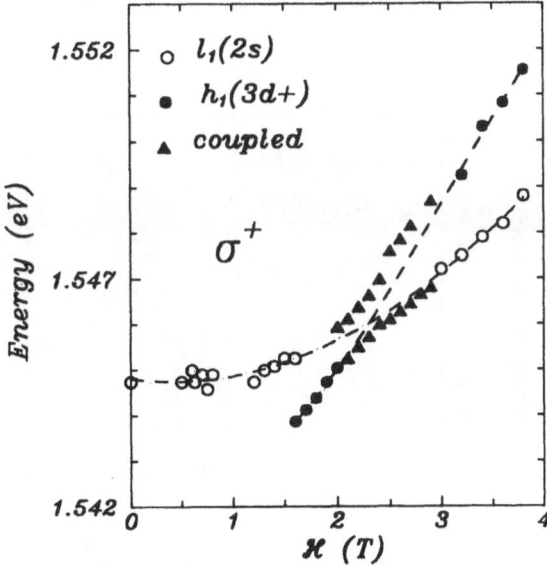

Fig. 7. Energy shifts versus magnetic field of $l_1(2s)$ and $h_1(3d+)$ measured with σ⁺ polarized light. This shows one of the many anticrossings observed in the fan-diagrams. The states are depicted by triangles in the region of strong interaction. The lines are a guide to the eye.

This figure shows many similarities with the absorption spectra of Rydberg atoms in the presence of a magnetic field, which display periodic structure extending above the ionization limit known as *quasi-Landau resonances*.[32] There is assumed, in the strong-field regime, that the states associated with the dominant lines are concentrated in the z=0 plane, and therefore the problem is reduced to a two-dimensional separable one, which can be treated within the WKB approach. This approximation, however, cannot predict oscillator strengths and prevents the explanation of many secondary lines present in the spectra.[33] From the analysis of our data, by comparison with the energies and oscillator strengths obtained from the calculations,[31] the main lines in these spectra correspond to s states of heavy-hole and light-hole excitons, while many other peaks are identified as excited states with higher angular quantum numbers. However, the theoretical and experimental fan diagrams show a large number of anticrossings between different excitons, which preclude the definitive identification of the states. For a given magnetic field and a photon-energy range many states are made conspicuous by a transfer of oscillator strength from neighbor levels, which suggests that the appearance of any particular resonance is governed by the optical excitation route employed. For these magnetic fields and energies, an analysis of the data taking into account stochastic processes, like in the case of atomic spectra,[34] could yield very valuable information.

Figure 7 depicts one of the many anticrossings observed in the experiments. We have selected from the fan diagram corresponding to right handed polarized light the energy shifts of $l_1(2s)$ [open circles] and $h_1(3d+)$ [full dots] for fields between 0 and 4T. Light-hole excitons show a smaller diamagnetic shift then heavy-hole ones, due to the larger in-plane effective mass of the light holes. Since these states belong to the same irreducible representation (Γ_6), they interact when they approach each other. The coupled states are shown as triangles in the region of strong interaction. The minimum separation between $l_1(2s)$ and $h_1(3d+)$ at 2.4T is 1±0.2meV, similar to that observed between $l_1(1s)$ and $h_1(2p)$ [Ref. 11]. The lines in the figure are only a visual aid and show the expected shifts in the absence of interaction.

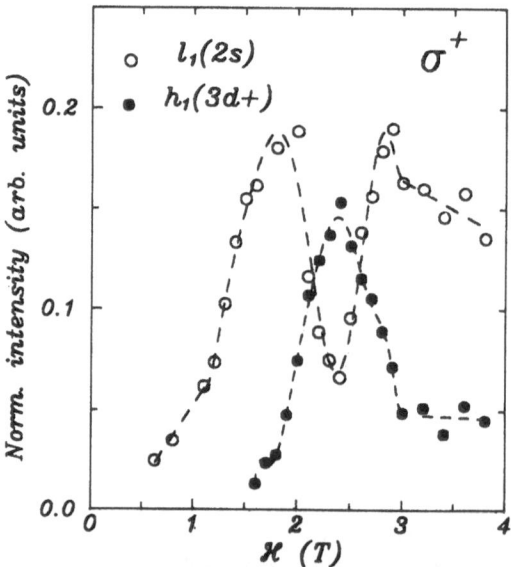

Fig. 8. Oscillator strengths of the exciton states shown in Fig.7 versus magnetic field. The intensities have been normalized to the corresponding intensity of $l_1(1s)$ at each field. The lines are only a guide to the eye.

The oscillator strengths of $l_1(2s)$ and $h_1(3d+)$, normalized to those of $l_1(1s)$, are shown in Fig.8 in the range from 0 to 4T. A large increase of the intensity of $l_1(2s)$ is observed between 0.6 and ~1.5T, as a consequence of the extra confinement by the magnetic field. At this field, $h_1(3d+)$ becomes close enough to $l_1(2s)$ so that the interaction becomes observable. They share their oscillator strengths and are completely mixed at 2.4T. As the field is increased, they separate again and $h_1(3d+)$ becomes progressively smaller; the asymmetry in the behavior of its oscillator strength is due to a new interaction with $h_{12}(2p-)$ at a field of ~4T. The $2s$ light-hole exciton increases again its strength at fields larger than 4T.

At larger fields the spectra become simpler in their low energy range, when the discrete excited states originating from different excitons are further apart. They resemble closer the spectra of Landau interband transitions, however many noticeable changes in the oscillator strength are still seen in the pseudo-absorption, which are characteristic of interacting excitons. We show in Fig.9 the excitation spectra at zero and 6.5T in a reduced energy range of ~45meV. We see again that the σ^- spectrum is more complicated than the σ^+ one. The peaks are identified by comparison of the energies and oscillator strengths with the results of the calculations.[31] Table II compiles the labeling

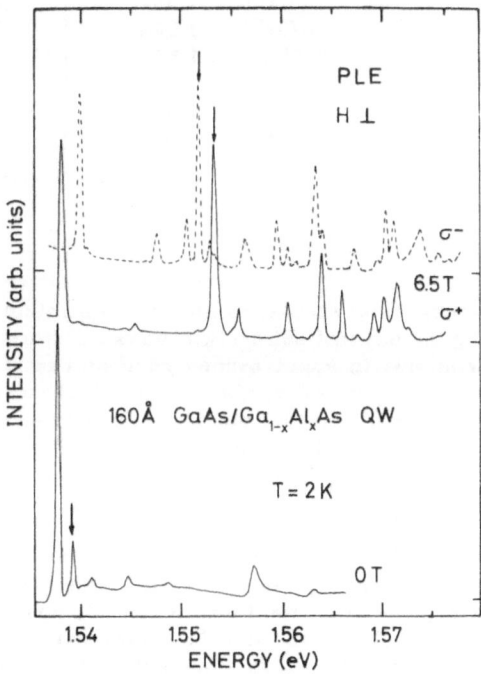

Fig. 9. Excitation spectra at zero field and 6.5T for σ^+ and σ^- polarized light over a reduced energy range. Note the considerable increase with field of the oscillator strength of $h_1(2s)$, marked by arrows. Table II compiles the labeling of the peaks.

and the energies of the lines. One should remember that in some cases this labeling represents only the main character of the transitions, since they are mixed with other excitons. At 6.5T we observe a strong enhancement of the lines in the spectra; the $h_1(2s)$ excited states (marked by arrows) are now comparable to the ground state of the light-hole exciton. Note that the spectra are completely polarized, except for some small structures like the small peak on the high energy side of $l_1(1s)$ in the σ^+ spectrum.

SUMMARY

We have studied the energy spectrum of excitons in GaAs/Ga$_{1-x}$Al$_x$As QW's in the presence of external electric and magnetic fields applied perpendicularly to the layers. The high quality of our sample has allowed us to resolve, for the first time, very sharp structure, which is assigned to ground and excited states of confined excitons. The labeling of the states is based on comparison with calculated oscillator strengths and energy shifts, taking into account the

Table II. Lines observed in the spectra of Fig. 9. The energies are in eV.

0T	E(eV)	6.5T(σ^+)	E(eV)	6.5T(σ^-)	E(eV)
l_1(1s)	1.538	l_1(1s)	1.538	l_1(1s)	1.540
h_1(2s)	1.539	h_1(2p-)	1.544	h_{12}(1s)	1.541
h_{12a}	1.541	h_{12}(1s)	1.545	l_1(3d-)	1.548
l_1(2s)	1.545	l_1(2p+)	1.552	h_1(3d-)	1.551
h_{12}(2p)	1.549	h_1(2s)	1.553	h_1(2s)	1.552
h_{13}(1s)	1.558	l_1(2s)	1.555	l_1(2s)	1.553
h_{13}(2s)	1.563	h_{13}(1s)	1.561	l_1(2p-)	1.553
		h_1(3s)	1.564	h_{13}(1s)	1.556
		l_1(3s)	1.566	h_{12}(2s)	1.560
		h_{12}(3p-)	1.569	l_1(5d-)	1.561
		h_1(4d+)	1.570	h_1(3s)	1.563
		h_1(4s)	1.572	l_1(3s)	1.564
				h_1(4s)	1.570
				l_1(4s)	1.571
				h_{13}(3d-)	1.574

complexity of the valence-band structure and exciton mixing in the presence of the fields. Our study goes beyond solid-state physics, since it is closely related to other important problems in spectroscopy such as the structure of atoms in very strong magnetic fields on the surface of neutron stars, the splitting and broadening of atomic spectral lines by electric and magnetic fields in a plasma, the Stark and Zeeman effects in the hydrogen atom, etc.

ACKNOWLEDGMENTS

The author wants to thank G.E.W. Bauer for the calculations, which were crucial for the identification of the lines in the spectra, W.I. Wang for the growing of the sample, J.C. Maan and M. Potemski for their help in the experiments performed in the Max-Planck-Institut, Hochfeld-Magnetlabor (Grenoble) and E.E. Mendez for his encouragement in the work. He also thanks C. Weisbuch for the availability of figure 1. This work was sponsored in part by CICYT Grant No. MAT-88-0116-C02-02.

References

1. R. Dingle, W. Wiegmann, and C.H. Henry, Phys. Rev. Lett. **33**, 827 (1974).
2. D.A.B. Miller, D.S. Chemla, T.C. Damen, A.C. Gossard, W. Wiegmann, T.H. Wood, and C.A. Burrus, Phys. Rev. Lett. **53**, 2173 (1984).
3. M. Potemski, J.C. Maan, K. Ploog, and G. Weimann, in *Proceedings of the 19th International Conference on the Physics of Semiconductors*, edited by W. Zawadzki (Institute of Physics, Polish Academy of Sciences, Warsaw, 1988), p. 119
4. R.C. Miller, D.A. Kleinman, W.T. Tsang, and A.C. Gossard, Phys. Rev. B **24**, 1134 (1981).
5. P. Dawson, K.J. Moore, G. Duggan, H.I. Ralph, and C.T. Foxon, Phys. Rev. B **34**, 6007 (1986); K.J. Moore, P. Dawson, and C.T. Foxon, *ibid.* **34**, 6022 (1986).
6. W.M. Theis, G.D. Sanders, C.E. Leak, D.C. Reynolds, Y.C. Chang, K. Alavi, C. Colvard, and I. Shidlovsky, Phys. Rev. B **39**, 1442 (1989).
7. J.E. Rowe and D.E. Aspnes, Phys. Rev. Lett. **25**, 162 (1970)
8. H.J. Polland, L. Schultheis, J. Kuhl, E.O. Göbel, and C.W. Tu, Phys. Rev. Lett. **55**, 2610 (1985).

9. G. Bastard, E.E. Mendez, L.L. Chang, and L. Esaki, Phys. Rev. B **28**, 3241 (1983)

10. R.T. Collins, K.v. Klitzing, and K. Ploog, Phys. Rev. B **33**, 4378 (1986)

11. L. Viña, R.T. Collins, E.E. Mendez, and W.I. Wang, Phys. Rev. Lett. **58**, 832 (1987).

12. D. Kleppner, M.G. Littman, and L. Zimmerman, in *Rydberg States of Atoms and Molecules*, edited by R.F. Stebbings and F.B. Dunning (Cambridge University Press, Cambridge, England, 1983), Chap. 3.

13. R. Dingle, Phys. Rev. B **8**, 4627 (1973).

14. M. Altarelli and N.O. Lipari, Phys. Rev. B **9**, 1733 (1974).

15. J.C. Maan, G. Belle, A. Fasolino, M. Altarelli, and K. Ploog, Phys. Rev. B **30**, 2253 (1984); J.C. Maan, A. Fasolino, G. Belle, M. Altarelli, and K. Ploog, Physica **127B**, 426 (1984).

16. N. Miura, Y.Iwasa, S. Tarucha, and H. Okamoto, in *Proceedings of the 17th International Conference on the Physics of Semiconductors*, edited by J.D. Chadi and W.A. Harrison (Springer, New York, 1984), p. 359; S. Tarucha, H. Okamoto, Y. Iwasa, and N. Miura, Solid State Commun. **52**, 815 (1984).

17. D.C. Rogers, J. Singleton, R.J. Nicholas, C.T. Foxon, and K. Woodbridge, Phys. Rev. B **34**, 4002 (1986).

18. W. Ossau, B. Jäkel, and E. Bangert, in *High magnetic fields in Semiconductor Physics*, edited by G. Landwehr (Springer, Berlin, 1987), p. 213; W. Ossau, B. Jäkel, E. Bangert, G. Landwehr, and G. Weimann, Surf. Sci. **174**, 188 (1986).

19. S.-R.E. Yang and L.J. Sham, Phys. Rev. Lett. **58**, 2598 (1987).

20. G.E.W. Bauer and T. Ando, Phys. Rev. B **38**, 6015 (1988), and references therein.

21. R.G. Ulbrich and C. Weisbuch, unpublished.

22. D.D. Sell, Phys. Rev. B **6**, 3750 (1972)

23. S.B. Nam, D.C. Reynolds, C.W. Litton, R.J. Almassy, T.C. Collins,and C. M. Wolfe, Phys. Rev. B **13**, 761 (1976).

24. L. Viña, G.E.W. Bauer, M. Potemski, J.C. Maan, E.E. Mendez, and W.I. Wang, Phys. Rev. B **38**, 10154 (1988).

25. Y. Shinozuka and M. Matsuura, Phys. Rev. B **28**, 4878 (1983).

26. R.T. Collins, L. Viña, W.I. Wang, L.L. Chang, L. Esaki, K.v. Klitzing and K. Ploog, in *Proceedings of the 18th International Conference on the Physics of Semiconductors*, edited by O. Engström (World Scientific, Singapore, 1987) p. 521.

27. L.C. Andreani and A. Pasquarello, Europhys. Lett. **6**, 259 (1988).

28. B. Zhu, Phys. Rev. B **38**, 13316 (1988).

29. G.E.W. Bauer and T. Ando, Phys. Rev. Lett. **59**, 601 (1987).

30. D.C. Rogers, J. Singleton, R.J. Nicholas, and C.T. Foxon, in *High magnetic fields in Semiconductor Physics*, edited by G. Landwehr (Springer, Berlin, 1987), p. 223.

31. L. Viña, G.E.W. Bauer, M. Potemski, J.C. Maan, E.E. Mendez, and W.I. Wang, to be published.

32. See, for example, D. Kleppner, M.G. Littman, and M.L. Zimmerman, in *Rydberg States of Atoms and Molecules*, edited by R.F. Stebbings and F.B. Dunning (Cambridge Univ. Press, Cambridge, England, 1983), Chap. 3.

33. See, for example, D. Delande, F. Biraben, and J.C. Gay, in *New Trends in Atomic Physics, Vol. 1*, edited by G. Grynberg and R. Stora. NATO ASI Series (North Holland, New York, U.S.A., 1984), Chap. 4.

34. V.S. Lisitsa, Sov. Phys. Usp. **30**, 927 (1987) [Usp. Fiz. Nauk **153**, 379 (1987)].

ANISOTROPY OF MAGNETO-OPTICAL PROPERTIES OF (Al,Ga)As QUANTUM WELLS

Gerrit E.W. Bauer

Philips Research Laboratories
5600 JA Eindhoven, The Netherlands

Magneto-optical spectra of undoped GaAs/Al_xGa_{1-x}As quantum wells grown in [001], [110] and [111] crystal directions are calculated in an effective mass approximation, taking into account exciton mixing interactions. Small but observable effects like different exciton diamagnetic shifts and spin-splittings as well as characteristic exciton fine structures are predicted.

INTRODUCTION

The emphasis in III-V semiconductor quantum well physics has traditionally been on samples grown on substrates oriented along the [001] axis of the single-crystal unit cell. The main reason for this choice is the superior crystal quality which can be achieved for this (or vicinal) orientations under conventional growth conditions. Only very recently it has been demonstrated that high-quality interfaces can be obtained by epitaxial growth on off-[001] oriented substrates also,[1-6] and a number of intriguing optical effects have been reported.[2-6] Since the conduction band is nearly parabolic, the anisotropic valence band structure can be held responsible for these effects.

Shanabrook et al.[2] interpreted photoreflectance spectra of [111]-oriented quantum wells in terms of an increase of the hole mass normal to the well. Hayakawa et al.[3] reported an enhanced optical oscillator strength in [111] quantum wells, which was thought to originate from a larger density of states. This result was consistent with the subsequent finding of the same group[4] that the ground state exciton binding energy in [111] quantum wells is increased by about 10 per cent. Hayakawa et al.'s conclusions are not supported by the present theory, however, as will be discussed later. Molenkamp and colleagues studied [001], [310] and [111] quantum wells by means of photoluminescence excitation spectroscopy.[5,6] They observed an anisotropy of the subband transition energies[5] which in conjunction with 8-band k.p calculations was used to derive a new effective mass parameter set for GaAs. Molenkamp *et al.*[6] were also able to measure the heavy-hole ground state exciton binding energy of [310]-oriented wells. In agreement with theoretical results[6,7] no significant changes compared with [001] quantum wells were found. Although not discussed here any further, the strain-induced electrooptical effects observed recently[8] in [111] GaInAs/AlAs quantum wells should be mentioned.

Besides the calculations of the **k.p** band structure[5] and of the exciton binding energies[6,7] mentioned above, there are only a few theoretical papers which deal with the effect of crystalline anisotropy on wide quantum wells and/or large-period superlattices. Mailhiot and Smith[9] calculated the electronic structure of [001] and [111] superlattices by means of a **k.p** theory (see also Ref. 8). Using a bond-orbital method Houng *et al.*[10] calculated band structures of off-[001] quantum wells, though their results differ conspicuously from the effective mass calculations presented here. Broido[11] also calculated ground state exciton binding energies in a few-subband model.

The magnetic field has a long record as a powerful tool in the study of the electronic structure and optical properties of solids. For intrinsic quantum wells too the magnetic field has proven its worth in exhibiting new effects.[12] A magnetic field furthermore enhances oscillator strengths and increases the energy differences between exciton states, which means that excited excitons can be studied even in less perfect samples. One might also hope that magnetic field effects will shed light on the origin of the conflicting results on exciton binding energies in Refs. 4 and 6.

Here, in anticipation of a magneto-optical study,[13] calculations of energies and wave functions of magnetoexcitons in differently oriented quantum wells are presented. From previous experiences on [001] quantum wells[12] one may be optimistic about the predictive power of the theory, and it is hoped that the results are of some use to experimenters.

EFFECTIVE-MASS HAMILTONIANS

For most purposes it is sufficient to use a parabolic conduction band and describe the hole band structure by the Luttinger Hamiltonian. Non-parabolicities cause small shifts of electron and light-hole subbands but qualitatively new effects do not occur. The exciton Hamiltonian for the most important crystal directions [001], [110] and [111] and a magnetic field parallel to the growth direction (always chosen to be parallel to the z-axis) can be derived directly from the expressions given by Suzuki and Hensel.[14] For future convenience the Hamiltonians are listed below. Using Rydberg atomic units the exciton Hamiltonians in the rest frame can be expressed in terms of electron and hole coordinates $\vec{r}_{e/h} = (x_{e/h}, y_{e/h}, z_{e/h})$, and the difference polar coordinates $\vec{r}_e - \vec{r}_h = (\rho, \phi, z_e - r_h)$ as:

$$\mathcal{H}_{ex} = \begin{bmatrix} A_{-3/2} & B & C & D \\ B^* & A_{-1/2} & 0 & C \\ C^* & 0 & A_{1/2} & -B \\ D^* & C^* & -B^* & A_{3/2} \end{bmatrix} + \left[V_{val}(z_h) - \frac{2}{\varepsilon |\vec{r}_e - \vec{r}_h|} + H_e \right] \mathbf{I}_4, \tag{1}$$

$$H_e = \frac{m_0}{m_e} \{ \frac{\partial^2}{\partial z_e^2} - \nabla_\rho^2 + \frac{\gamma^2 \rho^2}{4} + \gamma L_z \} + V_{cond}(z_e) \pm \frac{g_e}{2} \gamma \qquad (2)$$

where m_e and g_e are the effective mass and g-factor of the conduction band electron, I_4 denotes the 4×4 unit matrix, $L_z = -i\partial/\partial\phi$ is the operator for the angular momentum normal to the interfaces, γ denotes the magnetic field and ∇_ρ^2 is the 2-dimensional Laplacian. V_{val} and V_{cond} are the potential steps which represent the quantum well. The matrix character of Eq. (1) reflects the mixing of the spin-projected $J = 3/2$ spinor Bloch waves at the top of the valence band. The matrix in brackets follows from the Luttinger Hamiltonian and can be separated into an axially symmetric (superscript as) and a warping contribution (superscript wa). Introducing the Luttinger valence band parameters $\gamma_1, \gamma_2, \gamma_3, \kappa$ and q and the raising and lowering operators k^+ and k^-:

$$k^\pm = \frac{i}{\sqrt{2}} e^{\pm i\phi} \{ \frac{\partial}{\partial\rho} \pm \frac{i\partial}{\rho\partial\phi} \pm \frac{\gamma\rho}{2} \}, \qquad (3)$$

the non-zero matrix elements for a [001]-oriented quantum well read in the phase convention of Ref. 14:

$$A_{\pm 3/2}^{ax} = -(\gamma_1 - 2\gamma_2) \frac{\partial^2}{\partial z_h^2} + (\gamma_1 + \gamma_2)(-\nabla_\rho^2 + \frac{\gamma^2 \rho^2}{4} - \gamma L_z) \pm (3\kappa + \frac{27}{4} q)\gamma \qquad (4)$$

$$A_{\pm 1/2}^{ax} = -(\gamma_1 + 2\gamma_2) \frac{\partial^2}{\partial z_h^2} + (\gamma_1 - \gamma_2)(-\nabla_\rho^2 + \frac{\gamma^2 \rho^2}{4} - \gamma L_z) \pm (\kappa + \frac{q}{4})\gamma \qquad (5)$$

$$B^{ax} = -2\sqrt{6} \gamma_3 k^- \frac{\partial}{\partial z_h} \qquad (6)$$

$$C^{ax} = \sqrt{3}(\gamma_2 + \gamma_3)(k^-)^2 \qquad (7)$$

$$C^{wa} = \sqrt{3}(\gamma_2 - \gamma_3)(k^+)^2 \qquad (8)$$

For a [110]-oriented quantum well one obtains analogously:

$$A_{\pm 3/2}^{ax} = -(\gamma_1 - \frac{\gamma_2}{2} - \frac{3\gamma_3}{2}) \frac{\partial^2}{\partial z_h^2} + (\gamma_1 + \frac{\gamma_2}{4} + \frac{3\gamma_3}{4})(-\nabla_\rho^2 + \frac{\gamma^2 \rho^2}{4} - \gamma L_z) \pm (3\kappa + 6q)\gamma \qquad (9)$$

$$A_{\pm 1/2}^{ax} = -(\gamma_1 + \frac{\gamma_2}{2} + \frac{3\gamma_3}{2}) \frac{\partial^2}{\partial z_h^2} + (\gamma_1 - \frac{\gamma_2}{4} - \frac{3\gamma_3}{4})(-\nabla_\rho^2 + \frac{\gamma^2 \rho^2}{4} - \gamma L_z) \pm (\kappa + \frac{5}{2} q)\gamma \qquad (10)$$

$$A_{\pm 3/2}^{wa} = -(\gamma_2 - \gamma_3)((k^+)^2 + (k^-)^2) = -A_{\pm 1/2}^{wa} \qquad (11)$$

$$B^{ax} = -\sqrt{6}\,(\gamma_2 + \gamma_3) k^- \frac{\partial}{\partial z_h} \tag{12}$$

$$B^{wa} = -\sqrt{6}\,(\gamma_2 - \gamma_3) k^+ \frac{\partial}{\partial z_h} \tag{13}$$

$$C^{ax} = \frac{\sqrt{3}}{4}\,(3\gamma_2 + 5\gamma_3)\,(k^-)^2 \tag{14}$$

$$C^{wa} = \frac{\sqrt{3}}{4}\,(\gamma_2 - \gamma_3)\{3(k^+)^2 + \nabla_\rho^2 - \frac{\gamma^2 \rho^2}{4} + \gamma L_z - 2\frac{\partial^2}{\partial z_h^2}\} + \frac{3\sqrt{3}}{4}\,q\gamma, \tag{15}$$

whereas the [111] Hamiltonian contains the operators:

$$A^{ax}_{\pm 3/2} = -(\gamma_1 - 2\gamma_3)\frac{\partial^2}{\partial z_h^2} + (\gamma_1 + \gamma_3)\left(-\nabla_\rho^2 + \frac{\gamma^2 \rho^2}{4} - \gamma L_z\right) \pm (3\kappa + \frac{23}{4}q)\gamma \tag{16}$$

$$A^{ax}_{\pm 1/2} = -(\gamma_1 + 2\gamma_3)\frac{\partial^2}{\partial z_h^2} + (\gamma_1 - \gamma_3)\left(-\nabla_\rho^2 + \frac{\gamma^2 \rho^2}{4} - \gamma L_z\right) \pm (\kappa + \frac{13}{4}q)\gamma \tag{17}$$

$$B^{ax} = -2\sqrt{\frac{2}{3}}\,(2\gamma_2 + \gamma_3) k^- \frac{\partial}{\partial z_h} \tag{18}$$

$$B^{wa} = -2i\sqrt{\frac{2}{3}}\,(\gamma_2 - \gamma_3)\,(k^+)^2 \tag{19}$$

$$C^{ax} = \frac{2}{\sqrt{3}}\,(\gamma_2 + 3\gamma_3)\,(k^-)^2 \tag{20}$$

$$C^{wa} = \frac{4i}{\sqrt{3}}\,(\gamma_2 - \gamma_3) k^+ \frac{\partial}{\partial z_h} \tag{21}$$

$$D^{wa} = -2iq\gamma. \tag{22}$$

The symmetry properties of the effective mass Hamiltonian and the method of solving the exciton wave equation have been detailed in Ref. 7. Briefly, the exciton envelope function is expanded into a large, symmetry-adapted and non-orthogonal basis set and the exciton energies are obtained by solving the corresponding generalized eigenvalue problem, which also yields eigenvectors which determine the oscillator strengths of the optical transitions. The parameter set recommended in Ref. 5 is used throughout. This includes a 68/32 band-offset rule which is assumed to be independent of the crystal orientation.

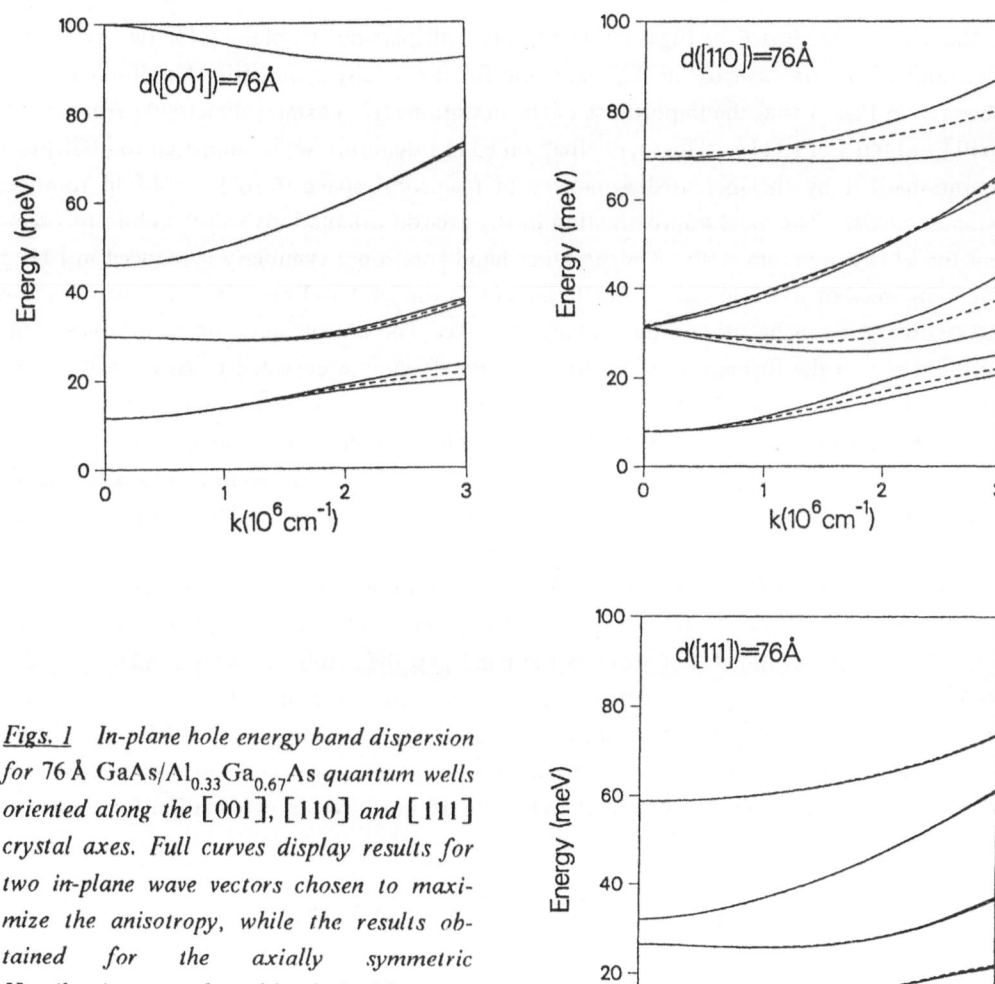

Figs. 1 *In-plane hole energy band dispersion for 76 Å GaAs/Al$_{0.33}$Ga$_{0.67}$As quantum wells oriented along the [001], [110] and [111] crystal axes. Full curves display results for two in-plane wave vectors chosen to maximize the anisotropy, while the results obtained for the axially symmetric Hamiltonians are plotted by dashed lines.*

WARPING DISTORTIONS AND HOLE BAND STRUCTURE

$\gamma_2 - \gamma_3$ and q are small for GaAs and additionally the above warping Hamiltonians do not contribute to first order to the exciton energies. Indeed the effect has been calculated to be very small for [001] quantum wells[7] and even in very high resolution spectra no trace of the warping interaction could be detected.[12] As discussed in Ref. 7 it is difficult to properly include warping in the exciton calculations for off-[001] quantum wells. The deviations from axial symmetry in off-[001] quantum wells can easily be appraised by inspection of the hole band structure, however. The hole Hamiltonian is obtained by omitting the Coulomb interaction and H_e in Eq. (1). The hole band structure at zero magnetic field is then easily calculated. The results for 76 Å quantum wells and an aluminum concentration of $x = 0.33$

in the barrier are plotted in Figs. 1. The in-plane dispersion is obtained for two symmetry directions of the wave vector in the plane and for the axially symmetric Hamiltonian. It is observed in Figs. 1 that the importance of the deviations from axial symmetry decreases from [110] to [001] to [111]. The small effect on [111] quantum wells compared to [001] can be rationalized by the increased symmetry in reciprocal space (6-fold vs. 4-fold rotation symmetry axis). The axial approximation in the exciton calculations will therefore be excellent for [111] quantum wells. On the other hand rotational symmetry is reduced in [110] quantum wells to a 2-fold axis. Small changes in the subband energies are caused by the interaction between heavy and light holes at k = 0. The anisotropies are to a large extent averaged out in the formation of excitons and its effects are expected to be quite small, as are the errors introduced by imposing axial symmetry. Still, [110] quantum wells are the most likely candidates to detect the effects of warping in magnetoexciton spectra.

It is observed in Figs. 1 that the effective mass parallel to the well, and thus the band-gap density of states, is decreased in [111] quantum wells relative to [001] quantum wells, which appears to be in conflict with the conclusions of Hayakawa et al..[3]

The band structure calculations in Ref. 5 are more advanced than the present ones in so far as non-parabolicities due to interactions between conduction band, light-hole and spin split-off band are included, but close to the band gap the results are very similar. Big differences exist on the other hand between the present results and those from Ref. 10. Houng et al.[10] claim for example enormous anisotropies in the in-plane dispersion of [110]-quantum wells. The zone center energies are different from the present ones and subband quantization seems to persist far above the potential well. We do not have an explanation for these discrepancies.

MAGNETOEXCITONS

The results of the calculations which have been carried out as described in Ref. 7 are summarized in Table I and Figs. 2 and 3 which will be discussed below. As explained in the previous section, warping effects are disregarded here.

The ground state properties of the excitons are listed in Table I (due to the omission of non-parabolicity corrections small differences exist between the present values and those in Ref. 6). It can be seen that the effect of crystal orientation on the ground state excitons is small. At zero field the largest effect is experienced by the light-hole exciton in [111] quantum wells due to the increased mixing with the p-type exciton in the second heavy-hole subband.[6] The differences are much smaller than the discrepancies between experimental and theoretical ground state exciton binding energies,[7] but it is hoped that the trends are less affected. Notable is the difference in the spin-splittings of excitons in [001] quantum wells vs. other orientations. Though small, this effect is quite significant, since small differences in well width, which in a comparative study can never be completely ruled out, have a negligible effect on this quantity,[15] which can be measured to an accuracy of 0.1 meV.[16] The absolute numbers cannot be trusted, however, since they depend quite strongly on the choice of the Luttinger parameters.[15]

Table I _Characteristic energies (in meV) of heavy (h) and light (ℓ) ground state excitons in 76 Å GaAs/Al$_{0.33}$Ga$_{0.67}$As quantum wells for different orientations. Listed are the exciton binding energy (E_1), exciton spin splittings, i.e. difference between exciton energies measured for σ$^-$ and σ$^+$ polarizations, for 10 tesla (E_2) and 20 tesla (E_3), and spin averaged energy (diamagnetic) shifts relative to zero field at 10 tesla (E_4) and 20 tesla (E_5)._

	E_1	E_2	E_3	E_4	E_5
h ([001])	-9.7	0.5	0.4	4.1	11.2
h ([110])	-9.4	1.2	1.4	4.5	12.1
h ([111])	-9.4	1.4	1.7	4.6	12.4
ℓ([001])	-11.6	1.7	4.0	1.4	7.0
ℓ([110])	-11.9	0.9	3.1	1.1	7.2
ℓ([111])	-12.6	0.7	2.7	1.5	7.9

The conclusion of Hayakawa _et al._ that the exciton binding energy is enhanced in [111] relative to [001] quantum wells by 10 per cent clearly contradicts the present theoretical results. The decrease of the parallel mass in [111] quantum wells leads here to a decrease of the binding energy, since there is no compensation due to enhanced mixings. The additional information as provided by magneto-optical spectroscopy might help to resolve this question.

Energies and oscillator strengths of ground and excited exciton transitions are plotted in Figs. 2-3 for σ$^\pm$ light polarization in the Faraday configuration. The ground state of the heavy-hole (light-hole) exciton and its excited states with the same symmetry are connected by full (dashed) lines. The two sets of states belong to different irreducible representations of the symmetry group and do not mix.[7] Transition oscillator strengths at integer magnetic fields are proportional to the area of the filled dots. Exciton states with zero (one-photon) oscillator strength are also plotted since they might become observable through unintentional electric fields or two-photon processes. The quasi-continuum region where exciton states are very closely spaced cannot be treated adequately by the present theory and is therefore shaded in Figs. 2-3. Basis set limitations can also cause numerical inaccuracies of several meV in the high-field and high-energy regime.

Open circles on the ordinate (zero magnetic field) give the location of the exciton resonance associated with the third heavy-hole subband. At high fields this exciton is easily identified due to its small diamagnetic shift. The strong dependence of the heavy-hole normal mass on the crystal orientation is reflected by the more or less parallel energy shift of this exciton, though the mixing with other excitons appears to be larger for [001].

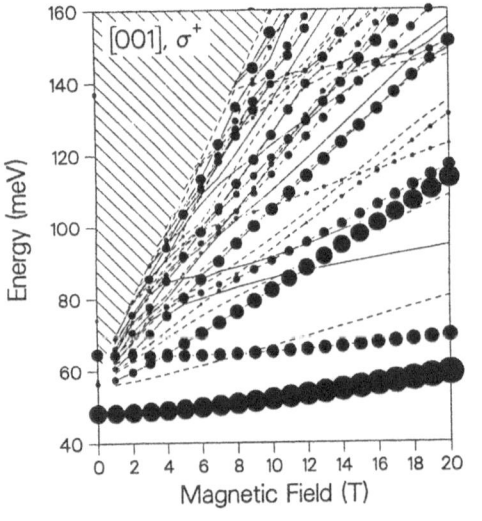

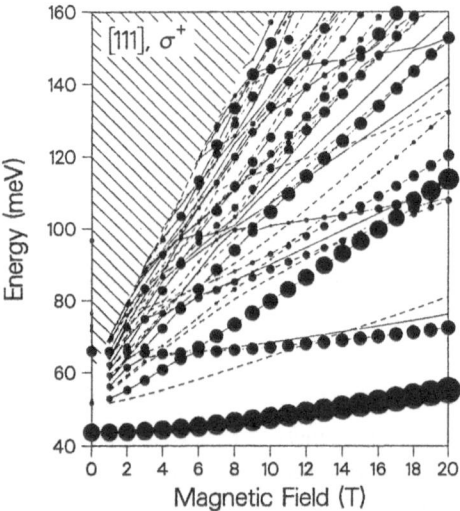

Figs. 2 σ^+ *polarized magnetooptical spectra of 76 Å GaAs/Al$_{.3}$Ga$_{.7}$As quantum wells oriented along the* [001], [110] *and* [111] *crystal axes. Dashed and full lines represent energies of excitons belonging to different symmetry representations. Transition oscillator strengths at integer magnetic fields are proportional to the area of the filled dots.*

While [111] and [110] spectra turn out to be quite similar, both display distinct differences compared to [001] especially in the σ^- polarization, which should be easily detectable. In the 15-20 tesla regime one observes for example that the p-type exciton transition associated with the second heavy hole band is weak for [001] (dashed line just beneath the third subband heavy-hole exciton). In the [111] spectra, on the other hand, this exciton is shifted to lower energies (but less than the third subband exciton) such that its oscillator strength is strongly increased at the expense of mainly the $2s$ exciton of the light hole.

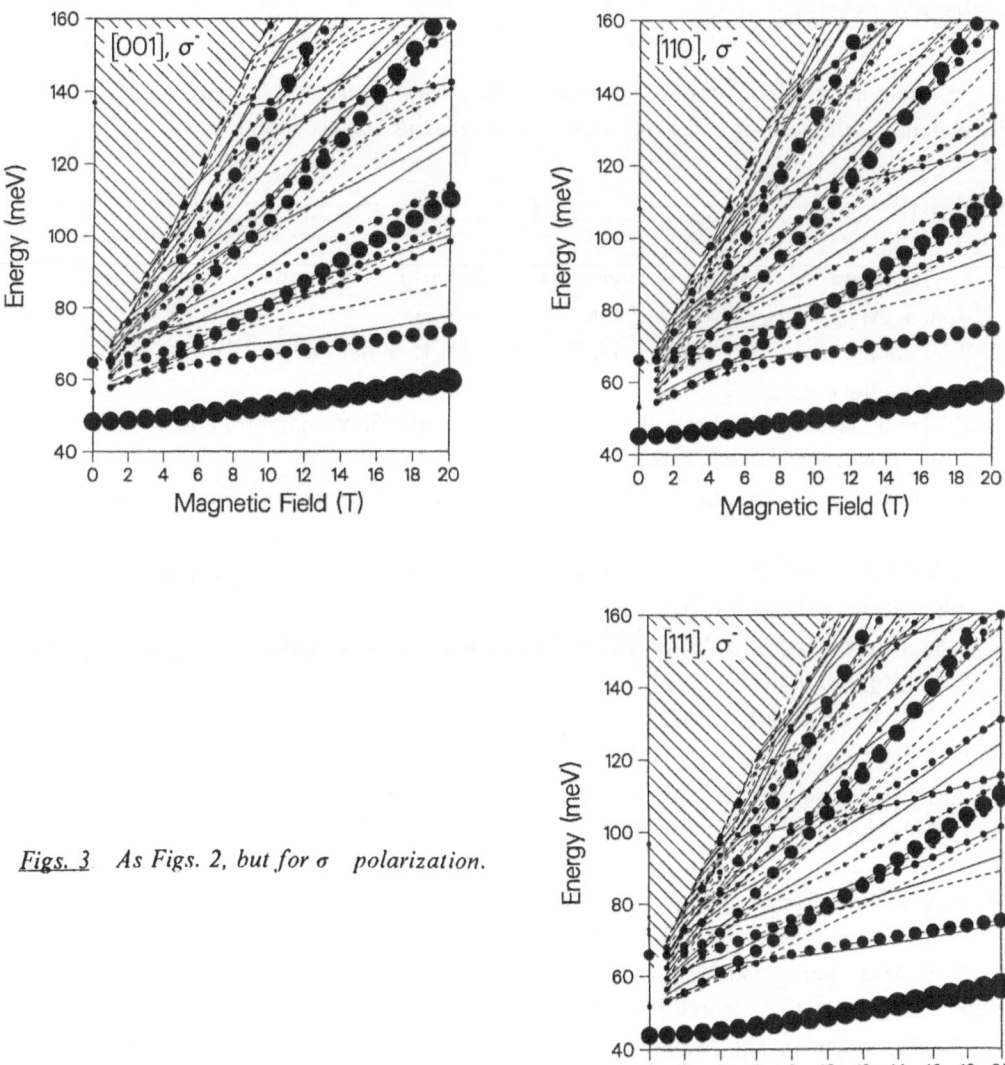

Figs. 3 _As Figs. 2, but for σ⁻ polarization._

CONCLUSIONS

An effective mass theory of magnetoexcitons is shown to reveal quite significant effects of the quantum well growth direction on magneto-optical spectra. Experimental (pseudo) absorption spectra in the presence of magnetic fields should provide for a sensitive test of the present calculations and help to reconcile controversies on binding energies in off-[001] quantum wells.

ACKNOWLEDGMENTS

I would like to thank drs. R. Eppenga, T. Hayakawa and L.W. Molenkamp for valuable discussions on the subject and Prof. M.F.H. Schuurmans for his support.

REFERENCES

1. T. Hayakawa, M. Kondo, T. Suyama, K. Takahashi, S. Yamamoto, and T. Hijikata, Jpn. J. Appl. Phys. 26:L302(1987).

2. B.V. Shanabrook, O.J. Glembocki, D.A. Broido, L. Viña, and W.I. Wang, J. Physique (Coll.) C5:235(1987).

3. T. Hayakawa, K. Takahashi, M. Kondo, T. Suyama, S. Yamamoto, and T. Hijikata, Phys. Rev. Lett. 60:349(1988).

4. T. Hayakawa, K. Takahashi, M. Kondo, T. Suyama, S. Yamamoto, and T. Hijikata, Phys. Rev. B38:1526(1988).

5. L.W. Molenkamp, R. Eppenga, G.W. 't Hooft, P. Dawson, C.T. Foxon, and K.J. Moore, Phys. Rev. B38:4314(1988).

6. L.W. Molenkamp, G.E.W. Bauer, R. Eppenga and C.T. Foxon, Phys. Rev. B38:6147(1988).

7. G.E.W. Bauer and T. Ando, Phys. Rev. B38:6015(1988).

8. B.K. Laurich, K. Elcess, C.G. Fonstad, J.G. Beery, C. Mailhiot, and D.L. Smith, Phys. Rev. Lett. 62:649(1989).

9. C. Mailhiot and D.L. Smith, Phys. Rev. B35:1242(1987).

10. M.-P. Houng, Y.-C. Chang, and W.I. Wang, J. Appl. Phys. 64:4609(1988).

11. D.A. Broido, Superl. Microstr. 5:471(1989).

12. L. Viña, these Proceedings and references therein.

13. L.W. Molenkamp and J. Singleton, private communication.

14. K. Suzuki and J.C. Hensel, Phys. Rev. B9:4184(1974).

15. G.E.W. Bauer, in The Application of High Magnetic Fields in Semiconductor Physics, edited by G. Landwehr, (Springer, Berlin, 1989).

16. W. Ossau, B. Jäkel, E. Bangert, G. Weimann: The Basic Properties of Impurity States in Superlattice Semiconductors, edited by C.Y. Fong, (Plenum, New York, 1988).

MAGNETO-OPTICAL STUDY OF MINIBAND DISPERSION AND

EXCITONIC EFFECTS IN GaAs/GaAlAs SUPERLATTICES

B. Deveaud, A. Chomette, F. Clérot, and A. Regreny
C.N.E.T., LAB/OCM, 22301 Lannion, France

R. Romestain
Laboratoire de Spectrométrie Physique, USMG, Grenoble

J.C. Maan
Max Planck Inst., Grenoble,   France

G. Bastard
E.N.S., 25 rue Lhomond, Paris,   France

Hanyou Chu and Yia Chung Chang
University of Illinois, Urbana, IL   61801

ABSTRACT

The luminescence excitation spectra of GaAs/GaAlAs superlattices show new structures when their period is such that the miniband width for electrons is of the order of 10 meV. The experimental spectra are remarkably reproduced by theoretical calculations, and we are then able to give an assignement to these structures. In particular, new excitonic transitions are observed which might be the equivalent of saddle-point excitons. We have performed excitation studies under magnetic field in both Faraday and Voigt configurations in order to check the different attributions. Confirmation of the excitonic character of the new peaks is obtained. When the superlattice period decreases, the exciton binding energy is observed to decrease and the miniband width to increase.

INTRODUCTION

In a crystal, the expansion of the optical transition energy in a Taylor series around one of the critical points of the band structure leads, in first approximation, to a parabolic development with three reduced mass parameters. Four types of critical points can be found, labelled M_j according to the number j of negative mass components. M_1 critical points are called saddle points due to the shape of the band structure around them. Superlattices (SLs) of type I such as GaAs/AlGaAs represent a very interesting case where the zone folding of the band structure keeps a M_0 singularity at the centre of the Brillouin minizone, and leads to the formation of saddle points at the Brillouin minizone edges ($k_z = \pm\pi/d$ where k_z is the wavevector along the growth direction z and d the SL period). By varying d, the miniband width for electrons can be tailored from less than 1 meV to more than 200 meV. Therefore, the miniband width can be tailored to be of the order, larger or smaller than the exciton binding energy, in contrast with bulk materials where the Coulomb energy cannot be adjusted and is generally quite small as compared to the band width. In the case of SLs of intermediate periods, the situation is particularly simple as in the two positive mass directions (x,y), the dispersion relation can be assumed to be parabolic, and in the other one (z), the dispersion relation is a cosine.

Coulomb effects lead to drastic changes in the absorption spectra of semiconductors. The most dramatical one is due to the formation of stable excitons which bring sharp resonances in the optical spectra[1]. Other effects such as Sommerfeld enhancement[2] or excitonic enhancement around the Fermi energy[3,4] are also observed and quite well understood theoretically. Coulomb interaction between electrons and holes associated with the saddle-point states may give rise to exciton resonances below the energy of the saddle point.

The possible existence and behaviour of saddle-point excitons (sometimes called "hyperbolic excitons") has been the subject of a controversy[5,6] as well as of numerous theoretical[7,8,9] and experimental[10,11] studies. In Alkali Halides[8], or in solid Xenon[12] for example, such excitons are observed as sharp structures in resonance with the band to band transitions.

In this paper, we describe the luminescence excitation (PLE) results obtained on a series of SLs with different periods. We compare our experimental results to theoretical calculations[13], and obtain a very good agreement without any adjustable parameter. The new structures observed in the PLE spectra of the SLs are found to derive from the existence of the miniband structure and of the saddle points in $k_z\pi/d$. Their excitonic character is confirmed by the experiments under a magnetic field. Other features of the PLE spectra such as high energy shoulder are observed and interpreted.

EXPERIMENTAL DETAILS

Our GaAs/GaAlAs samples are grown by Molecular Beam Epitaxy (MBE). Precise determination of the SL parameters is obtained by x-ray diffraction[14]: the result of this determination labels the SL (a/b means well width/barrier width in Å). In all samples, the barrier aluminium concentration is of the order of 27%. The growth conditions have been optimized for short-period structures, and high quality is achieved, characterized by a very good flatness of the interfaces, and a good reproducibility of the layer thicknesses[15,16]. The parameters of the samples used for the magnetooptic experiments are listed on Table I.

Magneto-excitation of the luminescence is performed in a superconducting magnet up to fields of 10 T and in a Bitter magnet up to fields of 20T. In order to obtain the photoluminescence excitation (PLE) spectra, we set the monochromator on the heavy-hole exciton peak and scan the excitation energy of a krypton pumped LD700 dye laser (with two possible incident polarisations, σ+ or σ- in the case of Faraday configuration, σ or π in the case of Voigt configuration). When the monolayer splitting of the luminescence peak is resolved[17], we set the detection on the high energy component of the luminescence in order to reduce the number of peaks observed in PLE[17]. If we instead detect on the low energy peak, all the observed transitions are split into two components approximately separated by the value of the monolayer splitting. The fan charts are thus very difficult to interpret due to the large number of points.

Table I Structural parameters of our samples, and calculated miniband widths in meV for the different carriers.

Sample N°	La	Lb	x	electrons	HH	LH
432	52	140	.23	.1	0	.4
433	52	65	.23	5	0	7
434	52	56	.23	8	.1	10
329	50	50	.24	10	.2	12
330	40	40	.23	30	1	26
682	30	30	.25	74	2.5	50

EXPERIMENTS AND THEORY WITHOUT MAGNETIC FIELD

A) EXPERIMENTS

PLE spectra of superlattices show a distinct behaviour from quantum well spectra at zero magnetic field already. Some of these differences have been described elsewhere and arise from the increase in three-dimensional character of the structure as the period decreases[18,19], (see also in [17] and [20] that the shape of the luminescence itself is affected). Basically, the spectrum gradually changes from the shape typical of MQWs to a shape and a behaviour similar to what is observed in bulk GaAs layers: as an example, the light hole excitonic structure disappears for very short SL periods[21,22]. In the case of short period SLs, as vertical transport becomes quite efficient, it is necessary to clad the superlattice between two thick GaAlAs barriers in order to prevent surface recombination[19]. If this is not done, the high energy side of the spectrum is washed out due to the escape of the carriers and phonon replica appear[23].

One new characteristic can be evidenced on the spectra of intermediate period superlattices (approximately 50/50), i.e. the appearance of new structures as shown in Fig.1 [24,25,26]. For sake of clarity, we plotted this figure in energy shift units with respect to the heavy-hole exciton energy. Going from the lower curve of the figure to the upper one corresponds mainly to a change in the barrier thickness. Calculated miniband widths are given in table I. The PLE spectrum for the MQW sample (see sample 52/140) shows, as usual, 2 excitonic resonance peaks (HH and LH) and 2 shoulders corresponding to the onset of band to band transitions[27].

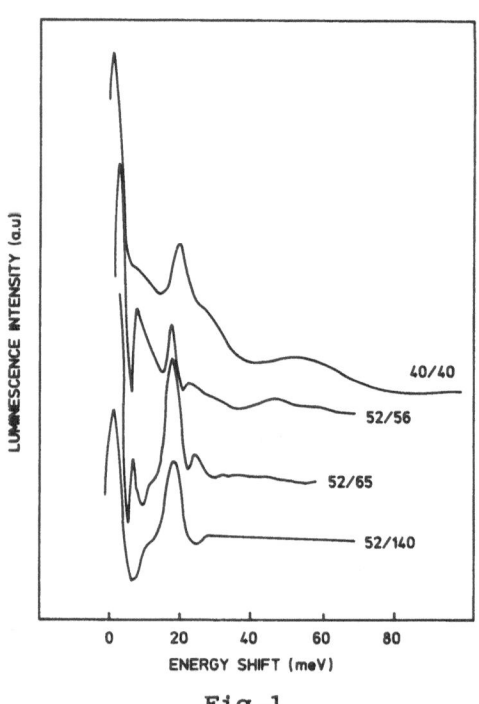

Fig.1

PLE spectra of a series of GaAs/GaAlAs superlattices. Note the new structures that appear with respect to the case of MQW systems represented by the bottom curve.

On the SL spectra shown in this figure two more sharp structures appear a few meV above each excitonic peak [25,26,28] (these structures are indicated by arrows 1 and 2 in Fig.2). Note that the width of these peaks increases as the superlattice period decreases, eventually the structure merges into the continuum[29]. Obviously, the shape of these structures relates to the corresponding miniband widths : the LH structure is broader than the HH one, due to the larger light-hole miniband width. Note the shoulder 1' for sample 52/65: although it looks very similar to the shoulder at 1.617 eV for sample 52/140, we will see in the following that they have different origins.

At smaller periods, further structures are observed in the PLE spectra. In samples which are not cladded between GaAlAs layers to avoid surface recombination they correspond to phonon emission[19,23] and are separated by about 42meV for the peaks corresponding to LO phonon emission by electrons, and by 36 meV for the ones corresponding to hot exciton or resonant Raman effects[30]. When surface recombination is not important (see sample 40/40 in fig.1), two new shoulders are observed at high energies (see structures 1' and 2' for 40/40 SL in Fig.2). The position of these structures roughly corresponds to the top of the minibands, however their shape is rather astonishing as a decrease of the oscillator

strength is observed rather than the increase that would be
expected in a simple minded picture. Equivalent shoulders are
also evidenced in different 30/30 SLs. These structures are
not observed any more when the period is further reduced
(20/20 and 10/10 for example).

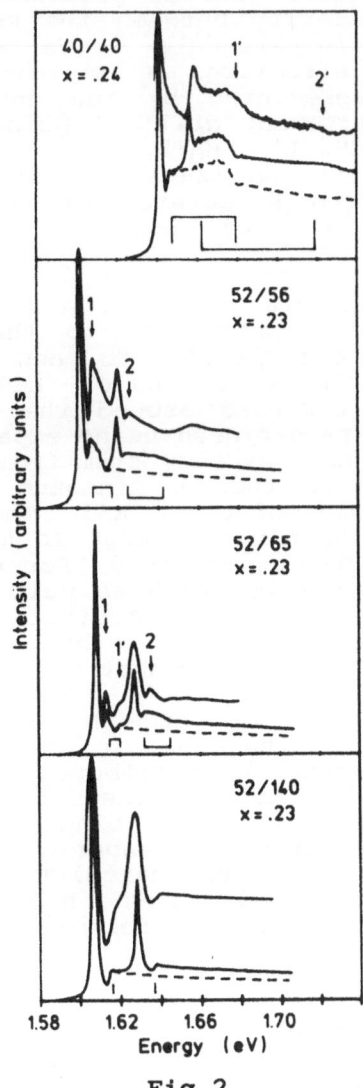

Fig.2

Comparison between experimental spectra (top
curves) and theoretical ones (bottom curves). There
is no adjustable parameter, we only allow for a small
rigid shift of the theoretical curve. Sample
parameters are given by x-ray measurements.

THEORY

We have reproduced the experimental results by computing
the theoretical absorption spectra shown in the lower curves

of Fig.2 with the method proposed by Chu and Chang[13], and using the x-ray parameters of the real samples[31]. The method has been improved by taking into account the effect of valence band mixing[32]. The absorption spectrum is obtained by computing the density of states for each heavy- or light-hole excitonic component, using the recursion method[33], and multiplying by the proper matrix element. Details of the theoretical calculations will be published elsewhere[34]. Note the very close similarity between the experimental and the theoretical curves, in particular as far as the shape and the position of the new structures is concerned. The contribution of HH alone is represented by the dashed line, and the positions of the different critical points are indicated on the figure (HH_0:1, HH_π:1', LH_0:2, LH_π:2' respectively meaning HH (resp. LH) to electron transition in $k_z=0$ (resp. π/d)). Theory predicts sharp structures around HH_0 and LH_0 and shoulders around HH_π and LH_π as is indeed observed experimentally.

The extra peaks are due partly to the usual excited states of the M_0 HH exciton or to the M_0 band to band transition onset, and partly to new resonances. These resonances are not the usual kind of "saddle-point excitons" as they cannot be expanded around the saddle point only; their wavefunctions rather need being expanded over the whole Brillouin minizone. However they stem from the existence of a dispersion relation in the SL miniband and thus from the presence of a saddle-point in π/d. Theory also predicts shoulders close to the minizone edge in π/d; such structures can be observed in our experiments. For example we evidence the HH_π shoulder in sample 52/65 as well as the HH_π and LH_π shoulders in sample 40/40. In the other samples, due to the particular value of the miniband width, these shoulders lye under the LH exciton peak and cannot be resolved.

Despite the similarity between the two structures, it should be stressed that the shoulder labelled 1' on the 52/65 SL spectrum can no longer be attributed, as in the MQW case, to the onset of M_0 band to band transitions. This would lead to a value of 13 meV for the binding energy of the HH exciton, much higher than both experimental observation and theoretical estimates[35]. Our calculations show that this structure rather corresponds to the heavy-hole M_1 point. We are going to confirm in the following with magneto PLE experiments that indeed such attributions were correct.

MAGNETIC FIELD EXPERIMENTS

FARADAY CONFIGURATION

When a magnetic field is applied parallel to the growth axis of a MQW, oscillations of the luminescence intensity are observed. The two excitonic transitions (HH and LH) show a small diamagnetic shift, and two series of Landau levels are evidenced that approximately extrapolate to their respective band-edges, corresponding to heavy-hole and light-hole related transitions (see Fig.3), (see for example[36,37,38], see also[39,40,41,42,43] for theoretical discussions). In a simple picture (neglecting excitonic effects and valence band mixing), one would expect for a 2D system a series of Landau transitions following:

$$E_n = (N+1/2)(heB/\mu) \text{ where } 1/\mu \text{ is the reduced mass}$$

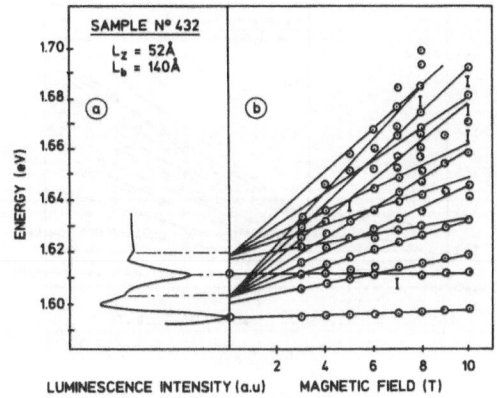

Fig.3

Fan chart of a MQW in Faraday configuration. Note the two excitonic lines with a very small shift, and the two series of Landau levels. Note that these series approximately extrapolate to the structure corresponding to the onset of band to band transitions.

$1/m_e + 1/m_h$ of the exciton.

It has been shown by Akimoto and Hasegawa[38] and others that, if excitonic effects were to be taken into account, the slope of the Landau transitions would be quite different:

$E_n = -1/(N+\frac{1}{2})^2 + 5/8(N+\frac{1}{2})^4(\hbar w_c)^2 + \ldots$ where w_c is the cyclotron frequency.

This has been observed by different authors, and is evidenced on Fig.3. For not too strong magnetic fields (below 10T), the slope of the series of levels roughly varies as $(n+1)(hw_c)$ and a strong excitonic effect is evidenced on the first two levels. However, the series of transition of N>1 still extrapolates very close to the band-edge energy. Oscillations corresponding to heavy-hole (resp. light-hole) transitions will extrapolate to the corresponding band edge of the MQW system and the distance of this point to the exciton will give the exciton binding energy[27,44]. Due to the non linearity in the magnetic behaviour just described (see also Plaut el al[45]), magneto-optical experiments are not perfect measure of the exciton binding energy. This technique is nevertheless one of the most accurate[46]. If valence band mixing is taken into account, the complexity of the spectra above 10T can be quite satisfactorily interpreted[47]

When a magnetic field is applied parallel to the growth axis of the superlattice, as in the case of MQWs, oscillations of the luminescence intensity are observed. These oscillations can be plotted on a fan chart (see Fig.4 for the 52/65 SL and Fig.5 for the 50/50 SL). On the series of SLs that we have studied, we observe the following features:

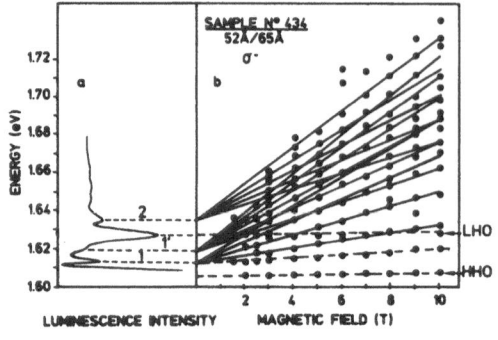

Fig.4

Fan chart of a 52/65 SL. Note the presence of 3
lines with a small diamagnetic shift, and the 3
series of Landau levels. Two of the series
extrapolate between HH and LH energies and are
interpreted as arising from the two edges of the
miniband in $k_z=0$ and π/d. This gives both an
approximate value of the miniband width and of the
exciton binding energy.

    -The two usual HH and LH excitonic transitions which
show a diamagnetic shift[48]. This shift is more easily
observed when the field is increased up to 20T.

    -As in the case of MQWs, we measure the exciton binding
energy as the distance between the extrapolation point of
the Landau transitions and the excitonic line. We observe a
very clear decrease of the exciton binding energy as the
period of the SL is reduced. The results of such an
estimation are reported on Fig.6 together with the result of
two theoretical calculations[13,18]: the decrease from about 9
meV for the MQW case to about 5meV is clearly evidenced from
our results and corresponds quite well to the reduction

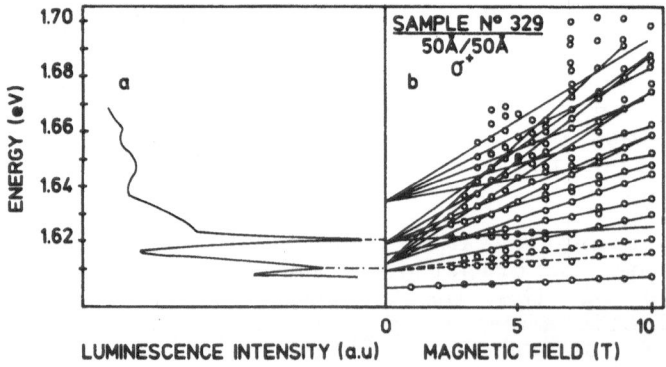

Fig.5

Same as Fig.4 but for a 50/50 SL. Note the increase
of the miniband width.

predicted by the theory. In order to interpret correctly all results on our SLs, it is then necessary to take this reduction of the exciton binding energy into account. Note that in our MQW sample the HH extrapolation point does coincide with the shoulder at 1.617 eV (see fig.2) which is generally interpreted as the onset of band to band transitions.

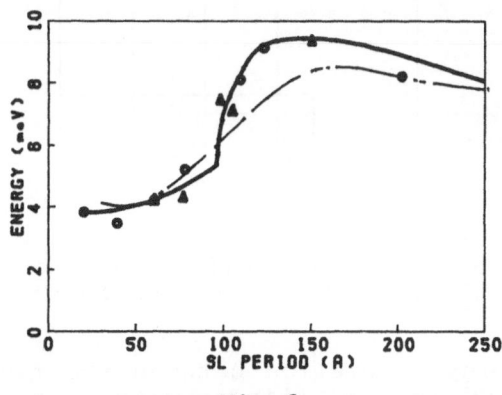

Fig.6

Exciton binding energy as a function of the SL period. The two curves correspond to theory (---)[13] and (—)[18]. Two series of experimental determinations are reported: this work(triangles) and [18] (circles).

-The new structures (labelled 1 on Fig.2), observed at zero magnetic field, show a small diamagnetic field (see the dotted lines in Figs.4 and 5). This evidences that at least some excitonic character is responsible for the occurrence of these new peaks and confirms the predictions of the theory. In some samples (see Fig.5), this structure eventually splits into two components at high magnetic fields.

-Instead of one series of Landau levels for heavy-holes we now observe two series of transitions corresponding to the peaks in the density of states at k_z = 0 and k_z = π/d[49]. In a quantum well, application of a magnetic field transforms the 2D density of states into a series of peaks (see Fig.7). In the case of a superlattice, the miniband in k_z has some extension and the application of a magnetic field induces a series of double peaked densities of states (see Fig.7). In the case of the 52/64 SL, these two series respectively extrapolate at 1.614 eV for the k_z = 0 transitions and 1.621 eV for the k_z = π/d transition in good agreement with the result of theoretical calculations[50].
The energy difference between the two extrapolated convergence points thus approximately gives the miniband width[22].

This additional series is not at all related to the monolayer splitting observed on the same samples for the following reasons :
i) we have detected the luminescence on the high energy component of the luminescence, so that only one HH transition is observed in PLE at zero magnetic field. If we set the detection on the low energy peak, all series of transitions are doubled due to this monolayer splitting.

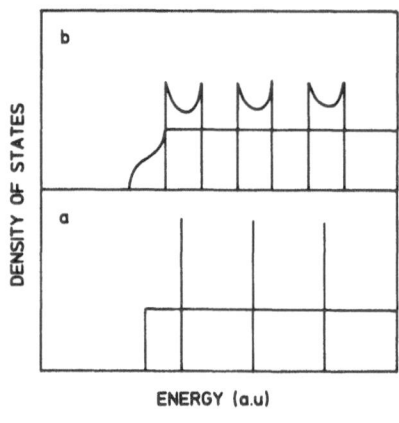

Fig.7

Effect of a strong magnetic field on: a) a 2D density of states and b) a quasi 3D density of states.

ii) The energy separation between the two series is not equal to the monolayer splitting energy, contrary to what we observe when we set the detection on the low energy line.

iii) We also observe the new series of transitions on samples that do not show any monolayer splitting.

-At high magnetic fields, the PLE spectrum is simplified and consists of a series of double peaks induced by the miniband dispersion in k_z (see Fig.8).

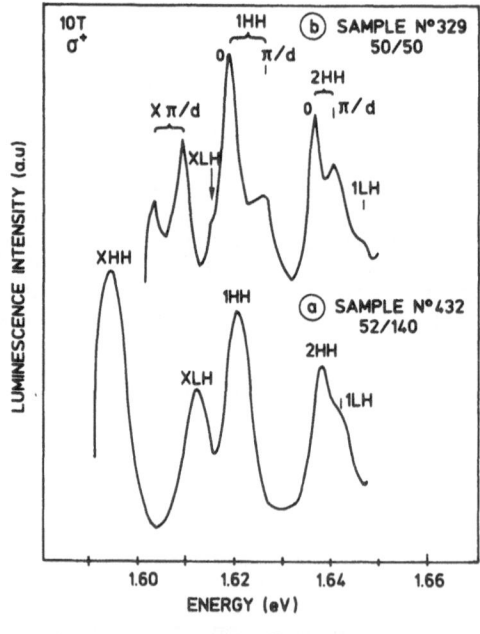

Fig.8

High magnetic field (10T) PLE spectrum of a 50/50 SL compared to a MQW. Note both the extra excitonic peaks labelled $X_{\pi/d}$ and the splitting of the HH Landau levels. The equivalent splitting of the LH Landau levels is not resolved in this experiment.

MAGNETIC FIELD ALONG THE LAYERS

When the magnetic field is applied in the plane of the layers, in the case of quantum wells, no oscillations are observed: the carriers are confined in the wells and cannot orbit[51]. In the case of superlattices, on the contrary, application of the magnetic field in the plane of the layers leads to oscillations; this effect has been used by Belle et al[52] for the measurement of the miniband width in a superlattice. When a magnetic field is applied, the carriers orbit with a radius given by[47]:

$$R_\mu^2 = (2N+1)L^2 \text{ where } L^2 = h/eB \text{ is the magnetic length.}$$

As pointed out by Maan[47] this is a unique situation. At 20T the N=1 electrons have an orbit radius of 5nm, of the order of the SL period. For larger N, the radius will be larger and larger and the electrons will test more and more periods of the SL during their orbit. However the carriers can only perform orbits if their energy is within the miniband of the superlattice[35,53,54]. As a result, the observation of Landau levels is only expected in a limited energy range corresponding to the coupled electron and hole miniband width.

We show on Fig.9 our results on 3 characteristic SLs at a magnetic field of 14T and on Fig.10 the fan chart of a 30/30 sample. As is obvious from these two figures:

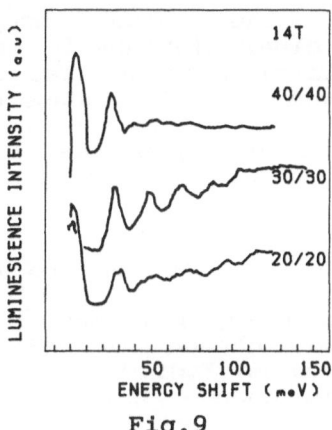

Fig.9

14T PLE spectra in Voigt configuration of 3 SLs of different periods. The electron miniband width of these samples are 20 meV for the 40/40, 60 meV for the 30/30 and 220 meV for the 20/20. Note the different energy positions at which the Landau transitions stop.

i) No Landau levels are observed above a certain energy. This confirms the experiment of Belle et al[52].

ii) The energy at which these oscillations stop depends on the SL period. In the case of the 20/20 SL oscillations are observed up to the top of the energy range allowed by the dye that we are using. When the SL period increases, a clear cutoff is observed closer and closer to the excitonic lines.

iii) It is possible to obtain informations on the coherence length of the electrons by the maximum number of observable transitions. For the case of the 30/30 sample shown on Fig.10, at 10T, we observe 7 light hole transitions which means that the corresponding orbit radius for the electrons is 273Å. If we make the same computation at 5 T we find 260Å. At higher fields, we are limited by the miniband width. We then conclude that in this sample the electrons keeps its coherence over at least 550Å.

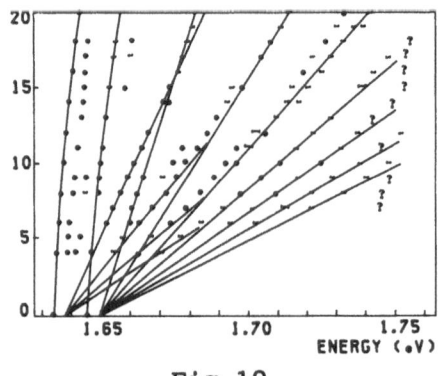

Fig.10

Fan chart of the 30/30 sample in Voigt configuration. A clear diamagnetic shift is observed for both excitonic transitions. Two series of Landau levels are observed. Note the change of slope on each transition around 1.68 eV. This energy is interpreted as corresponding to the top of the electron miniband.

The main difference between our results and the results previously published resides in the fact that we are able to differentiate between heavy-hole and light-hole associated transitions. A clear change of slope of the Landau transitions is observed on Fig.10 around the energy of 1.68 eV. We attribute this change of slope to the change from heavy-hole to light-hole character as the Landau levels reach the top of the miniband. This allows then to determine approximately both the electron and the light-hole miniband widths. If we compare our results to calculations for the case of the sample shown on Fig.10 (30/30 with x=.25), we find the following values:

30/30	Theory	experiment
electrons + heavy holes	76 meV	55±10
electrons + light-holes	124 meV	114±5

Although a complete comparison between the experimental spectra and calculations has not been performed yet[55], the indications given by magnetic-PLE in Voigt configuration do confirm that the shoulders observed in the PLE of SLs occur at energies close to the top of the miniband. The complete comparison with theory obviously needs the inclusion of band mixing effects in the valence band: as a matter of fact we observe that the slope of the LH related transitions is smaller than that of the HH ones, opposite to what is expected without mixing.

CONCLUSION

In summary, we have observed experimentally, studied in magnetic field and fitted theoretically the new structures that appear in the photoluminescence excitation spectra of superlattices with miniband width ranging from 5 to 30 meV. These structures might be called saddle-point excitons although they need expansion of the wavefunctions over the whole Brillouin minizone.

ACKNOWLEDGMENTS

We would like to thank P. Auvray and M. Baudet for the determination of sample parameters, G. Dupas and H. Krath for their technical assistance in sample growth and magnetic field measurements, A. Fasolino for useful discussions.

REFERENCES

1. R. Knox R., Theory of Excitons, Solid Sate Physics, Ed. F. Seitz, D. Turnbull, Supp. 5 (Academic press, NY 1963)
2. R.J. Elliott, Phys. Rev., 108, 1384 (1957)
3. S. Schmidt Rink, Phys. Rev., B33, 1183 (1986)
4. M.S. Skolnick, J.M. Rorison, K.J. Nash, D.J. Mowbray, P.R. Tapster, S.J. Bass, A.D.Pitt, Phys. Rev. Lett., 58, 2130 (1987)
5. J.C. Phillips, in Solid State Physics, Ed. F. Seitz and D. Turnbull (Academic Press, NY 1966) Vol.18, p55
6. C.B. Duke, B. Segall, Phys. Rev. Lett., 17, 19 (1966)
7. B. Velicky, J. Sak, Phys. Stat. Sol., 16, 147 (1966)
8. E.O. Kane, Phys. Rev.B, 180, 852 (1969)
9. I. Baslev, Solid State Commun., 52, 351 (1984)
10. J.C. Phillips, Phys. Rev. 136, 2949 (1964)
11. D.T.F. Marple, H. Ehrenreich, Phys. Rev. Lett., 8, 87 (1962)
12. G. Baldini, Phys. Rev., 128, 1562 (1962)
13. H. Chu, Y. C. Chang, Phys. Rev. B36, 2946 (1987)
14. J. Kervarrec, M. Baudet, J. Caulet, P. Auvray, J.Y. Emery, A. Regreny, J. Appl. Cryst.
15. B. Deveaud, J.Y. Emery, A. Chomette., B. Lambert, A. Regreny, Appl. Phys. Lett. 45, 1078 (1984)
16. A. Chomette, B. Deveaud, A. Regreny, G. Bastard, Phys. Rev. Lett. 57, 1464 (1986)
17. B. Deveaud, A. Regreny, J.Y. Emery, A. Chomette, J. Appl. Phys., 59, 1633 (1986)
18. A. Chomette, B. Lambert, B. Deveaud, F. Clérot, A. Regreny, G. Bastard, Europhys. Lett., 4, 461 (1987)
19. A. Chomette, B. Lambert, B. Clerjaud, F. Clérot, F. Liu, A. Regreny, J. Semicond. Science and Technol., 3, 351 (1988)
20. M. Krahl, J. Christen, D. Bimberg, *Appl. Phys. Lett.*,52, 798 (1988)

21. K. Ploog in *Physics and applications of quantum wells and superlattices*, Ed. E.E. Mendez, K. von Klitzing, NATO series Vol. 170, Plenum, NY, 1987, p43

22. A. Chomette, B. Deveaud, B. Lambert, F. Clérot, A. Regreny, to be published

23. K. Moore, G. Duggan, P. Dawson, C.T. Foxon, to be published in *Superlatt. Microstruc. Microdev.*

24. B. Deveaud, A. Regreny, M. Baudet, A. Chomette, J.C. Maan, R. Romestain, Proc. 18th Int. Conf. on the Physics of Semiconductors, Ed. Engstrom, World Scientific Pub., Singapore, 1987, p. 695

25. J.J. Song, P.S. Jung, Y.S. Yoon, H. Chu, Y.C. Chang, C.W. Tu, *Phys. Rev.*, **B39**, 5562 (1989)

26. B. Deveaud, A. Chomette, F. Clérot, A. Regreny, J.C. Maan, R. Romestain, G. Bastard, H. Chu, Y.C. Chang, (to be published)

27. R.C. Miller, D.A. Kleinman, W.T. Tsang, A.C. Gossard, *Phys. Rev.* B24, 1134 (1981)

28. Y.S. Yoon, P.S. Jung, J.J. Song, J.N. Schulman, C.W. Tu, H. Morkoc, Bull. Am. Phys. Soc., 33, 365 (1988)

29. We observed equivalent structures in all samples of equivalent period that we have studied: see for example Fig. in B. Deveaud, J. Shah, T.C. Damen, B. Lambert, A. Chomette, A. Regreny, IEEE, QE26, (1988)

30. F. Clérot, B. Deveaud, A. Chomette, B. Lambert, A. Regreny A., to be published

31. We only allow for a small rigid shift in order to superimpose the HH excitonic peaks.

32. A. Fasolino, M. Altarelli, in *Two-Dimensional Systems, Heterostructures and Superlattices, Ed. G. Bauer, F. Kuchar, H. Heinrich* (Springer Verlag, NY 1984)

33. D.M. Woodruff, S.M. Anlage, D.L. Smith, *Phys. Rev.*, **B36**, 1725 (1987)

34. H. Chu, Y.C. Chang, to be published.

35. A. Chomette, B. Lambert, B. Deveaud, F. Clérot, A. Regreny, G. Bastard, *Europhys. Lett.*, 4, 461 (1987)

36. J.C. Maan, A. Fasolino, G. Belle, M. Altarelli, K. Ploog, *Physica* **127B**, 426 (1984)

37. N. Miura, Y. Iwasa, S. Tarucha, H. Okamoto, Proc. 17th Int. Conf. Phys. Semicond., (1985)

38. P. Dawson, K.J. Moore, G. Duggan, H.I. Ralph, C.T.B. Foxon, *Phys. Rev.*, **B34**, 6007 (1986)

39. D.C Rogers, J. Singleton, R.J. Nicholas, C.T.B. Foxon, K. Woodbridge, *Phys. Rev.*, **B34**, 4002 (1986)

40. O. Akimoto, H. Hasegawa, *J. Phys. Soc. Japan*, **22**, 181 (1967)

41. F. Ancilotto, A. Fasolino, J.C. Maan, *Phys. Rev.*, **B38**, 1788 (1988)

42. S.R. Eric Yang, L.J. Sham, *Phys. Rev. Lett.*, 58, 2598 (1987)

43. G. Duggan, *Phys. Rev.*, **B37**, 2759 (1988)

44. J.C. Maan, G. Belle, A. Fasolino, M. Altarelli, K. Ploog, *Phys. Rev.*, **B30**, 2253 (1984)

45. A.S. Plaut, J. Singleton, R.J. Nicholas, R.T. Harley, S.R. Andrews, C.T.B. Foxon, *Phys. Rev.*, **B38**, 1323 (1988)

46. Let us note here that our results are in apparent contradiction with the value very recently obtained by two photon absorption: K. Tai, A. Mysyrowicz, R.J. Fisher, R.E. Slusher, A.Y. Cho, *Phys. Rev. Lett.*, **62**, 1784 (1989). We have no explanation yet for this discrepancy.

47. J.C. Maan in *Physics and applications of quantum wells and superlattices*, Ed. E.E. Mendez, K. von Klitzing, NATO series Vol. 170, Plenum, NY, 1987, p347.

48. Except for the very short period SLs where only one peak is resolved: as in bulk GaAs or GaAlAs there is no HH-LH splitting.

49. A new series is only observed clearly for HH transitions as the LH transitions are washed out in the high energy part of the spectrum.

50. The lines in the figure are not the result of a fit and we only consider them as an indication of the origin of the different transitions. As a consequence, we do not consider the extrapolation point energy to be very precise but rather to give an approximate energy range. However, we have taken care, in the drawing of the lines to change their slope by one cyclotron energy.

51. J.C. Maan, *Festkorperprobleme* **27**, Ed. P. Grosse, p137 (1987)

52. G. Belle, J.C. Maan, G. Weimann, G. Belle, J.C. Maan, G. Weimann, *Solid State Commun.*, **56**, 65 (1985) and *Surf. Sci.*, **170**, 611 (1986)

53. A. Fasolino, M. Altarelli, in *Band structure engineering in semiconductor microstructures*, Eds. R.A. Abram, M. Jaros, Plenum Press, in print

54. T. Duffield, R. Bhat, M. Koza, F. DeRosa, D.M. Hwang, P. Grabbe, S.J. Allen Jr, *Phys. Rev. Lett.*, **56**, 2724 (1986)

55. Calculations using the model proposed by Fasolino and Altarelli[36] are being performed.

MAGNETOTUNNELING IN SEMICONDUCTOR MICROSTRUCTURES

C.Tejedor, L.Brey, G.Platero and P.A.Schulz

Departamento de Fisica de la Materia Condensada
Universidad Autonoma. Cantoblanco, 28049 Madrid. Spain

1. INTRODUCTION

Tunneling is intrinsically a quantum mechanical effect which can be analyzed both by transport and spectroscopic experiments in potential barriers built up by means of semi-conductor microstructures. The main effort has been devoted to the study of tunneling between two media separated by a region where localized states can exist. In this case, there are two aspects of extraordinary interest: i) the appearance of negative differential resistance (NDR) and ii) the existence of two possible mechanisms, coherent and sequential, as responsible for resonant tunneling. The study of those problems has become more accessible with the development of growth techniques of semiconductor microstructures allowing the design of the required potential profiles. As in the case of any other spectroscopy, much insight on the phenomenon can be attained by modulation with some external agents. In particular, since tunneling is deeply connected with spatial localization or delocalization of quantum states, a magnetic field can be very useful as a modulator of the spatial shape of the wavefunctions.

Here, we will analyze the effect of a magnetic field on the current density j between two doped semiconductors separated by different potential profiles with an applied bias V_G, i.e. with the two semiconductors having uncommon chemical potentials. The best configuration to modulate the current is the one having B perpendicular to j (i.e. B parallel to the interfaces). Therefore, we will concentrate on such a case although some results in other configurations will be presented too. We start in section 2 by giving a brief sketch of the theoretical method used to analyze these kind of problems. Special emphasis will be put on the fact that when B is perpendicular to j, tunneling takes place only through a discrete set of channels. In section 3 the problem of a single barrier with B ⊥ j will be theoretically studied after a short revision of the experimental situation. Two phenomena will require our attention, the oscillatory dependence of j with B for a given V_G and the possible appearance of NDR in a non resonant system. Section 4 is devoted to the double barrier with B ⊥ j. As in the previous case we will start by presenting the available experimental information to follow with the analysis of the coherent and sequential contributions to the current. In section 5 we study the double barrier with B ∥ j. A general discussion and conclusions is contained in section 6.

2. THEORETICAL METHOD

2.1 Stationary spectrum

In order to study the motion of wavepackets, the first requirement is the knowledge of the eigenvalues and eigenstates of the system in presence of a magnetic field. Since we are dealing with doped semiconductors, the energy region of interest is small enough to make use of an effective mass approximation. We will consider the case in which the potential (barriers and bias with $B = 0$) is only a function of z, $(V(z))$ while in the plane x,y parallel to the interfaces carriers with a given effective mass move freely. As far as the magnetic field is concerned, we will treat only two limit cases: i) B parallel to the current density, i.e. $\vec{B} = B\vec{u}_z$ and ii) B perpendicular to the current density, i.e. $\vec{B} = B\vec{u}_x$. In the first case the eigenstates are in the xy plane Landau states completely independent of the wave function in the z direction. Due to this factorization, the tunneling becomes a one dimensional problem along z and the effect of the magnetic field reduces to impose conservation of the index of the Landau level instead of the conservation of $k_{x,y}$ occurring when $B = 0$. More complicated and interesting is the second case of $B\vec{u}_x$. By taking the gauge $\vec{A} = (0, - Bz, 0)$ the eigenstates have the form $e^{ik_x x + ik_y y}\phi$ (z) and, after some algebra, the Schrödinger equation becomes[1]:

$$[- \frac{\hbar^2}{2m^x m_0} \frac{\partial^2}{\partial z^2} + \frac{e^2 B^2}{2m^x m_0} (z + z_0)^2 + V(z) - E_n(k_y)]\phi(z) = 0 \qquad [1]$$

where m^x and m_0 are the effective and free electron mass respectively, z_0 is the magnetic orbit center

$$z_0 = \frac{\hbar k_y}{eB} = l_m^2 k_y \qquad [2]$$

and the energy of the magnetic level $E_n(k_y)$ is related to the total energy of the eigenstate by means of:

$$E = \frac{\hbar^2 k_x^2}{2m^x m_0} + E_n(k_y) \qquad [3]$$

For a given value of k_y (or z_0) the one-dimensional equation [1] can be numerically solved by a finite elements method. Figure 1 shows the band structures of single and double barriers with an applied bias and in the presence of a magnetic field. The main result is that states far from anticrossings have wave functions localized in only one side of the barriers, while just states close to the anticrossings have wave functions with significant weigth in more than one region. Therefore, the latter are the channels allowing the carrier flow between different parts of the system. This is the origin of the tunneling current theoretically analyzed in the next subsection as a phenomenon out of the equilibrium.

2.2 Generalized transfer Hamiltonian method for magnetotunneling

The Generalized transfer Hamiltonian (GTHM) method[2] is an extension of the usual transfer Hamiltonian formalism[3] for including the possibility of resonant processes. The total Hamiltonian H is separated in left and right terms by means of two auxiliary Hamiltonians $H_L \equiv H$ in the left side and $H_R \equiv H$ as shown in figure 2. To recover H, the required perturbations written as:

$$H \equiv H_L + V_L e^{\eta t} = H_R + V_R e^{\eta t} \qquad [4]$$

are switched on adiabatically ($\eta \to 0$). One starts with a time dependent wave function for the total system

$$|\Psi(t)> = f(t)e^{-i\omega_L t}|L> + \sum_R a_R(t)e^{-i\omega_R t}|R> \qquad [5]$$

where $\omega_L = E_L/\hbar$ and $\omega_R = E_R/\hbar$. Tunneling is related to very specific boundary conditions. At the initial time, $|\Psi(-\infty)>$ must describe a particle on the left side. This is fulfilled by imposing

$$f(-\infty) = 1; a_R(-\infty) = 0 \qquad [6]$$

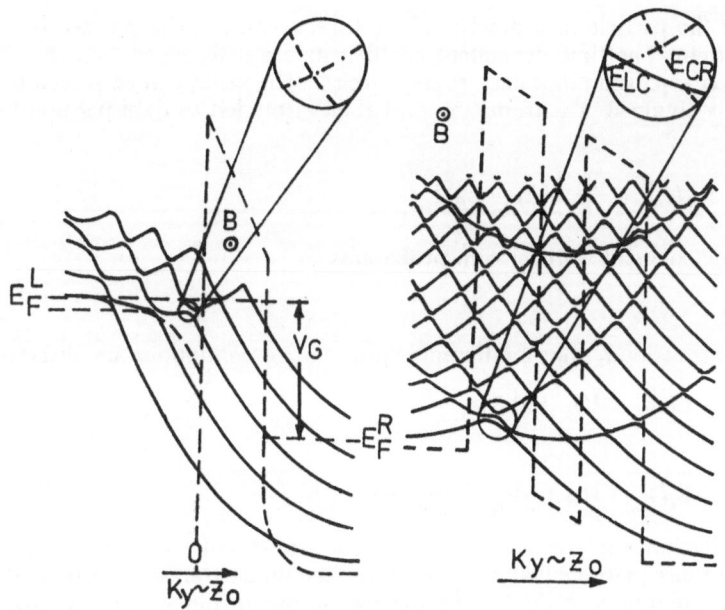

Fig. 1 Dispersion relation of the magnetic levels of simple and double barriers with B parallel to the interfaces.

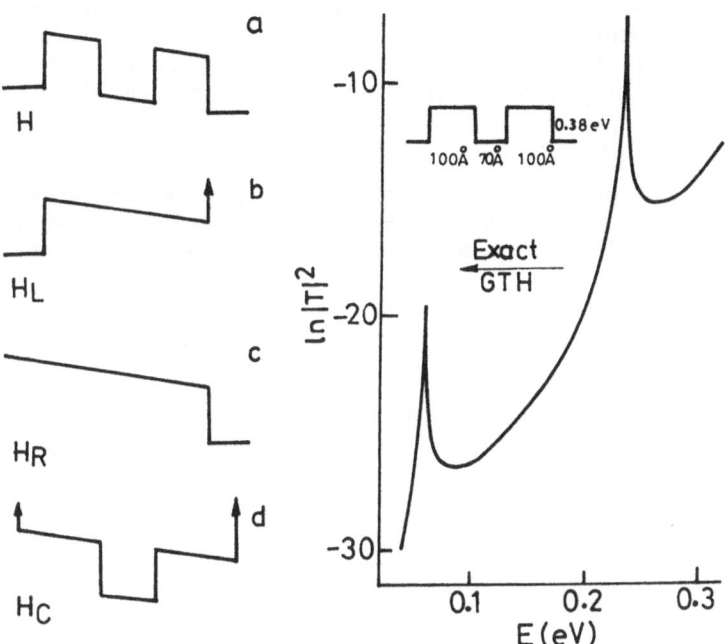

Fig. 2 Total(H), left(H_L), right(H_R) and center(H_C) Hamiltonians used in GTH method. To the right, we show the transmission probability obtained with GTH for a double barrier without bias.

After a while the particle in a precise left state $|L>$ can in principle evolve to any state $|R>$ to the right. The time dependent coefficients are determined from the Schrödinger equation by an expansion in a perturbation series. The series can be solved to any order[2] so that $a_R(t)$ is obtained. The transition probability from left to right per unit time is given by:

$$P_{LR} = \lim_{\eta \to 0} \frac{d}{dt} |a_R(t)|^2 = \frac{2\pi}{\hbar} \delta(E_L - E_R)T_C \qquad [7]$$

where T_C is the coherent transmission probability

$$T_C = |t_{LR}|^2 \qquad [8]$$

where the transmission t_{LR} has a different expresions for continuous and discrete problems:

$$t_{LR}^c = <L|V_L + V_L(E_L - H)^{-1}V_R|R> \qquad [9a]$$

and

$$t_{LR}^d = <L|V_L + V_L(E_L - H + \mathbb{P}_R V_R)^{-1}\mathbb{P}V_R|R> \qquad [9b]$$

$\mathbb{P}_R$ is the projection operator on the state $|R>$ and $\mathbb{P} = 1 - \mathbb{P}_R$. The first term in the matrix element of equations [9a] and [9b] gives the direct transition usually described by the transfer Hamiltonian method while the second one includes all the virtual transitions producing the resonance. The GTH method has proved to give good results for problems where the exact solution is known[2] as is shown in figure 2. This gives us confidence for using GTH in more difficult problems as the one of magnetotunneling. In our case, equation [9a] is the one required for B $\|$ j while [9b] is the one required for B $\perp$ j.

Let us now discuss how to compute the current density j from the transmission. We will concentrate on the case of B $\perp$ j because the existence of discrete tunneling channels reflects in interesting characteristics of j. Then, t_{LR} is computed from the wave functions obtained as described in section 2.1 from equations similar to [1] with V(z) corresponding to H_L and H_R respectively. The δ function in equation [7] implies that the current is produced only when $E_L = E_R$, i.e. at the crossings of the left and right dispersion relations which, as expected, corresponds to the discrete set of anticrossings in the spectrum of H. On top of that, all the potentials are only a function of z so that both k_x and k_y are good quantum numbers which must be conserved in the tunneling process. Then

$$t_{LR} \propto \delta(k_y^L - k_y^R)\delta(k_x^L - k_x^R) \qquad [10]$$

which means that only tunneling between very particular edge states at the two sides of the barriers is possible. Once t_{LR} is known, the current is computed by summing up all the transition probabilities between occupied states to the left and empty states to the right. For zero temperature, we have

$$j = \frac{2e}{h} \sum_{n,m} \int dk_x dk_y |t_{LR}(E,k_y)|^2 [\theta(E_L^F - E_{Ln}(k_y)) - \theta(E_R^F - E_{Rm}(k_y))]\delta(E_{Ln}(k_y) - E_{Rm}(k_y)) \qquad [11]$$

where E_L^F and E_R^F are the Fermi levels of the left and right contacts respectively and n, m are indices running over magnetic levels of H_L and H_R respectively. In equation [11] the integral in k_x is straightforward because for this variable the dispersion relations are simply parabolas. The integral in k_y is performed by using

$$\delta(E_{Ln}(k_y) - E_{Rm}(k_y)) = \frac{1}{\left| \dfrac{d(E_{Ln}(k_y) - E_{Rm}(k_y))}{dk_y} \right|_{k_y^c}} \delta(k_y - k_y^c) \qquad [12]$$

where k_y^c stands for the crossings n,m i.e. the value in which $E_{Ln}(k_y) = E_{Rm}(k_y)$. After some algebra, the current density is given by

$$j = \frac{2e\sqrt{2m_0 m^\times}}{\pi \hbar^2} \sum_c \frac{|t_{LR}(k_y^c)|^2}{|\frac{d(E_{Ln}(k_y) - E_{Rm}(k_y))}{dk_y}|_{k_y^c}} \times \qquad [13]$$

$$[\sqrt{E_L^F - E_c}\,(\theta(E_L^F - E_{Ln}(k_y)) - \theta(E_R^F - E_{Rm}(k_y))) + (\sqrt{E_L^F - E_c} - \sqrt{E_R^F - E_c}\,)(\theta(E_c)\theta(E_R^F - E_c))]$$

where the sum runs over all the crossings k_y^c with energy E_c above discussed. In equation [13] the origin of energies is in the bottom of the potential to the left and $t_{LR}(k_y^c)$ is calculated as discussed above. The derivative in equation [12] implies a contribution of the density of states in the y direction to the current along the z direction and is also evaluated from the dispersion relation of H_L and H_R. It must be pointed out that the current through the barriers in the z direction is essentially controlled by the k_y dependence because the magnetic field associates these two magnitudes as shown in equation [2].

3. MAGNETOTUNNELING THROUGH A SINGLE BARRIER

We concentrate here on the tunneling through a single barrier separating two GaAs semi infinite crystals with B parallel to the interfaces. Later, we will finish this section discussing briefly tunneling between two GaAs-GaAlAs superlattices (SL) where the energy minigaps introduce interesting consequences in the I-V characteristics as the appearance of NDR. It must be pointed out that, in the single barrier case no resonance due to virtual transitions is possible, so that the use of just the first term in the matrix element of equations [9] is a good approximation[1].

3.1 Semiconductor-barrier-semiconductor

Let us start with the case of GaAs-barrier-GaAs where a great amount of experimental information exists. The current, either as a function of B or as a function of V_G, presents quantum oscillations. Two typical examples are shown in figure 3[4,5]. In the case of B constant, these oscillations must be produced by the tunneling itself because nothing special happens in the bulk. However, in the case of fixed V_G, the variation of B produces variations of the bulk Fermi level that, as in the case of the Shubnikov-de Haas effect, manifest in current oscillations periodic in $1/B$. Nevertheless, in the tunneling experiment two periodicities are observed[5] being our intention to prove that one of the periods is a tunneling effect while the other is produced by the Fermi level variation. In other words, both when B varies or when V_G does, tunneling current presents oscillations which are typically mechanoquantical because no semiclassical approach is able to describe them. The physical origin of the oscillations is rather simple. When one of the fields (magnetic or bias) varies, the whole dispersion relation of H (and consequently of H_L and H_R) changes so that the tunneling channels, discussed in the previous section, move with respect to the left and right Fermi levels. When one of these channels enters in the energy region corresponding to occupied states to the left and empty to the right, a new contribution to the current is opened so that j varies abruptly. In this way, oscillations originated by tunneling are due to the opening of new channels. There is not such an intense effect when the channel moves below the bottom of the band of the left crystal because then the effective barrier seen by the carrier is higher and the transmission is rather small[6].

Once we have given the main physical idea behind all the features experimentally observed, we will discuss separately each case and its implications as well as some predictions about interesting but presently unobserved phenomena. Let us start by studying the case of fixed bias and variable B. In order to simplify the analysis we choose a system with a non degenerate semiconductor to the left. Then, only one magnetic level to the left is occupied and the different tunneling channels are due to anticrossings of that level with the possible ones coming from the right side, as shown in figure 1. In figure 4 we show the position of the tunneling channels as a function of $1/B$ for a system with a barrier of 230 Å of $Ga_{0.63}Al_{0.37}As$ between a nondegenerate substrate of GaAs ($N_S = 1.7 \times 10^{15} cm^{-3}$) and a degenerate gate of GaAs ($N_G = 9 \times 10^{17} cm^{-3}$) with a bias $V_G = 0.4V$. The dashed line shows the field B in which a crossing passes through the left Fermi level so that such channel

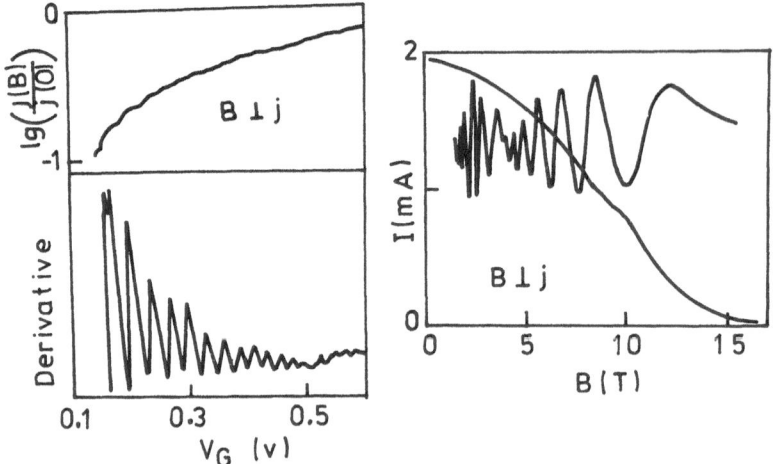

Fig.3 Current measured in a single barrier with B⊥j. To the
left, j is shown as a function of V_G for a given B(ref.
4) and to the right j is afunction of B for a given
V_G (ref. 5).

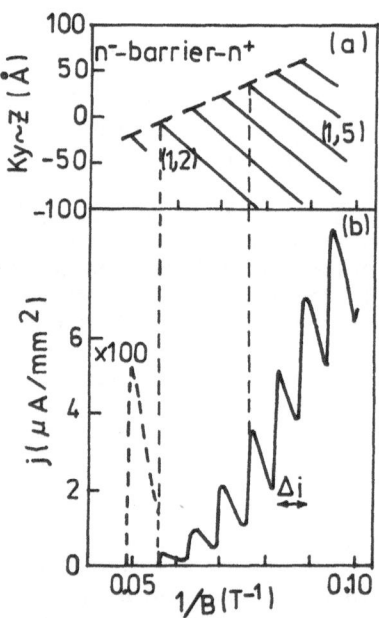

Fig.4 (a) Position of the
tunneling channels in a single
barrier(B⊥j). (b) Current density as a function of 1/B when
the left semiconductor is non-
degenerated.

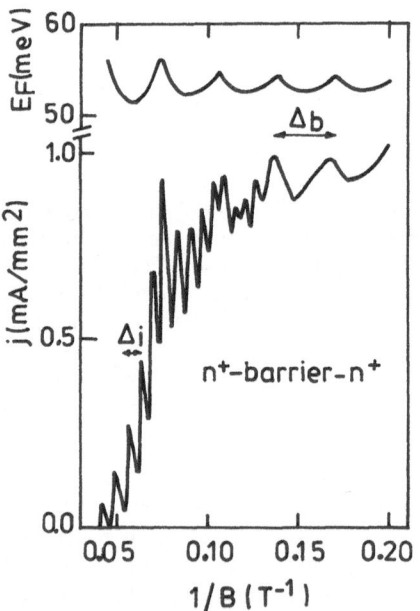

Fig.5 Current density through
a single barrier (B⊥j) separating two degenerated semiconductors as a function of
1/B.

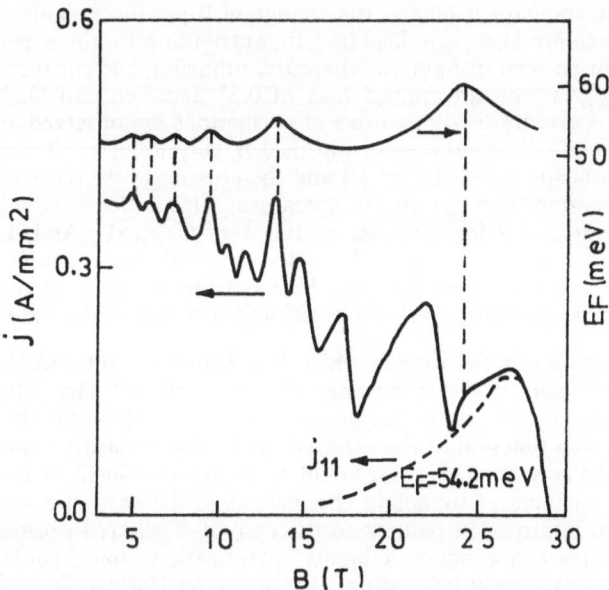

Fig.6 Fermi level oscillations and current density through a
single barrier with B⊥j. Dashed line shows the contribution of just one tunneling channel.

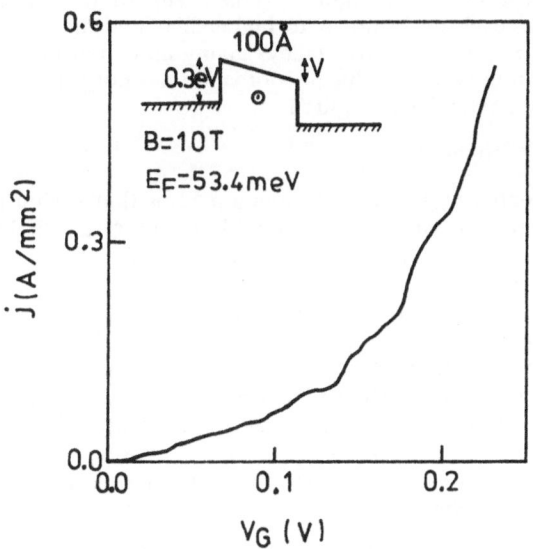

Fig.7 Current density through a single barrier with fixed B⊥j
as a function of the applied bias.

starts to carry current. Figure 4 shows that the cut off of successive channels is linear in $1/B$ so that the current presents oscillations in $1/B$ with a period Δ_i as is also shown in figure 4. This clearly proves the interfacial origin of the mechanism producing quantum oscillations in j. However there is another possible source of oscillations. When the density of the left semiconductor is higher, the change of B produces significant variations in the Fermi level which are also periodical in $1/B$ contributing to a new period Δ_b of j($1/B$). This bulk effect can be seen in figure 5 where the tunneling current through a barrier of 200 Å of $Ga_{0.68}Al_{0.32}As$ with an applied bias of 0.3V between two GaAs media with a doping of $10^{18}cm^{-3}$ is given. Clearly two sets of oscillations are observed as it has been experimentally measured[5]. When the magnetic field is very intense, all the magnetic levels suffer a rather high bending (see figure 1) and the crossings take place at energies higher than E_L^F so that no current flows at zero temperature. This is clearly seen in figure 6 where j is given as a function of B for a barrier of 100 Å of $Ga_{0.33}Al_{0.67}As$ between two GaAs crystals with a doping of $n = 10^{18}cm^{-3}$. In order to clarify the behaviour of the current density, we show (in dashed line) the contribution coming just from the channel corresponding to the lowest magnetic level of the left and that one of the right.

Even more interesting is the case in which B is kept constant and V_G varies. The reported experiments[4,5] indicate the appearance of steps in the I-V characteristics which are better detected in the derivatives. The origin is the same as before, i.e. the opening of new tunneling channels as a function of the external field. This is clearly shown in figure 7 for the same sample of figure 6. There, the oscillations are as evident as in the experiments and are due to the opening of tunneling channels as in the case of a varying B. The reported experimental information reduces to the case of $B \leq 13T$. In principle one can expect that higher fields can produce a higher quantization which could manifest more clearly. The result is extremely interesting. From a semiclassical viewpoint, B bends the electron trajectories so that a threshold is required to open the first channel. This manifest in a step-like increase of the current. The successive increase of the bias lowers the energy of the channel more rapidly than the lowering of the barrier. Therefore, it reflects in a higher effective barrier seen by the carrier. This produces a lowering of the transmission t_{LR} and consequently of the current density as shown in figure 8 for the same sample of figure 4 for several values of B. Such lowering produces the appearance of NDR with no resonant origin. When the bias is high enough, a second tunneling channel is opened and the process restarts as shown in the figure 8. This is superimposed to the increase of the current with V_G. Then, each peak moves to higher bias and increases exponentially with B. The same behaviour appears for the valleys giving an exponential increase of the peak to valley ratio as shown in figure 8. This phenomenon has not yet been observed, probably because of the high magnetic fields required.

3.2 Superlattice-barrier-superlattice

The novelty represented by the fact of having a SL is that such a perfect crystal has a periodicity in the z direction which is comparable to the magnetic length l_m. There are minigaps in the energy region of interest, meaning that the states of the system can be interpreted as coherent resonances. Then, the quantum effect of NDR is possible[7]. The effective mass approximation fails to be valid and the SL potential $V_{SL}(z)$ must be explicitly included in the Schrödinger equation [1]. Apart from this, the whole theoretical method above described can be directly applied. The existence of minibands and minigaps is well visualized in the density of states shown in figure 9 obtained from the perfect SL band structure by integration in k_x and k_y. Minibands and minigaps still exist although they are affected by the magnetic field. As in the case of a common semiconductor, B bends the dispersion relations for orbit centers z_0 close to the barrier interface. Tunneling gives information about such a bending.

Davies et al[7] have reported the I-V characteristics in SL-barrier-SL systems under a transverse magnetic field shown in figure 10. There it is shown that when B increases NDR regions move to higher biases and currents. The peak to valley ratio initially increases but later it decreases and eventually disappears. We show in figure 11 our results for the sample DB106 experimentally reported[7]. The lowest peak is produced by crossings

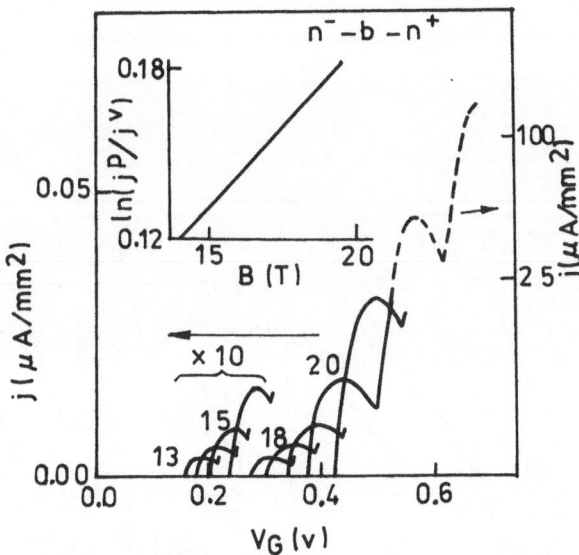

Fig. 8 Current density through a single barrier as a function
of the applied bias. NDR is shown for several magnetic
fields (in T). Inset shows the dependence on B of the
peak to valley ratio.

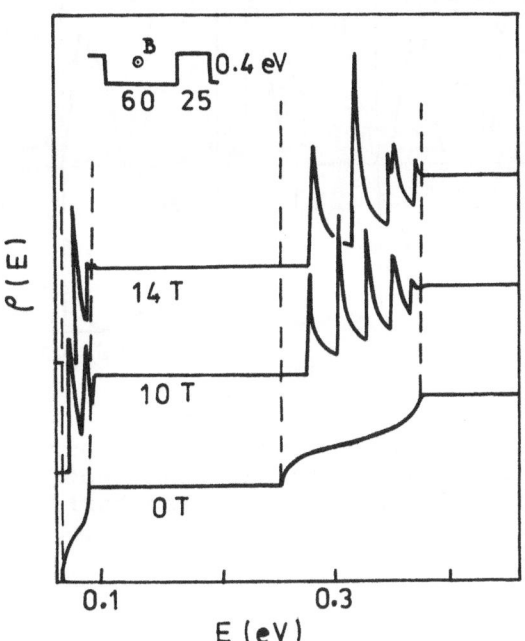

Fig. 9 Density of states of a SL for different values of the
magnetic field applied parallel to the barriers.

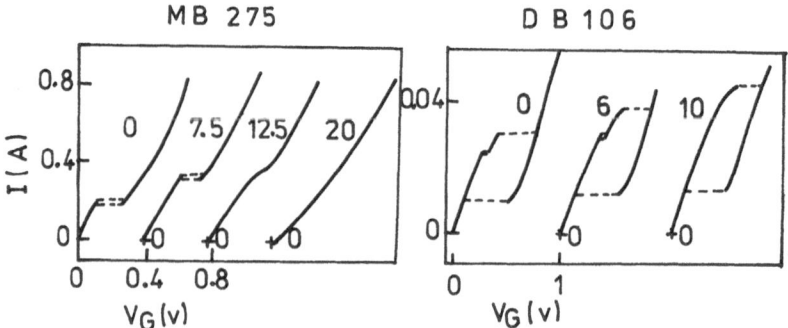

Fig. 10 Experimental current as a function of the bias applied
to a barrier separating two SL with different magnetic
fields parallel to the barriers (ref. 7).

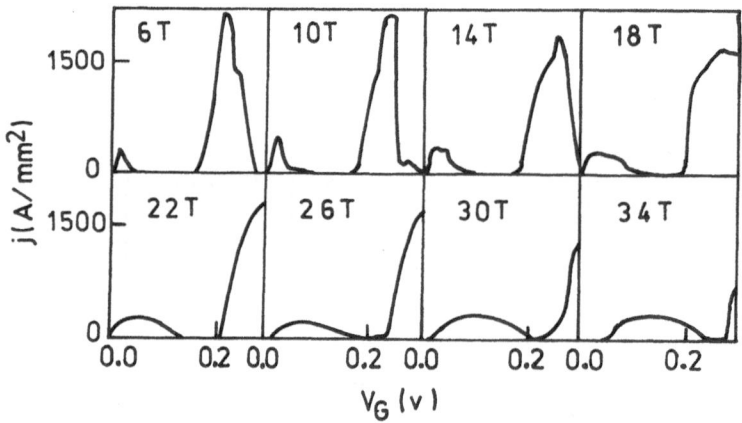

Fig. 11 Theoretical current density as a function of the bias
applied to a barrier separating two SL with different
magnetic fields parallel to the barriers.

between occupied magnetic levels coming from the first minibad of the left and empty levels from the first miniband of the right. When V_G increases these two energy regions are misaligned and no channels are available for tunneling giving no current. For higher bias, current is reestablished when the first miniband of the left becomes aligned with the second of the right. The agreement with the experiments is qualitatively satisfactory if one takes into account that these elastic features must be superimposed to an inelastic background. The NDR precludes the possibility of obtaining useful information from the representation of j as a function of B for fixed V_G[.]

4. TRANSVERSE MAGNETOTUNNELING TRHOUGH A DOUBLE BARRIER

The problem of a double barrier is significantly different from the one of the simple barrier discussed above because of the possibility of resonance through states localized in the well. From the theoretical point of view this reflects in the necessity of using the whole expression [9] in the calculation of transmission amplitudes. However there is a rather more important consequence. On top of the coherent tunneling hitherto discussed, now a sequential mechanism is possible in which a carrier elastically traverses the first barrier, then it spends some time in the well where it can suffer some inelastic scattering and finally it tunnels elastically through the second barrier. Since the two types of mechanisms are depending on the spatial localization, a magnetic field parallel to the barriers is an excellent tool to modulate the relative importance of the two mechanisms. However, the presently available experimental information throw little light on the problem[8,11]. The general trend is that the resonance peak occurring in j when $B = 0$ shifts to higher energies, decreases and is quenched when B increases as shown in figure 12. No particular features appear so that it is not straightforward to draw information on the possible tunneling mechanisms. That is the main problem we want to discuss in this section.

A double barrier where both coherent and sequential processes are possible, behaves like two channels in parallel. The resistance of each channel is proportional to the inverse of the corresponding transmission probabilities T_C or T_S respectively[12]. This gives for the total transmission probability

$$T = T_C + T_S \tag{14}$$

where all the magnitudes must be taken at the same energy which has some implications discussed below[13].

Let us start by seeing if just the coherent tunneling is able to explain the experimental information as it was for the single barrier. We can directly apply the procedure above discussed (equations [8], [9b] and [13]) to compute the coherent contribution j_c to the current density. We have checked that a good approximation is:

$$(E_L - H + \mathbb{P}_R V_R)^{-1} \simeq G_C(E_L, \Sigma(V_P)) \tag{15}$$

where G_C is the Green's function of the center Hamiltonian shown in figure 2 and the selfenergy is given by

$$\Sigma = V_P + V_P G_C V_P \tag{16}$$

with

$$V_P = H - H_C - \mathbb{P}_R V_R. \tag{17}$$

The virtual processes involved in the coherent tunneling are represented by energy differences $(E_L - E_C)$ appearing in the denominators of the Green's function. This denominators give very important contributions to the transmission so that when a state localized in the well with energy E_C is close to E_L a peak appears in the current. From figure 1 it is very simple to visualize that such fact is going to occur several times when varying the bias. This gives a rather complicated structure of j as a function of the bias. Figure 13(a) shows the coherent tunneling current through a double barrier $n^+GaAs - 100\text{Å}\,Ga_{0.6}Al_{0.4}As - 70\text{Å}\,GaAs - 100\text{Å}\,Ga_{0.6}Al_{0.4}As - n^+GaAs$ There the structure

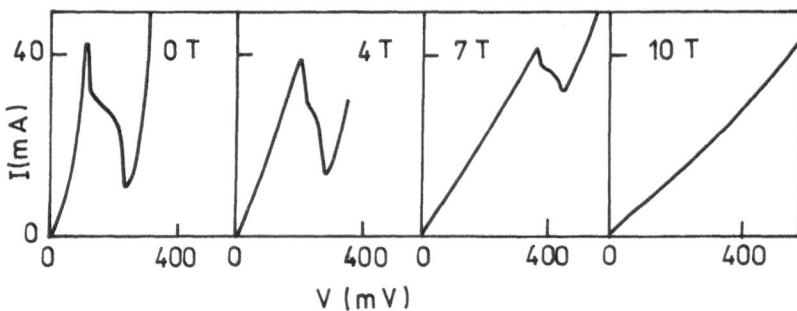

Fig.12 Experimental current through a double barrier with different magnetic fields applied parallel to the barriers
(ref. 8).

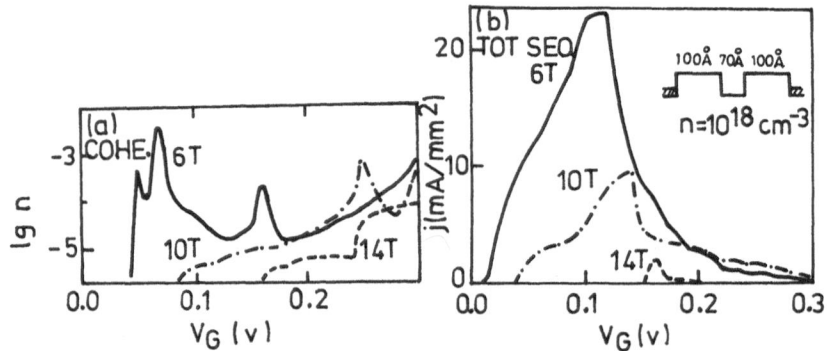

Fig.13 Coherent and sequential current densities through a
double barrier computed as a function of V_G (B ∥ j).

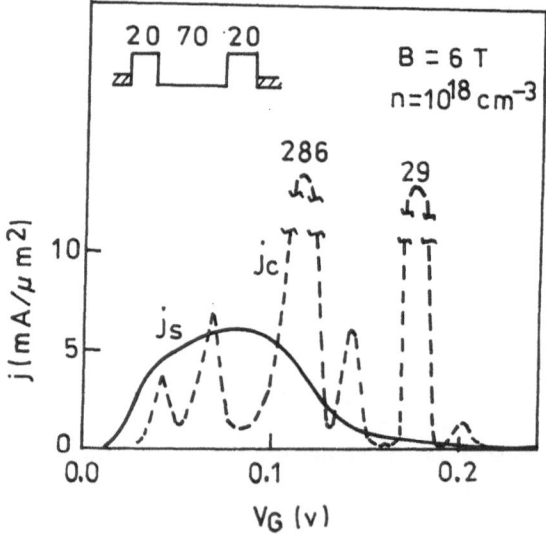

Fig.14 Coherent and sequential current densities through a
double barrier computed as a function of V_G. (B ∥ j).

above mentioned is clearly observed which is in clear disagreement with the experiments. This is an indication that sequential tunneling should be responsible for the experimental behaviour.

Sequential tunneling is a three step processes. First, the carrier traverses the left barrier, second it spends some time in the well suffering scattering processes that make it lose memory of its previous state and finally it tunnels through the second barrier. The first and third steps can be simply analyzed as it was done for the single barrier in sections 2 and 3. The argument of tunneling channels can be applied to these two steps so that it is quite clear that they occur at different values of k_y and energy. Therefore, some inelastic process in the well is necessary to connect the first and third steps. A microscopic evaluation of this memory loss should require some model on the mechanism producing it. This is not very simple to make so that we adopt the macroscopic point of view[12] which considers the well in contact with a reservoir where the memory loss takes place. The connection between first and third steps should have a small probability when the change of quantum numbers is large. This is similar to the impurity induced transitions between edge states at the two surfaces of a narrow channel with a transverse magnetic field where an exponential dependence on the change of k_y has been proposed[14,15]. Then we take for the transmission probability of the three step sequential process :

$$T_S^{-1} = exp[\alpha(l_m \Delta k_y)^2](T_{LC}^{-1} + T_{CR}^{-1}) \qquad [18]$$

where T_{LC} and T_{CR} are the transmission probabilities through the left and right barriers respectively and Δk_y is the variation of k_y between the first and third processes. α is of the order of unity[14,15], the exact value being not very crucial because in our case there are several channels with $\Delta k_y << 1/l_m$ so that the exponential factor is practically 1 for any reasonable value of α. It must be stressed that in equation [18] T_{LC} and T_{CR} are transmission probabilities taken at different energies (E_{LC} and E_{CR} in figure 1). This means an approximation for T_S, which is good when $\Delta k_y << 1/l_m$, becoming worse in converse limit case. This is not a problem because the latter limit has very small weight due to the exponential factor. So, the only processes playing an important role in sequential magnetotunneling are those corresponding to center states very close to a crossing of $E_{Ln}(k_y)$ and $E_{Rm}(k_y)$ i.e. to coherent channels. Moreover, only crossings close enough to the well give significant values of T_{LC} and T_{CR} because now there are no energy difference denominators connected with resonances. This gives a rather featureless current, compared with the coherent case. Figure 13b shows the total density current computed with an expression similar to [13] by changing $|t_{LR}|^2$ by the total T. From the comparison with the part (a) it is easily seen that the main contribution is due to the sequential part of tunneling while the coherent one is negligible in this case. The behaviour of j(V) for different values of the magnetic field is qualitatively that of the experiment so that one can conclude that sequential tunneling is that which is being experimentally observed.

In order to get a situation where coherent tunneling could be comparable with the sequential one, we have performed some calculations for other double barrier systems. Figure 14 shows the current density for $B = 6T$ (perpendicular to j) in a sample $n^+ GaAs - 20Å Ga_{0.6}Al_{0.4}As - 70Å GaAs - 20Å Ga_{0.6}Al_{0.4}As - n^+ GaAs$. Now the barriers are very narrow and the intensity of the virtual processes is high giving a coherent current comparable and even higher than the sequential one. To our knowledge no experimental information is presently available on samples with so narrow barriers and we hope that this type of experiments will confirm our prediction throwing much light on the comparison between coherent and sequential tunneling.

5. LONGITUDINAL MAGNETOTUNNELING IN DOUBLE BARRIERS

As we mentioned in the introduction, the case of B ∥ j is rather different from the one previously discussed. I-V characteristics measured for different magnetic fields in such a configuration[16,17] show the usual resonance peak. Figure 15 shows that, for incresing B, dj/dV_G develops a structure connected with tunneling between Landau states of the substrate and the gate with conservation of the index level. A previous calculation[18] re-

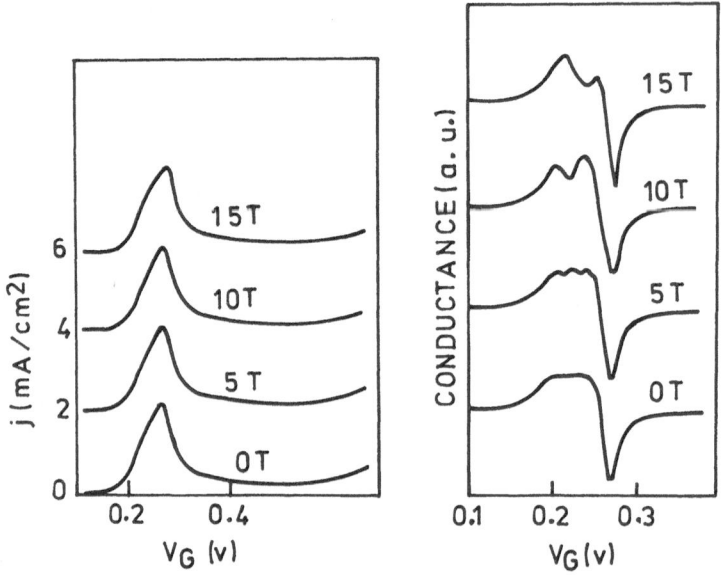

Fig.15 Experimental current and conductance throgh a double
       barrier with different B∥j (ref. 16).

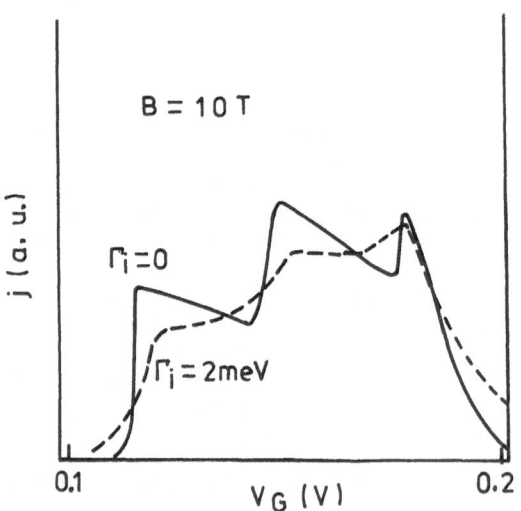

Fig. 16 Current density through a double barrier with B∥j
        computed as a function of V_G.

produces this experimental behaviour only in a qualitative way. Instead of the experimentally found smooth increase of j for low biases and the abrupt cutoff for higher bias, the theory seems to give the opposite behaviour. Such calculation only includes the coherent part of the current so that an appealing possibility is that the experiment could be showing the sequential part once again. In order to clarify this point one can apply the above discussed formalism to compute $j(V_G)$ in this case that, as mentioned in section 2, is only one dimensional[2]. Figure 16(a) shows our results for a double barrier $n^+ GaAs - 50\text{Å} Ga_{0.7}Al_{0.3}As - 40\text{Å} GaAs - 50\text{Å} Ga_{0.7}Al_{0.3}As - n^+ GaAs$ in complete agreement with the previously reported calculations[18]. In order to take into account sequential processes, one can include some inelastic scattering in the well just by introducing an imaginary part Γ_i in the energy of the well states. Since this violates the unitarity of the scattering matrix, the sum of the transmission and reflection coherent probabilities is different from 1. This can be interpreted[19] by seeing that there are sequential (inelastic) transmission and reflection probabilities that restore the unitarity.

$$T_S + R_S = 1 - T_C - R_C. \qquad [19]$$

This allows the calculation of T_S just by doing a guess on the relation between T_S and R_S. Figure 16(b) shows our results by taking[19] :

$$T_S = T_C \frac{\Gamma_i}{\Gamma_e} \qquad [20]$$

where Γ_e is the natural elastic width of the resonance. Now the results of the calculation show a better agreement with the shape of the experimental data. The consequence to be drawn is the importance of the sequential part in the total current density of the resonance.

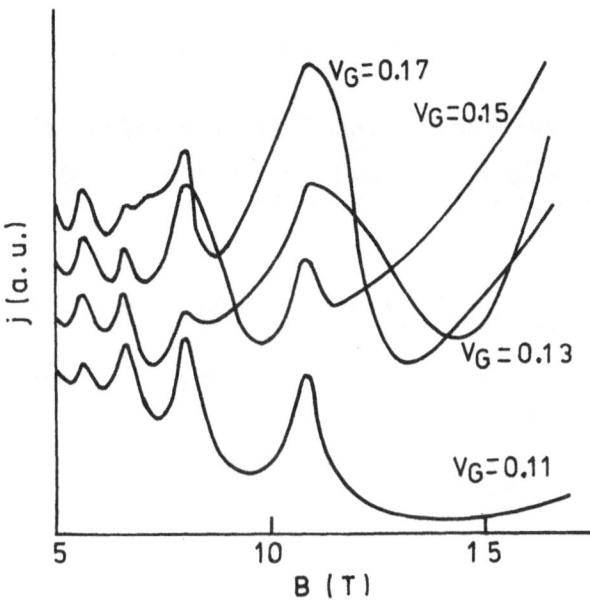

Fig. 17 Current density through a double barrier with B ∥ j computed as a function of B for different values of V_G.

Finally we show in figure 17 the dependence of j with B for different values of V_G. Two different type of oscillations appear which are periodic in 1/B. One of them is directly connected with the bulk Fermi level oscillations as can be easily checked out just by repeating the calculation with a given E_F independent of B. On top of this usual Shubnikov-de Haas bulk effect, the other oscillation is produced by the two dimensional electron gas within the well. The periodicity is not the same than the other one just because the different effective mass of the carriers at the barriers changes the cyclotron frequency of that two dimensional gas.

6. SUMMARY

We have reviewed our work on magnetotunneling in semiconductor microstructures. For the case of a single barrier with $B \perp j$, NDR and quantum oscillations as a function of either the magnetic field or an applied bias have been found as the most relevant results. In a double barrier system we have studied both the case of $B \perp j$ and $B \| j$. In both configurations, our main interest has been the analysis of the relation between the coherent and sequential mechanisms of the resonant tunneling. In all the cases our results present good agreement with the experiments when available. Moreover, some interesting predictions for other cases have been made. Then magnetotunneling shows clearly to be an interesting technique to get insight on the properties of semiconductor microstructures.

Acknowledgements

We are indebted to Drs. E.E.Mendez, J.C.Maan, F.Flores, C.E.T.Goncalves da Silva, C.Rossel, S.Ben Amor and T.C.Hickmott for helpful discussions or for providing us with experimental information before its publication. This work has been supported in part by the Comision Interministerial de Ciencia y Tecnologia of Spain under contract MAT88-0116-C02-01/02

REFERENCES

1. L.Brey, G.Platero and C.Tejedor, Phys. Rev. B, **38**, 9649 (1988)

2. L.Brey, G.Platero and C.Tejedor, Phys. Rev. B, **38**, 10507 (1988)

3. C.B.Duke, "Tunneling in solids" Solid State Physics. Supplement 10. (Academic Press. New York, 1969)

4. T.W.Hickmott, Solid State Commun., **38**, 371 (1988) and unpublished.

5. B.R.Snell, K.S.Chang, F.W.Sheard, L.Eaves, G.A.Toombs, D.K.Maude, J.C.Portal, S.J.Bass, P.Claxton, G.Hill and M.A.Pate, Phys. Rev. Lett. **59**, 2806(1987)

6. P.A.Schulz and C.Tejedor, Phys. Rev. B, to be published.

7. R.A.Davies, D.J.Newson, T.G.Powell, M.J.Kelly and H.W.Myron, Semicond. Sci. Technol. **2**, 61 (1987)

8. M.L.Leadbeater, L.Eaves, P.E.Simmonds, G.A.Toombs, F.W.Sheard, P.A.Claxton, G.Hill and M.A.Pate, Solid Stae Electronics, **31**, 707 (1988)

9. S.Ben Amor, K.P.Martin, J.J.L.Rascol, R.J.Higgins, A.Torabi, H.M.Harris and C.J.Summers, Appl. Phys. Lett. **53**, 2540 (1988)

10. E.E.Mendez, to be published.

11. P.Gueret, C.Rossel, E.Marclay and H.Meier, J. Appl. Phys. to be published.

12. M.Buttiker, IBM J. Res. Dev. **32**, 63 (1988)

13. G.Platero, L.Brey and C.Tejedor, to be published.

14. H.A.Fertig and B.I.Halperin, Phys. Rev. B, **36**, 7969 (1987)

15. J.K.Jain and S.A.Kivelson, Phys. Rev.Lett., **60**, 1542 (1988)

16. E.E.Mendez, L.Esaki and W.I.Wang, Phys. Rev. B, **33**, 2893 (1986)

17. M.L.Leadbeater, E.S.Alves, L.Eaves, M.Henini, O.H.Hughes, A.Celeste, J.C.Portal, G.Hill and M.A.Pate, Phys. Rev. B, **39**, 3438 (1989)

18. C.E.T.Goncalves da Silva and E.E.Mendez, Phys. Rev. B, **38**, 3994 (1988)

19. A.D.Stone and P.A.Lee, Phys. Rev, Lett. **54**, 1196 (1985)

PROPERTIES OF A DENSE QUASI TWO DIMENSIONAL ELECTRON-HOLE GAS AT HIGH MAGNETIC FIELDS

J.C. Maan, and M. Potemski*
Max Planck Institut für Festkörperforschung, HML, 166X F38042
Grenoble Cedex France

K. Ploog
Max Planck Institut für Festkörperforschung, Heisenbergstrasse 1
7000 Stuttgart 80, FRG

G. Weimann
Walter Schottky Institut, TU München, 8046 Garching, FRG

ABSTRACT

Luminescence spectra in GaAs/Ga$_{1-x}$Al$_x$As quantum wells in magnetic fields show the recombination from several Landau levels, filled with optically excited electrons and holes. A comparison of this luminescence with low power absorption spectra (involving empty states) shows in detail the changes of the optical properties with electron and hole density. For densities >10^{12}cm^{-2}, a decrease of the bandgap of twice the exciton binding energy, and which is independent on density is observed. Furthermore an excitonic behaviour of the ground state and a 20% enhancement of the effective mass is observed.

It is well known that the optical properties of semiconductors change when the density of free electron and holes, either excited optically or of extrinsic origin changes. This effect is caused by the additional Coulomb interaction between the charged particles (many-body effects) and this interaction may induce a correlation between them or lead to an exchange energy due the Pauli principle. These effects are actively studied because the implications in various experimental conditions are very different and therefore it is interesting to study them under new conditions. The advent of new layered semiconductor structures like quantum wells permits such studies in two dimensional systems, which may contribute further to their understanding. Modulation doped quantum wells, where a high mobility, almost purely two dimensional, electron gas is present have been actively studied, with luminescence[1-3] and Raman experiments [4] both with and without a magnetic field. At zero field there exist also several studies on undoped quantum wells, under intense optical excitation[5-8], but in a field one experiment with CW high excitation [9] and a few studies using time resolved luminescence have been reported [10,11]. In this paper we want to present a more detailed experimental study of the luminescence under high intensity excitation in GaAs/Ga$_{1-x}$Al$_x$As quantum wells in a magnetic field perpendicular to the layers.

There are both fundamental and practical reasons for the interest in such a study. From a fundamental point of view highly degenerate electron and hole Landau levels exist under these experimental conditions, each of these Landau levels having the same degeneracy, and separated from the others by gaps (the cyclotron energy) and occupied by the same number of electrons and holes. This is a fundamentally interesting situation, because it can easily be imagined that correlation and exchange energies are both strongly affected by this change in the density of states [12]. From a more practical point of view there are many advantages compared to high intensity luminescence studies at zero field and in bulk, namely; i) In quantum wells, the exciting light of a proper energy is absorbed in the barrier material, and the excited electrons and holes are accumulated in the wells leading to a carrier density, which can be higher and more homogenous than that obtained in bulk materials, ii) the exciton binding energy, which is a measure of electron-hole pair correlation may be varied with the well thickness [13] and finally iii) at high magnetic fields a rich energy spectrum is observed due to the fully discrete Landau levels[14-16]. This latter fact is very important because we will show that no lineshape analysis is necessary to obtain information from the data. At zero field where smooth, almost featureless spectra are observed, such an analysis is the only way to obtain more experimental information. Furthermore, since in a magnetic field the spectrum shows discrete peaks at all energies, not only information about the band edge but also from higher energies, can be obtained. In particular our data show a clear renormalization of the effective mass and a singular behaviour of the ground state; both these effects would have been impossible to see without a field. The importance of the rich spectra in fields becomes particularly clear from a comparison of the conventional low power absorption spectra -corresponding to the two-particle, excitonic absorption in empty states- with high intensity luminescence -resulting from the many-particle emission from filled states. This comparison shows in detail and quite dramatically the change in the optical properties due to many body effects.

The samples studied consisted of a thin GaAs cap layer, a 300nm thick Ga$_{1-x}$Al$_x$As top layer, three GaAs wells separated by 100nm Ga$_{1-x}$Al$_x$As barriers, on a GaAs substrate. Well thicknesses of the samples studied here were 3, 4.5, 9 and 13 nm. A thick barrier on the top of the samples was chosen to avoid the band bending due to the surface depletion and to minimize unwanted surface recombination. The energy of the exciting light (from a N$_2$ laser pumped Rhodamine 6G dye laser with 20 mJ pulses of 10ns at 620 nm wavelength, focused on the sample) in the high excitation luminescence was above the direct bandgap of the barrier material. This way an appreciable amount of the power was absorbed in the barriers and the excited carriers are collected in the wells. The Al content and thereby the barrier height, was adjusted with the well thickness so that the difference between the top of the barrier and the lowest confined state is always roughly about 250meV, giving a maximal density of the number of carriers that can be accomodated in the well of about 2 10^{13} cm^{-2}. On the other hand it was thought advantageous that the barrier should not be too high so that the excess kinetic energy that the carriers gain when they fall from the top of the barrier in the wells would be limited. One of the problems of high excitation intensity experiments is that the carrier temperature becomes very high, and this choice was thought to limit the problem, but as we shall see this turned out to be only partially true. Another problem in this type of experiments is to ensure the spatial homogeneity of the excited plasma. The two dimensional nature of the samples garanties already a homogeneous carrier density perpendicular to the surface. In order to maximize also the lateral homogeneity under high excitation, we analyzed only the central spot of the excited area by using a small pinhole in an intermediate image of the luminescence from the sample. We have verified that for a diameter of the spot size below a certain value and corresponding to an area smaller than the focus of the exciting beam on the sample, the shape of the luminescence does not depend on the spot size anymore. This focussing has turned out to be indeed very important, because the shape of the luminescence spectrum of the entire excited area (and not only the central part) is indeed different from that of the central area. We have not only performed high excitation luminescence but also conventional excitation spectra to characterize the samples under 'normal' conditions. These conventional, excitation spectra have been obtained with ion laser pumped dye lasers, using DCM and LD700 as dyes. All measurements were performed at 2K, and in magnetic fields up to 20T, using Bitter coils.

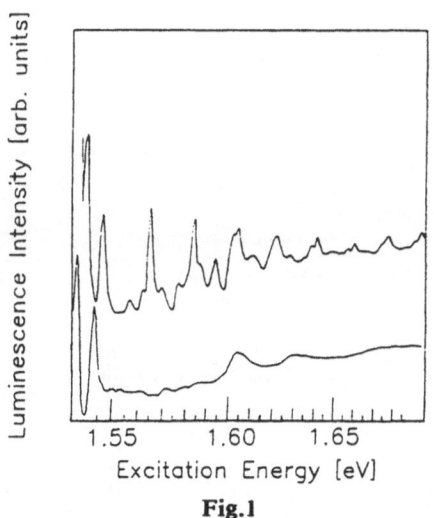

Fig.1

Conventional excitation spectrum of a 9 nm quantum well. Lower trace B=0T. Upper trace B=14T.

Fig. 1 shows on the bottom part a conventional low power excitation spectrum at zero field, and the corresponding spectrum at 10T. It can clearly be seen that the field leads to very pronounced features at higher energies. These spectra reflect basically the two particle excitonic excitations, which involve absorption between filled valence band states and empty conduction band states. These magneto-excitonic spectra have been extensively studied in the past [14-17]. In fig. 2a and b we show representative high intensity luminescence spectra at 0T and at 18T on a sample with 4.5nm well width. It is important to realize that all the spectra are taken on the same scale, and that therefore the saturation of the intensity directly demonstrates the filling of the states. It can be seen in fig. 2 that, analogous to the low power, pseudo-absorption spectra, the zero field luminescence is very broad and rather featureless, whereas in the magnetic field much more detail can be observed. However, note that these spectra correspond to emission coming from the recombination between filled electron states and empty valence band states.

Concentrating at the 0T spectra (fig. 2a), one first observes an increase and a broadening of the lowest luminescence peak with increasing laser power; upon further increase of the power this peak saturates and the higher energy tail begins to develop more rapidly than the low energy one. This demonstrates the filling and thereby the onset of recombination from higher occupied states. At even higher powers, the high energy tail reaches the top of the well, and at the highest power a rectangular shaped luminescence is observed. This spectrum directly reflects beautifully the rectangular shaped joint density of states in a two dimensional system. The high energy cut-off in these spectra corresponds to the top of the barriers, i.e. to the indirect GaAlAs gap in this sample. At these densities, no additional carriers can be accomodated in the well because carriers which are in the two dimensional subbands, but have an energy (high carrier temperature, or high Fermi energy) that exceeds the top of the well, will

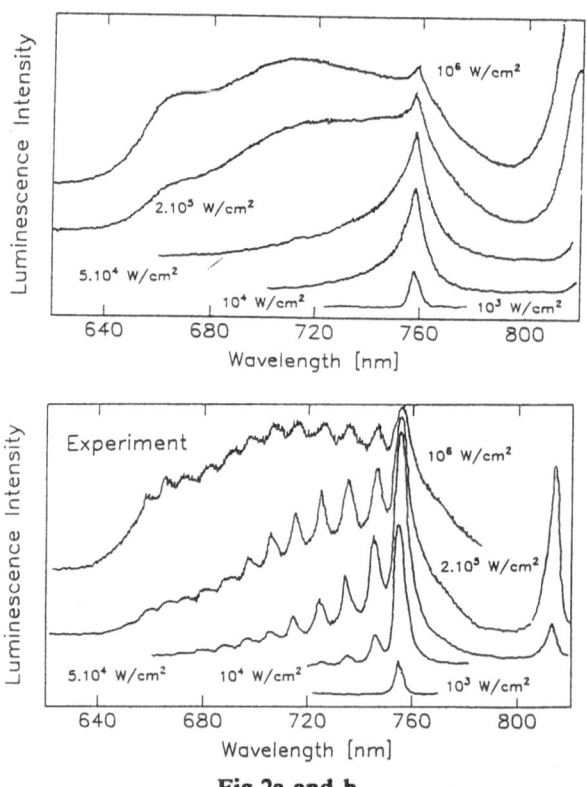

Fig.2a and b

Luminescence spectra for different intensities of the exciting light, at B=0T (above) and B=18T (below) of a 4.5nm quantum well. All spectra are recorded on the same scale and the saturation of the intensity as the states are filled can directly be observed. All spectra were taken at 1.8K He bath temperature.

escape into the bulk-like states in the GaAlAs. This band gap is indirect and therefore gives a very weak luminescence, not visible in our experiment. The escaping of the hottest particles effectively reduces the carrier temperature of the two-dimensional plasma. The evolution of the spectra with power in the presence of a 18T perpendicular magnetic field is shown in Fig. 2b and shows an analogous behaviour as the one observed at zero magnetic field. However, in this case a regular structure in the high energy side of the luminescence can be seen due to the Landau level quantisation by the field. This structure is much more regular than the complex magneto-excitonic transitions as observed in the pseudo absorption experiment. In fact all the peak positions have changed substantially. This difference is demonstrated more clearly in fig.3 where the energy position of the strongest pseudo-absorption peaks are compared with the high excitation luminescence peaks. The ground state has remained approximately at the same energy, but all other excited states are very different. This is demonstrated most clearly by the behaviour of the first excited state, which occurs in the high intensity luminescence at an energy where there is no corresponding absorption. However, the most surprising result is that once peaks have become visible in luminescence, their energy position does not change with increasing intensity, and thereby with increasing carrier density. The common wisdom is

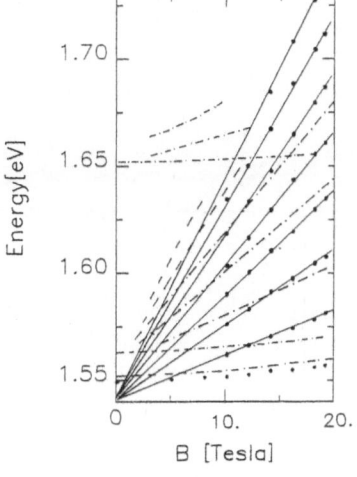

Fig.3

Position of the most pronounced extrema in the pseudo-absorption spectra (dashed lines) and of the luminescence peaks in the high excitation intensity experiment as a function of the magnetic field. The drawn lines in the latter case are a guide to the eye.

that due to many body effects, there is a reduction (renormalization) of the bandgap, which should keep on decreasing with increasing density. This would be visible in our spectra as a continuous shift to lower energy with increasing density of each of the peaks in the spectrum, which is not observed. As an example, the second peak remains at the same energy between $P=3\ 10^3 W/cm^2$, when it becomes just observable, to $P=10^6 W/cm^2$, when it is completely saturated. To be more quantitative we can estimate the carrier density that is excited at a given power, using the known Landau level degeneracy. The maximum intensity of an electron–hole Landau level transition occurs, when both levels are full. Knowing the Landau level degeneracy $2eB/h$, including spin, this corresponds to a density of $8.7\ 10^{11} cm^{-2}$ at 18T. At the lowest intensity shown in fig. 2b, the lowest peak is 20 of the maximum value, corresponding to a density of $2.2\ 10^{11} cm^{-2}$. At full power almost 13 Landau levels are full, i.e. the density of excited particles is $1.1\ ^{13} cm^{-2}$.

To demonstrate the filling of the states, we show in fig.4a and b the dependence of the total luminescence intensity and that of the first peak (which we isolated from the total luminescence higher intensities using a Lorentzian deconvolution) as a function of the power of the exciting light. It can clearly be seen that the total luminescence is close to linearly dependent on the power and that the intensity of the first peak saturates at a certain power. This demonstrates that this Landau level is full at that power, and that additional carriers are accomodated in higher Landau levels. As can be seen from fig.4b at the highest intensity the first peak contributes approximately only for 5 of the total intensity.

As was mentioned before, the remarkable feature of our data is that although we clearly see the change in the excitation spectrum between the low power, absorptive two particle spectrum and the high power many particle emission spectrum, there is no dramatic change in the position of the emission peaks with density. This experimental fact is illustrated in fig. 5 where we show the position of the emission peaks as a function of the excitation power, which is varied over many orders of magnitude. This figure demonstrates the remark-

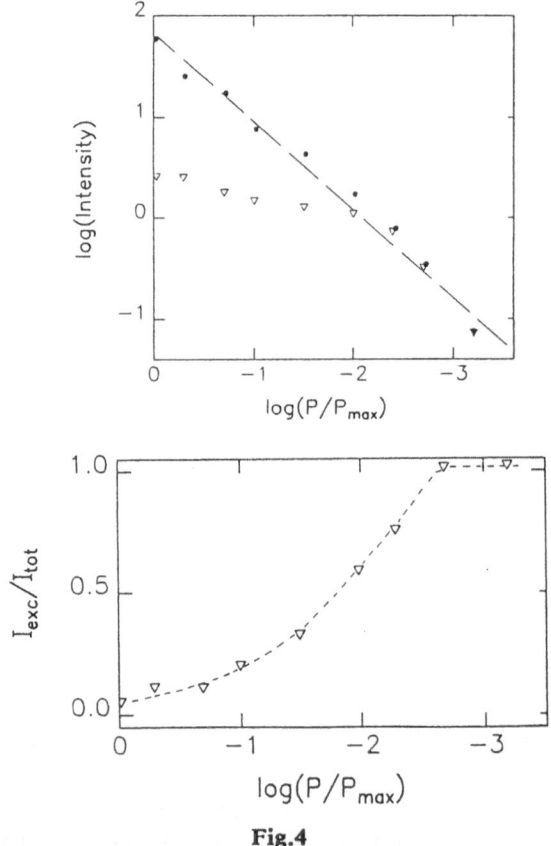

Fig.4

The dependence of the total luminescence intensity (full points) and of the intensity of the ground state (at 18 T) on the power density of the exciting beam. (above) At -2 on the horizontal scale the latter intensity saturates because the Landau level is full. Below, the intensity of the ground state luminescence as a fraction of the total luminescence, showing that at full power it contributes only up to 5.

able insensitivity of these peak positions on power and thereby on density. The conclusion from these results is that the energy of the maxima and thereby the level structure, remains unaffected, by a change in the carrier density of almost two orders of magnitude. On the other hand there is a clear change between the "pseudo absorption" spectra at vanishing densities and the emission spectra at densities $>10^{12}\mathrm{cm}^{-2}$.

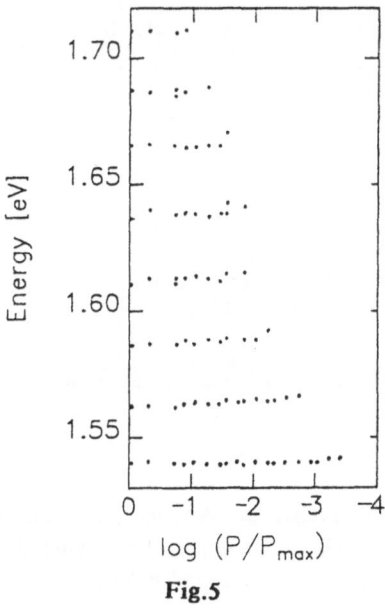

Fig.5

The dependence of the peak positions in the high intensity luminescence on the power density of the exciting light, showing the remarkable independence of the spectra on the density of excited carriers. The number of observable peaks increases of course, as the power is increased.

Comparing the pseudo-absorption spectra with the high intensity luminescence the renormalization of the fundamental gap can clearly been seen in our experimental results. In fig.3 we have drawn lines connecting the field dependence of the luminescence peaks other than the ground state. Concentrating on, for instance, the first excited state in high excitation, it can be seen that this transition can be associated with a corresponding –i.e. having the same slope- pseudo-absorption excitonic peak, but which is rigidly shifted to lower energy. Similarly, all higher excited states can be associated with their excitonic counterpart, equally shifted to lower energy by the same amount. This shift in energy directly demonstrates the band-gap renormalization. The only exception to this rule is the excitonic ground state, which in high excitation is almost at the same energy as the low power exciton peak. Apart from this ground state the lines through the transitions clearly show a Landau level fan, which originates at a certain energy at zero field. This energy is actually lower than the position of the ground state. This extrapolation point defines the position of the renormalized gap. We have observed the same behaviour in all investigated samples, and we have determined a bandgap renormalization as a function of the well thickness in this fashion. The result is shown in fig.6, together with the exciton binding energy as determined from the analysis of the low power spectra. The interesting conclusion is that we find that, within experimental uncertainty, the renormalization of the gap is exactly two times the exciton binding energy.

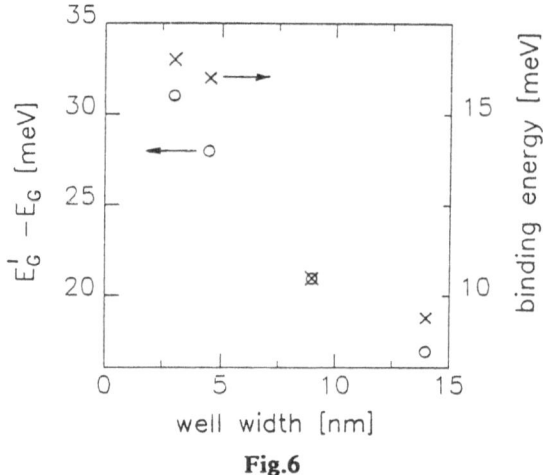

Fig.6

Dependence of the renormalization of the gap, at high densities (circles, left scale) and of the exciton binding energy at vanishing carrier densities (crosses, right scale) on well width. Note how the first quantity is exactly twice the other.

In the following we will analyze our data more quantitatively. Neglecting complexities of the valence band Landau levels for simplicity, and assuming a $\Delta N = 0$ selection rule, the transitions between Landau levels of electrons and holes are given by:

$$E_N{}^e - E_N{}^h = E_G - \frac{\hbar e B}{\mu} \qquad (1)$$

with $1/\mu = 1/m^*_e + 1/m^*_h$ the reduced mass. From a fit of the observed transition energies at high excitation we can therefore obtain the reduced mass and the renormalized gap. We have found that there is a notable non-parabolicity in our data which makes that the difference in energy between successive transitions is less at higher energies then at lower ones. We attribute this non-parabolicity to the conduction band which contributes mainly to the observed slopes because of the low electron mass. Transitions can be observed at higher energies either because one is considering a Landau level with high N or because, in thin wells there is an important size quantization of the subbands so that already the lower Landau levels are high in the GaAs band. A simple approximate formula that describes the effect of non-parabolicity in a coherent manner is [18]:

$$E_{N,n} = -\frac{E_G}{2} + \sqrt{\left(\frac{E_G}{2}\right)^2 + E_G D_N} \qquad (2)$$

with $D_N = (N+1/2)\hbar \omega_c + E_n(1 + E_n/E_G)$, and $\omega_c = eB/m_{0e}^*$ with m_{0e}^* the mass at the bulk GaAs bandedge. Fig. 7 shows a fit of eq. (2), i.e. neglecting entirely any contribution of the hole Landau levels, to the data, which is of course unrealistic. Already with this most conservative analysis we find that we have to use a 10% enhancement of the bulk GaAs band edge mass. If we assume more realistically some average hole mass of 0.6 (which describes reasonably the pseudoabsorption peaks) we find that we can still describe the excited states observed in all our samples in the same quasi-perfect manner but we then find a 20% heavier band edge mass. Our observations show therefore clearly that the reduced mass for transitions between states in the dense electron hole plasma is enhanced by at least 10%, but more realistically, by even 20%. This enhancement can also be derived more directly by comparing the slopes of the corresponding excitonic transitions at low excitation with those observed in the luminescence at high excitations. Such a comparison has been found to be completely consistent with the conclusions drawn from the from the fitting procedure.

It is important to realize that this quasi-perfect agreement is only obtained for the excited states. As has already been mentioned before, the ground state does not behave in the same way. In fact this ground state has a much weaker and quadratic field dependence contrary to the linear and steeper field dependence of a zeroth Landau level. This quadratic magnetic field dependence is what would have been expected for a bound electron-hole, excitonic state ('diamagnetic shift'). Indeed,as can be seen in fig 3., the field dependence of the ground state in high excitation is very analogous to that observed for the exciton in the pseudo-absorption spectra. The conclusion is therefore that the excited states behave like a free electron-hole plasma with a renormalized bandgap, which is twice the binding energy and a 20% renormalized mass. On the other hand there is a significant electron-hole pair correlation for the ground state, although this e-h pair is embedded in a dense, relatively hot, e-h plasma.

It is common practice in this type of high excitation experiments to obtain the information from the spectra using a lineshape analysis. We have shown in the previous paragraphs that clear conclusions can already be drawn from the results without such an analysis. However to be more complete, and in order to follow similar data analysis as other authors have performed before, we will show the results of a similar calculation applied to our data in a magnetic field. However, we want to describe first the ingredients that are generally used in such an analysis for data at zero field. Our zero field spectra show an analogous behaviour as reported in the literature before[5-8], except that our maximal densities are higher than reported before. The low energy tail, clearly visible in all high excitation spectra, is generally thought to be a many-body effect. It is often described by an energy dependent broadening of the states, which is due to the finite lifetime of the hole created in the Fermi sea when an electron and a hole recombine[19-20]. This broadening is proportional to the distance of the states involved in the recombination to the Fermi energy [20]. I.e. the states at the band-edge are broadened most, whereas states at the Fermi energy are not broadened. In the calculations the electron and hole states are filled with a certain number of particles distributed in the states according to Fermi statistics. Quasi-stationary conditions are generally assumed and the carrier distribution is characterized by the density, the effective carrier temperature (usually assumed to be equal for electrons and holes), and the quasi-Fermi energies for electrons and holes. The carrier density and renormalized bandgap are generally derived from the data by fitting the lineshape to calculations[7-8], taking a k-selection rule into account[7-8], or assuming that many body effects can be described by the absence of selection rules[21]. Fit parameters are then the (renormalized) bandgap, the broadening parameter, the carrier densities and the electron and hole effective temperatures. The final result is a plot of the renormalized gap against the carrier density, which is in fact the dependence of one fit parameter on another. Although it is difficult to do something better, the results of this type of analysis should nevertheless be taken with caution. A remarkable feature of the power dependence of the zero field spectra reported in the literature [7-8] is that the maximum of the luminescence peak does not change in energy with increasing carrier density. In the lineshape analysis this is explained by a compensation of a shift to higher energies due to the rising of the Fermi energy, and thereby the broadening, with a shift downward, namely the renormalization of the gap. Our data in a magnetic field show that in fact none of the luminescence peaks shifts in energy with increasing density beyond a certain density, and since we can observe singularities in the density of states, at each Landau level transition, we can see that these indeed hardly shift in energy.

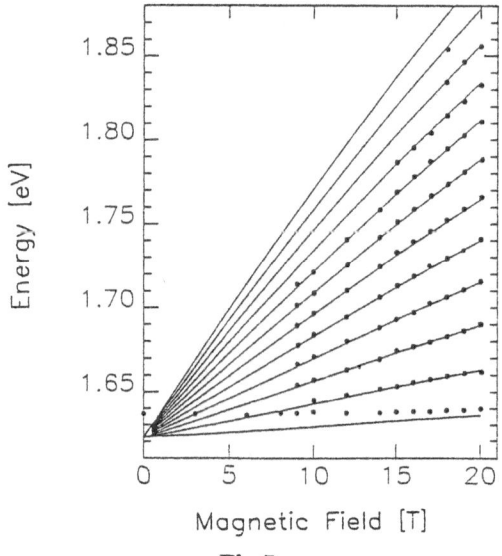

Fig.7

Magnetic field dependence of the position of the luminescence peaks in a 4.5 nm sample (data points). The drawn lines are the calculated interband transitions (eq. 2) including non-parabolicity and a 20% enhancement of the bulk GaAs band edge mass.

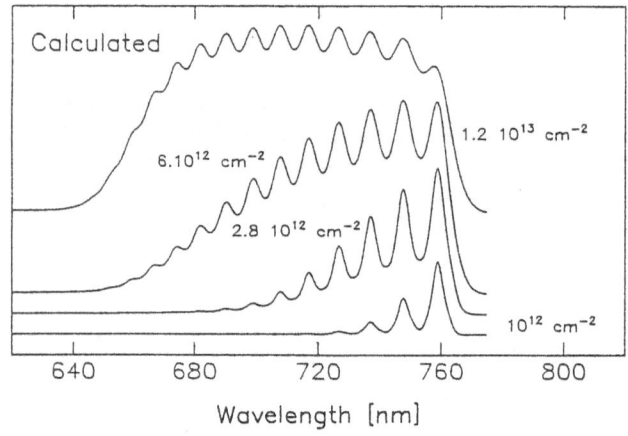

Fig.8

Calculated luminescence lineshapes using eq.3 and densities as indicated in the figure. Comparing with fig. 2b shows a good agreement except for the intensity and the position of the first peak, and the absence of the tail at low energies.

Following the conventional procedure we have calculated theoretical lineshapes more or less in a standard way [7,8,10,11]. The formula we have used is:

$$I(\hbar\omega) = \sum_{N=0}^{N_{max}} \int_0^{\hbar\omega+E'_G} D_{e,N}(E)f_e(E)D_{h,N}(\hbar\omega-E'_G-E)f_h(\hbar\omega-E'_G-E)dE \qquad (3)$$

where $D_{e,N}$ and $D_{h,N}$ are the density of states of the broadened Nth Landau level, f_e and f_h are the Fermi functions, $E'_G = E_G + E_n + H_n$, with E_n and H_n the respective confinement energies; where the subscripts e and h stand for electrons and holes. We used a Lorentzian broadened Landau level with equal width for conduction and valence band, having the proper degeneracy at each field and we assumed electron and hole temperatures to be equal. The effect of temperature and carrier density is included in the Fermi function. Formula (3) implies that we use a $\Delta N = 0$ selection rule, contrary to what has been done in previous work with time resolved luminescence [11,12]. Our data demonstrate that selection rules are still valid, because without selection rules and with the high densities and Fermi energies here, the transitions in the center of the spectra would be much more intense than those at the edges. This is because with increasing radiation energy more and more transitions become possible, and this would at zero field lead to a parabolic increase in intensity, which is incompatible with the more or less constant spectra we observe. Furthermore we have also found experimentally that the high intensity luminescence spectra clearly show a circular polarization dependence, which shows that there must be selection rules. The calculated intensities are shown in fig. 8. Fit parameters are the density, the temperature, the Landau level width and the bandgap. The densities we obtain are in reasonable ageement with those estimated from the saturation intensities as previously described. As also observed before [11,12] we find a clear increase in the Landau level width with carrier density, ranging from 4meV at the lowest density and increasing to 14 meV at the highest density. Carrier temperatures may be as high as 400K in the intermediate region of densities (4 10^{12} cm^{-2} range), at the highest intensities where carriers start spilling out of the well we observe the cooling effect as described in the introduction. Overall the agreement with the experiments is rather satisfactory. We should mention that we have of course used eq. (1) and (2) for the non-parabolicity, and the mass enhancement mentioned before. It is important to note that there is no explicit many body effect incorporated in our analysis. The bandgap is of course a fit parameter, but the values that we obtain are the same as those that can already directly be inferred from the extrapolation to zero field as shown in fig. 3 and 7. The most notable discrepancies are that the single particle picture cannot explain the tailing of the luminescence at low energies, which must therefore clearly be a many-body effect. Furthermore both the intensity and the position of the first peak, is consistently in disaccord with the experimental observations. We found already that the field dependence of this peak behaved differently with respect to the higher energy peaks; now we find from the calculated intensities that also its intensity dependence is anomalous. The fact that this peak is always more intense than calculated is consistent with the previous speculation about its excitonic character, because the oscillator strength of the exciton is always much stronger than that for band to band transitions. In summary, the standard analysis using lineshape fitting confirms the conclusions derived directly from the spectra.

We believe that our data have shown some interesting effects, which have not been observed before namely: a renormalization of the effective mass, a renormalization of the bandgap which is density independent at densities higher than 10^{12} cm^{-2}, and which scales exactly with the exciton binding energy, and finally strong indications that the ground state remains excitonic, both in its intensity as in its field dependence. Our data on the renormalized gap differ from those reported in the literature[7-8] because they are about 50% lower and furthermore we see a clear dependence on thickness, whereas previously the renormalization was claimed to be universal [8]. The thickness dependence we observe is in agreement with a recent calculation [22], and in fact the numbers we obtain agree reasonably with those calculated theoretically at a density of 10^{12} cm^{-2}. However this calculation does find a continuous decrease of the gap with increasing carrier density. Schmitt-Rink et al. have calculated the theoretical luminescence spectra as a function of density and they obtain an increasing band-gap renormalization with density, once more as opposed to our findings [23]. A mass renormalization has been calculated by Schmitt-Rink et al. in [24] but if we interpret their figure correctly they seem to find a decrease of the mass with density, opposite to the experiments. The excitonic behaviour that we observe suggests an important pair-correlation, which is surprising at these high densities where the average particle spacing is comparable or even less than the exciton Bohr radius. It is even more surprising because especially at intermediate densities the carrier temperature is rather high, so that $k_B T_{eff}$ is in fact higher than the exciton binding energy. We have no explanation for this effect, but can only say that the experimental results consistently show the importance of this pair correlation. We have verified the spatial homogeneity and the temporal homogeneity of the luminescence, in order to be sure that we do not see excitons at the end of the laser pulse after the recombination of the other electrons and holes, or excitons from the edge of the excitation spot where density and temperature of the carriers is less. The result of this check was that the entire luminescence from all energies is seen simultaneously within 1 ns. Furthermore, we have seen the importance of analysing only the central part of the excited luminescence spot using a pinhole in the intermediate image of the luminescence, because analysing the entire spot does indeed show the luminescence presence of cooler excitons coming from the edges of the spot. Nevertheless below a certain spot size, the spectra do not change anymore and the energy position of the ground state, for instance at 20T where there should be a marked difference between the free-electron hole and the excitonic transition , remains the same.

It has been considered theoretically [12] that under conditions of high magnetic fields in a 2D system there could be a coherent ground state. In fact, this was one of the starting ideas of this work. We do not believe that we have seen such effects, because the spectra evolve smoothly with magnetic field and it seems that the only effect is a splitting of the continuous density of states in Landau levels. Nevertheless, we believe that the use of a field has allowed us to draw precise conclusions from the experimental results because the experimental conditions are rather well defined.

It is a pleasure to thank H.Krath for the excellent technical assistance in this experiment and P.Wyder for the interest in this work.

* Permanent address: Institute of Physics, Polish Academy of Sciences, 02-668 Warsaw, Poland.

References

1.   C.Delalande, G.Bastard, J.Orgonasi, J.A.Brum, H.W.Liu, M.Voos, G.Weimann, W.Schlapp,Phys.Rev.Lett.59,2690,(1987).
2.   I.Kukushkin, V.Timofeev, K.von Klitzing, K.Ploog, Festkörperprobleme 28,21,(1988).

3.   A.Pinczuk, J.Shah, H.L.Störmer, R.C.Miller, A.C.Gossard and W.Wiegman, Surf.Sc.142,492,(1984).

4.   A.Pinczuk, J.P.Valladares, D.Heiman, A.C.Gossard, J.H.English, C.W.Tu, L.Pfeiffer and K.West, Phys.Rev.Lett.61,2701,(1988).

5.   R.C.Miller, D.A.Kleinman, O.Munteanu, W.T.Tsang, Appl.Phys.Lett.39,1,(1981).

6.   Z.Y.Xu, V.G.Kreismanis, C.L.Tang, Appl.Phys.Lett.43,415,(1983).

7.   L.S.Devine and W.T.Moore, Solid State Commun.65,177,(1988).

8.   G.Tränkle, H.Leier, A.Forchel, H.Haug, C.Ell and G.Weimann, Phys.Rev.Lett.58,419,(1987).

9.   M.Potemski, J.C.Maan, K.Ploog, G.Weimann, Proc. 19th Int. Conf. Phys. Semicon. 19, Warsaw 1988, p.119, ed.W.Zawadzki, IPPAS.

10.   T.T.J.M. Berendschot, H.A.J.M. Reinen and H.J.A. Bluyssen, Solid State Commun.63,873,(1987).

11.   R.W.J. Hollering, T.T.J.M. Berendschot, H.J.A. Bluyssen, H.A.J.M. Reinen, P.Wyder, Roozeboom, Phys.Rev.B38,13323,(1988).

12.   D.Paquet, T.M.Rice, K.Ueda, Phys.Rev.B32,5208,(1985).

13.   R.C.Miller, D.A.Kleinman, W.T.Tsang and A.C.Gossard, Phys.Rev.B24,1134,(1981). and E.S.Koteles and J.Y.Chi, Phys.Rev.B37,6332,(1987).

14.   J.C.Maan, G.Belle, A.Fasolino, M.Altarelli and K.Ploog, Phys.Rev.B30,2253,(1984).

15.   L.Viña, M.Potemski, J.C.Maan, G.E.W.Bauer, E.E.Mendez and W.I.Wang, Superlattices and Microstructures 5,371,(1989).

16.   D.C.Rogers, J.Singleton, R.J.Nicholas, C.T.Foxon and K.Woodbridge, Phys.Rev.B34,4002,(1984).

17.   S-R.E.Yang, and L.J.Sham, Phys.Rev.Lett.58,2598,(1987). G.E.W. Bauer, and T.Ando, Phys.Rev.Lett.50,601,(1987).

18.   H.Bluyssen, J.C.Maan, P.Wyder, L.L.Chang and L.Esaki, Solid State Commun.31,75,(1979).

19.   P.T.Landsberg, Phys.Stat.Sol.15,623,(1966).

20.   R.W.Martin and H.L.Störmer, Solid State Commun.22,523,(1977).

21.   S.Tanaka, H.Kobayashi, H.Saito and S.Shionoya, J.Phys.Soc.Jap.49,1051,(1980).

22.   G.E.W. Bauer, Proc. 19th Int. Conf. Phys. Sem. Warsaw 1988, p.143, ed.W.Zawadzki, IPPAS.

23.   S.Schmitt-Rink, C.Ell and H.Haug, Phys.Rev.B33,1183,(1986).

24.   S.Schmitt-Rink and C.Ell, J.Lum.30,585,(1985).

MAGNETO-ABSORPTION IN A $Hg_{1-x}Mn_xTe$-CdTe SUPERLATTICE: EVIDENCE OF THE EXCHANGE INTERACTION IN A SEMIMAGNETIC SEMICONDUCTING SUPERLATTICE

G. S. Boebinger

AT&T Bell Laboratories
600 Mountain Avenue
Murray Hill, NJ 07974 USA

Y. Guldner, J. M. Berroir, M. Voos, and J. P. Vieren

Groupe de Physique des Solides de l'Ecole Normale Superieure
24 rue Lhomond
75231 Paris, France

J. P. Faurie

Department of Physics
University of Illinois at Chicago
Chicago, IL 60680 USA

ABSTRACT

We discuss far-infrared magneto-absorption experiments performed on a $Hg_{0.96}Mn_{0.04}Te$-CdTe superlattice over a temperature range from 1.5 to 10 K. The results are interpreted from superlattice band-structure calculations which include the magnetic field. The temperature dependence of the cyclotron resonance and interband transitions show evidence of the exchange interaction between the localized Mn d-electrons and the conduction band electrons in this semimagnetic semiconducting superlattice. We also discuss very recent results which appear to resolve the valence-band offset dilemma in HgTe-CdTe superlattices and the impact of those results on our interpretation of the magneto-absorption data.

INTRODUCTION

In the field of semimagnetic semiconductors much work has been directed toward superlattice (SL) structures in search of lower-dimensionality electronic and magnetic effects.[1-4] The narrow-gap system, $Hg_{1-x}Mn_xTe$-CdTe, is unique because the two-dimensional electron confinement is in the layers which contain the dilute magnetic ions. In the future, this system should be particularly interesting for the experimental control available to probe the exchange interaction. In addition to controlling the magnetic ion concentration, as is done already in bulk samples, the growth of superlattices and quantum wells will allow control of the sample dimensionality, the magnitude of the band gap, and the electron concentration.

This paper does not attempt to review the field of semimagnetic semiconductor superlattices. Rather it discusses results on the narrow gap superlattice $Hg_{0.96}Mn_{0.04}Te$-CdTe which provide the first argued evidence in a semimagnetic superlattice of the spin-spin exchange interaction between the localized Mn magnetic moments and the conduction band electrons.[5] In the superlattice, there is a strong effect of the exchange interaction on cyclotron resonance transitions[5] which is not seen in bulk semimagnetic semiconductors.[6] We also discuss recent theoretical results which illuminate the source of the valence-band offset controversy in HgTe-CdTe superlattices[7] and subsequent experimental results[8] which provide compelling magneto-optical evidence of the large valence-band offset ($\Lambda \sim 300-350$meV) originally determined by X-ray photoemission spectroscopy.[9-11]

SAMPLE PARAMETERS AND BAND STRUCTURE CALCULATIONS

The superlattice was grown[14] by molecular beam epitaxy on a (100) GaAs substrate with a thick buffer layer of $\sim 2\mu$m thick CdTe(111) to accomodate the strain resulting from the lattice mismatch with the substrate. The superlattice itself consists of 100 periods of $d_1 = 168$ Å thick $Hg_{0.96}Mn_{0.04}Te$ quantum wells interspaced by $d_2 = 22$ Å thick CdTe barriers. The CdTe barriers contain up to 15% HgTe since the mercury shutter remains open during growth of the barrier layers.[15] The superlattice is n type with electron density $n = 6 \times 10^{16}$ cm^{-3} and mobility $\mu = 2.7 \times 10^4$ cm^{2}V^{-1}s^{-1} at T = 30 K.

The superlattice band structure calculations were carried out in the envelope function approximation.[12,13] To begin the calculation, the band structures of the $Hg_{0.96}Mn_{0.04}Te$ and CdTe layers near the Γ point under strong magnetic field are described by the 6×6 Pidgeon and Brown Hamiltonian[16]. This Hamiltonian includes the Γ_6 and Γ_8 bands and ignores the spin-orbit split-off Γ_7 band, due to the large spin-orbit splitting, $\Delta \sim 1$ eV. The mixing with higher bands is included up to second order through the modified Luttinger parameters γ_1, $\gamma_2 = \gamma_3 = \gamma$ (spherical approximation), and κ. The two other

TABLE 1   Band-parameter values utilized in the calculations.

	γ_1	γ	κ	E_g(meV)	E_p(eV)
$Hg_{0.96}Mn_{0.04}Te$	4.50	1.1	−0.85	−125	18
CdTe	1.54	0.015	0.6	1600	18

parameters which enter the calculation are E_g, the $\Gamma_6 - \Gamma_8$ energy gap, and E_P, related to the square of the Kane matrix element. The parameter values used in our calculations are given in Table I.[17]

For the $Hg_{0.96}Mn_{0.04}Te$ layers, the Pidgeon and Brown model is modified to take into account the exchange interactions between localized d electrons bound to the Mn^{2+} ions and the Γ_6 and Γ_8 s- and p-band electrons. The s-d and p-d interactions are introduced in the molecular field approximation through two additional parameters, r and A, where $r = \alpha/\beta$ is the ratio between the Γ_6 and Γ_8 exchange integrals, α and β, respectively.[18,19] Here A is the normalized magnetization defined by $A = (1/6)\beta x N_0 <S_z>$, where N_0 is the number of unit cells per unit volume of the crystal, x is the Mn composition (x = 0.04), and $<S_z>$ is the thermal average of the spin operator along the direction of the applied magnetic field, B. Extensive magneto-optical data [18,19] and direct magnetization measurements [20,21], obtained on bulk alloys with dilute Mn concentration, have shown that $<S_z>$ is well described by a modified spin-5/2 Brillouin function, where the temperature T is replaced with an empirical effective temperature $(T + T_0)$. T_0 is found to be ~ 5 K for x = 0.04.[21] The exchange integral $N_0\beta \sim 1$ eV and $r \sim -1$.[19]

For a superlattice in a magnetic field, the electron motion parallel to the layers is described by a six-component vector[12,13,22]: $\Psi_N = (C_1\phi_{N-1}, C_2\phi_{N-2}, C_3\phi_N, C_4\phi_N, C_5\phi_{N-1}, C_6\phi_{N+1})$, where ϕ_N is the Nth harmonic oscillator function and N = -1,0,1,2, ... For $N \leq 1$, the coefficients C_i corresponding to the negative oscillator index vanish. A system of six differential equations for the six-component envelope function is established from the Pidgeon and Brown Hamiltonian. The boundary conditions at the superlattice interfaces are satisfied by the continuity of the wave functions and the current density of the six coupled differential equations across an interface. Taking into account the superlattice periodicity through the Bloch theorem, one obtains a numerical solution for the Landau level energies. Figure 1 shows the results of the calculation at $k_z = 0$ for the lowest conduction band E_1 and the first heavy hole valence band HH_1. We have set the CdTe valence-band edge at zero energy and the $Hg_{0.96}Mn_{0.04}Te$ valence-band edge at 40 meV, following the early determination of a small valence-band offset in HgTe-CdTe superlattices from magneto-optical experiments.[22] [The value of the valence-band offset will be discussed in greater detail in the second part of the article.] The energy levels are identified by their harmonic oscillator index N. Note that for $N \geq 1$, there are two conduction-band solutions, denoted N and N'.

The resulting band structure, calculated for zero magnetic field, is shown in Figure 1. The position of the Fermi energy, E_F, is estimated from the large measured electron density and the calculated effective mass. The results of the Landau level calculations at k=0 are shown in Figure 2. The existence of a superlattice yields several visible effects in Figures 1 and 2. First, it lifts the Γ_8 degeneracy and opens a finite superlattice energy gap. Second, the superlattice causes additional mixing of the bulk electron states because there is a non-zero wave vector component normal to the superlattice layers. This destroys any simple semiclassical interpretation of the superlattice Landau levels. Note that as magnetization, A, is increased in Figure 2, all the conduction band levels increase in energy. There is no semiclassical assignment of the states as spin up and spin down states of purely orbital Landau levels. The experimentally determined magnitude of A which satisfies the modified Brillouin function for bulk $Hg_{0.96}Mn_{.04}Te$ is given in Table II for various magnetic fields and temperatures.[21,23]

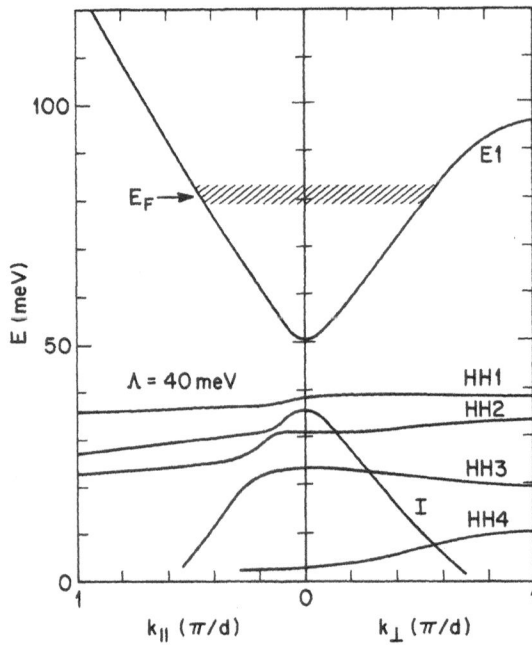

Figure 1   Calculated band structure of the
Hg$_{0.96}$Mn$_{0.04}$ Te-CdTe superlattice in zero
magnetic field. E_F is estimated from the
electron density.

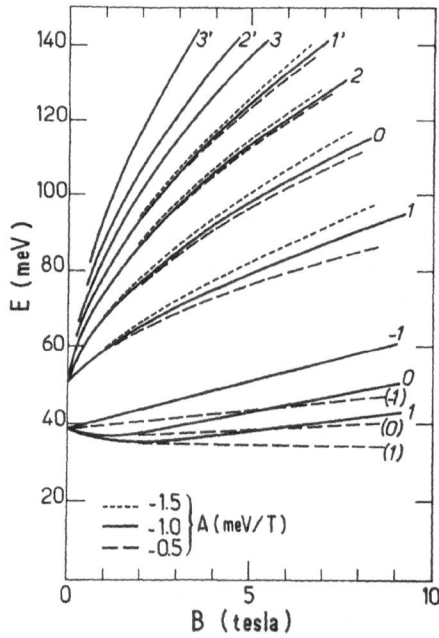

Figure 2   Calculated E_1 and HH$_1$ Landau levels at
k=0 corresponding to different values of the
harmonic oscillator index N=−1,0,1,1′....
Note the influence of magnetization, A, on
the Landau levels

EXPERIMENTAL RESULTS

The magneto-absorption experiments reported here utilized a CO_2 laser ($\lambda = 9.3 - 10.8\ \mu m$) and a molecular gas laser ($\lambda = 41-255\ \mu m$). The transmission signal, detected by a carbon bolometer, was measured at fixed infrared photon energies E, while the magnetic field was varied continuously to 12 T. Examples of the magneto-transmission spectra are shown in Figures 3 and 4 for $\lambda = 41\ \mu m$ and $\lambda = 9.32\ \mu m$. The cyclotron resonance or interband transitions correspond to minima in the transmission spectrum. The energy positions of the transmission minima as a function of B are given in Figures 5 and 6 as open circles.

The data of Figure 5 are interpreted as electron cyclotron resonance transitions since the superlattice is n-type and the data extrapolate to $E \sim 0$ at $B = 0$. The solid lines in Figure 5 correspond to all calculated cyclotron resonance transitions observable with a selection rule $\Delta N = +1$ and the estimated Fermi energy $E_F \sim 30\ meV$ above the conduction-band edge at $B = 0$. For simplicity, the solid lines in Figure 5 correspond to $A = -1\ meV/T$ at all magnetic fields. Figure 6 shows the energy positions of the transmission minima observed at several wavelengths of the CO_2 laser. The calculated interband transitions obeying $\Delta N = \pm 1$ are represented as solid lines and are in reasonable accord with the observed transitions. Each observed minimum probably results from the two unresolved allowed interband transitions. In the CO_2 laser energy range, transitions to lower conduction band levels N = 1, 0, and 2 would lie at magnetic fields above 12 T and are therefore not observed. The data and calculated transitions contained in Figs. 5 and 6 demonstrate that this superlattice is adequately modeled by the envelope function approximation.

We now turn attention to the temperature dependences of the observed transitions as evidence of the exchange interaction of the Mn ions. Figure 3 shows transmission spectra at $\lambda = 41\ \mu m$ between 1.5 and 10 K. The two identified minima shift to lower B with increasing temperature: the $1 \rightarrow 2$ transition by $\Delta B_r \sim -0.5\ T$ and the $0 \rightarrow 1'$ transition by $\lesssim -0.2\ T$. There are two contributions to these temperature-dependent shifts: the temperature dependence of the $Hg_{0.96}Mn_{0.04}Te$ energy gap and the temperature dependent Mn magnetization. In bulk HgTe, as T increases from 1.5 K to 10 K, the magnitude of the energy gap diminishes by less than 3 meV.[17] The effect in $Hg_{0.96}Mn_{0.04}Te$ is expected to be similar. Such a small shift in energy gap is calculated to have only a slight effect on the cyclotron resonance transitions.

TABLE II   Magnetization A given in meV/T at various magnetic fields and temperatures for bulk $Hg_{0.96}Mn_{0.04}Te$.

T(K)	B(T) 0.5	3	5	7	10
1.5	−2.3	−1.3	−1.0	−0.8	−0.6
4.2	−1.4	−1.1	−0.9	−0.7	−0.6
10.0	−0.5	−0.5	−0.45	−0.45	−0.4

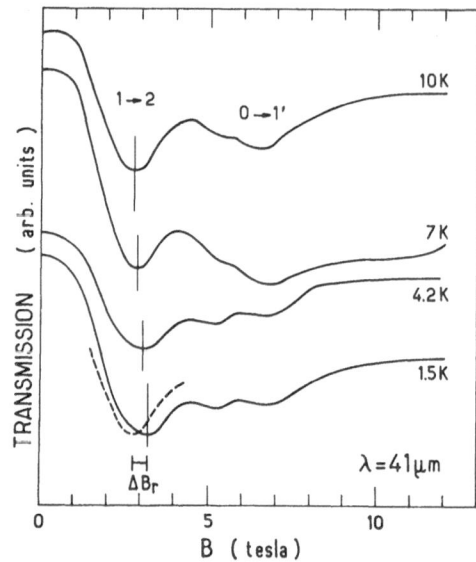

Figure 3   Far-infrared magnetotransmission spectra at
λ=41μm for temperatures from 1.5 to 10 K.
The dashed line reproduces the data at
10 K. The bracket ΔB_r indicates the
magnitude of the calculated shift in the
$1 \rightarrow 2$ cyclotron resonance transition.

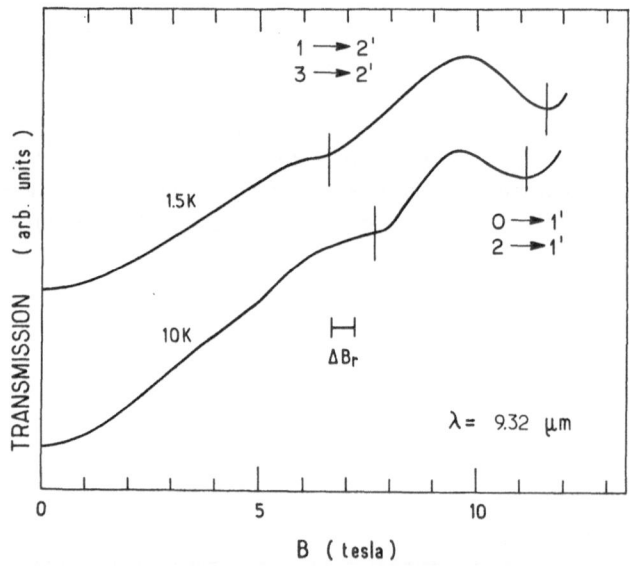

Figure 4   Infrared magnetotransmission spectra at
λ=9.32μm for temperatures 1.5 and 10 K.
The bracket ΔB_r indicates the magnitude of
the calculated shift in the $\rightarrow 2'$ interband
transition.

444

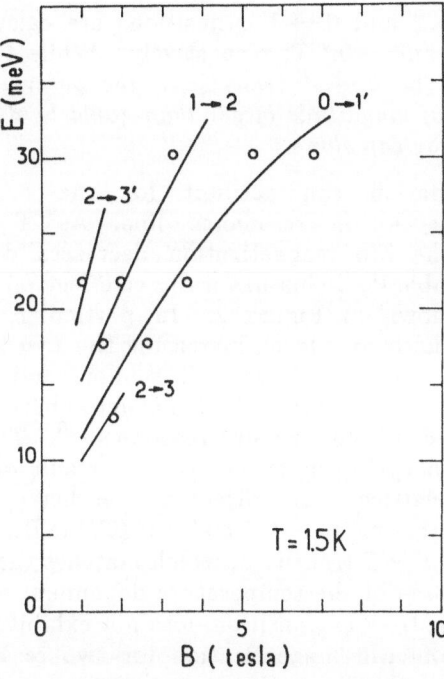

Figure 5   Observed far-infrared magnetotransmission
              minima (open circles) and calculated
              cyclotron resonance transitions (solid lines).

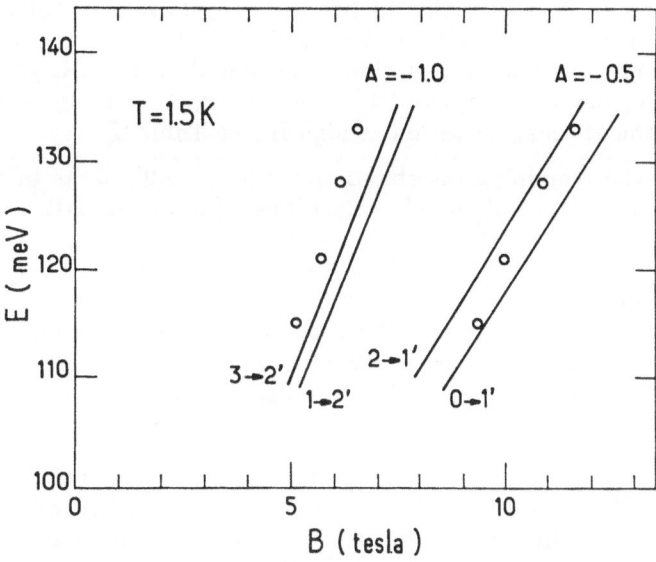

Figure 6   Observed infrared magnetotransmission
              minima (open circles) and calculated
              interband transitions (solid lines). The
              normalized magnetization, A, appropriate
              for T = 1.5 K is given for each transition.

In particular, the $1 \to 2$ and $0 \to 1'$ transitions are calculated to shift by at most $\Delta B_r \sim -0.05$ T and -0.1 T, respectively. While this accounts well for the observed shift of the $0 \to 1'$ transition, *the observed shift of the $1 \to 2$ transition is an order of magnitude larger than could be due to the temperature dependence of the energy gap alone.*

The Mn magnetization can account for the enhanced temperature dependence of the cyclotron resonance line. As T increases over our experimental range, the Mn magnetization decreases dramatically at lower magnetic fields (see Table II). This has a direct effect on the Landau levels in the superlattice as shown in Figure 2. In particular, the $N = 1$ level is depressed more than the $N = 2$ level, increasing the $1 \to 2$ cyclotron resonance energy as temperature is increased. In bulk semimagnetic semiconductors, the exchange interaction has been observed by magneto-absorption primarily in the spin and combined (cyclotron plus spin) resonances [6]. The observation of the exchange interaction in cyclotron resonance is a result of the increased band mixing in a superlattice, as discussed earlier. For magnetization $A = -1.3$ meV/T decreasing to -0.5 meV/T (Table II), we calculate a shift $\Delta B_r \sim -0.5$ T for the $1 \to 2$ transition, which matches the large observed shift. The calculated magnitude of the temperature-dependent shift ΔB_r is shown in Figure 3. Note that the $0 \to 1'$ transition does not exhibit a strong temperature dependence arising from Mn magnetization for two reasons. First, at higher magnetic fields ($B \sim 7$T) the magnetization changes less dramatically over the experimental temperature range (see Table II). Second, in Figure 2 we see that the $N = 0$ and $N = 1'$ levels shift by essentially the same amount for $A \sim -0.45 \to -0.8$ meV/T, corresponding to the experimental temperature range.

As would be expected, the experimental interband transitions (Figure 4) also exhibit temperature dependences. The higher magnetic field transition, $\to 1'$, shifts to lower field with increasing temperature ($\Delta B_r = -0.4$ T for $T = 1.5 \to 10$ K). As with the higher field cyclotron resonance transition ($0 \to 1'$), this temperature shift is attributable to the temperature dependence of the $Hg_{0.96}Mn_{0.04}Te$ energy gap alone (calculated to be $\Delta B_r \lesssim -0.5$ T). No temperature-dependent magnetization effect exists at $B \sim 10$T due to saturation of the Mn magnetization at high B (see Table II).

Note that the remaining interband transition, $\to 2'$, shifts in the direction contrary to all other observed transitions, (observed $\Delta B_r \sim +1.0$ T for $T = 1.5 \to 10$ K). Here, the effects of the band gap temperature dependence (calculated $\Delta B_r \lesssim -0.05$ T) are *both an order of magnitude too small and in the wrong direction.* The observed shift is dominated by a much larger shift in the proper direction which results from the change in Mn magnetization. This calculated shift, $\Delta B_r \sim +0.5$ T, is shown in Figure 4. The net calculated shift remains smaller than that observed, which suggests an even larger effect due to the presence of the Mn.

To summarize the results, we observe cyclotron resonance transitions and interband transitions in a $Hg_{0.96}Mn_{0.04}Te - CdTe$ superlattice which are in good agreement with band structure calculations in the envelope function approximation with small valence band offset ($\Lambda \sim 40$meV). The temperature dependence of cyclotron resonance transitions evidences not only the altered behavior of the Landau levels in a superlattice, but also the effect of the exchange interaction in the semimagnetic semiconductor superlattice. The magnetization observed in the superlattice is consistent with the bulk

$Hg_{0.96}Mn_{0.04}Te$ magnetization, due to the thick $Hg_{0.96}Mn_{0.04}Te$ layers and thin CdTe barriers in our superlattice.

THE VALENCE BAND OFFSET CONTROVERSY

Recently, much progress has been made toward the resolution of the dilemma posed by the two valence-band offsets reported from different experiments. Low-temperature magneto-absorption experiments have reported a small valence-band offset, $\Lambda \sim 40$ meV,[22,5] while room temperature X-ray photoemission spectroscopy has consistently shown a large valence-band offset, $\Lambda \sim 350$ meV. [9-11] The controversy is resolved upon realizing that both small and large valence band offsets can result in semiconducting superlattices with similar in-plane electron effective masses. This is illustrated in Figures 7 and 8 for the 100Å HgTe-36Å CdTe superlattice studied by magneto-absorption in Ref. 22. As Λ is increased from zero, the superlattice band gap decreases. Eventually the conduction band and heavy hole bands cross, yielding a semimetallic superlattice. As Λ continues to increase, an energy gap is opened at the Brillouin zone edge, $k_\perp = \pi/d$, yielding once again a semiconducting superlattice.

The in-plane band structure probed by the magneto-absorption experiments is coincidentally similar for $\Lambda \sim 40$ meV and $\Lambda \sim 350$ meV. This is shown in Figure 8 where the in-plane effective mass, $m_\parallel$, and band structure are illustrated as a function of Λ for the superlattice of Ref. 22. We have since completed detailed Landau level calculations for this superlattice and have concluded that the data of Ref. 22 are also consistent with $\Lambda \sim 300$ meV.

Very recent magneto-absorption experiments on HgTe-CdTe based superlattices have given compelling evidence that the valence-band offset must be large ($\Lambda \sim 300$ meV).[8] A large electron mass anisotropy ($m_\perp/m_\parallel$ as high as 40) has been observed in $Hg_{0.95}Zn_{0.05}Te$—CdTe superlattices. This anisotropy is consistent with band structure calculations only if there is a large valence-band offset resulting in a semimetallic superlattice.

Clearly, the bulk of experimental evidence is now behind a large valence-band offset $\Lambda \sim 300$-350 meV. As a result, efforts are now underway to perform the Landau level calculations with large valence-band offset for the $Hg_{0.96}Mn_{0.04}Te$—CdTe superlattice studied herein. There are important differences between this $Hg_{0.96}Mn_{0.04}Te$—CdTe superlattice and the HgTe-CdTe superlattice studied in Ref. 22 which make it unclear if coincidentally similar results will emerge for $\Lambda \sim 40$ meV and $\Lambda \sim 350$ meV. Among these are the smaller $\Gamma_6 - \Gamma_8$ energy gap due to the presence of Mn (-125 meV versus -300 meV) and the wider wells and narrower barriers for the $Hg_{0.96}Mn_{0.04}Te$—CdTe superlattice.

To date, these new calculations have been able to identify the cyclotron resonance transition labelled $1 \rightarrow 2$ in Figure 3 as the $-1 \rightarrow 0$ transition with $\Lambda \sim 300$ meV. Of greater importance with regard to this paper, only when the temperature dependence of the Mn magnetization is taken into account, as in the analysis presented earlier, is the observed shift of this cyclotron resonance line, ΔB_r, in good quantitative agreement with the calculations. Thus, although final interpretation of the experimental results reported herein must await the completion of these new calculations, preliminary indications suggest that the new calculations will not dramatically alter the conclusion: magneto-absorption

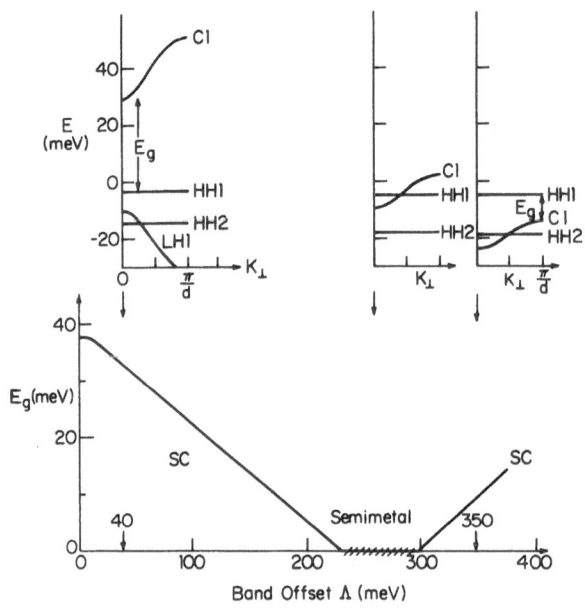

Figure 7   Band gap, E_g, versus band offset Λ for 100Å HgTe-36Å CdTe superlattice. Above are band structures for $k_{\parallel}=0$ for $\Lambda = 40, 260,$ and 350 meV. [Figure from Ref. 7]

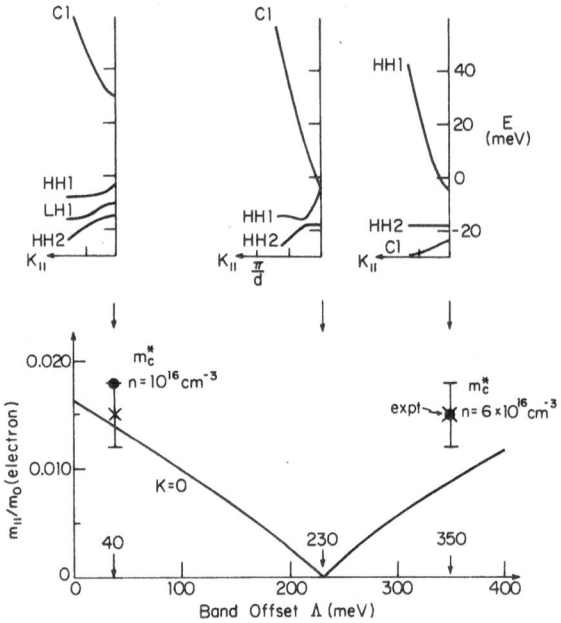

Figure 8   In-plane effective mass, $m_{\parallel}$, (solid line) versus band offset Λ for 100-Å HgTe/36Å CdTe superlattice. Experimental values from Ref. 21 are shown as crosses with error bars. Solid circles are calculated values for the given densities Above are band structures in the planes of the superlattice for $\Lambda = 40, 230,$ and 350 meV. [Figure from Ref. 7]

experiments on a $Hg_{0.96}Mn_{0.04}Te-CdTe$ superlattice reveal evidence of the exchange interaction between the localized Mn d-electrons and the conduction band electrons.

ACKNOWLEDGEMENTS

We wish to thank G. Bastard, G. Barilero, and C. Rigaux for very helpful discussions. This work was performed while GSB was at the Ecole Normale Superieure with support from the North-Atlantic Treaty Organization (NATO). This work was also partially supported by DREF. The Groupe de Physique des Solides de l'Ecole Normale Superieure is associated with the Centre National de la Recherche Scientifique.

REFERENCES

1. R. N. Bicknell, R. W. Yanka, N. C. Giles-Taylor, E. L. Buckland, and J. F. Schetzina, Appl. Phys. Lett. 45, 92 (1984)
2. L. A. Kolodziejski, T. C. Bonsett, R. L. Gunshor, S. Datta, R. B. Bylsma, W. M. Becker, and N. Otsuka, Appl. Phys. Lett. 45, 440 (1984)
3. K. A. Harris, S. Hwang, Y. Lansari, J. W. Cook, Jr., and J. F. Schetzina, Appl. Phys. Lett. 49, 713 (1986)
4. D.D. Awschalom, J.M. Hong, L.L. Chang, and G. Grinstein, Phys. Rev. Lett. 59, 1733 (1987).
5. G.S. Boebinger, Y. Guldner, J.M. Berroir, M. Voos, J.P. Vieren, and J.P. Faurie, Phys. Rev. B 36, 7930 (1987).
6. R. E. Kremer, A. M. Witowski, M. Jaczynski, and J. K. Furdyna, in Physics of Narrow Gap Semiconductors, Vol. 152 of Lecture Notes in Physics, edited by E. Gornik (Springer, Heidelberg, 1982), p. 307.
7. N.F. Johnson, P.M. Hui, and H. Ehrenreich, Phys. Rev. Lett. 61, 1993 (1988).
8. J.M. Berroir, Y. Guldner, J.P. Vieren, M. Voos, X. Chu, and J.P. Faurie, Phys. Rev. Lett. 62, 2024 (1989).
9. S.P. Kowalczyk, J.T. Cheung, E.A. Kraut, and R.W. Grant, Phys. Rev. Lett. 56, 1605 (1986).
10. T.M. Duc, C. Hsu, and J.P. Faurie, Phys. Rev. Lett. 58, 1127 (1987).
11. C.K. Shih and W.E. Spicer, Phys. Rev. Lett. 58, 2594 (1987).
12. G. Bastard, Phys. Rev. B 24, 5693 (1981) and Phys. Rev. B 25, 7584 (1982)
13. A. Fasolino and M. Altarelli, Surf. Sci. 142, 322 (1984)
14. X. Chu, S. Sivananthan, and J. P. Faurie, Appl. Phys. Lett. 50, 597 (1987).
15. C.G. Van de Walle and R.M. Martin, Phys. Rev. B 34 5621 (1986).
16. C. R. Pidgeon and R. N. Brown, Phys. Rev. 146, 575 (1966).
17. See M. H. Weiler, in Semiconductors and Semimetals, edited by R. K. Willardson and A. C. Beer (Academic, New York, 1981), Vol. 16.
18. G. Bastard, C. Rigaux, Y. Guldner, J. Mycielski, and A. Mycielski, J. Phys. (Paris) 39, 87 (1978).
19. G. Bastard, C. Rigaux, Y. Guldner, A. Mycielski, J. K. Furdyna, and D. P. Mullin, Phys. Rev. B 24, 1961 (1981).
20. W. Dobrowolski, M. von Ortenberg, A. M. Sandauer, R. R. Galazka, A. Mycielski, and R. Pauthenet, in Physics of Narrow Gap Semiconductors, Vol. 152 of Lecture Notes in Physics, edited by E. Gornik (Springer, Heidelberg, 1982), p. 302.

21. J. R. Anderson, M. Gorska, L. J. Azevedo, and E. L. Venturini, Phys. Rev. B 33, 4706 (1986).
22. J. M. Berroir, Y. Guldner, J. P. Vieren, M. Voos, and J. P. Faurie, Phys. Rev. B 34, 891 (1986).
23. G. Barilero and C. Rigaux (private communication).

POLARON COUPLING IN GaAs–GaAlAs HETEROSTRUCTURES OBSERVED BY CYCLOTRON AND MAGNETOPHONON RESONANCE

R.J. Nicholas*, D.J. Barnes*, D.R. Leadley*, C.J.G.M. Langerak[+], J. Singleton[+], P.J. van der Wel[+], J.A.A.J. Perenboom[+], J.J. Harris[#] and C.T. Foxon[#]

* Clarendon Laboratory, Parks Rd., Oxford, OX1 3PU, U.K.

[+] High Field Magnet Laboratory, Nijmegen University, Toernooieveld 65225 ED Nijmegen, NL

[#] Philips Research Laboratory, Redhill, Surrey, U.K.

Cyclotron resonance has been studied for 2-D electrons in GaAs–GaAlAs heterojunctions as a function of energy and electron concentration. The strength of the resonant polaron coupling is a strong function of the 2-D carrier density indicating the importance of screening and Landau level occupancy. For low density coupling is seen to the L.O. phonons which is slightly stronger than found in bulk GaAs, and mass enhancements as large as 10% have been observed. At high carrier concentrations the resonant coupling is almost totally suppressed by the exclusion principle, when the lowest Landau level approaches full occupancy at the polaron resonance condition. At higher temperatures the polaron coupling has different temperature variations depending on the energy and carrier concentration. For high concentrations second harmonic polaron coupling could be seen.

The temperature dependence of the cyclotron resonance linewidth at non-resonant fields can be used to study the phonon scattering, since it is relatively insensitive to long range order. For temperatures above 50K, the width is limited by optic phonon scattering, while at lower temperatures comparison with theory suggests that scattering from screened residual bulk acceptors in the GaAs is dominant.

Magnetophonon resonance due to the absorption of L.O. phonons has been studied in a range of heterojunctions. The amplitude and damping of the resonances are found to be a strong function of both carrier concentration and spacer layer thickness, with the damping determined mainly by the spacer layer. The resonance amplitude appears to increase at higher electron concentrations, despite some increase in damping. In tilted magnetic fields the heterojunctions show a very strong suppression of the resonance amplitudes, which does not occur for the quantum wells. This difference is presumably related to the assymetric form of the potential well binding the carriers in the heterojunctions.

1. INTRODUCTION

The interaction of electrons with optic phonons in semiconductors is one of the most fundamental problems in solid state physics. There is a strong coupling of electrons to the longitudinal optic (LO) phonons in ionic solids due to the long range electric field of the polarization wave. Probably the best known example of this is the formation of the polaron, in which the motion of an electron through an ionic solid is described as a composite particle, consisting of an electron dressed by a virtual phonon cloud. The strength of the coupling to phonon modes is defined by a dimensionless coupling constant, which for LO phonons is the Fröhlich constant, α, given by[1]

$$\alpha = \frac{e^2}{4\pi\epsilon_0\hbar} \left[\frac{1}{\epsilon_\infty} - \frac{1}{\epsilon_0} \right] \left[\frac{m^*}{2\hbar\omega_{LO}} \right]^{1/2} \qquad (1)$$

where ϵ_∞ and ϵ_0 are the high and low frequency dielectric constants, ω_{LO} the phonon frequency and m^* the effective mass. A typical value for α will be of order 0.07, using the parameters for GaAs for example. Added impetus has been given to the study of polaron effects by the increased control and flexibility available from the study of heterostructures which allow a variation in both the electronic and phonon properties.

Polaron coupling influences the electronic properties though coupling to both real phonons (scattering processes) and virtual phonons (mass renormalisation). At high temperatures LO phonon absorption plays a dominant role in limiting the momentum relaxation time and determining the carrier mobility. In magnetic fields it can be shown to be the dominant cyclotron resonance broadening factor above 50K, and resonant phonon absorption can be observed to cause oscillations in the magnetoresistance, known as the magnetophonon effect[2,3]. This will be the subject of the second half of this paper. Virtual phonons make up the lattice distortion which forms the renormalised particle known as the polaron. The main result of this is that the effective mass is altered, and becomes strongly frequency dependent.

In the weak coupling limit, the Frohlich polaron interaction leads to a renormalisation of the carrier effective mass at low frequencies, given by $m^*_{pol} = m^*(1 - \alpha/6)^{-1}$, and to the resonant polaron effect at higher frequencies, as the optic phonon frequencies are approached. This is often probed using a magnetic field, by varying the cyclotron frequency ω_c ($=eB/m^*$), as is shown schematically in fig. 1, where the unperturbed electron Landau levels are shown as dashed lines and the polaron states as solid lines. Also shown is the energy of the lowest Landau level ($n_L = 0$) plus one phonon. The point where this crosses the energy of the second Landau level defines the resonant polaron coupling, where the two states would become degenerate without the interaction. A similar coupling will occur at around a series of lower fields, as the higher Landau levels each hybridise successively with the one phonon (n_L = 0) state. Theoretical calculations of the resonant polaron regime are difficult, and have led to an extensive literature (3 – 6 and references therin).

In the case of 2–D systems it was at first thought that polaron effects should be stronger than in the corresponding bulk materials[7–10], but the finite wave functions in the third dimension[10–12], and both screening[11,13,14], and level occupany (Pauli principle) should reduce the coupling. Experiments on various 2–D systems have suggested both enhanced[15] and more often reduced[14,16–21] effects. Recently Langerak et al[21] have reported an extensive study of polaron coupling in GaAs–GaAlAs heterojunctions, which resolves some of the conflicting aspects of the data by examining in detail the influence of the carrier concentration and hence level occupany on the coupling. These results are now well described by theories which include the effects of degenerate statistics and hence level occupancy[22].

2. RESONANT POLARON COUPLING

A typical experiment to study the resonant polaron coupling in 2–D consists of a measurement of the frequency dependence of the cyclotron effective mass. Typical results are shown for this in fig. 2, which also shows the results of theoretical fits using a memory function approach in which the shift in the cyclotron resonance is calculated directly, including the effects of dynamic screening, Landau level occupancy and the finite thickness of the 2DEG[22]. It is also necessary to include the influence of band non–parabolicity, which causes the mass to increase linearly with energy, E, in the form

$$\frac{1}{m^*_b} = \frac{1}{m^*_o} \left[1 + \frac{2K_2}{E_g} (E + <T_z>) \right] \qquad (2)$$

where m^*_o is the band edge effective mass, $<T_z>$ is the kinetic energy from the confining potential and K_2 is a measure of the band non–parabolicity which is found to be −1.4 for bulk GaAs[23]. When all of the above factors have been taken into account, it can be seen that theory now provides a quite adequate description of the resonant polaron coupling. The overall result of the different factors is that at low energies the total energy dependence of the effective mass is actually higher in bulk GaAs[23] than for the 2DEG[20].

The understanding of the situation becomes much clearer when the carrier concentration is varied, and the results are extended into the region close to the

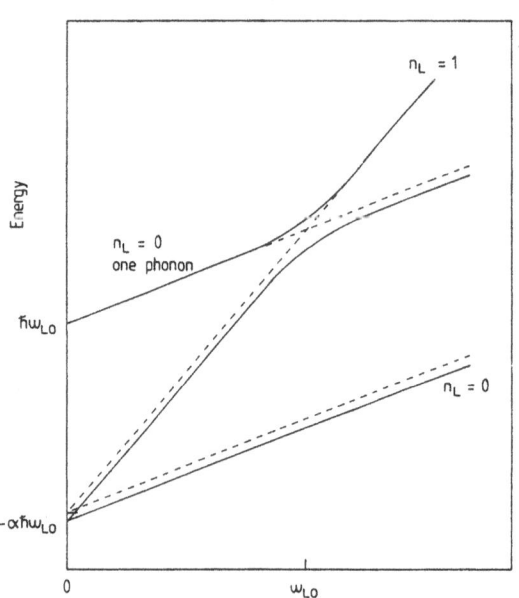

Fig. 1 A schematic view of the Landau levels in a polar optic solid.

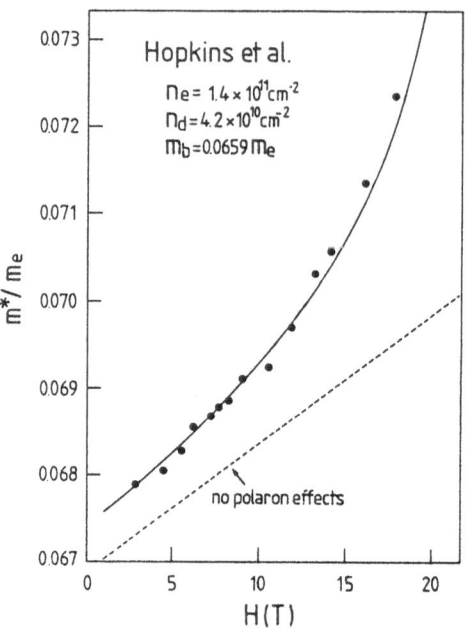

Fig. 2 The magnetic field, and hence frequency, dependence of the effective mass in a GaAs-AlGaAs heterojunction, compared with theory[22].

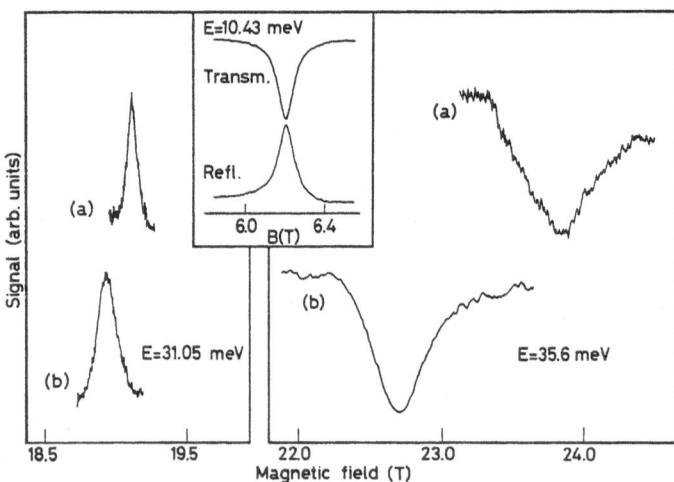

Fig. 3 Typical experimental recordings of the reflectivity of a GaAs-GaAlAs heterojunction. The Upper traces (a) have a carrier concentration $N_S = 3.4 \times 10^{11}$ cm^{-2}, while the lower (b) ones have $N_S = 5.2 \times 10^{11}$ cm^{-2}.

454

Reststrahlen band, where the photon energy becomes very close to the LO phonon. In this case the interpretation of the experimental data becomes more critical, due to the strong frequency dependence of the dielectric function. It is necessary to perform detailed fits of the optical response of the structures studied[24], and to use reflectivity to study thicker layers in the Reststrahlen band[21]. Some typical experimental results are shown in fig. 3, for a GaAs–GaAlAs heterojunction. It can be seen that at low energies the reflectivity produces a maximum at the same point as the transmission minimum, but for a photon energy above the T.O. phonon (33.9 meV), the resonance becomes a minimum in reflectivity. The figure also illustrates the interesting point that increasing the carrier concentration leads to a decrease in the resonance field, and hence the effective mass. This is shown quantitatively in fig. 4, where the effective mass shows a systematic decrease as a function of increasing carrier concentration, which becomes much stronger as the frequency approaches the LO phonon (36.7 meV). The origin of this behaviour is the suppression of the resonant polaron coupling due to the increasing occupancy of the lowest Landau level. Since the resonant polaron coupling is a second order effect it involves transitions in which a virtual optic phonon is absorbed and simultaneously emitted by an electron which must fall back into the lowest Landau level. If this level is almost filled the interaction will be strongly suppressed[22,24].

The importance of the carrier concentration to the energy dependence of the effective mass, and its relation to the level occupancy can be seen in fig. 5. This shows the energy dependence of the effective mass m^*/m^*_b, after normalising to the band mass m^*_b and thus removing the effects of non–parabolicity (with Eq. 2). The energy dependence is now entirely the result of the polaron coupling. At low densities (around 1×10^{11} cm^{-2}) the energy dependence is very similar in both the bulk and the 2–D systems, with some indications that the resonant interaction may be a little stronger in 2–D as the LO phonon is approached. As the carrier concentration rises the resonant coupling is rapidly suppressed. This is seen most dramatically for the data at 5.4×10^{11} cm^{-2}. This has an occupancy ($\nu = N_s h/eB$) of 2 at 11T (17 meV) and 1 at 22 T (34 meV), which will correspond to the total suppression of the polaron coupling when the spin splitting is either resolved ($\nu = 1$) or not ($\nu = 2$). The results suggest almost total removal of the polaron coupling up to 34 meV, and thus complete suppression due to a filled lowest spin state of the lowest Landau level. The theoretical curve, taken from the results of Peeters et al[22], shows less suppression of the coupling due partly to the lower carrier concentration (4×10^{11} cm^{-2}) and partly due to the neglect of spin splitting. The effects of screening were found theoretically[22] to be less important than the occupancy, even with a full treatment of dynamic screening, and to be only weakly frequency dependent.

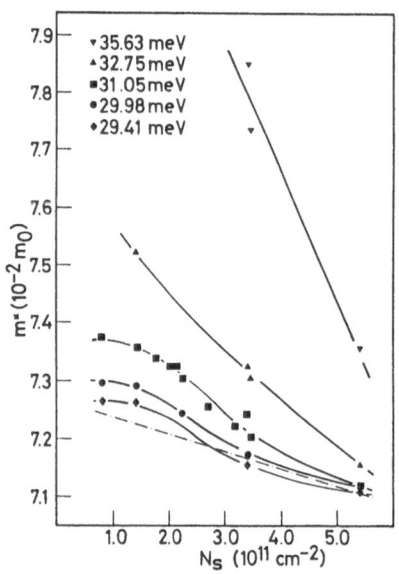

Fig. 4 The N_S dependence of the
effective mass at a variety of
different energies.

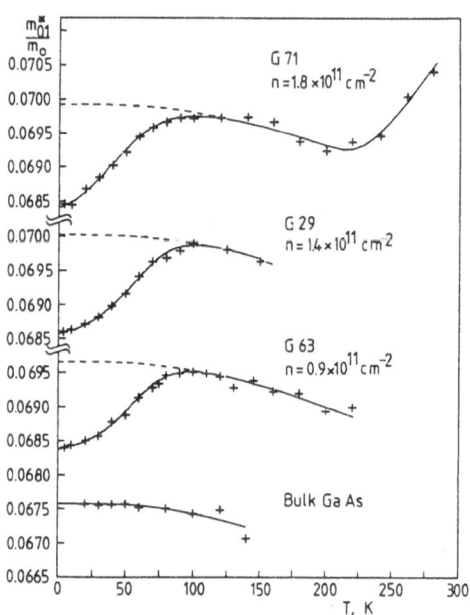

Fig. 6 The temperature dependence
of the effective mass in three
heterojunctions, compared
with results for bulk GaAs.

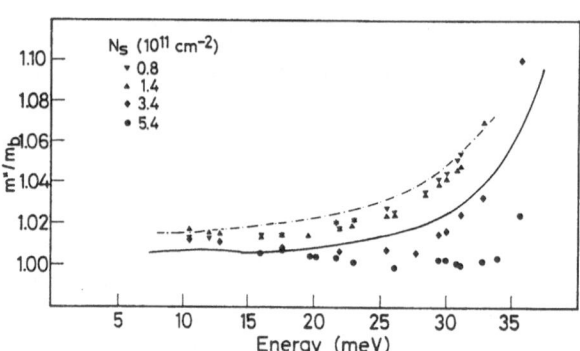

Fig. 5 The energy dependence of the effective mass for different
carrier concentrations. The dashed-dotted line shows results from bulk
material, and the solid lines shown the theoretical results[22].

The effects of temperature on the polaron coupling are not yet fully explained, with suggestions that both screening[19] and polaron non-parabolicity[25,26] may result in significant changes in the effective mass. Fig. 6 shows the low frequency effective mass as a function of temperature for a number of low density heterojunctions[19], again corrected for the change in band non-parabolicity. The results indicate that there is a rapid increase in the effective mass up to approximately 80K. The increase in mass, of order 2%, is quite close to the size of the total polaron contribution at low frequencies, as calculated for the finite 2-D system in the low density limit[11-14]. This led to the suggestion that the polaron mass enhancement was becoming larger at high temperatures due to the removal of the suppression effects caused by the occupancy and screening as described above, and that these are then removed once the carriers become spread over a wider range of states with broader levels, caused by the increasing temperature. Calculations suggest however that the intrinsic polaron non-parabolicity[25,26] should also play an important role, contributing of order 0.5 - 1% to the mass increase.

At higher temperatures (>10K) in samples with a higher carrier concentration (>3.5 x 10^{11} cm^{-2}), a prominent new anomaly can be seen in the energy dependence of the effective mass at energies corresponding to resonant polaron coupling between the N=0 and N=2 Landau levels. For cyclotron resonance energies just below half the LO phonon energy there is an enhanced increase in mass with temperature, while for energies just above half the LO phonon, a small reduction in mass is initially seen on raising the temperature from 4K. This behaviour is shown in fig. 7, which plots the mass change from 4 to 80K as a function of energy. The effect is visible because the increasing temperature removes the occupation effects in the N=0 Landau level suppressing the resonant interaction, and also populates the N=1 Landau level, allowing resonances involving transitions to the pinned N=2 level to be observed. At higher energies the fundamental polaron coupling takes over as the dominant effect, and the mass becomes increasingly temperature dependent. This is due again to the occupancy of the N=0 Landau level, which is decreasing as the statistics become less degenerate, thus allowing more resonant polaron coupling.

3. TEMPERATURE DEPENDENT LINEWIDTHS

The cyclotron resonance linewidth is of interest as a measure of the scattering time because, being a local transition, the linewidth is largely determined by potential fluctuations of range less than or of the order of the cyclotron radius (homogeneous broadening)[26]. Longer range fluctuations (inhomogeneous broadening) will have a similar effect on all the Landau levels and so will make little contribution to the lifetime broadening in the linewidth. Ando and Uemura[26] have calculated that the cyclotron resonance halfwidth for short-range scattering potentials $\Delta B_{\frac{1}{2}}(SR)$ can be expressed as

$$\Delta B_{\frac{1}{2}}(SR) = C \ (B_R/\mu_0)^{\frac{1}{2}} \tag{3}$$

where C is a constant of order $\sqrt{(2/\pi)}$. The $\sqrt{B_R}$ dependence of the linewidth is in contrast to classic theory, which predicts a field independent linewidth ($\Delta B_{\frac{1}{2}}/B_R = 1/\omega\tau$), and originates from the field dependence of the Landau level degeneracy and the self-consistency of the scattering rate.

The field dependence of the resonance linewidth at 4.2 K measured in samples with low mobilities[27,28-30] are in general agreement with Eq. (1), giving values for C of 0.65±0.03. Interface roughness or, in the case of GaInAs, alloy scattering have been suggested as the dominant scattering mechanisms. More recent studies of high mobility GaAlAs–GaAs heterojunctions have shown that the field dependence of the linewidth is more complex, involving both the level occupancy[31,32] and factors depending on the carrier concentration and the presence of impurities[33].

In figure 8 the true resonance linewidth ($\delta B_{\frac{1}{2}}$) is plotted as a function of temperature for four high mobility GaAs–GaAlAs heterojunctions with different spacer layer thicknesses. In order to deduce $\delta B_{\frac{1}{2}}$ it is necessary to correct for the finite carrier concentration, using a Drude formalism[34]. This gives an experimental half width $\Delta B_{\frac{1}{2}}$ which is related to the true linewidth $\delta B_{\frac{1}{2}}$ by the relation

$$\Delta B_{\frac{1}{2}} \ = \ \delta B_{\frac{1}{2}} \ + \ m^* \omega_p/e \tag{4}$$

where $\omega_p = Z_0 N_s e^2/m^*(1 + \sqrt{\epsilon})$, Z_0 is the impedance of free space and ϵ is the dielectric constant of GaAs. The result of this is that there is an additive finite linewidth even for a perfect system where τ, $\mu \rightarrow \infty$, and for higher carrier concentrations this can seriously limit the accuracy of the value of $\delta B_{\frac{1}{2}}$ (e.g. for $N_s = 2 \times 10^{11}$ cm^{-2}, $m^*\omega_p/e = 25$ mT). The right hand scale of the figure shows the cyclotron lifetime τ_c, deduced from $\tau_c = m^*/e\delta B_{\frac{1}{2}}$.

At high temperatures (> 50 K) the halfwidths of all four samples have similar values and temperature dependences, increasing approximately as T^2. Below 50 K $\delta B_{\frac{1}{2}}$ increased less rapidly and at a rate which was fastest for the thickest spacer layer samples. By ~1.5 K values as low as $\delta B_{\frac{1}{2}}/B_R \approx 1/2000$ were found. These results can be explained qualitatively as follows: at high temperatures the linewidth is dominated by optic-phonon scattering, which decreases rapidly with temperature, while

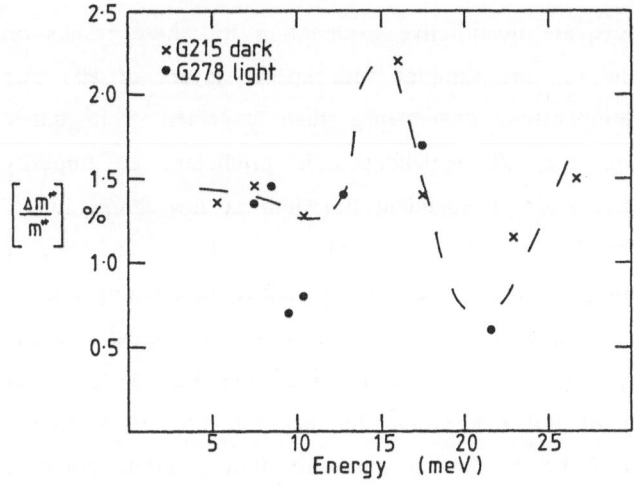

Fig. 7 The temperature dependent change of effective mass between 4.2K and 80K, as a function of energy, for two samples with carrier concentrations of 5.4×10^{11} cm^{-2}.

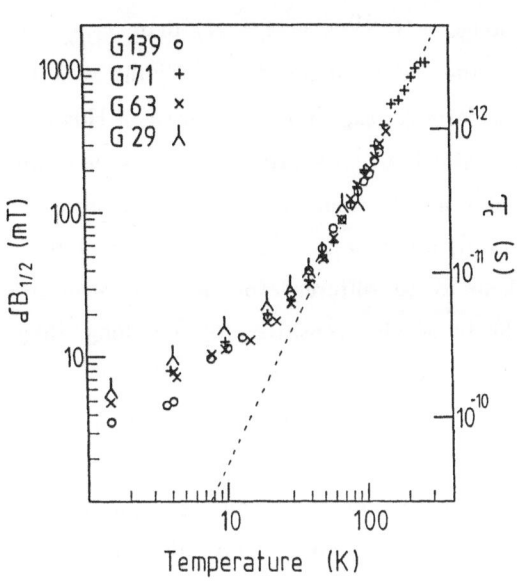

Fig. 8 The cyclotron linewidth as a function of temperature, for samples with spacer layer thicknesses of 1600 Å, (G139), 800 Å, (G63) and 400 Å, (G29, G71).

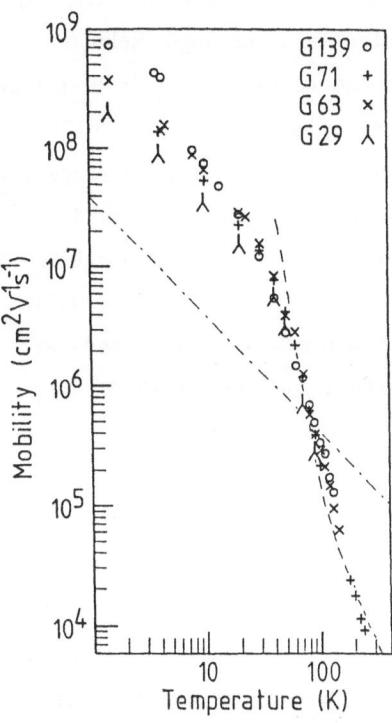

Fig. 9 The electron mobility as deduced from Cyclotron linewidth. Dashed line is theory[39].

at lower temperatures the scattering rate becomes increasingly limited by acoustic–phonon, ionized–impurity and possibly alloy–disorder scattering, which are less temperature dependent. However, linewidth theories in which screening is not treated self–consistently do not give an adequate quantitative explanation for these results on two points. Firstly, the linewidths for the samples with spacer layers of 800 and 1600 A have a rather greater temperature dependence than expected from either acoustic–phonon scattering, for which a $\sqrt{T}$ dependence is predicted, or impurity scattering, which should give a temperature independent linewidth at low temperatures. Secondly, the low–temperature linewidths of all the samples studied are considerably smaller than expected from theoretical calculations of the resonance broadening due to acoustic phonons[35,36]. For example, substitution of the measured low–field mobility (μ_0) at 4.2 K for a typical sample (0.75×10^6 cm^2/Vs) into the Eq. 3 for short range scatterers, gives a halfwidth of 0.059 T, while for acoustic phonon scattering theory[35] gives a halfwidth of 0.21 T in the low temperature limit. These compare with a measured halfwidth of 0.005 T.

As mentioned before, the relationship between the low–field D.C. mobility (and its associated D.C. scattering time $\tau_{D.C.}$ (= $\mu m^*/e$) and the resonance linewidth is not straightforward. The high field scattering time τ_c deduced from $\delta B_{\frac{1}{2}}$ and shown in fig. 10 can be over an order of magnitude longer (10^{-10} s at 1.5 K) than $\tau_{D.C.}$ for the same samples ($0.8 - 3 \times 10^{-11}$ s). In order to compare the linewidths with the low field mobilities the short range scattering formula Eq. 3 can be used. However, the linewidth in the quantum limit is determined by scattering processes involving wavevectors q fixed by the magnetic field, and not by the Fermi wavevector, as in the case of the low field mobility at low temperatures. Thus the constant of proportionality between $\delta B_{\frac{1}{2}}$ and $\mu_0^{-\frac{1}{2}}$ is likely to be different for different scattering mechanisms, and in particular $\delta B_{\frac{1}{2}}$ is unlikely to be sensitive to the long range potential fluctuations which limit $\tau_{D.C.}$ in wide spacer layer samples.

This is illustrated in Fig. 9, where C' $(\Delta B_{\frac{1}{2}})^{-2}$ is compared[37] with the low field mobilities μ_0 calculated by Walukiewicz et al[38] for a GaAs–GaAlAs heterojunction with $n_s = 2.2 \times 10^{11}$ cm^{-2} as a function of temperature. The constant C' was chosen to give a fit between μ_0 and the results around 200K. This gives a value of C' = 0.9 T, which corresponds to a value of C = 0.38 in Eq. 3. This is quite close to the value of 0.65 expected for short range scattering, and at temperatures above ~60 K there is good agreement between the temperature dependence of the mobility derived from the linewidths and the calculated optic–phonon limited mobility. Even closer agreement would in fact be given by a simple power law, with $\mu \propto T^4$. Below 50 K the mobility derived from the linewidth, $\mu(CR)$, falls below the calculated optic–phonon limited mobility but remains much larger than the acoustic–phonon limit.

The temperature dependence is however quite close to that expected for the acoustic–phonons ($\propto$ T). Finally, at around 2 K, the mobility and linewidth show signs of saturating, and there is little further narrowing when measurements are made at even lower temperatures. The values of μ(CR) at low temperature are much larger than the measured values of μ_0, or even those which would be deduced from the high field relaxation time τ_c.

The most obvious reasons for a discrepancy between the theoretical and experimental low–field mobilities and μ(CR) are the range of the scattering potentials and the enhanced effects of free–carrier screening in high magnetic fields. Numerical calculations of the cyclotron resonance linewidth where screening is included self–consistently have been made by Lassnig et al.[39] and Ando and Murayama[40], who calculated that when $N_A = 1 \times 10^{14}$ cm^{-3} the minimum linewidth is of order 0.01 T at 5 T. Given that for such samples N_A probably lies[41] in the range of $2 - 4.5 \times 10^{14}$ cm^{-3}, these calculated values are closer to the experimental measurements, although still somewhat bigger. The large temperature dependence of the linewidth is also consistent with the self–consistent screening calculations of Lassnig et al.[39], which predict a strongly temperature dependent linewidth even when limited by ionized impurity scattering. The conclusion that scattering by the background impurities in the GaAs is dominant, as suggested by these theories, is also consistent with the weak dependence of the linewidths on spacer layer thickness, as can be seen in Figs. 8 and 9, despite almost an order of magnitude change in carrier concentration. In a recent study of a slightly lower mobility heterojunction at low temperatures Chou et al[42] came to a similar conclusion, that scattering from the acceptors in the GaAs was dominant.

One process which may also be significant in limiting the cyclotron linewidth at high fields is alloy scattering due to the penetration of the wavefunction into the GaAlAs barrier. Calculations[38] show that the barrier penetration is almost linearly proportional to n* in the density range up to around 5×10^{11}cm^{-2}, where n* = (n_s + (32/11)n_{dep}); $\dot{n}_{dep}$ is the depletion charge in the GaAs. The fraction of the wavefunction inside the barrier is only of the order[38] of 1% and the alloy scattering limited low field mobilities are of order 10^8 cm^{2}V^{-1}s^{-1} and above for low n* values at low temperatures[38]. Since alloy scattering is a short range process we would expect the mobility values deduced from Equ. (5) to be a good estimate, and the data of fig. 9 suggest that the alloy scattering may well be responsible for limiting the linewidth and scattering times in the extreme case of high magnetic fields and low temperatures, despite its negligible contribution to the low field mobility. This would also account for the weak carrier concentration dependence of the linewidths.

4. THE MAGNETOPHONON EFFECT

The magnetophonon effect is the result of a resonant inelastic scattering process in high magnetic fields, usually involving LO phonons. It is characterised experimentally by an oscillatory component of the magnetoresistance, periodic in 1/B, whose amplitude reaches a few per cent of the monotonic component at maximum. The effect was first observed in the magnetoresistance of InSb, and has subsequently been observed in a number of bulk[3] and two dimensional[2] materials.

The normal magnetophonon effect arises from the absorption of optic phonons, which are essentially monoenergetic, causing transitions between well resolved Landau levels. In high fields the conductivity σ_{xx}, is proportional to the scattering rate, and hence to the joint density of states for transitions between two levels. The conductivity thus exhibits a maximum each time the LO phonon energy is equal to the separation of one or more Landau levels, giving the resonance condition:

$$N \; \hbar\omega_c \; = \; \hbar\omega_{LO}, \qquad \text{with } N = 1, \; 2, \; 3, \; .... \qquad (5)$$

The resistivity also shows a similar series of maxima since

$$\rho_{xx} \; = \; \frac{\sigma_{xx}}{\sigma_{xx}^2 \; + \; \sigma_{xx}^2} \qquad (6)$$

and usually $\mu B \gg 1$, $\sigma_{xy} \gg \sigma_{xx}$, so that $\rho_{xx} \propto \sigma_{xx}$. The resulting oscillations are periodic in 1/B, with the positions of the individual extrema given by

$$NB_N \; = \; m^*\omega_{LO}/e \qquad (7)$$

where the effective mass m^* must be adjusted from the band edge value to take account of non-parabolicity and also the resonant polaron coupling, which will produce a mass renormalisation at the resonance condition. This latter effect takes account of the fact that all the resonances occur, by definition, at the field values where resonant polaron coupling is strong. This has been shown empirically, in bulk material, to increase the mass[3] by a factor of $(1 + \alpha/4)$. The value for 2-D systems is not known, however the cyclotron resonance studies above suggest that the strength of resonant polaron coupling may be comparable in two and three dimensions, but may also be carrier concentration dependent.

The amplitude of the oscillations is limited by two factors. There must be a sufficient population of optic phonons to cause scattering, and so the amplitude decreases at low temperatures. The Landau levels must also be sufficiently well resolved to give a sharp resonance structure, and since there is increased broadening of the levels at higher temperatures (due to scattering by both optic and acoustic phonons) there is a maximum in amplitude, typically in the region 150K – 200K. The details of the broadening will depend upon the individual sample studied, however the main factors are thought to be due to remote impurity scattering[43] and optic phonon self–broadening, which is dominated by phonon <u>emission</u> in the temperature region below 300K.

The oscillatory part of the magnetoresistivity can be described as a damped oscillation given by the relation

$$\frac{\Delta\rho}{\rho_0} = A \sum_{r=1}^{\infty} \exp\left(-\gamma r\omega_{LO}/\omega_c\right) \cos(2\pi r\omega_{LO}/\omega_c) \tag{8}$$

where γ is the damping factor describing the decay of the oscillations at low field, and ρ_0 is the zero field resistivity. In practice there is little peak sharpening and only terms with r=1 need be considered in most cases. This expression was first introduced empirically, and is also found to come from theoretical treatments[43,44] which treat the broadening as Lorentzian with width Γ (giving $\gamma = 4\pi\Gamma/\hbar\omega_{LO}$). Deviations from this formula occur when γ (Γ) becomes magnetic field dependent.

Typical experimental recordings of magnetophonon oscillations in the resistivity of a high mobility GaAs–GaAlAs heterojunction are shown in fig. 10. These are taken by subtracting a voltage linearly proportional to the magnetic field, in order to remove the majority of the background magnetoresisitance, while still allowing quantitative measurements to be made of the oscillation amplitudes. The oscillations show the characteristics expected, with the amplitude peaking at around 200K, and the damping of the oscillations becoming less at lower temperatures. These features can be seen more directly in figures 11 and 12, which show the temperature dependence of the amplitude and damping factor explicitly. The higher harmonic number oscillations peak at a lower temperature as they are more sensitive to the amount of broadening present through the damping factor γ, and the value of γ falls with temperature as seen in fig. 12.

In an attempt to determine the systematic dependence of the magnetophonon amplitudes on factors such as electron concentration (n_e) and scattering processes, we have studied a range of GaAs–GaAlAs structures grown with varying spacer layer thicknesses, and we have also used photoexcitation to change the carrier concentration

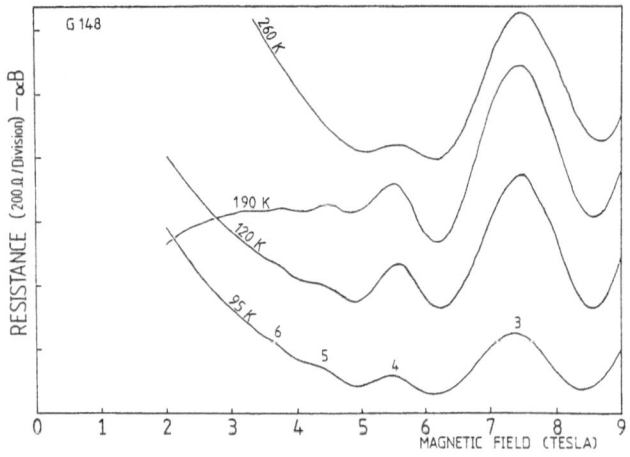

Fig. 10 Experimental recordings of the magneto-resistance of a GaAs-GaAlAs heterojunction, with a linear voltage subtracted to compensate the background magnetoresistance.

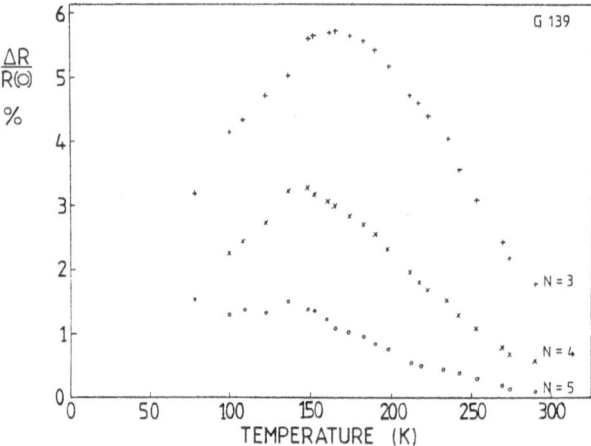

Fig. 11 The temperature dependence of the oscillation amplitudes for three of the magnetophonon resonances.

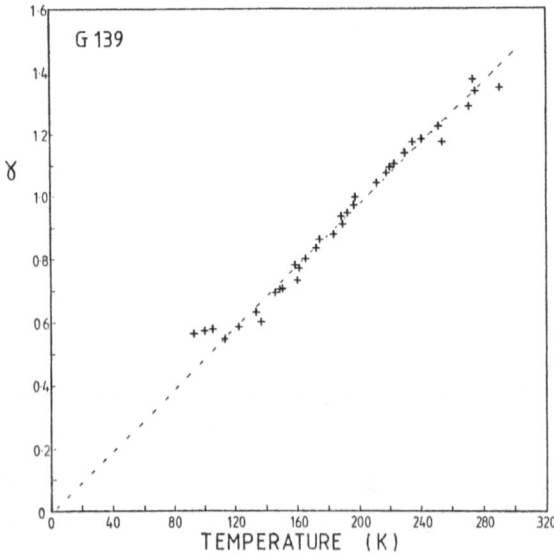

Fig. 12 The temperature dependence of the damping factor γ, deduced for the lower field resonances.

within one sample. The amplitude appears to follow a systematic dependence upon carrier concentration, as shown in fig. 13. This shows the amplitudes of the N=2 and N=3 harmonic resonances, together with the predicted strength of the fundamental N=1 resonance, deduced from the N=2 value and the measured damping parameter γ_{23}. The points show that at high carrier concentrations, with data taken from samples with different spacer thicknesses, the oscillation amplitudes fall, while for single samples with n_e varied by illumination the amplitudes rise with n_e (shown as pairs of points connected by a dashed line). At low carrier densities the amplitudes rise almost linearly with n_e, as can be seen best for the amplitude of the N=1 resonance, which is where the damping of the oscillations plays the least important role in determining the amplitude.

We can form a coherent view of these results if we distinguish two factors, the carrier concentration, and the remote impurity scattering which act essentially independently to influence the amplitude. The remote impurity scattering determines the damping of the oscillations, while the carrier concentration seems to determine the strength of the undamped coupling. The relation between the damping and the remote impurity scattering can be seen from fig. 14, in which the damping factor γ_{34} is shown to be a direct function of the spacer layer thickness in a series of samples with an essentially identical doping profile. (The one obvious exception being the sample at 100 Å, which had an n-type buffer layer). This conclusion is in good agreement with the suggestion and calculations of Kido et al[43], who concluded that this was the dominant effect in determining the amplitude of the N=3 oscillation, as reported in our earlier studies[45]. It should also be noted that the broadening is somewhat magnetic field dependent, and we find the approximate result that $\gamma_{34} - \gamma_{23} \approx 0.5$. Higher temperatures will then make a further additive contribution to γ, as shown in fig. 12. More surprising is the conclusion that the strength of the oscillations should be increased by increasing the carrier concentration, particularly when this occurs under conditions where the damping of the oscillations is increasing. This tendency does not seem to occur in the theoretical treatments published to date[43,44,46,47]. It might in fact be expected that an increase in carrier concentration would lead to a decrease in the strength of the coupling, through the action of screening and level occupancy, as found above for the resonant polaron coupling. It is conceivable, however, that both factors may act in the reverse manner by influencing the optic phonon level broadening coming from phonon emission. This process will be inhibited by high occupancies, thus leading to sharper levels and less broadening for a strongly coupled system. This would however be expected to lead also to a reduced damping of the oscillations. The measurements of the cyclotron resonance linewidth at high temperatures shown in fig. 8 demonstrate also that the

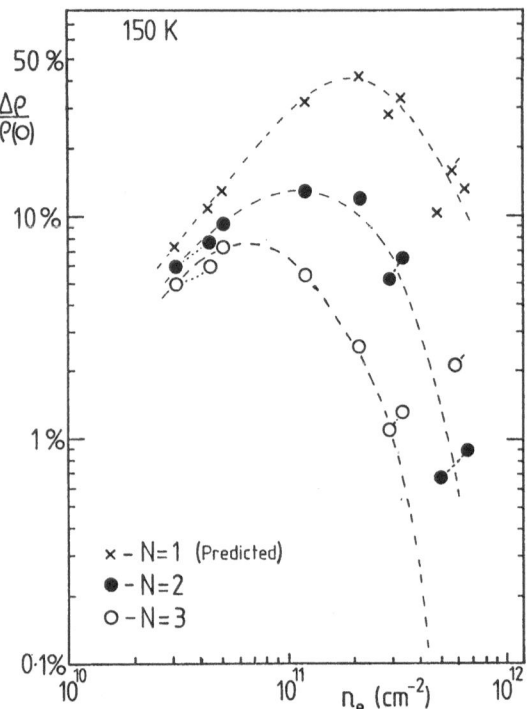

Fig. 13 The carrier concentration (n_e) dependence of the magnetophonon amplitudes in a range of different GaAs-GaAlAs heterojunctions. The value of n_e was varied by changing the spacer layer thickness, and by continuous illumination. Pairs of points connected by dashed lines show data from one sample. The amplitude of the N=1 resonance was predicted from that of N=2 and the assumption of the same damping factor as between N=2 and N=3.

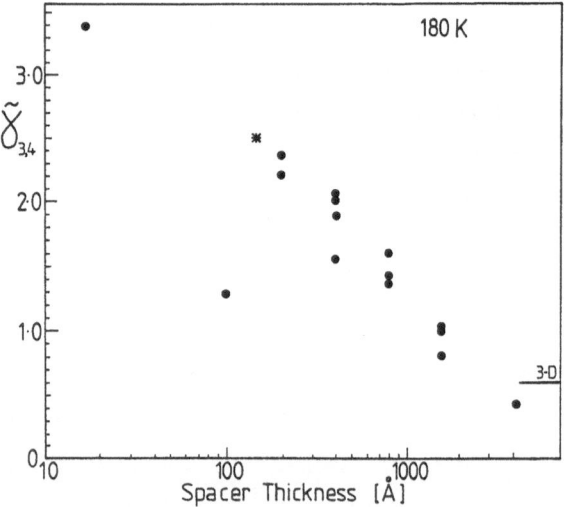

Fig. 14 The spacer thickness dependence of the damping parameter γ_{34} at 180K. The low point at 100Å is from a heterojunction with a slightly different structure.

non-resonant optic phonon scattering is independent of carrier concentration. The origin of the n_e dependence of the amplitudes is therefore not well understood.

An empirical relation to determine the amplitudes of the first three (N=1, 2, 3) oscillations in the structures studied is to a reasonable approximation

$$\frac{\Delta\rho}{\rho_0} = 0.02(n_e)^{3/2} \exp(-N\gamma)$$

(9)

where,

$$\gamma = 3.5 + T/200 - \tfrac{1}{2}\ln L_s$$

with n_e measured in units of 10^{10} cm^{-2} and L_s, the spacer thickness, in Å. In practice the best predictor and correlation between the magnetophonon amplitudes and any other single factor is the low temperature (4.2K) mobility. This is illustrated in fig. 15, which shows the magnetophonon amplitude (as predicted for the fundamental N=1 peak), as a function of 4K mobility, for samples covering the complete range of both carrier concentration and spacer layer thickness. This curve is particularly interesting, since it includes data at low mobilities with both low and high carrier concentrations. The factor of $n_e^{3/2}$ in equation (9), is thus highly reminiscent of the similar factor found in the low temperature mobility. It should however be remembered in this discussion that the low field transport mobility at the temperature where the oscillations are observed is almost completely independent of both the spacer layer thickness and carrier concentration, with the carriers exhibiting essentially non-degenerate statistics, despite the fact that it also is limited by optic phonon scattering.

One further factor found to influence the amplitude of the oscillations is the orientation of the magnetic field, relative to the sample normal. Fig. 16 shows just how sensitive the oscillations are to small tilt angles, with a rotation of around 10° sufficient to halve the amplitude[45]. This behaviour is almost certainly associated with the mixing of the electric subbands of the heterojunction, which may introduce additional broadening through mechanisms such as inter-subband scattering, or may lead to some additional screening of the phonon field[45]. Studies on quantum wells, mostly using GaInAs[48,49], show that tilting has little or no effect on the amplitude, due to the symmetric nature of the potential profile which makes subband coupling negligible.

In order to deduce the phonon frequencies from the resonance positions it is necessary to make accurate measurements of the effective mass, using cyclotron resonance, and then to apply a number of different corrections[2]. This has been

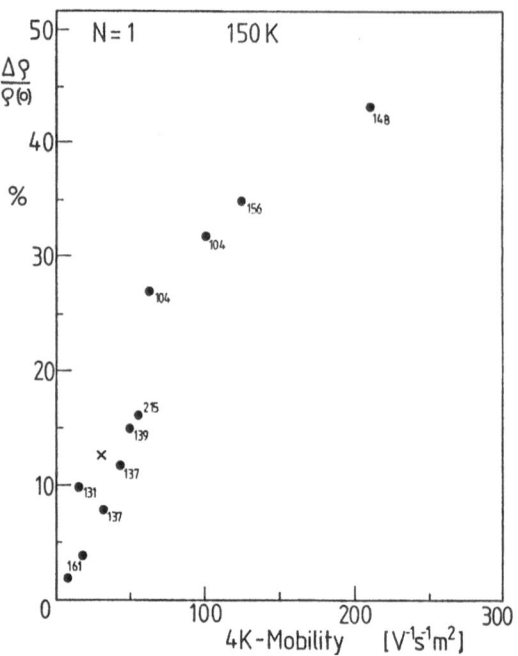

Fig. 15 The relation between the (predicted) amplitude of the N=1 magnetophonon resonance at 150K (deduced as described for fig. 13) and the measured 4.2K mobility for a range of GaAs heterojunctions.

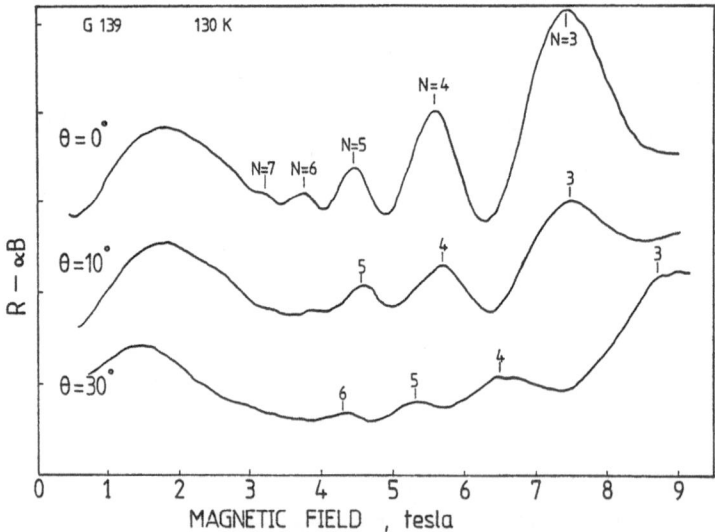

Fig. 16 Experimental recordings of the (compensated) magnetophonon oscillations, for various tilt angles, θ, relative to the magnetic field direction at 130K.

extensively discussed in our earlier work[19,45], and has led to the conclusion that the phonon frequencies lie intermediate between the LO and TO modes. It is not clear to what extent this corresponds to the presence of additional phonon scattering, or to a modification of the magnetophonon resonance condition in 2-D. Recently Warmenbol et al[44] have shown that the resonance positions may be shifted down in field by the presence of level broadening. This would be expected to lead to a systematic decrease in phonon frequency with spacer layer thickness, as the broadening becomes larger. We do not see any clear evidence for this behaviour in the samples studied to date.

5. ACKNOWLEDGEMENTS

We would like to acknowledge support from S.E.R.C.(U.K.), F.O.M., and N.W.O.(Netherlands).

6. REFERENCES

1. H. Frohlich, Adv. Phys. $\underline{3}$ 325 (1954)
2. R.J. Nicholas, Prog. Quantum Electron. $\underline{10}$ 1 (1985)
3. P.G. Harper, J.W. Hodby and R.A. Stradling, Rep. Prog. Phys. $\underline{36}$ 1 (1973)
4. D.M. Larsen, in "Polarons in Ionic Crystals and Polar Semiconductors", ed. J.T. Devreese, (North Holland, Amsterdam) (1971)
5. G. Lindemann, R. Lassnig, W. Seidenbusch and E. Gornik, Phys. Rev. $\underline{B28}$ 4693 (1983)
6. F.M. Peeters and J.T. Devreese, Phys. Rev. $\underline{B34}$ 7246 (1986)
7. S. Das Sarma and A. Madhukar, Phys. Rev. $\underline{B22}$ 2823 (1980)
8. S. Das Sarma, Phys. Rev. Lett. $\underline{52}$ 859 (1984)
9. F.M. Peeters and J.T. Devreese, Phys. Rev. $\underline{B31}$ 3689 (1985)
10. D.M. Larsen, Phys. Rev. $\underline{B30}$ 4595 (1984)
11. S. Das Sarma, Phys. Rev. $\underline{B27}$ 2590 (1983)
12. R. Lassnig and W. Zawadzki, Surf. Sci. $\underline{142}$ 361 (1984)
13. S. Das Sarma and B.A. Mason, Phys. Rev. $\underline{B31}$ 5536 (1985)
14. H. Sigg, P. Wyder, and J.A.A.J. Perenboom, Phys. Rev. $\underline{B31}$ 5253 (1985)
15. M. Horst, U. Merkt and J.P. Kotthaus, Phys. Rev. Lett. $\underline{50}$ 754 (1983)
16. W. Seidenbusch, B. Lindemann, R. Lassnig, J. Edlinger and E. Gornik, Surf. Sci. $\underline{142}$ 375 (1984)
17. J. Singleton, R.J. Nicholas and F. Nasir, Sol. State Commun. $\underline{58}$ 833 (1986)
18. M. Horst, U. Merkt, W. Zawadski, J.C. Maan and K. Ploog, Sol. State Commun. $\underline{53}$ 403 (1985)
19. M.A. Brummell, R.J. Nicholas, M.A. Hopkins, J.J. Harris and C.T. Foxon, Phys. Rev. Lett. $\underline{58}$ 77 (1987)
20. M.A. Hopins, R.J. Nicholas, M.A. Brummell, J.J. Harris and C.T. Foxon, Phys. Rev. $\underline{B36}$ 4789 (1987)
21. C.J.G.M. Langerak, J. Singleton, P.J. van der Wel, J.A.A.J. Perenboom, D.J. Barnes, R.J. Nicholas, M.A. Hopkins and C.T.B. Foxon, Phys. Rev. $\underline{B38}$ 13133 (1988)
22. F.M. Peeters, Wu. Xiaoguang, and J.T. Devreese, Surf. Sci. $\underline{196}$ 437 (1988); F.M. Peeters, Wu. Xiaoguang and J.T. Devreese, Phys. Scripta $\underline{T13}$ 282 (1986)
23. M.A. Hopkins, R.J. Nicholas, W. Zawadzki, P. Pfeffer, D. Gauthier, J.C. Portal and M.A. DiForte-Poisson, Semicond. Sci. and Technol. $\underline{2}$ 568 (1987)

24. K. Karrai, S. Huant, G. Martinez and L.C. Brunel, Solid State Commun. $\underline{66}$ 355 (1988)

25. Wu. Xiaoguang, F.M. Peeters and J.T. Devreese, Phys. Rev. $\underline{B36}$ 9765 (1987)

26. T. Ando and Y. Uemura, J. Phys. Soc. Jpn. $\underline{36}$, 959 (1974)

27. P. Voisin, Y. Guldner, J.P. Vieren, M. Voos, P. Delescluse and N.T. Linh, Physica $\underline{117B}$ & $\underline{118B}$ 634 (1983)

28. G. Abstreiter, J.P. Kotthaus, J.F. Koch and G. Dorda, Phys. Rev. B$\underline{14}$, 2480 (1976).

29. R.J. Wagner, T.A. Kennedy, B.D. McCombe and D.C. Tsui, Phys. Rev. B$\underline{22}$, 945 (1980).

30. M.A. Brummell, R.J. Nicholas, L.C. Brunel, S. Huant, M. Baj, J.C. Portal, M.Razeghi, M.A. DiForte–Poisson, K.Y. Cheng and A.Y. Cho, Surf. Sci. $\underline{142}$, 380 (1984).

31. R.J. Nicholas, M.A. Hopkins, D.J. Barnes, M.A. Brummell, H. Sigg, D. Heitmann, K. Ensslin, J.J. Harris, C.T. Foxon and G. Weimann. Phys. Rev. $\underline{B39}$ in press (1989)

32. H. Sigg, D. Weiss and K von Klitzing, Surf. Sci. $\underline{196}$ 293 (1988)

33. Z. Schlesinger, W.I. Wang and A.H. Macdonald, Phys. Rev. Lett. $\underline{58}$, 73 (1984)

34. K.W. Chiu, T.K. Lee and J.J. Quinn, Surf. Sci. $\underline{58}$ 182 (1976)

35. C.K. Sarkar and R.J. Nicholas, Surf. Sci. $\underline{113}$, 326 (1982).

36. M. Prasad and S. Fujita, Physica $\underline{91A}$, 1 (1978).

37. M.A. Hopkins, R.J. Nicholas, D.J. Barnes, M.A. Brummell, J.J. Harris and C.T. Foxon, Phys. Rev. $\underline{B39}$ in press (1989)

38. W. Walukiewicz, H.E. Ruda, J. Lagowski and H.C. Gatos, Phys. Rev. B$\underline{30}$, 4571 (1984)

39. R. Lassnig and E. Gornik, Solid State Commun. $\underline{47}$, 959 (1983).

40. T. Ando and Y. Murayama, J. Phys. Soc. Jpn. $\underline{54}$, 1519 (1985).

41. J.J. Harris, C.T. Foxon, D.E. Lacklison and K.W.J. Barnham, Superlatt. Microstruc. $\underline{2}$, 563 (1986).

42. V.J. Goldman, M. Shayegan and D.C. Tsui, Phys. Rev. Lett. $\underline{61}$ 881 (1988)

43. N. Mori, H. Murata, K. Taniguchi and C. Hamaguchi, Phys. Rev. $\underline{B38}$ 7622 (1988)

44. P. Warmenbol, F.M. Peeters and J.T. Devreese, Phys. Rev. $\underline{B37}$ 4694 (1988)

45. M.A. Brummell, D.R. Leadley, R.J. Nicholas, J.J. Harris and C.T. Foxon, Surf. Sci. $\underline{196}$ 451 (1988)

46. P. Vasilopoulos, Phys. Rev. $\underline{B33}$ 8587 (1986)

47. R. Lassnig and W. Zawadzki, J. Phys. C $\underline{16}$ 5435 (1983)

48. D.R. Leadley, R.J. Nicholas, M.S. Skolnick, S.J. Bass and L.L. Taylor, In Proc. Int. Conf. on the Application of High Magnetic Fields in Semiconductor Physics, Wurzburg, 1988, ed. G. Landwehr (Springer–Berlin, in press)

49. R.J. Nicholas, S. BenAmor, J.C. Portal, D.L. Sivco and A.Y. Cho, Semicond. Sci. and Technol. $\underline{4}$ 116 (1989)

CYCLOTRON RESONANCE OF 2D ELECTRON SYSTEMS IN INTENTIONALLY DOPED AlGaAs/GaAs QUANTUM WELLS AND HETEROSTRUCTURES

H. Sigg, J. Richter, K. von Klitzing, and K. Ploog

Max-Planck-Institut für Festkörperforschung
7000 Stuttgart 80
Federal Republic of Germany

ABSTRACT

An introduction is given to various aspects of the electron-impurity interaction that can be realized and studied in intentionally and selectively doped two-dimensional (2D) electron systems. We exemplify this by presenting transport and far infrared (FIR) measurements in differently doped AlGaAs/GaAs heterostructures and quantum well structures. The FIR response is studied for electrons bound to 2D confined impurities of donors in a quantum well. In the same system we have studied the transition from an insulator to metal behavior. We contrast the results on donor doped systems with our experiments on acceptor doped AlGaAs/GaAs heterostructures. The experimental findings for doping induced phenomena on the cyclotron resonance of 2D-electron systems lead us to discuss the validity of the single particle model of bound states.

1.   INTRODUCTION

In a homogeneously doped semiconductor sample the carrier concentration is strictly related to the doping concentration. However, for a two-dimensional (2D) electron system, for example the AlGaAs/GaAs heterostructure, carrier concentration, impurity concentration, scattering strength and charge of the impurities can all be set independently. Such systems can be realized using techniques like in situ intentional doping during molecular beam epitaxy (MBE). Starting from modulation doped heterostructures of high mobilities, an additional impurity doping at a given spatial separation from the 2D channel enables us to control and investigate the influence of the Coulombic scatterers on the 2D transport.

In the Si-MOS-system controlled intentional doping has been obtained by drifting Na^+-ions in an electrical field towards the SiO_2/Si interface [1]. Recent work on the dc-transport behavior of such Si-MOS-systems has been reported by Furneaux and Reineke [2]. A similar type of reversible doping that can be applied subsequently to the growth, is the electron-beam irradation of AlGaAs/GaAs heterostructures [3]. Experiments on the dc-and far infrared (FIR) transport properties of such irradiated systems have been reported by Sigg et al. [4]. Recently, the technique of δ-doping for MBE grown AlGaAs/GaAs systems has become very convenient [5]. It allows the realization of controlled doping with both, donors (by Si-atoms) and/or acceptors (by Be-atoms). The dc-magneto transport properties of such Si- and Be-doped AlGaAs/GaAs heterostructures has been investigated by Haug et al. [6], and the FIR response in such systems has been studied by Richter et al. [7]. In lightly doped AlGaAs/GaAs QW systems it was possible to realize the metal to insulator transition (Sigg et al. [8]).

Here, we review some recent results that we have obtained on intentionally doped AlGaAs/GaAs systems. It was found, that the influence of scatterers on the cyclotron resonance (CR) is dramatically changed for samples of different doping configuration. One can consider two extreme, well described cases (i) the impurity scatterers have nearly no influence on the resonance frequency and the FIR response is that of a free electron CR, (ii) the resonance is a "pure" single electron resonance and can be described by the so-called 1s-$2p^+$ transition of hydrogen atom with an effective Rydberg. In real systems we find, depending on the doping and electron concentration, a response which lies somewhere between the two models. We will discuss in the following our experimental results starting from these models and will in particular address the validity of the single-particle model of bound states.

The most elementary 2D system (with impurities) is to have one single donor with its electron confined in a quantum well. Since there is no further electron-electron interaction the energy levels and exitation energies can be treated rigorously (c.f. Bastard [9]). This system can be realized by weak doping with Si in the center of a QW (center-doping). We will discuss this system in chapter 3.1. As a next step we give a discussion in section 3.2 upon the same type of center-doped QW systems, where additional carriers are supplied by a modulation doping of the AlGaAs barrier. By this additional doping the transport behavior changes from an insulator to a metallic type of behavior which drastically influences the FIR response. In the final section of chapter 3 a very peculiar type of bound resonance is described which is observed in conventional modulation-doped heterostructures with an additional layer of acceptors close to the interface electron channel. Chapter 2 gives some details on the principle of the intentional doping of 2D electron systems. In chapter 4 we summarize the presented CR experiments and results and conclude that the FIR response in intentionally doped 2D-electron systems is a very interesting subject and still an open field.

2. INTENTIONAL DOPING OF 2D ELECTRON SYSTEMS

The principles and possibilities of intentional doping are sketched in Fig.1. In particular, it is possible to insert during the MBE process a δ-layer[5] of donors (Si) or acceptors (Be) at a well defined distance z_s from the interface (see inset.). For example, the elementary structure discussed above is realized if a sheet of donors is inserted at the center of the well.

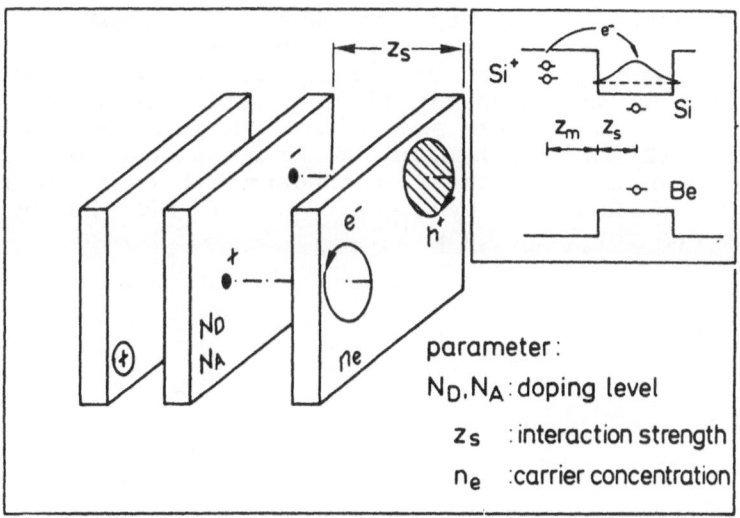

Fig. 1 Schematic illustration of the layered intentionally doped 2D electron systems. The inset shows an example of a doped quantum well structure. The layers to the left and to the right correspond to the modulation doped barrier and the 2D-electron channel respectively. The most relevant parameters that determine the interaction strength of the impurity with the electrons are N_A and N_B, the donor and acceptor concentration, and Z_s, the separation of the intentional doping layer from the 2D-electron channel.

Additionally, remote donors can be built into the AlGaAs barrier at a distance z_m, the standard modulation doping technique. For our discussion in the following it is essential, that we have negatively charged electron sheets with electron density n_e and at a distance z_s (which might be zero) positively or negatively charged acceptor or donor sheets. In this picture (see Fig.1) we can imagine that ionized donors form (possibly weak) bound states, and that, by symmetry, ionized acceptors may induce "quasi-holes" in the high density electron sheet. The latter situations will be discussed in section 3.3.

3. EXPERIMENTAL RESULTS AND DISCUSSION

The experiments are performed using a variable temperature inset (1.2K < T < 70K) in a 15 Tesla magnet cryostat. FIR sources are a CO_2 laser pumped CH_3OH gas laser system and a fast scanning Fourier transform spectrometer. The FIR response is measured in the Faraday configuration using transmission or photoconductivity.

3.1 The neutral 2D confined donor

Fig. 2 shows the magnetic field dependence of the resonance energies measured in a center-doped QW of width 15 nm. The intended doping level in the well is about 3.10^{10} cm^{-2}, the dopant atoms are Si donors. When the sample is tilted with respect to the magnetic field the resonance energies are shifted towards higher magnetic fields. The shift scales with the component of the magnetic field directed perpendicular to the layers and therefore proves the 2D character of these resonances. It is worthwhile to compare the observed 2D magneto resonance with the two fundamental exitations, i.e. the $1s-2p^+$ resonance of hydrogenic donors and the CR of the free electrons for bulk GaAs [10].

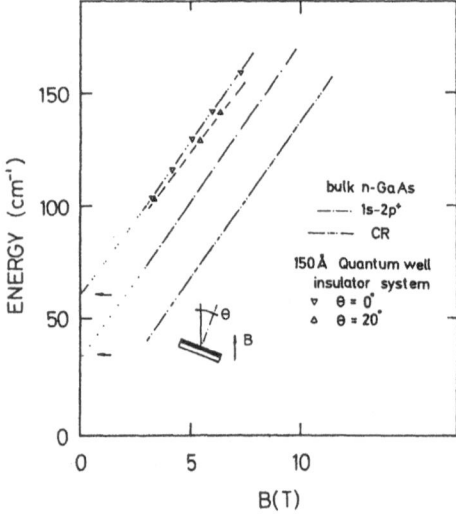

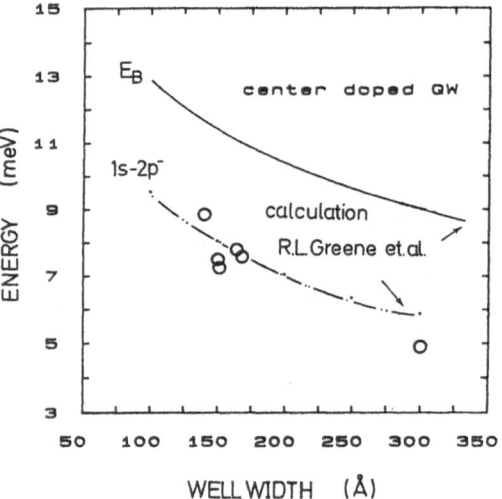

Fig. 2 Resonance energies plotted as a function of the magnetic field for a 15 nm QW (▽) are compared with the bulk n-GaAs system[10]. The shift of the resonance energies to higher magnetic fields (△) proves the 2D character of the QW confined donor impurities.

Fig. 3 Comparison of the experimentally deduced shift of the impurity resonance with the related $1s-2p^-$ transition energy (at B = 6 T) taken from the calculation by Greene and Bajaj [11].

As shown in Fig. 2, all these GaAs related resonances show up with a linear energy vs. magnetic field dependence of equal slope. The resonance with the highest energy (extrapolated offset at B = 0T is about 65 cm^{-1}) is found for the 2D system; next, with an offset of about 32 cm^{-1}, one has the bound resonance of the bulk GaAs and finally, with zero offset, the CR. The latter is the expected behavior for free electrons. Obviously, we do observe an impurity resonance in the QW system which, by analogy with the bulk impurity resonance, represents the 1s - $2p^+$ transition of a 2D confined shallow donor. The measured offset of the transition energy can be related to the binding energy. In Fig. 3 we compare the experimentally observed dependence of the energy offset on the well size with calculated values taken from the paper by Greene and Bajaj [11]. This calculation is performed for isolated donors in AlGaAs/GaAs QW systems. Note however, this calculation does not include any exchange interaction and screening effects. Therefore, the close agreement of experiment and theory shows, that in a dilute center-doped QW system, the FIR excitation indeed probes the effective binding strength of donors. This result, that has first been obtained by Jarosik et al. [12], is the basis for the following discussion of the FIR experiments on QW systems which are both center and modulation doped.

3.2 Quantum well system at the metal-insulator transition

Fig. 4 shows the dc-magneto transport properties of a QW system that is not only center-doped but also modulation doped in the barrier. The layered structure consists of a 15 nm center-doped well followed by an undoped AlGaAs barrier of width 5 nm. On top of these layers a 2 nm wide Si-doped GaAs well is grown followed by a 5 nm barrier and the next 15 nm well etc. A tenfold repetition is chosen in order to increase the FIR absorption strength.

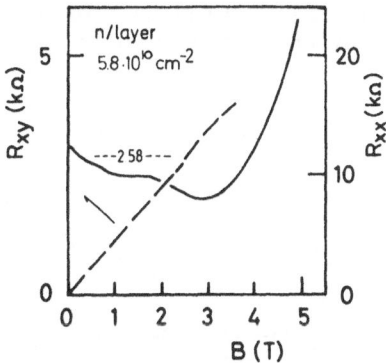

Fig. 4 DC-magneto transport properties R_{xy} (scale to the left hand side) and R_{xx} (right hand side) showing a magnetic field induced MI-transiton at B ≈ 3T.

Test specimens with undoped 2 nm wells show that such an additional well does not change the confinement energies in the main well: the same 2D hydrogenic type of bound state is found as has been described in the preceding section.

For samples where the 2 nm wells are slightly doped we expect a transfer of charge into the main well. Thereby we should reach the situation where the number of 2D carriers is in excess of the number of the 2D confined donors. This situation is obtained by doping the 2 nm well with Si at a level of the order of $3-10.10^{10}$ cm^{-2}. The free carrier concentration and conductivity of such samples remains now finite even at low temperature, i.e., the samples become metallic. For the sample shown in Fig. 4, carrier concentration and mobility, as obtained from Hall measurements, are 6.10^{10} cm^{-2} per layer and 2000 cm^2/Vs, respectively. On applying a weak magnetic field (directed perpendicular to the layers) a negative differential resistance is found. At somewhat larger fields, however, a sudden increase of the resistance is observed.

This behavior is attributed to a magnetic field induced metal-insulator (MI) transition. More about this transition can be learned from an analysis of the FIR response.

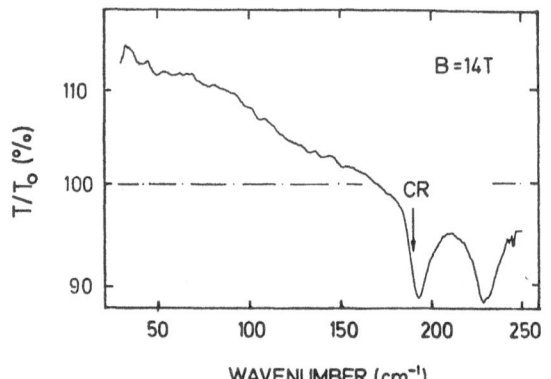

Fig. 5 Normalized Transmission spectrum T/T_o = T(14T)/T(B=0T) of the sample with MI-transition at 4 T. The excessive transmission strength at low frequency is due to a distinct absorption which appears in the zero magnetic field spectrum T_o = T(B=0T).

A typical spectrum obtained for a QW sample with MI-transition at about 4 Tesla is shown in Fig. 5. The spectrum taken at 14 Tesla resolves a double resonance. According to our simple approach the strength of binding of the two resonances is as follows. The high energy resonance corresponds to a bound state of rather high binding energy. The upshift in energy, measured with respect to the free CR, is about 40 cm^{-1} (see Fig. 5).

The low-energy resonance, however, is almost unshifted. The corresponding binding energy therefore is different from zero but very small. Intuitively, we may understand this behavior by considering the following arguments. Each of the donors which belong to either the 2 nm or the 15 nm well supplies a single electron. These electrons remain in the 15 nm well at the energetically most favorable positions. Such positons are the bound states at the center donors, and possibly, localized states at the interface boundary which are weakly bound to their host donor in the 2 nm well.

However, this explanation is somehow inconsistent with the observation that the high-energy resonance occurs at a too small energy. The upshift of the impurity resonance described in the preceding section was about 65 cm^{-1}. Here the upshift is about 40 cm^{-1} at 14 T and is further decreasing with decreasing magnetic field. The low-energy resonance, however, shows an increasing shift with decreasing magnetic field.

This is shown in Fig. 6, where the resonance positions are plotted as obtained from a line fitting procedure. Below 6 Tesla, due to a strong increase of the resonance linewidth, the fit could not produce any accurate values[8]. Towards even lower magnetic fields, where the MI-transition occurs, all features in the spectra get very broad. Nevertheless, we can definitely conclude that the shift of the impurity-like resonance does not approach zero, neither at B = 4T nor at zero magnetic field. Rather, an increasing absorption at the low-frequency side of the spectra is found for decreasing magnetic fields. This behavior is evident in Fig. 5 from the fictitious transmission of more than 100% which appears at energies below 150 cm^{-1} due to the normalization of the spectrum $T(B=14T)/T(B=0)$.

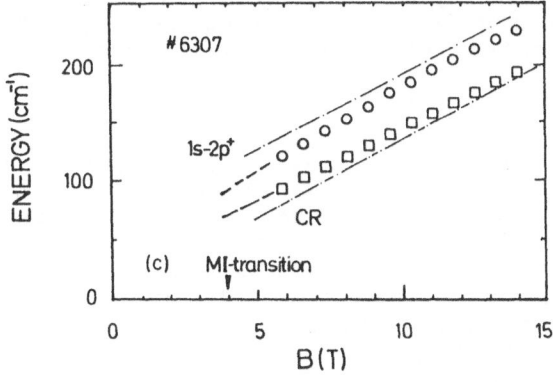

Fig. 6 Peak positions for the sample with MI-transition at 4 T obtained from a fit procedure assuming Lorentzian CR line profiles. For comparison, the energy vs. magnetic field dependence of the isolated 2D confined impurity resonance ($1s$-$2p^+$) and free CR is included.

We like to note that recent FIR investigations on systems where the MI-transmission has been realized in a different way than we have, show comparable results. In bulk GaAs systems, Ming-Way Lee et al.[13] reported an enhanced low-frequency absorption and an impurity-like resonance with a correspondingly reduced binding energy. Such a reduced binding energy is also reported for δ-doped GaAs systems by Qui-ye Ye et al. [14]. Although this is now a clear and experimentally verified trend, we do not have a convincing explanation for this weakening of the binding energy.

For the context of this paper, however, we like to point out the following. In deliberately doped systems the spectroscopically determined binding energy of an impurity resonance with in many respect normal behavior (e.g. linear energy vs. magnetic field dependence) can be smaller than the binding energy expected for the isolated case. It is the excess charge, supplied by the modulation doping, that leads to this reduction. Obviously, the transition from a bound to a free CR is very smooth. Consequently, at much higher doping level, we can hardly find any features of a bound resonance. Instead, one observes the CR of the free electrons which shows strong filling factor correlated oscillations of the linewidth and position [7,8]. A description of these oscillating CR features however, goes much beyond the model applied here of bound/unbound resonances. Therefore, we will switch the aspect of our discussion to another novel intentionally doped system: the acceptor doped 2D electron system, where the CR shows the behavior of a collective bound state.

3.3 Acceptor doped AlGaAs/GaAs heterostructures

Fig. 7 shows a series of CR spectra which have been obtained from a modulation doped AlGaAs/GaAs heterostructure with an additional δ-doped sheet of Be in the GaAs layer at 10 nm from the interface. The Be concentration is about 2.10^{10} cm^{-2}. The most spectacular finding is that the linewidth of the resonance strongly decreases with increasing magnetic fields. The CR splits into two resonances while the line position of the high-energy resonance shows a remarkable upshift in energy. From the quadratic plot, ω_c^2 versus B^2, in Fig. 8 we obtain that the line position is well approximated by $\omega_c^2 \approx (\omega_c^0)^2 + (\omega_s)^2$, where ω_c^0 is the free CR frequency and ω_s describes the upshift in frequency. In the foregoing, we dealt with a linear upshift of the CR that is characteristic of an electron bound by a Coulomb potential, V_c proportional to $1/r$. As has been shown by Mikeska and Schmidt [15], a quadratic upshift of the CR is characteristic of a harmonic type of binding; this is a smooth potential of the form $V_h = V_0(r/r_{max})^2$ where r_{max} is the radial extension of the potential. From the calculation given in Ref. 15 we obtain $V_0/r^2_{max} = m^* \cdot \omega_s^2/2$ that relates the quadratic upshift ω_s to the strength and extension of the potential.

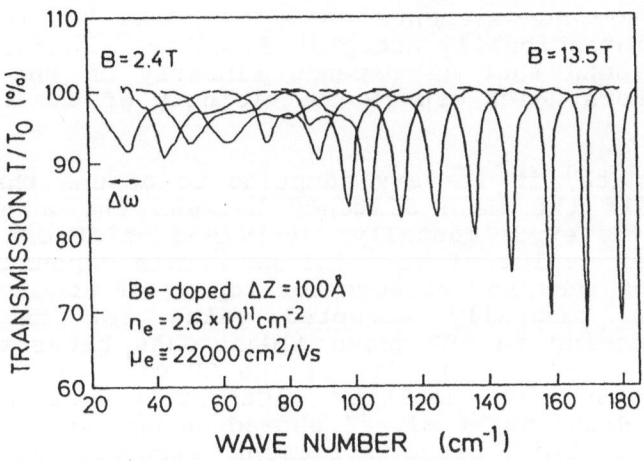

Fig. 7 Normalized CR transmission spectra, measured at various magnetic fields which have been increased in constant increment within the indicated range for an acceptor doped AlGaAs/ GaAs heterostructure (T = 2.2 K).

Next, the question arises about the origin of the smooth potential and about the microscopic interpretation for V_o and r_{max}. In order to answer this question the experimental fact should be noted, that the shift ω_S depends on the density of the acceptors which interact with the 2D electron gas. For increasing acceptor concentration a monotonically increasing shift ω_S is observed [7].

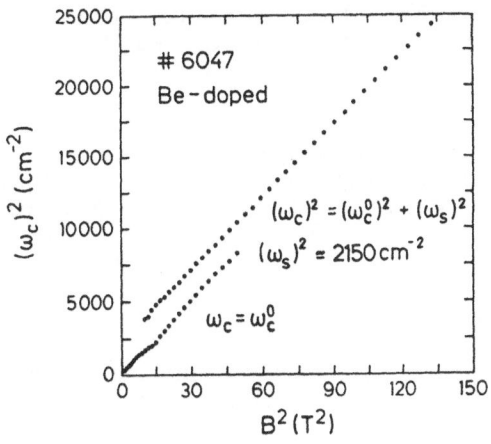

Fig. 8 Double quadratic plot of the resonance energies squared ω_c^2 versus B^2 for an acceptor doped AlGaAs/GaAs heterostructure. Note that the resonances obey the relation $\omega_c^2 \approx (\omega_c^0)^2 + (\omega_S)^2$ which correspond to electrons confined in a harmonic oscillator type of potential.

Moreover, from measurements on AlGaAs/GaAs heterostructures which were intentionally acceptor doped by e-beam irradiation [4], it was found that ω_s^2 depends linearly on the number of those acceptors which are in the vicinity of the 2D electron gas [16].

From these data, it is very tempting to assume that r_{max} is given by half the mean distance between the acceptors. In this case, the experimentally determined shift can be fitted by a constant value of $V_o = 1.5$ meV. This dependence of the shift implies that for an acceptor doping of about 10^{14} cm^{-3}, which is a generally accepted value for the residual background doping in MBE-grown AlGaAs/GaAs heterostructures, we expect a quadratic upshift of the CR of $\omega_s^2 \approx (1.5$ cm$^{-1})^2$. Experiments on high mobility AlGaAs/GaAs heterostructures reported by Nicholas et al. [17] showed an upshift of typically $(1.85$ cm$^{-1})^2$. This close agreement strongly supports the proposed harmonic potential model. However, it should be noted that in our model r_{max} is identified with the <u>mean</u> scatter distance. This means that a <u>collective</u> property of the random scatterer distribution causes the observed shift of the CR frequency.

A deeper understanding of this experimental fact, however, goes beyond our rather simple picture of the CR that is essentially a single particle description. Although the mechanism which leads to this spectacular averaging is not yet fully understood, we interpret the observation of the quadratically upshifted CR as follows [7,16]. The smooth harmonic type of potential is related to potential minima formed by the tails of the repulsive centers of the ionized acceptor ions. In a more rigorous approach one should consider the effect on the FIR response of the "quasi-holes" which form those bound states with the ionized acceptors (see Fig. 1).

4.   SUMMARY AND CONCLUSION

Intentionally doped 2D electron systems open the possibility to study the transport behavior of the 2D carriers which interact in a controllable way with impurities, like ionized shallow donors and acceptors. The FIR response has been investigated and discussed for differently designed structures. Examples have been given where, induced by the dopants, the observed resonance shows a frequency upshift that is characteristic for a bound resonance. We have stressed the experimental observation that the thereby obtained binding energies do not necessarily correspond to the single particle excitation of the isolated bound state. Collective phenomena are important as is shown for the case of acceptor doped 2D electron systems.

In conclusion, FIR experiments were presented showing the possibilities and limitations of a single particle model in explaining doping induced phenomena of the CR in 2D electron systems.

Acknowledgement   Part of this work has been supported by the "Bundesministerium für Forschung und Technologie" (BMFT).

REFERENCES

1. A. Hartstein and A.B. Fowler, Phys. Rev. Lett. **34**, 1435 (1975)
2. J.E. Furneaux and T.L. Reinecke, Phys. Rev. **B33**, 6897 (1986)
3. D. Pons, and J.C. Bourgoin, J. Phys. C: Solid State Phys. **18**, 3839 (1985)
4. H. Sigg, D. Weiss, and K. von Klitzing, Surf. Sci. **196**, 293 (1988)
5. K. Ploog, J. Cryst. Growth **81**, 304 (1987)
6. R.J. Haug, R.R. Gerhardts, K. von Klitzing, and K. Ploog Phys. Rev. Lett. **59**, 1349 (1987)
7. J. Richter, H. Sigg, K. von Klitzing, and K. Ploog Phys. Rev. **B39**, 6268 (1989).
8. H. Sigg, K. von Klitzing, M. Hauser, and K. Ploog, in: "Shallow Impurities in Semiconductors 1988", B. Monemar, ed.; IOP Publishing Ltd.; p11
9. G. Bastard, Phys. Rev. **B24**, 4714 (1981)
10. H. Sigg, H.J.A. Bluyssen, and P. Wyder, Solid State Commun., **48**, 897 (1983)
11. R.L. Greene and K.K. Bajaj, Solid State Commun. **45**, 825 (1983).
12. N.C. Jarosik, B.D. McCombe, B.V. Shanabrook, I. Comas, I. Ralston, and G. Wicks, Phys. Rev. Lett. **54**, 1238 (1985)
13. Ming-Way Lee, D. Romero, H.D. Drew, M. Shayegan, and B.S. Elman, Solid State Commun. **66**, 23 (1988)
14. Qui-Yi Ye, A. Zrenner, F. Koch, H. Sigg, D. Heitmann, and K. Ploog, Proc. of MSS (1989), to be published.
15. H.J. Mikeska and H. Schmidt, Z. Physik **B20**, 43 (1975)
16. H. Sigg, J. Richter, K. von Klitzing, and K. Ploog, in: "The Application of High Magnetic Fields in Semi conductor Physics", G. Landwehr,ed.; Springer Series in Solid State Sciences (Springer, Berlin 1989) in press
17. R.J. Nicholas, M.A. Hopkins, D.J. Barnes, M.A. Brummel, H. Sigg, D. Heitmann, K. Ensslin, J.J. Harris, C.T. Foxon, and G. Weimann, Phys. Rev. **B39**, 10955 (1989)

MAGNETOPHONON RESONANCE CONDITION IN QUASI

ONE- AND TWO-DIMENSIONAL ELECTRON SYSTEMS

Peter Warmenbol[*]

Alcatel-Bell Advanced Research
F. Wellesplein 1
B-2018 Antwerpen, Belgium

INTRODUCTION

Measurements of the magneto-resistance in polar semiconductors at sufficiently high temperature ($T > 100K$) show oscillations at magnetic fields for which $N\omega_c = \omega_{LO}, N = 1, 2, 3, ...$ with $\omega_c = eB/m^*$ the cyclotron resonance frequency and ω_{LO} the LO-phonon frequency. These oscillations in the magneto-resistance ρ_{xx} (and in the derivative to the magnetic field) are referred[1] to as magneto-phonon resonances (MPR). This resonance also appears in the magnetic field dependence of e.g. thermopower and warm electron coefficient. The magnetophonon effect is a powerful spectroscopic tool.

In the last years there has been considerable interest in systems in which the electron motion is strongly confined such as single and multiple heterostructures, quantum wires, quantum dots, etc. Since in these systems one, two or three of the dimensions become of the order of the de Broglie wavelength for electrons, new quantum effects are expected[2]. Also, such systems are expected to have high electron mobilities with possible device applications. So far there has been a variety of experimental observations: the oscillatory behavior of the conductivity of quantum wires as function of the impurity density[3], the quantization of the resistance of ballistic point contacts[4], the quenching of the Hall effect[5], etc. ... The magnetophonon effect in two-dimensional (2D) systems has attracted a considerable amount of interest. The review paper of Nicholas[1] includes recent experimental results (see also the contribution of Nicholas in this volume).

Shift of the MPR peak positions

Recent measurements of the magnetophonon effect in the linear regime on $GaAs/Al_xGa_{1-x}As$ heterostructures[6] and $Ga_{0.47}In_{0.53}As/InP$ heterostructures[7,8] reveal the surprising result that the resonant condition is altered to $N\omega_c = \omega^*$ where the electrons appear to interact with the LO-phonons at an effective frequency $\omega^* < \omega_{LO}$. Furthermore this frequency ω^* decreases with increasing temperature, much faster than the bulk LO-phonon frequency. This frequency ω^* is also sensitive to the electron density and, in fact decreases with increasing electron density. Also in the non-linear regime (high electric fields), the MPR peak positions were found[9] to be shifted to lower magnetic fields. Here the shift was observed to increase with increasing applied electric field (or euivalently with increasing electrical current density). Independently from experiment, this non-linear effect was described theoretically in Refs. 10 and 11. It is analogous to the three-dimensional case[12].

Until now there is no agreement on the physical interpretation of the shift in the linear regime and neither in the non-linear regime. The aim of this contribution is to analyse different possible contributions to the shift of the MPR peak position in quasi-2D (Q2D) and Q1D electron systems. First the 2D case is treated in detail and subsequently a brief discussion of the 1D case is given. The main approximation in this paper is the assumption that the electron subsystem of the single heterostructures interacts with bulk LO-phonons, which is supported by recent results[13]. Multiple heterostructures (e.g. superlattices), in which dimensional confinement of the phonons is expected to play a role[14], fall out of the scope of the present paper.

Density of states

The density of states (DOS) of a 2D electron gas (2D EG) can not be observed directly. Nevertheless on the basis of magneto-transport[15], electronic specific heat[16,17], magnetic susceptibilty[18], optical measurements[19] it was found that there exist many more states at the Fermi energy than would be expected from the independent particle picture calculation of Ando et al[20]. In Refs. 15-18 the results could be explained by modelling the DOS by a series of Gaussians on top of a positive background. In more recent calculations of the DOS of a 2D EG it was found that the DOS varies much more smoothly than was first anticipated. Also much more DOS was found in between the Landau levels. Early calculations in which electron screening was included[21] resulted in a Fermi level dependent broadening of the DOS. This broadening of the Landau levels is much larger when the Fermi energy is between two Landau levels. Recently Gudmundsson and Gerhardts[22] stressed the importance of statistical averaging over inhomogeneities and calculated the DOS explicitly. Das Sarma and Xie[23] also included the Landau level coupling effects together with the non-linear screening and found that the DOS of a 2D EG in a magnetic field is surprisingly smooth as a function of the energy.

For our purposes it is sufficient to include the essential features of the DOS. Therefore the DOS will be parametrized by a series of Gaussians, which is characterized by a broadening parameter Γ_0, on top of a constant background which contains a fraction x of the total DOS. This parametrized form for the electronic DOS was employed in Ref. 10 (hereafter referred to as I) in a study of the linear and non-linear magnetophonon effect within a momentum balance approach. It was found that the broadening of the DOS induces a shift of the MPR peaks. From Ref. 23 it is found that values for x up to .5(and even higher) are not unrealistic. From a Shubnikov de Haas measurement[12] on a GaAs-heterostructure, in which two electric subbands are occupied, x-values are obtained around .2 − .5.

Damping of the MPR oscillations

Experimental data[1] for the damping of the magnetophonon oscillations show a complicated dependence on the applied electric field E, the magnetic field B (harmonic number N), the lattice temperature T and the 2D electron density N_e: various contributions to the broadening interplay. The model of Ref. 10 agrees qualitatively with experimental data for the linear regime, except for the density dependence, and with the theory of Ref. 24, which introduced a Lorentzian broadened DOS. However both models tend to overestimate the MPR amplitude at low magnetic fields, independently from the way of introducing broadening. In Ref. 25 a relation between the Gaussian broadening parameter Γ_0 and N_e was extracted by fitting the experimental electron density dependence of the $N = 3$ MPR amplitude at $T = 180\,K$ of Ref. 26 to the theoretical Γ_0 dependence of the MPR amplitude. The experimental MPR amplitudes decreases monotonically with increasing electron density for $N_e > 6 \times 10^{10} \text{cm}^{-2}$. For this range of electron densities the fit results in $\Gamma_0 \sim N_e$, which agrees with recent findings of Ref. 27, who included Thomas-Fermi static screening of the remote impurities. However, further data of the Oxford group[8] reveal a maximum in the $N = 3$ MPR amplitude at $N_e \approx 6 \times 10^{10} \text{cm}^{-2}$. The occurence of a maximum in the electron density dependence of the MPR amplitude cannot be explained by either of the arguments; moreover these arguments are far from self-consistent.

Selfconsistent model

A full selfconsistent model for the interacting electron-impurity-phonon system is available for 3D in the linear regime (e.g. see Ref. 28). On the other hand, the standard approximations that are made for the calculation of 1) the 2D electronic density of states (self-energy effects, many body effects, screening, ...), 2) for the wavefunction of the electron in a confining potential; and 3) the transport parameters (e.g. the contribution of electron-phonon and -impurity interaction to the resistivity) are not consistent. Numerical calculation of the full selfconsistent model has, to our knowledge, not been attempted: it requires formidable computer time. Therefore one is forced to make simplifying assumptions for the transport problem, especially in the non-linear regime.

HOT ELECTRON EFFECT

The non-linear momentum balance equation

In I a drifted-Maxwellian form of the electron momentum distribution function is emloyed, which has the advantage of leading to tractable expressions. This ansatz is used as a starting point and, although its applicability to the full hot electron regime is limited, we believe it leads to the correct physical trends (for a discussion see e.g. Refs. 10 and 29). For clarity the approximations are listed
   - an effective mass approximation for the electrons,
   - a parabolic conduction band is assumed,
   - only the first electric subband is populated (Fang-Howard wavefunction), which is valid for not too high electron densities.
   - screening is neglected,
   - the LO-phonons are taken to be the bulk phonons (3D) of GaAs, and
   - the first order Born approximation is used (weak electron-phonon interaction), which is valid for GaAs with $\alpha = 0.06$.
Within these approximations the momentum balance equation can be written as

$$e[\vec{E} + (\vec{v} \times \vec{B})] = \vec{F}(\vec{v}) \tag{1}$$

where $\vec{E}$ is the total electric field, $\vec{B}$ is the magnetic field, $\vec{v}$ is the average electron velocity, $-e$ is the electron charge and $\vec{F}(\vec{v})$ is the average force exerted on the electron by the interaction with LO-phonons

$$\vec{F}(\vec{v}) = \frac{2}{N_e} \sum_{\vec{q}} \frac{\vec{q}}{\hbar} |V_{\vec{q}}|^2 [n(\omega_{\vec{q}}) - n(\omega_{\vec{q}} - \vec{q}.\vec{v})] \, \mathrm{Im}\{D^r(\vec{q}, \omega_{\vec{q}} - \vec{q}.\vec{v})\} \tag{2}$$

with $\mathrm{Im}\{D^r(\vec{q}, \omega)\}$ the imaginary part of the retarded density-density correlation function. $V_{\vec{q}}$ is the electron-phonon interaction Fourier coefficient, N_e the electron density, $\omega_{\vec{q}}$ the phonon frequency with phonon wavevector $\vec{q}$ and $n(\omega_{\vec{q}})$ is the phonon occupation number.

From the momentum balance equation (we choose $\vec{v}$ along the x-axis) the resistivity components are obtained as follows

$$\rho_{xx} = \vec{v}.\vec{F}(\vec{v})/[N_e(ve)^2] \,, \tag{3a}$$

$$\rho_{xy} = B/(e \, N_e) \,. \tag{3b}$$

Shift versus splitting of the MPR peaks

From numerical solution of Eqs. (1) and (2) it was found in I that the maxima in the magnetoresistance at low fields (linear regime) become minima at the same position at higher fields (hot electron or non-linear regime). The peak position shifts gradually to lower magnetic fields as the electric field is increased. This was confirmed by the measurements of Leadley et al.[9] on $GaAs/Al_xGa_{1-x}As$ heterojunctions. In the 3D case a similar conversion of the magnetophonon resonance maxima into minima with increasing electric field was established in the experiments of Eaves et al.[12] on a $n^+ - n^- - n^+$ GaAs

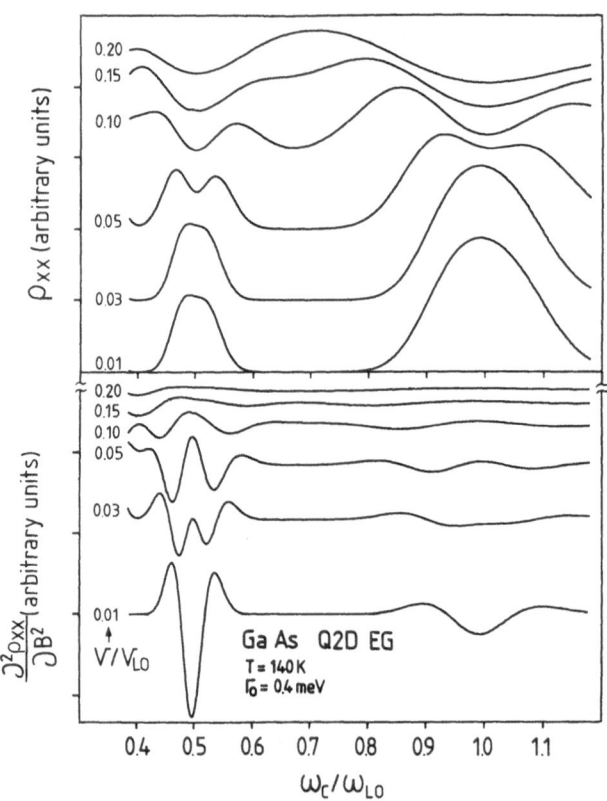

Figure 1. The non-linear resistivity ρ_{xx} (top) and its second derivative with respect to the magnetic field (bottom) as a function of the cyclotron frequency at T= 140 K and Γ_0 = 0.4 meV. Results are displayed in arbitrary units for different values of the average electron velocity as indicated on the left.

structure. In Ref. 30 this case was studied and a splitting was found in the second derivative of the magnetoresistance, in agreement with experiment. In contrast to this splitting for a 3D EG our theoretical results in I for a 2D EG, the model of Ref. 11 and the experimental results of Ref. 9 give only evidence for a shift (not a splitting) of the magnetophonon resonance peaks in the 2D EG. This apparently different behaviour of the 2D case and the 3D case, puts us before the question: what is the basic physical mechanism behind the shift/splitting of the magnetophonon resonance peaks?

To that purpose we will be concerned with the case of small broadening, as opposed to I where Eq. (1) was solved for $\vec{v}$, for rather large broadening ($\Gamma_0 = 3.7\,\text{meV}$) and only a shift of the MPR peaks was found. Fig. 1 shows the non-linear resistivity ρ_{xx} (top) and its second derivative with respect to the magnetic field (bottom) as a function of the cyclotron frequency for T= 140 K and $\Gamma_0 = 0.4$ meV. Peaks are found in the resistivity for $\omega_c = \omega_{LO}/N$ where $N = n - m$. Let us concentrate on the $N = 2$ peak which occurs at $\omega_c/\omega_{LO} \approx 0.5$. For $v/v_{LO} \geq 0.05$ this peak splits into two peaks. This splitting is more pronounced in the second derivative as shown in the bottom part of Fig. 1.

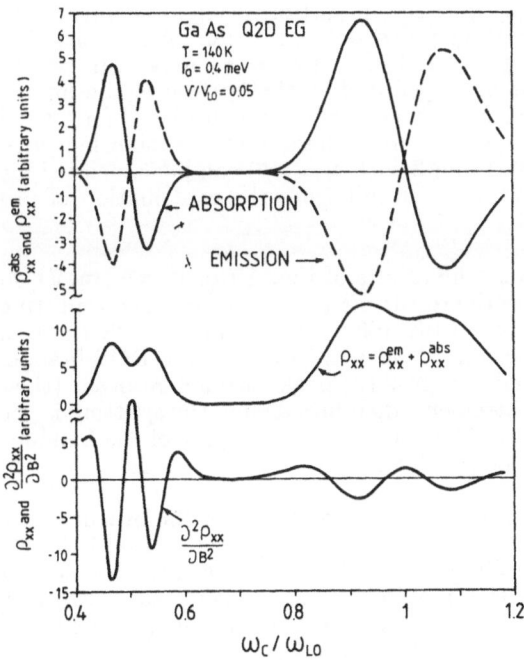

Figure 2. The separate contribution to the non-linear resistivity ρ_{xx} of emission and absorption processes (top) as a function of cyclotron frequency at T= 140 K and $\Gamma_0 = 0.4$ meV. The total non-linear resistivity $\rho_{xx} = \rho_{xx}^{em} + \rho_{xx}^{abs}$ and its second derivative with respect to the magnetic field (bottom) are also displayed. Results are given in arbitrary units for a fixed value of the average electron velocity $v/v_{LO} = 0.05$.

Next the physical origin of the splitting of the MPR-peaks will be discussed. For the purpose of the discussion it is instructive to write Eq. (2) such that the contributions from LO-phonon emission and absorption processes to the force $\vec{F}(\vec{v})$ are separated

$$\vec{F}(\vec{v}) \sim \sum_{\vec{q}} \sum_{n,m=0}^{\infty} B(\vec{q}, n, m) I(\vec{v}, \vec{q}, n, m) \qquad (4a)$$

$$I(\vec{v}, \vec{q}, n, m) = \int_{-\infty}^{\infty} dE \, e^{-\beta E} \, Im[\mathcal{G}_m(E)] \left\{ n(\omega_{LO}) \, Im[\mathcal{G}_n(E + \hbar\omega_{LO} - \hbar\vec{q}.\vec{v})] \right.$$

$$\left. - \left(1 + n(\omega_{LO})\right) Im[\mathcal{G}_n(E - \hbar\omega_{LO} + \hbar\vec{q}.\vec{v})] \right\} \tag{4b}$$

where $\beta = 1/k_B T$ and $Im\mathcal{G}_n(E)$ is the imaginary part of the Greens function for an electron in the n-th Landau Level. The first term between curly brackets in Eq. (4b) represents the process of absorption of an LO-phonon by the electron while the second term represents the corresponding emission process. For Gaussian broadening without a background one has: $Im[\mathcal{G}_n(E)] = G(E + \hbar\omega_{LO} - \epsilon_n)$ with $\epsilon_n = \hbar\omega_c(n + 1/2)$. Details of the calculation can be found in Ref. 25.

The numerical results for the magnetoresistivity as obtained from Eqs. (1) and (4) are displayed in Fig. 2. The top part of this figure depicts the separate contribution of emission and absorption processes to the non-linear resistivity ρ_{xx} as a function of cyclotron frequency at T= 140 K and $\Gamma_0 = 0.4$ meV. The total non-linear resistivity $\rho_{xx} = \rho_{xx}^{em} + \rho_{xx}^{abs}$ is also displayed and at the bottom of the figure its second derivative with respect to the magnetic field. Results are displayed for a fixed value of the average electron velocity $v/v_{LO} = 0.05$. From these data the following physical picture emerges: the N^- peaks (in Figs. 1 and 2 with $N = 1, 2$) can be associated mainly with absorption processes while the N^+ peaks result from the contribution of processes in which LO-phonons are emitted.

In contrast with the results of Fig. 1 (for $\Gamma_0 = 0.4$ meV), for much larger values of Γ_0 the MPR peaks in the resistivity shift, but there is no splitting. The reason for this is the competition between two different contributions to the broadening of the MPR peaks: 1) due to the electric field and 2) due to the DOS braodening. For large DOS broadening the N^+ and the $(N + 1)^-$ peaks are not resolved (they have merged for all velocities), while for extremely small broadening the splitting is very clearly resolved for small average velocities. This means that the value of the Landau level broadening can lead to different hot-electron MPR behaviour.

Note that even for rather large values of the broadening parameter (for which the N^+ and the $(N - 1)^-$ peaks are seen as only one peak) the $N = 1^+$ peak should be observable experimentally, since for the $N = 1^+$ peak there is no corresponding $(N-1)^- = 0^-$ peak. In practise this requires 1) high quality samples in order to minimise the DOS broadening and 2) measurements in high magnetic fields: e.g. for GaAs systems $20T < B < 30T$.

In our momentum balance model it was assumed so far that the effect of an electron temperature T_e different from the lattice temperature T is small on the MPR effect. It was checked that this does not lead to an additional shift of the peak positions. Only the MPR amplitudes are affected.

<u>Simple model for the physical interpretation of the splitting</u>

In the following a simple argument based on Eq. (4) will be given which contains most of the above results. In the case of zero broadening the density of states factors $Im G_n(E)$ are Dirac delta-functions $\delta(E - \epsilon_n)$. After performing the energy integral in this case, Eq. (4b) takes the very simple form

$$I_E \sim (n(\omega_{LO}) + \frac{1}{2} \mp \frac{1}{2})\delta[\epsilon_m - (\epsilon_n \mp \hbar\omega_{LO} + \hbar\vec{q}.\vec{v})] \tag{5}$$

where the minus sign corresponds to the absorption and the plus sign to the emission process. Eq. (5) expresses the conservation of energy for the process of absorption or emission of an LO-phonon. Introducing the approximate average of the recoil term $\langle \vec{q}.\vec{v} \rangle \approx q_{LO}v$, with $\hbar q_{LO} = m^* v_{LO}$ in Eq. (5) then leads to the following MPR resonance condition

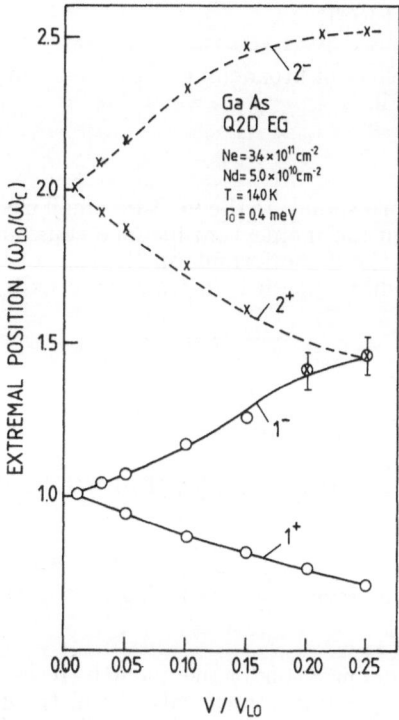

Figure 3. The position of the split $N = 1^-, 1^+$ and $N = 2^-, 2^+$ magnetophonon resonance maxima in units ω_{LO}/ω_c as a function of the average electron velocity at $T = 140\,K$ and $\Gamma_0 = 0.4\,\text{meV}$. The curves are a guide to the eye.

$$\omega_c = \frac{\omega_{LO}}{N} \mp \frac{q_{LO}v}{N} \qquad (6)$$

where $N = n - m$. In this way the N^- peaks are again associated with absorption and the N^+ peaks with emission processes. Note the v/N dependence of the shift of the resonant magnetic field.

Numerical results for the position of the non-linear MPR-peaks

The position of the $1^+, 1^-$ and $2^+, 2^-$ maximum are displayed in Fig. 3 as a function of the average electron velocity. The qualitative trend of these results can be explained on the basis of our simple model (Eq. 6). However the shift of the position of the maxima is linear in the electron velocity only for small velocities. For larger velocities (e.g. $v/v_{LO} > 0.15$ for $N = 2$) a saturation of this shift sets in and the N^+ and $(N + 1)^-$ peaks start to merge.

A critical average electron velocity v_{crit} may be defined as the velocity for which the Nth MPR-maximum in the resistivity occurs at the magnetic field for which a minimum occurs in the linear regime ($v = 0.$). Within our simple model (Eq. (6)) it follows that $v_{crit} \sim 1/N$, which agrees with the qualitative trend of the numerical results. Temperature does not affect the value of the critical velocity significantly.

DENSITY OF STATES EFFECT

In this section the effect of broadening of the electronic DOS on the MPR peak position is examined in detail.

Linear regime

The linear regime corresponds to the limit for small electron velocities of Eq. (2). The result, as obtained from the momentum balance equation, leads to an expression for ρ_{xx} which is identical to the Kubo-formula in the case of high magnetic fields (Ref. 28). For the purpose of the discussion it is instructive to write out Eq. (2) in the linear response limit as follows

$$F \sim \sum_{\vec{q}} \sum_{n,m=0}^{\infty} A(\vec{q},n,m) I_E(n,m) \tag{7}$$

$$I_E(n,m) = \int_{-\infty}^{\infty} dE \, f(E)[1 - f(E + \hbar\omega_{\text{LO}})] \, [G_m(E) + BG_m] \, [G_n(E + \hbar\omega_{\text{LO}}) + BG_n] \tag{8}$$

where we have introduced Gaussian (G) broadening on top of a background (BG) part: $Im[\mathcal{G}_n(E)] = [G(E - \epsilon_n) + BG_n]$.

The DOS enters our expression for the magneto-resistivity through the convolution of the broadened initial and final states, with the distribution function as a weight. In contrast with this, Vasilopoulos[11] and Mori et al.[27] have introduced a phenomenological broadening after performing the energy integration; this is equivalent with a collisional broadening (the delta-function for the conservation of energy is broadened to a Lorentzian function) and is different from our approach were the broadening is introduced on the level of the DOS. They do not find any shift in the resonant fields. The present approach has a more sound theoretical basis[28] and is closer to the actual experimental situation.

In Eq. (8) four different terms contribute, corresponding to the different contributions of the DOS : G-G, B-G, G-B, B-B where B stands for background and G for Gaussian. In Fig. 4 the relative contributions of these four terms to the resistivity are displayed for a background of 50%, which is taken large in order to show the effect in a pronounced way. It is clear that i) the resonances are shifted down in magnetic field from the unperturbed values and ii) the terms which include DOS from the background, result in the largest shift.

Additionally the temperature dependence of the resonant magnetic field value for a $GaAs$-heterostructure was calculated[31] for different values of the background. Following the work of Ref. 27 the following temperature dependence was assumed: $\Gamma(meV) = (T/T_0)^{1.7}$, with $T_0 = 60K$. The results indicate that i) the shift grows with increasing background contribution and ii) without fitting the width of the Landau levels and the background, the experimental results can be well represented by a background in the range $25 - 50\%$.

Fig. 5 shows the electron density dependence of the effective resonant frequency. As discussed in the introduction, one may assume a linear dependence of Γ on the density of electrons: here $\Gamma(meV) = 4N_e(10^{11}cm^{-2})$. The experimental results are from Ref. 8 on a $GaInAs/InP$ heterostructure. Only the magneto-phonon resonance corresponding to the $GaAs$ LO-phonon mode is shown. Note that here the calculated shift in the MPR resonances is a consequence of: 1) the broad DOS with background, in combination with 2) relatively high temperatures at which the electron distribution function is broadened. For example one can easily shown that for very low temperature, where the electron distribution is given by a sharp Fermi-Dirac function, Eq.(8) would not give any shift in the resonant field. This is the reason why in cyclotron resonance experiments such a shift was not observed: those experiments are typically done at liquid helium temperature.

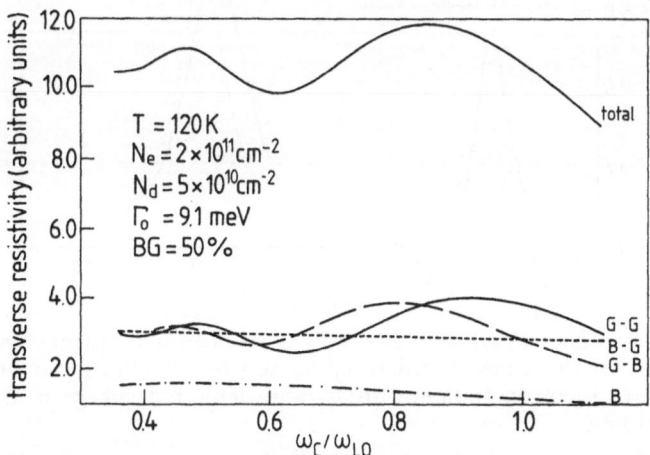

Figure 4. The magneto-resistance as function of the magnetic field. The different contributions resulting from the Gaussian (G) and background (B) part of the density of states are displayed.

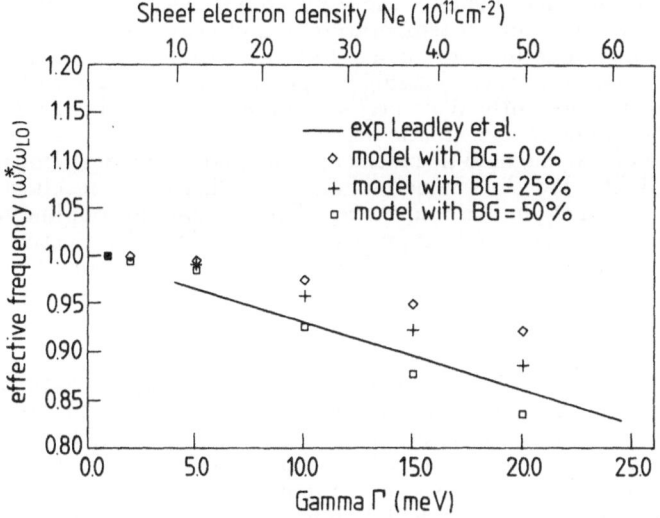

Figure 5. The resonant magnetic field as function of the broadening of the DOS. The Gaussian broadening parameter Γ is assumed to be proportional to the electron density.

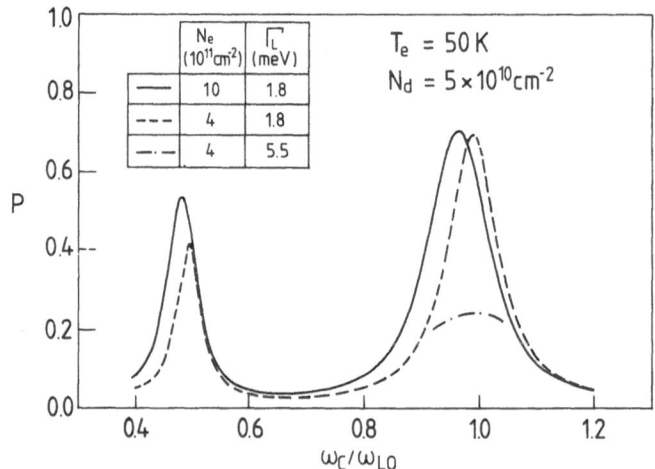

Figure 6. P (from Eq. (14)) as a function of cyclotron frequency for different values of sheet electron density N_e and broadening parameter Γ_L (see table). Electron temperature T_e and depletion charge density N_d are fixed here.

SCREENING OF THE ELECTRON-PHONON INTERACTION

In general, many-particle phenomena[32] result in a blocking effect due to the Pauli exclusion principle and furthermore lead to screening of the electron-phonon interaction. The blocking effect modifies the MPR amplitudes, but not its peak positions. Screening is expected to result in temperature and electron density dependent effects. So far this mechanism has been discussed only in the low temperature limit in connection with cyclotron resonance. Lassnig[33] suggested the possibility of pinning to the TO-phonon frequency. Recently Wu[34] performed a more detailed calculation of the 2D plasmon-phonon coupling in a magnetic field (at T = 0 K) and found that indeed there is a strong mixing of the magneto-plasmons and the phonon modes. But he was able to rule out the possibility that the magneto-phonon coupling would result in a splitting of the cyclotron mass at TO-phonon frequencies. Furthermore no coupling to a frequency below the TO-phonon frequency is expected. However, in order to study the effect of screening on the MPR peak positions one has to adopt a temperature dependent model. In that case it will turn out that screening of the electron-LO-phonon interaction may lead to an electron density dependence of the MPR peak position.

In this section the magnetophonon effect in the energy relaxation rate for a 2D EG will be investigated. Since i) the averaged correlation function is largely independent of temperature (see below), and ii) correlation effects are expected to be similar for the energy relaxation and for the conductivity of a 2D EG, we believe that screening will have qualitatively the same effect on the MPR peak position in the energy relaxation rate as in the resistivity. A model calculation is performed, which includes the effect of dynamical screening of the 2D EG and electron LO-phonon interaction within the well known RPA approximation

$$W = \frac{\hbar\omega_{LO}}{\tau_0}\left[n(\omega_{LO}) - n_e(\omega_{LO})\right]P \tag{9}$$

with

$$1/\tau_0 = \frac{2\alpha\hbar\epsilon_\infty}{e^2} \left(\frac{\hbar}{2m^*\omega_{LO}}\right)^{1/2} (\omega_{LO})^2 \tag{10}$$

$$P = \frac{1}{2\pi N_e} \int_0^\infty dq\, q\, Im \frac{-1}{\epsilon(q,\omega_{LO})} \tag{11}$$

and where $\epsilon(q,\omega)$ is the dielectric function of the 2D EG, ϵ_∞ is the high frequency dielectric constant and α is the Fröhlich coupling constant. The dielectric function as calculated in the RPA approximation is expressed as $\epsilon(q,\omega) = 1 - v(q)\Pi^0(q,\omega)$ where $v(q)$ is the Fourier transformed electron-electron interaction matrix element and $\Pi^0(q,\omega)$ is the polarizability of the unperturbed 2D EG[35]

$$\Pi^0(q,\omega) = \frac{m^*\omega_c}{2\pi\hbar} \sum_{n,m=0}^\infty J_{n,m}(l_B^2 q^2/2)\, \Pi_{n,m}^0(\omega) \tag{12}$$

with

$$\Pi_{n,m}^0(\omega) = -2 \frac{f(\epsilon_m) - f(\epsilon_n)}{\hbar\omega - \epsilon_m + \epsilon_n + i\Gamma_L} \tag{13}$$

where $f(x)$ is the electron energy distribution function, taken here as a Boltzmann distribution function, and where Lorentz type broadening was introduced by adding a nonzero imaginary part (Γ_L) in the denominator. Eq. (13) involves transitions between Landau levels with index n and m. The summation over Landau level indices can be exactly converted into an integral (see e.g. Glasser[35]) and this results in

$$\Pi^0(q,\omega) = \frac{-2}{\pi\hbar} \left[1 - e^{-2\pi\Gamma_L/\hbar\omega_c + i2\pi\omega/\omega_c}\right]^{-1}$$
$$\times \int_0^{2\pi} dt\, \sin(a\sin t)\, e^{a(\cos t - 1)}\, G_j[2a(1 - \cos t)]\, e^{-t\Gamma_L/\hbar\omega_c + it\omega/\omega_c} \tag{14}$$

where $a = q^2/2l_B^2$, j is the largest integer contained in $\mu/\hbar\omega_c$ and $G_n(x) = L_n(x) + 2L_{n-1}^1(x)$ (see Eq. (27) of Ref. 35). The temperature dependence of the dielectric function is included along the lines of Ref. 36.

From the numerical results[37] it appears that all the temperature dependence is contained in the phonon occupation numbers, in agreement with the results of I. The Pauli exclusion principle does not affect the MPR peak positions and only starts to diminish the MPR amplitude for filling factors larger than 2. The dependence of P on the sheet electron density N_e and broadening parameter Γ_L (see table) is shown in Fig. 6. Electron temperature T_e and depletion charge density N_d are fixed here. The effect of screening on the magnitude of the energy relaxation rate is clearly not very large. Even for a sheet electron density $N_e = 1.4 \times 10^{12}\,cm^{-2}$ (for fixed $\Gamma_0 = 1.8\,meV$) the magnitude of the $N = 1$ MPR peak is still 76 % of the magnitude of the corresponding peak for $N_e = 0.4 \times 10^{12}\,cm^{-2}$. The peak positions of all the curves are clearly shifted down in magnetic field from the normal MPR peak positions at $\omega_c/\omega_{LO} = 1/n$ with $n = 1, 2$. One observes that this shift increases with increasing N_e, if the broadening parameter is fixed. On increasing the broadening parameter Γ_L with a factor of three while keeping N_e fixed, the relative shift of the peak position is smaller than 0.1 %, but the magnitude of P is strongly reduced (with a factor three).

1D WIRES - EFFECT OF ADDITIONAL ELECTRON CONFINEMENT

In quantum wires the electron states are localized in the z direction and additionally in the y direction, but free along the axis of the wire (x direction). This,

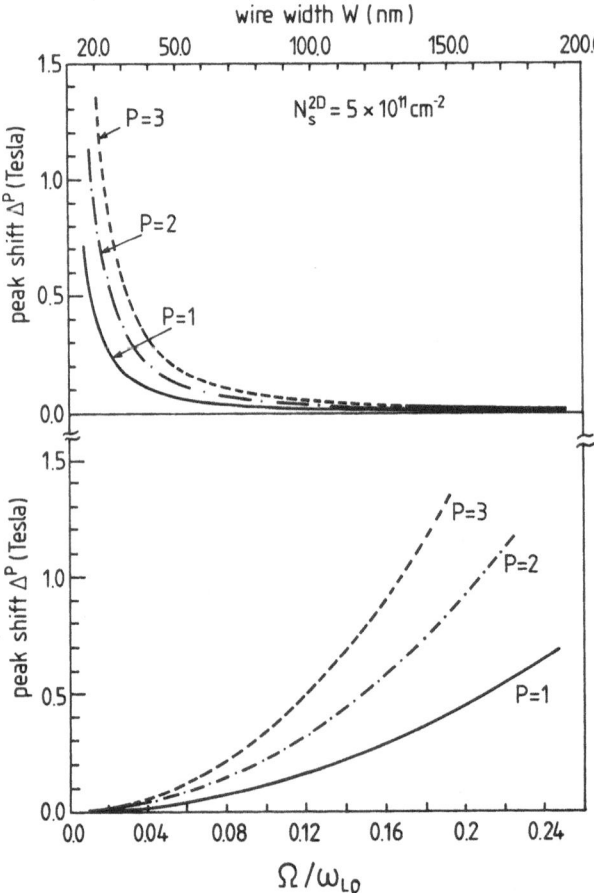

Figure 7. Magnetophonon resonance peak shift for a Q1D EG versus confining frequency (bottom) and versus wire width $W = a[N_e^{1D}]^{1/3}\Omega^{-2/3}$ (top) for three values of the harmonic number P. The 2D electron density is kept constant.

together with the quantization of the electron energy, modifies the physical properties of the system. Here the influence of the confinement on the MPR peak positions will be studied.

Consider a quasi-one-dimensional electron gas (Q1D EG) confined in a wire of dimensions $L_x : L, L_y : W$ and L_z such that $W, L_z \ll L$, i.e $W \sim 0.1\mu m$ or less, and $L_z \sim 50 - 100\mathring{A}$. The vertical confinement (z axis) is modelled with a triangular well, as it occurs in heterostructures, and the lateral one (y axis) with a parabolic potential of frequency Ω. The magnetic field B is applied along the z axis. In the Landau gauge $\vec{A} = (-yB, 0, 0)$ where $\vec{A}$ is the vector potential, the one-electron Hamiltonian H, eigenstates $|\zeta>$ and eigenvalues E_ζ read

$$H = \frac{(\vec{P} + q\vec{A})^2}{2m^*} + \frac{m^*\Omega^2 y^2}{2} + H_0(z), \tag{15}$$

$$|\zeta> = \Psi_{n,k_x,0} = \frac{1}{\sqrt{L}}\Phi_n(y - y_0)e^{ik_x x}\Psi_0(z), \tag{16}$$

494

$$E_\zeta = E_{n,k_x,0} = (n + \frac{1}{2})\hbar\tilde{\omega} + \frac{\hbar^2 k_x^2}{2\tilde{m}} + E_0(z) \qquad (17)$$

where $\vec{P}$ is the momentum operator, q is the electron charge, n is the Landau level index, $\tilde{\omega} = \sqrt{(\omega_c^2 + \Omega^2)}$, $\tilde{m} = m^*\omega_c^2/\Omega^2$, $y_0 = \tilde{b}\tilde{l}_B^2 k_x$ with $\tilde{b} = \omega_c/\tilde{\omega}$, $\tilde{l}_B^2 = \hbar/m^*\tilde{\omega}$ is the magnetic length and k_x is the wave vector in the x direction. The main difference of Eqs. (15-17) from the corresponding ones in the 2D case, in the absence of the confining potential $m^*\Omega^2 y^2/2$, is that the k_x degeneracy of the energy levels, for $\Omega = 0$, is now lifted and the electrons appear heavier since $\tilde{m} > m^*$ for $\Omega > 0$. This particular form of the confining potential was chosen because it is analytically tractable and because theoretical self-consistent calculations[38] have indicated that it is appropriate for the narrow wires we are considering.

For the evaluation of the linear dc magnetoconductivity σ_{xx} when an electric field $E = E_x$ is applied along the wire, the formalism of Ref. 39 is applied (which is equivalent to the well-known Kubo formula). The same formalism was applied in Ref. 40 to model the low temperature oscillations of the conductivity of a quantum wire as a function of the impurity density observed in Ref. 3. The approximations listed above for the momentum balance equation are also made here. In addition, the impurity system is assumed to be one dimensional and a range of magnetic fields and wire widths is considered such that the cyclotron orbits are closed: $l_B < W < L_z$ (no skipping orbits). In order to get tractable expressions, some minor but physically well supported approximations were made (for details see Ref. 41). Assuming a Lorentzian broadening (Γ_L) of the delta-functions with shift zero, the result for the conductivity is

$$\sigma_{xx} \sim \frac{e^2}{h}\beta(1 + 2n(\omega_{LO}))e^{-\beta\mu}\frac{\tilde{b}^2\tilde{l}_B^2}{\hbar\tilde{\omega}}\frac{1}{sh(\beta\hbar\tilde{\omega}/2)}$$
$$\times \frac{sh(2\gamma)}{ch(2\gamma) - cos(2\pi\delta)} \qquad (18)$$

where $\delta = \omega_{LO}/\tilde{\omega}$, $\gamma = \pi\Gamma_L$ and μ is the chemical potential. Γ_L is the total level width obtained from the sum of the widths for interaction of the electrons with phonons and impurities, respectively. It was found to be largely independent of Landau level index N.

At resonance $\delta = \omega_{LO}/\tilde{\omega} = P$, an integer, and the last line of Eq. (18) is equal to $coth(\gamma)$. In this way Eq. (18) exhibits the magnetophonon effect with a "renormalized" cyclotron frequency: $\tilde{\omega}_c = (\omega_c^2 + \Omega^2)1/2$. Since $\omega_c < \tilde{\omega}$, for $\Omega > 0$, the resonances are shifted to smaller magnetic fields. For $\Omega = 0$, i.e. in the absence of confinement, $\tilde{\omega} = \omega_c$ and the usual resonance condition $\omega_{LO} = P\omega_c$ is recovered. The monotonic part of σ_{xx} is given by the first line of Eq. 18; the oscillatory part of σ_{xx} is given by Eq. 18 with $sh(2\gamma)$ replaced by $cos(2\pi\delta) - exp(-2\gamma)$. The general trends of the Q1D magnetoconductivity agree qualitatively with the corresponding two- or three-dimensional results as far as the temperature and the magnetic field dependences are concerned, e.g. at high temperatures σ_{xx} decreases with increasing T.

Through the impurity level width, σ_{xx} depends on the impurity concentration. Another dependence, i.e. on the 1D electron density N_e^{1D} enters through the factor $exp(\beta\mu)$. Using Eq. (17), the unperturbed quasi-one-dimensional density of states $N^{1D}(E)$ reads (spin included)

$$N^{1D}(E) = \sum_n \sqrt{\frac{2m^*}{E - E_{n,0}}} \qquad E > E_{n,0} \qquad (19)$$

where $E_{n,0} = (n + 1/2)\hbar\tilde{\omega} + E_0(z)$. At high temperatures, the Fermi function f(E) is approximately $exp[-(E - \mu)]$ and this readily gives for N_e^{1D} the result

$$N_e^{1D} = \sqrt{\frac{2\tilde{m}}{\beta\hbar^2\pi}} \frac{e^{\beta\mu}}{2sh(\hbar\tilde{\omega}/2)} \qquad (20)$$

through the definition $N_e^{1D} = \int N^{1D}(E)f(E)\,dE$. The main dependence of σ_{xx} on Ω comes from the term N_e^{1D}, since the other terms involve $\tilde{\omega} \approx \omega_c + \Omega^2/2\omega_c = \omega_c(1+\epsilon)$ with $\epsilon \ll 1$ for $\Omega < 0.3\omega_c$. As in Ref. 42, N_e^{1D} is taken equal to $N_e^{2D}W$ and the 2D electron density is assumed to remain constant. The width is related to the confining frequency here through the approximate relation[42]: $W(\text{in } m) = a[N_e^{1D}(\text{in } m^{-1})]^{1/3}\Omega^{-2/3}$ with $a = 0.021$. This only serves as a qualitative indication, since the width of the parabolic potential is not well defined and the confinement is not rigid since the wave functions do not vanish at the edges of the sample. Related to that is, of course, the fact that no rigorous relation exists between the confining frequency Ω and the width W of the sample (corresponding values of W and Ω can in principle be obtained from a selfconsistent calculation[38]). In this respect, the relation between W and Ω, as given in Ref. 42 at low temperatures, is far from being rigorous but it is used here in order to understand qualitatively the influence of the confinement on the magnetophonon effect. Note that the width of the wires has not been expermentally determined with satisfactory precision so far.

In Fig. 7 the magnetophonon resonance (MPR) peak shift $\Delta^P = B_P - \tilde{B}_P$ for a GaAs Q1D wire is displayed as function of the confining frequency Ω in the lower part of the figure, and as a function of the 1D width in the upper part. The shift Δ^P is the difference between the normal resonant magnetic field $B_P = (m^*/e)(\omega_{LO}/P)$, in the case of a Q2D EG (no confinement in the y-direction) and the resonant field for a Q1D EG $\tilde{B}_P = (m^*/e)((\omega_{LO}/P)^2 - \Omega^2)1/2$. For $\Omega \ll \omega_{LO}$, one has $\Delta^P \approx (m^*/e)P\Omega^2/2\omega_{LO}$. For fixed Ω, the peak shift increases linearly with increasing harmonic number P, i.e. with decreasing magnetic field. This means that, for fixed width W or Ω, the relative peak shift $\Delta^P/B_P = 1-(1-P^2(\Omega/\omega_{LO})^2)^{1/2}$ increases strongly with increasing P or decreasing magnetic field. For the $P = 1, 2$ and 3 peaks at $\Omega/\omega_{LO} = 0.08$, corresponding to approximately $W=50$nm, one has $\Delta^P = 0.071, 0.143$ and 0.216 Tesla, respectively.

CONCLUSIONS

The present analysis of the shift in the position of the MPR peaks for single heterostructures is consistent with an interaction of the electrons with bulk LO-phonons. This is in agreement with cyclotron resonance experiments on $GaAs/Al_xGa_{1-x}As$ heterostructures done at low temperature where the presence of coupling with bulk LO-phonons was demonstrated[13], irrespective of the electron density. It was shown that

1. with increasing electron velocity (or equivalently applied electric field) the magnetophonon resonance peak splits up into two peaks when the broadening of the Landau levels is small. Considering the separate contributions of LO-phonon absorption and emission processes to the non-linear momentum balance equation a physical interpretation for this splitting is proposed. A shift to lower magnetic fields (no splitting) of the magnetophonon peaks is found when the broadening of the Landau levels is large,

2. the convolution of the broadened initial and final density of states (DOS) of the two-dimensional electron gas (2D EG) in combination with a thermally broadened electron distribution function can lead to substantial shifts in the position of the MPR resonances,

3. the inclusion of screening on the magnetophonon effect in the energy relaxation rate within the RPA approximation may lead to a shift of the MPR peak positions to lower magnetic fields with increasing electron sheet density; and

4. the confinement of electrons in quantum wires of width $W \sim 0.1\,\mu m$ much larger than the magnetic length and much smaller than the mean free path, modelled with a parabolic potential of frequency Ω, may lead to an observable shift of the magnetophonon resonance peaks to smaller magnetic fields.

For the 2D EG the relatively most important contribution to the shift of the

MPR peak positions comes from the hot electron effect in the non-linear regime, while in the linear regime the DOS effect may dominate. In order to check on the latter or equivalently to correctly describe the dependence of the linear MPR oscillations (both position and broadening of the peaks) on all parameters (B, N_e, T), at least the effect of screening and the effect of a broad DOS will have to be treated consistently. For the higher electron densities the population of a second electric subband will have to be accounted for[43]. A microscopic model as well as experimental data for the effect of temperature on the DOS of a 2D EG are lacking and consequently it is not clear what are realistic values for the parameters of phenomenological broadening models. A thorough understanding of this topic is of importance for other phenomena too, like cyclotron resonance, breakdown of the quantum Hall effect.

ACKNOWLEDGMENTS

Part of this work was supported by I.I.K.W. (Interuniversitair Instituut voor Kernwetenschappen), Project No. 4.0002.83, Belgium. I would like to thank all the people with whom I have had the pleasure to work on the magnetophonon effect and in particular F.M. Peeters, X. Wu, P. Vasilopoulos and J.T. Devreese.

REFERENCES

* Formerly at: Physics Department, University of Antwerp (U.I.A.), B-2610 Wilrijk, Belgium.

1. R.J. Nicholas, Prog. Quant. Electr. 10, 1 (1985); in "Landau Level Spectroscopy," E.I. Rashba and G. Landwehr, eds., North-Holland, Amsterdam, in press.
2. M.J. Kearny and P.N. Butcher, J. Phys. C20, 47 (1987).
3. T. Oshima, M. Okada, M. Matsuda, N. Yokoyama and A. Shibatomi, Proc. 4th Int. Conf. on Superlattices, Microdevices and Microstructures (Trieste, 1988), Superlattices and Microstructures, in press.
4. B.J. Van Wees, H. van Houten, C.W.J. Beenacker, J.G. Williamson, L.P. Kouwenhoven, D. van der Marel and C.T. Foxon, Phys. Rev. Lett. 60, 848 (1988).
5. M.L. Roukes, A. Scherer, S.J. Allen, Jr.H.G. Craighead, R.M. Ruthen, E.D. Beebe and J.P. Harbison, Phys. Rev. Lett. 59, 3011 (1987).
6. M.A. Brummell, R.J. Nicholas, M.A. Hopkins, J.J. Harris and C.T. Foxon, Phys. Rev. Lett. 58, 77 (1987).
7. M.A. Brummell, R.J. Nicholas, J.C. Portal, K.Y.Cheng and A.Y. Cho, J. Phys. C16, L579 (1983).
8. D.R. Leadley, R.J. Nicholas, L.L. Taylor, S.J. Bass and M.S. Skolnich, (1988), to be published.
9. D.R. Leadley, M.A. Brummell, R.J. Nicholas, J.J. Harris and C.T. Foxon, Proc. 5th Int. Conf. on Hot Carriers in Semiconductors, (Boston, 1987), Solid State Electr. 31, 781 (1988).
10. P. Warmenbol, F.M. Peeters and J.T. Devreese, Phys. Rev. B37, 4694 (1988).
11. P. Vasilopoulos, M. Charbonneau and C.M. Van Vliet, Phys. Rev. B35, 1334 (1987).
12. L. Eaves, P.S.S. Guimaraes, J.C. Portal, T.P. Pearsall and G. Hill, Phys. Rev. Lett. 53, 608 (1984).
13. C.J.G.M. Langerak, J. Singleton, P.J. van der Wel, J.A.A.J. Perenboom, D.J. Barnes, R.J. Nicholas, M.A. Hopkins and C.T. Foxon, Phys. Rev. B38, 13133 (1988).
14. M.V. Klein, IEEE J. Quantum Electronics QE-22, 1760 (1986).
15. G. Gobsch, D. Schulze and G. Paasch, Phys. Rev. B38, 10943 (1988).
16. E. Gornik, R. Lassnig, G. Strasser, H.L. Störmer, A.C. Gossard and W. Wiegmann, Phys. Rev. Lett. 54, 1820 (1985).
17. H.P. Wei, A.M. Chang, D.C. Tsui and M. Razeghi, Phys. Rev. B32, 7016 (1985).
18. J.P. Eisenstein, H.L. Störmer, V. Narayanamurti, A.Y. Cho, A.C. Gossard, and C.W. Tsu, Phys. Rev. Lett. 55, 875 (1985).
19. I.V. Kukushkin and V.B. Timofeev, Surface Science 196, 196 (1988).
20. T. Ando and Y. Uemura, J. Phys. Soc. Jpn. 36, 959 (1974).

21. S. Das Sarma, Solid Stat. Commun. 36, 357 (1980); Phys. Rev. B23, 4529 (1981); R. Lassnig and E. Gornik, Solid, Stat. Commun. 47, 959 (1983); T. Ando and Y. Murayama, J. Phys. Soc. Jpn. 54, 1519 (1986); Phys. Rev. B35, 2252 (1987); W. Cai and C.S. Ting, Phys. Rev. B33, 3967 (1986).
22. R.R. Gerhardts and V. Gudmundsson, Phys. Rev. B34, 2999 (1986); ibid. B35, 8005 (1987).
23. S. Das Sarma and X.C. Xie, Phys. Rev. Lett. 61, 738 (1988).
24. R. Lassnig and W. Zawadski, J. Phys. C16, 5435 (1983).
25. P. Warmenbol, F.M. Peeters and J.T. Devreese, Phys. Rev. B39 (1988), in press.
26. M.A. Brummell, D.R. Leadley, R.J. Nicholas, J.J. Harris and C.T. Foxon, Surface Science 196, 451 (1988).
27. N. Mori, H. Murata, K. Taniguchi and C. Hamaguchi, Phys. Rev. B38, 7622 (1988).
28. R. Kubo, S.J. Miyake and N. Hatshitsume, Solid State Phys. 17, 269 (1965).
29. R.L. Peterson, Vol.10, Ch. 4, in "Semiconductors and Semimetals," R.K. Williardson and A.C. Beer, eds., Academic, N.Y. (1975).
30. N. Mori, N. Nakamura, K. Taniguchi and C. Hamaguchi, Semicond. Sci. Technol. 2, 542 (1987).
31. F. M. Peeters, P. Warmenbol and J.T. Devreese, (1989), to be published.
32. X. Wu, F.M. Peeters and J.T. Devreese, Phys. Rev. B 36, 9760 (1987).
33. R. Lassnig, Surface Science 170, 549 (1986).
34. X. Wu, Phys. Rev. B38, 4212 (1988).
35. M.L. Glasser, Phys. Rev. B28, 4387 (1983).
36. P.F. Maldague, Surface Science 73, 246 (1978).
37. P.Warmenbol, F.M. Peeters, X. Wu and J.T. Devreese, (1989), to be published.
38. S.E. Laux and F. Stern, Appl. Phys. Lett. 49, 91 (1986).
39. P. Vasilopoulos, Phys. Rev. B33, 8587 (1986).
40. P. Vasilopoulos and F.M. Peeters, (1989), to be published.
41. P. Vasilopoulos, P. Warmenbol, F.M. Peeters and J.T. Devreese, Phys. Rev. B (1989), in press.
42. K.F. Berggren, G. Roos, and H. van Houten, Phys. Rev. B37, 10118 (1988).
43. T.P. Smith III, F.F. Fang, U. Meirav and M. Heiblum, Phys. Rev. B38, 12744 (1988).

INTERACTION OF SURFACE ACOUSTIC WAVES WITH TWO-DIMENSIONAL ELECTRON SYSTEMS IN GaAs / AlGaAs HETEROJUNCTIONS

Achim Wixforth

Univers. Hamburg, Institut f. Angewandte Physik, D-2000 Hamburg 36
F.R.Germany
present address

Materials Dept., Univers. of California, Santa Barbara, CA 93106, U.S.A.

I. Introduction

The properties of quasi two-dimensional electron systems (2DES) as realized in space charge layers on semiconductors have been studied extensively during the last twenty years using various spectroscopic methods [1]. A lot of the recent work on this field   has only become possible due to the tremendous progress and success in the evaluation of growth techniques such as molecular beam epitaxy (MBE)[2] and the improvements in material processing and lithography. Needless to mention, that also the physics and technology of semiconductor devices profits very strongly of these efforts. The discovery of the quantum Hall effect (QHE)[3] as well as the fractional quantum Hall effect (FQHE)[4] has led to a strong interest in the properties of a 2DES in high magnetic fields and at low temperatures. The amount of publications on this subject during the last few years demonstrate that there are still open questions and continuosly new questions and answers arise.

Here, I wish to review recent investigations [5..8,25] of a magnetically quantized 2DES at low temperatures using surface acoustic waves (SAW) in the radio frequency regime (50..500 MHz). Corresponding to the relatively low sound velocity the wavelengths of these coherent and monochromatic lattice vibrations lie in the 10 μm range. The effect of ultrasonics on electrons in semiconductors has first been pointed out by Keldysh [9] in 1962. He considered a superlattice effect on the electron spectrum due to the spatially periodic quasi-stationary field of a sound wave. For sufficiently short wavelengths and large amplitude of the wave this should lead to the occurence of minibands and gaps in the electron spectrum. Our experiments can be regarded as the long-wavelenght limit of this interaction and thus can mostly be described in terms of classical models. As I will show, on piezoelectric semiconductors such as GaAs the interaction between a SAW and a 2DES in this frequency range is dominated by the piezoelectric field accompanying the SAW with speed of sound. As it turns out, the electron-SAW interaction is of relaxation type and most pronounced in the regime of very small conductivities σ of the 2DES. This is the case at high magnetic fields under QHE conditions i.e. near integer filling factors ν of Landau levels or at B=0 near inversion threshold.

II. Surface acoustic waves on piezoelectric heterostructures

Surface acoustic waves [10] are modes of elastic energy propagating along the surface of an elastic medium. The particle displacement amplitudes decay in an exponential fashion with depth into the material. Hence, the main energy flow is concentrated within a distance of the order of a wavelenght λ beneath the surface. The particle displacement at and beneath the surface is elliptic, leading to an elliptical polarization of the wave. On piezoelectric materials the mechanical surface wave can be accompanied by electric fields which in turn interacts with mobile carriers in the medium. Here, I will focus on the propagation of a SAW on a piezoelectric GaAs/AlGaAs heterostructure, which is modelled by a semiinsulating, piezoelectric substrate, having near its surface a high mobility 2DES. The discussion is limited to a crystal cut and sound propagation direction which is piezoelectrically active, represented by a large electro-mechanical coupling coefficient.

For sound waves in a piezoelectric material the mechanical stress T and the electrical displacement D are connected with the mechanical strain S and the electric field E. In a simplified one-dimensional form the basic equations of state are :

$$T = cS - eE \qquad \text{and} \qquad D = eS + \varepsilon E. \tag{1}$$

Here, c denotes the elastic constant, e the piezoelectric constant and ε the electric permittivity of the material. Generally, both equations have to be written in tensor form, taking into account the anisotropy and the symmetry of the crystal as well as boundary conditions like in the case of a SAW, the surface. For the sake of legibility and simplicity the discussion here is restricted to the one-dimensional case. Stress T and strain S are related to the displacement amplitude u in x direction using

$$\frac{\partial S}{\partial x} = \frac{\partial^2 u}{\partial x^2} \qquad \text{and} \qquad \frac{\partial T}{\partial x} = \rho \frac{\partial^2 u}{\partial t^2} \tag{2}$$

where ρ is the density of the material under consideration. Combining (1) and (2) leads to a wave equation for the problem, again written in one-dimensional form :

$$\rho \frac{\partial^2 u}{\partial t^2} = c \left(1 + \frac{e^2}{c\varepsilon}\right) \frac{\partial^2 u}{\partial x^2} - \frac{e}{\varepsilon} \frac{\partial D}{\partial x} \tag{3}$$

Considering this wave equation, we can easily discuss two limiting cases. If the piezoelectric crystal is a perfect conductor, i.e. $\sigma = \infty$, the internal electric field E in (1) has to vanish. Equation (3) then describes longitudinal sound waves in a medium, which appears non-piezoelectric and has a sound velocity $v_0 = (c / \rho)^{1/2}$. If, on the other hand, the material is an insulator, i.e. $\sigma = 0$, Poisson's equation requires the last term in eq. (3) to cancel and thus describing a medium with increased elastic constants $c' = c (1 + e^2 / c\varepsilon)$. The sound velocity in this case is $v' = (c' / \rho)^{1/2} > v_0$. This effect is called piezoelectric stiffening and is used to define an electro-mechanical coupling coefficient

$$K^2 = \frac{e^2}{c\varepsilon} \approx \frac{2 (v' - v_0)}{v_0} \tag{4}$$

as a measure of the strenght of the piezoeffect for a given material and geometry. It usually is a small quantity, for the best piezoelectric it is of the order of 0.05. For surface waves one obtains an effective coupling coefficient that differs slightly from the one derived above, since here, too, boundary conditions have to be taken into account. For a SAW on a (100)GaAs surface propagating in a [011] or equivalent direction, one finds [11] $K_{eff}^2 = 6.4 \cdot 10^{-4}$. The intermediate case of finite conductivity in the medium has been treated in detail for the bulk case by different authors [12,13]. The electric field accompanying the SAW now is non-zero and can couple to the mobile carriers in the material. This leads to induced currents and resulting Ohmic loss σE^2, resulting in an attenuation of the wave. Simultaneously, the sound velocity is changed by piezoelectric stiffening of the elastic constants. The attenuation Γ and the change in sound velocity $\Delta v/v_0$ as a function of the conductivity σ in this case turn out to be [12]:

$$\Gamma = \frac{\omega}{v_0} \frac{K_{eff}^2}{2} \frac{(\omega_c/\omega)}{1 + (\omega_c/\omega)^2} \tag{5}$$

and

$$\frac{\Delta v}{v_0} = \frac{v - v_0}{v_0} = \frac{K_{eff}^2}{2} \frac{1}{1 + (\omega_c/\omega)^2} \tag{6}$$

Here, v denotes the SAW velocity at finite conductivities σ, $\omega_c = \sigma / (\varepsilon_1 + \varepsilon_2)$ is the Maxwell conductivity relaxation frequency and ε_1 and ε_2 are the dielectric constants of the substrate and the medium above it, respectively. If the SAW frequency ω is much smaller than the conductivity relaxation frequency ω_c, the carriers can redistribute themselves rapidly enough to screen the electric field of the SAW. If ω is comparable or larger than ω_c, this screening will be less perfect and finally if $\omega >> \omega_c$, the piezoelectric field will be nearly the same as for the corresponding insulator. Maximum attenuation occurs at $\omega = \omega_c$ and piezoelectric stiffening occurs for $\omega \geq \omega_c$.

This description has to be slightly changed, if the substrate is not a homogenous conductive medium, but if the conductivity is distributed inhomogenously across it. Such is the case if one considers, e.g., a semiinsulating, piezoelectric substrate with a conductive layer at or near its surface. If the thickness d of this layer, which might be a metallization [14,15] of the surface or a 2DES [16..18] as we consider it, is much smaller than the wavelenght λ, the longitudinal electric field of the SAW can only be screened at or near the surface, i.e. z $\cong 0$. With depth into the material it will recover, until at a certain distance from the surface, the influence of the thin conducting sheet is not longer important and the field is the same as for the insulating case. This is demonstrated in Fig. 1, where the absolute value of the SAW potential $|\varphi_{SAW}|$ as a function of the depth into the piezoelectric substrate is depicted. Both cases for the metallized as well as for the free surface are shown.

The description of the attenuation Γ and the change in sound velocity $\Delta v/v_0$ as represented in eq. (5) and (6) in principle apply here, too. The fact that only a thin layer of thickness d<<λ is able to screen the electric field of the SAW, results effectively in a modification of the relaxation frequency ω_c. This frequency now becomes dependent on the wave vector $k = 2\pi/\lambda$. It turns out to be given by $\omega_c = \sigma_\square \cdot k / (\varepsilon_1 + \varepsilon_2)$, where $\sigma_\square = \sigma \cdot d$ is the sheet conductivity of the conducting layer at the surface of the insulating substrate [14]. Thus for the two-dimensional case, the ratio (ω / ω_c) in eq. (5) and (6) has to be replaced by the frequency independent ratio $(\sigma_\square / \sigma_M)$, where $\sigma_M = v_0 (\varepsilon_1 + \varepsilon_2)$ is a characteristic sheet conductivity, at which maximum attenuation occurs. We now have

$$\Gamma = \frac{\omega}{v_0} \frac{K_{eff}^2}{2} \frac{(\sigma_\square / \sigma_M)}{1 + (\sigma_\square / \sigma_M)^2} \qquad \text{and} \qquad \frac{\Delta v}{v_0} = \frac{K_{eff}^2}{2} \frac{1}{1 + (\sigma_\square / \sigma_M)^2} \tag{7, 8}$$

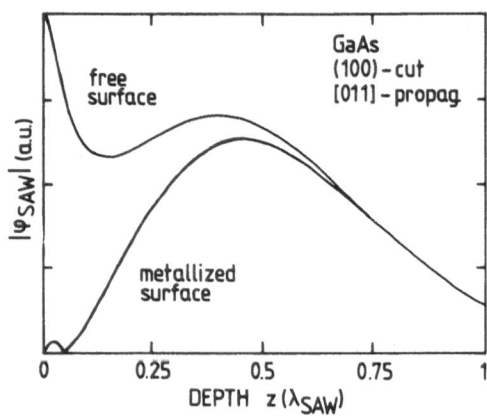

Fig. 1. Normalized absolute value of the SAW potential for a (100) GaAs surface with SAW propagation in [011] or equivalent direction as a function of the depth into the substrate. Both cases for a metallized as well as for a free surface are shown.

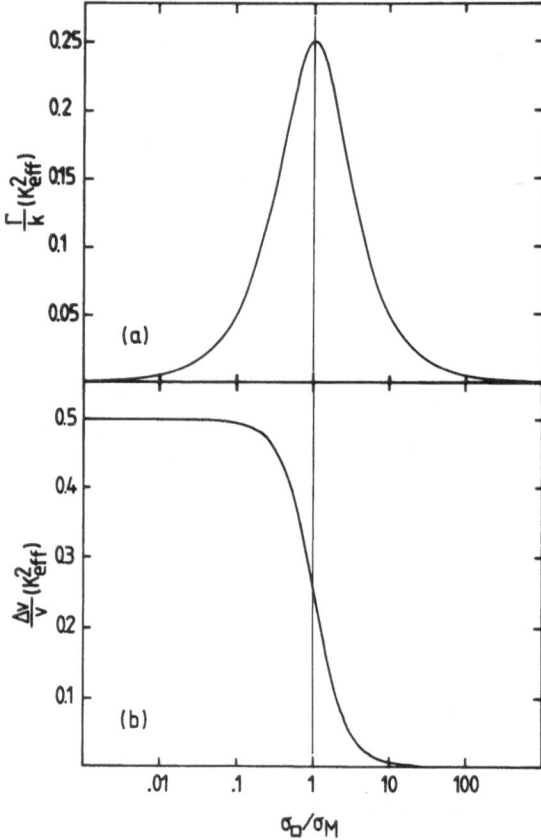

Fig. 2. Attenuation Γ per unit length (a) and change in sound velocity $\Delta v/v_0$ (b) according to eqs. (7) and (8) as a function of the sheet conductivity $\sigma_\Box$ in units of σ_M as defined in the text.

Equations (7) and (8) are illustrated in Fig. 2, where the attenuation per unit length as well as the relative change of the sound velocity as a function of the sheet conductivity $\sigma_\square$ of the 2D layer at the surface is depicted. The attenuation Γ shows a maximum at $\sigma = \sigma_M$, whereas the sound velocity changes rather steplike around this characteristic conductivity, demonstrating the effect of piezoelectric stiffening. For the special case of a (100) GaAs surface with the SAW propagating in [011] or equivalent direction the characteristic conductivity σ_M is about $3.3 \cdot 10^{-7} \Omega_\square^{-1}$, indicating the sensitivity of SAW experiments on very small conductivities. Since most of our experiments are carried out in magnetic fields, we have to consider the anisotropy of the conductivity in a magnetic field and $\sigma_\square$ in eqs.(7) and (8) has to be replaced by the diagonal component of the magnetoconductivity tensor, i.e. by σ_{xx} (B).

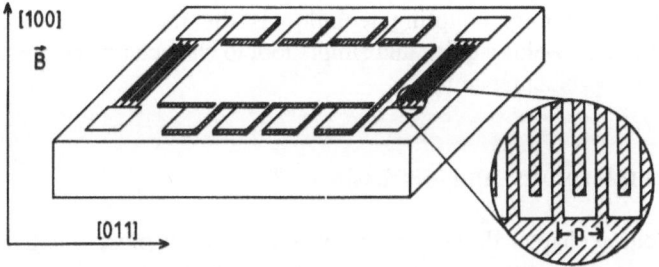

Fig. 3.   Sketch of a typical sample for use in SAW experiments. The heterojunction containing a high mobility 2DES is partly removed to form a Hall bar mesa. Indium diffused contacts allow the measurement of the magnetoresistance components ρ_{xx} and ρ_{xy}. To the left and the right of the active sample area a pair of interdigital transducers with periodicity p provide the launch and the detection of a SAW .

III. Experiment

A sketch of a typical sample arrangement used to study the interaction of SAW with a 2DES is given in Fig. 3. The sample is prepared on a conventional GaAs/AlGaAs heterojunction grown by MBE on a semiinsulating GaAs substrate. A Hall bar shaped mesa is defined by chemical wet etching. Indium diffused contacts allow us to simultaneously measure the magnetoresistance components ρ_{xx}(B) and ρ_{xy}(B) of the 2DES in a magnetic field. On both sides of this mesa we photolithografically define interdigital transducers with periodicities p in the range of four to fourty microns to excite and detect SAW. In standard heterojunctions, the total thickness of the cap layer, the doped AlGaAs layer and the spacer layer is about 100nm, much less than the wavelenght of a SAW and thus much less than its penetration depth. Our experiments are carried out at low temperatures (T > 2K) and in high magnetic fields (B<12T) applied perpendicularly to the sample surface. The SAW delay line operates at a fundamental frequency f = v / p , p again being the periodicity of the transducer structure. Short RF pulses (0.1..1 μsec) of this frequency are fed into one of the transducers, leading to the emission of a SAW packet. This SAW packet thus is transmitted through the mesa containing the high mobility 2DES and after typical delay times of 1.5 .. 2.5 μsec a SAW signal is detected at the other transducer, where it is analyzed in intensity and phase using conventional boxcar integration techniques.

Another useful method to investigate the interaction of a SAW with a 2DES is the proximity coupling technique [7,19]. In this case the SAW is excited not on the heterostructure sample itself, but on a separate piezoelectric substrate which is brought into close contact with the sample under investigation. For this purpose we use Y-cut, Z-propagating $LiNbO_3$ SAW delay lines, which have a very high electromechanical coupling coefficient ($K^2_{eff} = 0.048$). The $LiNbO_3$ delay line and the heterostructure thus form a sandwich structure, with the heterojunction face down on the surface of the $LiNbO_3$ on which the SAW is propagating. The system is acoustically mismatched, so that no SAW is excited on the GaAs surface. If the gap between the SAW delay line and the heterostructure surface, however, is considerably smaller than the wavelength of the SAW, the electric field can penetrate into the sample and interact with the carriers in the 2DES. For this sample arrangement the expression for the attenuation and the change in sound velocity as represented by eqs. (7) and (8) have to take into account the gap between the two materials as well as the fact, that the SAW only is excited on the $LiNbO_3$ sample [19]. Nevertheless, with a reasonable airgap of a few microns and SAW wavelengths of some ten microns, the observed interaction between SAW and 2DES turns out to be of the same order as for the direct coupling method. The clear advantage of the proximity coupling techniqe is the fact, that it is contactless and nondestructive and thus yields a quick and simple tool to characterize heterostructure samples without the need of special preparation.

IV. Experimental results and discussion

As pointed out in chapter II, electron-SAW interaction is most sensitive in the range of very small sheet conductivities $\sigma_\square \approx \sigma_M$. Therefore, SAW experiments are appropriate to study the regimes of low conductivity of a 2DES that occur at low temperatures in a quantizing magnetic field B or at B=0 near inversion threshold. In high magnetic fields and at low temperatures the energy spectrum in the plane of the 2DES condenses in a series of Landau levels, leading to strong Shubnikov-de Haas oscillations in the magnetoconductivity σ_{xx}, when either the carrier density or the magnetic field is varied. Deep minima of σ_{xx} are observed, whenever the Fermi level lies between two Landau levels, i.e. if the Landau level filling factor $v = N_s h/eB$ is close to an integer i (i = 1,2,3 ...). N_s in this expression is the areal carrier density of the 2DES, h is Planck's constant and e the electron charge. Simultaneously the Hall resistance becomes quantized, i.e. $\rho_{xy} = h/e^2 i$. Exactly in this regime SAW experiments are most sensitive and thus may yield a better understanding of the QHE.

1. Quantum oscillations in the SAW - propagation parameters

Under this headline, I want to discuss the influence of a completely quantized 2DES in a strong magnetic field on the propagation parameters of a SAW, transmitted through the heterojunction. These propagation parameters are the transmitted intensity and the sound velocity of the SAW. The measured magnetic field dependencies of the transmitted SAW intensity I(B) and the change in sound velocity $\Delta v/v_0$ as extracted from measurements of the phase shift of the SAW signal for a typical GaAs/AlGaAs heterojunction are shown in Fig. 4(a) and (b), respectively. The experiment is done with a 39 μm SAW excited at sufficiently low amplitudes to avoid nonlinear effects. Fig. 4(c) for comparison shows the magnetoconductivity of the sample, exhibiting very pronounced Shubnikov-de Haas oscillations. The data for σ_{xx} (B) are calculated from ρ_{xx} (B) and ρ_{xy} (B) as measured in Hall bar geometry, using

$$\sigma_{xx} = \frac{\rho_{xx}}{\rho_{xx}^2 + \rho_{xy}^2} \qquad (9)$$

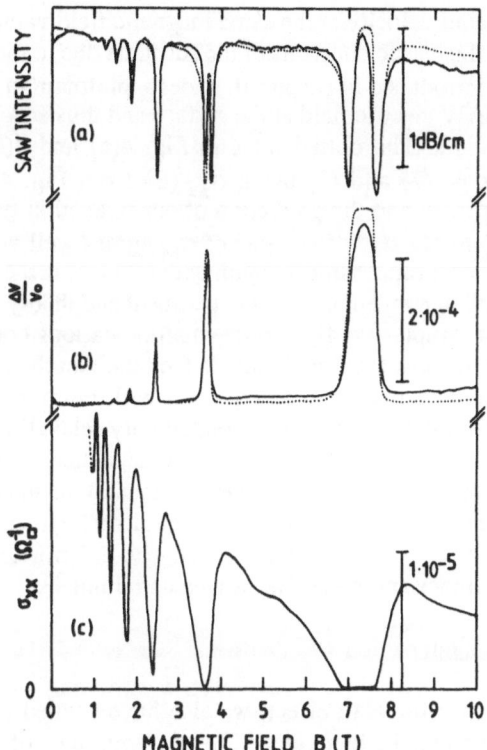

Fig. 4. Experimental results for (a) the transmitted SAW intensity and (b) the change in sound velocity for a typical sample as a function of the magnetic field B. In(c) for comparison the magnetoconductivity σ_{xx} as extracted from measurementsof ρ_{xx} and ρ_{xy} in Hall bar geometry is shown. Both the transmitted intensity as well as the sound velocity exhibit quantum oscillations which coincide in position with minima in σ_{xx}. The dotted lines in (a)and (b) represent the corresponding calculated quantities using eqs. (7) and (8) and σ_{xx} from (c).

Due to the high mobility of the 2DES the magnetoconductivity σ_{xx} (B) changes at low magnetic fields by orders of magnitude. For clarity this part has been omitted in the figure. In the same magnetic field range, both the transmitted SAW intensity as well as the relative change in sound velocity are nearly unaffected. At higher magnetic fields quantum oscillations appear in the SAW intensity I(B) and in $\Delta v/v_0$, which coincide in position on the B scale with minima in the magnetoconductivity σ_{xx} (B), i.e. they appear around integer filling factor v. Each time, the magnetoconductivity drops to low values, the SAW intensity decreases, i.e. the SAW becomes attenuated, whereas the sound velocity increases. For even higher magnetic fields the quantum oscillations in the SAW intensity reveal a very characteristic feature. They split into two distinct minima, which are not observed in the sound velocity, nor in the magnetoconductivity. As it turns out, this splitting is a kind of a fingerprint for electron-SAW interaction in the regime of the quantum Hall effect. It occurs, whenever σ_{xx} (B) drops to very low values below σ_M. In such a deep minimum, σ_{xx} (B) passes σ_M twice, once as it falls and once as it rises. According to eq.(7), this leads to the occurance of two maxima in the attenuation Γ or, in turn, to the occurance of two minima in the transmitted intensity I(B) = $I_0 \cdot \exp(-\Gamma \cdot L)$, where L is the length of the interaction region, i.e. the length of the Hall bar mesa in propagation direction. The center maximum in I(B) thus corresponds to the point of lowest magnetoconductivity.

The increase in the sound velocity at the same magnetic field values is explained by the piezoelectric stiffening of the elastic constants of the substrate due to the reduced screening of the SAW electric field as introduced in chapter II. A deep minimum in σ_{xx} (B) results in less perfect screening of the SAW electric field at the surface and thus according to eq. (3) stiffening of the elastic constants. The dotted lines in Fig. 4(a) and 4(b) are the calculated quantities according to eqs. (7) and (8) using σ_{xx} (B) from Fig. 4(c). To achieve best agreement between experiment and the prediction of our relaxation type model, we choose $\sigma_M = 4 \cdot 10^{-7} \, \Omega_\square^{-1}$ and $K_{eff}^2 = 6.4 \cdot 10^{-4}$. The value of σ_M agrees well with the one predicted, whereas the value for the electromechanical coupling coefficient is the same as measured by other authors[11]. The overall agreement between experiment and theory is clearly satisfactory, but nevertheless for most samples we find some small deviations compared to the simple model discussed above. The most striking feature is the fact, that the double minima in I(B) do not have equal amplitude as predicted by eq. (7). I will return to the discussion of this deviation below. Apart from that, the classical conductivity relaxation model, as presented above, provides a good description of the influence of a magnetically quantized 2DES on the SAW propagation parameters. It should be emphasized, that all the interaction of a SAW and a 2DES in this frequency range can be quantitatively described by the interaction of the SAW electric field and the mobile carriers alone. Interaction due to deformation potential effects can be shown to be much less important in the long-wavelength limit [20].

2. Power dependence of the interaction

In the regime of the quantum Hall effect, we also have studied the dependence of the electron-SAW interaction on the incident power, i.e. the amplitude of the surface wave [5]. A representative result of such an experiment is given in Fig.5, where we plotted the transmitted SAW intensity of a split quantum oscillation ($\nu \approx 4$) as a function of the magnetic field B for four different power levels of the SAW. With increasing SAW amplitude, the splitting of the oscillation disappears. This occurs by a decrease of the amplitude of the center maximum in I(B) and a simultanous decrease of the separation of the both minima on the B scale. At the highest power level, both minima have merged. Further increase of the power then leads to a decrease in the absorption depth of the no longer split quantum oscillation. For very high power levels a complete suppression of a given quantum oscillation may appear. The same behavior is observed for all split quantum oscillations that occur at low temperatures. Oscillations corresponding to lower magnetic fields, or for a given carrier density to higher filling factor ν, however, begin to disappear at lower power levels. We interpret the observed behavior as a heating of the 2DES within the electric field of the SAW. As shown before, the center maximum in a split quantum oscillation in I(B) corresponds to a deep minimum in σ_{xx} (B), occuring at integer filling factor ν. Under these conditions, the Fermi level lies in the center of a mobility gap between two adjacent Landau levels. Increasing the temperature of the electron system thermally activates carriers from the energetically lower lying level to the higher one [1]. This activation in turn leads to an increase in the minimum of the conductivity, which in our model manifests in a decrease of the center maximum of a split oscillation. However, electron heating in the field of a piezoactive surface wave is significantly different to the heating of the electrons by means of an external electric field, like, e.g., a DC drift field. As demonstrated above, the electric field of a SAW depends strongly on the conductivity of the material, on which it propagates. That means, for a given mechanical amplitude of the SAW, the electric field is partially screened by the mobile carriers and decreases with increasing conductivity σ. It turns out to be given by [6]

$$E(\sigma_\square) = \frac{E_o}{\sqrt{1+(\sigma_\square / \sigma_M)^2}} \tag{10}$$

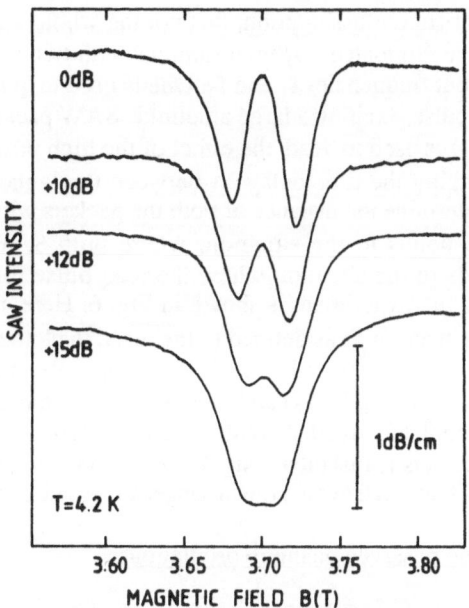

Fig. 5. Power dependence of a split quantum oscillation in I(B). With increasing SAW amplitude the splitting disappears, while the double minima merge on the B scale. The disappearance of the splitting is interpreted in terms of electron heating in the electric field of the SAW .

Thus the electric field in the SAW adjusts itself selfconsistently to the activated conductivity in an intensity oscillation. This fact has to be included in a quantitative analysis, and though we performed such an analysis, it momentarly contains too many unknown parameters to satisfactorily explain the observations. However, SAW traces measured at different lattice temperatures reveal the same behavior and show, that heating of $\Delta T \approx 10$ K is necessary to explain the lineshape of the quantum oscillation at the highest power level in Fig.5.

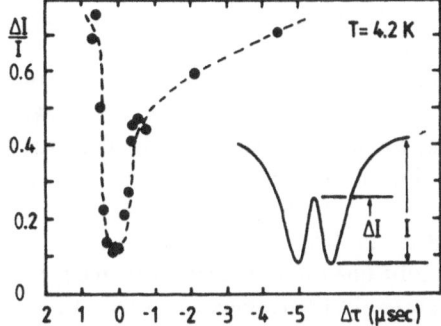

Fig. 6. Amount of splitting $\Delta I/I$ of a split quantum oscillation in I(B) as defined in the inset as function of the time delay between 'pumping' SAW packet and 'read' SAW packet. $\Delta \tau = 0$ corresponds to the situation where both SAW completely overlap. Only under these conditions $\Delta I/I$ tends towards zero, indicating a quasi instantaneous heating process .

To experimentally confirm our assumption, that not the whole sample, but only the free carriers are heated, we have fabricated a special sample[6] with two sets of interdigital transducers, operating at different frequencies f_1 and f_2. One is used to 'pump' the 2DES, i.e. it is fed with a high power RF pulse, exciting a large amplitude SAW packet at frequency f_1. The other set of transducers (f_2) is used to 'read' the effect of the high power pulse, operated at a low power level. By changing the time delay $\Delta\tau$ between the launching of the two SAW packets we actually can determine the distance of both the packets with respect to each other. A time delay $\Delta\tau=0$ corresponds to the situation, where both SAW packets completely overlap, $\Delta\tau <0$ corresponds to the situation where the read-pulse follows the pump-pulse, respectively. The result of the experiment is shown in Fig. 6. Here, the amount of splitting $\Delta I/I$ of an oscillation in the read-pulse as defined in the inset, is plotted versus the time delay $\Delta\tau$ between the two pulses. The power level of the pump-pulse in this experiment was comparable to the highest one in Fig.5. As can be seen, a substantial decrease in $\Delta I/I$ is only observed, if both pulses completely overlap. Within the time resolution of our experiment ($\approx$ 50 nsec) the heating effect thus is instantaneous. We therefore can exclude a heating of the whole sample, which would thermalize on a much longer timescale in the range of msec.

3. Line shape analysis of the observed quantum oscillations

As pointed out in chapter IV.1, the observed lineshape of the quantum oscillations in the transmitted SAW intensity I(B) reveal a characteristic fingerprint of the electron-SAW interaction. In high magnetic fields or for a given carrier density at low filling factors ν they split into two distinct minima. This splitting could be explained by our simple relaxation type model and yields a sensitive test for the description of the experimental results. However, even at low SAW power level, there were minor differences between the prediction of the model and the observation on most samples, namely that both minima in a split oscillation do not reach the same depth. According to eq.(7) both minima in a split quantum oscillation should be equal in amplitude, since the maximum attenuation at $\sigma_{xx}(B)=\sigma_M$ is given by

$$\Gamma_{max}= k \cdot K_{eff}^2/4 \tag{11}$$

leading to a minimal transmitted intensity

$$I_{min}=I_0 \cdot exp(-\Gamma_{max} \cdot L) \tag{12}$$

Here, L is again the geometrical length of the interaction region, i.e. the length of the Hall bar mesa in sound propagation direction. This value for I_{min} should be the same for all double minima in I(B). As we could show [8], this is only true if one assumes the conductivity $\sigma_{xx}(B)$ to be perfectly homogeneous across the whole heterojunction area $F = W \cdot L$. W denotes the geometrical width of the interaction region, i.e. the width of the soundpath, or the width of the hall bar mesa, whichever is smaller. The detected SAW signal I(B) is proportional to this width W, since the detecting transducer integrates over its aperture. In the small signal approximation, which applies here since $\Gamma \cdot L$ is a small quantity, the detected signal is also proportional to the length L, namely $I(B) \approx I_0 (1 - \Gamma \cdot L)$. To explain the different amplitudes in a split quantum oscillation in I(B), we now define an *effective* interaction length. For a perfectly homogeneous sample this effective l equals the geometrical L. For an inhomogeneous sample, we model the inhomogeneity by the assumption of several independent homogenous domains of area $F_i=W \cdot l_i$, each of which has a somewhat different conductivity $\sigma_{xx}^i(B)$ and thus a slightly different $\Gamma_i(B)$ as given by eq.(7). The total transmitted SAW intensity then will be

$$I_{tot} = I_0 (1 - \Sigma \Gamma_i \cdot l_i) \tag{13}$$

Here, the effective interaction length is $l_i = F_i/W$ and the sum of the domain areas $\Sigma F_i = F$ is the geometrical area of the Hall bar mesa. The most obvious way to achieve an inhomogeneous conductivity in a heterostructure sample is to have slight inhomogeneities in the carrier concentration across the sample area. In a given magnetic field, different carrier densities on different part of the sample lead to different Landau level filling factors $v_i = N_{s,i} h / eB$.

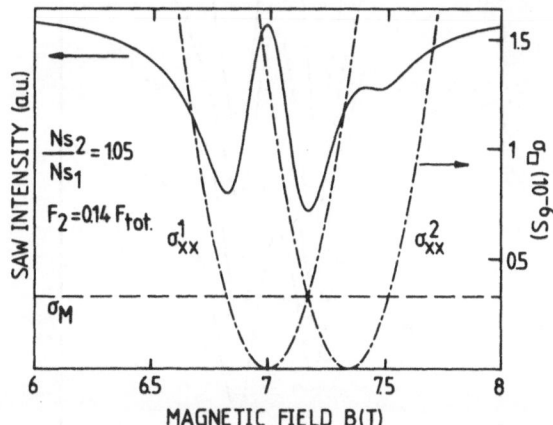

Fig. 7. Simulation of the influence of a small spatial inhomogeneity in the 2D carrier density on the lineshape of a split quantum oscillation in I(B). For simplicity, only two domains of different carrier density are taken into account. The higher density domain ($N_{s,2}$, $\sigma_{xx,2}$) covers 14% of the total sample area. The dot-dashed lines represent the magnetic field dependencies of the $\sigma_{xx,i}$, modeled by a parabola. The horizontal dashed line indicates the position of σ_M.

A simulation of the influence of such small inhomogeneities on the transmitted SAW intensity is given in Fig.7, where we only assume two domains of different density $N_{s,1}$ and $N_{s,2}=1.05\ N_{s,1}$ and model the minimum in σ_{xx}^i(B) by a parabola. Even this simplest case results in an asymmetry in the depth of the double minima, indicating a variation of the effective interaction length with the magnetic field. The higher density domain in this case manifests in a small shoulder in the I(B) oscillation on the high magnetic field side. To experimentally verify our assumption that the lineshape anomalies in I(B) are due to slight inhomogeneities in the carrier density of the samples under investigation, we have studied a sample with an intentionally created inhomogeneity domain [8]. Here, we illuminated a small area ($\approx 15\%$) of a standard sample by means of a focussed LED, thus creating a domain of higher carrier density by the persistent photoeffect. Since the LED was pulsed for very short periods of time, we were able to gradually increase the density in the domain. The result of this experiment is shown in Fig.8, where we depict the development of a split oscillation in I(B) under partial illumination of the sample. Parameter is the total time of illumination,

somehow a measure for the difference in densities in the domain and the sample area. With increasing density in the domain, the character of the lineshape completely changes, leading to a strong asymmetry in the depth of the double minima. At the highest density after $t_{LED}=16$ msec the shoulder on the high field side appears as predicted. It should be noted, that such small spatial inhomogeneities in the carrier density of a 2DES are difficult to extract from DC transport measurements. Though one observes small changes in the magneto-resistance ρ_{xx} (B), they depend crucially on the choice of the potential probes of the Hall bar mesa. The Hall plateaus in ρ_{xy} (B) are broadened by inhomogeneity of the carrier density

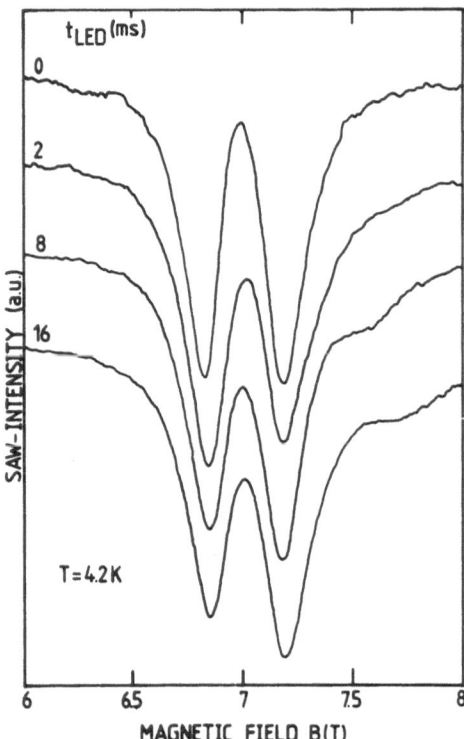

Fig.8. Experimentally determined influence of an intentionally induced inhomogeneity on the lineshape of a split quantum oscillation in I(B). A domain of higher electron density is created by partial illumination of the sample area (15%). Parameter is the total time of illumination t_{LED}. An increase of the carrier density in the domain is accompanied by a change in the symmetry of the quantum oscillation.

which can be seen in Fig.9. Here we show two quantum oscillations in I(B) together with the corresponding Hall plateaus in ρ_{xy}(B) for both a rather homogeneous situation (a) and the situation where at least two domains of different carrier densities (b) are present on the sample. This is nicely indicated by the occurence of a second set of quantum oscillations in I(B), the one with higher density having a smaller area leading to a smaller absorption depth. Since a SAW integrates over the whole interaction region, it proves advantageous over conventional DC transport experiments, where the averaging process is not yet know. On some samples, inhomogeneities in the carrier density seem to be connected with the occurence of a small parallel conductance in the doped AlGaAs layer in certain regions of the sample area. Here, it is necessary to include this bypass conductivities in our model to explain the observed lineshapes. Since SAW experiments are very sensitive to small conductivities, they provide a useful tool to investigate this parallel conductance and its influence on the QHE. If we neglect the small spacing between the 2DES and the bypass layer, we can model the bypass influence by replacing $\sigma_{\square}$ in eq.(7) by $\sigma_{\square} = \sigma_{xx} + \sigma_{by}$ and assume σ_{by} to be independent of the magnetic field for the range of a single quantum oscillation. The result of such a calculation is shown in Fig.10, where we compare it with an actual measured lineshape. Again, we only assume two different domains on the sample area. It shows, that even very asymmetric double minima can be successfully modeled by the assumption of small spatial inhomogeneities in the 2D carrier concentration.

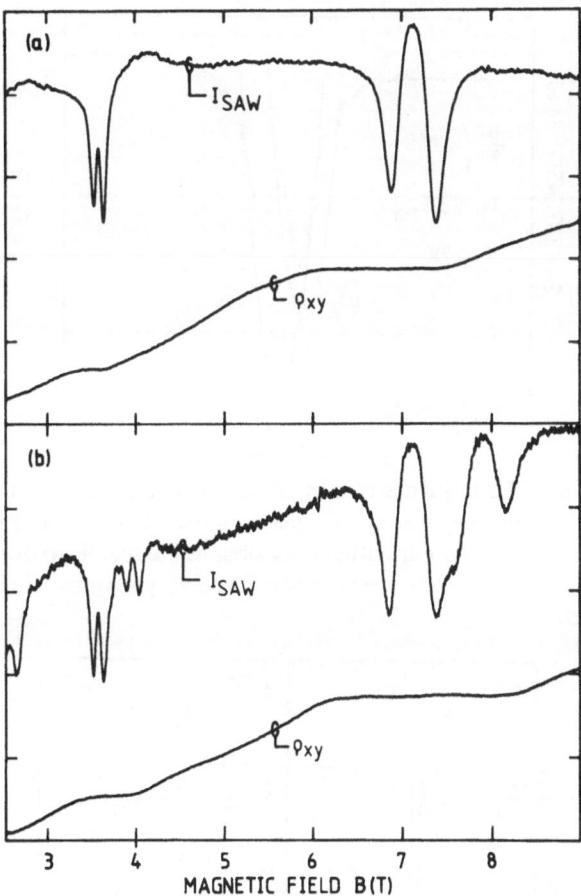

Fig. 9. Influence of a small spatial inhomogeneity in the 2D carrier density on the quantum Hall effect. In (a), the situation corresponds to a quite homogenous sample, indicated by only one set of quantum oscillations in I(B). In (b) the sample exhibits two distinct domains of different carrier densities, as indicated by the occurence of a second set of quantum oscillations in I(B). The Hall plateaus in ρ_{xy} broaden with increasing inhomogeneity.

Using this technique, we could also show [6] that domains of slightly different carrier densities can be generated in a heterostructure sample by the application of drift field pulses between the source and drain contacts. Our results might be of importance in interpreting so called 'hot electron' experiments, where the 2DES is heated in DC drift fields. We studied the homogeneity of the 2D carrier density by use of a low power SAW probing beam, while a DC pulse was applied to the contacts. The result of such an experiment is depicted in Fig. 11. Here, the development of a split quantum oscillation in I(B) is shown during the application of short (5 μsec) DC pulses at low repetition rate (10 kHz). Parameter is the electric field strength of these drift pulses. It should be mentioned, however, that although the pulses strongly affect the 2DES, the electric field strengths in our experiment are very moderate in comparison to those in experiments, where hot electron phenomena are investigated. In (a), the pulse height is varied, whereas in (b) the pulse height is kept constant and parameter is the total time of their application to the sample, which is proportional to the total number of pulses. As can be seen in (a), a second well defined density domain begins to appear on the sample, growing in area as indicated by an increase in absorption depth of the additional quantum oscillation. The growth of this area is at expense of the remaining, original density

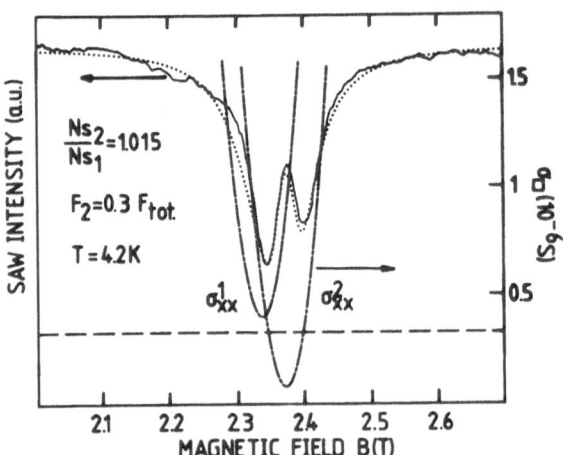

Fig.10. Simulation of the influence of a small parallel conductivity in the doped AlGaAs layer on the lineshape of a split quantum oscillation in I(B) (dotted line) and comparison with an experimentally obtained trace. Two domains of slightly different carrier density are assumed. Even very asymmetric lineshapes can be simulated.

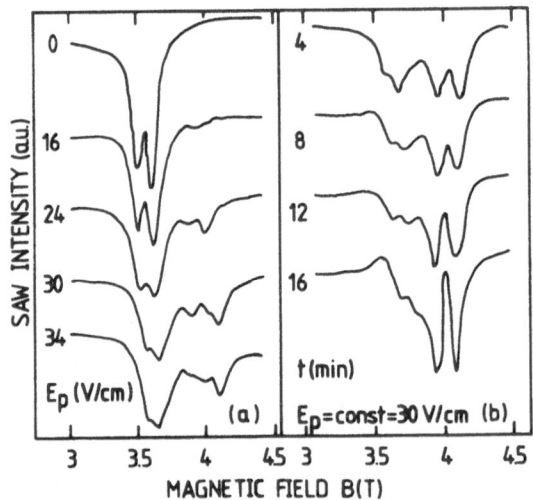

FIG.11. Influence of small DC drift pulses on the homogeneity of the 2DES as observed in the lineshape of the transmitted SAW intensity. In (a) the electric field E_p in the pulse is varied, whereas in (b) the pulse height is kept constant for a certain time. A second, well defined domain of higher carrier density appears on the sample, growing in area, as indicated by an increase in absorption depth.

on the sample, indicated by a decrease of the absorption depth of the original quantum oscillation. The effect is accumulative, as shown in (b). The longer drift pulses are applied to the 2DES, the larger the area of the higher density domain. At the end of this experiment, nearly all the sample area showed this higher carrier density. After switching off the drift pulse, this higher density remains persistent, similar to the effect of illumination of the sample with band gap radiation (persistent photoeffect). Similar results are obtained by replacing the electrical drift pulse by a high power SAW pulse, comparable to the experiment shown in Fig.6. Here, we find that in this case the magnetic field strength at which the SAW pulse is applied is important for the amount of inhomogeneity induced. This is consistent with the fact that the SAW electric field is screened by the mobile carriers except for magnetic field ranges around integer filling factors ν, indicating the principal difference of the selfconsistent SAW electric field from external electric fields.

The study of small spatial inhomogeneities in the carrier density of a 2DES thus is shown to be very powerful by use of SAW experiments. The influence of such inhomogeneities on the quantum Hall effect has recently attracted much attention [21]. The apparent density of states (DOS) between Landau levels seems to be related to a certain inhomogeneity in the carrier distribution, and many quantities related to the DOS such as capacitance [22] and specific heat [23] have been successfully modeled by the assumption of a finite DOS between Landau levels. There have also been suggestions [24] to explain the integral QHE only on bases of small spatial inhomogeneities in the carrier density, explaining many experimental results in a very simple manner.

4. Electron-SAW interaction on gated HEMT structures

The electron-SAW interaction is very sensitive in the regime of small sheet conductivities $\sigma_\square$ of the 2DES. Such small conductivities also occur in the absence of a magnetic field near inversion threshold on gated heterostructure samples. Also, gated heterostructures provide the possibility of changing the carrier concentration in a given sample, while the mobility remains essentially unaffected. The use of a gate, however, is somehow in contradiction to the applicability of SAW experiments. As has been pointed out in chapter II., the metallization of the sample surface results in a screening of the SAW electric field, leaving the crystal surface appear to be non-piezoelectric. The use of a backgate, or gating the 2DES in the proximity coupling technique by a gate on a thinned $LiNbO_3$ SAW delay line have not proven to be very successful for this purpose, since a homogeneous depletion of the electron channel is hardly achievable. As can be seen in Fig.1, however, in a certain distance from the surface the SAW electric field recovers even for a metallized surface. Using this fact, we deposit between the surface of the heterostructure sample and the gate electrode a thin insulating photoresist layer to increase the distance between gate and 2DES. Although the potential distribution with depth here is more complicated as shown in Fig.1, we expect the electric field of the SAW at the position of the 2DES to be finite in this case.

First experimental results [25,26] for the interaction of surface acoustic waves with a 2DES in gated HEMT structures are given in Fig.12. In (a), the transmitted SAW intensity as a function of the magnetic field is plotted for various gate voltages V_g. The sample in this case has SAW transducers operating at $\lambda = 16$ μm and the thickness of the insulating photoresist layer between sample surface and gate electrode is d = 4600Å. The 2D electron channel is depleted at a threshold voltage of about V_{th}=-8V. As can be seen by the shift of corresponding quantum oscillations in the SAW intensity towards lower magnetic fields, the carrier density in the 2DES can be decreased quite homogeneously and linear with V_g. In Fig.12(b) the detected phase signal, being proportional to the change in the sound velocity at different magnetic fields is plotted versus the gate voltage for an other sample. Here, the SAW wavelength is $\lambda = 8$ μm and the insulator thickness d = 3000Å, leading to a threshold voltage V_{th}=-3.8V. At this voltage, the electron channel becomes depleted, resulting in a large drop in the conductivity and according to eq.(8), the sound velocity changes steplike due to the piezoelectric stiffening. The pronounced quantum oscillations in the B=2T and B=3T traces correspond to the change in the sound velocity caused by the deep minimum in σ_{xx} (B) around filling factor ν=2. The results presented here nicely demonstrate that also on gated heterostructures SAW experiments have become possible. An alternative method to increase the distance between gate electrode and the 2DES is the use of a sample, in which the channel is buried in a quite large depth, i.e. the spacer consisting of a MBE grown GaAs cap layer. A first sample of this kind has recently been fabricated and is presently under investigation.

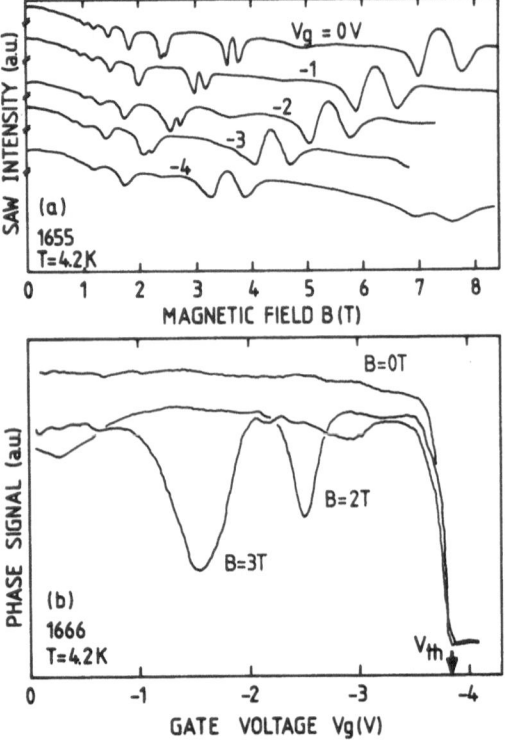

Fig.12.   Interaction of SAW with a 2DES in gated heterostructure samples. In (a) the quantum oscillations in I(B) shift towards lower magnetic fields as the gate voltage V_g is varied, indicating a decrease in the 2D carrier density. In (b) the measured phase signal, being proportional to the change in sound velocity, exhibits quantum oscillations as a function of the gate voltage V_g, if a magnetic field is applied. Around $V_g = -3.8$ V the 2D channel becomes depleted, resulting in piezoelectric stiffening of the elastic constants.

V. Finally...

In this review, I presented some recent investigations on the interaction of surface acoustic waves with a two dimensional electron system in GaAs/AlGaAs heterostructures. This work represents a new and different experimental approach to the fascinating properties of low-dimensional systems in high magnetic fields and at low temperatures. Surface acoustic waves prove several advantages over conventional quasistatic methods, since the mechanisms of interaction exhibit some unique features. The electron-SAW interaction has been shown to be most sensitive in regions of very low sheet conductivities of the 2DES, e.g. in the regime of the quantum Hall effect or near inversion threshold of gated HEMT structures. In high magnetic fields and at low temperatures, we observe quantum oscillations in the transmitted sound intensity as well as in the sound velocity, both of which could be quantitatively explained in a quite simple relaxation-type model. The study of the power dependence of the interaction led to the conclusion, that the 2DES can be uniformly heated by the selfconsistent electric field of large amplitude SAW pulses. In a 'pump and read' experiment we could show, that this 'phonoconductive' response is instantaneous within the time resolution of our technique.

The advantage of SAW experiments in comparison to conventional DC transport experiments is besides the high sensitivity at low conductivities the integrative character of the interaction. This makes them very sensitive to changes in the spatial homogeneity of the carrier density of the 2DES. We could use this to semi-quantitatively describe the observed lineshapes of the quantum oscillations in the transmitted SAW intensity by the assumption of small inhomogeneities in the carrier density of our samples. Such small inhomogeneities of only a few percent are difficult to detect using conventional experimental techniques, although they seem to be important for the interpretation of many experimental results. We could also show, that spatial inhomogeneities can show certain dynamics on the sample area, if the electron system is heated by electric drift field pulses. The use of more sophisticated transducer structures eventually will allow us to also study the spatial distribution of these domains.

As has been pointed out in chapter III., electron-SAW interaction is also observable in a seperate media proximity coupling technique. Here, the SAW is not excited on the heterostructure sample itself, but on a strongly piezoelectric $LiNbO_3$ delay line which is brought into close contact to the heterojunction. The SAW electric field then penetrates the heterostructure and interacts with the 2DES. This technique provides, besides much higher electric fields for long wavelengths a very simple, nondestructive and contactless method to characterize heterostructure samples without the need of special preparation.

Finally, I presented first experimental results on the interaction of a SAW with a 2DES in gated heterostructure samples. Here, the 2D carrier density can be controlled by the application of a gate voltage between channel and a front gate electrode. We believe, that this way a new field of investigation is accessible. We now can investigate the properties of high mobility samples with very low carrier density ($N_S < 10^{11}$ cm^{-2}). In high magnetic fields and at low temperatures we thus should be able to enter the regime of the fractional quantum Hall effect. Also, experiments on samples with microstructured gate electrodes become accessible. Here, the investigation of the interaction of SAW with quasi one dimensional electron systems [27,18] would be the challenge. At very low carrier densities also acoustic charge transport in the potential wells of the SAW and acoustic amplification for conductivities $\sigma_\Box \approx \sigma_M$ in the presence of electric drift pulses are interesting aspects for the future. Both up to now has only been observed in bulk semiconductor materials and besides the scientific interest, the use of low dimensional structures promises interesting applications in RF signal processing techniques.

Acknowledgement

This work was done in intimate collaboration with my coworkers J.P.Kotthaus, J.Scriba, and M.Wassermeier at the University of Hamburg. I greatfully acknowledge their cooperation and enthusiasm, which only made possible many of the results that I presented. Special thanks to A.V.Chaplik for many fruitful discussions and hints. The GaAs/AlGaAs heterostructure samples have been grown by G.Weimann and W.Schlapp, the $LiNbO_3$ substrates have been provided by Ch. Grabmeier (Siemens Forschungslabor). Further I would like to thank A.C.Gossard for his hospitality and for the possibility to grow specially designed samples for SAW experiments. This work has been financed by support of the Deutsche Forschungsgemeinschaft .

References

1. For a comprehensive review see, e.g., Ando T., Fowler A.B., and Stern F., Rev. Mod. Phys **54**, 437-672 (1982).

2. Weimann G., in Festkörperprobleme / Advances in Solid State Physics, Grosse P., ed., Vieweg, Braunschweig (1986), p.231.
3. v.Klitzing K., Dorda G., and Pepper M., Phys. Rev. Lett. **45**, 494 (1980).
4. Tsui D.C., Störmer H.L., and Gossard A.C., Phys. Rev. Lett. **48**, 1559 (1982).
5. Wixforth A., Kotthaus J.P., and Weimann G., Phys. Rev. Lett. **56**, 2104 (1986).
6. Wixforth A., Kotthaus J.P. in The Application of High Magnetic Fields in Semiconductor Physics, Landwehr G., ed., Springer, Berlin, in press.
7. Wixforth A., Scriba J., Wassermeier M., Kotthaus J.P., Weimann G., and Schlapp W., Journ. Appl. Phys. **64**, 2213 (1988).
8. Wixforth A., Scriba J., Wassermeier M., Kotthaus J.P.,Weimann G., and Schlapp W., Proceedings of the 19th ICPS , Warsaw (1988), to be published.
9. Keldysh L.V., Sov. Phys. Sol. State **4**, 1658 (1962).
10. For a review see, e.g., Acoustic Surface Waves (Topics of Applied Physics) Vol. **24**, Oliner A.A., ed., Springer, Berlin (1978).
11. Webster R.T., and Carr P.H., in Rayleigh Waves, Theory and Applications, Ash E.A., and Paige E.G.S., ed., Springer series in wave phenomena, Springer, Berlin (1985), pp. 122-130.
12. Hutson A.R., and White D.L., Journ. Appl. Phys. **33**, 40 (1969).
13. Ingebrigsten K.A., Journ. Appl. Phys. **41**, 454 (1971).
14. Adler R., IEEE Trans. on Sonics and Ultrasonics, **SU-18**, 115 (1971).
15. Bierbaum P., Appl. Phys. Lett. **21**, 595 (1972).
16. Chaplik A.V., Sov. Tech. Phys. Lett. **10**, 584 (1984).
17. Ridley B.K., Semicond. Sci. Techn. **3**, 542 (1988).
18. Magarill L.I., and Chaplik A.V., Sov. Phys. JETP, **66**, 1289 (1987).
19. Schenstrøm A., Quian Y.J., Xu M.F., Baum H.P., Levy H., and Sarma B.K., Solid State Commun. **65**, 739 (1988).
20. McFee J.H., in Physical Acoustics,Vol. **4A**, Mason W., ed., Academic Press, New York (1966), pp. 1-47.
21. Gudmundsson V., and Gerhardts R.R., Phys. Rev. **B 35**, 8005 (1987).
22. Weiss D., v.Klitzing K., and Mosser V., in Two Dimensional Systems: Physics and New Devices, Bauer G., Kuchar F., and Heinrich H., ed., Springer, Berlin, (1986), 204.
23. Lassnig R., and Gornik E., in Two Dimensional Systems: Physics and New Devices, Bauer G., Kuchar F., and Heinrich H., ed., Springer, Berlin (1986), 218.
24. Woltjer R., Eppenga R., Mooren J., Timmering C.E., and Andre J.P., Europhys. Lett. **2**, 149 (1986).
25. Wassermeier M., Wixforth A., Kotthaus J.P., Weimann G., Nickel H., and Schlapp W., submitted to the II. International Conference on Surface Waves in Solids and Layered Structures, Varna (1989).
26. Wixforth A., Scriba J., Wassermeier M., Kotthaus J.P., Weimann G., and Schlapp W., to be published.
27. Hansen W., Horst M., Merkt U., Sikorski Ch., Kotthaus J.P., and Ploog K., Phys. Rev. Lett. **58**, 2586 (1987).

SCREENING OF IMPURITIES IN THE
QUANTUM HALL REGIME

Vidar Gudmundsson

Science Institute
University of Iceland
Dunhaga 3
IS-107 Reykjavik
Iceland
EARN: vidar@ raunvis.hi.is

Abstract

The equilibrium screening of a single impurity by a two-dimensional electron gas (2DEG) residing on a finite disk in a quantizing perpendicular magnetic field at low temperature is investigated. The electron-electron interaction is included in the Hartree approximation and the impurity is represented by the Coulomb potential of a negatively or positively charged point particle situated in the plain of the 2DEG. We observe how the binding energy of the impurity oscillates with the filling factor of the Landau bands (Lb's), reflecting the dependence of the screening on the location of the chemical potential with respect to the Landau bands. Implications for the optical properties of the 2DEG are discussed.

1. Introduction

In this paper we investigate the interactions of a Coulomb impurity (donor or acceptor) and a two-dimensional electron gas (2DEG) in a perpendicular quantizing magnetic field B at low temperature. This theoretical investigation is spured by recent developements in the measurements of the cyclotron resonance in GaAs heterostructures [1]. Experimental results [2] have confirmed, that the line width Γ of the cyclotron resonance is governed by two different mechanisms. The component of the line width that depends on $\sqrt{B}$ is caused by short-range randomly distributed scatterers [3], but the component oscillating with the filling factor ν of the Landau bands is found to result from the interaction of the 2DEG with the coherent superposition of the long-range tails of the impurity potentials [4]. The oscillations of Γ are thus caused by the nonlinear screening properties of the 2DEG. Further evidence has gathered that other properties of the cyclotron resonance, such as the mass discontinuity etc. [1], may be caused by the interactions of the 2DEG with impurities or potential fluctuations located near, or in, the 2D-system. Interactions here do not refer to the simple scattering by short-range potentials, but rather the Coulomb interaction between the impurity and the interacting 2DEG, so that screening effects are included. These ideas have been further strengthened by results comming from the ever increasing experimental activities on intentionally doped two dimensional electron systems [1], especially, the dilutely doped once, where the impurities do not lead to impurity bands. In these experiments on heterostructures, where a δ-doping layer of donors or acceptors has been situated in the 2DEG, effects on the optical properties and the cyclotron resonance have been observed, some of which confirm the above mentioned ideas and some that are only partially understood [5].

The screening of impurities by a 2DEG in no magnetic field has been studied in simple strictly 2D models [6] that show clearly the difference between screening in two and three dimensions, and in models that reflect the surrounding heterostructure [7] allowing therefore a direct quantitative comparison with experiments. In the presence of a magnetic field, conductivity measurements (and thereby the quantum Hall effect) have led to greater interest in considering the effects of randomly distributed impurities on the properties of the 2DEG [4,8-12]. On the other hand, the binding energy of electrons to impurities in quantum wells has been measured by far infrared spectroscopy as a function of both the well width and the location of the impurity [13]. Several researchers have also calculated the dependence of the impurity location on the energy spectrum by applying variational [14] or numerical [15] methods with or without a magnetic field. How the energy spectrum or the impurity potential change as functions of the magnetic field when a 2DEG is also present in the well has not been considered to the best of our knowledge.

Therefore, this paper is devoted to the study of the screening and the energy spectrum of the impurity potential in a quantizing magnetic field. In this paper we take the first steps towards understanding the interactions between the 2DEG and the impurities by considering the equilibrium ground state properties of the system, such as the energy spectrum and the screened impurity potential, which are the necessary foundations for the optical response calculation to be published later. In order to see qualitatively what might be expected in a real heterostructure where the 2DEG is confined to the lowest electrical subband, we consider a Coulomb center (positive or negative unit point charge) placed in a disk of strictly 2D electrons. Perpendicular to the disk is a homogeneous magnetic field and the total system is kept electrically neutral by a homogeneous positive background charge also located in the disk. Here we are therefore concerned with the screening of a long-range potential, that can vary strongly on the scale of the magnetic length $l = (\hbar c/eB)^{\frac{1}{2}}$, in contrast to the earlier study of the screening of long-range potential fluctuations of small amplitude [16]. Further details of the model will be presented in section 2 with results and discussion following in section 3 and 4.

2. The Model

Using the symmetry of the impurity potential we select the circular gauge for the vector potential $\vec{A}(\vec{x}) = (-\frac{1}{2}By, \frac{1}{2}Bx, 0)$. The Schrödinger equation for the Hartree interacting 2DEG then assumes the form

$$(2.1) \quad [-\frac{\hbar^2}{2m}\{\partial_r^2 + \frac{1}{r}\partial_r + \frac{1}{r^2}\partial_\varphi^2 + \frac{i}{l^2}\partial_\varphi\} + \frac{m\omega_c^2}{8}r^2 + V_{imp}(r) + V_H(r) + V_c(r)]\psi_\alpha = E_\alpha \psi_\alpha$$

with the cyclotron frequency $\omega_c = \frac{eB}{mc}$, and the quantum numbers $\alpha = (n, M)$ to be introduced later.

518

The impurity potential in the confining plane of the 2DEG is a simple Coulomb potential in a medium with a dielectric constant κ ($\kappa = 12.4$ for GaAs)

$$(2.2) \qquad\qquad V_{imp}(r) = \pm \frac{e^2}{\kappa r}.$$

The Hartree potential $V_H(r)$ is derived via the Poisson's equation from the local charge density $-en_s(r)$ of the 2DEG and the neutralizing background charge density $en_b(r)$

$$(2.3) \qquad\qquad V_H(r) = \frac{e^2}{\kappa} \int d\tilde{r}' \frac{n_s(r') - n_b(r')}{|\tilde{r} - \tilde{r}'|},$$

where $\tilde{r} = (x, y)$. The electron density $n_s(r)$ can be found knowing the energy spectrum E_α and the eigenfunctions ψ_α

$$(2.4) \qquad\qquad n_s(r) = \sum_\alpha |\psi_\alpha(r)|^2 f(E_\alpha - \mu),$$

where f is the equilibrium Fermi distribution and μ is the chemical potential to be determined from the constraint of a constant number of electrons in the system.

In order to facilitate numerical analysis, to be described later, the system size has to be limited. This has to be done in conjunction with the choice of material parameters. The experiments have been carried out in GaAs heterostructures so we select $m = 0.067m_0$ and $\kappa = 12.4$. Then the effective Bohr radius a_0 and the Rydberg energy $E_{Ryd.}$ are

$$(2.5) \qquad a_0 = \frac{\hbar^2}{m(e^2/\kappa)} \approx 97.9\text{Å}, \quad E_{Ryd.} = \frac{\hbar^2}{2ma_0^2} \approx 5.92meV.$$

The system resides on a disk of radius R, which has to be much larger than a_0 in order to avoid the edge from influencing the energy spectrum of the impurity, we are thus examining a 2D "bulk" impurity. The positive background charge density $en_b(r)$ is now assumed to be homogeneous on a disk of radius R. For convenience, in the numerical calculations to follow, $n_b(r)$ is represented by a smooth function

$$(2.6) \qquad\qquad n_b(r) = \bar{n}_s[\exp(\frac{r - R}{0.25 \cdot \Delta r}) + 1]^{-1},$$

where $\bar{n}_s$ is the average electron density inside the disk and $\Delta r = 22.0\text{Å}$. As has been pointed out [17], $en_b(r)$ cannot confine the 2DEG. Thus the confinement has to be guaranteed with an external potential $V_c(r)$, which we select to be a smooth step function rising up around $r \sim R$ to the value of U_0

$$(2.7) \qquad\qquad V_c(r) = U_0[\exp(\frac{R - r}{0.25 \cdot \Delta r}) + 1]^{-1}.$$

In view of the magnetic field range and the average electron density used in section 3 we have chosen $U_0 = 30meV$ and $R = 1000\text{Å}$. The cyclotron resonance energy $\hbar\omega_c$ will be of the same order of magnitude as $E_{Ryd.}$.

Due to the circular symmetry of the problem the angular integration in (2.3) can be carried out leading to:

$$(2.8) \quad V_H(r) = \frac{4e^2}{\kappa} \int_0^\infty dr' \{\theta(r - r')\frac{r'}{r}K(\frac{r'}{r}) + \theta(r' - r)K(\frac{r}{r'})\}\{n_s(r') - n_b(r')\},$$

where $\theta(x)$ is the Heaviside unit step function and $K(x)$ is the complete elliptic integral of the first kind

$$(2.9) \qquad K(x) = \int_0^{\pi/2} \frac{d\alpha}{\sqrt{1 - x^2 sin^2\alpha}}.$$

The elliptical integral can be approximated, to a high degree of accuracy, by a product of a polynomial and a logarithm [18]. The integral in (2.8) can therefore be done piecewise analytically around expansion points for the electron density, thus avoiding numerical integration around the singular points of the integrand. The effects of the finite thickness of the 2DEG, caused by including the wavefunction of the lowest electric subband (in the z-direction perpendicular to the 2DEG disk), would be a slight quantitative modifications of $V_H(r)$ as long as only the lowest subband was occupied. On the other hand this modification might cost considerable complications in the numerical methods to calculate $V_H(r)$, and would be device dependent. In a similar manner, moving the neutralizing background charge outside the 2DEG plane, so the system would resemble more closely a heterostructure, would only lead to slight quantitative modifications of $V_H(r)$. Both of these modifications of $V_H(r)$, caused by the finite thickness and the location of the neutralizing background, would leave the qualitative properties of the 2DEG, which we are investigating, unchanged.

Due to the choice of the confining potential $V_c(r)$ we can solve the self-consistent Hartree problem (2.1 - 2.4) using the eigenfunctions of the non-interacting 2DEG as a functional basis, which can be found by setting $V_{imp} = V_H = V_c = 0$ in (2.1) with the boundary condition that they vanish as $r \to \infty$. These eigenfunctions (of the Landau system) are in polar coordinates

$$(2.10) \qquad \phi_{n,M}(r,\varphi) = \{\frac{n_r!}{\pi(|M| + n_r)!(2l^2)^{|M|+1}}\}^{1/2}r^{|M|}e^{-r^2/4l^2} L_{n_r}^{|M|}(\frac{r^2}{2l^2})e^{iM\varphi},$$

where $n_r = 0, 1, 2, \cdots$ is the radial quantum number, $M = 0, \pm1, \pm2, \cdots$ is the angular momentum quantum number and $L_{n_r}^{|M|}$ are the Laguerre polynomials.

The eigenenergies represent the usual Landau levels in an infinite Landau system

$$(2.11) \qquad E_n = \hbar\omega_c(n + 1/2),$$

where the Landau level index $n = 0, 1, 2, ...$ is determined from

$$(2.12) \qquad n = \frac{|M| - M}{2} + n_r,$$

meaning that in the nth Landau level M can take on integer values in the range $-n \leq M < \infty$.

The spherical symmetry of all the potential terms in (2.1) results in, that the Hamiltonian leading to (2.1) has only to be diagonalized with respect to the quantum number n_r but not M in each iteration of the Hartree system of equations (2.1-2.4). That is, the angular symmetry is conserved but the interaction mixes different Landau levels forming Landau bands. The number of basis functions used in the diagonalization of the interacting Hamiltonian is then chosen such that a further increase of the subset results in unchanged electron density $n_s(r)$.

3 Results and discussion

For the magnetic field range used to calculate the ground state properties of the system we have $R \gg l$, and thus it is still possible to use the average filling factor $\nu = 2\pi l^2 n_s$ to describe the occupation of the lower Landau bands in the interior of the system. Here the spin has been entirely neglected and is left for a later study in connection with exchange and higher order effects. The filling factor ν can be varied as usually with either the magnetic field B or the average electron density $\bar{n}_s$. But if $\bar{n}_s$ is strongly reduced to attain low ν, then the approximation used to derive the Hartree potential V_H (2.8) from the total electron density will fail. Therefore we consider $\bar{n}_s$ fixed at a reasonable experimental value, $\bar{n}_s = 1.59 \cdot 10^{11} cm^{-2}$, and vary the filling factor ν through the magnetic field B.

The energy spectrum of the interacting 2DEG without an impurity is presented in Fig.1a, where the Landau bands are shown vs. the quantum number M. The filling factor $\nu = 2$ as can be verified from the position of the chemical potential μ between the second and the third Landau band. All the Lb's and especially the lower ones approach simple flat Landau levels in the interior of the system. The corresponding electron and background densities are seen in Fig.1b. In the interior of the system the electron density assumes the average value, but shows two humps near the edge corresponding to the crossing points of the chemical potential μ and the two lowest Lb's [16,19]. Fig.2 shows the same system with the addition of one donor i.e. a positive point charge producing a Coulomb potential at the center of the coordinate system. In Fig.3 the same system is seen, but there $\nu = 0.8$. The main change in the energy spectrum due to the presence of the donor like impurity is that the energy of the states around $M = 0$ decreases, and by the largest amount for the lowest Landau band. The reason why the $M = 0$ states suffer the largest energy shift is that the wavefunctions for these states are the only ones which do not vanish at the center point $r = 0$, in the neighbourhood of which the potential is strongest. The energy shift for the various Lb's as a function of the quantum number M can be seen in Fig.4. The electron bound to the impurity can be seen in Fig.2b and 3b as the density peak near the origin, that in both cases (though differently) is caused by the shifting of an $M = 0$ state below μ compared with the neighbouring M-states in the same Landau band. The decrease in the energy shift, or the "binding energy" of the electron, with increasing Lb number n can be partially explained by a smaller influence of the Coulomb potential on higher Lb's, but is also due to the mixing of different Landau levels (Ll's) in the calculation of the Lb's. This is clear from the case, if the impurity potential was taken to be a δ-function, then the first order perturbation theory would predict the same shift for all the $M = 0$ states, but exact diagonalization of the Hamiltonian for any number of Ll's would confirm that the mixing of Ll's causes the energy shift to decrease with increasing n (For the δ-impurity an energy cut off has to be used in order for to attain physical results that do not depend on the size of the functional basis used).

In Fig.5 we see the energy spectrum and the electron density for a 2DEG when the impurity is a negatively charged acceptor, rather than a donor. Now the $M = 0$

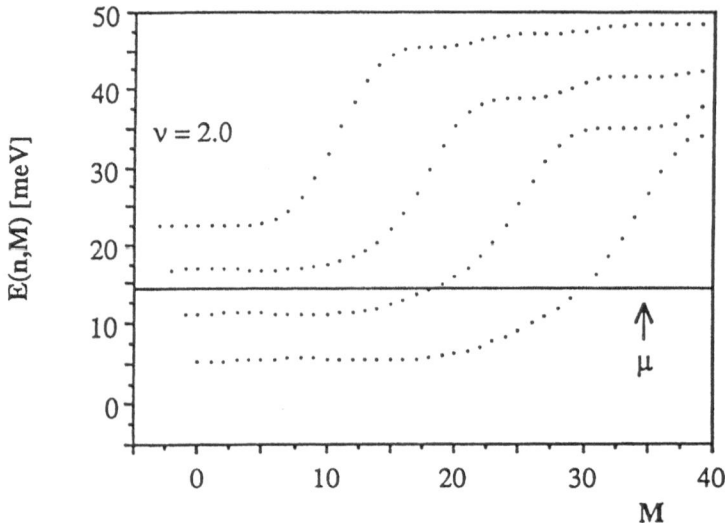

Fig.1a The energy spectrum (dotts) and the chemical potential μ (horizontal line) when no impurity is in the system and $\nu = 2$. $R = 1000\text{Å}$, $U_0 = 30.0 meV$, $\bar{n}_s = 1.59 \cdot 10^{11} cm^{-2}$, $\kappa = 12.4$, and $m = 0.067 m_0$.

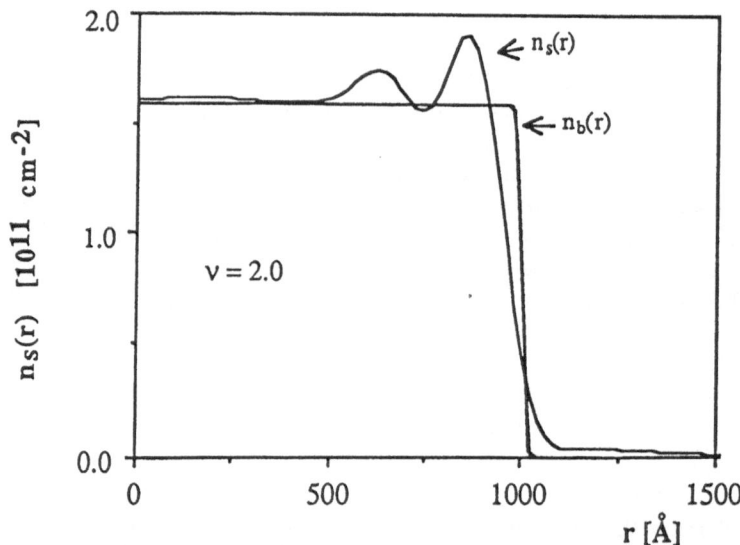

Fig.1b The electron $n_s(r)$ and the neutralizing background density $n_b(r)$ as functions of the radius r. Same parameters as in Fig.1a.

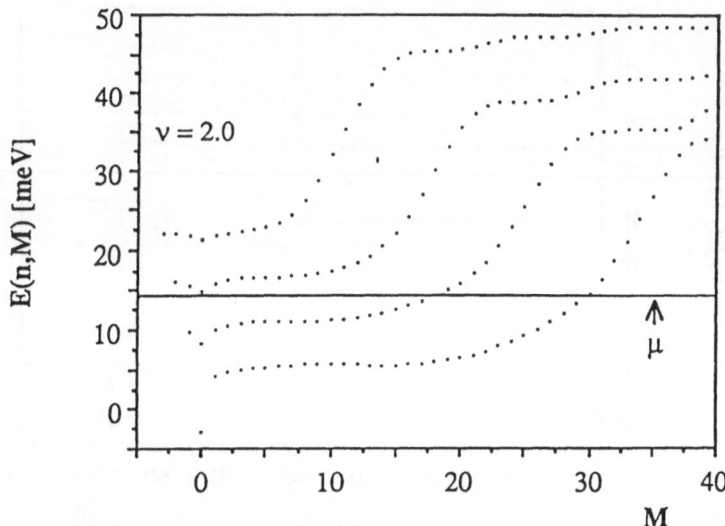

Fig.2a The energy spectrum (dotts) and the chemical potential μ (horizontal line) when a donor like impurity is present in the system (a positive point charge of unit size at $r = 0$) and $\nu = 2$. The parameters are the same as in Fig.1 except the background charge density has been reduced by one unit to compensate for the impurity.

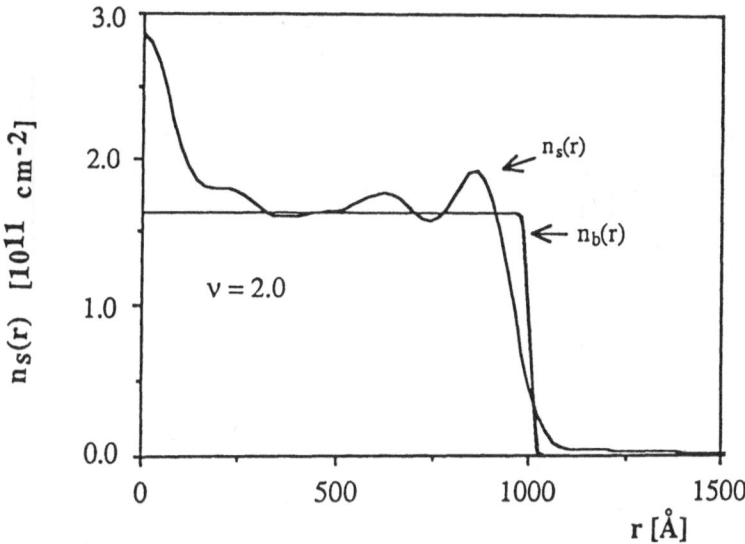

Fig.2b The electron $n_s(r)$ and the neutralizing background density $n_b(r)$ as functions of the radius r. Same parameters as in Fig.2a.

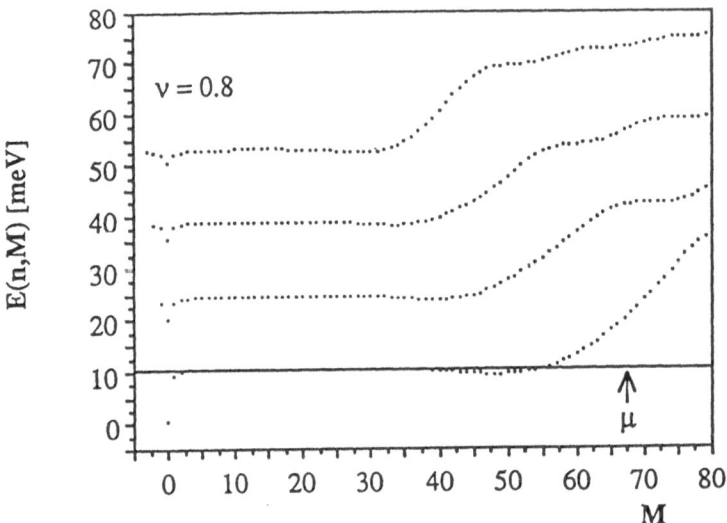

Fig.3a The energy spectrum (dotts) and the chemical potential μ (horizontal line) when a donor like impurity is present in the system (a positive point charge of unit size at $r = 0$) and $\nu = 0.8$. The parameters are the same as in Fig.1 except the background charge density has been reduced by one unit to compensate for the impurity.

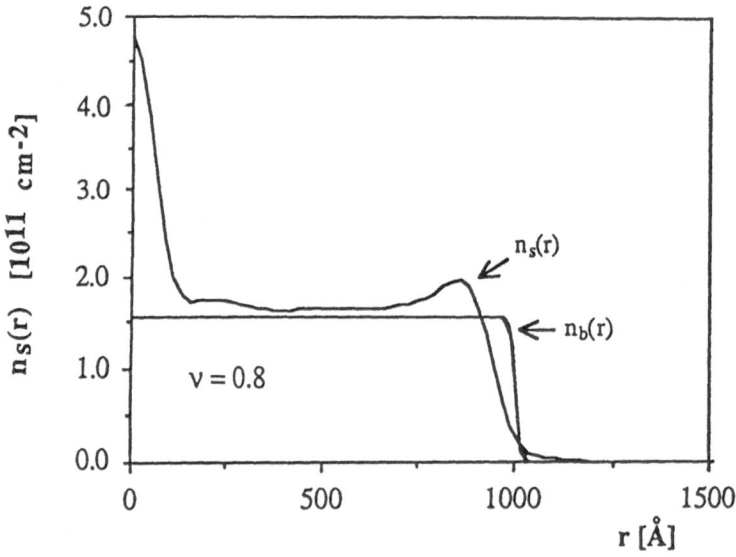

Fig.3b The electron $n_s(r)$ and the neutralizing background density $n_b(r)$ as functions of the radius r. Same parameters as in Fig.3a.

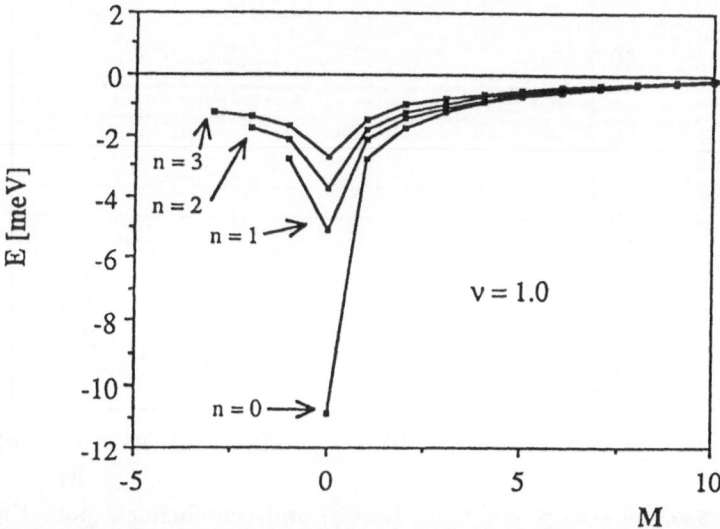

Fig.4a The difference in the energy of the Landau bands with and without an impurity present in the system as a function of the quantum number M. Here $\nu = 1.0$.

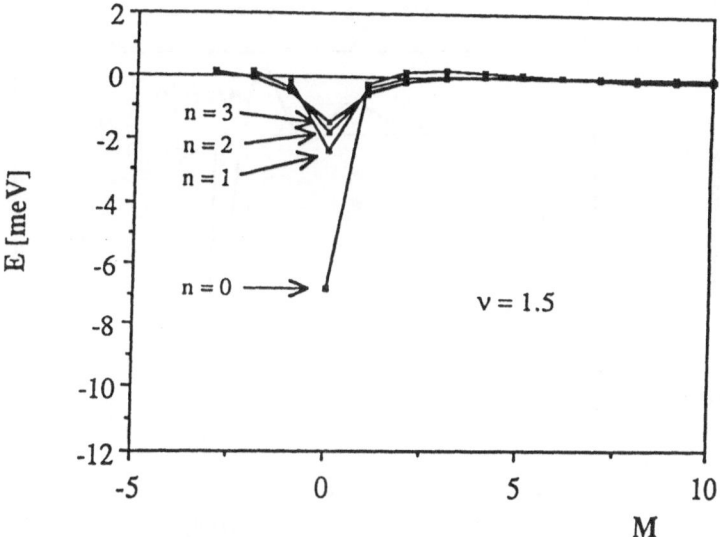

Fig.4b The difference in the energy of the Landau bands with and without an impurity present in the system as a function of the quantum number M. Here $\nu = 1.5$.

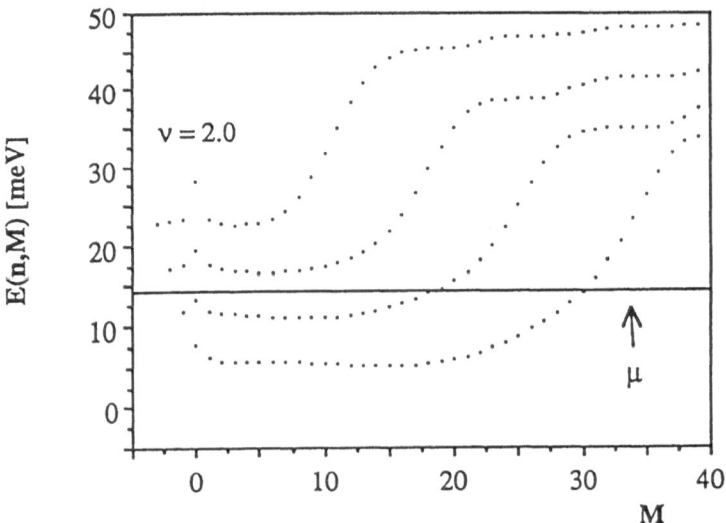

Fig.5a The energy spectrum (dotts) and the chemical potential μ (horizontal line) when an acceptor like impurity is present in the system (a negative point charge of unit size at $r = 0$) and $\nu = 2$. The parameters are the same as in Fig.1 except the background charge density has been increased by one unit to compensate for the impurity.

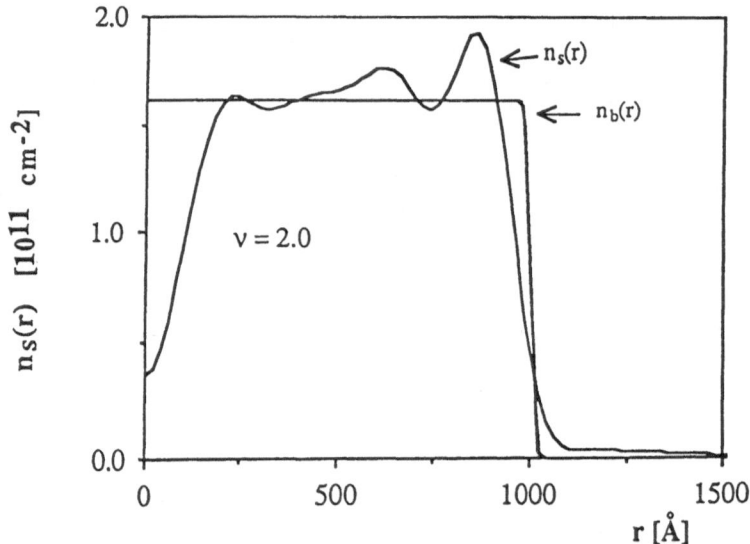

Fig.5b The electron $n_s(r)$ and the neutralizing background density $n_b(r)$ as functions of the radius r. Same parameters as in Fig.5a.

states are shifted upwards on the energy scale, and the mixing of the Ll's causes the shift to be constant or increase slightly with increasing n. The large shift of the $M = 0$ state in the top Lb is artificial and is damped if higher Ll's are included in the calculation (it shows the electron-hole symmetry in a system with a finite number of Lb's). Fig.5b shows the "hole" that is bound to the acceptor.

Now we shall concentrate on how the screening properties of the 2DEG manifest themselves, starting with the energy spectrum. Fig.4 shows the difference between the energy shifts for an integer ν (a), where μ is situated between two Lb's, and a half integer ν (b), where μ is pinned to a Lb and therefore more states are available for the electrons to redistribute themself in order to screen the impurity potential [16,20-21]. Not only is there a large difference in the energy shift for the $M = 0$ states in these cases, but for the integer filling factor the neighbouring M-states are also all shifted downwards, whereas for the $\nu = 1.5$ case the other M-states are hardly affected at all, reflecting the strong screening in the non-integer case. The origin of this difference in the screening strength is ofcourse, as was just stated, the different position of μ in the highly degenerate and discrete energy spectrum of a 2DEG in a perpendicular magnetic field. When $\nu = 1.5$ the neighbouring M-states to the $M = 0$ state are partially filled and can therefor be slightly rearranged to maximize the screening, opposed to the $\nu = 1.0$ case, where the neighbouring M-states are either empty or filled and far enough from the chemical potential that a slight rearrangement of the Lb does not alter their occupation.

The electron binding energy in a 2D hydrogenic atom is $4 \cdot E_{Ryd.} \approx 23.7 meV$ without a magnetic field [6], and in the magentic field range used here one can expect the binding energy to be around $23 meV$. The $M = 0$ energy shift in Fig.4a is only about half this value, due mainly to the nonlinear screening properties of the system, i.e. the screening does not entirely vanish when ν assumes an integer value [16,22-23]. How the energy shift for the $M = 0$ and the $M = 1$ states behaves as a function of ν can be seen in Fig.6 for a donor, and in Fig.7 for an acceptor like impurity, for all the included Lb.'s in each case. As expected, the energy shift is maximum for both types of impurities for integer filling factors, reflecting the weakened screening in that case. The screening is much weaker for the $\nu = 1$ case, than when $\nu = 2$, since $\hbar\omega_c$, the distance between Ll's in the former case is twice the one for the latter case, and all interaction effects in the angular symmetric system come around from the mixing of different Ll's with M unchanged. A partial electron hole symmetry, that was visible in the energy spectra (Fig.2 and 4) is also apparent in Fig.6 and 7, for example, the steeper side of the peaks or dips is on the left for the acceptor, but is on the right side for the donor.

As a consequence of the energy spectra we can expect that a calculation of the cyclotron resonance as a function of ν yields a broadening of the resonance line, shifting one side of it away from the center at $\hbar\omega_c$. This splitting of the resonance, caused by the impurity, will oscillate in strength with ν and take on maximum values for ν being an integer. The shift will also either grow abrubtly or slowly when the integer ν is approached from below or above, depending on whether the impurity is donor or acceptor like. The discontinuity seen in the "cyclotron mass" $m = eB/\alpha\omega_c$ at integer ν in experiments might be of a similar origin, as has been suggested elsewhere [1,5]. Care has ofcourse to be taken here when a model system with one impurity is compared with a real system with many impurities. In a system where the impurities are not all exactly in the 2D-plane those outside the plane would be much more strongly screened due to their weakened bare potential seen by the 2DEG [1]. Then,

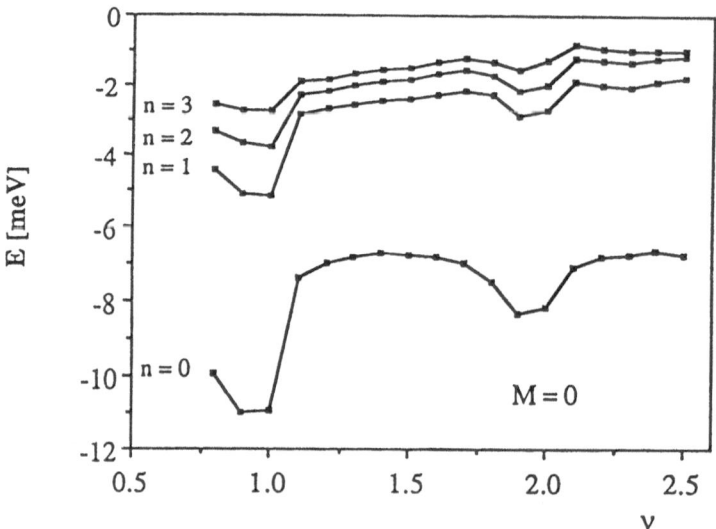

Fig.6a The energy shift of the $M = 0$ state as a function of the filling factor ν, when a donor like impurity is present in the system. Parameter choice as in Fig.1.

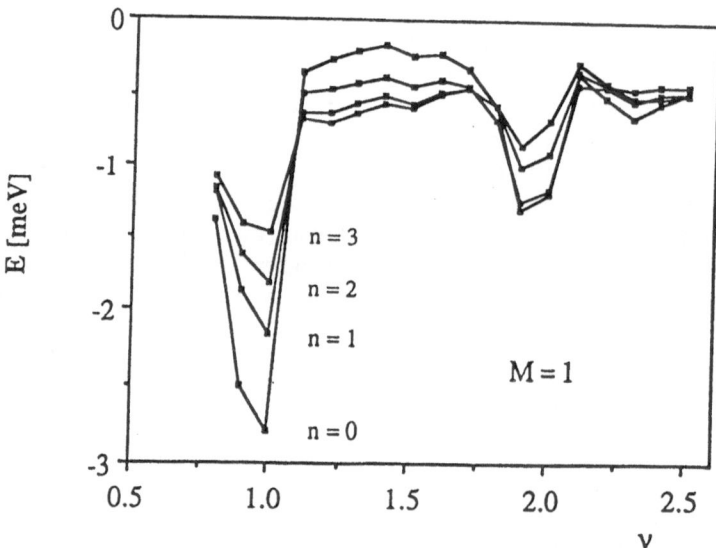

Fig.6b The energy shift of the $M = 1$ state as a function of the filling factor ν, when a donor like impurity is present in the system. Parameter choice as in Fig.1.

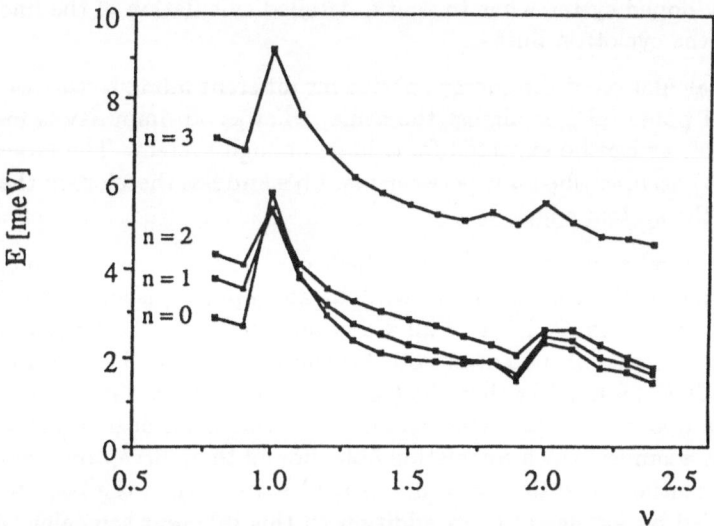

Fig.7a The energy shift of the $M = 0$ state as a function of the filling factor ν, when an acceptor like impurity is present in the system. Parameter choice as in Fig.1.

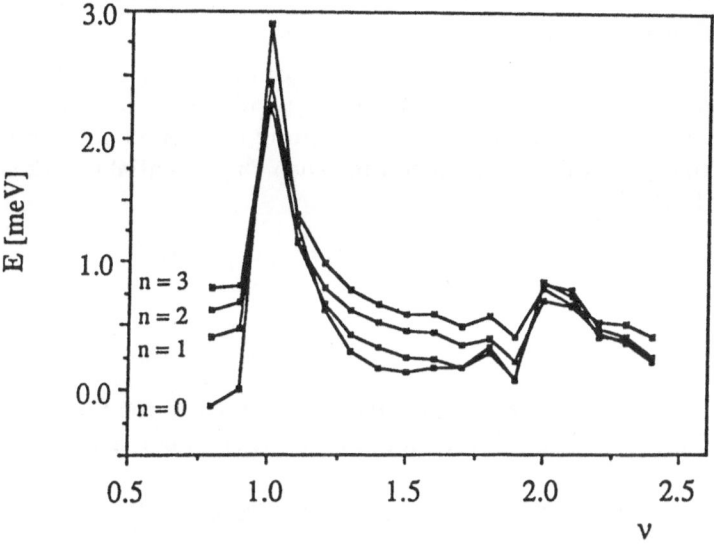

Fig.7b The energy shift of the $M = 1$ state as a function of the filling factor ν, when an acceptor like impurity is present in the system. Parameter choice as in Fig.1.

instead of a component being split off from the main cyclotron resonance line, one would have a resonance line that would be asymmetrically broadened to one side, in the case of one type of dominant impurities. Comparison with experiments in very dilutely doped systems has to wait a detailed calculation of the line shapes and positions of the cyclotron lines.

In the calculation of the energy spectra for different filling factors we notice, that the chemical potential μ is almost the same, whether an impurity is present in the system or not, as can be expected for a large enough system. The largest deviation ($\Delta\mu \ll \hbar\omega_c$) occures when μ is between two Lb's and lies therefore in the low density of states at the system edge.

The screened impurity potential, which can be found as the difference in the total potential for the case with and without an impurity, is shown in Fig.8 for two different filling factors ν. First of all we notice in Fig.8b the large deviation from the unscreened Coulomb potential both for the integer and the half-integer ν. This can be explained by the fact that the $(M = 0, \quad n = 0)$ state is always filled first in the case of a donor, and this state bound to the impurity, does screen the Coulomb potential. In a similar way a correlation hole, bound to an acceptor, would screen the Coulomb potential, since the $(M = 0, \quad n = 0)$ state is then the last state in the Lb to be occupied by an electron. In addition to this inherent screening of the bound state $(M = 0, \quad n = 0)$, the neighbouring partially filled M-states in the $\nu = 1.5$ case screen the rest of the potential quite effectively compared to the $\nu = 1.0$ case, where the neighbouring states are either almost filled or empty and can therefore not participate in the screening. The strong screening at non-integer ν is always characterized by oscillations of the impurity potential within the boundaries of the system, analogous to Friedel oscillations. These oscillations vanish entirely for integer filling factors, where the potential approaches the bare Coulomb potential more closely than for other ν, except for a curious crossing of the r-axis at large r-values, which we shall illuminate as a pure 2D behavior.

A glance at Fig.2b and 3b suggests the following simple model. Assume a positive unit point charge (the impurity) surrounded by a uniform disk of a negative charge of same quantity, but with radius r_0, representing the bound state of the impurity. The rest of the system is assumed neutral. Then the potential of this simple model is according to (2.8)

(3.1) $\qquad V(r) = \begin{cases} -\frac{e^2}{\kappa r} + \frac{4e^2}{\pi r_0^2 \kappa} r [E(\frac{r_0}{r}) - (1 - (\frac{r_0}{r})^2) K(\frac{r_0}{r})], & r > r_0 \\[2mm] -\frac{e^2}{\kappa r} + \frac{4e^2}{\pi r_0^2 \kappa} r_0 E(\frac{r}{r_0}), & r \leq r_0, \end{cases}$

where $K(x)$ and $E(x)$ are the complete elliptic integrals of the first (2.9) and second kind respectively [18]. The following limiting forms can be derived for the potential

$$V(r) \to -\frac{e^2}{\kappa r}, \qquad\qquad r < r_0, \quad r \to 0$$

(3.2) $\qquad V(r) > 0, \qquad\qquad\qquad r = r_0$

$$V(r) \to \frac{e^2}{8\kappa}(\frac{r_0^2}{r^3}) + \cdots \geq 0, \quad r > r_0, \quad r \to \infty.$$

So as expected, the asymptotic form of the potential for small r is the unscreened Coulomb potential, but as r increases the potential crosses the r-axis and approaches

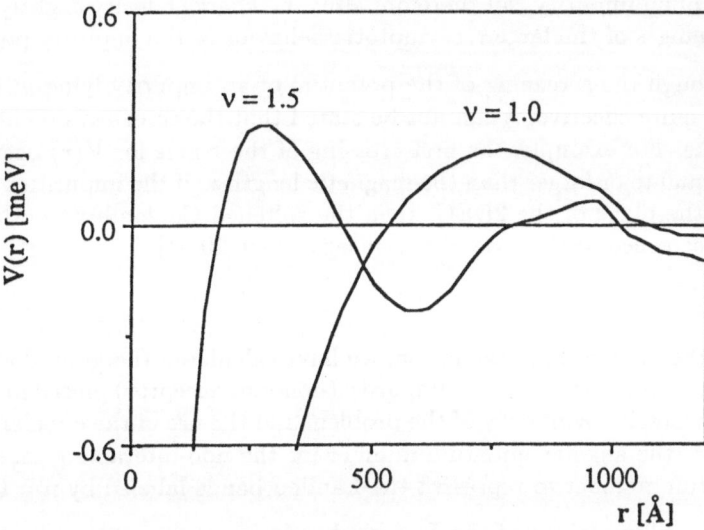

Fig.8a The screened impurity potential $V(r)$ for two different filling factors, $\nu = 1.0$ and $\nu = 1.5$.

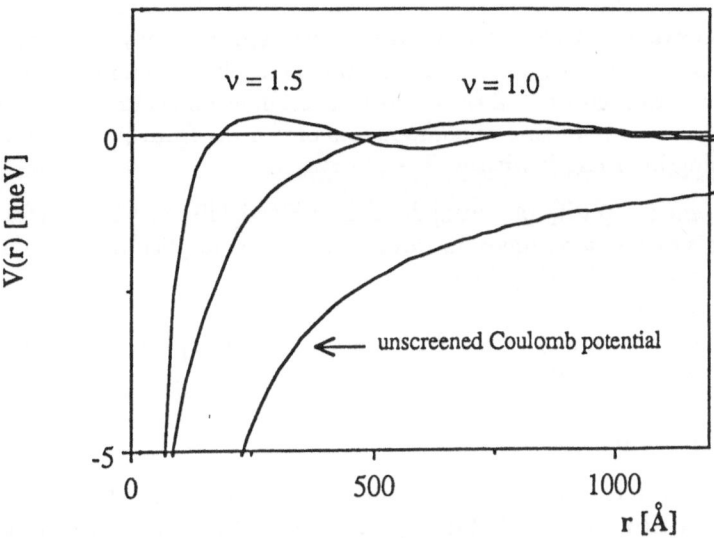

Fig.8b The screened impurity potential $V(r)$ for two different filling factors, $\nu = 1.0$ and $\nu = 1.5$ compared with the unscreened Coulomb potential.

the axis again from the *positive* side. This simple model shows therefore similarily this 2D property of overscreening. The difference being though, that the simple model reflects better what happens in a infinite system, since in a finite system the chemical potential μ crosses the Lb's near the system edge. Any slight change in μ due to the addition of an impurity can therefore alter the charge density slightly at the edge leading to changes of the large r asymptotic behavior of the impurity potential.

Even though the screening of the potential of an impurity lying in the plane of the 2DEG is quite effective, it can not be stated that the screened potential becomes δ-function like. For example, the first crossing of the r-axis for $V(r)$ occures usually for lengths equal to or larger than the magnetic length l. If the impurities are situated well outside the plane of the 2DEG, then the softened Coulomb potential may even be perfectly screened in the case of non-integer ν [16,20-21]

4 Summary

Within the Hartree approximation, we have calculated the ground state properties of the compound system of one impurity (donor or acceptor) placed in the plane of a 2DEG. The angular symmetry of the problem and the use of the circular gauge have resulted in M, the angular quantum number for the non-interacting case, still being a good quantum number to represent the Landau bands labeled by $n = 0, 1, 2, \cdots$.

When the filling factor of the Landau bands is not an integer, i.e. the highest occupied Lb is not fully occupied, then almost only the $M = 0$ states are affected, being shifted up or downwards in energy, depending on whether the impurity is donor or acceptor like respectively. The energy shift of the $M = 0$ state does oscillate with ν, taking on its largest values for ν being an integer, reflecting the dependence of the screening abilities of the 2DEG on the position of the chemical potential with respect to the Lb's.

The important consequences for the optical spectroscopy, in samples with one dominant type of impurities, are therefore the facts, that the ionization energy of the impurities oscillates with ν and the cyclotron resonance lines broaden asymmetrically to one side due to the shifting of components of it away from the value $\hbar\omega_c$, with the shift oscillating in strength with ν.

The screened impurity potential does show Friedel like oscillations for non-integer ν, which vanish when ν assumes an integer value, leaving behind a potential that for small r-values resembles the Coulomb potential, but crosses the axis for a large r, and then approaching it from same side again asymptotically. The screened potential thus shows the tendency of a 2DEG to overscreen the impurity potential, even when the screening itself is weak.

Acknowledgement

I would like to express my sincere thanks to Prof. Dr. Rolf R. Gerhardts at the Max-Planck Institute for solid state research in Stuttgart for suggesting this model to study the screening of impurities in two-dimensional electron systems.

References

[1] K. Ensslin, *Doctoral Theses*, Max-Planck Institute, Stuttgart (1989).
[2] D. Heitmann, M. Ziesmann and L.L. Chang, *Phys. Rev.* **B34**, 7463 (1986).

[3] T. Ando, *J. Phys. Soc. Jpn.* **38**, 989 (1975).

[4] V. Gudmundsson and R.R. Gerhardts, *Phys. Rev.* **B35**, 8005 (1987).

[5] J. Richter, H. Sigg, K.v. Klitzing and K. Ploog, *Phys. Rev.* **B**, March (1989).

[6] F. Stern and W.E. Howard, *Phys. Rev.* **163**, 816 (1969).

[7] Martin and Wallis, *Phys. Rev.* **B18**, 5644 (1978).

[8] T. Ando and Y. Uemura, *J. Phys. Soc. Jpn.* **36**, 959 (1974).

[9] R.R. Gerhardts, *Z. Phys.* **B21**, 275 (1975).

[10] S. Luryi, *High Magnetic Fields in Semiconductor Physics*, Vol 71 p.16 of *Springer Series in Solid-State Sciences*, edited by G. Landwehr (Springer, Berlin 1987).

[11] T. Ando and Y. Murayama, *J. Phys. Soc. Jpn.* **54**, 1519 (1985).

[12] W. Cai and T.S. Ting, *Phys. Rev.* **B33**, 3967 (1986).

[13] B.D. McCombe, N.C. Jarosik and J.M. Mercy, in *High Magnetic Fields in Semiconductor Physics*, Ref.8, p.238.

[14] G. Bastard, *Phys. Rev.* **B24**, 4714 (1981).

[15] R.L. Greene and K.K. Bajaj, *Phys. Rev.* **B31**, 913 (1985).

[16] U. Wulf, V. Gudmundsson and R.R. Gerhardts, *Phys. Rev.* **B38**,4218 (1988).

[17] O. Heinonen and P.L. Taylor, *Phys. Rev.* **B32**, 633 (1985).

[18] M. Abramowitz and I.A. Stegun, *Handbook of Mathematical Functions*, Dover (1972).

[19] V. Gudmundsson, R.R. Gerhardts, R. Johnston and L. Schweitzer, *Z. Phys.* **B70**, 453 (1988)

[20] V. Gudmundsson and R.R. Gerhardts, *The Application of High Magnetic Fields in Semiconductor Physics*, editor G. Landwehr, *Springer Series in Solid-State Sciences*, (Springer, Berlin 1989) in press.

[21] R.R. Gerhardts and V. Gudmundsson, *Proceedings of* **ICPS-19**, (Warszawa, 1988) in press.

[22] J. Labbé, *Phys. Rev.* **B35**,1373 (1988).

[23] R.R. Gerhardts and V. Gudmundsson, *Solid State Commun.* (1988).

ULTRAFAST LUMINESCENCE STUDIES OF TUNNELING IN SEMICONDUCTOR MICROSTRUCTURES

Jagdeep Shah
AT&T Bell Laboratories
Holmdel, N. J. 07733, U. S. A.

Tunneling in semiconductor microstructures such as double barrier diodes and double quantum wells is of fundamental as well as practical interest and has received intense attention in recent years. These phenomena have been investigated primarily using electrical techniques such as current-voltage measurements. Ultrafast optical techniques provide a powerful means of investigating the dynamics of carrier transport and tunneling in semiconductor microstructures. We present a brief review of the basic concepts behind all-optical techniques and then discuss some recent results on the dynamics of tunneling using luminescence spectroscopy.

I. INTRODUCTION

Transport of carriers in the direction perpendicular to the planes of a semiconductor superlattice was first considered by Esaki and Tsu [1], who predicted many interesting properties, including negative conductance and Bloch oscillations. The considerable activity in the field of superlattices in the early and mid 1970's has been reviewed recently by Esaki [2]. After this initial flurry of activity, the emphasis shifted towards the study of quantum wells, and the quasi-two-dimensional electron gas (2DEG) at hetero-interfaces and in quantum wells. These investigations have led to a number of exciting discoveries in the transport and optical properties of 2DEG.

In recent years, there has been a dramatic increase in research on resonant tunneling and perpendicular transport in semiconductor microstructures, spurred by advances in the growth of high quality semiconductor microstructures. The microwave experiments of Sollner and coworkers [3] were followed by intense experimental and theoretical activities in the field. Many aspects of this research have been discussed in excellent review articles by Esaki [2] and Capasso et at [4] in 1986.

The vast majority of studies on tunneling and perpendicular transport in superlattices have used current-voltage measurements as their primary technique. These studies have provided much valuable information and most of the static aspects of these phenomena are reasonable well understood. In contrast, the dynamical aspects of tunneling and perpendicular transport are only now being explored. Transient photocurrent measurements have been made on multiple quantum well structures [5-7]. The time resolution of transient measurements can be improved considerably by exciting the microstructures with the recently developed picosecond and femtosecond laser sources and using *optical detection techniques*. Such ultrafast optical techniques provide a valuable and powerful means of investigating the dynamics of these phenomena in semiconductor microstructures. Many such investigations using *all-optical techniques* have been reported in the past two years. These optical studies provide information that can not be obtained from electrical measurements and therefore complement the electrical studies. The purpose of this article is to discuss these optical techniques briefly (Sec. II.A), discuss some basic concepts behind tunneling (Sec. II.B) and then review some recent measurements that have directly determined dynamics of tunneling in double barrier diodes (Sec. III.A) as well as in coupled quantum well structures (Sec. III.B).

II. BASIC CONCEPTS

We briefly discuss in this section some basic concepts related to all-optical techniques and to tunneling in semiconductor microstructures.

II.A ALL-OPTICAL TECHNIQUES

Recently developed picosecond and femtosecond lasers have been used to investigate tunneling and transport in a number of different ways. The most direct method is to use an ultrafast laser to generate free carriers and measure the time evolution of the electrical current to determine the transport of carriers. This hybrid time-of-flight technique has been used to study transport in bulk semiconductors [8], parallel transport in inversion layers and quantum wells [9,10], and also perpendicular transport in GaAs/AlGaAs superlattices [5-7,11]. While this technique is very useful, the time resolution available in measuring electrical transients directly is much longer than the pulsewidths of the ultrashort laser pulses and insufficient to investigate tunneling processes.

The time resolution in the measurement of the electrical transients can be improved considerably by using one of the optical sampling techniques (e.g. electro-optic or photoconductive sampling). This is a promising approach and has been recently applied to investigate resonant tunneling diodes [12] and other semiconductor devices [13]. The strength of this technique is that it allows a direct measurement of devices.

The physics of carrier transport can be more directly explored by applying all-optical techniques to specially designed microstructures. A technique that deserves mention in this context makes use of the fact that the electric field in a semiconductor following excitation by an ultrashort pulse changes as a function of time because of the motion of photoexcited electrons and/or holes. The time evolution of the field can be measured by monitoring a field-sensitive optical property such as absorption. *Since the change in the field is related to the motion of the carriers, information about carrier transport can be obtained from such measurements*. This electroabsorption technique was first used by [14] to measure velocity overshoot effects in GaAs and has been recently applied to study the transport of carriers in multiple quantum well structures [15].

We will concentrate on a different all-optical technique which provides the means for a *direct measurement of* **transport and tunneling of carriers** *in semiconductor microstructures*. This technique involves the use of an optical "marker" [16] i.e. a thin region of the sample with optical properties different from the rest of the sample. The marker can be a different semiconductor, the same semiconductor with different doping characteristics, or a more complicated microstructure. Such a structure is schematically illustrated in Fig. 1. If carriers are created near the surface by photoexcitation, then their transport to the interior of

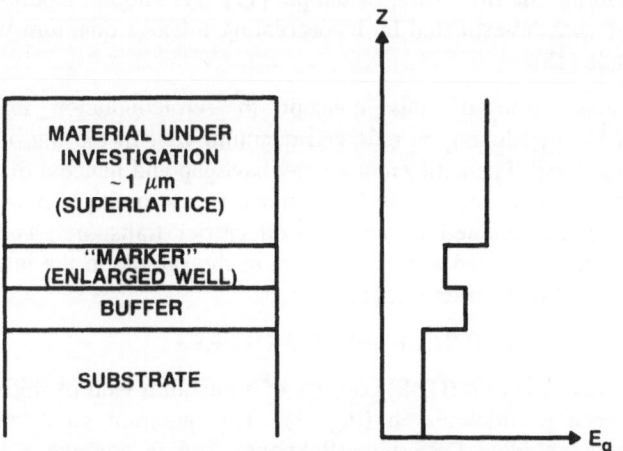

Fig. 1   Schematic diagram illustrating the concept of an optical "marker."

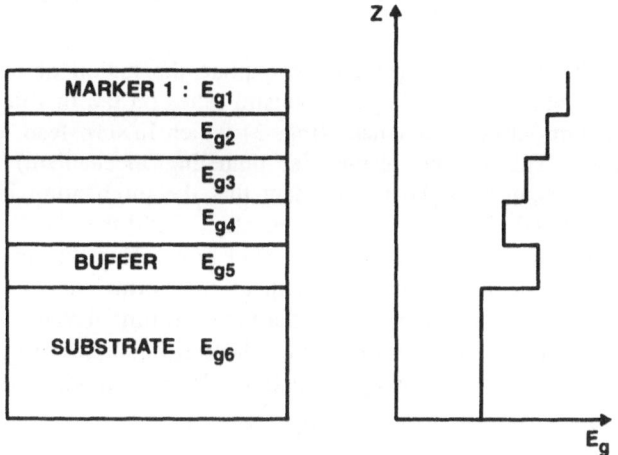

Fig. 2   Schematic diagram illustrating the concept of "multiple optical markers."

the sample can be monitored by measuring some optical property of the marker region that is modified when the photoexcited carriers arrive in the marker region. *The basic idea is that information about a specific **spatial** region, and hence about dynamics of transport and tunneling, can be directly obtained by studying optical properties in a specific **spectral** region.* For example, transport in Si was investigated by monitoring the luminescence from a thin doped region in the interior of the Si sample [17]. As another example, the transport of carriers in bulk InP was investigated by incorporating InGaAs quantum well markers in the interior of the sample [18].

The first application of this concept to semiconductor microstructures was demonstrated [19] by introducing an enlarged quantum well in the interior of a superlattice sample. The enlarged well has a different energy bandgap and hence a distinct luminescence spectrum. This technique can obviously be extended to use multiple markers (Fig. 2) in the sample to provide more detailed information on carrier transport. The dynamics of the transport of carriers photoexcited near the surface of the sample to the interior of the sample has been explored using such structures [16].

II.B TUNNELING IN DOUBLE BARRIER STRUCTURES

A double barrier structure (DBS) consists of a quantum well of thickness a surrounded by two barriers, each of thickness b (Fig. 3). The quantum mechanical description of transmission through an ideal DBS is well known and is analogous to the Fabry-Perot resonance in optics. The dependence of transmission on electron energy E is shown schematically in Fig. 3. The transmission goes through a strong resonance at the quantum well energy, and has a Full Width Half Maximum (FWHM) of Γ which can be related to the barrier height and the barrier thickness.

A double barrier diode (DBD) consists of a double barrier structure surrounded by two n^+ contacts (Fig. 4). As a function of an applied voltage, the current through a DBD initially increases, goes through a maximum and then decreases, giving rise to a region of negative differential resistance. This property has led to the use of these diodes as high frequency oscillators [3]. Interesting physics related to this negative differential resistance continues to be investigated [20].

Two mechanisms have been proposed to explain the region of negative differential resistance. The first (coherent tunneling mechanism) relies on the fact that the transmission through such a structure shows resonances (Fig. 3) which in turn lead to current peaks at certain bias voltages. The second (sequential tunneling mechanism) leads to negative differential resistance because of the requirement that the momentum in the plane of the interface must be conserved [21]. It was shown recently [22,23] that the DC current in a DBD is identical in the two cases, provided that the level broadening due to collisions is small compared to other energies in the problem such as the Fermi energy and the separation between resonant levels. The determination of the traversal time through such a structures is of interest and has been discussed earlier [24-26]. Theoretical studies of quantum mechanical tunneling based on the Wigner function approach have also been reported [27,28].

No experimental studies of the transmission of an electron wavepacket incident on such a structure from the left have been reported. However, a number of studies on how fast an electron optically created in the quantum well escapes from the well have been reported. For comparison with such studies one may argue that the escape time is related to Γ of the calculated transmission coefficient as a function of energy (Fig. 3). Calculation of the transmission of a Gaussian wavepacket through such structures [29,30] gives a reasonable agreement with this simple approach.

The transmission of an electron of energy E through a barrier of height V_0 and thickness b is given by

DOUBLE BARRIER STRUCTURE

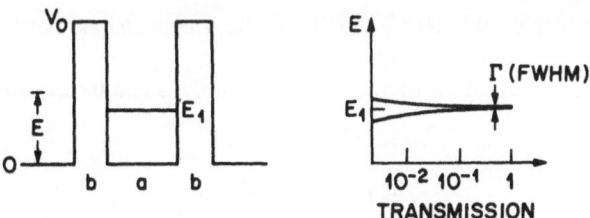

Fig. 3   Schematic of a Double Barrier Structure (DBS) and the transmission as a function of electron energy.

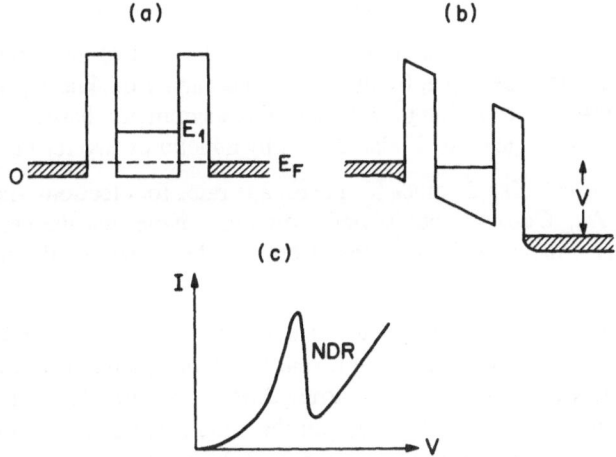

Fig. 4   Schematic of a Double Barrier Diode (DBD) under two bias conditions and of the I-V curve for a DBD.

$$T(E) = \frac{4rE(V_0-E)}{((r-1)E + V_0)^2 \sinh^2(2\pi b/\lambda) + 4rE(V_0-E)}$$

where r is the ratio (m_B/m_W) of the electron mass in the barrier to that outside the barrier. Since sinh (x) approaches exp (x) / 2 for large x, the transmission coefficient becomes proportional to exp ($-4\pi b/\lambda$). λ is the de Broglie wavelength given by

$$\lambda = \frac{h}{(2m_B(V_0-E))^{1/2}}$$

For the double barrier case let T_1 and T_r be the transmission coefficients (at energy E) of the left and the right barrier respectively and let T_{av} be the average of these two transmission coefficients. Then the escape time from a DBS can be shown to be [23]

$$\tau_{es} = 1/(vT_{av})$$

$$v = (2E_R/m_W^*)^{1/2}/2a$$

Here v is the frequency at which the electron attempts to escape **one** of the barriers (2a/(velocity of the electron)), and E_R is the energy of a resonant level in the structure. If Γ is the FWHM of the transmission resonance of a DBS:

$$T(E) = T(E_R)\frac{(\Gamma/2)^2}{(E-E_R)^2 + (\Gamma/2)^2}$$

then it can be shown that [23]

$$\Gamma = (2\hbar/\tau_{es})$$

or

$$\tau_{es} = (\hbar/(\Gamma/2))$$

Note that if one starts with a symmetric barrier ($T_1 = T_r$) and decreases one of the transmission coefficient (for example, by increasing the barrier thickness), then τ_{es} increases. The escape time becomes twice as long for the strongly asymmetric barrier case (e. g. $T_1 \ll T_r$) compared with the symmetric case. Γ also differs by a factor of two for these two cases.

Fig. 5 shows the calculated values for the escape rates for electrons and heavy holes for the case of GaAs / $Al_{0.3}Ga_{0.7}As$ double well structures. Note that the escape rates for the heavy holes are much smaller than those for electrons. The rates for the light holes can be higher than those for electrons.

One final point that needs to be discussed is the case when the conduction band in the barrier has a minimum at a different point in k-space than the quantum well material. Such is the case for the AlAs/GaAs/AlAs double barrier diodes. A number of discussions of this point have appeared in the literature [31-33] and the conclusion appears to be that tunneling through the indirect transitions are important for thick barriers but not for thin barriers [31]. It is important to note in this regard that even if tunneling proceeds through the Γ valley in the barrier, the effective mass of the electron may not be the electron mass at the bottom of the conduction band; it must be calculated appropriately using a two-band or a more sophisticated model.

II.C TUNNELING IN DOUBLE WELL STRUCTURES

A double quantum well structure (DQWS) consists of two quantum wells separated by a barrier of thickness b. In a symmetric DQWS, the two quantum wells are identical, whereas in an asymmetric double quantum well structure (a-DQWS), the two quantum wells have

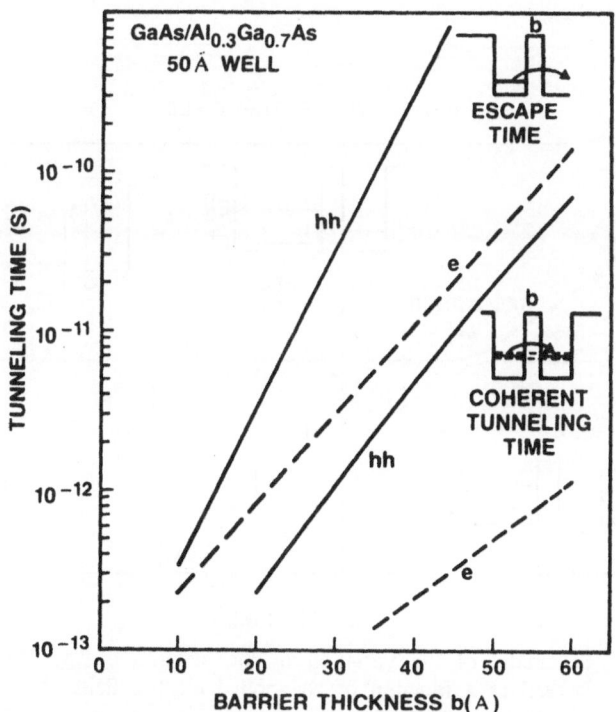

Fig. 5   Calculated variation for the escape times from a Double Barrier
Structure and the Coherent Tunneling times for a Double Quantum
Well Structure for both electrons and heavy holes.

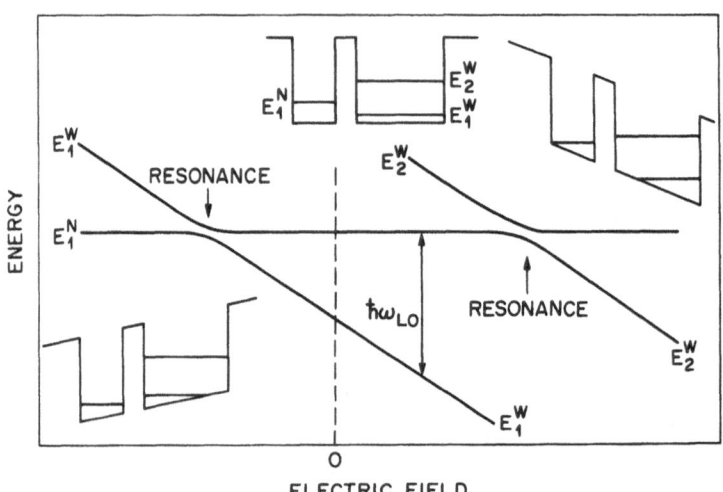

Fig. 6   Schematic of the variation of the electron energy levels in a-
         DQWS as a function of an applied electric field. Note the two
         anti-crossings indicating resonances.

unequal thicknesses (a_1 and a_2) and may be composed of different semiconductors (e.g. GaAs and AlGaAs). Double quantum wells are the simplest structures that exhibit resonant coupling between two quantum states and field induced localization. They form the building block for the multiple quantum well structures and superlattices in which many interesting transport phenomena have been investigated. For this reason, there has been a growing interest in both symmetric and asymmetric double quantum well structures [34-37]. The advantage of an asymmetric DQWS is that the energy bandgaps for the two wells are different so that there are two optical "markers" in the system.

Consider an a-DQWS (Fig. 6) such that the narrow well has only one confined electron subband whereas the wide well has two. Under the condition of flat bands, the energy minima of the two conduction subbands in the wide well (E_1^W and E_2^W) straddle the energy minimum of the conduction subband in the narrow well (E_1^N). Fig. 6 schematically illustrates the dependence of these energy levels on an applied electric field, with the center of the narrow well as the zero of energy. There are several regions of interest as the relative energy separation is tuned by varying the electric field. We discuss these in the next subsections.

II.C.1 NON-RESONANT TUNNELING

When the energy levels are well separated, the electron wavefunctions are primarily localized in one well or the other, and the tunneling between the two wells requires a momentum conserving collision (*non-resonant tunneling* process). As shown in Fig. 7, this can be an elastic collision (with an impurity or an interface defect) or an inelastic collision with an acoustic or optical phonon. As a result of such a transition, an electron whose wavefunction is primarily localized in one well transfers or tunnels into a state whose wavefunction is localized primarily in the other well.

For such non-resonant tunneling, it is interesting to explore the dependence of the tunneling rate on the barrier thickness. In calculating this transition rate, one must use the eigenstates of the entire double well system as the initial and final states and use the appropriate matrix elements for each process under consideration. This calculation is in principle similar to the calculation of inter-subband transition rates, except that most of the wavefunction overlap comes from the barrier region separating the two wells. For this reason, one must consider phonons related to both the wells and the barrier. An approximate calculation of such rates for optical phonon emission has been reported by Oberli et al [38]. Since the overlap of the wavefunction occurs primarily in the barrier region where the wavefunctions are decaying exponentially, the tunneling rate depends exponentially on the barrier thickness. More detailed calculations for all the above processes have also been reported recently [39], and Monte Carlo simulations of optical phonon assisted tunneling rates have also been reported [40].

Another interesting case for non-resonant tunneling arises when the difference $E_1^N - E_1^W$ is tuned through an optical phonon energy. Since the optical-phonon assisted tunneling rate is expected to be larger than those associated with the other processes discussed above, an increase in tunneling rate is expected as the energy difference between the levels increases and equals the phonon energy. Beyond this threshold, a gradual decrease in the tunneling rate is expected because the wavevector of the emitted phonon increases (leading to a decrease in the Frohlich matrix element) and the overlap between the wavefunctions decreases. Detailed calculations for the symmetric double well case have been reported recently [39].

II.C.2 RESONANT TUNNELING AND COHERENT OSCILLATIONS

For the case of the asymmetric double well structure shown in Fig. 6, there are two values of the applied field where the narrow well level is in resonance with one of the wide well levels. At the field where the levels would have crossed, the strong resonant coupling leads to an anti-crossing and a splitting of the level into two levels. In contrast to the case

NON-RESONANT TUNNELING

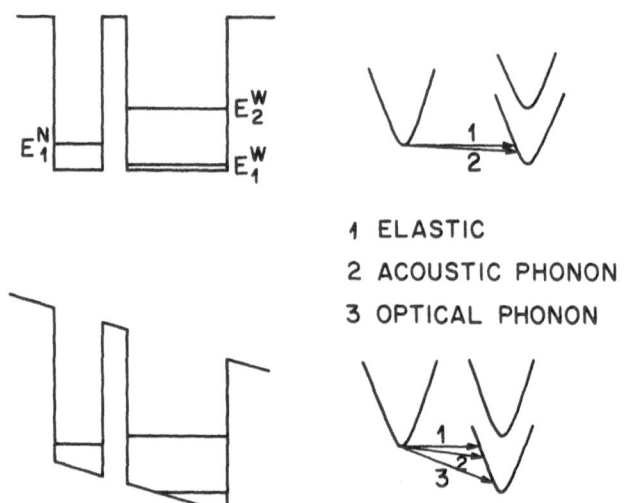

1 ELASTIC
2 ACOUSTIC PHONON
3 OPTICAL PHONON

Fig. 7   Schematic of processes involved in non-resonant tunneling in a-DQWS.

RESONANT TUNNELING

(a)   (b)

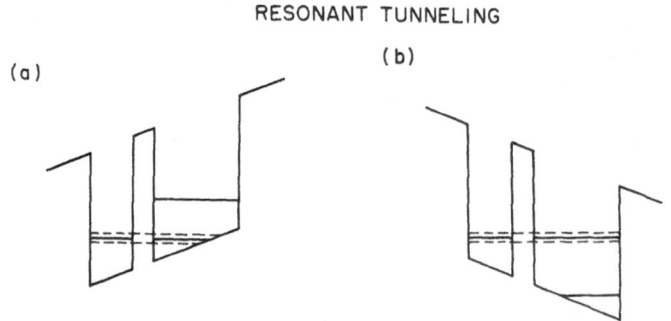

Fig. 8   Schematic of resonant tunneling in a-DQWS.

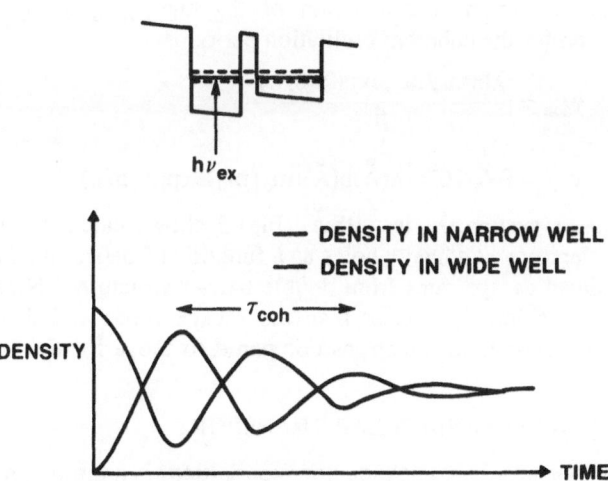

Fig. 9   Coherent oscillations resulting from resonant tunneling process
indicated in Fig. 8.

discussed above, no change in the momentum or the energy of the electron is required for tunneling (Fig. 8); this is the case of *resonant tunneling*.

Under the condition of tunnel coupling, it is well known that the true eigenstates of the system are given by the symmetric and the anti-symmetric combinations of the individual quantum well states. The wavefunctions corresponding to these eigenstates are *delocalized* and have equal amplitude in each well. If the electrons are *initially* prepared in a linear superposition of these two eigenstates so that the electron wavepacket is localized in a single quantum well, then the electron density in each well will oscillate with a period ($\tau_{coh} = h/\Delta E$) where ΔE is the energy splitting between the two resonantly coupled energy levels. These *coherent oscillations*, and the role of collisions in damping these oscillations, have been considered in detail by Luryi [41,42]. The expected behavior of the electron density in each well is shown schematically in Fig. 9. As pointed out by Kane [43], the oscillation period τ_{coh} represents one of the few cases in which a tunneling time is defined unambiguously. Observation of these oscillations, and the measurement of the period would therefore be of great scientific interest.

Since the coherent oscillation time is related to the energy splitting at resonance, τ_{coh} is best calculated from a numerical calculation of the energy splitting ΔE. However, an approximate expression for the coherent oscillation period is

$$\tau_{coh} = \frac{\lambda b(m_e/m_0)\exp(2\pi b/\lambda)}{\hbar}$$

$$\tau_{coh} = 8.6 \times 10^{-17} \lambda(\text{\AA}) b(\text{\AA})(m_e/m_0)\exp(2\pi b/\lambda)$$

For GaAs/Al$_{0.3}$Ga$_{0.7}$As system, λ_e is ≈ 95 Å. Fig. 5 shows calculated coherent tunneling times ($\tau_{coh}/2$) for electrons and heavy holes as a function of barrier thickness and compares them with the calculated escape rates from double barrier structures. Note that the coherent tunneling times are considerably shorter than the escape times and that the characteristic times for electrons are shorter for electrons compared to those for the heavy holes in both cases.

II.D TUNNELING AND PERPENDICULAR TRANSPORT

Perpendicular transport in semiconductor superlattices is a fascinating field of study. A number of electrical study of such transport have been reported [4-7,15]. An understanding of the perpendicular transport is not only of fundamental importance but also of particular interest in optimizing the performance of many devices based on perpendicular transport in superlattices.

It has been established that perpendicular transport in superlattices is influenced by tunneling between adjacent wells [4] as well as by other phenomena. The interplay between tunneling and these other phenomena are best investigated by directly studying perpendicular transport in superlattices. All-optical studies of perpendicular transport in superlattices using luminescence [16,44] techniques have been reported recently. Electro-absorption technique [15] and four-wave mixing techniques [45] have also been applied recently to this problem.

While it is interesting to explore transport in superlattices directly, it is clear that additional complications arising from transport in multiple quantum wells or superlattices should be avoided, if one is interested in understanding tunneling. Therefore, it is important to investigate *isolated structures*, such as DQWS and DBS, for obtaining insights into tunneling.

III. ULTRAFAST OPTICAL STUDIES OF TUNNELING

Tunneling in semiconductor microstructures such as double barrier diodes and multiple quantum well structures have been investigated primarily by cw current-voltage

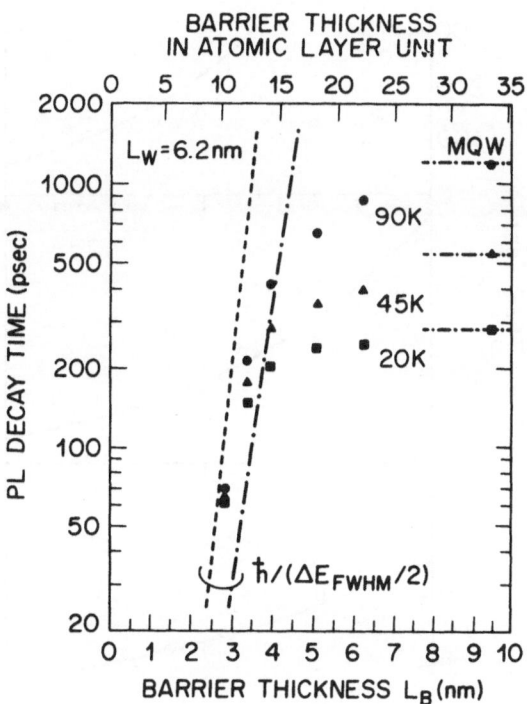

Fig. 10   Escape times from a DBS obtained by time-resolved
photoluminescence spectroscopy (from Tsuchiya et al [46]).

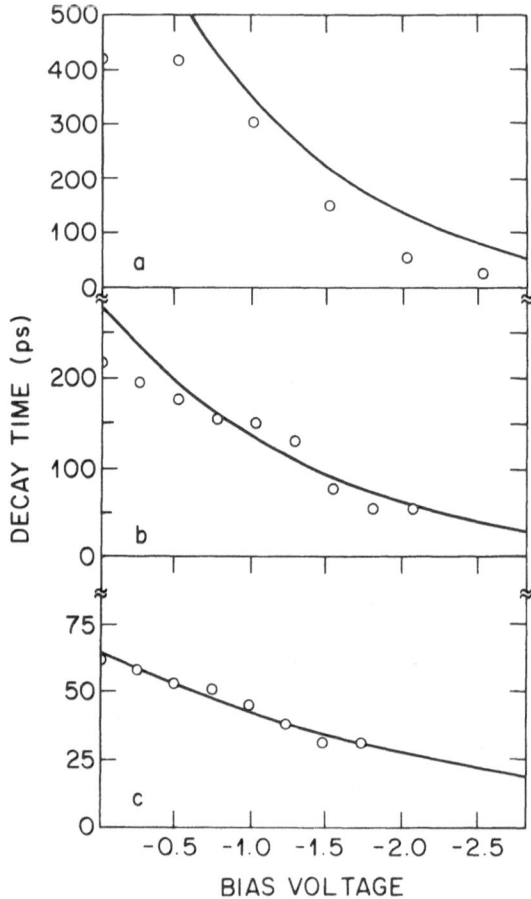

Fig. 11   Escape times from a DBS under the application of an electric field
(from Norris et al [48]).

measurements [4]. These studies have clearly established the existence of tunneling in such structures and perpendicular transport over large distances ($\approx \mu m$). Transient current-voltage measurements have been performed on multiple quantum well structures to investigate the dynamics of carrier transport [5-7]. In this section we review some of the experimental results on tunneling obtained by using all-optical techniques.

III.A DOUBLE BARRIER DIODES

The first study of escape rates from a DBS were performed [46] in an elegant experiment in which the quantum well in a GaAs/AlAs DBS was photoexcited directly at the n=1 light hole exciton energy. The decay time of the n=1 heavy hole exciton luminescence was investigated as a function of barrier thickness and temperature using a streak camera. Their data on the variation of the decay time versus the barrier thickness are presented in Fig. 10. For thin barriers, they found that the escape rate varies exponentially with b. The escape times calculated from the FWHM of transmission curves for two different values of the conduction band offsets (Fig. 10) give good agreement with the data. The authors also found good agreement when they compared their results to the product of the attempt frequency and the transmission through a single barrier.

The shortest time measured in the above studies was ≈ 60 ps. Similar measurements using the technique of luminescence correlation spectroscopy with an excitation source near 620 nm were performed down to 10 ps time resolution [47]. They found escape times which were about 4 times longer than in the earlier study [46]. They reported that a calculation using two-band model for the effective mass in the barrier gives a good agreement with their data. Use of different techniques and much higher excitation energy must also be considered to account for the differences in the two sets of results.

The escape times from DBS as a function of an applied electric field were also measured using streak camera [48]. The escape times decreased with increasing electric field in a manner consistent with the change in the effective barrier with electric field. Their results for three different barrier thicknesses are shown in Fig. 11.

One question that has not been resolved satisfactorily in these studies is what happens to the holes when electrons tunnel out. Jackson et al [47] argue that holes also escape at the same rate. I find this conclusion rather surprising and believe that a further investigation of this phenomenon would be interesting. It would also be interesting to inject carrier on the emitter side of a DBD and measure their arrival time on the collector side of the DBD. This would simulate the actual operation of a DBD; however, no such measurements have been reported to my knowledge.

III.B ASYMMETRIC DOUBLE QUANTUM WELLS

Investigation of DQWS using cw spectroscopy have provided interesting information about various aspects of the coupled well system [34-37]. Oberli et al [38,49,52] have performed time resolved studies on such structures and directly determined the *resonant and non-resonant tunneling times for electrons in asymmetric double quantum well structures*. Matsusue et al [50] have also reported investigations of tunneling in double quantum wells. We review these results in this section.

As discussed above, information about tunneling can be best obtained by investigating an isolated tunneling structure rather than by investigating transport across a superlattice. Oberli et al [38,49,52] accomplished this by studying electron tunneling from the narrow to the wide quantum well in an a-DQWS with eight periods with a thick AlGaAs barrier layer isolating each period (Fig. 12). The eight period structure was embedded in the i-region of a p-i-n structure so that an electric field may be applied to change the relative alignment of energy levels in the two wells (see Fig. 6). Three different samples with well thickness of 60 Å for the narrow well and 88 Å for the wide well, and different barrier thicknesses, were investigated. The thicknesses were checked by TEM as well as by extensive optical studies.

p⁺	GaAs		2000 Å
p⁺	AlGaAs		600 Å
i	AlGaAs		600 Å
i	GaAs		a_1
i	AlGaAs		b_1
i	GaAs		a_2
i	AlGaAs		150 Å
i	AlGaAs		2000 Å
n⁺	AlGaAs		5000 Å
n⁺	GaAs	BUFFER	2000 Å
n⁺	GaAs	SUBSTRATE	

Fig. 12   Schmatic of the sample structure used in the study of tunneling in a-DQWS by subpicosecond time resolved luminecence spectroscopy (Oberli et al [38]).

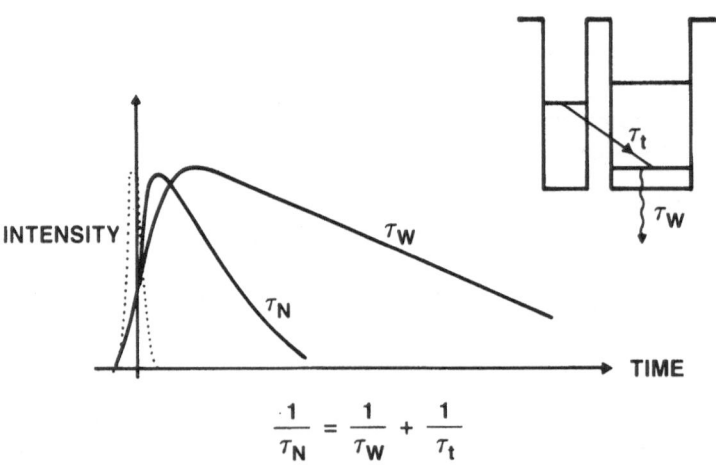

$$\frac{1}{\tau_N} = \frac{1}{\tau_W} + \frac{1}{\tau_t}$$

Fig. 13   Schematic of how tunneling times can be obtained from luminescence decay measurements.

The basic idea behind the experiment is to investigate the decay and rise times of luminescence from each well. Since tunneling provides an additional decay channel, the luminescence decay time of the well from which electrons tunnel out is smaller than the decay time of the other well. Therefore, a comparison of the decay times of the two wells gives a direct measure of the tunneling time. This is shown schematically in Fig. 13.

The recently developed luminescence upconversion technique [51] was used to measure time resolved luminescence spectra with sub-picosecond time resolution. Near resonant excitation with an infrared dye laser was used in these experiments. The time resolved luminescence spectra from a 60 Å barrier sample for two different delay are shown in Fig. 14. Both the narrow well and the wide well luminescence are clearly resolved. The narrow well luminescence decreases quickly with time whereas the wide well luminescence increases during this time. The tunneling times were determined as discussed above.

III.B.1 Non-Resonant Tunneling in a-DQWS

Consider first the case of zero applied bias so that there is no resonance between the electron levels of the two wells. The luminescence decay times for the wide well are longer than 500 ps for all samples under these conditions. However, the luminescence decay times for the narrow well are considerably shorter. This is a result of tunneling of electrons from the narrow well to the wide well. Note that the hole tunneling times are expected to be much longer and hole tunneling is not expected to play any part in these experiments (Fig. 5).

The dependence of non-resonant tunneling times of electrons on the barrier thickness are shown in Fig. 15. The non-resonant tunneling times increase rapidly as the barrier thickness increases. Under the conditions of the experiments, the separation between the two lowest electron levels was larger than an optical phonon energy so that optical phonon assisted tunneling processes are expected to dominate. Fig. 15 also shows the tunneling times for optical phonon (LO) assisted transitions between the two wells calculated by Oberli et al [38] using a simple model. Although the general trend is similar, the measured values are somewhat shorter than those calculated. The values calculated using Monte Carlo simulations [40] are shorter than these values. Clearly more investigations of this process are required to understand this tunneling in more detail.

Oberli et al [52] have also investigated the dependence of the tunneling rate when the energy separation between the two electronic states is tuned through an optical phonon resonance. Their results show a clear onset of optical phonon assisted tunneling. Optical phonon assisted tunneling processes have received considerable attention in recent years [53,54] and these experiments, being the first attempts to investigate these processes quantitatively, are very interesting. Further work is continuing in this direction.

III.B.2 Resonant Tunneling in a-DQWS

Oberli et al [38,49,52] also investigated resonant tunneling of electrons under the conditions when the n=2 level in the wide well is in resonance with the n=1 level in the narrow well. The variation of the luminescence decay time with applied bias is shown in Fig. 16. The decay time decreases strongly as the bias voltage is increased. This results from the increase in the tunneling rate as the n=1 level of the narrow well and the n=2 level of the wide well are brought into resonance. These results are summarized in Fig. 17 for the sample with 50 Å barrier width. The tunneling time remains nearly constant at low bias voltages; this is the *non-resonant tunneling time*. With increase in the reverse bias, there is first a *sharp reduction* in the tunneling time, followed by an *increase* in the tunneling time. This *non-monotonic* behavior of the tunneling time is a clear evidence for the tunneling resonance. Additional evidence for the resonance has been discussed earlier [49]. We note that the system time resolution for these measurements was ≈ 0.7 ps, so that the measured resonant tunneling time of 7 ps is not limited by the instrument response.

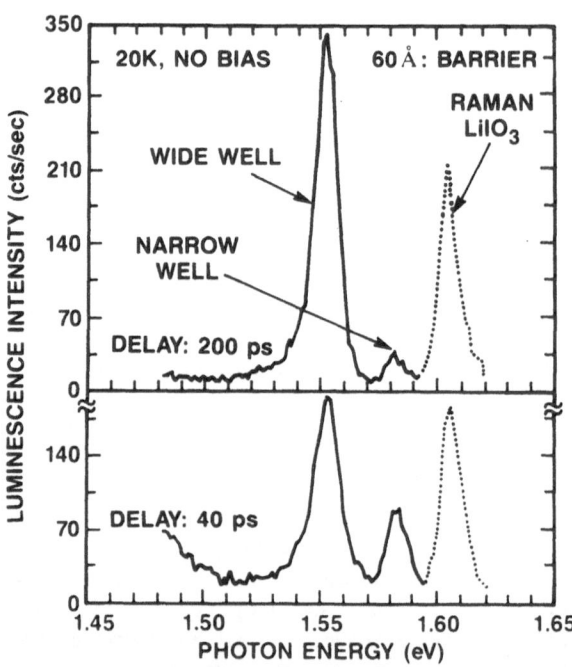

Fig. 14   Time resolved luminescence spectra from a-DQWS with a 60 Å barrier (from Oberli et al [38,49]).

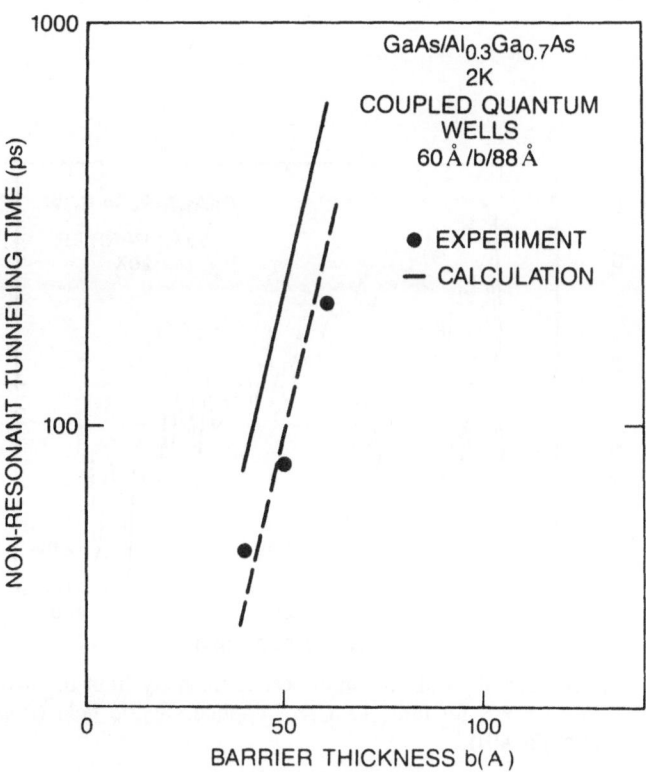

Fig. 15   The dependence of non-resonant tunneling times in a-DQWS as a
function of the barrier thickness. Also shown is an approximate
calculation of the phonon-assisted tunneling times (from Oberli et
al [38,49]).

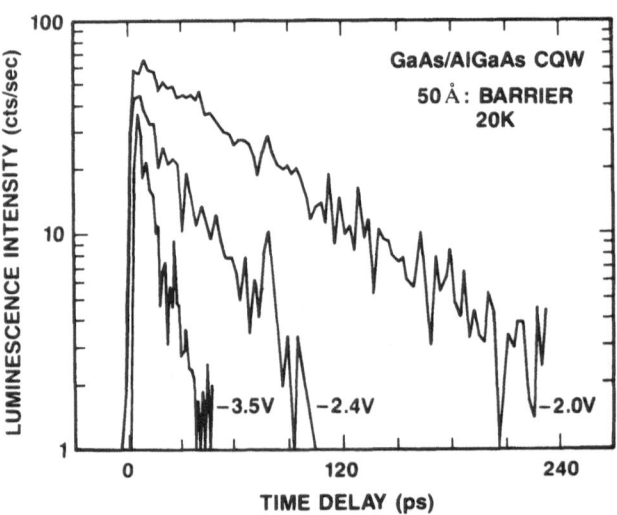

Fig. 16   Time evolution of the luminescence intensity from the narrow well in a-DQWS for three values of applied electric field (from Oberli et al [38,49]).

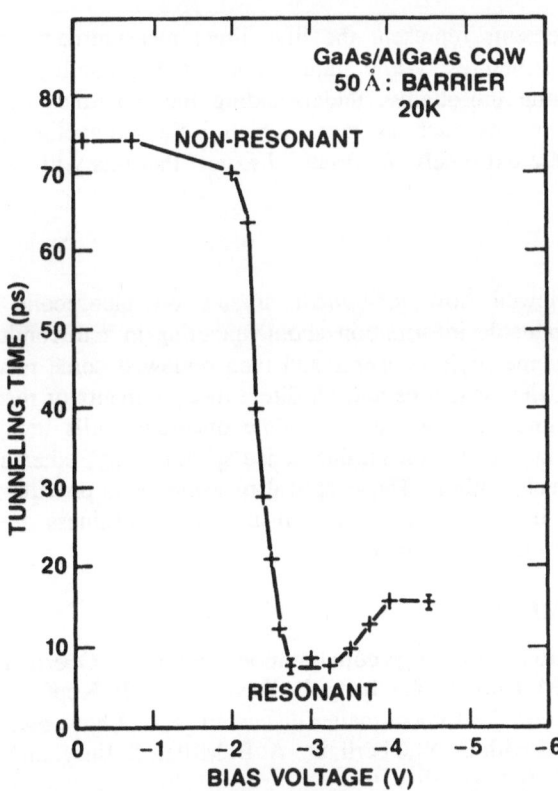

Fig. 17   The variation of the tunneling time from the narrow well to the the
           wide well in an a-DQWS as a function of an applied electric field.
           The non-monotonic behavior of the tunneling times demonstrates
           that the decrease is caused by a resonant process. The time
           resolution of the measurement system was ≈ 0. 7 ps (from Oberli
           et al [38,49]).

The significance of the tunneling time measured at resonance and its comparison with the coherent tunneling time been discussed at length Oberli et al [49]. That discussion may be summarized as follows: inhomogeneities in the well widths as well as in the electric field in the multiple double well structure contribute to the width of the resonance and may restrict the shortest tunneling time measured. However, the coherent tunneling time is expected to be ≈ 1 ps, considerably shorter than the measured resonant tunneling time. We believe that inter-subband scattering in the wide well may make an important contribution to the measured resonant tunneling time. If this is indeed correct, this technique may also shed some light on the currently controversial question of inter-subband scattering rates in quantum wells.

These measurements represent the first direct measurements of tunneling times in semiconductor microstructures. While static properties of tunneling in microstructures have been investigated quite thoroughly, understanding the dynamics of tunneling is equally important and experiments such as these which isolate tunneling and directly measure tunneling dynamics are extremely important. I expect that this will be an active and fruitful field of research.

IV. SUMMARY

We have reviewed how *all-optical* studies on picosecond and sub-picosecond timescales provide valuable information about tunneling in semiconductor microstructures. We have discussed some basic concepts and then reviewed some recent results on escape rates from double barrier structures and on direct measurements of resonant as well as non-resonant tunneling times in asymmetric double quantum wells using optical techniques. Although we have concentrated on luminescence spectroscopy, other optical techniques can also be applied for such studies. These optical measurements provide new insights into the physics of tunneling in such systems and illustrate the usefulness of optical techniques in investigating tunneling in microstructures.

V. ACKNOWLEDGMENTS

It is a pleasure to acknowledge collaborations with D. Y. Oberli, T. C. Damen, D. A. B. Miller, C. W. Tu, J. M. Kuo, T. Y. Chang, B. Deveaud, R. F. Kopf, N. J. Sauer, J. E. Henry and A. E. DiGiovanni on the subject matter discussed here. I have also benefited from many stimulating discussions with D. Y. Oberli, D. A. B. Miller, S. Luryi and F. Capasso.

REFERENCES

[1] L. Esaki, and R. Tsu, IBM J. Res. Dev. **14**, 61 (1970).

[2] L. Esaki, IEEE J. of Quantum Electronics **QE-22**, 1611-1624 (1986).

[3] T. C. L. G. Sollner, W. D. Goodhue, P. E. Tannenwald, C. D. Parker and D. D. Peck, Appl. Phys. Lett. **43**, 588-590 (1983); and T. C. L. G. Sollner, P. E. Tannenwald, D. D. Peck and W. D. Goodhue, Appl. Phys. Lett. **45**, 1319-1321 (1984).

[4] Federico Capasso, Khalid Mohammed and Alfred Y. Cho, IEEE J. of Quantum Electronics **QE-22**, 1853-1869 (1986).

[5] S. Tarucha, K. Ploog and K. von Klitzing, Phys. Rev. **36**, 4558 (1987).

[6] H. Schneider, K. von Klitzing and K. Ploog, J. de Physique **48**, Colloque C5-431 (1987).

[7] H. Schneider, K. von Klitzing and K. Ploog, presented at the Fourth Int'l conf. on Superlattices, Microstructures and Microdevices, Trieste, Italy (1988).

[8] A. G. R. Evans and P. N. Robson, Solid State Electronics **17**, 805 (1974)

[9] D. F. Nelson, J. A. Cooper, Jr. and A. R. Tretola, Appl. Phys. Lett. **41**, 857-859 (1982).

[10] R. A. Höpfel, Jagdeep Shah, D. Block and A. C. Gossard, Appl. Phys. Lett. **48**, 148-150 (1986).

[11] C. Minot, H. Le Person, F. Alexandre and J. F. Palmier, Physica **134B**, 514-518 (1985).

[12] J. F. Whittaker, G. A. Mourou, T. C. G. Sollner and W. D. Goodhue, Appl. Phys. Letters **53**, 385 (1988).

[13] K. J. Weingarten, M. J. Rodwell and D. M. Bloom, IEEE J. of Quantum Electronics **QE-24**, 198 (1988).

[14] C. V. Shank, R. L. Fork, B. I. Greene, F. K. Reinhart, and R. A. Logan, Appl. Phys. Letters **38**, 104 (1981).

[15] G. Livescu, D. A. B. Miller, T. Sizer, D. J. Burrows, J. E. Cunningham, A. C. Gossard and J. H. English, Appl. Phys. Letters **54**, 748 (1989). see also, Topical Meeting on Quantum Wells for Optics and Optoelectronics, 1989 Technical Digest Series, Vol. 10, 254 (1989).

[16] B. Deveaud, Jagdeep Shah, T. C. Damen, B. Lambert, A. Chomette and A. Regreny, IEEE J. of Quan. Electronics **QE-24**, 1641 (1985).

[17] A. Forchel, B. Laurich, H. Hillmer, G. Tränkle and M. Pilkuhn, J. of Luminescence **30**, 67-81 (1985).

[18] D. J. Westland, D. Mihailovic, J. F. Ryan and M. D. Scott, Appl. Phys. Lett. **51**, 590-592 (1987).

[19] A. Chomette, B. Deveaud, J. Y. Emery, A. Regreny and B. Lambert, Solid State Commun. **54**, 75-78 (1985).

[20] E. R. Brown, C. D. Parker, T. C. L. G. Sollner, C. I. Huang, and C. E. Stutz, Proceedings of Topical Meeting on Picosecond Electronics and Optoelectronics, Salt Lake City, 1986.

[21] S. Luryi, Appl. Phys. Letters **47**, 490 (1985).

[22] T. Weil and B. Vinter, Appl. Phys. Letters **50**, 1281 (1987).

[23] M. Jonson and Anna Grincwajg, Appl. Phys. Letters **50**, November 1987.

[24] K. W. H. Stevens, J. Phys. **C16**, 3649 (1983).

[25] M. Buttiker and R. Landauer, Physica Scripta **32**, 429 (1985).

[26] S. Collins, D. Lowe, and J. R. Barker, J. Phys. C: Solid State Phys. **20**, 6213 (1987).

[27] W. R. Frensely, Proceedings of the Fifth International Conference on Hot Carriers in Semiconductors, Boston (1987); edited by Jagdeep Shah and G. J. Iafrate, Pergammon Press (1988); p. 739.

[28] N. C. Kluksdahl, A. N. Kriman, C. Ringhofer, and D. K. Ferry, Proceedings of the Fifth International Conference on Hot Carriers in Semiconductors, Boston (1987); edited by Jagdeep Shah and G. J. Iafrate, Pergammon Press (1988); p.743.

[29] H. Guo, K. Diff, G. Neofotistos, and J. D. Gunton, Appl. Phys. Letters **53**, 131 (1988).

[30] H. C. Liu, Appl. Phys. Letters **52**, 453 (1988).

[31] E. E. Mendez, E. Calleja and W. I. Wang, Appl. Phys. Letters **53**, 977 (1988).

[32] H. C. Liu, Appl. Phys. Letters **51**, 1019 (1988).

[33] A. C. Marsh, Semicond. Sci. and Technol. **1**, 320 (1986).

[34] H. Q. Le, J. J. Zayhowski, and W. D. Goodhue, Appl. Phys. Letters **50**, 1518 (1987).

[35] S. Charbonneau, M. L. W. Thewalt, E. Koteles and B. Elman, Phys. Rev. **B38**, 6287 (9188).

[36] M. N. Islam, R. L. Hillman, D. A. B. Miller, D. S. Chemla, A. C. Gossard and J. H. English, Appl. Phys. Letters **50**, 1098 (1987).

[37] R. Sauer, K. Thonke and W. T. Tsang, Phys. Rev. Letters **61**, 609 (1988).

[38] D. Y. Oberli, Jagdeep Shah, T. C. Damen, C. W. Tu and D. A. B. Miller, Topical Meeting on Quantum Wells for Optics and Optoelectronics, 1989 Technical Digest Series, Vol. 10, 272 (1989).

[39] R. Ferreira and G. Bastard, to be published.

[40] S. Goodnick and P. Lugli, to be published.

[41] S. Luryi, Sol. St. Commun. **65**, 787 (1988).

[42] S. Luryi, presented at the Fourth Int'l conf. on Superlattices, Microstructures and Microdevices, Trieste, Italy (1988).

[43] E. O. Kane, in Tunneling in Solids, edited by C. B. Duke, Supplement 10 of Solid State Physics: advances in research and applications, Academic Press New York (1969).

[44] B. Deveaud, Jagdeep Shah, T. C. Damen, B. Lambert and A. Regreny, Phys. Rev. Lett. **58**, 2582-2585 (1987).

[45] A. Miller, R. J. Manning, and P. J. Bradley, presented at the Topical Meeting on Quantum Wells for Optics and Optoelectronics, 1989 Technical Digest Series, Vol. 10, 38 (1989).

[46] M. Tsuchiya, T. Matsusue, and H. Sakaki, Phys. Rev. Letters **59**, 2356 (1987).

[47] M. K. Jackson, M. B. Johnson, D. H. Chow, T. C. McGill, and C. W. Nieh, Appl. Phys. Letters **54**, 552 (1989).

[48] T. B. Norris X. J. Song, W. J. Schaff, L. F. Eastman, G. Wicks and G. A. Mourou, Appl. Phys. Letters **54**, 60 (1989).

[49] D. Y. Oberli, Jagdeep Shah, T. C. Damen, C. W. Tu, T. Y. Chang, D. A. B. Miller, J. E. Henry, R. F. Kopf, N. Sauer, and A. E. DiGiovanni, to be published. see also Proceedings of Topical Meeting on Picosecond Electronics and Optoelectronics, Salt Lake City (1989), to be published.

[50] M. Tsuchiya, T. Matsusue, and H. Sakaki, Proceedings of the 6th International conference on Ultrafast Phenomena, edited by T. Yajima, K. Yoshihara, C. B. Harris and S. Shionoya, Springer-Verlag, Berlin (1988); p. 304. see also, T. Matsusue, M. Tsuchiya, and H. Sakaki, Topical Meeting on Quantum Wells for Optics and Optoelectronics, 1989 Technical Digest Series, Vol. 10, 266 (1989).

[51] Jagdeep Shah, IEEE J. of Quan. Electronics **QE-24**, 276 (1988).

[52] D. Oberli, Jagdeep Shah, T. C. Damen, and J. M. Kuo, in the Proceedings of Topical Meeting on Picosecond Electronics and Optoelectronics, Salt Lake City (1989), to be published.

[53] N. S. Wingreen, K. W. Jacobsen and J. W. Wilkins, Phys. Rev. Letters **61**, 1396 (1988).

[54] B. Y. Gelfand, S. Schmitt-Rink, and A. F. J. Levi, Phys. Rev. Letters **62**, 1683 (1989).

MONTE-CARLO SIMULATION OF FEMTOSECOND CARRIER RELAXATION IN SEMICONDUCTOR

QUANTUM WELLS

Stephen M. Goodnick

Center for Advanced Materials Research
Oregon State University
Corvallis, Oregon 97331

INTRODUCTION

Ultra-fast optical studies in bulk semiconductors and quantum well systems have provided a great deal of insight into the dynamics of nonequilibrium phenomena on the picosecond and subpicosecond time scale. Typically such experiments are variants of the pump and probe type in which a short pump pulse is used to generate a nonequilibrium electron-hole plasma in the solid, and time delayed probe pulses are used to characterize the decay back to equilibrium. Quantum well systems in particular have proved quite interesting due to the high quality of samples grown using epitaxial growth techniques and the effects of carrier confinement which enhance certain optical phenomena such as excitonic absorption (Chemla et al., 1988). Quantum well systems have now been extensively studied using time resolved photoluminescence, Raman, and absorption spectroscopy which have increased our understanding of the effects of nonequilibrium phonons, intercarrier scattering, intersubband scattering, and tunnelling phenomena in quasi-two-dimensional systems (see for example the review by Shah, 1986).

Simulation of ultra-fast optical excitation experiments using numerical Monte Carlo particle methods has provided a rigorous theoretical tool with which to understand the complicated carrier dynamics occurring during photoexcitation. In ultra-fast optical studies of semiconductor systems, the principle experimental observables depend strongly on the distribution functions of electrons and holes, and hence provide a sensitive indicator of nonequilibrium behavior. The strength of the Monte Carlo technique lies in the fact that the carrier distribution functions are calculated directly for the nonequilibrium state with a minimum of assumptions regarding the form of the distributions. The drawback to this technique is that in order to include the proper physics into the simulation, extensive computer resources are required which usually precludes an understanding of the experimental results in terms of a simple analytical model. Indeed, Monte Carlo simulation is most frequently employed in a deductive, experimental fashion. That is, given a model that correctly incorporates all the relevant physics of the system, one may selectively 'turn off' various physical mechanisms to deduce which one is most important in reproducing actual experimental results. Examples of this approach will be discussed further in relation to specific ultra-fast experiments.

In the present article, we discuss the use of Monte Carlo simulation in gaining a deeper insight into the results of ultra-fast spectroscopic studies of quantum well systems. We first begin with a discussion of the numerical technique and the relevant scattering mechanisms which are necessary in order to compare to photoexcitation experiments in the following section. This discussion will be followed by simulated results for different types of relaxation observed experimentally in quantum well systems. First we consider energy relaxation and the role that hot phonons play in the long time constant energy loss rate observed experimentally in such systems (Ryan et al., 1984). We then will consider the role of intercarrier scattering in the apparent thermalization of electrons observed using low energy excitation in quantum wells (Knox et al., 1986, 1988). Finally, we simulate relaxation via intersubband scattering and the influence of the envelope function overlap in controlling the relaxation time.

MONTE CARLO SIMULATION

The Monte Carlo method itself is quite general and has found wide-spread use in a variety of applications ranging from theoretical physics to the design of integrated circuit interconnections (Kalas and Whitlock, 1986). This technique has been successfully applied to problems related to semiconductor transport for over two decades, and has been the subject of several excellent reviews (Jacoboni and Reggiani, 1983; Price, 1979). We will only outline the basic algorithm in the following.

The idea of a Monte Carlo simulation is to simulate the random walk of particles in a crystal subject to external forces and scattering events due to imperfections, phonons, etc. The trajectory of each particle is broken up into free flights during which the motion is only influenced by external forces, and collisions in which the momentum and energy of the particle are instantaneously changed. This Ansatz for the particle dynamics is fundamentally the same as that employed in the derivation of the Boltzmann transport equation. Hence the Monte Carlo method is often referred to as a numerical solution of this equation, although it is in fact more general.

The probability, $P(t)dt$, of a particle making a free flight for a time t after the previous collision and then scattering in a small time interval dt about time t, may be written

$$P(t)dt = S(k(t))e^{-\int S(k(t'))dt'} dt \qquad (1)$$

where $\hbar k(t)$ is the particle crystal momentum, and $S(k(t))$ is the scattering rate. Here $S(k(t))dt$ is the probability of scattering in dt around t and the exponential is the probability of not having suffered a collision after time t. The introduction of self-scattering (Boardman et al., 1968) allows this expression to be expressed analytically so that the probability in (1) may be written

$$P(t)dt = Se^{-St}dt . \qquad (2)$$

where S is the sum of the actual scattering rate, $S(k(t))$, and the self-scattering rate, $S_o(k(t))$, which adjusts itself so that S is independent of time. Random flight times may be generated according to the probability distribution in (2) using uniformly distributed random numbers between 0 and 1 by inversion of (2) to give

$$t_r = 1/S \ln(r) \tag{3}$$

where r (which is a uniform random number between 0 and 1) is usually
generated using the intrinsic random number generator of the computer.

The physics of the system under consideration are included through
the scattering rate, $S(k(t))$, which is calculated for each scattering
mechanism using time dependent perturbation theory. Once the scattering
cross-section as a function of energy is calculated, the flight times and
final state after each collision are generated stochastically. For
transient problems such as that of carrier relaxation in quantum well
systems, an ensemble of particles must be simultaneously simulated in
which measured quantities such as the occupancy of states or the average
energy are calculated as averages over the ensemble of particles. In
order to synchronize the motion of these particles during the simulation,
the time scale is discretized and a time step is introduced into the
simulation which is usually short enough that the time evolution of the
system over this interval is small.

For quantum well systems, the scattering rates for various mechanisms
such as polar optical phonons and intercarrier scattering must be calculated between initial and final quantized states in the well. Thus it is
necessary to make some approximation for the envelope functions representing the effect of the band offsets on the underlying Bloch functions of
the crystal. As will be discussed later, these envelope functions are of
particular importance in determining the intersubband relaxation rate.

Quantum Well States

For the GaAs/Al$_x$Ga$_{1-x}$As system, the conduction band states in a
quantum well system are adequately described in the effective mass
approximation (see for example Weisbuch, 1987). We assume the envelope
functions are separable as

$$\Psi(R) = \zeta_i(z)e^{ik \cdot r}/A \ , \tag{4}$$

where R is the position vector, r and k are the position and wave vectors
in the plane parallel to the well, z is the normal direction, and A is the
normalization area. The envelope function in the z direction satisfies a
one-dimensional Schroedinger equation with the potential given by the
position dependent conduction band edge and electrostatic contributions
due to ionized impurities and free carriers(in the Hartree approximation).
For many situations, particularly in lightly doped systems, solutions to a
finite square well potential are adequate. However, for modulation doped
samples in which the ionized impurities are spatially separated from the
confined carriers, self-consistent solutions are necessary. Fig. 1 shows
the comparison of the solutions for a finite square well and the same
system including self-consistent solutions of Poisson's equation. The
wells in Fig. 1 are 50Å wide separated by a 400Å modulation doped barrier.
Two things are important to note. For scattering processes involving
electrons residing in the lowest subband, the scattering rates are only
weakly influenced by inclusion of the self-consistent potential due to the
modulation doping. However, if excitation to higher subbands occurs such
as in the infra-red absorption experiments of Seilmeier (1987) discussed
later, dramatic effects in the calculated relaxation time occur depending
on the model used due to the spatial localization of the wavefunction in
the barrier material.

For valence band states in a quantum well system, even the effective
mass approximation is complicated due to band mixing effects (Weisbuch,
1987). In particular, there is not a clear distinction between light-hole

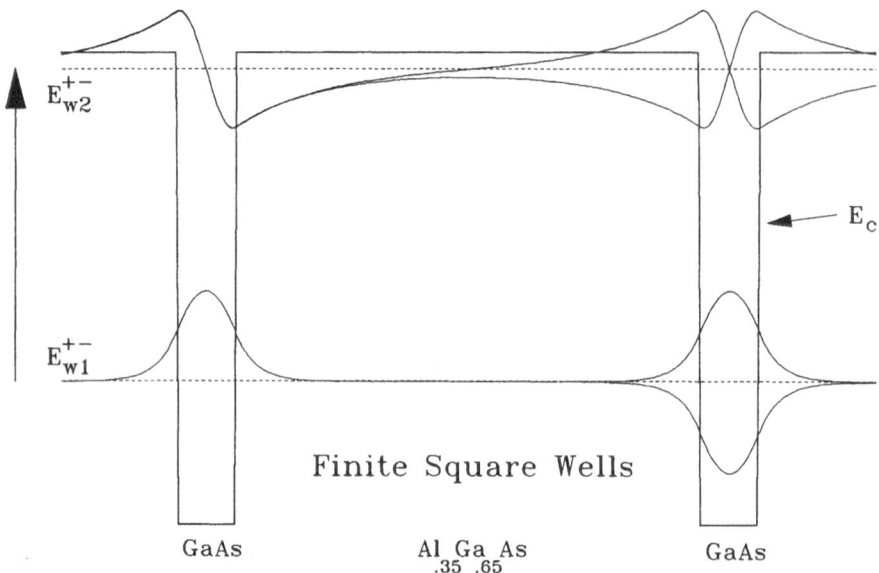

Fig. 1a. Envelope functions, potential variation and subband energies for two 50Å GaAs wells separated by a 400Å Al$_{.35}$Ga$_{.65}$As barrier.

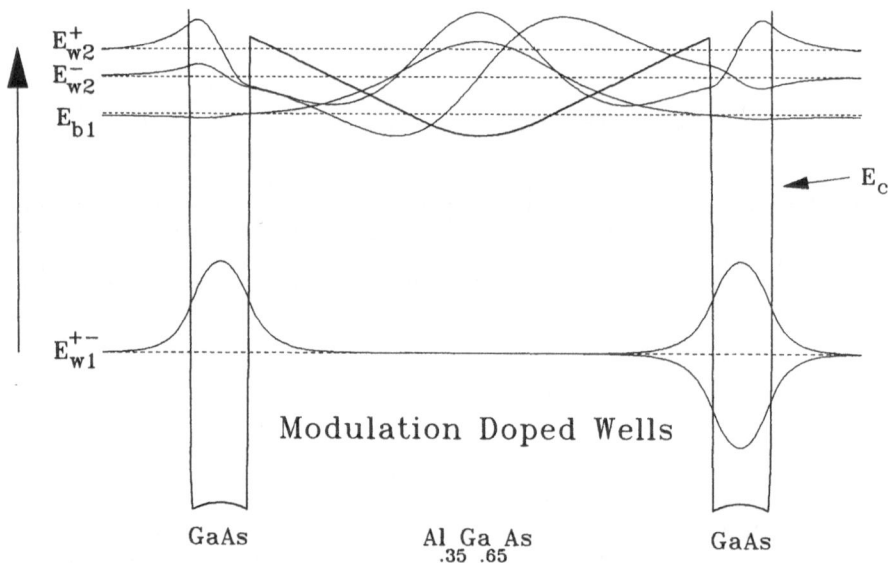

Fig. 1b. Same barrier and well thicknesses as Fig. 1a but modulation doped such that a density of 5x10^{11}/cm^2 electrons reside in the wells.

and heavy-hole states as in the bulk case. Although the influence of
holes is often neglected in the interpretation of optical experiments,
equal numbers of electrons and holes are generated during the initial
pulse, and thus their effect should properly be included. As a first
approximation to their effect in quantum well systems for some simulations, we include holes assuming parabolic bands.

Scattering Mechanisms

As mentioned earlier, the scattering rates calculated using time
dependent perturbation theory provide the primary connection between the
Monte Carlo simulation and the physical system being modeled. For carrier
relaxation in III-V semiconductors, the primary mechanisms which we
consider are polar optical(LO) and deformation potential optical(TO)
scattering, intervalley scattering, and intercarrier scattering. We have
included impurity, acoustic phonon, and surface roughness scattering in
various simulations, but their effect has proven unimportant in comparing
to photoexcitation experiments.

Polar Optical Scattering. The polar optical interaction is one of
the most important scattering mechanisms in polar materials such as GaAs
and is calculated via the Fröhlich Hamiltonian. In quantum well systems,
this mechanism has been calculated assuming that phonons are bulk-like
(Riddoch and Ridley, 1983) or are characterized by slab modes (Riddoch and
Ridley, 1985). Here we treat the phonons as bulk-like (Goodnick and
Lugli, 1988a), although little difference has been found in the carrier
scattering rates when slab modes were included for relatively wide
wells(greater than 150Å).

If one assumes only bulk screening of longitudinal modes, then the
matrix element for scattering is given by (Riddach and Ridley, 1983)

$$|<k \pm q, j|H|k, i>|^2 = \frac{2\pi\hbar^2 e E_0 (q^2 + q_z^2)}{V m^* (q^2 + q_z^2 + q_s^2)^2} \times \left(n_{\omega_0} + \frac{1}{2} \mp \frac{1}{2}\right) |G_{ij}(q_z)|^2 , \tag{5}$$

where q and q_z are the parallel and normal components with respect to the
well of the phonon wave vector, q_s, is the 3D inverse screening length, m^*
is the effective mass, i and j denote the initial and final subband
indices, and the effective field $e E_0$ is given by

$$e E_0 = \frac{m^* e^2 \hbar \omega_0}{\hbar^2} \left(\frac{1}{\kappa_\infty} - \frac{1}{\kappa_0}\right) \tag{6}$$

with κ_0 and κ_∞ the low- and high-energy dielectric constants, and $\hbar\omega_0$ is
the phonon energy. The overlap integral $|G_{ij}(q_z)|^2$ is given by

$$|G_{ij}(q_z)|^2 = \int_{-\infty}^{\infty} dz \int_{-\infty}^{\infty} dz' \rho_{ij}(z) \rho_{ij}^*(z') e^{iq_z(z-z')} , \tag{7a}$$

$$\rho_{ij}(z) = \zeta_i^*(z)\zeta_j(z) , \tag{7b}$$

between initial and final subbands. The overlap integral is controlled by
the orthogonality of the subband envelope functions and the degree of
overlap of these functions. If the overlap is small, as in the case of
intersubband scattering in coupled wells, the intersubband rate may be
quite small.

The calculated scattering rate due to bulk–like polar optical phonons at 300K for a 150Å well is shown in Fig. 2. The three dimensional scattering rate is included for reference. As seen in Fig. 2, the 2D scattering rate is a piece–wise approximation for the 3D scattering rate, with the discontinuities representing the onset of intersubband emission and absorption when the carrier energy is an LO phonon energy away. The magnitude of the intersubband scattering rate compared to the intrasubband rate is indicated by the relative height of the discontinuity. The largest rate in Fig. 2 corresponds to intersubband emission occurring for subband 1 to subband 2 which is seen to be about 25% of the intrasubband rate for this particular geometry.

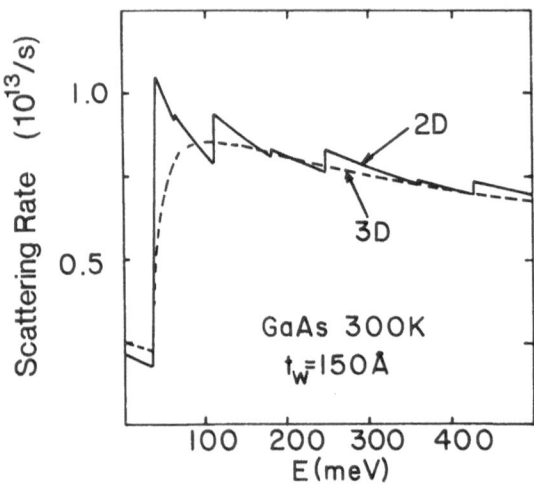

Fig. 2.   Total scattering rate due to polar optical phonons (bulk) for a 150Å well for electrons in the first subband (solid line). The dashed line is the 3D scattering rate (Goodnick and Lugli, 1988a).

Strong emission of polar optical phonons by heated electrons in a semiconductor, particularly during optical excitation, drives the phonon population out of equilibrium. Since the electron and hole emission and absorption rates for such phonons depend strongly on the phonon occupancy, buildup of a nonequilibrium phonon distribution feeds back into the electron system modifying the scattering rates which in turn affects the phonon population. In order to simulate this coupled behavior of elec-trons and phonons, we simulate the phonon dynamics simultaneously with that of the electrons (Lugli and Goodnick, 1987). For each emission and absorption event during the simulation we modify the phonon population at the particular crystal momentum of the event. The phonon population is assumed to decay due to anharmonic processes via a phenomenological decay time

$$\frac{\partial N_q}{\partial t}\bigg|_{ph-ph} = -\frac{N_q - N_L}{\tau_{op}} \tag{8}$$

where τ_{op} is the phonon lifetime taken as 7ps from bulk GaAs Raman studies (Von der Linde et al., 1980), N_q is the phonon occupancy for a parallel momentum $\hbar q$, and N_L is the equilibrium occupancy given by

$$N_L = (e^{\hbar\omega_0/k_B T_L} - 1)^{-1} \qquad (9)$$

where ω_0 is the optical phonon frequency and T_L the lattice temperature. Numerically, Eq. (9) is implemented using a finite difference scheme with the discretization occurring over the simulation time step. Sufficiently small time steps are required so that the phonon population does not change to rapidly over one timestep.

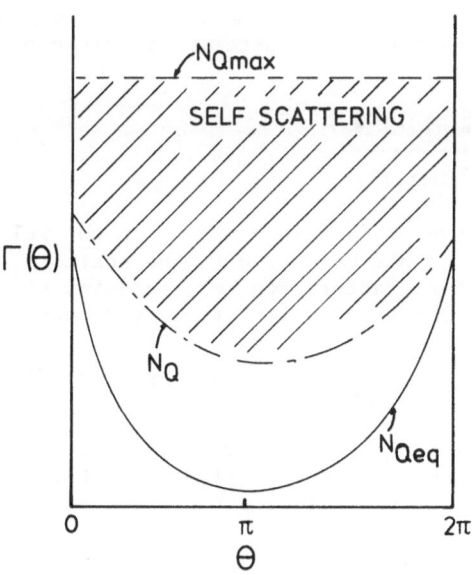

Fig. 3. Final state selection for nonequilibrium polar optical phonon scattering.

To account for the temporal modification of the electron–phonon scattering rate as the simulation proceeds, a self scattering technique is employed as shown in Fig. 3. A maximum phonon occupancy is assumed at the start of the program which is greater than the maximum value that the actual distribution function ever reaches. The total scattering rate (integrated over the scattering angle in Fig. 3) is calculated with the maximum occupancy, and when the final state is chosen during the simulation, a random number is chosen between the maximum value and zero. If this number lies below the curve of the instantaneous scattering rate in Fig. 3, the scattering event is accepted as valid, otherwise it is treated as self–scattering.

<u>Intercarrier Scattering</u>. Carrier–carrier (CC) scattering plays an important role in the relaxation of a nonequilibrium distribution back to

equilibrium. In fact, it is precisely the two particle interaction between carriers in an isolated system which is included in the collision integral of the Boltzmann transport equation in order to prove the now famous H-theorem (see for example Tolman, 1938) governing relaxation of an arbitrary distribution to an equilibrium state. Since this mechanism is inelastic, energy is exchanged between different carriers in the system which broadens the initial photoexcited distribution, and allows a redistribution of energy between electrons and holes which are initially created with unequal excess energies due to the differences in effective mass. The CC scattering rate is usually calculated using the Born approximation assuming a two-particle screened Coulomb interaction (Goodnick and Lugli, 1988a). For quasi-two-dimensional systems, the CC scattering rate (electron-electron, electron-heavy hole, etc.) may be generalized from the form given for electron-electron (Goodnick and Lugli, 1988a) with the inclusion of the relative mass between different carrier types. For two-particle interactions, we consider the first particle in the well with wave vector k in subband i and band u, scattering from a second particle of wave vector k_0 in subband j in band v. Neglecting interband transitions, the final states of these two particles are k,m and u for the first particle, and k'_0,n and v for the second particle. The matrix element squared for this scattering is given by (Goodnick and Lugli, 1988a)

$$|H^{CC}|^2 = \frac{4\pi^2 e^4 \delta(\mathbf{k'}+\mathbf{k'_0}-\mathbf{k}-\mathbf{k_0})}{A^2 \kappa^2 (q+q_0)^2} \ |F^{uv}_{ijmn}(q)|^2 \tag{10}$$

where κ is the relative dielectric constant in the well, q_0 is the inverse screening length in two dimensions, q is the scattered wave vector, $q = |\mathbf{k}-\mathbf{k'}|$, and the form factor, $F^{uv}_{ijmn}(q)$, contains the overlap integral of the subband envelope function for the respective bands

$$F^{uv}_{ijmn}(q) = \int_{-\infty}^{\infty} dz \int_{-\infty}^{\infty} dz' \zeta^u_i(z)\zeta^v_j(z') \ \zeta^{*u}_m(z)\zeta^{*v}_n(z') e^{-q|z-z'|} \ . \tag{11}$$

The inverse screening length, q_0, is obtained from the matrix dielectric function in the 2D random phase approximation (RPA) by assuming diagonal static screening in the long wavelength limit (Goodnick and Lugli, 1988b)

$$q_0 = \sum_{i,n} q_{in} \ f_{in}(0) \ , \tag{12}$$

where n is the band index (electrons, heavy holes, etc.) and i is the subband index within each band. The constant q_{in} is given by

$$q_{in} = m^*_{in}/\pi h^2 (2\pi e^2/\kappa) \ , \tag{13}$$

with κ the semiconductor dielectric constant. The occupancy, $f_{in}(0)$, of the bottom of the band is calculated at every time step (5 fs) during the simulation and used to update the CC rate which changes dynamically during photoexcitation. The total scattering rate for the first particle may be written (assuming no correlation with the position of the other particles)

$$\Gamma^{uv}_{im}(k) = \frac{4\pi e^4 m_r}{\hbar^3 A \kappa^2} \sum_{k_0,j,n} f^v_j(k_0) \int_0^{2\pi} d\theta \ \frac{|F^{uv}_{ijmn}(q)|^2}{(q+q_0)^2} \ , \tag{14}$$

where m_r is the relative mass, $m_r = 2m_u m_v/(m_u+m_v)$, and $f^v_j(k_0)$ is the distribution function for carriers in subband j and band v. The form factor

(11) appearing in (14) greatly diminishes the intersubband scattering rate due to the orthogonality of the wave functions.

We include CC scattering into the QW Monte Carlo simulation using a generalization of the self-scattering technique proposed by Brunetti et al. (1985) for the bulk. Here the partner electron for scattering is chosen at random from the ensemble, and the self-scattering rejection technique is used to account for the variable scattering rate in (14) due to changes in $f_j^v(k_0)$ and q_0 as the simulation progresses.

Other Scattering Mechanisms. As mentioned earlier, we have included other scattering mechanisms into the simulation such as intervalley, acoustic and impurity scattering. Apart from intervalley scattering (and TO phonon scattering for holes), other mechanisms, particularly elastic scattering, play little role in relaxation. Intervalley and TO phonon scattering are calculated assuming initial and final quantized states in the well and assuming a deformation potential interaction.

The influence of final state occupancy is included in the Monte Carlo code using a self-scattering technique (Bosi and Jacoboni, 1976) in which the occupancy of states in k space is tabulated from the ensemble. This effect becomes particularly important at low temperatures and high densities.

Laser Excitation Model. In order to simulate the effect of band to band optical transitions during laser excitation, electrons and holes are added to the simulation according to the temporal generation rate of the pulse. Various degrees of sophistication are possible for the interaction of light with the solid. Here we simply assume that the temporal generation of carriers during ultra-fast laser excitation may be given experimentally by

$$G(t) = I_0 \cosh^{-2}(2.634t/t_p) \tag{15}$$

where I_0 is the incident intensity which we choose to match the total number of injected carriers, and t_p is the half-width of the pulse. Extra particles are added to the simulation at each time step according to (15). To account for spectral broadening of the pulse, the carriers are introduced according to a Gaussian distribution 20 meV around the injection energy.

SIMULATION OF ULTRA-FAST EXPERIMENTS

We have used the Monte Carlo technique discussed in the previous section to elucidate the results of various ultra-fast optical experiments in semiconductor quantum well systems. Some or all of the scattering mechanisms discussed previously are included for a given simulation depending on what physical mechanisms are relevant for understanding the experimental results. The number of particles used in the simulation varies according to the accuracy that is needed. Calculations in which only simple averages are required such as the mean energy or particle number as a function of time need several thousand particles in order to give an acceptable statistical error. More detailed calculations of the carrier distribution functions may require as many as 50,000 particles. Material parameters for GaAs and AlGaAs are essentially those given by Adachi (1986).

We are primarily interested in the physics of carrier relaxation, so that experimental estimates of the injected carrier density, spectral width, and the temporal evolution of the laser pulse are necessary inputs

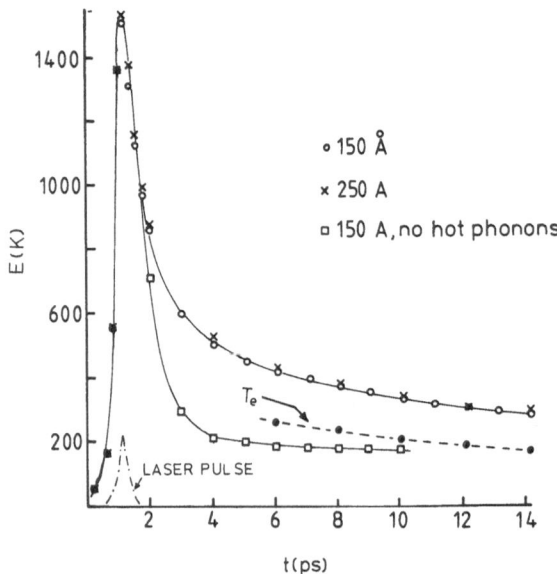

Fig. 4.   Average total kinetic energy versus time for $2.5 \times 10^{11}/cm^2$
electrons injected at 0.25 eV excess energy for two different
well widths with (o,x) and without (□) hot phonons included.
Additional curve (•) shows the variation of the electron
'temperature' fit assuming a heated Fermi—Dirac distribution.

to the simulation. In many cases, these quantities are only known within
50–100% error, so that quantitative comparison of Monte Carlo results to
experiment is often poor at best.

Nonequilibrium Phonon Effects

Results from time—resolved studies of hot—carrier relaxation in GaAs—
AlGaAs quantum wells have shown that the cooling rates for photoexcited
carriers are much longer than are expected from simple consideration of
the bulk electron—LO phonon interaction (Shah et al., 1985; Ryan et al.,
1986). Several possibilities exist for a reduced electron—phonon coupling
including confinement effects by the quantum well, screening of the
electron—LO phonon interaction and nonequilibrium phonons.

To elucidate the effect of a nonequilibrium phonon distribution, we
simulated high energy electron injection into a square quantum well (Lugli
and Goodnick, 1987) using parameters similar to those used experimentally
in studying carrier cooling. Fig. 4 shows the result for 150Å and 250Å
wide wells at a lattice temperature of 5K and an injection energy 0.25 eV
above the lowest subband energy in the conduction band. Holes were not
considered in these simulations. A cold background density of
$2.5 \times 10^{11}/cm^2$ electrons is assumed and an injected density of $2.5 \times 10^{11}/cm^2$.
For the simulation which includes nonequilibrium phonons, the energy (in
units of kT) decays rapidly after the initial pulse (shown in Fig. 4) but
then develops a long time tail. In contrast, without hot phonons, the
distribution decays after 5 ps to the equilibrium energy associated with a
cold electron gas of $5 \times 10^{11}/cm^2$ carriers at 5K. Since the electrons are
fermions, the average energy (kT in two dimensions) is not the same as the
electron temperature; rather it approaches a value which is half the Fermi

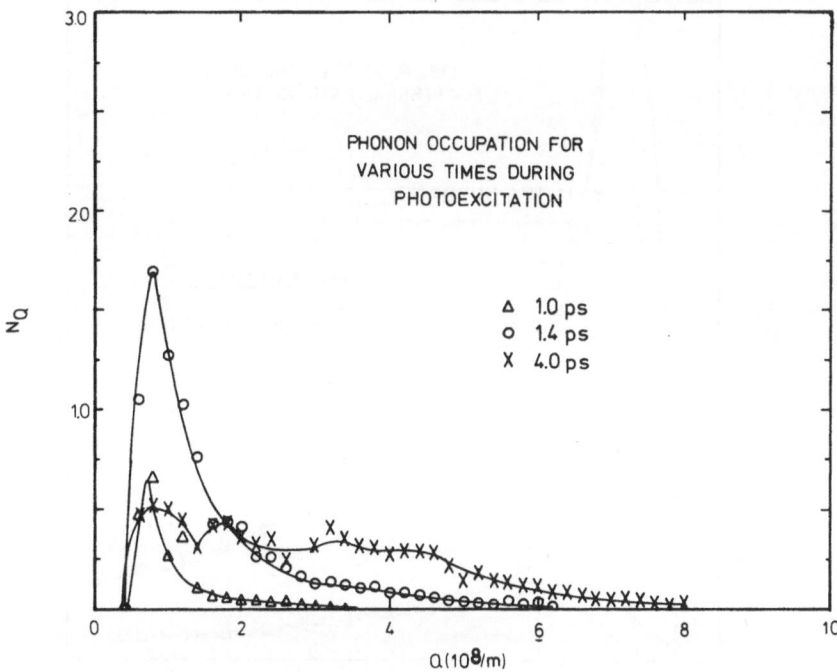

Fig. 5.   Density of occupied phonon modes as a function of total parallel
           momentum for times during and after laser excitation (Lugli and
           Goodnick, 1987).

energy at OK. Shown for comparison in Fig. 4 is the electron 'temperature' as a function of time corresponding to the upper curve (i.e. with
hot phonons) which is smaller than the average energy due to the effects
of degeneracy.

    In Fig. 5 we show the phonon distribution as a function of the
parallel wavevector ($q_z = 0$) for various times during and after the laser
pulse. As shown in Fig. 5, the initial buildup at small q of the nonequilibrium phonon distribution occurs due to phonon emission during the
initial relaxation of high energy carriers. At later times, reabsorption
and phonon decay reduce the distribution at low q while at larger values
of the phonon wavevector, the population continues to grow due to phonon
emission by cooler carriers.

    The effect of the phonon distribution on the electron polar optical
phonon scattering rate is shown in Fig. 6. Phonon absorption is proportional to N_q while emission depends on N_q+1 due to contributions of both
stimulated and spontaneous emission. Normally spontaneous emission
dominates since $N_q \approx 0$ at low temperatures. Fig. 6 illustrates the effect
which is responsible for the long time constant decay in energy observed
experimentally. As N_q becomes large, the emission and absorption rates
become nearly equal due to the increase in N_q. A simple increase in N_q
itself does not reduce the energy loss rate. However, after the distribution has cooled somewhat, most of the carriers reside with energies below
the emission threshold (36 meV) and hence absorption of hot phonons is
enhanced over emission. Since these two rates are now nearly equal, the
energy loss rate from the electron system is severely restricted, and the
primary loss mechanism from the coupled system is from phonon decay.

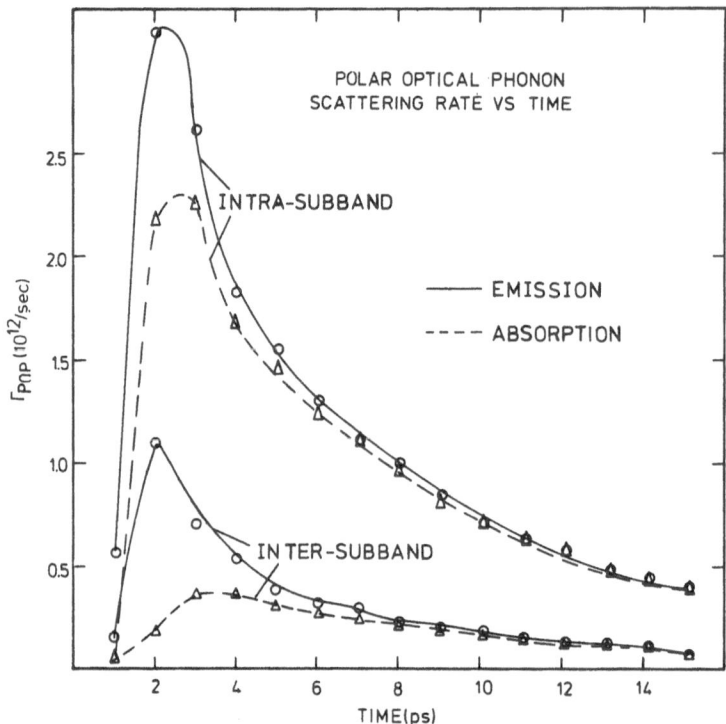

Fig. 6. Polar optical scattering rate versus time tabulated from
scattering events during the simulation. The rates due to both
intra- and intersubband emission and absorption are shown.

Intercarrier Relaxation

In the experiments of Knox and co-workers (1986; 1988), pump and
probe absorption spectroscopy was used to study the relaxation kinetics of
optically excited carriers in undoped and modulation doped quantum wells.
For nonexcitonic band-to-band absorption, the absorption coefficient
depends on the number of occupied initial states and the number of
unoccupied final states. For low-level excitation, the differential
transmission, $\Delta T/T_0$, is equal to the sum of the electron and hole distributions

$$\Delta T(h\nu, t)/T_0 = f_e(\mathbf{k}, t) + f_h(\mathbf{k}, t) \ , \tag{16}$$

where $f_e(\mathbf{k}, t)$ and $f_h(\mathbf{k}, t)$ are the time-dependent electron and hole distribution functions for a wave vector $\mathbf{k}$ corresponding to a photon energy $h\nu$.
In the experiments of Knox and co-workers (1986), a peak in the differential transmission was observed centered at the pump energy at t=0 which
evolved into a Maxwellian shape after about 200 fs. Based on equation
(16) above, the evolution of the differential transmission spectrum
corresponds to the relaxation of electrons and holes from their initial
nonequilibrium distribution to that of a thermalized Maxwell-Boltzmann
distribution. In these experiments, the pump energy was chosen such that
the excess electron and hole energies lie below the threshold for optical
phonon emisssion. Hence, the observed relaxation in these studies was
argued to result from intercarrier scattering since no other mechanism is
effective on a 200 fs time scale. If this argument is valid, the system
essentially behaves as an isolated system of interacting particles, which

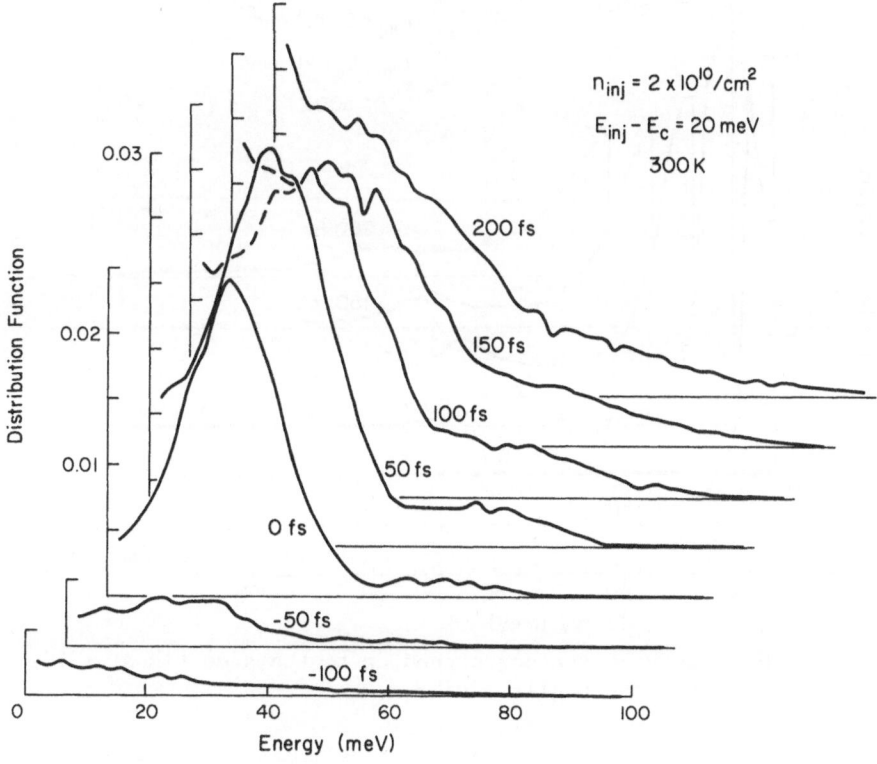

Fig. 7. Calculated electron occupancy versus energy for various times before and after the pump pulse which peaks at time t = 0 (Goodnick and Lugli, 1988b).

according to Boltzmann's H-theorem (Tolman, 1938) evolves from an arbitrary nonequilibrium state to the state of maximum entropy over a characteristic time related to the intercollisional time.

We have simulated these experiments (Goodnick and Lugli, 1988b) using the Monte Carlo technique described earlier including both electron-electron and electron-hole scattering. Corresponding to the samples studied by Knox et al. (1986), we model optical injection of electrons and holes in a 95Å wide undoped quantum well for an injected density of $2 \times 10^{10}/cm^2$ carriers. The excess energy of the electrons in the conduction band is taken as 20 meV and a 100 fs laser pulse duration with spectral width of 20 meV is assumed. Electrons and holes are added simultaneously to the simulation with the same k vector in order to simulate the effect of band to band optical generation.

The calculated electron distribution function before and after photoexcitation is shown in Fig. 7. The maximum of the pulse occurs at time t = 0. The observed time evolution of the electron distribution closely resembles that which is found experimentally in the differential transmission experiments discussed above; that is, an athermal carrier distribution develops at the peak of the pulse which subsequently relaxes to a Maxwellian shaped distribution after 200 fs. Optical phonon absorption accounts for the slight bump in the distribution at higher energies which eventually merges into the Maxwellian tail after 200 fs due to carrier-carrier scattering.

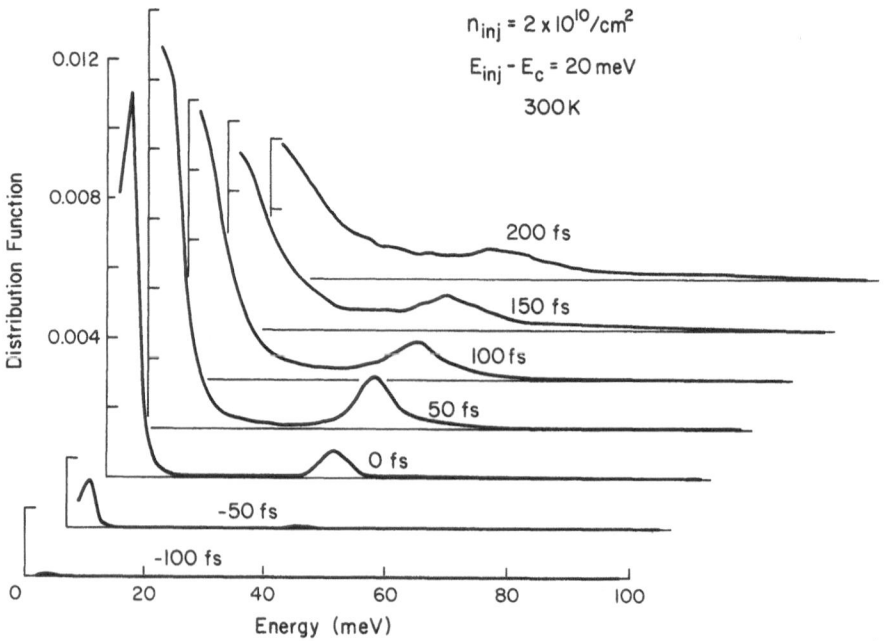

Fig. 8.   Hole distribution corresponding to the same times as Fig. 7
(Goodnick and Lugli, 1988b).

The evolution of the hole distribution during and after the pulse is
illustrated in Fig. 8.   Due to the large effective mass of the holes in
comparison to the electrons (the bulk value of the heavy hole mass was
taken as 0.51 m_o; light holes were neglected due to the low injection
energy and short time scales involved), the photoexcited holes are excited
with a very small excess energy, much less than the thermal energy at
300K.   Two effects are observed in Fig. 8:   the athermal portion of the
photogenerated hole distribution close to the injection energy thermalizes
after only 50 fs due to hole-hole scattering.   However, a second athermal
distribution in the hole population develops after the peak of the pulse
at one optical phonon energy above the injection energy (at approximately
40 meV in Fig. 8) due to absorption of phonons by the initially cold hole
gas.   This absorption results in a rapid heating of the hole system over a
time scale of 200 fs.   The athermal phonon replica thermalizes on a longer
time scale with the low energy distribution forming a broad 300K distribution after 200 fs as shown in Fig. 8.   Thus the dynamics of the coupled
electron-hole system are much more complicated than that of a simple
relaxation of electrons due to intercarrier interaction.

In Fig. 9 we show the differential transmission spectrum calculated
from the sum of the electron and hole distributions of Figs. 7 and 8 using
(16).   The differential transmission spectrum seems to follow closely the
time evolution of the electron distribution, with relatively little
contribution to the features of the optical spectrum due to the complicated hole dynamics.   In fact, the athermal phonon replica in the hole
distribution shown in Fig. 8 results in spectral features well outside the
optically coupled experimental region which is shown in Fig. 9.   The large
excess hole energy of the phonon replica should result in spectral
features at much higher energies, approximately at 1.7 eV photon energies

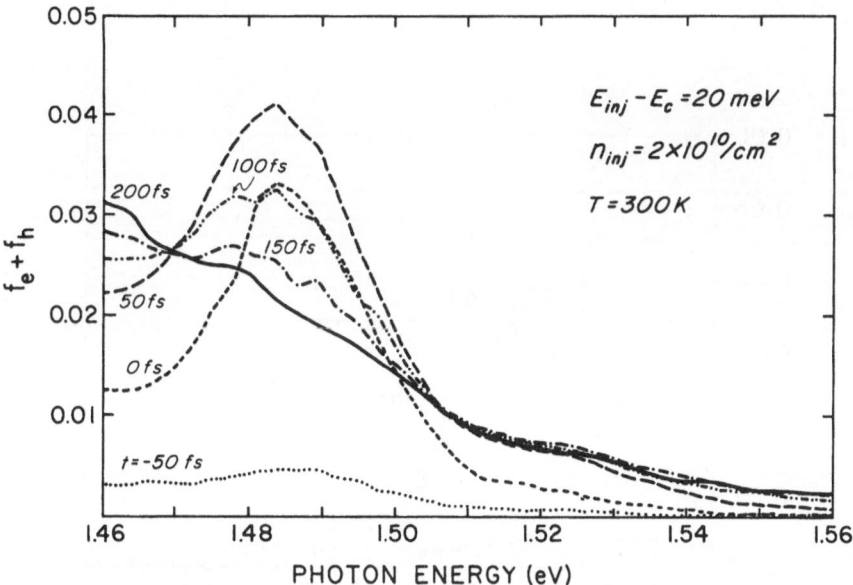

$E_{inj} - E_c = 20\,meV$

$n_{inj} = 2 \times 10^{10}/cm^2$

$T = 300\,K$

Fig. 9.   Calculated differential transmission spectrum corresponding to
the electron and hole distributions of Figs. 7 and 8 (Goodnick
and Lugli, 1988a).

if the effective mass of the holes is constant (which in fact it is not).
The relative magnitude of the effect, however, is almost an order of
magnitude smaller than the signal arising from the electron distribution,
and hence may be difficult to observe experimentally. The close agreement
between the simulated electron distribution, the calculated differential
transmission spectrum and the experimental differential transmission
verify the assumption that what was observed experimentally was in fact
the time evolution of the electron distribution.

To verify that this time evolution is mainly due to intercarrier
scattering, we selectively turn-off the intercarrier interaction during
the simulation and observe the time evolution of the differential transmission spectrum. The calculated spectrum without intercarrier scattering
is shown in Fig. 10. As can be seen, the spectrum is essentially frozen
in its nonequilibrium shape for the entire timescale of the experiment.
This figure rather dramatically illustrates that intercarrier scattering
is the essential mechanism responsible for the experimentally observed
relaxation.

Finally, in order to ascertain which type of carrier-carrier scattering influences the relaxation, we tabulate the rate of energy exchange for
various scattering processes in Fig. 11. For LO phonons and electron-hole
scattering, we show the net power gain at different times during the
simulation averaged over several time steps. Electron-electron and
electron-hole exchange are calculated from the magnitude of the energy
exchanged for each collision. It is rather simple to show from the
collision integral of the Boltzmann equation that the time evolution of
the system, particularly that of the H-function, depends upon the rate of
energy exchanged (i.e. inelastic processes) rather than simply the
scattering rate. Thus in order for intercarrier scattering to be

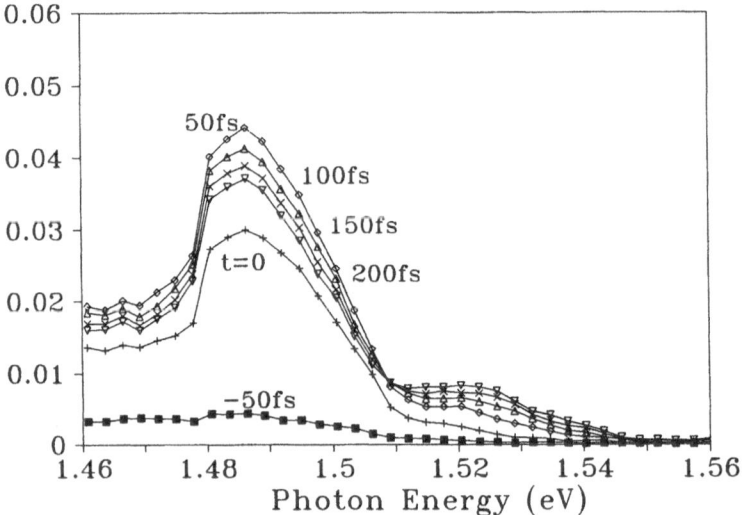

Fig. 10. Differential transmission spectrum corresponding to the same parameters as Fig. 10 but excluding intercarrier scattering during the simulation.

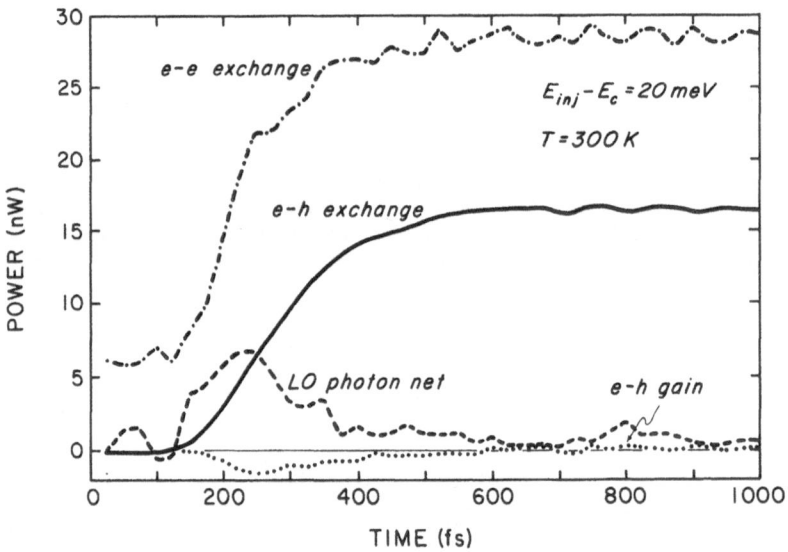

Fig. 11. Net power gain and exchange for LO phonons, electron–hole and electron–electron scattering for the simulated results of Figs. 7–9.

effective in driving the distribution function towards a Maxwellian distribution, carriers must scatter from their initial constant energy surface. Electron–hole scattering is inherently more elastic than electron–electron due to the large difference in effective masses, although the actual scattering rate is higher. In Fig. 11 we see that the rate of energy exchange due to electron–electron scattering is almost a factor of two greater than that of electron–hole scattering. Thus the dominant relaxation mechanism appears to be that due to electron–electron scattering. However, screening plays an important role in determining the exact scattering rate, and by necessity a rather simple form has been employed in the present set of simulations. It is not known at present to what extent dynamical screening and plasmon interactions modify the above picture.

Intersubband Relaxation in Modulation Doped Structures

Recent time resolved experiments in semiconductor quantum wells have focused on the relaxation of hot carriers via intersubband transitions (Oberli et al., 1987; Seilmeier et al., 1987; Oberli et al., 1989). Usually, intersubband transitions involve a large change in wave vector due to the change in kinetic energy associated with the change in poten- tial energy going from one subband to another. Hence this scattering rate is reduced with respect to the intrasubband scattering rate except in certain special cases. In addition, intersubband processes contain an overlap integral such as appears in equations (7) and (11) which attenu- ates the scattering matrix element due to the orthogonality of the subband wavefunctions. This attenuation becomes severe when the subbands are spatially separated as occurs when for instance two quantum wells are considered (Oberli et al., 1989) or when real space transfer into the barrier is important as discussed in connection with Fig. 1. In GaAs quantum wells, the dominant intersubband scattering mechanism is usually polar optical phonon scattering. Other processes may become important under certain conditions such as intercarrier intersubband scattering at high carrier densities.

In the experiments of Seilmeier and coworkers (1987), long time constants were observed when carriers were excited using far infrared intersubband excitation in narrow quantum wells. There carriers were resonantly excited from the ground subband to the upper subband where the decay from the upper to lower subband was monitored from the intersubband absorption. For a 47Å well, they found an intersubband relaxation time of 14 ps while for wider wells, this time decreased to 8 ps (Seilmeier et al., 1987; 1988). Such times are in excess of what one expects for intersubband relaxation time based on optical phonon emission in a single narrow quantum well.

To simulate these experiments, we compared the effect of including self–consistent solutions to Poisson's and Schroedinger's equations into the Monte Carlo program to those obtained using a simple square well potential. The samples used in their studies consisted of multiple quantum wells which were modulation doped $5 \times 10^{11}/cm^2$ with donors in barriers that were 400Å thick. The differences between square well solutions and self–consistent potential solutions were discussed earlier in regard to Figs. 1a and 1b. In Fig. 1a, the eigenfunctions and eigen- values are illustrated for two 50Å wells and a 400Å barrier. For both the lower and upper states, the wavefunction decays exponentially into the barrier, and thus is primarily localized in the well. Intersubband scattering in this case is not much different than the case of an infinite square well. In contrast, when the effects of ionized donors and the self–consistent potential due to electrons in the well are considered, a potential minima exists in the barrier which localizes the wavefunction

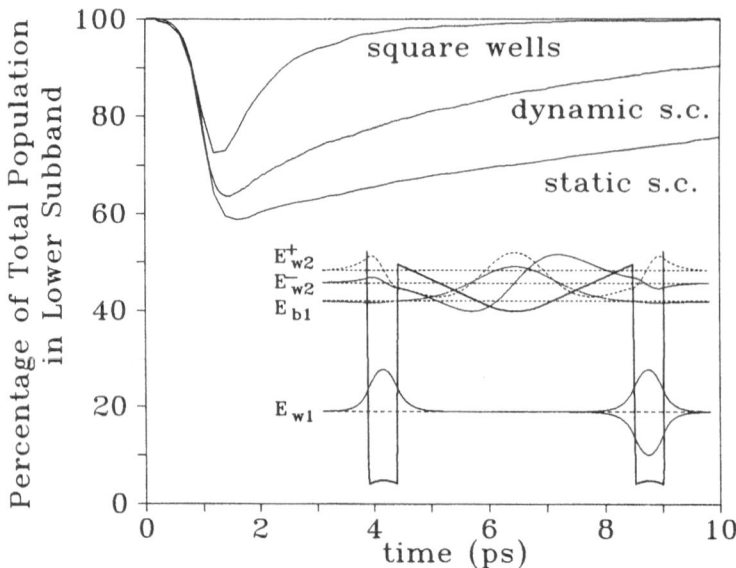

Fig. 12. Population of the lowest subband after intersubband excitation
using finite square wells versus self—consistent solutions. The
inset shows the self—consistent conduction band profile and
wavefunctions for 50Å GaAs wells separated by 400Å of modulation
doped $Al_{.35}Ga_{.65}As$.

for certain states outside of the wells. Two of the states, E_{w2}^+ and E_{w2}^-
contain some of the symmetry of the well, and therefore we associate them
with the excited states of the well even though they exist primarily in
the barrier. The third state, E_{b1}, looks purely like the ground bound
state of the barrier. Electrons which reside in this state have a very
small intersubband scattering cross section as the overlap integral (7) is
non—zero only near the boundary. The barriers on the left and right sides
of the wells were treated as quasi—infinite so that only bound states
existed within the system.

To compare the Monte Carlo simulation with the experimental absorption decay, we assume a 1 ps pump pulse which peaks 1 ps into the simulation. Experimentally it was reported that the intersubband absorption was
saturated by the pump pulse. We simulate this effect by taking half of
the carriers from the state E_{w1} to E_{w2}^+ vertically (i.e. conserving the
parallel wavevector) according to the phenomenological generation rate of
the pulse given by (16). The results are shown in Fig. 12 where we plot
the fraction of the population in the ground subband as a function of time
before and after the pulse for three different approximations. The first
approximation corresponds to assuming square wells as shown in Fig. 1a.
The relaxation time in this case is rather short, about 2 ps as shown in
Fig. 12 because both the upper and lower subband wavefunctions are
localized in the same space, enhancing the intersubband scattering rate.
The next approximation is to calculate the self—consistent wavefunctions
and associated carrier scattering rates only once at the start of the
simulation and to keep these rates fixed throughout the run. This
procedure reduces the amount of cpu time required, but neglects the
dynamic change in the self—consistent potential induced by the transfer of
charge from the well to the barrier. As shown in Fig. 12 for the curve
labeled "static s.c.", the return time for electrons is now quite long.

Carriers first thermalize in the barrier where they drop into the lowest barrier state E_{b1}. The decay out of this state is quite slow with some carriers returning via phonon absorption to higher barrier states. The calculated 1/e time is found to be 17 ps which is longer than the 14 ps time measured experimentally. A further refinement in the calculation is obtained by recalculating the subband energy levels, eigenfunctions, and carrier scattering rates at various time steps throughout the simulation. The result of this calculation is shown in Fig. 12 by the curve labeled "dynamic s.c." in which the scattering rates were updated whenever the population in the ground subband changed by more than 10%. For this calculation, the return time is decreased because the localization of photoexcited carriers in the barrier screens the potential of the ionized impurities hence reducing the potential well which exists there. This fact in turn reduces the localization of the wavefunction in the barrier increasing the intersubband scattering rate from the barrier to the well. The time constant obtained in this manner was found to be 6.6 ps which is less than the experimental return time.

Qualitatively, our results show that the mechanism responsible for the experimentally measured slow rate of return is the localization of the upper subband wavefunction in the AlGaAs barrier which reduces the overlap integral for LO phonon intersubband scattering. As the wells become wider, the upper subbands shift downwards in energy and eventually become fully localized in the well which results in shorter relaxation times, more similar to those obtained for a finite square well. Quantitatively, the agreement is not as good. The dynamic self-consistent calculation, which should be more accurate, in fact gives times which are too short. On the other hand, the times given by a square well potential are much shorter than those found for even wide wells (Seilmeier et al., 1988). However, several problems exist with the present model which might influence the quantitative agreement. In particular, we have used bulk GaAs LO phonon modes to describe the intersubband scattering rate, when in fact confined GaAs and AlAs modes will play an important role, especially in a 50Å well. Educato and coworkers (Educato et al., 1989) have included slab mode phonons in Monte Carlo calculations of these experiments, and find the same qualitative trends as here although quantitative agreement with experiment was also not found.

Intersubband Scattering in Coupled Quantum Wells

Recently, time resolved photoluminescence (PL) has been used in the study of intersubband relaxation in coupled asymmetric quantum wells (Norris et al., 1989; Oberli et al., 1989). The insert of Fig. 13 shows the well configuration used by Oberli and coworkers (1989) and the calculated envelope wave functions for a particular barrier thickness, b, separating the two wells. We can distinguish two types of subband states in this calculation, states which are localized in the wide well (88Å) labeled by the unprimed coordinates and states which are localized in the narrower well (60Å) labeled by the primed coordinates. Experimentally, carriers are excited in both wells by a picosecond pump pulse, and the luminescence intensity from each well is measured at subsequent times thereafter. The wells are situated in the intrinsic region of a p-i-n diode, and thus a field is also applied across the wells which shifts the relative energies of the subband states. Carriers were injected at low density in order to avoid charging effects during the experiment.

We have simulated the dynamics of electrons injected into the double barrier structure shown in Fig. 13 neglecting the effect of holes in order to focus on the dynamics of intersubband transfer in the electron system. The driving force for carrier transfer from the narrow to the wide well

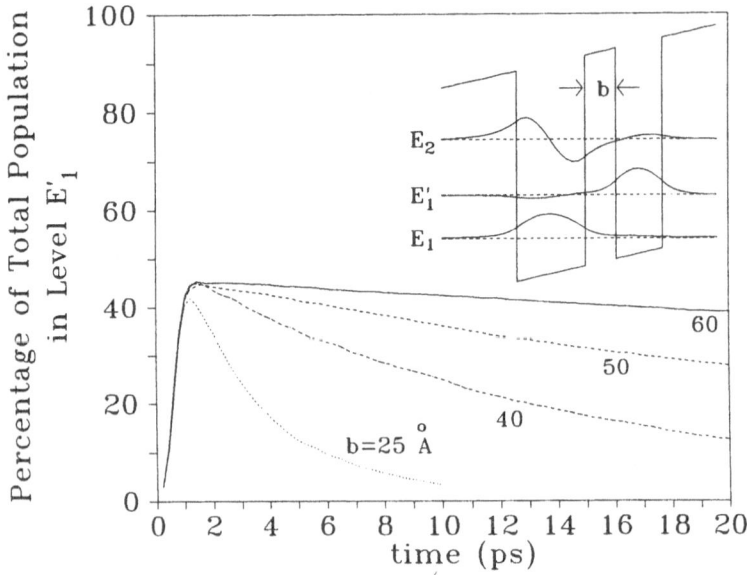

Fig. 13. Population in the first excited level E_1' for coupled 88Å and 60Å undoped GaAs wells under applied bias for various $Al_{.30}Ga_{.70}As$ barrier widths after a pulse peaked at 1 ps.

after photoexcitation is that of intersubband scattering, particularly from the E_1' to the E_1 state in Fig. 13. As in the case of modulation doping discussed in the previous section, the rate of decay from the upper to the lower subband is strongly dependent on the overlap of the subband envelope functions, which are spatially separated by the barrier. The decay time increases exponentially as the barrier thickness is increased as shown in Fig. 13 where the population in the narrow well ground subband as a function of time is calculated during and after the pump pulse for various barrier thicknesses. A constant field of 20 kV/cm is assumed corresponding to the built-in field in the p–i–n structure. The calculated decay times are found to be slightly lower than those measured experimentally (Shah, 1989). Here we have again used bulk GaAs modes, as the description of confined phonon modes in coupled wells such as those in Fig. 13 is beyond the scope of the present work. This fact may influence the absolute value of the intersubband transition time measured as the dominant contribution to the scattering time is due to the overlap integral in the narrow AlGaAs barrier.

Oberli and coworkers (1989) were able to bias the structure in Fig. 13 so that the E_1' and E_2 levels were in resonance. In this case, carriers tunnel from the narrow to the wide well in some characteristic time which depends on the details of the initial state formation. However, the PL decay time depends on the loss of carriers from the narrow well, and hence depends on the intersubband scattering rate from the upper to the lower state. We have simulated the effect of bias on the tunneling time as shown in Fig. 14 where we plot the calculated decay time (found from an exponential fit to the density versus time curves from the narrow well state) versus the field across the double well structure for a fixed barrier width of 55Å. For small fields, the decay time is basically that of the two weakly coupled wells as in the case of Fig. 13. As the upper state of the wide well and the lower state of the narrow well come into resonance, the decay time decreases rapidly and approaches that of the

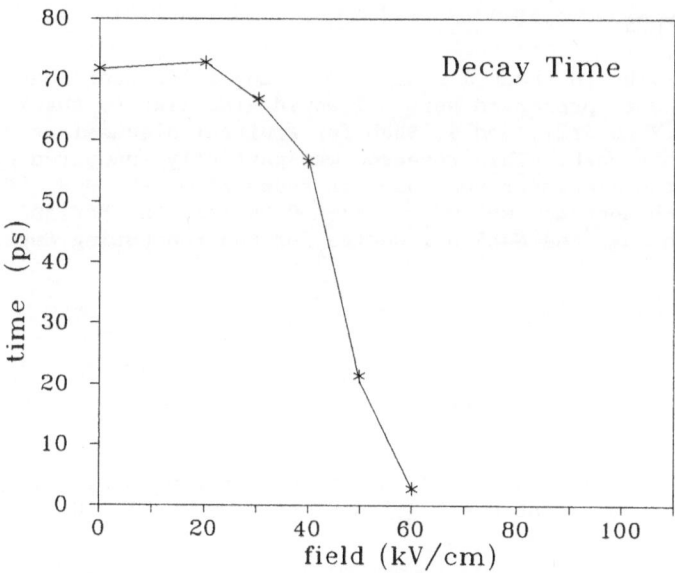

Fig. 14.   Intersubband relaxation time versus electric field for the
structure shown in Fig. 13 with b = 55Å.

intersubband scattering time in a single quantum well. These results seem
to agree both qualitatively and quantitatively with those found experimentally (Oberli et al., 1989) except at resonance where we find shorter
relaxation times. For the experimental barrier heights determined by the
Al mole fraction, our calculation breaks down after 60 kV/cm due to the
fact that the electron envelope functions become unbound so that field
emission out of both wells becomes very probable.

SUMMARY

    In the present manuscript, we have reported on some of the applica-
tions of Monte Carlo simulation in comparing and understanding ultrafast
optical experiments in quantum well systems. Here we have used this
technique to show the effect of hot phonons in the slow energy loss rate
by carriers in quasi-two-dimensional systems, the effect of electron-
electron and electron-hole scattering in short time relaxation, and
finally the role that spatial separation of the subband wavefunctions on
the intersubband scattering rate in modulation doped and coupled quantum
wells. In all cases, the qualitative agreement is good. However,
quantitative agreement is not always found which to a large extent still
reflects differences between what is measured experimentally, which is an
optical coupling such as the absorption coefficient or emission intensity,
and what is calculated in the Monte Carlo, which is usually some average
over the carrier ensemble reflecting the carrier distribution functions.
By combining the information contained in the electron and hole distribution functions, direct comparison may be obtained in some cases such as
the case of the intercarrier scattering experiments discussed above.
However, even in these experiments, complicated contributions to the
optical spectrum exist such as resonant absorption due to excitons which
make a quantitative comparison difficult. Future work in this area will
seek to address these issues so that quantitative information may be
obtained form detailed comparisons between theory and experiment.

ACKNOWLEDGEMENTS

    I would like to thank J. Lary and P. Lugli for contributions to the
technical results presented here. I would also like to thank D.S. Chemla,
W.H. Knox, D.Y. Oberli, and J. Shah for fruitful discussions and collaborations in this work. This research was partially supported by funds
provided by the Alexander von Humboldt foundation, the U.S. Office of
Naval Research Contract No. N00014-87-K-0686, and the National Science
Foundation through the National Center for Supercomputing Applications.

REFERENCES

Adachi, S., "GaAs, AlAs, and Al$_x$Ga$_{1-x}$As: Material Parameters for use in
Research and Device Applications," J. Appl. Phys., vol. 58, pp. R1-R29,
1985.

Bacchelli, L. and Jacoboni, C., "Electron-Electron Interactions in Monte
Carlo Transport Calculations," Solid State Comm., vol. 10, pp. 71-74,
1971.

Boardman, A.D., Fawcett, W., and Rees, H.D., "Monte Carlo Calculation of
the Velocity-Field Relationship for Gallium Arsenide," Solid State Comm.,
vol. 6, pp. 305-307, 1968.

Brunetti, R., Jacoboni, C., Matulionis, A., and Dienys, V., "Effect of
Interparticle Collisions on Energy Relaxation of Carriers in Semiconductors," Physica, vol 134B, pp. 369-373, 1985.

Chemla, D.S., Miller, D.A.B., and Schmitt-Rink, S., "Nonlinear Optical
Properties of Semiconductor Quantum Wells," in Optical Nonlinearities and
Instabilities in Semiconductors, Academic Press, New York, pp. 83-120,
1988.

Educato, J.L., Leburton, J.P., Dailey, D., and Hess, K., "Intersubband
Scattering in Modulation Doped Quantum Wells," presented at the Conference
on Quantum Wells and Superlattices, Salt Lake City, March 1989.

Goodnick, S.M., and Lugli, P., "Influence of Electron-Hole Scattering on
Subpicosecond Carrier Relaxation in Al$_x$Ga$_{1-x}$As/GaAs Quantum Wells", Phys.
Rev. B, vol. 38, no. 14, pp. 10135-10138, 1988a.

Goodnick, S.M. and Lugli, P., Effect of Electron-electron Scattering on
Nonequilibrium Transport in Quantum-well Systems," Phys. Rev. B, vol. 37,
no. 5, pp. 2578-2588,.1988b.

Jacoboni, C. and Reggiani, L., "The Monte Carlo Method for the Solution of
Charge Transport in Semiconductors with Applications to Covalent Materials," Rev. Mod. Phys., vol. 55, no. 3, pp. 645-705, 1983.

Kalos, M.H. and Whitlock, P.A., Monte Carlo Methods, John Wiley & Sons,
New York, 1986.

Knox, W.H., Hirlimann, C., Miller, D.A.B., Shah, J., Chemla, D.S., and
Shank, C.V., "Femtosecond Excitation of Nonthermal Carrier Populations in
GaAs Quantum Wells," Phys. Rev. Lett., vol. 56, no. 11, pp. 1191-1193,
1986.

Knox, W.H., Chemla, D.S., and Livescu, G., "High Density Femtosecond
Excitation of Nonthermal Carrier Distributions in Intrinsic and Modulation

Doped GaAs Quantum Wells," Solid-State Electronics, vol. 31, no. 3/4, pp. 425–430, 1988.

Lugli, P. and Goodnick, S.M., "Nonequilibrium Longitudinal-Optical Phonon Effects in GaAs-AlGaAs Quantum Wells," Phys. Rev. Lett., vol. 59, no. 6, pp. 716–719, 1987.

Lugli, P. and Ferry, D.K., "Investigation of Plasmon-Induced Losses in Quasi-Ballistic Transport," IEEE Elec. Dev. Lett., vol. EDL-6, no. 1, pp. 25–27, 1985.

Norris, T.B., Vodjdani, N., Vinter, B., Weisbuch, C., and Mourou, G.A., "Electron Tunneling Times in Coupled Quantum Wells," to be published.

Oberli, D.Y., Wake, D.R., Klein, M.V., Klem, J., Henderson, T., and Morkoc, H., "Time Resolved raman Scattering in GaAs Quantum Wells," Phys. Rev. Lett., vol. 59, no. 6, pp. 696–699, 1987.

Oberli, D.Y., Shah, J., Damen, T.C., Tu, C.W., Miller, D.A.B., "Electron Tunneling Times in Coupled Quantum Wells," in Optical Society of America Technical Digest Series, vol. 10, pp. 272–275, 1989.

Price, P., "Monte Carlo Calculation of Electron Transport in Solids," in Semiconductor and Semimetals, R.K. Willardson and A.C. Beer, eds., Academic Press, New York, pp. 249–308, 1979.

Riddoch, F.A., and Ridley, B.K., "On the Scattering of Electrons by Polar Optical Phonons in Quasi-2D Quantum Wells," J. Phys. C, vol. 16, pp. 6971–6982, 1983

Riddoch, F.A., and Ridley, B.K., "Electron Scattering Rates Associated with the Polar Optical Phonon Interaction in a Thin Ionic Slab," Physica, vol. 134B, pp 342–346, 1985.

Ryan, J.F., Taylor, R.A., Turberfield, A.J., Maciel, A., Worlock, J.M., Gossard, A.C., and Weigmann, W., "Time-Resolved Photoluminescence of Two-Dimensional Hot Carriers in GaAs-AlGaAs Heterostructures," Phys. Rev. Lett., vol. 53, no. 19, pp. 1841–1844, 1984.

Seilmeier, A., Hübner, H.J., Abstreiter, G., Weimann, G., and Schlapp, W., "Intersubband Relaxation in GaAs-Al$_x$Ga$_{1-x}$As Quantum Well Structures Observed Directly by an Infrared Bleaching Technique," Phys. Rev. Lett., vol. 59, no 12, pp. 1345–1349, 1987.

Sielmeier, A. Hübner, H.J., Wörner, M., Abstreiter, G., Weimann, G., and Schlapp, W., "Direct Observation of Intersubband Relaxation in Narrow Multiple Quantum Well Structurds," Sol. State Elec., vol. 31, no. 3, pp. 767–770, 1988.

Shah, J., Pinczuk, A., Gossard, A.C., and Wiegmann, W., "Energy-Loss Rates for Hot Electrons and Holes in GaAs Quantum Wells," Phys. Rev. Lett., vol. 54, no. 18, pp. 2045–2048, 1985.

Shah, J., "Hot Carriers in Quasi-2-D Polar Semiconductors," IEEE Journal of Quantum Electronics, vol. QE-22, no. 4, pp. 1728–1743, 1986.

Shah, J., private communication.

Tolman, R.C., The Principles of Statistical Mechanics, Oxford University Press, London, 1938.

von der Linde, D., Kuhl, J., and Klingenberg, H., Raman Scattering from Nonequilibrium LO Phonons with Picosecond Resolution," <u>Phys. Rev. Lett.</u>, vol. 44, no. 23, pp. 1505-1508, 1980.

Weisbuch, C., "Fundamental Properties of III-V Semiconductor Two-Dimensional Quantized Structures: The Basis for Optical and Electronic Device Applications," in <u>Semiconductors and Semiconductors</u>, Vol. 24, R.K. Willardson and A.C. Beer, eds., Academic Press, New York, pp.1-133, 1987.

ULTRAFAST TRANSPORT MEASUREMENTS IN BULK SEMICONDUCTORS AND TUNNELING DEVICES USING ELECTRO-OPTIC SAMPLING

Kevin E. Meyer

Cavendish Laboratory
Madingley Road
University of Cambridge
Cambridge, England CB3 OHE

1. INTRODUCTION

Electro-optic sampling (EOS) is a technique which has been developed for measuring transient electrical fields with subpicosecond resolution. It takes advantage of the inherent speed of the linear electro-optic effect (Pockels effect) to convert a fast electrical transient into a fast optical transient. The optical transient is then probed using the conventional pump/probe approach in conjunction with a short-pulse dye laser. Two applications of this technique will be discussed which are relevant to the study of transport physics on small time and dimension scales.

The first application is the characterization of transient photoconductivity in bulk GaAs. Fast photoconductivity is induced by optically exciting a simple photoconductive switch, and the photocurrent creates a transient voltage in the external circuit which is measured electro-optically. An overshoot in the photocurrent has been observed which is directly related to velocity overshoot in the bulk semiconductor. These measurements constitute the first fully time-resolved observation of velocity overshoot in these materials. They are in qualitatively good agreement with Monte Carlo simulations of the transport and the circuit response carried out at Arizona State University.

The second application is the measurement of switching speeds of resonant-tunneling diodes. In this application, a fast sinusoidal or stepped voltage pulse is applied to the input of the device while it is biased near resonance. The transient voltage exceeds the device switching threshold, and the device switches with a speed which is determined by the tunneling time and the device parasitics. The transient switching event is measured by monitoring the voltage at the output of the device electro-optically. This measurement yields an upper limit to the tunneling time through the quantum well in the device. Two groups have recently reported measurements of similar devices on a picosecond time scale. Their results will be reviewed, and future applications of EOS will be discussed.

2. DESCRIPTION OF THE ELECTRO-OPTIC SAMPLING (EOS) METHOD

The electro-optic sampling technique was introduced in 1982[1] as a way of utilizing short laser pulses and the inherent speed of the linear electro-optic effect to probe very fast electrical transients. Since then the technique has been used extensively in the study of pulse propagation on transmission lines[2,3], discrete device testing[4,5], on-chip wafer probing[6-8], and measurements of subpicosecond photoconductivity in GaAs[9,10]. This paper will concentrate on those applications, namely device characterization and photoconductivity measurements, which are relevant to the transport physics of short time and dimension scales.

A schematic of a typical measurements system is shown in Fig.1[1]. The laser source in the figure is a colliding pulse mode-locked laser (CPM), but may just as easily be a synchronously-pumped dye laser or a diode laser[11,12]. The source is split into two beams, a pump and a probe beam. The pump beam is used to trigger the electrical signal of interest. The electrical signal is coupled onto a traveling wave Pockels cell, where it induces a transient birefringence in the modulator. This birefringence is detected as a change in polarization of the probe beam focussed through the modulator. If the probe beam is optically biased at the modulator's quarter-wave point then the optical change is a linear function of the electric field. The time-dependent waveform is recovered by using a lock-in amplifier and sweeping the pump/probe delay.

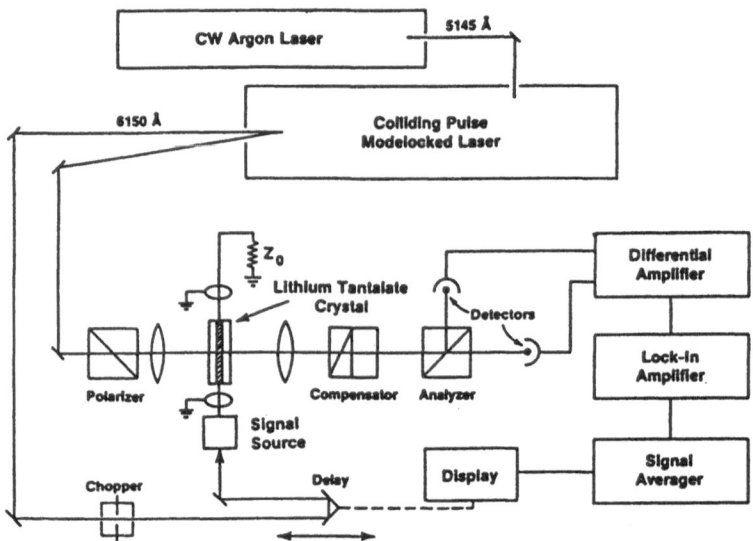

Fig. 1 Schematic of a typical EOS measurement system[1].

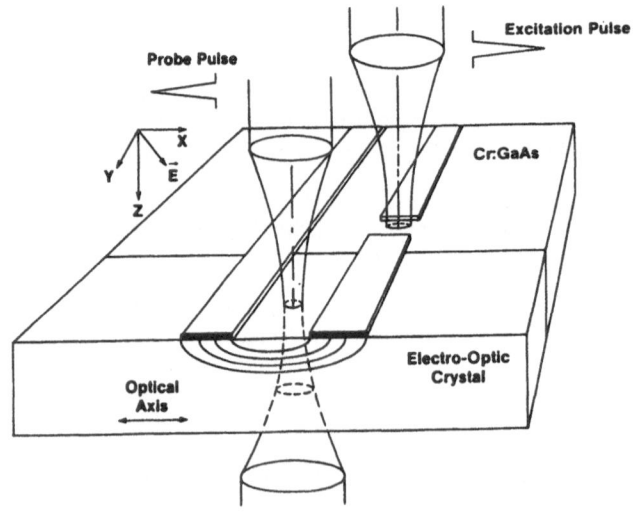

Fig. 2 Details of a coplanar EOS sampling geometry[13].

Details of a particular coplanar sampling geometry are shown in Fig. 2[13]. In this example the signal source was a Cr:GaAs photoconductive switch. The 50μm coplanar stripline provided the necessary bandwidth to couple the signal with minimum distortion onto the modulator. The electrical waveform obtained with this geometry is shown in Fig.3[13],

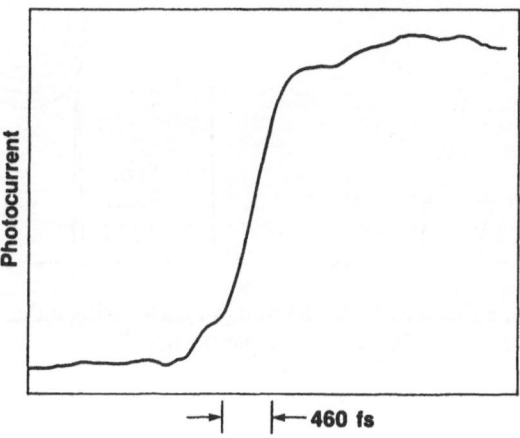

Fig. 3 Electrical waveform obtained with the coplanar sampling geometry of figure 2[13].

demonstrating the subpicosecond temporal resolution that is possible with this technique. The temporal resolution is determined by a convolution of (1) the laser pulsewidth, (2) the transit time of the optical probe pulse through the electric field region, and (3) the transit time of the electrical pulse across the diameter of the focussed probe beam. The latter two contributions may be minimized by utilizing small transmission lines, because the electric field depth scales with the separation of the transmission lines, and by focussing the probe beam as tightly as possible. With an optimized geometry a risetime as short as 300fs has been resolved[5].

For subpicosecond transients the constraint of very high-bandwidth coupling into the modulator can be very restrictive. Fortunately, it is not necessary to use a modulator at all; all that is required is that some fringing field lines from the device under test be coupled into an electro-optic medium. Several different demonstrated sampling geometries are shown in Fig. 4. In (a), (b), and (c) the sampling media consist of transmission line modulators. In (d), (e), and (f) an electro-optic medium has been placed in proximity to the device under test, in this case a transmission line, and the fringing fields above the device extend into the sampling crystal. The sampling medium may be a plate with a high-reflectivity coating[14], as in (d), or a "finger probe" as in (e) and (f)[10,15]. In short, there are a number of different geometries available to the experimentalist, all providing subpicosecond resolution.

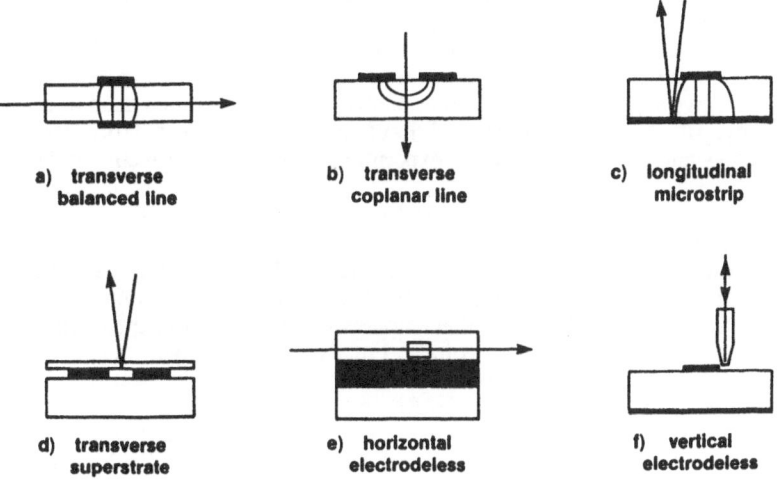

Fig. 4 Examples of various sampling geometries.

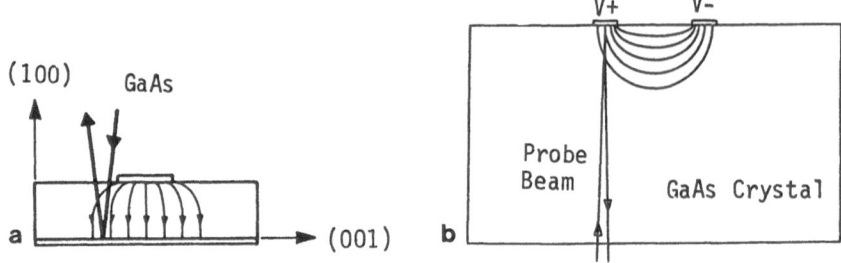

Fig. 5 EOS geometries for in-situ sampling in GaAs using either (a) a microstrip or (b) a coplanar design [16].

A different EOS application has been developed which allows direct on-chip probing of transient voltages without the need of an external sampling crystal, provided that the semiconductor under test is electro-optic[16]. GaAs is a cubic crystal which does not possess a center of inversion symmetry and is therefore electro-optic. There are two requirements which must be met for in-situ EOS. The first is that an optical source must be available at a sub-bandgap energy with a sufficiently short pulsewidth. Kollner and Bloom[16] used a CW mode-locked Nd:YAG laser at 1.064um whose output was compressed with a fiber-grating arrangement[17] to yield 1.8pS pulses. Since then, other short-pulse sources in the near-IR have been developed, including dye lasers[18-20] and color-center lasers[21]. The second requirement is that the standard crystal orientations and geometries used for GaAs devices must also be useful for electro-optic sampling. This is the case for the industry standard orientation (100). For this orientation EOS is achieved by passing the probe beam through the substrate normal to its surface, with the probe beam focussed on a metallization whose potential is to be measured. The metallization reflects the probe beam back out to the detection electronics. Two different geometries may be used, illustrated in Fig. 5, depending on whether the device under test is laid out in a microstrip or coplanar geometry[16]. This approach has demonstrated a temporal resolution of better than 2 picoseconds.

3. THEORETICAL AND EXPERIMENTAL ASPECTS OF TRANSIENT PHOTOCONDUCTIVITY IN GaAs

During the last 15 years an extensive literature has developed in which various techniques were applied to the prediction of transport transients in nonstationary, nonequilibrium charge carrier transport in a semiconductor[23-25]. The central dilemma in this field has been a significant lack of properly time-resolved experimental data. In recent years several groups have suggested that transient photoconductivity can be used to experimentally measure some of these transients and have presented results from first-generation experiments[26-28] and more recently from experiments utilizing improved MBE samples[29].

In this section, the detailed Ensemble Monte Carlo model of the transient transport, which has been developed at Arizona State University by R. Grondin, will be described. A survey of Monte Carlo parameters has been carried out[10] which demonstrates the need for discriminating experiments in this field. A circuit model of the photoconductivity experiment will be discussed which clearly shows that an overshoot in the carrier drift velocity will result in an overshoot in the device terminal voltage. Finally, the EOS experiments performed at the University of Rochester[9,10] will be described which constitute the first fully time-resolved measurements of velocity overshoot in GaAs. In addition, ongoing experiments which will include p-i-n structures and studies of transport in AlGaAs will be discussed.
A parameter study has been carried out to investigate the sensitivity of calculated velocity transients on the input parameters. Particular interest has been paid to the degree of velocity overshoot and the threshold field for velocity overshoot for a particular excitation

3.1   Ensemble Monte Carlo (EMC) Simulation of the Transient Transport

The GaAs transient response of the photoexcited electron-hole plasma in a uniform field has been studied with a Monte Carlo simulation[10,29]. A three valley electron and three band hole model has been used. Initial optical generation and distribution of the carriers in k-space takes into account anisotropic distributions. Carrier degeneracy has been suitably included through a rejection technique proposed by Lugli et. al.[30].

The bipolar EMC includes all the relevant carrier-phonon and electron-hole interactions. Only single mode LO and TO coupling have been considered and all plasmon-phonon interactions have been ignored for the present. A static but time-evolving screening model, proposed by Ferry et. al.[31], has been used for all of the polar interactions. Only intraband electron-hole processes have been included, leaving out possible multiple band scattering as discussed by D'yakonov et. al[32].

Hot phonon effects are treated using the EMC algorithm proposed by Lugli et. al.[33]. Both PO and intervalley phonon populations have been modified since the photoexcitation levels and electric field strengths cause large Γ-L transfer. Simulations have also been performed for AlGaAs. This large band gap material is used to investigate the Jones-Rees effect[34] and the bias dependence of the initial velocity rise[35]. For this case, two LO phonon modes are used and the relative interaction strengths chosen according to the hot phonon data of Kash et. al.[36].

TABLE I   Parameters for GaAs Monte Carlo Program

Parameter		Γ	L	X
Density (g/cm^3)	5.36			
Energy-band gap at 300 K (eV)	1.43			
High-frequency dielectric constant	10.92			
Static dielectric constant	12.9			
Velocity of sound (cm/sec)	5.24×10^5			
Number of valleys		1	4	3
Effective mass ratio				
m_l		0.063	1.5	'1.5
m_t		0.063	0.10	0.25
Nonparabolicity factor (eV^{-1})		0.69	0.64	0.55
Valley separation from Γ valley (eV)			0.33	0.52
Polar optic-phonon energy (eV)		0.035	0.0343	0.0343
Acoustic deformation potential (eV)		8.0	8.0	8.0
Coupling constant (10^8eV/cm)				
from Γ valley to			10	10
from L valley to		10	10	9
from X valley to		10	9	9
Intervalley phonon energy (eV)	0.026			
Heavy-hole band m^* (hh)	0.70			
Light-hole band m^* (lh)	0.082			
Split-off band m^* (so)	0.20			

wavelength, and to the determination of which parameters are most critical for velocity overshoot. The four parameter sets are as follows. The first set is shown in Table I. The second set are the parameters of Wysin et. al.[37]. Their deformation potential parameters are identical to those of Brennan and Hess[38] while their effective masses were based on a pseudopotential calculation. The third set, which differs in its valence band parameters, is that of Taylor et. al.[39]. The fourth set is that of Shah et. al.[40].

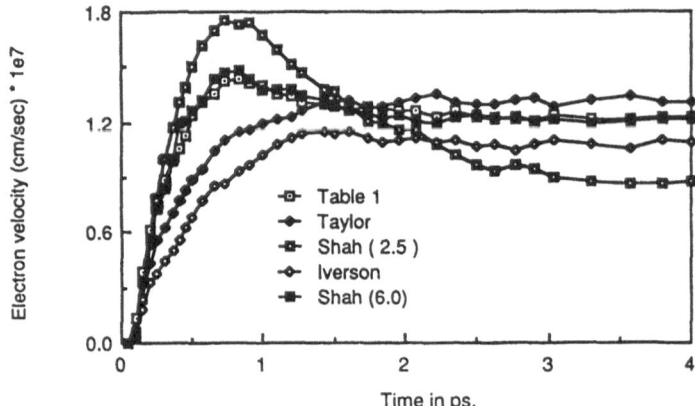

Fig. 6 Calculated electron velocity transients for the four parameter sets discussed in the text. The electric field is 10kV/cm and the excitation energy is 2.0eV[10].

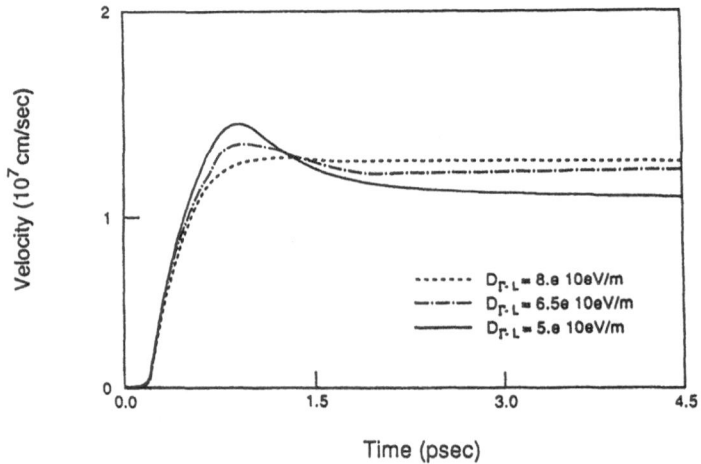

Fig. 7 Calculated transient electron velocities for Γ-L deformation potential values of 5, 6.5, and 8x10[8] eV/cm, as given by Shah et. al.[40]. The applied field is 10kV/cm and the excitation wavelength is 620nm [10].

Figure 6 shows calculated velocity transients for the four parameters sets assuming an excitation wavelength of 620nm, pulsewidth of 100fs, and an applied electric field of 10kV/cm. All four calculations are in reasonably good agreement in the predictions of the peak velocity but differ considerably in the details of the transients. The parameters used by Wysin et. al. predict that no velocity overshoot should occur at this excitation wavelength. Chamoun et. al.[10] have also shown that the four parameters sets differ considerably in their predictions of the threshold field for onset of velocity overshoot. Clearly there is a need for time-resolved experiments in this area to determine the proper choice of Monte Carlo parameters.

Velocity overshoots and other transient phenomena are sensitive to the Γ to L deformation potential $D_{\Gamma L}$ of the conduction band. The role of $D_{\Gamma L}$ in influencing the velocity overshoot phenomena has been investigated for E=10kV/cm and 2.0eV photoexcitation. For concreteness the recent experimental data of Shah et. al. has been used[40]. In that experiment, GaAs and InP samples were excited by a subpicosecond pulse and the luminescence intensity was measured. The luminescence intensity for GaAs increased very slowly in contrast with that of InP. Since there was no significant intervalley scattering in InP at the excitation energy used, the slow GaAs luminescence rise was attributed to the return of L-valley electrons to the Γ-valley. There experimental results were compared with an ensemble Monte Carlo calculation and the Γ-L deformation potential was determined to be $(6.5\pm1.5) \times 10^8$ eV/cm.

The uncertainty in the experimentally determined value of $D_{\Gamma L}$ translates into a rather large deviation of the transient velocities. Keeping within the error of Shah et. al. the computation of the transient electron velocities was carried out. Three values of $D_{\Gamma L}$ were used (5×10^8, 6.5×10^8, and 8×10^8 eV/cm). The results are shown in Fig. 7 for E=10kV/cm and an excitation wavelength of 620nm. The velocity curve for $D_{\Gamma L} = 5 \times 10^8$ eV/cm shows a significant overshoot compared to the other two values of $D_{\Gamma L}$. The velocity curve for $D_{GL} = 6.5 \times 10^8$ eV/cm shows a slight peak while that for $D_{\Gamma L} = 8 \times 10^8$ eV/cm just increases to a steady state velocity. It therefore seems that the existence of an overshoot at E=10kV/cm for 2.0eV excitation can only be determined experimentally. Such an experiment would help to

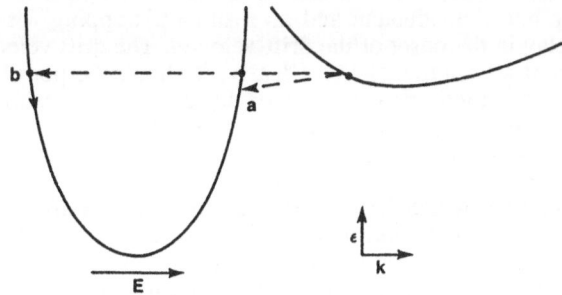

a) return to satellite valley
b) remain in central valley
     but velocity is negative

Fig. 8 Schematic of the Γ and L valleys in GaAs, illustrating the dynamics of transport for high-energy excitation.

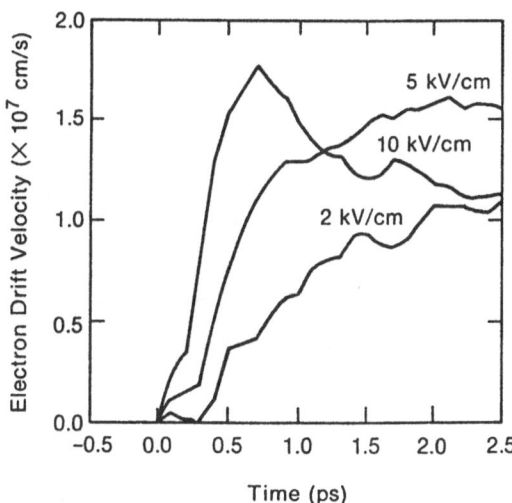

Fig. 9 Calculated electron velocity transients in GaAs for 2.0eV photoexcitation.

determine the $D_{\Gamma L}$ value. Since it would look at the departure of electrons from the Γ to the L valley, it would complement the experiment of Shah et. al. which observed the L to Γ transition.

Another phenomenon of interest on subpicosecond time scales in photoconductive experiments is the Jones-Rees effect, which is apparent as a bias-dependent delay in the onset of the drift velocity. The dynamics for high-energy photoexcitation are shown in Fig. 8. Unlike the case for electrons injected through contacts, photoexcited carriers have no initial net momentum and are distributed uniformly in the Γ valley. Consider what happens when a low electric field is applied. Electrons originally at point (a) will scatter into the L valley and will immediately slow down (and will scatter back into the Γ valley on a longer time scale). Electrons at point (b) have negative velocities and due to the presence of the field immediately fall below the threshold for intervalley transfer and are "trapped" in the Γ valley. The net drift drift velocity at short times is therefore very low. This is demonstrated in Fig. 9, which is a calculation of the electron velocity transients for photoexcitation at 2.0eV. For E=2kV/cm the net drift velocity is nearly zero for the first 300fs of transport. At high fields the electrons are accelerated much more rapidly and the negative-velocity electrons do not play as large a role, with the result that the drift velocity increases immediately after photoexcitation.

This interpretation may be tested by considering what happens when electrons are excited closer to the bandedge, below the threshold for Γ-L transfer. This may be simulated by reducing the excitation energy .This is shown in Fig. 10. Because there is no immediate intervalley scattering into the sideband and no resultant "trapping" of electrons in the Γ valley, there is no delay in the onset of the drift velocity. The drift velocity rises smoothly from the time origin at all applied fields. Clearly, it should be possible to observe this phenomenon in photoconductivity experiments by comparing results at two different excitation energies, above and below the threshold for Γ-L transfer.Such experiments will be discussed in the following section.

So far the simulations which have been discussed have only been of carrier drift velocities. However, the transient photoconductivity experiment measures transient voltages on the device terminals induced by the transient photocurrent. A circuit model of the photoconductive switch has been developed at Arizona State University[10] which uses the Monte Carlo velocity transients as an input. The equivalent circuit is shown in Fig. 11. The model is an extension of Auston's photoconductive switch analysis[41].

In the new model, the conductance is replaced by a photoconductive element and the characteristic impedance of the transmission line is assumed to be frequency dependent. The Monte Carlo velocity calculation simulates the photoconductive element. For simplicity a spatially uniform field was assumed and therefore space charge effects will not be considered

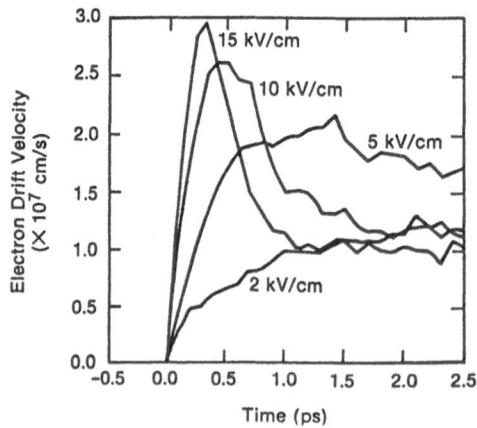

Fig. 10 Calculated electron velocity transients for 1.63eV photoexcitation in GaAs. Note that there is no evidence for a bias-dependent delay in the onset.

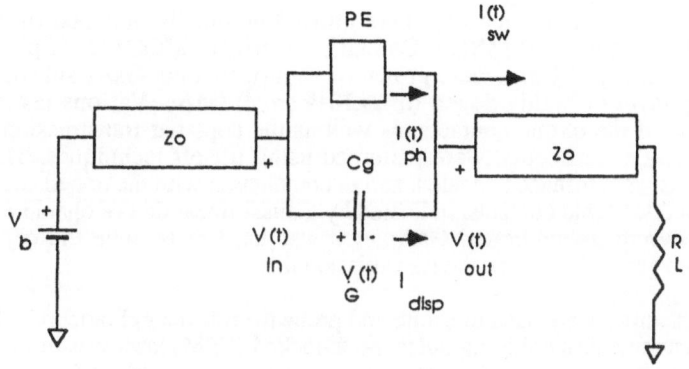

Fig. 11 Equivalent circuit of the photoconductive switch[10].

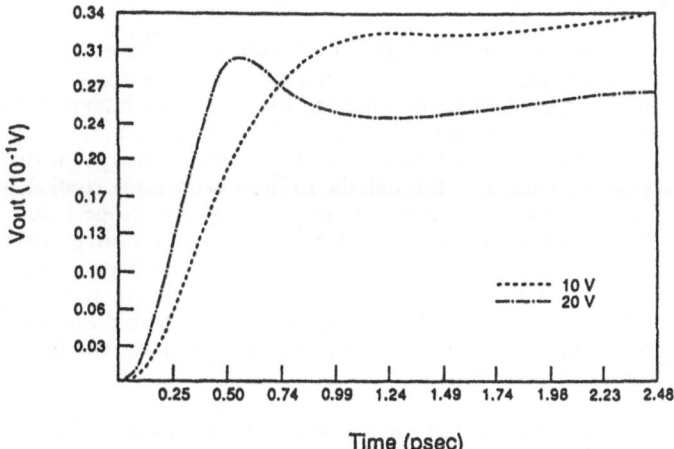

Fig. 12 The transient voltage waveforms obtained from the circuit simulation described in the text [10].

here. The transmission line model of Whitaker et. al.[42] was used to calculate the frequency-dependent line impedance. The photocurrent was determined using the continuity equation. The photocurrent was folded into an appropriate convolution integral which was solved numerically. For further details of the methodology of the calculation refer to Chamoun et. al.[10].

A representative calculated voltage transient is shown in Fig. 12. The modeled geometry is as follows: the photoconductive gap length is 10μm and the transmission line width is 50μm. The separation between the coplanar transmission lines is 50μm and the propagation distance is 20um. The average photoexcitation power is 5mW at a wavelength of 620nm. The background carrier concentration was $5 \times 10^{15} cm^{-3}$ and the gap capacitance was 1fF. The conclusion to be drawn from Fig. 12 is that electron velocity overshoot, as illustrated in Fig. 9, will result in a clearly observable overshoot in the terminal voltage in photoconductive switches, and the time dependence of the transient velocities will be reflected in the transient voltages.

3.2   Experimental Results of Transient Photoconductivity Measurements

This section will describe the transient photoconductivity experiments which have been performed at the University of Rochester [9,10] and the ongoing studies at the University of Cambridge.

All of the measurements reported here utilized nominally undoped (n=5x10^{15} cm^{-3}) high-mobility GaAs grown via MBE at Cornell University or MOCVD at Spire Corporation. Two microns of undoped GaAs were grown on semi-insulating GaAs substrates, followed by a 500Å cap layer of highly doped (n=2x10^{18} cm^{-3}) GaAs. Various test structures for characterization of the ohmic contacts, as well as the coplanar transmission lines for the photoconductivity experiments, were patterned using lift-off techniques. The evaporated NiAuGe contacts were furnace annealed, and in conjunction with the doped cap layer formed highly reproducible ohmic contacts, indicated by a quasi-linear dc I-V characteristic to fields as high as 10kV/cm. A calibrated GaAs etch was used to remove the cap layer in the photoconductive gap and between the transmission lines.

Two laser sources were used to excite and probe the transient photoconductivity. For the first set of measurements a colliding pulse mode-locked (CPM) laser was utilized, which had a wavelength of 620nm, a repetition rate of 100MHz, a pulsewidth of 60fs, and an average power of 5mW per beam. In subsequent experiments a synchronously-pumped linear cavity dye laser was used, which generated 300fs pulses at 760nm at a repetition rate of 100MHz and an average power of 10mW.

General aspects of electro-optic sampling were discussed in Section 2. The specific EOS geometry used for these experiments was the reflection-mode geometry, which is shown in detail in Fig. 13. In this embodiment a thin plate of LiTaO$_3$ with a high reflectivity dielectric coating on one surface was placed on top of the GaAs sample with the dielectric in contact with the GaAs. A small window, not shown in the figure, was etched in the dielectric coating to allow for passage of the excitation through the LiTaO$_3$ to the photoconductive switch. The excitation beam was focussed symmetrically on the gap. The probe beam was focussed between the transmission lines a short distance from the gap. Voltage calibration of the measurements was carried out by applying a known voltage at the lock-in frequency to the transmission line and recording the dc lock-in output. During measurements of the photoconductive transients, one side of the gap was biased with a known dc voltage, the transmission line on the opposite side was terminated into 50 ohms, and the excitation beam was chopped at the lock-in frequency.

The sample geometry consisted of a coplanar transmission line with a gap in one line to form the photoconductive switch. The dimensions of the stripline were 50μm wide lines separated by 50μm, and the gap length used was 10um. The pump and probe beams were each focussed separately to a spot size of 10μm, and the propagation distance from the gap to

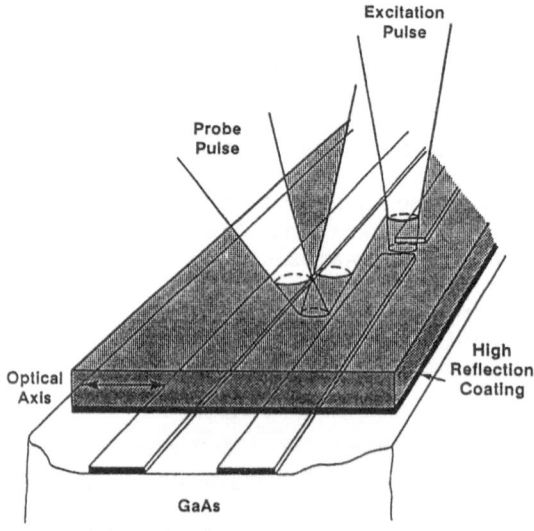

Fig. 13 Reflection-mode sampling geometry used in the transient photoconductivity measurements[9,10].

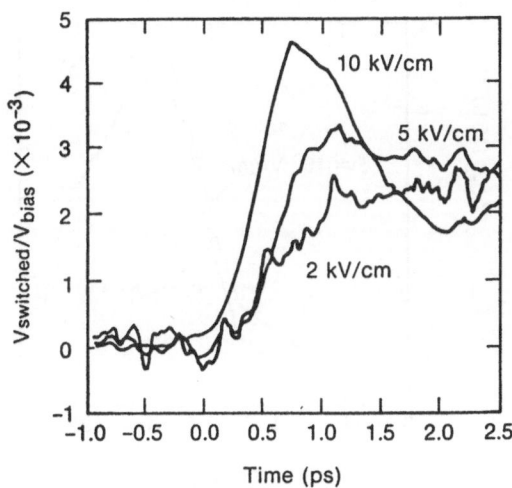

Fig. 14 Measured photoconductive voltage transients obtained for 2.0eV excitation [9,10].

the probe point was 20μm.

As discussed in Section 2, the temporal resolution of a particular sampling geometry is determined by the pump and probe pulsewidths, the probe spot size, and the extent of the electric field lines traversed by the probe pulse. For the geometry described above, this results in a temporal resolution of approximately 200fs for the 620nm excitation experiment and 500fs for the 760nm excitation measurements, the difference being due solely to the difference in laser pulsewidths.

Results obtained with 620nm excitation are shown in Fig. 14, plotted as the transient voltage measured across the transmission line normalized to the dc bias voltage. Note that the measured transient voltage was only 0.01% of the bias voltage and therefore the associated displacement current was much smaller than the particle current. Two features are clearly present in the data: a significant voltage overshoot that occurs at high bias voltage but not at the intermediate or low bias, and a much faster risetime of the transient for high bias. The overshoot is interpreted as evidence of electron velocity overshoot in the GaAs. The degree of overshoot and the temporal behavior is very similar to the velocity overshoot calculated for the same conditions in Fig. 9. Furthermore, the delay in the onset of the transient at low field is interpreted as evidence of the bias-dependent Jones-Rees effect discussed in the previous section, which is also apparent in Fig.9. These experimental results constitute the first fully time-resolved measurements of velocity overshoot and the Jones-Rees effect in GaAs.

In order to further investigate the interpretation of this data the experiment was repeated with an excitation wavelength of 760nm. Recall from the previous section that in this case there should be no bias-dependent onset of the transient. This is because the electrons are excited below the threshold for Γ-L scattering and therefore there is no bias-dependent Jones-Rees effect (see, for example, Fig. 10). The experimental results are shown in Fig. 15. A definite voltage overshoot was observed once again at high bias but not at low bias, exhibiting the expected velocity overshoot behavior. In addition, within the noise level of the experiment, no bias-dependent delay in the onset was observed, consistent with the interpretation that no Jones-Rees effect occurred under these conditions. However, this result is not conclusive since the temporal resolution in this case was only 500fs, which may not have been sufficient to resolve a bias-dependent delay.

The experimental results are in qualitative agreement with the Monte Carlo simulations, but are not in qualitative agreement with the circuit model predictions, which may be seen by comparing Figs. 12 and 14. The discrepancy may be due to the existence of a spatially nonuniform field in the n^+-i-n^+ structures investigated. It is possible that nonuniform fields existed in the samples due either to Gunn domain formation[43,44] or space-charge effects. The latter may be included in the device simulation by introducing a Poisson solver into the

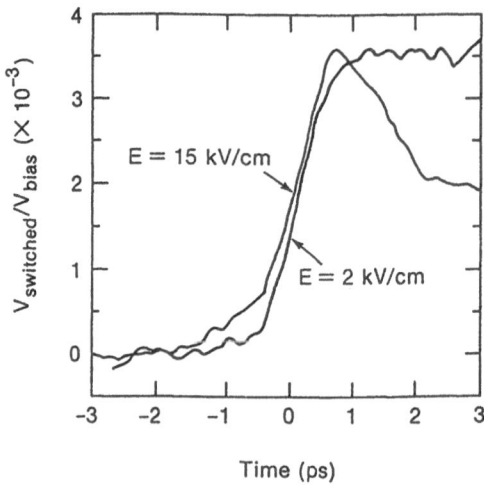

Fig. 15 Measured photoconductive voltage transients obtained for 1.63eV excitation[9,10].

model which will properly take into account field nonuniformities caused by the flow of the photocurrent.

Ongoing experiments at the University of Cambridge will utilize reverse-biased horizontal p-i-n structures fabricated on MBE GaAs to insure a uniform field in the samples and defeat the formation of Gunn domains. The experiments will be performed with both 620nm and 800nm excitation sources with 100fs pulsewidths. In addition, measurements will be carried out in AlGaAs samples to study in detail the Jones-Rees phenomenon. Finally, the measurements will also be repeated at cryogenic temperatures where velocity overshoot phenomena will be optimized. These measurements, in conjunction with the Monte Carlo simulations being performed at Arizona State University, will reduce the uncertainty in the range of parameters necessary for accurate simulation of device physics and transport processes.

4. TIME DOMAIN CHARACTERIZATION OF THE RESONANT TUNNELING DIODE (RTD)

4.1 Transport Physics of the RTD

The concept of devices based on quantum mechanical tunneling through double-barrier heterostructures was introduced by Tsu and Esaki[44]. The advantage of such a structure is that, in principle, the speed of the device could be determined by the tunneling time of electrons through the device, which can be very fast. Evidence for such high speed transport was presented by Sollner et. al.[45], who showed that the negative resistance feature in the conductance-voltage curve was evident to frequencies as high as 2.5THz. More recently RTD's have demonstrated fundamental oscillation frequencies in excess of 200GHz [46]. A full quantum mechanical simulation of the transient response of a RTD has indicated that current oscillations occur within the device on a time scale of 100fs[47].

A simplified band diagram of a typical RTD is shown in Fig. 16. Discrete electronic levels are formed in the thin GaAs layer sandwiched between the higher-bandgap AlGaAs barriers. Also shown in the figure is a representative I-V curve of the device. When the applied voltage is sufficient to align the Fermi level in the emitter with the quantized well level then resonant tunneling occurs and the current through the device increases sharply. At higher voltages the Fermi level is no longer in resonance and the current decreases. Devices are characterized by the slope of the negative resistance region and ratio between the peak and valley currents.

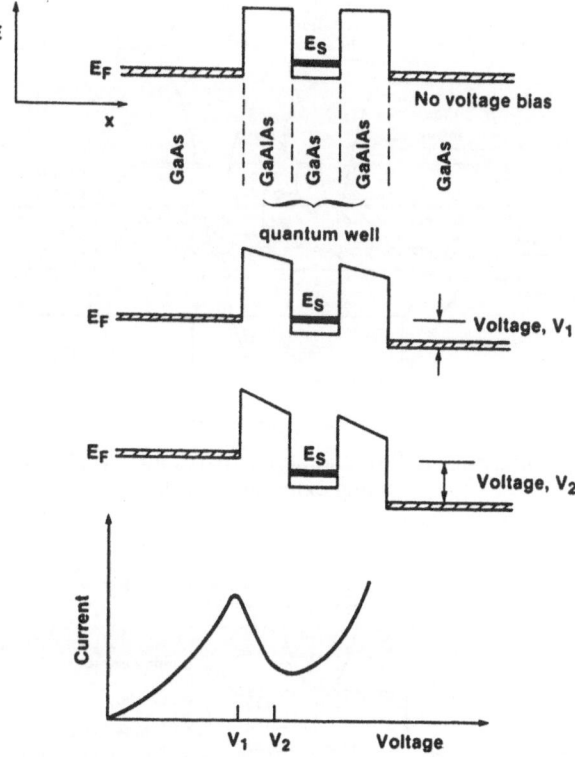

Fig. 16 Band diagram and I-V curve of a Resonant Tunneling Diode.

The transient behaviour may be described as follows. The region of negative resistance is unstable and can give rise to oscillations. If the RTD is biased just below resonance and a step in voltage is applied to take the device above threshold, the device will switch. The device current will change significantly and the new operating point will be to the right of the local I-V minimum. The load line of the device, determined by the external circuit and the device fixture, will govern the post-switching operating point. The switching time will be a function of the tunneling time through the device and the device parasitics.

It is important to study the response of RTD's in the time and frequency domain for two reasons. First, measurements will yield information about the tunneling times into and out of the quantum well, particularly if the well parameters, applied field, and temperature are varied. Most predicted tunneling times fall within the range of 10^{-13}-10^{-11} seconds, depending critically on these parameters. Recently a measured tunneling time of 7.5pS between two coupled wells separated by a 55Å barrier was reported[48]. Secondly, these measurements will address practical questions concerning the usefulness of such devices as switching components, oscillators, and mixers.

4.2 Picosecond Switching Times Measured Using External EOS

This section will discuss the results of Whitaker et. al.[49], who were the first researchers to study the transient response of a resonant tunneling diode on a picosecond time scale. They used a coplanar EOS arrangement with $LiTaO_3$ as the sampling medium, as shown in Fig. 17. The input to the device consisted of a dc bias combined with a fast voltage step generated by a GaAs photoconductive switch. The amplitude of the input voltage step was variable and could be made as large as 174mV. The RTD was mounted on edge and electrical connection to the emitter mesa was achieved with a gold-coated phosphor-bronze wire that had been etched to a 1μm tip. The device output was connected via silver epoxy to a coplanar

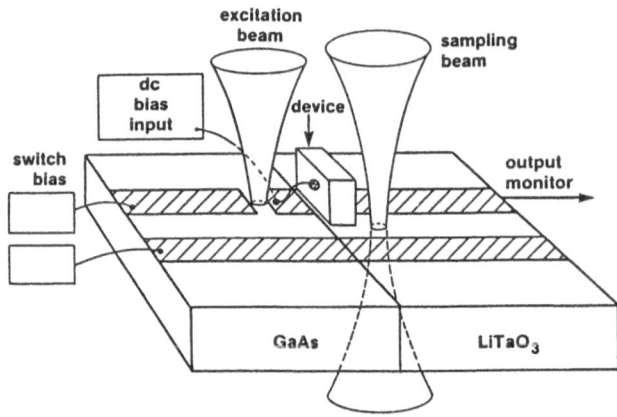

Fig. 17 Experimental configuration for time-resolved RTD characterization using external EOS[49].

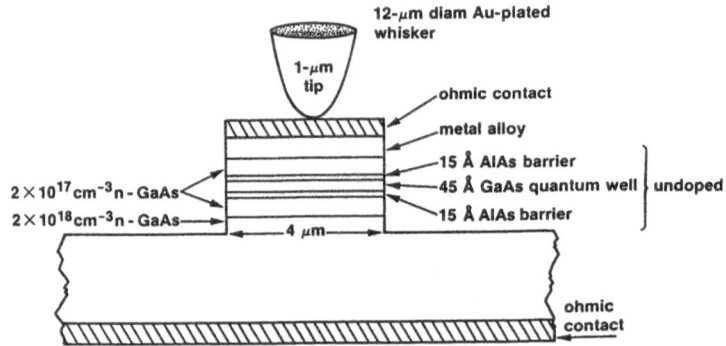

Fig. 18 Cross-section of the RTD characterized by Whitaker et. al[49].

transmission line fabricated on LiTaO$_3$. Synchronized 80fs pump and probe pulses from a CPM dye laser were used to activate the GaAs switch and probe the transient device output voltage across the LiTaO$_3$ modulator. The transient response was measured as a function of the dc bias voltage at room temperature.

The device under test in this case, shown in cross-section in Fig. 18, was grown via MBE and consisted of a 45Å GaAs well sandwiched between two 15Å AlAs barriers[50]. The outer regions of the device were doped to $2 \times 10^{17} cm^{-3}$ and the current density was $4 \times 10^4 A/cm^2$. A 4μm diameter RTD was tested which had a calculated capacitance under bias of 20fF and a series resistance of 15ohms.

Experimental results are shown in Fig. 19. In the first column (a) are shown the dc I-V curves and the load lines for different bias conditions. The second column (b) shows an analytical representation of the corresponding output voltage transients, and column (c) displays the experimental results. The quantity that was measured was the <u>difference</u> between the responses at different bias levels. The "reference waveform" in each case is curve 2, the transient response under low applied bias. At that operating point the I-V curve is approximately linear and the device response is passive, the shape of the output transient reflecting the shape of the incident voltage transient from the GaAs switch. Row 1 in the figure compares the response of two transients with the device biased well below resonance. Curve 1 is at a higher dc bias than curve 2, but still below resonance, and the I-V curve remains quasi-linear. The difference signal in column 3 shows no major features and reflects only the difference in curvature of the I-V curve at the two biases.

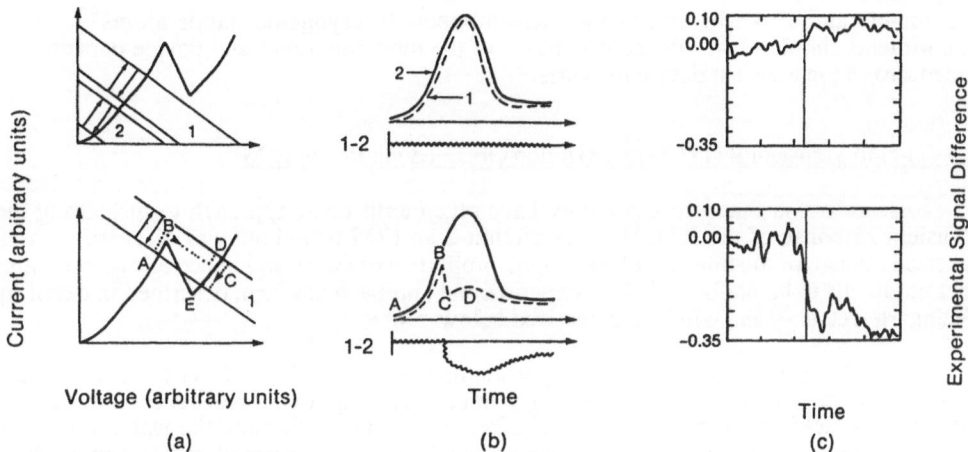

Fig. 19 RTD switching versus bias conditions: (a) I-V curves with load lines; (b) representations of output voltage waveforms resulting from shift in load lines; (c) experimental difference signal[49].

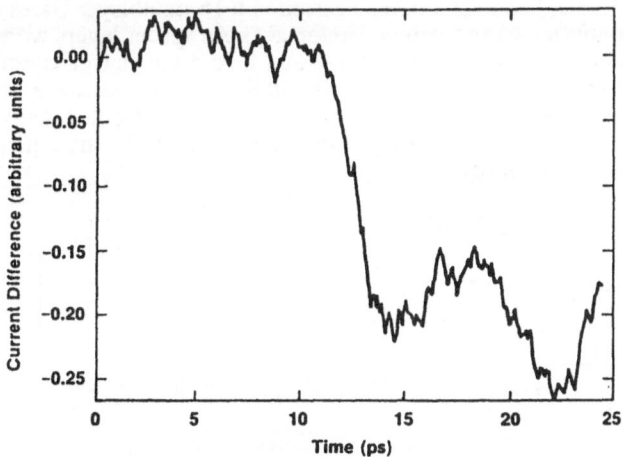

Fig. 20 Expanded plot of the difference signal with the device biased near resonance. The 10%-90% risetime of the switching transient is 1.9ps[49].

Now consider what happens when the bias point of curve 1 is moved up to just below the switching threshold. The additional voltage provided by the input step drives the device from points A to B, raising it above threshold. There is subsequently a sharp switching transient from points B to C, and as the input voltage continues to rise the output voltage follows briefly to point D, before settling back to the stable point E. On a longer time scale the device is assumed to switch back to its original operating point A.

Fig. 20 shows an expanded plot of the experimental difference signal obtained with the RTD biased near resonance. The 10%-90% switching time was 1.9ps. Whitaker et. al. compared this result with a calculation of the charging time of the well[50,51], assuming a voltage difference of 440mV, a current difference of 3mA, and a calculated device capacitance of 20fF. The calculated result was 2.9ps which compared favorably with the experimental result.

It is not clear how the tunneling time is associated with the measured switching between device states, and further experiments are needed. Whitaker and Mourou are carrying out studies at the University of Michigan at Ann Arbor of this device response for different well parameters and also will extend the measurements to cryogenic temperatures[53]. These experiments should clarify the contributions of the tunneling times and device parasitics to the macroscopic transient device response.

4.3 On-chip Transient RTD Characterization Obtained via In-situ EOS

Researchers at Stanford University have taken a different approach to measuring the transient response of the RTD[54,55]. In section 2 an EOS technique was discussed which takes advantage of the inherent electro-optic properties of GaAs and hence allows on-wafer measurements to be performed. This experimental approach has been described in detail by Weingarten et. al.[56] and will be summarized below.

The EOS system used for these experiments is shown in Fig. 21. The source of sub-bandgap probe pulses is the fiber-grating compressed output of a Nd:YAG mode-locked laser which yields 1.5ps pulses. The probe beam is focussed through the wafer substrate, reflects off of the device metallization whose potential is being measured, and is detected by a slow photodiode and vector receiver. The wafer and probe beam are integrated into a cascade wafer probe station to provide microwave excitation of the device and simultaneous S-parameter measurements.

The device under test in this case was a planarized version of the RTD, shown in Fig. 22[57]. The double-barrier heterostructure consisted of a 16-monolayer GaAs well sandwiched between two 6-monolayer AlAs barriers. Undoped GaAs spacer layers were grown adjacent to the barriers and doped GaAs layers above and below the spacer layers were grown to provide the top and bottom ohmic contacts. The individual devices were isolated via proton implantation. This device was designed for maximum current density and minimum capacitance, which according to the analysis of Diamond et. al.[54] are critical to achieve the shortest possible switching times.

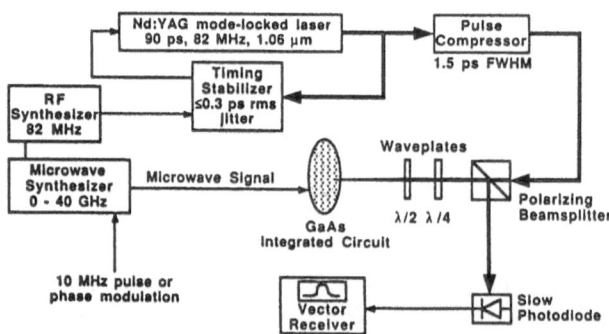

Fig. 21 EOS system used for on-wafer RTD testing[56].

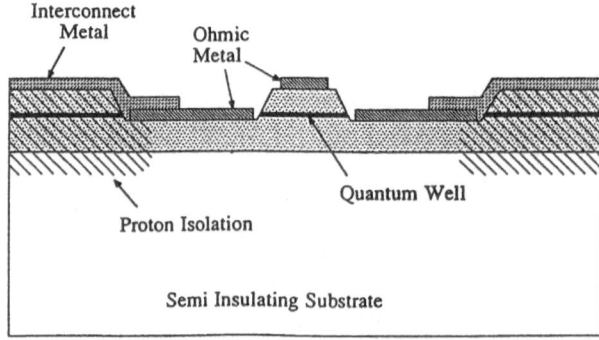

Fig. 22 Cross section of the planarized RTD[57].

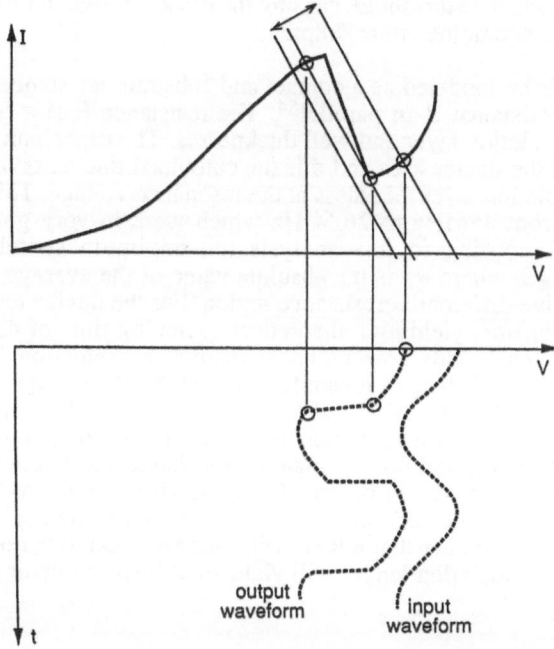

Fig. 23 Schematic representation of the expected output waveform of the planarized RTD when the input is a sinusoid plus a dc bias[55].

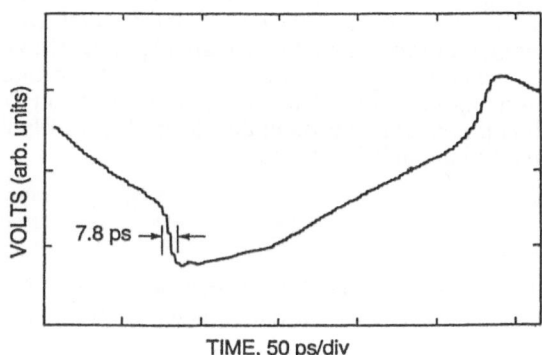

Fig. 24 Output switching waveform measured with on-wafer EOS. The risetime of the fast step is 7.8ps.

A graphical representation of the expected transient response is shown in Fig. 23[55]. The electrical input to the device was a variable dc bias superimposed on a 2GHz sine wave. The device response is similar to that discussed in the previous section, with the difference that in this case the input transient was a sine wave instead of a step function. When the device was biased just below resonance and the input sine wave drove it above threshold, the device switched rapidly, indicated by a step in the output waveform. At a later time the falling input waveform caused the device to be switched back to its original state.

A representative measured output waveform is shown in Fig. 24[55]. The measured response indicates that the device followed the sinusoidal input except when the sinusoid fell above or below the resonance threshold, causing the device to switch rapidly. The switching time for this RTD was determined to be 7.8ps.

The device has been modeled as a contact and substrate resistance R_s in series with a capacitance C and resistance R in parallel[54]. The resistance R is a function of the well impedance and the depletion layer and well thicknesses. The capacitance was estimated by $C = \varepsilon A/d$ where A is the device area and d is the combined thickness of the double barrier thickness and the depletion layer thickness at the resonance voltage. This model resulted in predictions of S_{11} from 45MHz to 26.5GHz which were in very good agreement with measured values. According to this analysis the minimum switching time may be approximated by $5R_nC$, where R_n is the absolute value of the average negative resistance throughout the negative differential resistance region. For the device tested Rn was 650Ω–μm^2 and C was $1.3fF/\mu m^2$, yielding a theoretical switching time of 4.2ps. The measured risetime was several picoseconds slower than the theoretical prediction. This may have been due to jitter in the device switching or nonuniformities in the device capacitance.

Note that according to this interpretation the device speed is determined solely by the RC time constants of the device and not by quantum mechanical time constants. Diamond et. al.[54,55] argue that the resonant build-up time for this device is of the order of 150fs, and the depletion layer transit times are also subpicosecond; these times are negligible relative to the RC time constants. They predict that a RTD with increased cathode doping levels, thinner barriers, and optimized depletion length will yield an ultimate switching speed of less than 5ps.

ACKNOWLEDGEMENTS

It is a pleasure to acknowledge the continued collaboration with Robert Grondin at Arizona State University, who has developed the Monte Carlo simulations of the electron transport and circuit response and has been the source of many useful discussions. The RTD testing using external EOS was performed by John Whitaker at the University of Rochester under the supervision of Gerard Mourou (both are currently at the University of Michigan at Ann Arbor); they have generously provided extensive data from their experiments. The device characterization using in-situ EOS was carried out by Scott Diamond at Stanford University under the direction of David Bloom; they have also readily provided figures and data from both published and unpublished work.

REFERENCES

1. J. A. Valdmanis, G. Mourou, and C. W. Gabel, Appl. Phys. Lett. **41**, 211 (1982).
2. J. F. Whitaker, T. B. Norris, G. A. Mourou, and T. Y. Hsiang, IEEE Trans. Microwave Theory and Tech. **MTT-35**, 42 (1983).
3. Charles J. Cryjak, Kevin E. Meyer, and Gerard A. Mourou, Proceedings of the Topical Meeting on Picosecond Electronics and Optoelectronics, Lake Tahoe, Nevada, March 13-15, 1985 (Springer-Verlag, New York) pp.244-8.
4. K. E. Meyer, D. R. Dykaar, and G. A. Mourou, ibid, pp.54-7.
5. Douglass R. Dykaar, Ph.D thesis, Dept. of Electrical Engineering, University of Rochester, New York, May, 1987.
6. U. Keller, S. K. Diamond, B. A, Auld, and D. M. Bloom, Appl. Phys. Lett. **53**, 388 (1988).
7. K. J. Weingarten, M. J. Rodwell, H. K. Heinrich, B. H. Kollner, and B. M. Bloom. Electron. Lett. **21**, 765 (1985).
8. X. C. Zhang and R. K. Jain, Electron. Lett. **22**, 264 (1986).
9. Kevin Meyer, Maurice Pessot, Gerard Mourou, Robert Grondin, and Sleiman Chamoun, Appl. Phys. Lett. **53**, 2254 (1988).
10. S. N. Chamoun, R. Joshi, E. N. Arnold, R. O. Grondin, K. E. Meyer, M. Pessot, and G. A. Mourou, to be published in the Journal of Applied Physics, July 1989.

11. J. Nees and G. Mourou, Electron. Lett. **22**, 918 (1986).
12. J. M. Weisenfeld, R. S. Tucker, A. Antreasyan, C. A. Burrus, A. J. Taylor, V. D. Mattera jr., and P. A. Garbinsk, Appl. Phys. Lett. **50**, 1310 (1987).
13. G. A. Mourou and K. E. Meyer, Appl. Phys. Lett. **45**, 492 (1984).
14. K. E. Meyer and G. A. Mourou, Electron. Lett. **21**, 568 (1985).
15. J. A. Valdmanis, Electron. Lett. **23**, 1308 (1987).
16. B. H. Kollner and D. M. Bloom, IEEE J. Quantum Electron. **JQE-22**, 79 (1986).
17. D. Grischkowsky and A. C. Balant, Appl. Phys. Lett. **41**, 1 (1982).
18. Hartmut Roskos, S. Optiz, Alois Seilmeir, and W. Kaiser, IEEE J. Quantum Electron.**JQE-22**, 697 (1986).
19. Martin P. Dawson, Thomas F. Boggess, and Arthur L. Smirl, Optics Lett. **12**, 590 ((1987).
20. P. Beaud, B. Zysset, A. P. Schwarzenbach, and H. P. Weber, Optics Lett. **11**, 24 (1986).
21. N. Langford, K. Smith, and W. Sibbet, Opt. Commun. **64**, 274 (1988).
22. B. H. Kollner, D. M. Bloom, and P. S. Cross, Electron. Lett. **19**, 574 (1983).
23. R. O. Grondin, P. Lugli, D. K. Ferry, and H. L. Grubin, in Picosecond Optoelectronics, Proc. SPIE **439**, 18 (1983).
24. D. K. Ferry, H. L. Grubin, and G. J. Iafrate, in Semiconductors Probed by Ultrafast Laser Spectroscopy, ed. by R. R. Alfano (Academic Press, NY, 1984).
25. K. Hess and G. J. Iafrate, Proc. IEEE **76**, 519 (1988).
26. R. B. Hammond, Physica **B134**, 475 (1985).
27. G. Mourou, K. Meyer, J. Whitaker, M. Pessot, R. Grondin, and C. Caruso, in Picosecond Electronics and Optoelectronics II, Springer Series in Electronics and Photonics **24**, 40 (1987).
28. M. C. Nuss, D. H. Auston, and F. Capasso, Phys. Rev. Lett. **58**, 2355 (1987).
29. R. Joshi, S. Chamoun, and R. O. Grondin, Proceedings of the Picosecond Electronics and Optoelectronics Topical Meeting, Salt Lake City, Utah, March 8-10, 1989 (Optical Society of America, 1989).
30. P. Lugli and D. K. Ferry, IEEE Trans. Electron. Dev. **ED-32**, 431 (1985).
31. M. A. Osman and D. K. Ferry, J. Appl. Phys. **61**, 5330 (1987).
32. M. D'yakonov, I. Perel, and I. N. Yassievich, Sov. Phys. Semicond. **11**, 801 (1987).
33. P. Lugli, C. Jacobani, L. Reggiani, and P. Kocevar, Appl. Phys. Lett. **50**, 1521 (1987).
34. D. Jones and H. D. Rees, J. Phys. **C6**, 1781 (1973).
35. R. O. Grondin and M. J. Kann, Solid State Electron. **31**, 567 (1988).
36. J. A. Kash, S. S. Jha, and J. C. Tsang, Phys. Rev. Lett. **58**, 1869 (1987).
37. G. M. Wysin, D. L. Smith, and A. Redondo (unpublished).
38. R. Brennan and K. Hess, Phys. Rev. **B29**, 5581 (1984).
39. A. J. Taylor, D. J. Erskine, and C. L. Tang, J. Opt. Soc. Am. **B2**, 663 (1985).
40. J. Shah, B. Deveaud, T. C. Damen, W. T. Tsang, A, C. Gossard, and P. Lugli, Phys. Rev. Lett. **59**, 2222 (1987).
41. D. H. Auston, IEEE J. Quantum Electron. **19**, 639 (1983).
42. J. F. Whitaker, R. Sobolewski, D. Dykaar, T. Hsiang, and G. Mourou, IEEE Trans. Microwave Theory and Techniques **MTT-36** (1988).
42. Kevin E. Meyer, Ph.D thesis, Department of Physics and Astronomy, University of Rochester, New York, May 1988.
43. H. L. Grubin, private communication.
44. R. Tsu and L. Esaki, Appl. Phys. Lett. **22**, 562 (1973).
45. T. C. L. G. Sollner, W. D. Goodhue, P. E. Tannerwald, C. D. Parker, and D. D. Peck, Appl. Phys. Lett. **43**, 588 (1983).
46. E. R. Brown, W. D. Goodhue, and T. C. L. G. Sollner, J. Appl. Phys. **64**, 1519 (1988).
47. N. C. Kluksdahl, A. M. Kriman, and David K. Ferry, IEEE Electron Dev. Lett. **9**, 457 (1988).
48. D. Y. Oberli, J. Shah, T. C. Damen, C. W. Tu, and D. A. B. Miller, Proceedings of the Quantum Wells for Optics and Optoelectronics Topical Meeting, March 6-8, 1989, Salt Lake City, Utah (Optical Society of America, Washington, DC).

49. J. F. Whitaker, G. A. Mourou, T. C. L. G. Sollner, and W. D. Goodhue, Appl. Phys. Lett. **53**, 385 (1988).

50. W. D. Goodhue, T. C. L. G. Sollner, H. Q. Le, E. R. Brown, and B. A. Vojak, Appl. Phys. Lett. **49**, 1086 (1986).

51. D. D. Coon and H. C. Liu, Appl. Phys. Lett. **49**, 94 (1986).

52. H. C. Liu and D. D. Coon, Appl. Phys. Lett. **50**, 1246 (1987).

53. John Whitaker, private communication.

54. S. K. Diamond, E. Ozbay, M. J. W. Rodwell, D. M. Bloom, Y. C. Pao, and J. S. Harris, Appl. Phys. Lett. **54**, 153 (1989).

55. S. K. Diamond, E. Ozbay, M. J. W. Rodwell, D. M. Bloom, Y. C. Pao, E. Wolak, and J. S. Harris, Proceedings of the 1989 Picosecond Electronics and Optoelectronics Conference, March 8-10, 1989, Salt Lake City, Utah (Optical Society of America, Washington, DC, 1989).

56. Kurt J. Weingarten, Mark J. W. Rodwell, and David Bloom, IEEE J. Quant. Electron. **QE-24**, 198 (1988).

57. S. K. Diamond, E. Ozbay, M. J. W. Rodwell, David M. Bloom, Y. C. Pao, E. Wolak, and James H. Harris, IEEE Electron Dev. Lett. **10**, 104 (1989).

OPTICAL TRANSPORT EXPERIMENTS IN HETEROSTRUCTURES

R.A. Höpfel, R. Christanell, S. Juen, and N. Sawaki[*]

Institut für Experimentalphysik, Universität Innsbruck
A–6020 Innsbruck, Austria

ABSTRACT

Various transport processes in GaAs/AlGaAs heterostructures are investigated by optical techniques using ultrashort laser pulses.

(1) The transfer of optically injected electron–hole populations in single heterostructures from AlGaAs to GaAs is observed by studying the luminescence above the AlGaAs bandgap. The technique of "population correlation" allows time–resolved luminescence experiments also at low emission intensities. We find that the "thermal emission" of carriers from AlGaAs to GaAs is limited by ambipolar diffusion, leading to transfer times of 5 to 11 ps depending on the layer thickness.

(2) In double–quantum–well structures consisting of quantum wells with two different well widths the tunneling times from the narrow well (higher ground state) to the wide well are quantitatively measured. The tunneling times are of the order of 200 ps, due to the barrier width and the nonresonant process which requires additional phonon emission or absorption for momentum conservation.

(3) Electron–hole scattering in quantum wells leads to negative mobility of minority carriers in modulation–doped structures due to the "carrier drag" effect. Recently we could observe also the negative photoconductivity associated with the negative mobility. Quantitative evaluation gives more exact values on electron–hole momentum scattering times.

(4) Photoluminescence from materials for subpicosecond photoconductors (radiation–damaged $Ga_{0.47}In_{0.53}As$) gives direct information on the ultrafast recombination processes. Decay times of 0.9 ps are observed, as well as evidence for extreme nonequilibrium carrier distributions ("inverted" luminescence spectra).

1. TIME–RESOLVED LUMINESCENCE IN SINGLE HETEROSTRUCTURES

The interest of the present paper is to study the transport of photoinjected carriers in the AlGaAs layer of single heterostructures, i.e. the "real space transfer" from AlGaAs to GaAs, by means of time–resolved luminescence experiments. Time–resolved photoluminescence can be measured directly with (sub)picosecond time resolution by a streak camera[1] or by optical gating techniques using frequency up–conversion[2]. We use the "population correlation" technique[3-5] for measuring the time dependence of the photoinjected carrier population in AlGaAs. This technique is based on the non–linear dependence of the luminescence intensity on the excitation intensity: The time dependent luminescence intensity $I(\omega,t)$ in a given wavelength interval is proportional to

---

[*] Permanent address: Department of Electronics, Nagoya University, Chikusa–ku, Nagoya 464, Japan.

the product of the electron and hole concentrations $n(\mathbf{k},\mathbf{x},t)$, $p(\mathbf{k},\mathbf{x},t)$ in the corresponding momentum interval:

$$I(\omega,t) \propto n(\mathbf{k},\mathbf{x},t) \cdot p(\mathbf{k},\mathbf{x},t) \tag{1}$$

The luminescence (at various wavelengths) above the bandgap of the AlGaAs is measured as a function of the delay time of two laser pulses using a heterodyne technique[3]: The two parts of a splitted laser beam are chopped with different frequencies (ω_1, ω_2). One of the beams is time–delayed (τ) in respect to the other and both beams are focussed on the sample (see Fig. 1). The electron and hole densities generated by the first pulse (chopped with ω_1) at time t are $n_1(t)$ and $p_1(t)$, and the photogenerated densities generated by the delayed second pulse (chopped with ω_2) are $n_2(t-\tau)$ and $p_2(t-\tau)$. The luminescence signal is measured with a photomultiplier and integrated in a lock–in amplifier triggered with the difference frequency $\omega_1-\omega_2$. Since only the mixed terms with the difference frequency $\omega_1-\omega_2$ contribute to the measured signal, the measured time integrated luminescence signal is given by:

$$\int I(t,\tau) \cdot dt \propto \int \left[n_1(\mathbf{k},\mathbf{x},t) \cdot p_2(\mathbf{k},\mathbf{x},t-\tau) + p_1(\mathbf{k},\mathbf{x},t) \cdot n_2(\mathbf{k},\mathbf{x},t-\tau) \right] \cdot dt \tag{2}$$

Thus, the recorded correlation signal is proportional to the cross correlation of the electron concentration injected by one beam and the hole concentration injected by the other beam. If we assume an exponential decay of the electron and hole concentration with a relaxation time τ_n for electrons and τ_p for holes, the physical meaning of the cross correlation becomes clear $[n_1(0) = p_1(0), n_2(0) = p_2(0)]$:

$$\int I(t,\tau) \cdot dt \propto n_1(0) \cdot p_2(0) \cdot \left[\exp(-\tau/\tau_n) + \exp(-\tau/\tau_p) \right] \tag{3}$$

The dependence of the correlation signal on the delay time τ directly gives the time dependence of the electron and hole concentrations. This technique allows us to study the time–resolved electron and hole populations even at low luminescence intensities.

In our experiments we use a "balanced colliding pulse modelocked dye laser"[6] (CPM: $\lambda \approx 620$ nm, repetition rate 100 MHz). The time resolution is in the order of 100 fs due to the small pulse width. Our experimental set–up is shown in figure 1.

We investigated three different single heterostructure (SH) samples with different AlGaAs layer thickness. All samples have some general structure of a thick undoped GaAs layer followed by undoped and doped AlGaAs layers (different in each sample) and a thin GaAs top–layer.

Sample I (d = 1750 Å AlGaAs layer thickness, insert top left Fig. 2) is a p–i–i–n structure[7] with the following dimensions: On top of a semi–insulating substrate (Cr–doped GaAs), a Be–doped GaAs layer is grown (thickness 1 μm, $p = 10^{18}$ cm^{-3}), followed by undoped GaAs (500 Å), undoped $Al_{0.3}Ga_{0.7}As$ (1000 Å), Si–doped $Al_{0.3}Ga_{0.7}As$ (750 Å, $n = 10^{18}$ cm^{-3}), and a top layer of 100 Å Si–doped GaAs ($n = 10^{18}$ cm^{-3}).

Sample II (d = 1145 Å) has following layers: a Cr–doped GaAs substrate, about 1 μm undoped GaAs, then four $Al_{0.26}Ga_{0.74}As$ layers with 107 Å undoped, 214 Å Si–doped, 642 Å undoped, 182 Å Si–doped, and as top layer 220 Å Si–doped GaAs.

The third sample (d = 495 Å) has a 4 μm GaAs substrate, 215 Å $Al_{0.34}Ga_{0.66}As$ undoped and 280 Å Si–doped ($n = 3.5 \times 10^{18}$ cm^{-3}) layer, and an undoped 180 Å ($p < 10^{14}$ cm^{-3}) GaAs top layer.

In figure 2 the correlation signals as a function of the delay time τ at room temperature are shown. The insert top right in figure 2 shows the luminescence spectra. The arrow indicates the maximum of the AlGaAs emission where the data have been obtained. In the insert top left the typical bandedge of the heterostructure is plotted. The correlation time τ_r decreases with decreasing AlGaAs layer thickness from 11 ps for d = 1750 Å, to 9 ps (d = 1145 Å) and to 5 ps for d = 495 Å. This suggests an interpretation by a transport process. Figure 2 shows that an exponential decay with the relaxation time τ_r is a good representation for the curves.

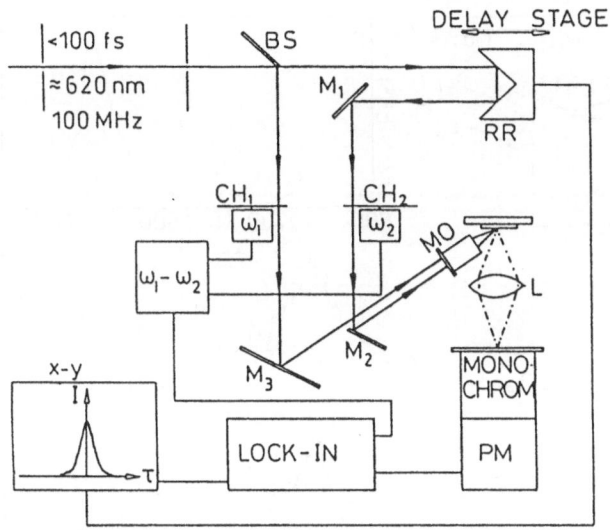

Fig. 1. Experimental setup. The outcoming beam of a CPM laser is split into two equal parts by a beamsplitter (BS). One part is chopped with frequency ω_1 ($\omega_1 = 250$ Hz), the second part is delayed by a retroreflector (RR) and chopped with ω_2 ($\omega_2 = 225$ Hz). The two beams are brought in a parallel arrangement via mirrors (M$_1$,M$_2$,M$_3$) and focussed on the sample through a microscope objective (MO). The luminescence signal is imaged with a lens (L) into a monochromator. The measured luminescence signal out of the photomultiplier (PM) is integrated in a lock–in amplifier triggered at $\omega_1-\omega_2$ ($\omega_1-\omega_2 = 25$ Hz). The correlation signal as a function of the delay is recorded.

Before injecting the carriers with the laser pulses the situation in the AlGaAs layers is as shown in the insert of figure 2: The AlGaAs layer exists of three parts: The depleted top layer (1), the doped region with the minimum in the band edge (2) and the spacer region (3). After exciting the carriers with the laser, the electric field in the depletion region and in the spacer separates the photoinjected electron–hole–pairs. Due to the high photoinjected carrier densities (about 1×10^{18} cm^{-3}), the space charge of the electrons and holes screen the internal electric field (after a displacement of 15 Å) and, therefore, the internal field can be neglected. This screening is a very fast process compared with the measured correlation times. The screening time τ_{screen} can be estimated if we assume that carriers move in the high internal field with an average velocity of 2×10^7 cms^{-1} through the distance needed to screen the field ($\tau_{screen} = s/v_d = 1.5 \cdot 10^{-9}$ m / $2 \cdot 10^5$ ms^{-1} = 7,5 fs). This time is much shorter than the times observed in the experiments (several picoseconds). Therefore, after injecting the carriers into the heterostructure there are flat bandedges in the AlGaAs and GaAs layers with a potential step at the interfaces. In this flatband situation (no electric fields present) the dynamics of the carriers is determined by energy relaxation, recombination and ambipolar diffusion.

The data of figure 2 suggest that the dominating process is transport of carriers out of the AlGaAs layers, because of the thickness dependence of the correlation times. The electric fields in the AlGaAs are screened and, caused by the strong space charge coupling between the carriers, ambipolar diffusion has to be taken into account, with an ambipolar diffusion constant $D = (n + p)/(n/D_P + p/D_n)$. We use the following model to describe the dominating processes for the evolution after the initial phase of thermalization and screening, i.e. the time after 1 ps: The carriers near the interface of the AlGaAs layer are transferred into the GaAs layer with thermal velocity (="thermal emission"). In GaAs the carriers lose energy by emitting optical phonons and by

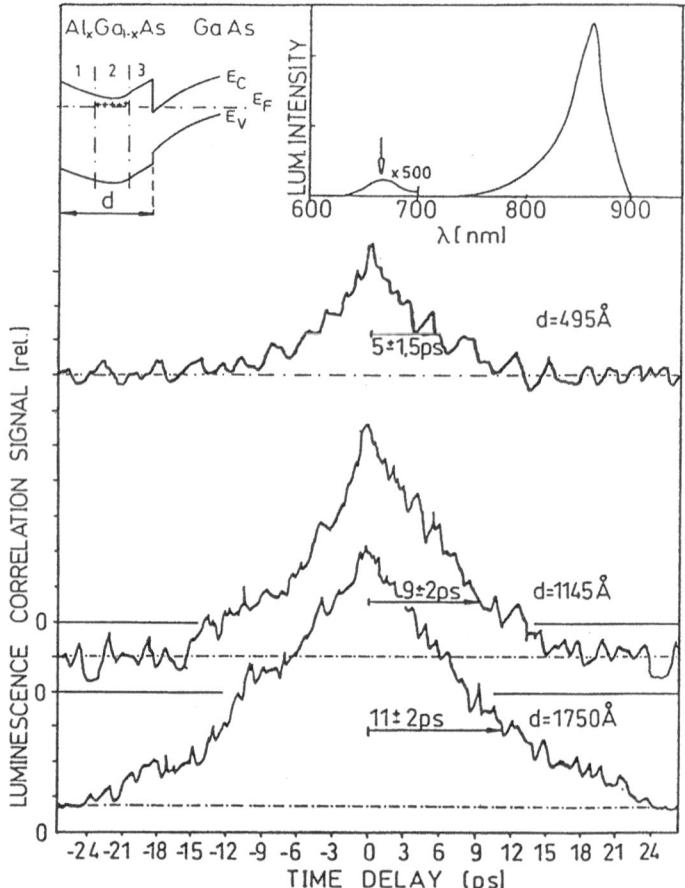

Fig. 2. Luminescence correlation signal as a function of delay for three different samples
with different AlGaAs layer thickness. Insert top right: Luminescence spectra of
the sample with d = 1365 Å. Insert top left: typical band edge of a AlGaAs/GaAs
heterostructure with the three described parts: 1 depleted top region, 2 doped
region, 3 spacer region.

carrier–carrier interaction and they relax <u>below</u> the band edge of the AlGaAs. Therefore,
at the interface the diffusion is only into one direction, which can be written as

$$j_z(\pm d/2, t) = \rho(\pm d/2, t) \cdot v_{th} \tag{4}$$

Equation (4) represents a boundary condition for the diffusion equation

$$\frac{d}{dt}\rho(z,t) = D \cdot \frac{d^2}{dz^2}\rho(z,t) \tag{5}$$

Numerical solution of equation (5) gives values of 6.4 ps for the thickest sample
(d = 1750 Å), 3.0 ps for d = 1145 Å (sample II) and for the thinnest one 0.5 ps
(d = 495 Å). These short times indicate that this process (thermal emission and
ambipolar diffusion) is the fastest process leading to a decay of the combined
electron–hole population, since recombination is not expected to be in the range below
10 ps. Therefore we interpret this process to be the dominating one.

The difference between the theoretical model and the experimental data are due
to the following reasons: Backscattering of carriers at the interfaces increases the
transfer time, if we take into account the quantum mechanical reflection of an electron

wave on a potential step. Classical backscattering by phonon scattering in GaAs has the same influence. Furthermore, valley transfer ($\approx 2,5$ ps for electrons from L to Γ valley in GaAs[9]) causes an initial rise of the luminescence which also leads to longer correlation times, as observed in our experiments.

2. TUNNELING IN DOUBLE QUANTUM WELL STRUCTURES

The double quantum well (DQW) structure with two different quantum wells of different width separated by a thin potential barrier, is an important structure for device applications. The dynamics of carriers, such as cooling of hot electrons and tunneling through the potential barrier has been the scope of several works[10]. Special interest is on the mechanism and the speed of the tunneling transfer of electrons from one quantum well into the other well.

Several authors have shown that negative differential resistance (NDR) devices can be obtained by utilizing the DQW structure, where the transfer of electrons from a quantum well with high mobility into the other quantum well with lower mobility plays an essential role (real space transfer)[11-14]. One of the present authors (N.S.) and Akasaki[15] claimed that one can treat these phenomena in the framework of the intersubband transitions. The intersubband transitions require momentum relaxation such as ionized–impurity scattering or phonon emission/absorption. The tunneling transfer in DQW systems can be calculated using the same concepts as for intersubband transitions. If the two quantized levels are not at the same energy (not resonant), the transfer of electrons is due to "indirect" tunneling assisted by impurity or phonon scattering[11].

We have investigated the carrier dynamics of photoexcited hot electrons by picosecond luminescence spectroscopy. The samples were made by molecular beam epitaxy on semiinsulating (001) GaAs. The unit cell of the DQW consists of two different GaAs quantum wells separated by thin $Al_xGa_{1-x}As$ ($x \approx 0.3$) barrier layer. The parameters of the structures are listed in Table 1. Ten unit cells of the structure are separated by thick layers of $Al_xGa_{1-x}As$ ($500 - 700$ Å).

Table 1. Parameters of DQW structures.

Sample	Doping	L_{w1} (GaAs)	$L_b(Al_xGa_{1-x}As)$	L_{w2} (GaAs)
378	$p_0=2\times10^{11}$ cm^{-2}	140 Å	60 Å (x=0.50)	60 Å
211	$n_0=4\times10^{11}$ cm^{-2}	175 Å	35 Å (x=0.26)	75 Å
209	$n_0=6\times10^9$ cm^{-2}	87 Å	82 Å (x=0.26)	——

Time–resolved photoluminescence spectroscopy was performed by using the two–beam correlation method as described above in Chapter 1. For the excitation we used again the CPM dye laser at $\lambda = 620$ nm (pulse width 150 fs, repetition rate 100 MHz). By measuring the correlation signal at various wavelengths as a function of the delay time τ, we can determine the decay time constant of the photoexcited hot carriers at various kinetic energies.

Figure 3 shows schematically the processes involved. Since the excitation energy (620 nm, 2.0 eV) is larger than the energy gap of the AlGaAs barrier layer, the carriers are excited in the whole region. After the capture time into the QW's, which is less than 1 ps [16], the hot electrons in the QW lose the energy down to the bottom or the lowest states in the QW by emitting phonons (B → C). Those carriers captured in the narrow QW escape into the wide QW by phonon–assisted tunneling (C → D → E). Because of the energy difference $\Delta\varepsilon_{12}$ between the lowest two subbands, the direct tunneling is forbidden due to momentum conservation. The same applies for the holes, except that the energy relaxation time is shorter than for the electrons. Due to the short capture times, the decay time of the photoluminescence longer than 1 ps from the narrow well is due to the energy relaxation of carriers within the quantum wells as well as the tunneling time between the wells.

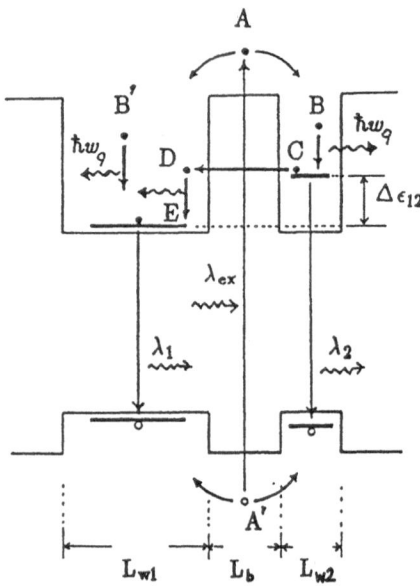

Fig. 3. Schematic illustration of the DQW structure and the processes concerned.

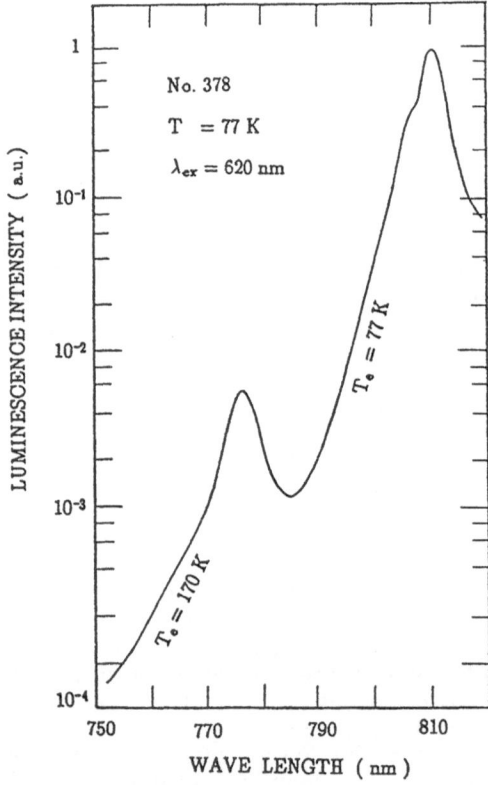

Fig. 4. Time–integrated photoluminescence spectrum of sample #378.

Figure 4 shows the typical photoluminescence spectra at 77 K. The main peak at $\lambda_1 = 812$ nm is by the electron–hole recombination in the wide quantum well, and the small peak at $\lambda_2 = 776$ nm is that in the narrow quantum well. The electron temperature determined by the high energy tail is $T_e = 77$ K for the wide quantum well and $T_e = 170$ K for the narrow well, which shows that the electrons in the wide well are in equilibrium with the lattice, in contrast to those in the narrow well. Since the sample is highly p–doped (with Be), the lifetime determined by the photoluminescence decay should reflect that of the electrons in the conduction band.

The correlation signals as a function of the delay time (not shown in this paper, see ref. 17) are described well by a single exponential decay $\exp(-\tau/\tau_e)$, from which we can determine the decay time constant τ_e of the electron population as a function of the luminescence wavelength. The results are shown in fig. 5: At the peak wavelength λ_1, the time constant was too long to be measured (longer than 1 ns). As the wavelength decreases, or as the kinetic energy of electrons and holes increases, the time constant decreases rapidly and reaches a value around 100 ps with a maximum at $\lambda = \lambda_2$. At the shortest wavelength measured, the correlation signal becomes very weak and shows a time constant as short as 15 ps.

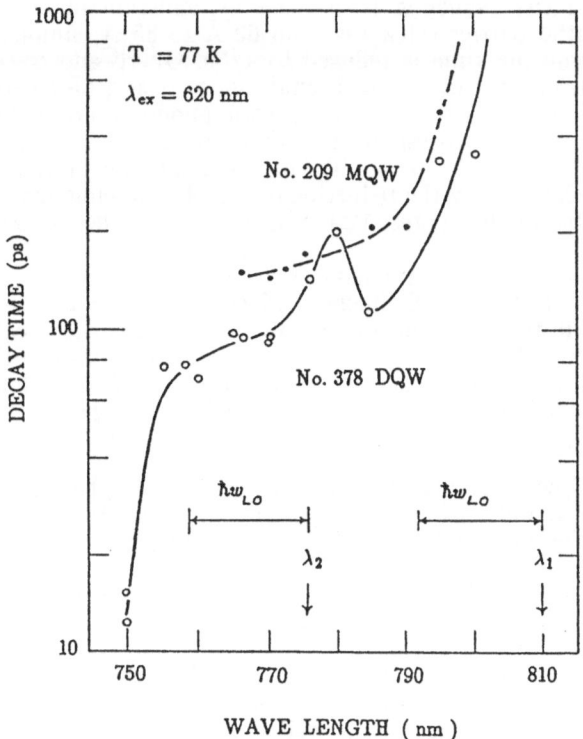

Fig. 5. Correlation time or the decay time of photoexcited hot electrons in sample No.378. For comparison, the data for a conventional multi–quantum well structure is also shown (#209). The decay time at $\lambda = 776$ nm gives the tunneling time of electrons from the bottom of the narrow well into the wide well. At the bottom of the wide quantum well the decay time is longer than 1 ns.

Since the relaxation time of hot holes to relax to a thermally steady state is much less than that of the electrons, the time constant obtained above is mainly determined by the relaxation time of electrons. The time constant longer than 1 ns at λ_1 should be the lifetime of electrons at the bottom of the wide well. Similarly, the time constant at λ_2 should be the lifetime of electrons at the bottom of the narrow quantum well. We cannot explain the tremendous difference in the magnitude by the quantum well width dependence of the lifetime[18]. An additional relaxation process for the electrons in the narrow quantum well is the *tunneling escape* into the wide quantum well. From the experimental data, the tunneling transition time in this sample is of the order of 150 ps. This time constant is nearly equal to the tunneling escape time in a double barrier (DB) structure. Using the results by Tsuchiya and coworkers[19-20] and assuming WKB approximation[21], we estimated the tunneling escape time in a DB structure having 60 Å $Al_{0.3}Ga_{0.7}As$ barriers. We obtained a value around 100 ps. The time constant obtained here is a little bit longer than the estimated one. This is probably due to the fact that the tunneling escape in DQW structures occurs into a quantized level in a wide well, while in DB structures there is a three–dimensional continuum outside the barriers.

The energy separation of the two quantized levels (see Fig. 4) is estimated to be 40 meV for the sample #378. Since this is larger than the LO phonon energy of GaAs, the tunneling is assisted by the emission of LO phonons. We performed similiar measurements about a sample #211 (see table 1). In this sample we also obtained a similar wavelength dependence of the decay time constant. But, in spite of the narrower potential barrier (L_b = 35 Å), the tunneling time constant is 300 ps, which is longer than that of sample #378. This is because the energy separation is as small as 24 meV for this sample, and the tunneling is assisted by emission of acoustic phonons. By decreasing the potential barrier thickness from 60 Å to 35 Å, simple WKB calculation suggests that the tunneling time is reduced to 1/50. Oberli and coworkers[22] measured the intersubband relaxation time τ_{12} in a quantum well and obtained 8 ps for the LO phonon emission process and 500 ps for the acoustic phonon process. Thus the relaxation time by the acoustic phonon process in more than 60 times longer than the LO phonon process. If this is the case, the possibility of the reduction of the tunneling time in sample #211 is cancelled out by the reduction of the phonon emission rate. Thus the role of LO phonon emission for the nonresonant tunneling process becomes obvious.

As pointed out earlier, the tunneling transition in DQW can be considered as a kind of intersubband transition. Seilmeier and coworkers[23] measured the decay time constant of electrons in the upper level in MQW's. They obtained a value of the order of 8 ps for ε_{12} = 100 meV. If we consider the overlap integral between the two associated subband states, the longer time constant of the order of 150 ps for sample #378 is attributed to the reduction of the overlap integral in the DQW structure. The precise estimation including phonon localization will be the subject of future work.

Next let us focus our attention to the relaxation phenomena in the narrow quantum well. The sample #378 is p–type doped with Be. Therefore we have a negligibly small number of equilibrium electrons in the DQW. Since the total number of photoexcited carriers is less than 5×10^{10} cm^{-2}, the number density in the narrow quantum well is estimated to be of the order of 5×10^8 cm^{-2}. Therefore the carrier–carrier scattering in the narrow quantum well is expected to be weaker than the electron–phonon interaction[24]. As shown in Fig.5, in the wavelength range of 755 − 770 nm, the time constant is rather insensitive to the wavelength. At even shorter wavelengths, we observe time constants as short as 15 ps. At these wavelengths the kinetic energy of electrons in the narrow well is larger than the LO phonon energy of GaAs. The sharp reduction of the decay time constant is attributed to the emission of LO phonons within the narrow quantum well.

3. ELECTRON–HOLE SCATTERING IN QUANTUM WELLS

In modulation–doped quantum well structures of GaAs/AlGaAs majority carriers from ionized acceptors or donors in AlGaAs form a high–mobility and high–density hole or electron plasma in the GaAs quantum wells at low temperatures[25]. It has been shown that in these high–mobility plasmas electron–hole scattering is the dominating scattering process for injected minority carriers[26-27]. For the parallel transport of minority carriers at low electric fields electron–hole scattering leads to the recently observed effect of "negative absolute mobility"[26-27]: The high–mobility majority carriers "drag" the injected minority carriers in the "wrong" field direction (e.g., minority electrons to the negative electrode). Negative drift of minority carriers has been studied in n– and p–modulation doped quantum wells by spatially resolved luminescence[26-27]. In the present work we describe a *different* experimental manifestation of the negative absolute mobility.

Injection of electron–hole pairs in the quantum wells also changes the total in–plane conductivity. In the regime of negative absolute mobility, a *decrease* of the conductivity due to carrier drag has been postulated theoretically in Ref.27: The negative drift of the minority carriers and the reduction of the majority carrier mobility by electron–hole scattering both contribute to a conductivity decrease, whereas the increase of the majority carrier concentration contributes to a conductivity increase. The analysis shows that the total photoconductivity becomes negative *exactly* when the mobility of minority carriers becomes negative.

In our experiments we studied negative photoconductivity due to "carrier drag" of minority electrons in p–modulation doped GaAs/AlGaAs quantum wells. In these structures long–lived carrier separation is the dominating photoconductivity process[28-29]. By spectral resolution (selective excitation in the quantum wells) and simultaneous saturation of trapping effects, it is possible – as it will be shown – to identify the intrinsic negative photoconductivity due to negative absolute mobility.

In the experiments, we used the identical p–modulation doped quantum well sample as in previous photoconductivity and photoluminescence measurements[27,29], consisting of 20 periods of undoped GaAs (d=112 Å), Be–doped $Al_{0.48}Ga_{0.52}As$ (d=49 Å), and undoped $Al_{0.48}Ga_{0.52}As$ "spacer" layers (d=294 Å) between the doped regions and the GaAs quantum wells. The holes from ionized acceptors in AlGaAs form a high mobility hole plasma in GaAs with a concentration of $p_s = 1.5x10^{11}$ cm^{-2} (per layer) and a mobility $\mu_h = 53800$ cm^2/Vs (at 4.2 K). The values at 77 K are $p_s = 1.83x10^{11}$ cm^{-2} and $\mu_h = 3750$ cm^2/Vs. The photoconductivity experiments were performed using a 3 mW AlGaAs diode laser with $\lambda=750$ nm as light source, operated by current modulation with a repetition frequency of 1kHz and a duty cycle of 1/4. The wavelength of 750 nm was chosen in order to (a) only excite electrons in the GaAs quantum wells and (b) achieve enough absorption in the quantum wells and minimize the influence of the substrate. The light intensity was controlled by the diode current and filters. The laser light was focussed on the four–point geometry sample with a constant current flowing through the outer contacts. The voltage at the inner contacts was measured and lock–in correlated with the diode laser current. In addition, the sample was illuminated with white light from a tungsten halogen lamp in order to saturate trapping effects. Photoconductivity was determined from the change of sheet conductivity of the whole system of 20 quantum wells and normalized to unit light intensity of 1 W/cm^2, designated as normalized photoconductivity $\Delta\sigma/I_0$.

The photoconductivity signal is positive at high temperatures, according to an increase of the sheet conductivity with injection of minority electrons, and slowly varying as the temperature is decreased down to about 100 K. At a temperature T_0 between 95 K and 85 K (depending on the laser light intensity) a transition from positive to negative photoconductivity occurs. With decreasing temperature there is a strong increase of the negative signal. At a given temperature the absolute value of the normalized photoconductivity $\Delta\sigma/I_0$ decreases both with increasing laser light intensity I_0 and background white light intensity I_{wl} [30]. *Saturation* of the normalized photoconductivity occurs for a laser and white light intensity of 150 W/cm² and 0.5 W/cm², respectively, in the whole temperature range. At the lowest sample temperature a transition from negative to positive photoconductivity is also observed as a function of electric field, which leads to heating of the majority hole plasma[30].

The observed intensity dependence of the normalized photoconductivity at low temperatures and low light intensities is attributed to the presence of trapping mechanisms in GaAs, such as described in Ref.28 and Ref.29. A sufficient background illumination and a "background" minority electron concentration is required to saturate these effects. The intrinsic photoconductivity is seen only for additional injected minority electrons. In the experiment this is achieved by measuring the conductivity change between two high injection levels. The conductivity change divided by the intensity difference yields the *differential photoconductivity*, designated as $d(\Delta\sigma)/dI_0$.

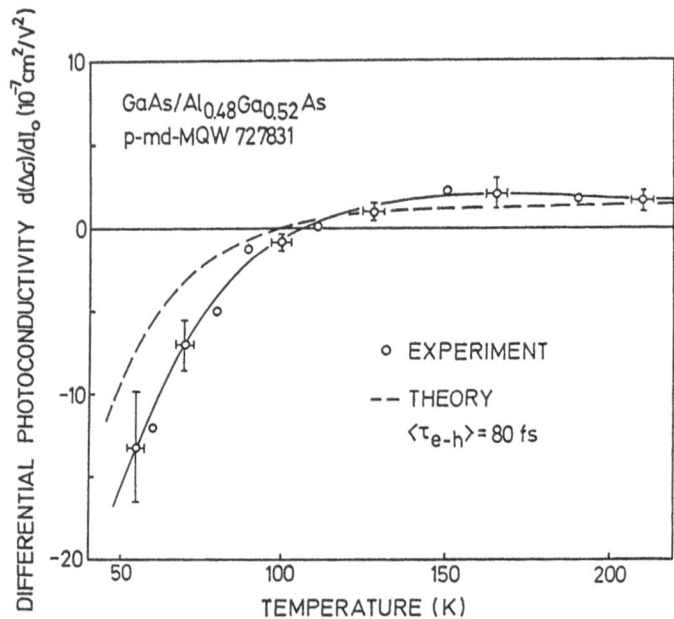

Fig. 6 . *Differential* photoconductivity $d(\Delta\sigma)/dI_0$ as a function of temperature. The experimental values are evaluated from the signal difference between laser light intensities $I_1=64$ W/cm² and $I_2=130$ W/cm². Background white light intensity: $I_{wl} = 0.82$ W/cm². Dashed line: Theoretically expected behaviour due to "carrier drag" with a momentum relaxation time $<\tau_{e-h}> = 80$ fs.

In Fig.6 the differential photoconductivity, determined from the signal difference at light intensities of 64 W/cm² and 130 W/cm², divided by the intensity change of 66 W/cm², is plotted as a function of temperature. The value of the background intensity is 0.82 W/cm². The transition from negative to positive photoconductivity occurs at a temperature of 105 K. The experimental values in Fig.6 are in good agreement with the theoretically predicted behaviour (dashed curve). The theoretical curve is obtained from the following derivation: In a p–doped semiconductor with equilibrium hole concentration p_0 ($n_0 \approx 0$) the total change of conductivity by injection of $p-p_0$ (= n) electron–hole pairs is given by

$$\Delta\sigma = n\cdot e\cdot\mu_e + p\cdot e\cdot\mu_h - p_0\cdot e\cdot\mu_{h-1} \qquad (6)$$

(μ_e, μ_h ... mobilities of electrons and holes under photoexcitation, including the influence of electron–hole scattering, μ_{h-1} ... hole mobility without photoinjection of carriers, determined only by hole–lattice scattering). With the expressions for the mobilities μ_e and μ_h from Ref.27 ("hydrodynamic" approximation with averaged relaxation times) we obtain from equation (1)

$$\Delta\sigma = p \cdot e^2 \cdot \left[1 + \frac{p \cdot m_h \cdot <\tau_{e-l}>}{n \cdot m_e \cdot <\tau_{h-l}>} \cdot \left(1 + \frac{<\tau_{e-h}>}{<\tau_{e-l}>} \right) \right]^{-1} \cdot$$

$$\cdot \left[\frac{<\tau_{e-h}>}{m_e} \cdot \left(1 + \frac{m_h \cdot <\tau_{e-l}>}{m_e \cdot <\tau_{h-l}>} \right) - \frac{<\tau_{e-l}>}{m_e} \cdot \left(1 - \frac{n}{p} \right) - \frac{p_0}{p} \cdot \frac{<\tau_{h-l}>}{m_h} \right] \qquad (7)$$

(m_e, m_h ... effective masses, $<\tau_{e-h}>$ ... momentum relaxation time of electrons relative to the hole plasma by electron–hole scattering, $<\tau_{e-l}>$ and $<\tau_{h-l}>$ ... momentum relaxation times of electrons and holes by lattice scattering). In Fig.6 , $\Delta\sigma$ according to the equation above, multiplied by the number of quantum wells (N=20), and normalized to unit light intensity, is plotted as a function of temperature (dashed curve). The values for the lattice scattering times are known from temperature dependent Hall measurements (for majority holes) and from data of comparable high–quality samples (for minority electrons). The values for $<\tau_{e-h}>$ are known from optical transport experiments and are below 10^{-13} s (<100 fs) for all temperatures[26-27]. The minority electron concentration is calculated from the light intensity and a recombination lifetime of 1.0 ns. This value was determined from time–resolved photoluminescence[27] of the identical sample at minority injection levels of 3×10^{10} cm^{-2}, and is almost constant over a wide temperature range (20 K to 300 K). The relation between $\Delta\sigma$ according to eq.(7) and the intensity I_0 is linear for n $\ll$ p (assuming that the lifetime is independent of the minority electron concentration). The differential photoconductivity is equal to the normalized photoconductivity in this approximation. From the good agreement of the experimental data with the theoretically expected behaviour we conclude therefore *that the observed negative photoconductivity is due to carrier drag in the negative mobility regime.*

A simple rate equation with a carrier lifetime of 1.0 ns yields a minority electron concentration of 4.0×10^9 cm^{-2} for a laser intensity of 130 W/cm^2 (λ=750 nm) at a temperature of 100 K. This concentration is much smaller than the majority hole concentration of 1.8×10^{11} cm^{-2}, which justifies the approximation n $\ll$ p .

The experimental value for $<\tau_{e-h}>$, the momentum relaxation time of electrons in a resting hole plasma, is 80 fs $\pm$ 20 fs at a temperature of 100 K (from Ref.27). The dashed curve in Fig.6 was calculated for a value of $<\tau_{e-h}>$ = 80 fs. The reduction of this value by 10 fs would result in a shift of the whole curve to temperatures 10 K higher as in Fig.6 and vice versa. Therefore the difference of the measured photoconductivity and the theoretical values suggests slightly stronger electron–hole scattering at low temperatures, which is within the accuracy of the experiments of Ref.27, and is also expected from theoretical calculations[27,31]. For $<\tau_{e-h}>$ = 30 fs at a temperature of 50 K, increasing to 80 fs at 115 K the theoretical values agree with the measured photoconductivity within the experimental errors.

A further important point of discussion is the influence of background illumination on both majority hole mobility and concentration. The relative change of the sample conductivity under illumination with white light of intensities as given above is positive and about 10 % for T > 150 K, decreases with temperature to zero at 90 K, and becomes negative for lower temperatures. The reason is the reduction of the hole concentration in GaAs under illumination with photon energies above the bandgap of AlGaAs due to spatial separation of electron–hole pairs (generated in AlGaAs) by the junction electric field. The holes are trapped in AlGaAs, whereas the electrons transferred to the GaAs recombine subsequently with the hole plasma. This process has been investigated in detail in Ref.29. With decreasing hole concentration also the hole mobility is reduced[32]. Therefore the decrease of the minority electron mobility by electron–hole scattering is weaker, and the expected photoconductivity for additional injected minority carriers should be slightly decreased. The relative change of the expected photoconductivity is less than 10% for temperatures higher than 65 K.

The normalized photoconductivity is in first order proportional to the recombination lifetime $<\tau_{rec}>$ at a given temperature. The value for $<\tau_{rec}>$ was measured at minority electron concentrations of 3×10^{10} cm^{-2}. The lifetime at lower electron concentrations as in the photoconductivity experiments might be longer due to trapping or assymmetry of the quantum wells. This would lead to an increase of the normalized photoconductivity at low light intensities, which is observed in the expriment.

In conclusion, we have observed negative photoconductivity in p—doped GaAs/AlGaAs quantum wells with excitation at 750 nm at low temperatures. The observed temperature and field dependences agree with the expected behaviour due to "negative absolute mobility" of injected minority electrons in the 2D—hole plasma. The observation of the intrinsic photoconductivity due to "carrier drag" was only possible after saturating trapping effects by a high intensity background illumination.

Finally we emphasize that photo—Hall measurements would give additional information on the majority carrier concentration and mobility, however, the dramatic effect of the photoconductivity going through zero at $\mu_e = 0$ is theoretically not expected for $\Delta \sigma_{xy}$ (Hall conductivity).

4. SUBPICOSECOND LUMINESCENCE FROM Ga$_{0.47}$In$_{0.53}$As

Radiation damaged III—V semiconductors, such as GaAs, InP and recently InGaAs, are important new materials for fast optoelectronic switches[33]. Extremely short carrier lifetimes are achieved by the high density of recombination centers induced by radiation damage, usually high energy ion bombardment, such as He$^+$ and Be^{++} with energies exceeding 100 keV. Main applications are ultrafast optoelectronic devices, such as pico— and subpicosecond photodetectors[33] and optically bistable devices, where short carrier lifetimes are essential for fast optical switching in absorbing materials. Ga$_{0.47}$In$_{0.53}$As lattice matched to InP is an important material for detectors in the visible to infrared region (up to 1.6 μm)[34-35]. Downey and coworkers[36] investigated Ga$_{0.47}$In$_{0.53}$As radiation damaged by Be^{++} ions: Photoconductivity decay times down to 40 ps were observed. In addition to time—resolved photoconductivity, an all—optical technique has been applied to characterize materials with extremely short lifetimes: The relaxation of optically injected carriers in radiation—damaged silicon—on—sapphire was measured by time—resolved optical reflectivity with femtosecond optical pulses[37].

In the present work we study the *photoluminescence* in such radiation—damaged material. Luminescence is a very useful tool to study carrier relaxation in semiconductors[38], since the instantanous light emission intensity is directly proportional to the product of electron and hole distributions at a given wavevector k, as described in Chapter 1. We investigate the luminescence far above the bandgap ($E_g = 0.75$ eV) in the range of 1.4 to 1.9 eV (laser photon energy 2.0 eV). The time dependence of the luminescence intensity is measured with the nonlinear correlation technique as in the experiments described in Chapters 1 and 2. The excitation level is of the order of 10^{18} cm^{-3}. Since the luminescence is detected by a photomultiplier with a GaAs photocathode, only the luminescence between 1.4 eV and 1.9 eV, far above the bandedge of Ga$_{0.47}$In$_{0.53}$As is investigated. The spectral resolution of the spectrometer (single 1/4 m) is set to > 5 nm (large slit width), in order to allow good time resolution (better than 100 fs). All measurements are performed at room temperature.

The samples have the following structure: On top of semi—insulating InP single crystals an epitaxial layer of lattice matched Ga$_{0.47}$In$_{0.53}$As (thickness 1.0 μm) has been grown by vapour phase epitaxy. The doping level is estimated as $N_d - N_a \approx 1 \times 10^{16}$ cm^{-3}. On the surface a passivation layer of 1000 Å silicon nitride has been deposited, which does not play any role for our experiments, due to its transparency throughout the visible and near infrared region. Some of the samples were bombarded by He$^+$ ions, with the following doses and kinetic energies, respectivly: 1×10^{14} cm^{-2} with 320 keV, 7×10^{13} cm^{-2} with 200 keV, 4.7×10^{13} cm^{-2} with 110 keV, 2.85×10^{13} cm^{-2} with 50 keV, and

1.76×10^{13} cm^{-2} with 30 keV. The average penetration depth is estimated as about 2 μm, which is larger than the thickness of the epitaxial layer and is by an order of magnitude larger than the absorption length of the laser excitation at $\lambda = 620$ nm ($\alpha^{-1} < 10^{-5}$ cm). Therefore the optical excitation occurs only near the surface of the radiation damaged (or undamaged) Ga$_{0.47}$In$_{0.53}$As.

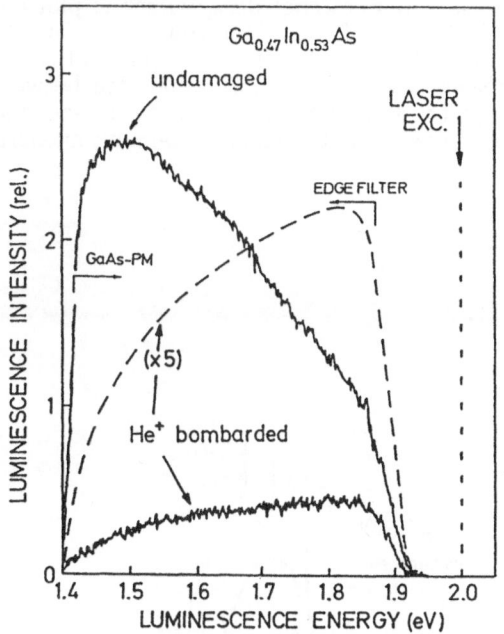

Fig. 7. Luminescence spectra (time integrated) of undamaged and radiation damaged (He$^+$ bombarded) Ga$_{0.47}$In$_{0.53}$As, excited by 100–fs laser pulses at 2.0 eV. Dashed line: Spectra of the He$^+$ bombarded sample, expanded by a factor of 5, for comparison.

In Fig. 7 the luminescence spectra of the undamaged and radiation damaged Ga$_{0.47}$In$_{0.53}$As are shown. The excitation level is of the order of 10^{18} cm^{-3}. The sharp decrease of the luminescence signal above 1.85 eV is due to the longpass filter placed between sample and spectrometer in order to cut off scattered laser light. The decrease below 1.45 eV is due to the cut–off of the photomultiplier sensitivity below the bandgap of GaAs. In the range between 1.45 and 1.85 eV the sensitivity of the system is almost constant ($\pm$ 10 %). Therefore the curves in Fig. 1 represent the real luminescence spectra in this energy range. The luminescence intensity of the undamaged material increases at lower photon energies, which is the usual "hot luminescence" characteristics of the radiative band–band recombination at energies far above the bandedge[40]. The spectra of the radiation damaged material show a completely different behaviour: The intensity is by a factor of about 5 lower, and *decreases with decreasing photon energies*. The luminescence intensity is lower at energies closer to the bandgap and higher for energies closer to the excitation laser energy (2.0 eV).

In Figure 8 the results of the time–resolved luminescence measurements are shown. The correlation signal is plotted as a function of the delay τ between the two 100–femtosecond laser pulses, for several different photon energies. All curves show a sharp maximum at $\tau = 0$, with a width in the order of the autocorrelation of the laser pulses (< 200 fs). This "coherent spike" is caused by interference of the two pulses, causing higher local carrier concentrations. This coherent artifact is only used for determining the zero delay. The width of the total luminescence correlation signals, however, is of the order of a few picoseconds and changes as a function of the photon energy, for the undamaged material (upper curves). The correlation signals are larger and broader at lower photon energies. The luminescence correlation signals of the radiation–damaged samples (lower curves) have a much smaller width indicating a more rapid decay. The shape and size of the correlation signals is roughly independent of the photon energy. In Fig. 9, the "luminescence decay time" is plotted as a function of the photon energy: This decay time is defined as the time delay, at which the luminescence correlation signal decreases to $1/e$. (The coherent spike at $\tau = 0$ is not taken into account.) For the undamaged sample the decay times are between 4 ps (at a photon energy of 1.48 eV) and 2 ps (at 1.85 eV). The radiation damaged sample shows a much shorter decay time (0.9 ps), roughly independent of the photon energy.

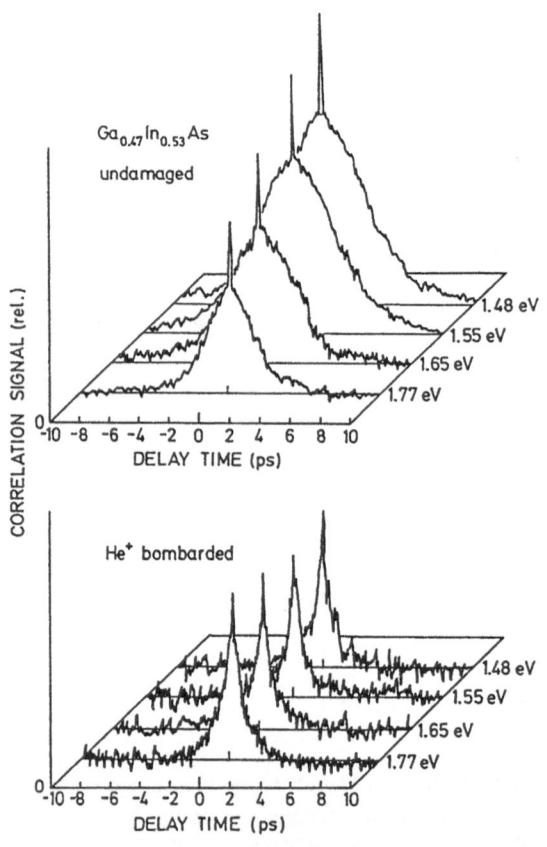

Fig. 8. Luminescence correlation signal of undamaged (upper curves) and He+ bombarded (lower curves) $Ga_{0.47}In_{0.53}As$ as a function of the delay τ between the two laser pulses, for several different luminescence photon energies.

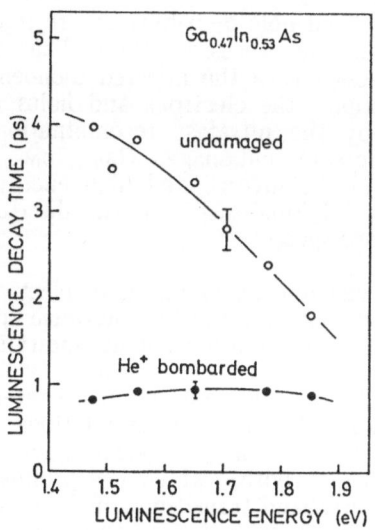

Fig. 9. Decay time of the luminescence (decay to 1/e, coherent spike not taken into account), as a function of the luminescence energy.

The large difference between the radiation—damaged and the undamaged sample is a direct consequence of the recombination centers induced by the ion bombardment. The luminescence decay time of 0.9 ps is due to the drastically shortened recombination lifetime of the photoexcited carriers. Before discussing these results in detail, we comment on the results of the *undamaged* material. In the undamaged material, where the recombination lifetime of thermalized carriers is of the order of 100 ps to 1 nanosecond, the luminescence decay in the high energy range is due to energy relaxation of the hot injected carriers. Qualitatively, energy relaxation and thermalization from the initially excited monoenergetic carrier distributions to a room temperature Fermi—Dirac distribution, explains the longer correlation time at lower energies and the shorter correlation time (more rapid decay) at higher energies. Quantitatively, the observed relaxation times of 2–4 ps seem longer than expected for LO phonon emission[41-42], which may be due to several effects, such as nonequilibrium phonons, weaker LO—phonon coupling in InGaAs as compared to GaAs, as well as screening of the electron—phonon interaction. In addition, in the high energy range of our experiments, which is closer to the laser energy than to the band gap, and within the short time scale of $\leq$ 4 ps, the carrier dynamics also contain the relaxation by carrier—carrier scattering and — most important — the transitions to and from the satellite valleys (both Γ–X and Γ–L transitions) as well as the energy relaxation therein. The details of the band structure in InGaAs are not completely known, but the L—valley is about 0.55 eV higher than the Γ—valley[35]. At this energy, the transition from the Γ—valley to the heavy hole valence band gives a photon energy of 1.32–1.35 eV. Therefore the energies of the electrons in the Γ—valley, which contribute to the observed luninescence, are *above* the minimum of the L—valley. The time constant for the L–Γ transition in GaAs was measured and reported as $\approx$ 2 ps [43], which would be comparable to the present results. Thus, for a detailed analysis of our results we would have to take into account the Γ–L as well as Γ–X transitions.

Now let us focus our discussion on the results of the radiation damaged sample: The doses and energies of the ion bombardment obviously resulted in extremely short lifetimes of 0.9 ps. This time is considerably shorter than the energy relaxation time in this energy range (2 to 4 ps). Therefore the luminescence spectra are the result of the average carrier distributions *within less than one picosecond after photoexcitation.* This explains the "inverted" shape of the luminescence spectra of the radiation damaged sample: Monoenergetic electron and hole distributions are injected by the 2.0 eV laser pulses. *Before* the energy relaxation processes "thermalize" the distribution function, i.e., *before* the scattering processes turn the injected monoenergetic distributions into thermal Fermi–Dirac distributions, the electrons and holes *leave* the conduction and valence bands, respectively, by the ultrafast recombination via the recombination centers. From the measurements in undamaged $Ga_{0.47}In_{0.53}As$ the energy relaxation times are known (2 to 4 ps in the investigated high energy range). Thus the short lifetime of 0.9 ps prevents the thermalization of the distributions, and explains the observed "inverted" luminescence spectra.

In order to obtain the distribution functions of electrons and holes, separately, additional information is necessary, since the luminescence spectra are proportional to the *product* of electron and hole distributions. This additional information might be obtained from nonlinear transmission experiments (yielding the sum of the distributions), photoconductivity measurements (sum weighted with mobilities) and/or theoretical considerations including Monte Carlo simulations. In any case, however, the observed "inverted" luminescence spectra give evidence for inverted *intraband* carrier distributions (of electrons, holes or both), which has promising implications for light amplifying devices: Radiative intraband transitions might be stimulated, e.g., cyclotron resonance[44], free carrier intraband optical transitions by carrier scattering[45], and radiative subband transitions[46] in confined systems. Such novel, optically pumped intraband laser devices would have the advantages of room temperature operation and wide tunability in the infrared and far infrared wavelength range.

ACKNOWLEDGMENTS

This work has been supported by the "Fonds zur Förderung der wissenschaft–lichen Forschung" (FWF), Austria, under project # P 6184. The samples were provided by scientists from several institutions: We thank A.C. Gossard and T.Y. Chang (AT&T Bell Laboratories), G. Weimann (Walter–Schottky–Institut, München), H. Kano (Toyota R&D) and E. Wintner (TU Vienna) for the collaboration. The partial support by the Murata Science Foundation for the stay of N.S. at the University of Innsbruck is acknowledged.

REFERENCES

1 M. Tsuchiya, T. Matsusue, and H. Sakaki, Phys. Rev. Lett. 59, 2356 (1987).
2 J. Shah, IEEE Journal of Quantum Electronics 24, 276 (1988).
3 D. Rosen, A.G. Doukas, Y. Budansky, A. Katz, and R.R. Alfano, Appl. Phys. Lett. 39, 935 (1981).
4 D. von der Linde, J. Kuhl, and E. Rosengart, J. Luminescence 24/25, 675 (1981).
5 A. Von Lehmen, J.M. Ballantyne, Appl. Phys. Lett. 44, 87 (1984).
6 J.A. Valdmanis, R.L. Fork, and J.P. Gordon, Optics Lett. 10, 131 (1985).
7 R.A. Höpfel, J. Shah, T.Y. Chang, and N.J. Sauer, Appl. Phys. Lett. 51, 1815 (1987).
8 R. Christanell, R.A. Höpfel, J. Appl. Phys. (submitted).
9 J. Shah, B. Deveaud, T.C. Damen, and W.T. Tsang, Phys. Rev. Lett. 59, 2222 (1987).
10 T. Tada, A. Yamaguchi, T. Ninomiya, H. Uchiki, T. Kobayashi, and T. Yao, J. Appl. Phys. 63, 5491 (1988).
11 N. Sawaki, M. Suzuki, Y. Takagaki, H. Goto, I. Akasaki, H. Kano, Y. Tanaka, and M. Hashimoto, Superlattices and Microstructures 2, 281 (1986); N. Sawaki, M. Suzuki, E. Okuno, H. Goto, I. Akasaki, H. Kano, Y. Tanaka, and M. Hashimoto, Solid State Electronics 31, 351 (1988).
12 H. Kano, Y. Tanaka, N. Sawaki, M. Hashimoto, and I. Igarashi, J. Cryst. Growth 81, 144 (1987).
13 S.W. Kirchoefer, R. Magno, and J. Comas, Appl. Phys. Lett. 44, 1054 (1984).
14 J.M. Pond, S.W. Kirchhoefer, and E.J. Cukauskas, Appl. Phys. Lett. 47, 1175 (1985).
15 N. Sawaki, and I. Akasaki, Physica 134B, 494 (1985).

[16] B. Deveaud, J. Shah, T.C. Damen, W.T. Tsang, Appl. Phys. Lett. 52, 1886 (1988).

[17] N. Sawaki, R.A. Höpfel, E. Gornik, and H. Kano, Appl. Phys. Lett. (submitted).

[18] U. Cebulla, G. Bacher, G. Mayer, A. Forchel, W.T. Tsang, and M. Razeghi, Superlattices and Microstructures 5, 227 (1989).

[19] M. Tsuchiya, T. Matsusue, and H. Sakaki, Phys. Rev. Lett. 59, 2356 (1987).

[20] M.K. Jackson, M.B. Johnson, D.H. Chow, T.C. McGill, and C.W. Nieh, Appl. Phys. Lett. 54, 552 (1989).

[21] M.G.W. Alexander, W.W. Ruehle, R. Sauer, and W.T. Tsang (preprint).

[22] D.Y. Oberli, D.R. Wake, M.V. Klein, J. Klem, T. Henderson, and H. Morkoc, Phys. Rev. Lett. 59, 696 (1987).

[23] A. Seilmeier, M. Wörner, G. Abstreiter, G. Weimann, and W. Schlapp, Superlattices and Microstructures 5, (1989).

[24] K. Leo, W.W. Ruehle, and K. Ploog, Phys. Rev. B38, 1947 (1988).

[25] J.H. English, A.C. Gossard, H.L. Störmer, and K.W. Baldwin, Appl. Phys. Lett. 50, 1826 (1987).

[26] R.A. Höpfel, J. Shah, P.A. Wolff, and A.C. Gossard, Phys. Rev. Lett. 56, 2736 (1986).

[27] R.A. Höpfel, J. Shah, P.A. Wolff, and A.C. Gossard, Phys. Rev. B37, 6941 (1988).

[28] M.J. Chou, D.C. Tsui, and G. Weimann, Appl. Phys. Lett. 47, 609 (1985).

[29] R.A. Höpfel, Appl. Phys. Lett. 52, 801 (1988).

[30] R.A. Höpfel, S. Juen, J. Shah, and A.C. Gossard, Superlattices and Microstructures 5, 15 (1989).

[31] W. Cai, T.F. Zheng, and M. Lax, Phys. Rev. B37, 8205 (1988).

[32] D.C. Tsui, A.C. Gossard, G. Kaminsky, and W. Wiegmann, Surf. Sci. 113, 464 (1982).

[33] For a review, see D.H. Auston in "Ultrashort Laser Pulses and Applications" (Editor: W. Kaiser), Springer–Verlag, Berlin 1988, p.183–233.

[34] "GaInAsP Alloy Semiconductors", (Editor: T.P. Pearsall), John Wiley & Sons, New York 1982.

[35] K.Y. Cheng, A.Y. Cho, S.B. Christman, T.P. Pearsall, and J.E. Rowe, Appl Phys. Lett. 40, 423 (1982).

[36] P.M. Downey, R.J. Martin, R.E. Nahory, O.G. Lorimor, Appl. Phys. Lett. 46, 396 (1985).

[37] F.E. Doany, D. Grischkowsky, and C.C. Chi, in "Picosecond Electronics and Optoelectronics II" (Editors: F.J. Leonberger, C.H. Lee, F. Capasso and H. Morkoc), Springer Series in Electronics and Photonics 24, 228 (1987).

[38] J. Shah, IEEE J. Quantum Electron. QE–24, 276 (1988).

[39] For an arbitrary time–dependence of the distribution functions, the correlation curves I(τ) must be fitted numerically, with certain assumptions on the rising onset of the luminescence.

[40] J. Shah, Solid State Electron. 21, 43 (1978).

[41] J.C. Kash, J.C. Tsang, and J.M. Hvam, Phys. Rev. Lett. 54, 2151 (1985).

[42] K. Kash and J. Shah, Appl. Phys. Lett. 45, 401 (1984).

[43] J. Shah, B. Deveaud, T.C. Damen, W.T. Tsang, A.C. Gossard, and P. Lugli, Phys. Rev. Lett. 59, 2222 (1987).

[44] E. Gornik, Phys. Rev. Lett. 29, 595 (1972).

[45] R.A. Höpfel, G. Weimann, Appl. Phys. Lett. 46, 291 (1985).

[46] E. Gornik, D.C. Tsui, Phys. Rev. Lett. 37, 1425 (1976).

OPTICAL PROPERTIES OF GaAs/AlAs SUPERLATTICES

E. O. Göbel, J. Feldmann, R. Fischer, G. Peter, and
R. Sattmann

Philipps-Universität, Fachbereich Physik
Renthof 5, 3550 Marburg, F.R.G.

J. Hebling, J. Kuhl, R. Muralidharan, and K. Ploog

Max-Planck-Institut für Festkörperforschung
Heisenbergstr. 1, 7000 Stuttgart 80, F.R.G.

P. Dawson, and C.T. Foxon

Philips Res. Lab., Redhill, Surrey RH1 5HA, U.K.

ABSTRACT

The optical properties of GaAs/AlAs short period superlattices
(SPS) are investigated by photoluminescence and photoluminescence
excitation spectroscopy, picosecond photoluminescence, and
subpicosecond excite and probe experiments. The transition from a type
I to a staggered type II SPS for GaAs layer thicknesses below $\simeq$ 30A is
clearly revealed in the stationary as well as time resolved optical
experiments. The characteristic time constants for scattering between
electron states originating from the Γ-conduction band of the GaAs into
X-states of the AlAs are measured.

INTRODUCTION

The possibility of tailoring the band structure of semiconductors
by growing short period superlattice (SPS) structures has always
fascinated material scientists. Already in 1974, the possibility of
obtaining direct optical transitions in indirect gap materials using a
SPS has been discussed[1]. Semiconductor technology now is capable of
realizing some of these dreams employing modern hetero-epitaxy
processes. High quality quantum wells (QW) and superlattices with
different band alignment can be fabricated in II-VI, III-V and element
IV material systems. The most common examples of different band gap
alignment are schematically depicted in Fig. 1, showing the type I,
staggered type II, and misaligned type II band alignments. In the type
I structures (Fig. 1 (a)), the valence band and conduction band of the
material with the smaller band gap (material A in Fig. 1) lies within
the band gap of the larger band gap material (B in Fig. 1). In the
staggered type II systems (Fig. 1 (b)) either the conduction or valence
band of material A lies outside the band gap of material B. Finally,
none of the bands of material A lies within the gap of material B in

the misaligned type II structures (Fig. 1 (c)). A more specific description of the different structures of course has to take into account the details of the band structure of the respective materials as well as possible built in tension or stress, which in fact can provide an additional degree of freedom to tailor the band structure. Examples for the different structures and their physical properties can be found in recent review articles[2-5].

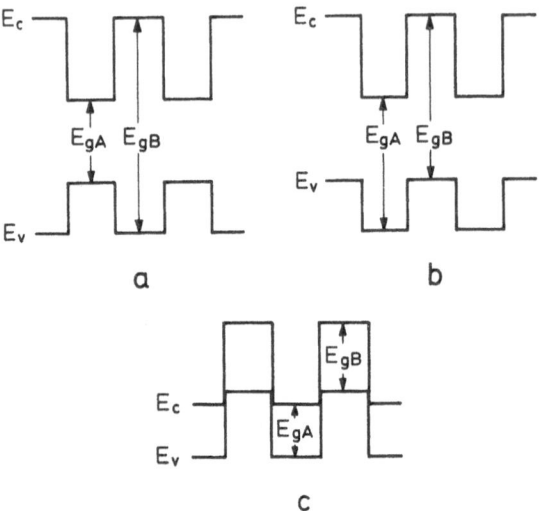

Fig. 1   Band alignments of a type I (a), type II (b), and misaligned type II (c), semiconductor heterostructure.

Among the different material systems, the GaAs/AlAs quantum wells and SPS are of particular interest, because a continuous transition from a type I to a staggered type II can be realized by decreasing the layer thicknesses (see ref. 4,5 for a review). The optical properties of (GaAs/AlAs) QW and SPS thus can be drastically different depending on whether the respective structure is type I or type II. In this paper we shall illustrate some of the characteristic differences of the optical properties of type I and type II GaAs/AlAs QW and SPS. Special emphasis is paid on the dynamics of photoexcited carriers including recent results on the time constants for the real space charge transfer in type II SPS.

Band Alignment of the GaAs/AlAs System

A direct gap (GaAs) and an indirect gap (AlAs) semiconductor are combined in the GaAs/AlAs system to form the QW or SPS. The valence band structure of bulk GaAs and AlAs is principally similar with a heavy- and light hole band being degenerated at the Γ-point of the Brillouin zone (BZ). The lowest conduction band minimum in GaAs is at the Γ-point. The energy of the band gap amounts to $E_g = 1.519$ eV at low temperatures[6]. The lowest conduction band minimum of AlAs is close to the X-point of the BZ with a band gap of about 2.24 eV. The X-point conduction band minimum is highly anisotropic with a longitudinal mass, i.e. in (001) direction of about 1.1 m_0 and a transverse mass of about 0.19 m_0. The direct gap of AlAs has an energy above 3eV. On the base of presently accepted values for the band offsets a band alignment of the GaAs/AlAs system as depicted in Fig. 2 is obtained. The confined energy levels are also shown schematically in Fig. 2. The Γ-electron states in the GaAs exhibit the largest confinement energy due to their small effective mass. The lowest electron states are no longer in the GaAs but in the AlAs for confinement energies larger than about 200 meV

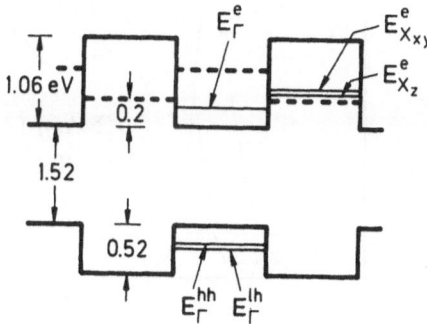

Fig. 2   Band alignment of a type I(GaAs) (AlAs) superlattice.

(GaAs thickness smaller than $\simeq$ 35Å) and the structure then represents a staggered type II QW or SPS. The confined energies associated with the X-minimum in AlAs are different for states with a component of momentum parallel to the growth direction (001) (X_z-states) and those with momentum in the plane (X_{xy}-states) due to the different confinement mass. Since the confinement mass for the X_z-states is given by the longitudinal mass, one would expect the confined X_z-states being always lowest. In this simple consideration, however, mixing of the states due to the superlattice potential as well as effects due to strain are neglected. Recent experimental[7,8,9] as well as theoretical work[9,10,11] support the assignment of the X_z-states being lowest in SPS with thin AlAs layer thickness. This topic, however, will be discussed in more detail in separate contributions to this book.[12,13]

Optical Properties

The spatial separation between electrons and holes in type II QW and SPS naturally will result in a reduction of the optical transition strength similar to the trends observed in nipi-structures[14]. In addition, the lowest excitation in type II QW and SPS corresponds to an optical transition from Γ to X electronic states which is forbidden due to k-conservation. It may become (partly) allowed by additional scattering with phonons like e.g. in bulk GaP, scattering at the interface due to disorder[15] or mixing of Γ and X_z-states by the superlattice potential[7]. Both effects, the spatial separation of electrons and holes (often referred to as an indirect transition in real space) as well as the fact that the transition connects states with at least partly different symmetry (partly indirect in k-space), will result in a pronounced reduction of the strength of the lowest optical transitions in type II SPS as compared to type I structures. This difference in transition strength can be demonstrated in stationary as well as time resolved optical experiments and may be used to identify the type of the SPS.

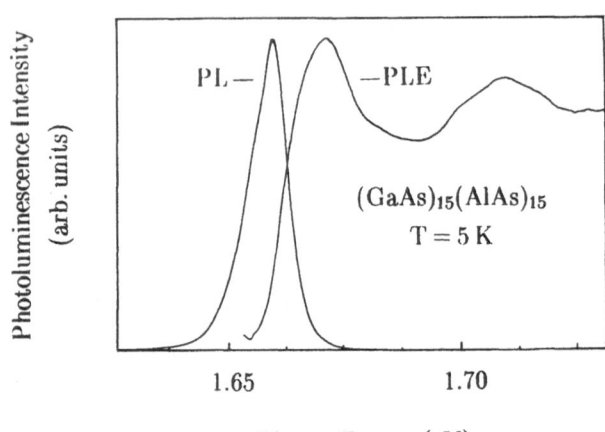

Fig. 3   Low temperature (T = 5K) photoluminescence (PL) and photoluminescence excitation (PLE) spectrum of a (GaAs)[15] superlattice.

a) Photoluminescence and Excitation Spectroscopy

A low temperature photoluminescence excitation (PLE) and photoluminescence (PL) spectrum of a type I GaAs/AlAs SPS with 15 layers of GaAs and AlAs, respectively, is depicted in Fig. 3. The heavy and light hole exciton feature at 1.675eV and 1.701eV are clearly resolved in the PLE spectrum.The PL is close to the heavy hole exciton peak. The low energy shift is attributed to relaxation of excitons within localized states due to inter- and intralayer well width fluctuations. The PLE and PL spectra of all the different type I SPS are qualitatively similar to isolated quantum wells, nevertheless, miniband formation may affect e.g. the absolute numbers for the exciton binding energy[15].

The optical behaviour of type II SPS is significantly different. Figure 4 shows the PLE and PL spectrum of a type II GaAs/AlAs with 6 layers of GaAs and AlAs. The most striking difference compared to type

I SPS is the pronounced shift between the onset of the dominant absorption and the PL reflecting the separation between the Γ- and X-conduction band states: The dominant absorption corresponds to transitions involving Γ-valence and conduction band states, while the luminescence arises from zero phonon- and phonon assisted recombination of electrons within X-like states of the AlAs with holes in the GaAs[8,9,14,15,17]. It follows from the PLE spectrum that electrons are efficiently scattered from Γ-conduction band states of the GaAs into X-like AlAs states for excitation energies higher than the corresponding Γ-Γ transition. This scattering process will be discussed below. The weak PLE signal inbetween the PL feature at about 1.9eV and the onset of the strong absorption at 2.1eV corresponds to type II transition, which are weaker by more than one order of magnitude. It may be concluded on the base of optical[14,17] as well as photoacoustic[18] spectroscopy that the transition from a type I to a type II SPS in symmetic structures occurs at 11 or 12 monolayers.

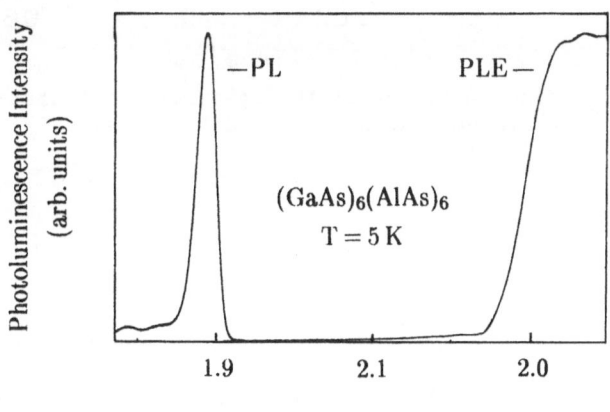

Fig. 4   Low temperature (T = 5K) photoluminescence (PL) and photoluminescence excitation (PLE) spectrum of a (GaAs)$_6$ (AlAs)$_6$ superlattice.

b) Photoluminescence Decay

The photoluminescence decay times of type I SPS are of the same order than for isolated QW. Figure 5 depicts as an example the decay of the spectrally integrated PL of two SPS with 15 monolayers of GaAs and respectively 15 and 5 monolayers of AlAs. Both samples had barrier layers on either side of the SPS to prevent carrier diffusion. The photoluminescence decay times are about 400 ps, in either case, which is comparable to the data reported for isolated QW[19]. We thus conclude that the radiative decay is still enhanced in these SPS compared to bulk exciton recombination due to the exciton confinement and miniband formation is not yet very effective, which is in accordance with simple Kronig-Penney calculations. An increase of the photoluminescence decays together with a decrease of the exciton binding energy has been reported recently for (GaAs)(AlGaAs) SPS and was attributed to loss of exciton confinement in narrow SPS[20],[21].

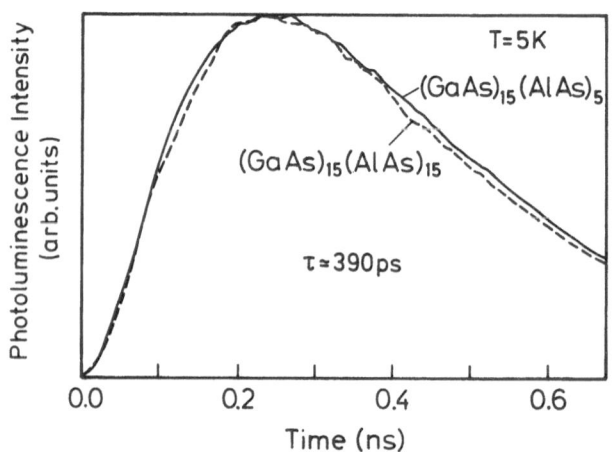

Fig. 5 Time behaviour of the spectrally integrated (type I) photoluminescence of a (GaAs)$_{15}$ (AlAs)$_{15}$ and (GaAs)$_{15}$ (AlAs)$_5$ superlattice at T = 5K.

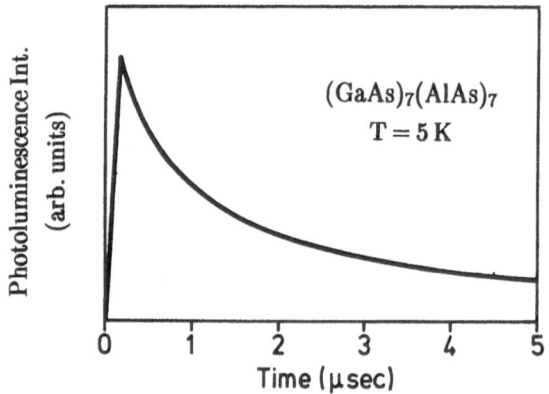

Fig. 6 Time behaviour of the spectrally integrated (type II) photoluminescence of a (GaAs)$_7$ (AlAs)$_7$ superlattice at T = 5K.

The luminescence decay in type II GaAs/AlAs SPS has been studied by several groups (for a comprehensive summary see ref. 4). The most striking result is the much slower decay as compared to type I or bulk GaAs recombination. An experimental result for the PL-decay of a type

II (GaAs)$_7$ (AlAs)$_7$ SPS at T = 5K is shown in Fig. 6 [22]. The slow
PL-decay reflects again the weaker transition strength of the type II
excitons. The detailed mechanism for this originally forbidden optical
transition has been discussed controversely in the literature[4].
According to the presently available data it seems likely that both,
mixing between the X_z (AlAs) electron states and the Γ (GaAs) electron
states by the superlattice potential[7] or scattering at the
heterointerface[23] as well as scattering by disorder[16] provides the
oscillator strength for the type II exciton recombination with relative
magnitude depending on SPS parameters and possibly also quality, e.g.
interface roughness.

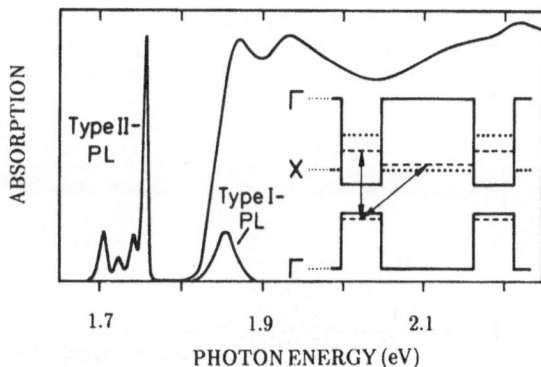

ABSORPTION

Type II-
PL

Type I-
PL

Γ

X

Γ

1.7   1.9   2.1

PHOTON ENERGY (eV)

Fig. 7   Low temperature (T = 5K) absorption and photoluminescence
          of a (GaAs)$_{11}$ (AlAs)$_{24}$ superlattice. The band alignment
          together with the confined energy states is also
          indicated.

c) Γ-X Scattering In Type II SPS

Electrons excited in the Γ-conduction band states of the GaAs in
type II SPS are scattered to the lower lying X (AlAs) electronic states
as revealed e.g. by the excitation spectra (see e.g.Fig. 4). This
intervalley scattering thus involves a spatial charge transfer. The
characteristic time constants for this process have been determined
recently for different (GaAs)(AlAs) SPS by means of femtosecond excite
and probe experiments[24],[25]. The respective time constants for the Γ-X
scattering are obtained from the partial recovery of the bleaching of
the type I exciton transition after excitation with a short ($\simeq$ 100 fs)
laser pulse with photon energy higher than the effective type I band
gap. This will be illustrated in somewhat more detail for a (GaAs)$_{11}$
(AlAs)$_{24}$ SPS as an example. The absorption and PL spectrum of this
particular SPS are shown in Fig. 7 together with a schematic drawing of
the band alignment (full line for the Γ-bands, dotted line for the
X-conduction band) and the confined electron and hole states (dashed
lines). The PLE spectrum reveals the heavy- and light hole exciton
transition at about 1.87eV and 1.93eV, respectively, as well as the

exciton transition involving holes in the split-off valence band at
about 2.2eV. The photoluminescence bands correspond to the type I
exciton recombination at about 1.87eV and the zero phonon and phonon
assisted type II recombination at lower energies. The type I PL decays
with a time constant which is below the temporal resolution ($\simeq$ 20 ps)
of our present set up revealing the fast transfer of carriers into the
X-states of the AlAs.

The result of the femtosecond excite and probe experiment is
depicted in Fig. 8, where time resolved differential transmission
spectra ($\Delta T/T_0$) are plotted. The photon energy of the 100 fs excitation
pulses of 2eV is indicated by a line. An increase of the transmission
after the excitation is observed at energies corresponding to the heavy
hole, light hole and split-off exciton transition reflecting the
bleaching of the respective transitions by the excited carriers. State
filling as well as screening may contribute to the bleaching of the
exciton peaks with state filling being the dominant contribution in
quantum wells[26]. It should be noted that in case of the split-off
transition at about 2.2eV state filling only of electrons can
contribute since holes are not excited in the split-off valence band.

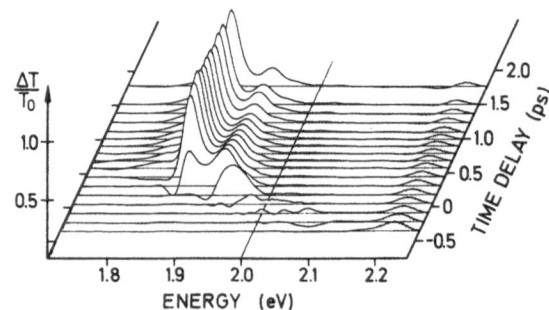

Fig. 8   Differential transmission spectra ($\Delta T/T$) at different
delay times of a (GaAs)$_{11}$ (AlAs)$_{24}$ superlattice.

The time behaviour of the ($\Delta T/T_0$)-signals at the respective
exciton peaks is summarized in Fig. 9. An initial partial recovery of
the bleaching signal within about 2 ps is found for all exciton
transitions, opposite to the behaviour found in GaAs quantum
wells[27,28]. The remaining bleaching after about 2 ps is strongest for
the heavy hole exciton and less than one third of the initial value for
the light hole and split-off exciton. The initial partial recovery of
the bleaching signal can definitely be attributed to the scattering of
electrons out of the Γ-like GaAs states into the X-states of the AlAs.
This transfer of electrons removes the contribution of conduction band
states filling to the exciton bleaching and in addition screening might
be different for electrons being in Γ-like states within the GaAs slabs
or X-like states in the AlAs. It should be noted that the remaining
bleaching of the split-off transition (full points in Fig. 9) can be
solely caused by screening of the exciton transition by heavy and light

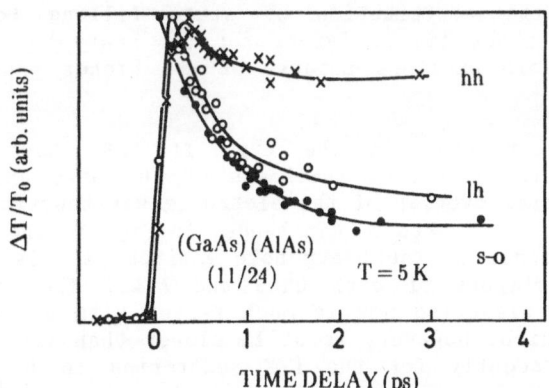

Fig. 9   Time behaviour of the differential transmission signal
         ($\Delta T/T_o$) of the $(GaAs)_{11}$ $(AlAs)_{24}$ superlattice at photon
         energies corresponding to the heavy hole exciton (hh), light
         hole exciton (lh) and split-off transistion (s-o).

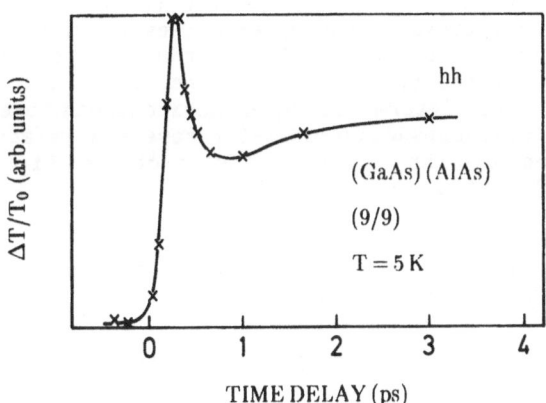

Fig. 10   Time behaviour of the differential transmission signal
         ($\Delta T/T_o$) at the energy corresponding to the heavy hole
         exciton (hh) transition.

holes and electrons within X-states of the AlAs. We thus are able to separate the contribution of electron state filling from screening. The results show that the contribution of electron state filling to the bleaching of the split-off exciton transition is comparable to the screening by electrons (in the AlAs) <u>and</u> holes. The difference in the time behaviour of the DTS signal of the heavy- and light hole exciton transition can be attributed to thermalization of light and heavy holes taking place on the time scale of the order of 1ps[29]. Due to this thermalization the contribution of state filling to the exciton bleaching after about 1ps is larger for the heavy hole excitons than for the light hole excitons because of the higher population of the heavy hole band.

The Γ-X scattering in the type II SPS shows a pronounced dependence on the respective thickness of the GaAs and AlAs layers, i.e. on the spatial overlap of the elctronic wavefunctions of the GaAs (Γ) and the AlAs (X) states or minibands. In Fig. 10 the time course of the bleaching signal at the heavy hole exciton peak is depicted for a SPS with 9 monolayers of each GaAs and AlAs. The initial partial recovery of the bleaching now is much faster with a time constant of about 100 fs, which, however, is still slower than the value of about 55fs reported recently for the Γ-X scattering in bulk GaAs[30]. The values for the overlap of the envelope wavefunctions ($|\langle\Psi_x|\Psi_\Gamma\rangle|$), which have been calculated within a simple Kronig-Penney model amount to about 0.34 and 0.08 for the (9,9) and (11,24) SPS, respectively. The difference in the Γ-X transfer times thus can at least be partly attributed to the different overlap of the envelope wavefunctions.

SUMMARY

The possibilities of band structure engineering with SPS and the consequences on optical properties and in particular on the dynamics of photoexcited, nonequilibrium carriers are demonstrated for the (GaAs) (AlAs) system for example. We find that the time behaviour of the photoluminescence of type I SPS is similar to GaAs quantum wells with characteristic decay times on the order of several 100 ps to ns. The PL decay times are longer in type II SPS by order of magnitudes reflecting the weak strength of the type II transitions. The Γ-X scattering times in type II SPS are determined by subspicosecond excite and probe experiments and their dependence on SPS parameters is attributed to the different overlap of the envelope wavefunctions of the respective initial and final states.

Acknowledgement: The work at the University of Marburg is supported by the Deutsche Forschungsgemeinschaft.

References

1. U. Gnutzmann, K. Clausecker, Theory of direct optical transitions in an optical indirect semiconductor with a superlattice structure, Appl. Phys. 3: 9 (1974)
2. L. Esaki, A bird's-eve view on the evolution of semiconductor superlattices and quantum wells, IEEE J. Quantum Electron., QE-22: 1611 (1986)
3. L.L. Chang, Semiconductor quantum-well heterostructures, in "Layered Structures and Epitaxy", M.Gibson, G.C. Osbourn, R.M. Tromp, eds., Mat. Res. Soc. Symp. Proc., vol. 56, pp. 267 (1986)

4. B. Wilson, Carrier dynamics and recombination mechanisms in staggered-alignment heterostructures, IEEE J. Quantum Electron., QE-24: 1763 (1988)

5. K. Ploog, Molecular beam expitaxy of artificially layered III-V semiconductors on an atomic scale, in: "Physics and Applications of Quantum Wells and Superlattices", E.E. Mendez, K. von Klitzing, eds., Plenum Press, New York, pp. 43, 1988

6. A collection of the material and band structure parameters of the AlGaAs system can be found in: S. Adachi, GaAs, AlAs and $Al_xGa_{1-x}As$: material parameters for use in research and device application, J. Appl. Phys., 58: R1, (1985)
    see also in: "Landolt-Börnstein, vol. 22", O. Madelung, ed., Springer-Verlag, Berlin, New York, 1987

7. M.-H. Meynadier, R.E. Nahory, J.M.Worlock, M.C. Tamargo, J.L. de Miguel, M.D. Sturge, Indirect-direct anticrossing in GaAs/AlAs superlattices induced by quantum confined Stark effect: evidence for Γ-X mixing, Phys. Rev. Lett. 60: 1338 (1988)

8. P. Dawson, K.J. Moore, C.T. Foxon, Photoluminescence studies of type II GaAs/AlAs quantum wells grown by MBE, Proc. SPIE 792: 208 (1987)

9. K.J. Moore, G. Duggan, P. Dawson, C.T. Foxon, Short period GaAs-AlAs superlattices: optical properties and electronic structure, Phys. Rev. B38: 5535 (1988)

10. D.Z.-Y. Ting, Y.C. Chang, Γ-X mixing in GaAs/AlGaAs and AlGaAs/AlAs superlattices, Phys. Rev. B 36: 4359 (1987)

11. S. Gopalan, N.E. Christensen, M. Cardona, Band edge states in short-period $(GaAs)_m (AlAs)_n$ superlattices, Phys. Rev. B 39: 5165 (1989)

12. K. Moore et al., this book

13. M.-H. Meynadier, this book

14. G. Peter, E.O. Göbel, W.W. Rühle, J. Nagle, K. Ploog, Carrier dynamcis in $(GaAs)_m (AlAs)_n$ superlattices, Superlatt. and Microstruct. 5: 197 (1989)

15. A. Chomette, B. Lambert, B. Deveaud, F. Clerot, A. Regreny, G. Bastard, Exciton binding energy in small period GaAs/GaAlAs superlattices, Europhys. Lett. 4: 461 (1987)

16. E.Finkmann, M.D. Sturge, M.-H. Meynadier, R.E. Nahory, M.C. Tamargo, D.M. Hwang, C.C. Chang, Optical properties and band structure of short period GaAs/AlAs superlattices, Journ. Luminesc. 39: 57 (1987)

17. J. Nagle, M. Garriga, W. Stolz, T. Isu, K. Ploog, Position and character (Γ or X) of energy states in short-period $(GaAs)_m(AlAs)_n$ superlattices, Journ. de Physique 48: C5-495 (1987)

18. R. Cingolani, M. Ferrara, L. Baldassare, M. Lugara, K. Ploog, Type I - type II transition in ultra short period GaAs/AlAs superlattices, Phys. Rev. B, in press

19. J.Feldmann, G. Peter, E.O. Göbel, P. Dawson, K. Moore, C. Foxon, R.J. Elliott, Linewidth dependence of radiative exciton lifetimes in quantum wells, Phys. Rev. Lett. 59: 2337 (1987)

20. B. Deveaud, B. Lambert, A. Chomette, F. Clerot, A. Regreny, J. Shah, T. Damen, B. Sermage, in: "Optical Switching in Low Dimensional Systems", H. Haug, L. Banyai, eds., Plenum Press, 1989

21. B. Lambert, B. Deveaud, A. Chouette, A. Regreny, B. Sermage, Density-dependent transition from electron to ambipolar vertical transport in short period GaAs-AlGaAs superlattices, Semicond. Science and Techn., in press

22. E.O. Göbel, R. Fischer, G. Peter, W.W. Rühle, J. Nagle, K. Ploog, Carrier relaxation and recombination in (GaAs) (AlAs) short period superlattices, in: "Optical switching in low dimensional systems", H. Haug, L. Banyai, eds., Plenum Press, 1989

23. P. Dawson, K. J. Moore, C.T. Foxon, G.W't Hooft, R.P. M. van Hal, Photoluminescence decay time of type II GaAs/AlAs quantum well structures, Journ. Appl. Phys. 65: 3606 (1989)

24. J. Feldmann, R. Sattmann, E.O. Göbel, J. Kuhl, J. Hebling, K. Ploog, R. Muralidharan, Subpicosecond real space charge transfer in type II GaAs/AlAs superlattices, Phys. Rev. Lett. 62: 1892 (1989)

25. P. Saeta, J.F. Federici, R.J. Fischer, B.I. Greene, L. Pfeiffer, R.C. Spitzer, B.A. Wilson, Γ to X Transport of photoexcited electrons in type II GaAs/AlAs multi-quantum well structures, Appl. Phys. Lett. 54: 1681 (1989)

26. D.S. Chemla, D.A.B. Miller, S. Schmitt-Rink, Nonlinear Optical Properties of Semiconductor Quantum Wells, in: "Optical Nonlinearities and Instabilities in Semiconductors", H. Haug, ed., Academic Press, San Diego, 1988, pp. 83

27. W. Knox, C. Hirlimann, D.A.B. Miller, J. Shah, D.S. Chemla, C.V. Shank, Femtosecond excitation of nonthermal carrier populations in GaAs quantum wells, Phys. Rev. Lett. 56: 1191 (1986)

28. W.Z. Lin, R.W. Schoenlein, J.G. Fujimoto, E.P. Ippen, Femtosecond absorption saturation studies of hot carriers in GaAs and AlGaAs, IEEE J. Quantum Electr. QE-24: 267 (1988)

29. R.P. Joshin, D.K. Ferry, Hot-phonon effects and interband relaxation processes in photoexcited GaAs quantum wells, Phys. Rev. B39: 1180 (1989)

30. R.C. Becker, H.L. Fragnito, C.H.Brito Crutz, J. Shah, R.L. Fork, J.E. Cunningham, J.E. Henry, C.V. Shank, Femtosecond intervalley scattering in GaAs, Appl. Phys. Lett. 53: 2089 (1988)

HOT PHONONS IN MICROSTRUCTURES

P. Kocevar

Institut für Theoretische Physik, Universität Graz
Universitätsplatz 5, A-8010 Graz, Austria

ABSTRACT

The transient as well as steady-state dynamics of highly energetic carriers in semi-conductors is of great interest for both basic research and technological applications in electrical and electrooptical devices. In recent years hot-carrier spectroscopy in bulk and microstructured III-V compounds has given strong evidence for a substantial, nonequilibrium-LO-phonon induced reduction of the electronic energy transfer into the lattice. The present report gives a general introduction into the concept of "hot phonons", discusses their possible consequences for the response of photo-and/or d.c.field-excited hot carriers in GaAs-based quantum-well and multiquantum-well structures, and summarizes the experimental and theoretical support for these expectations.

INTRODUCTION

The energy relaxation rates of highly energetic charge carriers in semiconductors has since the early times of hot-electron physics been the subject of intensive experimental and theoretical studies. In this context the last years have seen an increasing number of spectroscopic investigations of possible effects of nonequilibrium optical phonons on the steady-state and on the ultrafast nonlinear response of electrically or optically excited hot electrons and holes in bulk semiconductors and semiconductor microstructures.

The present volume contains a number of contributions, which address the problematics of phonons in layered semiconductor microstructures ("MS"). Together with a rapidly growing number of publications on this subject they give ample theoretical and experimental evidence for the formation of optical slab modes, of confined modes as their descendants in quantum-well (QW) and multi-quantum-well (MQW) structures, of interface modes and of the zone folding of the broad-band bulk acoustic branches. The determination of these modes is presently one of the most revealing fields of phonon spectroscopy, as is well documented by several contributions to this workshop.

This type of research determines so to say the kinematical properties of phonons in microstructures, whereas their dynamical properties, especially in connection with so-called "hot-phonon" effects, have been addressed in the foregoing contributions of J.Shah and S.Goodnick and concern the interpretation of abnormally low energy-relaxation rates of highly photo excited carriers generally found in the picosecond laser-pulse spectroscopy of QW and MQW structures of polar semiconductors and their alloys.

After some years of intense dispute about the question whether free-carrier screening of the dominant polar-optical (po) carrier-LO phonon couplings alone could explain this retardation, there seems now to be a general agreement about the much more decisive role of LO-phonon "heating" for the reduced power dissipation of highly excited and dense carrier plasmas within the picosecond timescale.

THE CONCEPT OF HOT PHONONS

Let us introduce the concept of "hot phonons" in such a way as to distinguish between nonequilibrium-phonon effects in general and the specific phenomena connected with highly photo- or d.c.field-excited charge carriers. So we shall speak of hot-phonon effects, whenever nonequilibrium phonon distributions induce, *in addition to possible drag effects*, noticeable modifications of the nonlinear electronic response of a solid.

Here we have distinguished between the many well-known cases of phonon disturbances in the form of a mean displacement of the phonon distribution function $N(\vec{q})$ in $\vec{q}$-space ("phonon drift") and disturbances involving an overall amplification of the distribution in the conventional sense of particle heating. In analogy to the hot-electron concept this distinction does by no means require the existence of an elevated "phonon temperature" T_{ph} in a diffusion approximation $N(\vec{q}) = N_0(q) + N_1(\vec{q})$ with the small drift term N_1 superimposed on an isotropic thermal Planck distribution N_0 at T_{ph} or, more generally, does not imply the realization of a drifted and heated Planck distribution.

636

For the strongly coupled carrier-phonon (c-ph) systems of our present concern the functional form of the disturbed phonon distributions will for each mode depend on the dominant c-ph interaction —and thereby on the carrier distribution functions $f(\vec{k})$— and quite decively also on the dominant nonelectronic phonon-thermalization processes, as established by ph-ph and phonon-boundary (ph-b) scatterings. These dependences are the direct counterpart of the dependences of the carrier distributions on the external driving forces (d.c. or laser fields) and on the dominant carrier scatterings.

To introduce into this hot-phonon problematics we first outline the historical development of the subject. For this purpose we begin with the older question of possible effects of acoustical-phonon disturbances on semiconductor mobilities.

Acoustic Phonons

Quite generally one has to expect for strongly interacting carrier-phonon systems in external fields deviations of the distribution function of the phonons from their thermal-equilibrium Planck distribution, whenever the rate of momentum and/or energy transfer from the excited carriers to the phonons is comparable or greater than the thermalization rate of these phonons through ph-ph or ph-b scatterings. The resulting mutual drag and heating effects between the carriers and the electronically active phonons had for a long time been studied in connection with thermoelectric effects in semiconductors like the Peltier or the Seebeck effects and were in fact found to be a transport-theoretical prerequisite for the validity of the Kelvin relation.

Successful attempts to perform a quantitative transport analysis of these phenomena and of the related question of stability criteria for the charge transport were only possible after the advent of experimental techniques for the spectroscopy of highly nonequilibrium phonons and the resulting determination of phonon-thermalization rates. Thereafter one discovered the decisive difference between the slow relaxation of nonequilibrium phonons within a similarly disturbed phonon reservoir (as in the case of the totally streaming phonon population in lattice heat conduction) and the much faster relaxation of disturbed phonons belonging to a small nonequilibrium frequency band into the "thermal reservoir" of the remaining equilibrium modes (as in the case of ultrasonic attenuation).

I stress these historical facts because exactly the same sort of considerations not only concerns the case of hot optical phonons, but is also just coming up in connection with the recent studies of subpicosecond thermalization of photoexcited carriers among themselves or within a larger or comparable reservoir of cold carriers (see the contributions of J.Shah and S.Goodnick to this volume).

In the early seventies quite substantial difficulties arose from this relaxation problematics for the transport formulations of coupled carrier-acoustic phonon systems [1]. These difficulties were connected with the attempt to apply the longstanding practice of carrier-temperature approaches to nonohmic transport (for phonon equilibrium) to use rate equations for the energy and momentum balance of the carriers together with separate relaxation times for momentum and energy. It turned out that a similar treatment of the nonequilibrium acoustic phonons is rather questionable mainly due to the following complications:

(i) The electronically amplified modes relax by 3-phonon processes with partner phonons belonging themselves to electronically active and therefore perturbed modes. This has two severe consequences. Firstly one cannot use the "single-mode" thermalization rates as measured by ultrasonic attenuation or the spectroscopy of heat-pulse or Josephson-junction generated ballistic phonons. These rates would be inadequate for the description of the relaxation within a broad band of disturbed lattice modes. Secondly, because of their coupling to the carriers, the decay products of the initially excited phonons will in case of their absorption by carriers feed energy and momentum back to the carrier system. So the use of purely lattice-dynamical phonon decay rates as energy or momentum relaxation rates for the disturbed phonons in one of the usual kinetic models for the phonon transport is highly questionable.

(ii) For small samples, the energy and the momentum relaxation of the phonons at the crystal boundaries are strongly dependent on the (experimentally not easily controllable) microscopic surface properties and on the outside cooling agent, and are in any case badly understood.

Acoustic-phonon disturbances are restricted to very low lattice temperatures (below typically 20 K) because of the very rapid increase of the phonon thermalization rate with temperature (typically proportional to T^3 or T^4) and because of the rather slow rates ($\approx$ of the order of $10^6 sec^{-1}$ or lower) for the build up of acoustic phonons by hot and/or strongly drifting carriers. Transient effects of such acoustic-phonon disturbances would occur within time scales of at least microseconds. These times are characteristic of the onset of acoustoelectric phenomena (and some of their accompanying GHz emissions) and, for longer times, of the heating or eventual thermal damage of the crystal for too long high-field pulses. This brings us to the recent low-temperature studies of ref.[2], where the finding of MHz oscillations resembles the macroscopic, collective phenomena of bulk acoustoelectrics, but where the discovery of high-frequency emission might, like in the bulk case, be a candidate for a "kinetic" (i.e. individual-particle) heating of a kinematically favoured subsystem of acoustical phonons [3].

Optical Phonons

Most of the difficulties encountered for acoustic-phonon disturbances are practically nonexistent for the theoretical description of optical-phonon disturbances by d.c.field- or photo-excited hot carriers, at least in *bulk* materials. Here we are in the position to assert the validity of using the experimentally measured thermalization rate of LO and TO phonons as the rate of energy and momentum dissipation out of the coupled system of carriers and optical phonons. This favourable situation follows from the fact that the electronically most strongly coupling near-zone-center optical phonons decay into pairs of electronically inactive near-zone-boundary phonons, each of which rapidly further decays into still lower-lying phonons of the acoustic branches, initiating in this way the phonon cascade of an eventual lattice heating.

The frequencies of scattering of electrons and holes by optical phonons are of the order of 10^{12} to $10^{13} sec^{-1}$ and therefore in general noticeably larger than the nonelectronic thermalization rates of optical phonons (of the order of 0.5 to 1.0 $\cdot$ $10^{11} sec^{-1}$ at temperature $T = 0$, increasing only by about a factor of three between $T = 0$ and room temperature. So the very short mean free times and the very weak frequency dispersion of the optical modes allows one to neglect spatial diffusion of the amplified phonons out of their excitation region during their lifetime.

For the QW or the MQW systems of our present concern we should be aware of possible size effects on the relaxation of (confined or delocalised) optical phonons due to the modification in the number of their kinematically allowed decay modes. But a restriction of nonequilibrium-phonon effects due to a strong increase in the number of phonon-decay channels cannot be expected, because the number of decay partners (as e.g. in case of a folding of the zone-boundary LA modes) allowed by energy and momentum conservation should, with the exception of a possible overlap interaction with modes of the nearby barrier layers, remain practically the same as in the bulk [4][5]. Moreover, besides these phase-space considerations the finding of an almost universal decay rate of optical phonons in all standard semiconductors indicates a rather weak dependence of the anharmonic ph-ph couplings on the chemical composition of the lattice. This insensitivity to the details of the local interatomic bonds leads us to expect minor size effects on the strength of the ph-ph couplings, at least in lattice-matched microstructured samples.

In any case, as the degree of phonon amplification is determined by the interplay between phonon generation by the carriers and nonelectronic phonon thermalization, we stress the lack of even a semiquantitative theory of the decay dynamics of nonequilibrium phonons in layered microstructures. This deficiency, like the early difficulties with our quantitative understanding of acoustical-phonon relaxation stands in the way of fully reliable numerical estimates of nonequilibrium-phonon effects in semiconductor microstructures. It also illustrates the urgent present need for a detailed spectroscopy not only of the mode structure of the electronically most active phonons, but also of their decay or thermalization rates as functions of frequency, temperature, excitation conditions (carrier density), and sample geometry.

In view of these uncertainties our following discussion of certain presently debated aspects of hot optical phonons in heterojunctions and QW or MQW structures will necessarily rely on (and in most cases also justify) our general expectation of a close similarity between bulk and microscopic materials regarding the dominant relaxation mechanisms within and out of highly excited c-ph systems.

CARRIER RELAXATION IN THE PRESENCE OF HOT OPTICAL PHONONS

Hot-Phonon Effects (general)

Figure 1 shows in a (hopefully) self-explanatory representation how the amplification of phonons which dominate the energy and/or momentum relaxation of the carriers can strongly modify the d.c.field or optical response of hot carriers.

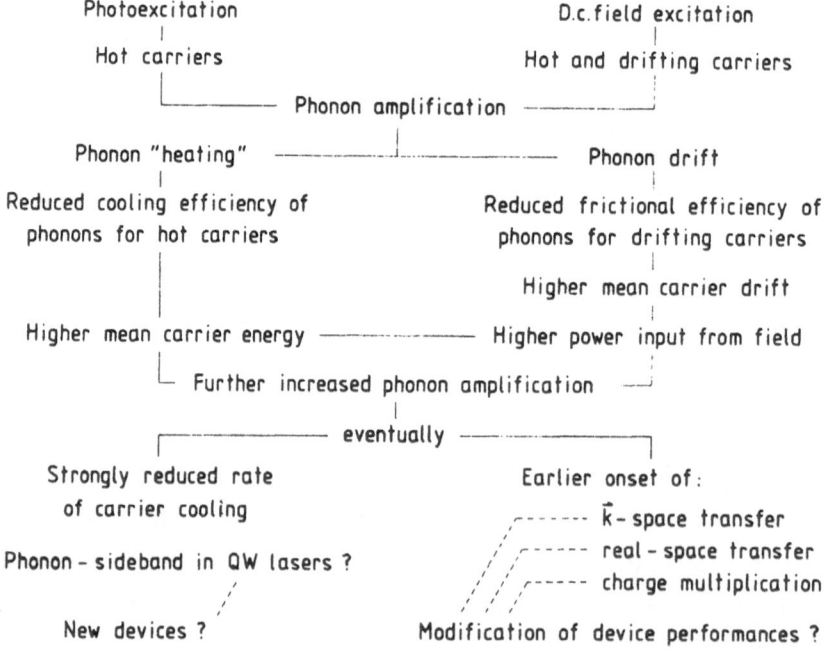

Figure 1. Hot-Phonon Scenario

Figures 2 and 3 contain the basic ingredients of the hot-LO-phonon effect. Figure 2 shows the statistical weights associated with the equilibrium situation (a), the standard hot-carrier picture for phonon equilibrium (b), and the hot carrier—hot phonon situation of our present concern (c). For the equilibrium c-ph system before the onset of the excitation the total rates for phonon emission and absorption are of course equal. For the "active" carriers with energies $E \geq \hbar\omega_{LO}$ the net emission rate is determined by the statistical phonon weight 1 for spontaneous emission, since the absorption and induced emission, both of weight $N^{(0)}$, cancel each other. The total absorption rate is therefore determined solely by the phonon absorptions, of weight $N^{(0)}$, by the "passive" carriers with energies below the emission threshold $\hbar\omega_{LO}$. In spite of the fact that for all lattice temperatures of practical interest $N^{(0)} \ll 1$, the emission and absorption rates are equal because of the higher carrier occupation of the passive energy regime $E < \hbar\omega_{LO}$.

Figure 2b shows the situation shortly after the carrier excitation with already internally thermalized hot carriers but with still cold phonons. The statistical weights of the phonons are still the same as before, but now many carriers have been transferred from the passive to the active energy regime. This results in a strong relative

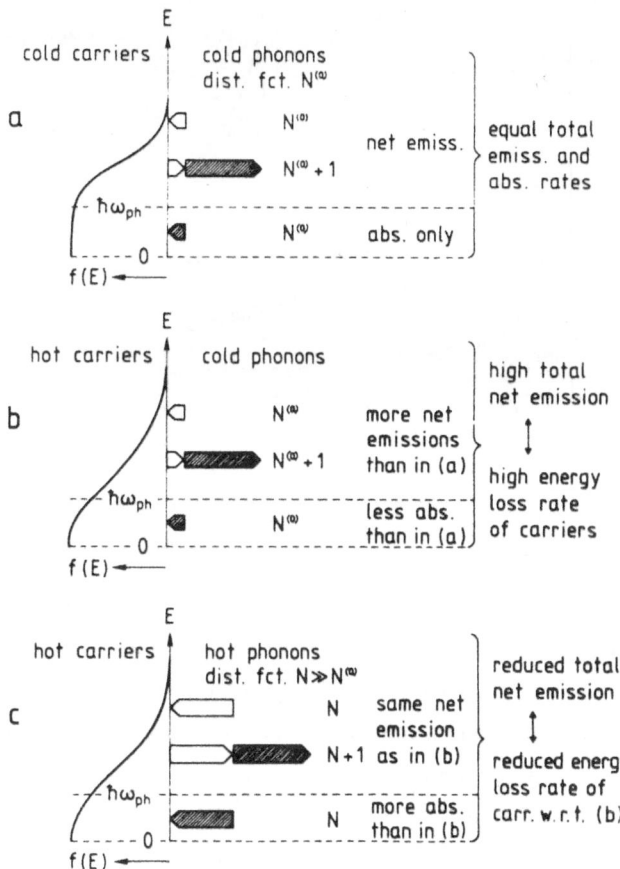

Figure 2. Hot-Phonon Effect: Statistical Weights

increase of net phonon emission, a correspondingly reduced phonon reabsorption and therefore in a large energy (momentum) relaxation rate of the carriers.

As the next stage of the c-ph evolution Figure 2c shows the situation after the eventual phonon amplification resulting from the fast energy relaxation phase of Figure 2b. Now the phonon distribution function has strongly increased, $N \gg N^{(0)}$. Still the phonon weight for the active carriers remains 1 because of the cancellation of their now strongly enhanced absorption and induced emission. But the weight N for the only remaining phonon absorption in the passive carrier region is substantially increased. As a consequence the large initial net phonon emission during the transient stage (b) is now strongly reduced by the increased phonon reabsorption, resulting in a much smaller energy (momentum) dissipation of the carriers to the LO phonons and therefore in a higher mean carrier energy (drift).

So the reduction of the cooling and viscuous action of the phonons originates

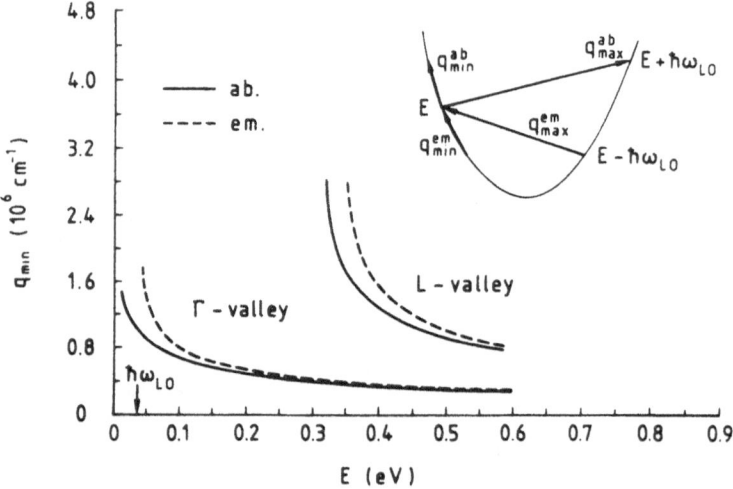

Figure 3. Phase-Space Analysis of Electron-LO-Phonon Scattering in GaAs

from the phonon amplification and the ensuing relative increase of the hot-phonon reabsorption by carriers with energies $E \leq \hbar\omega_{LO}$. For a completely isolated c-LO-ph system the LO phonons would finally become completely inefficient as cooling and frictional agent. Due to the nonelectronic phonon processes, however, the LO-ph losses through the anharmonic LO-phonon decays will eventually keep N below its limiting "detailed-balance" value N_c for mutual hot carrier—hot phonon equilibrium [6][7].

The foregoing discussion of Figure 2 was rather crude, because it neglected the important additional phase-space restrictions from energy and wavevector conservation per c-ph process as depicted in Figure 3 for Γ and L electrons in bulk GaAs. Consider e.g. the excitation of electrons by a short intense laser pulse to kinetic energies far above the LO-emission threshold and follow their cooling through LO-ph emission. In Figures 3 are plotted the magnitude of the minimum phonon wavevectors q_{min} for emission and absorption of LO phonons by a carrier of energy E. As the dominant polar optical c-LO-ph coupling favours forward scattering, most of the emitted or absorbed LO phonons will have wavenumbers q just slightly above the q_{min}-curves. So as long as there are still enough carriers with sufficiently high energies, the initially emitted phonons are likely to be reabsorbed before decaying through the anharmonic ph-ph interaction. However, after a cooling period of typically a few picoseconds most of the electrons have relaxed to low energies and cannot reabsorb the phonons they had initially emitted. Thereafter these electronically inactive phonons can only decay through ph-ph processes [8].

We have for simplicity discussed the bulk situation, where these strong phase-space restrictions indeed have important consequences e.g. on the interpretation of time-resolved Raman spectra in terms of lattice-dynamical phonon lifetimes [6][8]. In QW systems and heterojunctions, even for unconfined modes the phonon interaction with the size-quantized carriers will also introduce size effects on the disturbed phonon distributions due to the smeared-out restrictions for the components $q_\perp$ of the phonon wavevectors perpendicular to the layers. Some wavevector-conserving and -nonconserving formulations of the c-ph interactions and some of the corresponding difficulties in describing the wave coupling of 2d-carriers and 3d-phonons have been extensively discussed in the literature [9] [10] [11] [12] [13]. The most promising approach seems to be that of refs.[14] and [15], where the problem of the unavoidable size effects and for some cases of spurious dependences of the hot-phonon effects on the well or layer width even for unconfined modes is circumvented by the use of appropriately constructed phonon wavepackets and of a corresponding phonon density matrix.

It is of some practical advantage to have a rough estimate of the conditions, i.e. mainly of the carrier densities n_c, for which noticeable hot-LO-ph effects have to be expected during or after d.c.field or laser excitation. There are two factors governing the degree of LO-phonon amplification. First the phonon disturbance is naturally proportional to the density of hot carriers. Second for the free-carrier screening of the po couplings the transition propabilities are changed from their proportionality to q^{-2} for no screening to roughly (i.e. for simple Thomas-Fermi screening) a proportionality to $q^2/(q^2 + q_s^2)^2$, with the screening parameter $q_s \propto n_c^{1/2}$. So for small n_c (typically below several $10^{16} cm^{-3}$) the phonon disturbances are negligible and reach a maximum at densities between several 10^{17} and several $10^{18} cm^{-3}$. They finally again disappear for very large densities above several $10^{19} cm^{-3}$ due to the eventual dominance of the screening parameter $q_s^4 \propto n_c^2$ in the denominator of the transition probabilities, which simply means the complete screening of the po couplings and the total quenching of the polar optical c-LO-ph interaction.

For the frequently encountered experimental situations with favourable amplification conditions quite large LO-ph disturbances have been predicted by various theoretical methods: at room temperature (with a thermal equilibrium LO-ph Planck distribution $N^{(0)}$ of 0.34) hot and strongly forward-peaked distributions have been found for bulk n-GaAs for d.c.transport (with $N \approx 5$ and more for moderate doping and fields of several kV/cm)[16]. A similar LO-ph amplification is also predicted for photoexcitation (with maximal N values of ≈ 2.5 for bulk GaAs and InP [8] for frequently encountered experimental excitation conditions and a lattice temperature of 77K (at which $N^{(0)} = 0.005$). For GaAs-based QW structures the theoretically predicted phonon amplification is similarly pronounced; for low temperatures of a few degrees K (i.e. for practically vanishing $N^{(0)}$) a Monte-Carlo analysis again finds N values around 2 for standard photoexcitation conditions [17][18][19] and the already mentioned density-matrix approach finds for similar experimental situations maximal N values larger than 1 [20].

In all the cited calculations of the cooling rates of highly laser-excited carrier plasmas a drastic reduction of the relaxation rates of the hot carriers by more than one order of magnitude is found, in excellent agreement with our general scheme of Figure 1 and with the experimental data of 2D systems [21] [22] [23]. In contrast to these decisive hot-phonon effects for high photoexcitation the influence of nonequilibrium phonons on the nonohmic response in both bulk and microstructured semiconductors is rather modest, with typical modifications of the hot-carrier mobilities of a few ten percent for the fields and moderate dopings of practical interest [15] [16], even in cases of very strong LO-phonon amplification. At least for the more transparent bulk situation the reason for this insensitivity of the electrical carrier response to the strong phonon build up can be traced to the phase-space restrictions discussed in connection with Figures 2 and 3, which for instance obstruct an interference of the LO-phonon amplification with the velocity-overshoot phenomenon in n-GaAs.

For our following comments on some special aspects of hot LO phonons we should, in view of the foregoing contributions of J.Shah and S.Goodnick, keep in mind that for the ultrashort carrier-relaxation phenomena presently investigated in femtosecond laser spectroscopy the picosecond nonequilibrium-phonon dynamics is of no immediate relevance.

D.C.Fields

The scheme in Figure 1 of mutual drag and heating effects within a highly d.c.field-excited c-ph system has been recently invoked and analysed in terms of a wave-packet description of nonequilibrium LO phonons interacting with quasi-twodimensional displaced Fermian (or Maxwellian) electrons in GaAs-based heterojunctions [15].

In complete accordance with our declared strategy to expect great similarities between bulk and MS materials, the main hot-phonon effects predicted by these authors just duplicate completely analogous predictions from much older calculations for bulk materials. Quite typically for many presently published results for quasi-twodimensional systems the authors seem to be unaware of this older work, giving the impression that some phenomena might be a consequence of the lower dimensionality of the carrier system. For brevity the following list of the hot-phonon effects (listed as D1—D4 below) predicted and discussed for quasi-twodimensional (2D) systems in [15] contains only a short description of the effects, together with some references to earlier predictions of the same effects for bulk (3D) semiconductors.

D1) Dominance of carrier drag by drifting nonequilibrium LO phonons at low fields (including the ohmic limit) and transition to a frictional effect of LO-phonon heating with reduced carrier mobility at higher fields. 3D: for the analogous single-valley model of GaAs see [7][16][24].

D2) Carrier *cooling* below the lattice temperature due to the transfer of random thermal energy into directed kinetic energy of the drift motion for neglible ionized impurity scattering. This effect should only be realizable for situations of remote ionized-impurity scattering in modulation-doped heterostructures. 3D: thoroughly discussed in [25] for phonon equilibrium, but ref.[15] cites [26] and [27] as primary literature. The further increase of this effect in the presence of LO-phonon disturbances as found by [15] is of course the consequence of the mechanism D1 above.

D3) Nonequilibrium LO-ph-induced collective breakdown in a single-valley model of GaAs as described in Figure 1. Ref.[15] cites [24] as an earlier prediction for the 3D case; meanwhile, as already conjectured in [24], the quenching of this effect by intervalley transfers of the carriers in the actual many-valley conduction band of GaAs has been demonstrated for 3D in analogous heated and displaced maxwellian (HDM) carrier models [7] and in Monte-Carlo simulations [16]. The earlist prediction (for n-InSb) of the nonequilibrium-LO-ph-induced acceleration of the single-valley breakdown would be [3].

D4) Possibility of a compensation of hot-LO-ph effects by antiscreening of the polar-optical (po) couplings for the case of comparable carrier-plasma and LO-phonon frequencies, as mentioned by [15], referring to [28]. In earlier 3D work for both d.c.field and laser excitation it was shown, that *in the presence of noticeable hot-LO-ph effects* the latter almost exactly cancel any change in the c-ph couplings; e.g.   increasing the coupling by antiscreening increases on one hand the number of c-ph processes, on the other hand the resulting increase of the phonon amplification reduces the average frictional and cooling effectivity of each individual c-LO-ph process. This was demonstrated in both HDM-carrier [24] and Monte-Carlo calculations [16]. It should be mentioned that the possibility of phonon-plasmon effects and of a corresponding antiscreening of the po couplings in 3D was originally predicted by [29].

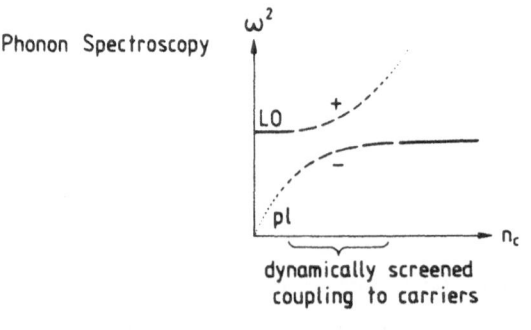

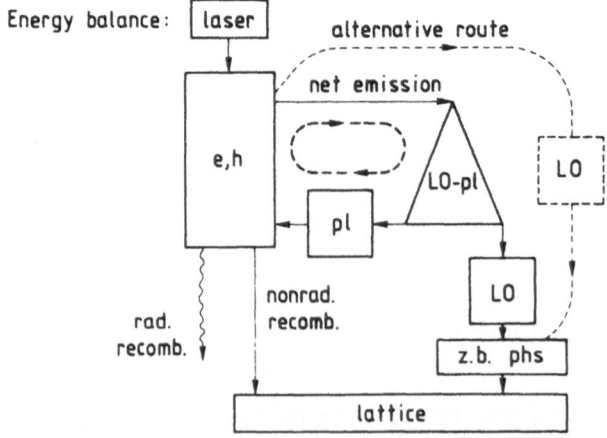

Figure 4. Role of LO-Phonon—Plasmon Hybridisation

D5) Hot-phonon effects in QW systems are also found in the presence of quantizing magnetic fields [30], in complete analogy to the bulk case [31].

D6) The possibility of a hot-LO-ph-induced enhancement of the thermionic emission of hot electrons from single GaAs QWs over the barriers has been found in Monte-Carlo simulations [32].

D7) An adverse effect of LO-ph heating on the (still contoversial [33]) concept of an intrinsic negative differential resistance due to the abrupt LO-ph emission threshold in polar 2D systems is being predicted from Monte-Carlo calculations [34].

Photoexcitation

P1) The modification of the energy relaxation of hot carriers through the formation of mixed LO-ph-plasmon modes: such effects have been mainly discussed in connection with photoexcitation, for 3D among others by [35] and for 2D by [36]. The basic features of this effect are shown in Figure 4. The upper part of the figure shows the hybridisation of the LO-phonon and the plasmon modes at carrier concentrations, for which $\omega_{LO} \approx \omega_{pl}$. For such cases the appropriate theoretical description is to introduce the two resulting mixed modes (+ and -) and their corresponding dynamically screened polar couplings to the carriers. But this refinement might rise more problems than it is supposed to solve, as long as the nonelectronic lifetimes of these modes are taken as the bare LO-phonon decay times. The error in this treatment is to neglect that the plasmon part of each mode will always decay back into single-particle excitations of the carrier system, whereas only the phonon part will, through its decay into the zone-boundary phonons, act as net energy dissipation channel for the initially photoexcited carrier system. As two hybrid modes just share the weight of one LO-ph mode, the much simpler alternative route of a properly free-carrier screened "bare" c-LO-ph dynamics should provide a sufficiently accurate description of the picosecond power transfer into the lattice. The introduction of mixed modes would become decisive for the description of the intermediate reshuffling of electronic energy between free-particle and plasma degrees of freedom and consequently for the analysis of the presently studied femtosecond single-particle carrier dynamics. These transient plasma effects somehow resemble the intermediate storage of electrons in the secondary L valleys and the corresponding retardation of the energy relaxation of the central-valley electrons, a transitory mechanism adding to the reduction of the carrier cooling by LO-phonon heating.

P2) The conjecture in [15] of a cancellation of hot-LO-ph effects by antiscreening in the experiments of [37]: however, as already mentioned above, the lower threshold for noticeable LO-ph amplification is roughly at carrier densities n_c of several $10^{16} cm^{-3}$, trivially explaining the finding in [37] of no hot-LO-ph effects for $n_c = 10^{16}$.

P3 The longstanding question of a possible hot-phonon-induced LO-ph sideband emission in GaAs QW lasers has to our knowledge never been fully clarified [38][39].

SUMMARY

Practically all of the meanwhile experimentally and theoretically established hot-phonon effects in highly d.c.field- or laser-excited bulk tetrahedral semiconductors are found or can be theoretically expected to occur also in heterostructures and in quantum-well or multiquantum-well structures of these materials. A completely reliable theoretical treatment and quantitative numerical estimates of these phenomena will still have to wait until a detailed spectroscopy of the nonelectronic dissipation of microstructured phonons or mixed phonon-plasmon modes becomes available. But nevertheless carrier-temperature models and Monte-Carlo simulations are at present in satisfactory agreement with most of the experimental data.

REFERENCES

[1] *P.Kocevar*, Phys.Stat.Sol.(b) **84**, 681 (1977); *P.Kocevar and E.Fitz*, Phys.Stat.Sol.(b) **89**, 225 (1978).

[2] *N.Balkan, B.K.Ridley, and J.S.Roberts*, Proc. 4th Int.Conf.on Superlattices, Microstructures and Microdevices (Trieste 1988) Ed. *J. Dow*, Superlattices and Microstructures (1989), to be publ.

[3] *P. Kocevar*, J.Phys.C5, 3349 (1972); Acta phys. Austriaca **37**, 259 (1973).

[4] *K.T.Tsen and H.Morkoc*, Phys.Rev.B **34**, 4412 (1986); Phys.Rev.B **37**, 7137 (1988); Phys.Rev.B **38**, 5615 (1988).

[5] *J.A.Kash, J.M.Hvam, J.C.Tsang, and T.F.Kuech*, Phys.Rev. **B 38**, 5776 (1988).

[6] *W.Pötz and P.Kocevar*, Phys.Rev.B **28**, 7040 (1983).

[7] *P.Kocevar*, Festkörperprobleme (*Advances in Solid State Physics*) Ed. *P.Grosse* (Vieweg, Braunschweig 1987), Vol.**27**, p.197.

[8] *P.Lugli, C.Jacoboni, L.Reggiani,* and *P.Kocevar,* Appl.Phys.Lett.**50**, 1251 (1987); see also *P. Lugli, C.Jacoboni, L.Reggiani,* and *P.Kocevar,* SPIE Vol.793 on "Ultrafast Lasers Probe Phenomena in Bulk and Microstructure Semiconductors" (1987), p.102; *P.Lugli, P.Bordone, L.Reggiani, M.Rieger, P.Kocevar, and S.M.Goodnick,* Phys.Rev.**B** (1989) to be publ.

[9] *P.Price,* Ann.Phys.**133**, 217 (1981).

[10] *B.K.Ridley,* J.Phys.**C 15**, 5899 (1982).

[11] *F.A.Riddoch and B.K.Ridley,* Physica **134 B**, 342 (1985).

[12] *P.Price,* Physica **134 B**, 164 (1985).

[13] *S.M.Goodnick and P.Lugli,* Phys.Rev.**B 37**, 2578 (1988).

[14] *W.Cai, M.C.Marchetti, and M.Lax,* Phys.Rev.**B 34**, 8573 (1986).

[15] *W.Cai, M.C.Marchetti, and M.Lax,* Phys.Rev.**B 37**, 2636 (1988).

[16] *M.Rieger, P.Kocevar, P.Bordone, P.Lugli, and L.Reggiani,* Solid State Electron.**31**, 687 (1988); *M.Rieger, P.Kocevar, P.Lugli, P.Bordone, L.Reggiani, and S.M.Goodnick,* Phys.Rev.**B** (1989) to be publ.

[17] *S.M.Goodnick and P.Lugli,* Proc.18th Int.Conf.Phys.Semicond., Ed. *O.Engström* (World Scientific Press, Singapore 1987) p.1335.

[18] *P.Lugli and S.M.Goodnick,* Phys.Rev.Lett.**59**, 716 (1987).

[19] *P.Lugli, P.Bordone, L.Reggiani, M.Rieger, P.Kocevar, and S.M.Goodnick,* Phys.Rev.**B** (1989) to be publ.

[20] *W.Pötz and M.C.Marchetti,* SPIE Proc.**942**, 100 (1988).

[21] *J.F.Ryan, R.A.Taylor, A.J.Turberfield, A.Maciel, J.M.Worlock, A.C.Gossard, and W.Wiegmann,* Phys.Rev.Lett.**53**, 1841 (1984).

[22] *J.Shah, A.Pinczuk, A.C.Gossard, and W.Wiegmann,* Phys. Rev.Lett.**54**, 2045 (1985).

[23] *K.Leo, W.W.Rühle, H.J.Queisser, and K.Ploog,* Phys.Rev. **B 37**, 7121 (1988); *K.Leo, W.W.Rühle, and K.Ploog,* Phys.Rev. **B 38**, 1947 (1988).

[24] *P.Kocevar,* Physica **134B**, 401 (1985).

[25] *V.V.Paranjape and E.de Alba,* Proc.Phys.Soc.**85**, 945 (1965).

[26] *X.L.Lei and C.S.Ting,* Phys.Rev.**B 32**, 1112 (1985).

[27] *G.Mahan,* Phys.Rev.**B 35**, 4365 (1987).

[28] *S.Das Sarma, J.K.Jain, and R.Jalabert,* Phys.Rev.**B 37**, 1228 (1988).

[29] *S.Doniach,* Proc.Phys.Soc.**73**, 849 (1959).

[30] *H.A.J.M.Reinen, T.T.J.M.Berendschot, R.J.H.Kappert, and H.J.A.Bluyssen,* Solid State Commun.**65**,1495 (1988); *R.W.J.Hollering, T.T.J.M.Berendschot, H.J.A.Bluyssen, H.A.J.M.Reinen, P.Wyder, and F.Roozeboom,* Phys.Rev.**B 38**, 13323 (1988).

[31] *G.Bauer, H.Kahlert, and P.Kocevar*, Phys.Rev.**B 11**, 968 (1975).

[32] *K.Kim, K.Hess, and F.Capasso*, Appl.Phys.Lett. **52**, 1167 (1988).

[33] *M.Artaki and K.Hess*, Phys.Rev.**B 37**, 2933 (1988).

[34] *B.K.Ridley and M.Al-Mudares*, Solid State Electron.**31**, 683 (1988).

[35] *J.Collet, A.Cornet, M.Pugnet, and T.Amand*, Solid State Commun. **42**, 883 (1982).

[36] *S.Das Sarma, J.K.Jain, and R.Jalabert*, Phys.Rev.**B 37**, 4560 (1988).

[37] *C.H.Yang, J.M.Carlson-Swindle, S.A.Lyon, and J.M. Worlock*, Phys.Rev.Lett. **54**, 2045 (1986).

[38] *N.Holonyak, R.M.Kolbas, W.D.Laidig, B.A.Vojak, K.Hess, R.D.Dupuis, and P.D.Dapkus*, J.Appl.Phys. **51**, 1328 (1980).

[39] *K.Hess, N.Holonyak, W.D.Laidig, B.A.Vojak, J.J.Coleman, and P.D.Dapkus*, Solid State Commun. **34**, 749 (1980).

CONTRIBUTORS

Abstreiter G.

Ahmed H., see Wharam

Andrews D. A., see Davies

Baldereschi A., see Baroni

Balslev I.

Baribeau J.-M., see Gant

Barnes D. J., see Nicholas

Baroni S.

Bastard G.

  see also Devaud

Bauer G. E. W.

Berroir J. M., see Boebinger

Boebinger G. S.

Brey L., see Tejedor

Cardona M.

Chomette A., see Devaud

Christianell R., see Höpfel

Ciccacci F.

Clérot F., see Devaud

Davies G. J.

Dawson P., see Göbel

Devaud B.

Eberl K., see Abstreiter

Ekenberg U., see Wharam

English J. H., see Menéndez

Ettienne B., see Paquet

Fasol G.

Fasolino A., see Fasol

  see also Molinari

Faurie J. P., see Boebinger

Feldmann J., see Göbel

Ferreira R., see Bastard

Fischer R., see Göbel

Foxon C. T., see Nicholas

  see also Göbel

Friess E., see Abstreiter

Frost J. E. F., see Wharam

Gant T. A.

Gérard J. M., see Marzin

Göbel E. O.

Goodnick S. M.

Gossard A. C., see Menéndez

Grambow P., see Demel

Gudmundsson V.

Guldner Y., see Boebinger

Hanyou Chu , see Devaud

Harris J. J., see Nicholas

Hasko D. G., see Wharam

Hebling J., see Göbel

Heitmann D., see Demel

Höpfel R. A.

Ishibashi A.

Joncour M. C., see Paquet

Jones G. A. C., see Wharam

Juen S., see Höpfel

Kanellis G.

Kocevar P.

Kuhl J., see Göbel

Langerak C. J. G. M., see Nicholas

Laruelle F., see Paquet

Leadley D. R., see Nicholas

Lockwood D. J., see Gant

Lugli P., see Fasol

PARTICIPANTS

Gerhard Abstreiter	Walter-Schottky-Institut der Technischen Universität München James-Franck-Straße D-8046 Garching Federal Republic of Germany
Ivar Balslev	Fysisk Institut Odense Universitet DK-5230 Odense M Denmark
Stefano Baroni	SISSA Strada Costiera 11 I-34100 Trieste Italy
Gerald Bastard	Groupe de Physique des Solides Ecole Normale Supérieure 24, rue Lhomond F-75231 Paris Cedex 05 France
Gerrit Bauer	Philips Research Laboratories P. O. Box 80.000 NL-5600 JA Eindhoven The Netherlands
Greg Boebinger	AT&T Bell Laboratories Room 1D-208 600 Mountain Avenue Murray Hill, New Jersey, 07974 USA
Manuel Cardona	Max-Planck-Institut für Festkörperforschung Heisenbergstraße 1 D-7000 Stuttgart 80 Federal Republic of Germany
Franco Ciccacci	Istituto di Fisica del Politecnico Piazza Leonardo da Vinci 32 I-20133 Milano Italy
Edouard Da Silva	DILOR SA 244ter, rue des Bois Blancs F-59000 Lille France

Graham Davies	British Telecom Research Centre
	Martlesham Heath
	Ipswich, Suffolk, IP5 7RE
	United Kingdom
Benoît Devaud	Centre National d'Etudes de Télécommunications
	LAB/OCM
	F-23301 Lannion
	France
Gerhard Fasol (Director)	Cavendish Laboratory
	Madingley Road
	Cambridge CB3 0HE
	United Kingdom
Annalisa Fasolino (Co-Director)	
	SISSA
	Strada Costiera 11
	I-34100 Trieste
	Italy
T. A. Gant	National Research Council
	Ottawa K1A 0R6
	Canada
Ernst Göbel	Philipps Universität
	Fachbereich Physik
	Renthof 5
	D-3550 Marburg
	Federal Republic of Germany
Stephen Goodnick	Dept. of Electrical and Computer Engineering
	Oregon State University
	1320 SE Park Avenue
	Corvallis, OR 97331
	USA
Vidar Gudmundsson	Science Institute
	Dunhaga 3
	IS-103 Reykjavìk
	Iceland
Detlev Heitmann	Max-Planck-Institut für Festkörperforschung
	Heisenbergstraße 1
	D-7000 Stuttgart 80
	Federal Republic of Germany
Ralph Höpfel	Institut für Experimentalphysik
	Universität Innsbruck
	Technikerstraße 25/4
	A-6020 Innsbruck
	Austria
Akira Ishibashi	SONY Corporation Research Centre
	174 Fujitsuka-cho
	Hodogaya-ku
	Yokohama 240
	Japan

George Kanellis

Department of Physics
University of Thessaloniki
GR-54006 Thessaloniki
Greece

Peter Kocevar

Institut für Theoretische Physik
Universität Graz
A-8010 Graz
Austria

Paolo Lugli (Co-Director)

Dipartimento di Ingegneria Meccanica
Università di Roma II
"Tor Vergata"
I-00173 Roma
Italy

Jan-Kees Maan

Max-Planck-Institut
Hochfeldmagnetlabor
B. P. 166X
F-38042 Grenoble Cedex
France

Jean - Yves Marzin

Centre National d'Etudes de Télécommunications
196, avenue Henri Ravéra
F-92220 Bagneux
France

José Menendez

Physics Department
Arizona State University
Tempe, AZ 85287
USA

Ulrich Merkt

Institut für Angewandte Physik
Universität Hamburg
Jungius-Straße 11
D-2000 Hamburg 36
Federal Republik of Germany

Roberto Merlin

University of Michigan
Harrison M. Randall Laboratory of Physics
Ann Arbor, Michigan, 48109-1120
USA

Kevin Meyer

Cavendish Laboratory
Madingley Road
Cambridge CB3 0HE
United Kingdom

Marie - Hélène Meynadier

AT&T Bell Laboratories
Room C335
600 Mountain Avenue
Murray Hill, New Jersey, 07974
USA

Elisa Molinari

CNR Istituto di Acoustico "Corbino"
Via Cassia 1216
I-00189 Roma
Italy

Karen Moore   Philips Research Laboratories
Cross Oak Lane
Redhill RH1 5HA
United Kingdom

Robin Nicholas   Clarendon Laboratory
Parks Road
Oxford OX1 3PU
United Kingdom

Daniel Paquet   Centre d'Etudes de Télécommunications
196, avenue Henri Ravéra
F-92220 Bagneux
France

Klaus Ploog   Max-Planck-Institut für Festkörperforschung
Heisenbergstraße 1
D-7000 Stuttgart 80
Federal Republic of Germany

Hiroyuki Sakaki   Institute of Industrial Science
University of Tokyo
7-22-1 Roppongi
Minato-ku
Tokyo 106
Japan

Jagdeep Shah   AT&T Bell Laboratories
Holmdel, New Jersey 07733
USA

Hans Sigg   Max-Planck-Institut für Festkörperforschung
Heisenbergstraße 1
D-7000 Stuttgart 80
Federal Republic of Germany

C. Tejedor   Departamento de Fisica de la Materia Condensada
Universidad Autonoma
Cantoblanco
E-28049 Madrid
Spain

Jim Tsang   IBM T. J. Watson Research Center
Mail 29-161
P. O. Box 218
Yorktown Heights, New York 10598
USA

Luis Viña   Instituto de Ciencia de Materiales
Dept. Fìsica Aplicada C IV
Universidad Autónoma de Madrid
Cantoblanco
E-28049 Madrid
Spain

Peter Warmenbol   ALCATEL - Bell Advanced Research
F. Wellesplein 1
B-2018 Antwerpen
Belgium

David Wharam

Cavendish Laboratory
Madingley Road
Cambridge CB3 0HE
England

Achim Wixford

University of California
Dept. of Electrical and Computer Engineering 4113
Santa Barbara, CA 93106
USA

INDEX